14th International Conference on Hypernuclear and Strange Particle Physics (HYP2022)

EPJ Web of Conferences Volume 271 (2022)

Prague, Czech Republic
27 June – 1 July 2022

Editors:

J. Bielčíková
P. Bydžovský
D. Gazda
F. Křížek
J. Mareš
J. Óbertová

ISBN: 978-1-7138-6619-0

Printed from e-media with permission by:

Curran Associates, Inc.
57 Morehouse Lane
Red Hook, NY 12571

Some format issues inherent in the e-media version may also appear in this print version.

For additional information, please contact EDP Sciences – Web of Conferences at the address below.

EDP Sciences – Web of Conferences
17, Avenue du Hoggar
Parc d'Activité de Courtabœuf
BP 112
F-91944 Les Ulis Cedex A
France

Phone: +33 (0) 1 69 18 75 75
Fax: +33 (0) 1 69 28 84 91

contact-edps@webofconferences.org

Additional copies of this publication are available from:

Curran Associates, Inc.
57 Morehouse Lane
Red Hook, NY 12571 USA
Phone: 845-758-0400
Fax: 845-758-2633
Email: curran@proceedings.com
Web: www.proceedings.com

TABLE OF CONTENTS

High Accuracy Spectroscopy of 3- And 4-Body Λ Hypernuclei at Jefferson Lab 1
Gogami T., Achenbach P., Akiyama T., Androic D., Asaturyan A., Brash E., Bukhari M. H., Camsonne A., Covrig Dusa S., Ebata K., Elaasar M. A., Fujii Y., Fujiwara T., Furic M., Garibaldi F., Gueye P., Higinbotham D. W., Ishige T., Itabashi K., Kaneta M., Kino R., Lashley N., Markowitz P., Meekins D., Mizuno M., Mkrtchyan H. G., Mkrtchyan A. H., Nagafusa S., Nagano S., Nagao S., Nakamura S. N., Nakamura Y. R., Niculescu G., Niculescu I., Okuyama K., Pandey B., Pochodzalla J., Reinhold J., Rodriguez V. M., Samanta C., Sawatzky B., Shabestari M. H., Shahinyan A., Sirca S., Suzuki K. N., Tachibana K., Tang L., Toyama Y., Tsutsumi K., Uehara K., Umezaki E., Urciuoli G. M., Watanabe D., Wood S. A.

Lifetime of the Hypertriton 7
Gazda Daniel, Pérez-Obiol Axel, Gal Avraham, Friedman Eliahu

Comparison of 3H/4H Production Cross-Section Via (K-,pi0) Reaction at J-PARC.. 15
Akaishi Takaya, Asano Hidemitsu, Chen Xurong, Clozza Alberto, Curceanu Catalina, Del Grande Raffaele, Guaraldo Carlo, Han Chengdong, Hashimoto Tadashi, Iliescu Mihai, Inoue Kentaro, Ishimoto Shigeru, Itahashi Kenta, Iwasaki Masahiko, Ma Yue, Murayama Rie, Miliucci Marco, Noumi Hiroyuki, Ohnishi Hiroyuki, Okada Shinji, Outa Haruhiko, Piscicchia Kristian, Sada Yuta, Sakaguchi Atsushi, Sakuma Fuminori, Sato Masaharu, Scordo Alessandro, Sirghi Diana, Sirghi Florin, Shirotori Kotaro, Suzuki Shoji, Tanida Kiyoshi, Toda Taiga, Tokuda Megumu, Tsuruta Masato, Yamaga Takumi, Yuan Xiaohua, Zhang Hongqiang, Zhang Pengming, Zhang Yapeng

Single- & Double-Strangeness Hypernuclei Up to a = 8 Within ÷EFT.. 18
Le Hoai

Baryonic Effective Field Theory for Light Hypernuclei 26
Barnea Nir, Bazak Betzalel, Contessi Lorenzo, Gal Avraham, Mareš Jiøí, Schäfer Martin

Commissioning of the Hypertriton Binding Energy Measurement at MAMI 32
Eckert Philipp, Achenbach Patrick, Akiyama Takeru, Chen Jinhui, Distler Michael O., Esser Anselm, Geratz Julian, Helmel Christian, Hoek Michael, Itabashi Kosuke, Kaneta Masashi, Kino Ryoko, Klag Pascal, Ma Yu-Gang, Merkel Harald, Mizuno Masaya, Müller Ulrich, Nagao Sho, Nakamura Satoshi N., Nakamura Yuki R., Okuyama Kazuki, Pochodzalla Josef, Schlimme Björn Sören, Sfienti Concettina, Shao Tianhao, Steinen Marcell, Tachibana Koga, Thiel Michaela, Toyama Yuichi, Uehara Keita

Studying ΛN Interactions Through the 208Pb(e,e'K+)208ËTl Reaction 38
Garibaldi Franco, Benhar Omar, Bydzovsky Petr, Gogami Toshiyuki, Covrig Silviu, Markowitz Pete E.C., Millener John, Motoba Toshio, Nagao Sho, Nakamura Satoshi N., Reinhold Joerg, Tang Liguang, Urciuoli Guido Maria, Vidana Isaac

Hypernuclear Gamma-Ray Spectroscopy: Summary and Future Prospect 44
Ukai Mifuyu

Feasibility Study for Measurement of Beta-Decay Rates of Ë Hypernuclei 50
Kamada Kento, Fujita Manami, Tamura Hirokazu

Hypernuclear Production Spectroscopy with an Extended Shell Model 54
Umeya Atsushi, Motoba Toshio, Itonaga Kazunori

Spectrum of Few-Body Systems in a Finite Volume 58
Bazak Betzalel, Schäfer Martin, Yaron Roee, Barnea Nir

Mean Field and Beyond Mean Field Calculations of Hypernuclei for Study of Electroproduction 63
Veselý P., De Gregorio G., Knapp F., Petrellis D., Bydžovský P., Denisova D., Pokorný J., Skoupil D., Lo Iudice N.

Kinematical Effects in Electroproduction of Hypernuclei 66
Denisova Daria

The Strangeness Program at GlueX 69
Pauli Peter

Cross-Section Measurement of Virtual Photoproduction of Iso-Triplet Three-Body Hypernucleus, Ěnn 77
Gogami T., Suzuki K. N., Pandey B., Itabashi K., Nagao S., Okuyama K., Nakamura S. N., Tang L., Abrams D., Akiyama T., Androic D., Aniol K., Ayerbe Gayoso C., Bane J., Barcus S., Barrow J., Bellini V., Bhatt H., Bhetuwal D., Biswas D., Camsonne A., Castellanos J., Chen J-P., Chen J., Covrig S., Chrisman D., Cruz-Torres R., Das R., Fuchey E., Gnanvo K., Garibaldi F., Gautam T., Gomez J., Gueye P., Hague T. J., Hansen O., Henry W., Hauenstein F., Higinbotham D. W., Hyde C. E., Kaneta M., Keppel C., Kutz T., Lashley-Colthirst N., Li S., Liu H., Mammei J., Markowitz P., McClellan R. E., Meddi F., Meekins D., Michaels R., Mihovilovic M., Moyer A., Nguyen D., Nycz M., Owen V., Palatchi C., Park S., Petkovic T., Premathilake S., Reimer P. E., Reinhold J., Riordan S., Rodriguez V., Samanta C., Santiesteban S. N., Sawatzky B., Širca S., Slifer K., Su T., Tian Y., Toyama Y., Uehara K., Urciuoli G. M., Votaw D., Williamson J., Wojtsekhowski B., Wood S. A., Yale B., Ye Z., Zhang J., Zheng X.

Study of the A/E0 Electroproduction in the low-Q2 Region at JLab 83
Okuyama K., Itabashi K., Nagao S., Nakamura S. N., Suzuki K. N., Gogami T., Pandey B., Tang L., Abrams D., Akiyama T., Androic D., Aniol K., Ayerbe Gayoso C., Bane J., Barcus S., Barrow J., Bellini V., Bhatt H., Bhetuwal D., Biswas D., Camsonne A., Castellanos J., Chen J-P., Chen J., Covrig S., Chrisman D., Cruz-Torres R., Das R., Fuchey E., Gnanvo K., Garibaldi F., Gautam T., Gomez J., Gueye P., Hague T. J., Hansen O., Henry W., Hauenstein F., Higinbotham D. W., Hyde C. E., Kaneta M., Keppel C., Kutz T., Lashley-Colthirst N., Li S., Liu H., Mammei J., Markowitz P., McClellan R. E., Meddi F., Meekins D., Michaels R., Mihovilovic M., Moyer A., Nguyen D., Nycz M., Owen V., Palatchi C., Park S., Petkovic T., Premathilake S., Reimer P. E., Reinhold J., Riordan S., Rodriguez V., Samanta C., Santiesteban S. N., Sawatzky B., Širca S., Slifer K., Su T., Tian Y., Toyama Y., Uehara K., Urciuoli G. M., Votaw D., Williamson J., Wojtsekhowski B., Wood S. A., Yale B., Ye Z., Zhang J., Zheng X.

Possible ENN Resonances in the 3H(e,e'K+) Reaction 89
Gibson Benjamin F., Afnan Iraj R.

The Impact of Final-State Interaction on the piE Photoproduction in the A(1405) Region 92
Ciepl ý Aleš, Bruns Peter C.

Study of ƒ©n FSI with ƒ© Quasi-Free Productions on the 3H(e, e□ŒK+)X Reaction at JLab 96
Itabashi K., Suzuki K.N., Pandey B., Okuyama K., Gogami T., Nagao S., Nakamura S.N., Tang L., Abrams D., Akiyama T., Androic D., Aniol K., Ayerbe Gayoso C., Bane J., Barcus S., Barrow J., Bellini V., Bhatt H., Bhetuwal D., Biswas D., Camsonne A., Castellanos J., Chen J-P., Chen J., Covrig S., Chrisman D., Cruz-Torres R., Das R., Fuchey E., Gnanvo K., Garibaldi F., Gautam T., Gomez J., Gueye P., Hague T.J., Hansen O., Henry W., Hauenstein F., Higinbotham D.W., Hyde C.E., Kaneta M., Keppel C., Kutz T., Lashley-Colthirst N., Li S., Liu H., Mammei J., Markowitz P., McClellan R. E., Meddi F., Meekins D., Michaels R., Mihovilovic M., Moyer A., Nguyen D., Nycz M., Owen V., Palatchi C., Park S., Petkovic T., Premathilake S., Reimer P.E., Reinhold J., Riordan S., Rodriguez V., Samanta C., Santiesteban S.N., Sawatzky B., Širca S., Slifer K., Su T., Tian Y., Toyama Y., Uehara K., Urciuoli G.M., Votaw D., Williamson J., Wojtsekhowski B., Wood S.A., Yale B., Ye Z., Zhang J., Zheng X.

The BGOOD Experiment at ELSA Exotic Structures in the Strange Quark Sector? 101
Jude Thomas, Alef Stefan, Beck Reinhard, Braghieri Alessandro, Cole Philip, Di Salvo Rachele, Elsner Daniel, Fantini Alessia, Freyermuth Oliver, Fromberger Frank, Ghio Francesco, Groß Johannes, Kohl Katrin, Levi Sandri Paolo, Mandaglio Guiseppe, Pedroni Paolo, Romaniuk Mariia, Scheluchin Georg, Schmieden Hartmut

A Polarization Measurement of the pi-p-K0A Reaction in J-PARC E40 Experiment 105
Sakao Tamao, Honda Ryotaro, Miwa Koji, Nakada Yoshiyuki, Nanamura Takuya, Takahashi Toshiyuki, Ukai Mifuyu, Yamamoto Takeshi

K+E-. Photoproduction Within an Isobar Model 109
Skoupil Dalibor, Bydžovský Petr, Cieplý Aleš, Petrellis Dimitrios, Zachariou Nicholas

Application of Model Selection Criteria for K+Ó. Photoproduction Within an Isobar Approach 113
Petrellis Dimitrios, Bydžovský Petr, Cieplý Aleš, Skoupil Dalibor, Zachariou Nicholas

Exploring the Neutron Skin by Hyperon–antihyperon Production in Antiproton–nucleus Interactions 117
Christiansen Martin, Schupp Falk, Achenbach Patrick, Bölting Michael, Pochodzalla Josef, Steinen Marcell

High Resolution Spectroscopy of the "EN Cusp" by Using the d(K-,pi-) Reaction 121
Ichikawa Yudai, Ahn Jungkeun, Choi Sungwook, Fujita Manami, Gogami Toshiyuki, Hasegawa Shoichi, Harada Takeshi, Hayakawa Shuhei, Ishimoto Shigeru, Jung Woose-ung, Kang Byungmin, Kim Shinhyung, Miwa Koji, Nagae Tomofumi, Nanamura Takuya, Sako Hiroyuki, Sato Susumu, Suzuki Shoji, Tamura Hirokazu, Tanida Kiyoshi, Ukai Mifuyu, Yamamoto Takeshi, Yang Seongbae

Strangeness Production in the NA61/SHINE Experiment at the CERN SPS Energy Range 126
Balkova Yuliia

X Ray Spectroscopy on ... Atoms (J-PARC E03, E07 and Future) 130
Yamamoto Takeshi. O., Fujita M., Gogami T., Harada T. K., Hayakawa S. H., Hosomi K., Ichikawa Y., Ishikawa Y., Kamada K., Kanauchi H., Koike T., Miwa K., Nagae T., Oura F., Takahashi T., Tamura H., Tanida K., Ukai M.

ƒ¬-Nuclear Constraints from ... Emulsion Capture Events 135
Friedman Eliahu, Gal Avraham

A Skyrme Force for ... Hypernuclei 141
Schulze Hans-Josef, Guo Jing, Zhou Xian-Rong

Production of Light ..-Hypernuclei, ..7H 147
Fujioka Hiroyuki

Results of the ... Atomic X-Ray Measurement in J-PARC E07 150
Fujita Manami, Ishikawa Y., Ukai M., Kanauchi H., Koike T., Tamura H., Hosomi K., Yamamoto T.O., Ekawa H., Hayakawa S.H., Nakazawa K., Yoshida J., Yoshimoto M., Kasagi A., Nishimura N., Hayashi K.

High Resolution Spectroscopy of .. Hypernuclei with Active Fiber Target 154
Harada Takeshi K., Ebata K., Gogami T., Hayakawa S. H., Honda R., Ichikawa Y., Miwa K., Nagae T., Nanamura T., Takahashi T., Ukai M., Yamamoto T. O.

A Spectroscopic Study of ..10Li Hypernucleus Via the 10B(K.,K+)X Reaction 158
Tokiyasu Atsushi

Preparation Status of Missing-Mass Spectroscopy for .. Hypernuclei with S-2S Magnetic Spectrometer 161
Ebata Kengo, Fujioka H., Fujita M., Gogami T., Harada T. K., Hayakawa S. H., Honda R., Ichikawa Y., Kamada K., Kobori T., Miwa K., Nagae T., Nanamura T., Negishi R., Oura F., Sakao T., Son C., Takahashi T., Takahashi H., Tamura H., Tokiyasu A. O., Ukai M., Yamamoto T. O.

Recent Progress and Future Prospects of Hyperon Nucleon Scattering Experiment 164
Miwa Koji, Nanamura Takuya, Sakao Tamao, Ahn J. K., Akazawa Y., Aramaki T., Ashikaga S., Callier S., Chiga N., Chiga N., Choi S. W., Ekawa H., Evtoukhovitch P., Fujioka N., Fujita M., Gogami T., Harada T., Hasegawa S., Hayakawa S. H., Honda R., Hoshino S., Hosomi K., Ichikawa M., Ichikawa Y., Ieiri M., Ikedai M., Imai K., Ishikawa Y., Ishimoto S., Jung W. S., Kajikawa S., Kanauchi H., Kanda H., Kitaoka T., Kang B. M., Kawai H., Kim S. H., Kobayashi K., Koike T., Matsuda K., Matsumoto Y., Nagao S., Nagatomi R., Nakada Y., Nakagawa M., Nakamura I., Naruki M., Ozawa S., Raux L., Rogers T. G., Sakaguchi A., Sako H., Sato S., Shiozaki T., Shirotori K., Suzuki K. N., Suzuki S., Tabata M., Taille C. d. L., Takahashi H., Takahashi T., Takahashi T. N., Tamura H., Tanaka M., Tanida K., Tsamalaidze Z., Ukai M., Umetsu H., Wada S., Yamamoto T. O., Yoshida J., Yoshimura K.

Results of Analysis of Ó+p Scattering Events in J-PARC E40 Experiment: Differential Cross Sections and Phase Shifts of 3S1 and 1P1 States 171
Nanamura Takuya, Miwa Koji, Hayakawa Shuhei H., Honda Ryotaro, Nakada Yoshiyuki, Takahashi Toshiyuki, Ukai Mifuyu, Yamamoto Takeshi O.

Status of the Hyperon-Nucleon Interaction in Chiral Effective Field Theory 176
Haidenbauer Johann, Meißner Ulf-G.

Study on Hadron-Hadron Interaction with Femtoscopic Technique 184
Kamiya Yuki

SU(3)f Constraints on Hypernuclear Energy Density Functionals 190
Lenske Horst, Dhar Madhumita

Exploration of In-Medium Hyperon-Nucleon Interactions 197
Dhar Madhumita, Lenske Horst

ΛΛ Pairing Effects in multi-Λ Hyperisotopes Within a Skyrme-Hartree-Fock Model 201
Guo Jing, Zhou Xian-Rong

How Can We Solve the Hyperon Puzzle? —Introduction to "topical Session on ËNN Three-Body Force" 205
Tamura Hirokazu

ËNN Content of A-Nucleus Potential 209
Friedman Eliahu, Gal Avraham

Equation of State and Strangeness in Neutron Stars - Role of Hyperon-Nuclear Three-Body Forces - 217
Weise Wolfram

Experimental Study of K.NN and Future Experiments for Kaonic Nuclei 227
Yamaga Takumi

K.N Interaction, P-Wave Terms 235
Feijoo Albert, Gazda Daniel, Magas Volodymyr, Ramos Àngels

On K -Nuclear Interaction, K -Nuclear Quasibound States and K Atoms 243
Obertova Jaroslava, Friedman Eliahu, Mare. Ji.i, Ramos Angels

A Novel Approach to the Measurement of the Hyperon Nucleon/S Interaction by AMADEUS 251
Piscicchia Kristian, Cargnelli Michael, Del Grande Raffaele, Fabbietti Laura, Marton Johann, Moskal Pawel, Ramos Àngels, Scordo Alessandro, Sirghi Diana, Skurzok Magdalena, Vazquez Doce Oton, Wycech Slavomir, Zmeskal Johann, Branchini Paolo, Czerwinski Eryk, De Leo Veronica, De Lucia Erika, Di Cicco Alessandro, Diociaiuti Eleonora, Fermani Paolo, Fiore Salvatore, Martini Matteo, Perez Del Rio Elena, Selce Andrea, Silarski Michal, Curceanu Catalina

Decay of the A(1405) Hyperon to E0pi0 Measured at GlueX 258
Wickramaarachchi Nilanga, Schumacher Reinhard A., Kalicy Grzegorz

Triangle Singularity in the K-d-pE-Reaction and Its Relevance to Find the K.N Amplitude Below Threshold 264
Oset Eulogio, Molina Raquel, Dai Lian Rong, Feijoo Albert

Four-Body Faddeev-Type Calculation of the K.NNN System 269
Shevchenko Nina V.

Meson-Baryon Scattering and A(1405) in Chiral Effective Field Theory 273
Ren Xiu-Lei

Study of Strange Matter in STAR with Express Analysis 277
Kisel Ivan

Recent Studies on Hypernuclei Lifetimes from STAR 284
Leung Yue Hang

Studies of Hypernuclei with Heavy-Ion Beams, Nuclear Emulsions and Machine Learning 289
Saito Takehiko R.

New Results on Light Nuclei, Hyperons and Hypernuclei from HADES (HADES Collaboration) 298
Abou Yassine Rayane, Adamczewski-Musch Jörn, Becker Marten, Bergmann Philip, Blanco Alberto, Blume Christoph, Chlad Lukas, Chudoba Petr, Ciepa³ Izabela, Cordts Malte, Dreyer Jörn, Esmail Waleed Ahmed, Firlej Miroslaw, Fiutowski Tomasz, Floersheimer Henrik, Fonte Paulo, Friese Jürgen, Fröhlich Ingo, Förtsch Jörg, Galatyuk Tetyana, Gniazdowski Tomasz, Greifenhagen Robert, Grunwald Mateusz, Grzonka Dieter, Gumberidze Malgorzata, Harabasz Szymon, Heinz Thorsten, Höhne Claudia, Hojeij Fatima, Holzmann Romain, Huck Holger, Idzik Marek, Kämpfer Burkhard, Kampert Karl-Heinz, Kardan Behruz, Kedych Vadym, Koenig Ilse, Koenig Wolfgang, Kohls Marvin, Kolas Jedrzej, Korcyl Grzegorz, Kornakov Georgy, Kornas Frederic, Kotte Roland, Krueger Wilhelm, Kugler Andrej, Kulessa Pawel, Lalik Rafal, Lebedev Semen, Linev Sergey, Lopes Luís, Lorenz Manuel, Malige Akshay, Markert Jochen, Matulewicz Tomasz, Messchendorp Johan, Metag Volker, Michel Jan, Molenda Aleksandra, Moron Jakub, Müntz Christian, Nabroth Marvin, Naumann Lothar, Orliñski Jan, Otto Jan-Hendrik, Parpottas Yannis, Parschau Mirco, Pauly Christian, Pechenov Vladimir, Pechenova Olga, Perez Andrade Gabriela, Pfeifer Dennis, Piasecki Krzysztof, Pietraszko Jerzy, Povar Tetiana, Prozorov Alexandr, Przygoda Witold, Pysz Krzysztof, Ramstein Béatrice, Rathod Narendra, Ritman James, Rost Adrian, Rustamov Anar, Salabura Piotr, Saraiva Joao, Schadmand Susan, Schild Niklas, Schwab Erwin, Seck Florian, Selyuzhenkov Ilya, Singh Udai, Skorpil Leon, Smyrski Jerzy, Sobiella Manfred, Spataro Stefano, Spies Simon, Stefaniak Maria, Ströbele Herbert, Stroth Joachim, Sumara Konrad, Svoboda Ondøej, Swientek Krzysztof, Szala Melanie, Tlusty Pavel, Traxler Michael, Tsertos Haralabos, Wagner Vladimir, Wasiluk Mateusz, Weber Adrian Amatus, Wendisch Christian, Wintz Peter, Zbroszczyk Hanna, Zherebtsova Elizaveta, Zielinski Marcin, Zumbruch Peter

Strange Hadron Correlations in Heavy-Ion Collisions at RHIC Energies and Below 306
Zbroszczyk Hanna

Directed Flow of Ë from Heavy-Ion Collisions and Hyperon Puzzle of Neutron Stars 314
Ohnishi Akira, Jinno Asanosuke, Murase Koichi, Nara Yasushi

Centrality Dependence of Multistrangeness Production in High-Energy Heavy-Ion Collisions..................... 320
Arakelyan Gevorg H., Merino Carlos, Shabelski Yuli M.

Recent Progress on Global and Local Polarization of Hyperons in Heavy-Ion Collisions.............................. 326
Niida Takafumi

Femtoscopic Measurements of Two-Kaon Combinations in Au+Au Collisions at the RHIC Energies 330
Paw owska Diana

Study of Charge Symmetry Breaking in a = 4 Hypernuclei in NN = 3 GeV Au+Au Collisions at RHIC 335
Shao Tianhao

Kaon Flow in Au+Au Collisions at 1.23AGeV Measured with HADES 339
Chlad Lukáš

WASA-FRS HypHI Experiment at GSI for Studying Light Hypernuclei 342
Ekawa Hiroyuki

Investigation of the Ó0 Hyperon Production in p(3.5GeV)+p Collisions 347
Esmail Waleed

Neutron Stars and the Hyperon Puzzle... 352
Vidaña Isaac

Strange Mesons in Nuclei and Neutron Stars 360
Tolos Laura

Exploring the Role of Hyperons in High Density Matter in the Quark-Meson-Coupling Model 366
Stone Jirina R.

Studying . Meson Properties in Nuclear Matter from Dilepton and K+K. Decays 372
Gubler Philipp, Bratkovskaya Elena

Hyperonic Equation of State at Finite Temperature for Neutron Stars 376
Kochankovski Hristijan, Ramos Angels, Tolos Laura

The T+cc Exotic State and Its Doubly Bottom B(*)(s)B(*)(s) Counterparts 380
Oset Eulogio, Liang Wei Hong, Dai Lian Rong, Molina Raquel, Roca Luís, Martínez Torres Alberto, Khemchandani Kanchan, Feijoo Albert

Nature of Excited .. Baryons with Threshold Effects 384
Nishibuchi Takuma, Hyodo Tetsuo

Estimation of Compositeness with Correction Terms 388
Kinugawa Tomona, Hyodo Tetsuo

Analysis of Coupled-Channel Potentials with Quark and Hadron Degrees of Freedom 392
Terashima Ibuki, Hyodo Tetsuo

J-PARC Hadron Experimental Facility Extension Project 395
Sakuma Fuminori, Aoki Kazuya, Fujioka Hiroyuki, Gogami Toshiyuki, Hidaka Yoshimasa, Hiyama Emiko, Honda Ryotaro, Hosaka Atsushi, Ichikawa Yudai, Ieiri Masaharu, Isaka Masahiro, Ishii Noriyoshi, Ishikawa Takatsugu, Komatsu Yusuke, Komatsubara Takeshi, Lim GeiYoub, Miwa Koji, Morino Yuhei, Nagae Tomofumi, Nagao Sho, Nakamura Satoshi N., Nanjo Hajime, Naruki Megumi, Nemura Hidekatsu, Nomura Tadashi, Noumi Hiroyuki, Ohnishi Hiroaki, Ozawa Kyoichiro, Sawada Shinya, Sekihara Takayasu, Shim Sang-In, Shiomi Koji, Shirotori Kotaro, Tajima Yasuhisa, Takahashi Hitoshi, Takahashi Toshiyuki, Takeuchi Sachiko, Takizawa Makoto, Tamura Hirokazu, Tanida Kiyoshi, Ukai Mifuyu, Yamamoto Takeshi O., Yamamoto Yasuo

Strangeness Physics Programs by S-2S at J-PARC 402
Gogami Toshiyuki, Achenbach Patrick, Ahn Jung Keun, Androiæ Darko, Aoki Kanae, Asaturyan Arshak, Botta Elena, Bukhari Masroor. H., Camsonne Alexandre, Covrig Silviu. C., Ebata Kengo, Ekawa Hiroyuki, Evtoukhovitch Petr, Feliciello Alessandro, Fujioka Hiroyuki, Fujita Manami, Garibaldi Franco, Harada Takeshi. K., Hasegawa Shoichi, Hasegawa Tomoyuki, Hayakawa Shuhei. H., Hirose Erina, Honda Ryotaro, Hosomi Kenji, Ichikawa Yudai, Imai Kenichi, Ishige Tatsuhiro, Jung Wooseung, Kamada Kento, Kanatsuki Shunsuke, Kato Seigo, Kim Shinhyung. H., Kobori Takumi, Marcello Simonetta, Markowitz Pete, Miwa Koji, Mkrtchyan Arthur, Mkrtchyan Hamlet, Moritsu Manabu, Nagae Tomofumi, Nagao Sho, Nakagawa Manami, Nakamura Satoshi. N., Nanamura Takuya, Naruki Megumi, Negishi Ryosuke, Okuyama Kazuki, Oura Fumiya, Pandey Bishnu, Pochodzalla Josef, Sakaguchi Atsushi, Sakao Tamao, Sako Hiroyuki, Samanta Chhanda, Sato Susumu, Shabestari Mitra. H., Shahinyan Albert, Shirotori Kotaro, Širca Simon, Son C., Sugimura Hitoshi, Takahashi Hitoshi, Takahashi Shuji, Takahashi Tomonori, Takahashi Toshiyuki, Tamura Hirokazu, Tanida Kiyoshi, Tokiyasu Atsushi, Tsamalaidze Zviadi, Uchida Makoto, Ukai Mifuyu, Une Chiaki, Urciuoli Guido. M., Yamamoto Takeshi. O.

Future Prospects of Spectroscopic Study of Lambda Hypernuclei at JLab and J-PARC HIHR 409
Nakamura Satoshi N.

The J-PARC Heavy Ion Project 420
Ozawa Kyoichiro, Aoki Kazuya, Esumi Shin-ichi, Gunji Taku, Hachiya Takashi, Harada Hiroyuki, Ichikawa Yudai, Kitazawa Masakiyo, Miake Yasuo, Morino Yuhei, Nagamiya Shoji, Nara Yasushi, Naruki Megumi, Niida Takafumi, Nonaka Toshihiro, Saito Takehiko R., Sakaguchi Takao, Sako Hiroyuki, Sato Susumu, Takahashi Hitoshi, Tamura Hirokazu, Tanaka Kazuhiro, Yokkaichi Satoshi

Information and Perspective for Doubly-Strange Hypernuclei with Nuclear Emulsion Detector 426
Nakazawa Kazuma

Unique Approach for Precise Determination of Binding Energies of Hypernuclei with Nuclear Emulsion and Machine Learning 432
Nakagawa Manami, Kasagi Ayumi, Liu Enqiang, Ekawa Hiroyuki, Yoshida Junya, Dou Wenbo, He Yan, Muneem Abdul, Nakazawa Kazuma, Rappold Christophe, Saito Nami, Saito Takehiko R., Sugimoto Shohei, Taki Masato, Tanaka Yoshiki K., Wang He, Gao Yiming, Yanai Ayari, Yoshimoto Masahiro

Development of a Triple Coincidence Method of Reaction, Gamma-Ray, and Weak Decay in the Hypernuclear Gamma-Ray Spectroscopy at J-PARC 436
Oura Fumiya, Fujita Manami, Ishikawa Yuji, Rogers Tatyana G., Tamura Hirokazu, Ukai Mifuyu, Yamamoto Takeshi O.

Overview of Hypernuclear and Strange Particle Physics –Experimental Summary of HYP2022 440
Tamura Hirokazu

Author Index

HYP 2022 PRAGUE

PROCEEDINGS OF THE

14th INTERNATIONAL CONFERENCE ON HYPERNUCLEAR AND STRANGE PARTICLE PHYSICS

June 27 – July 1, 2022, Prague, Czech Republic

Organized by

Nuclear Physics Institute
Czech Academy of Sciences
and
Faculty of Nuclear Sciences and Physical Engineering
Czech Technical University in Prague

Edited by

J. Bielčíková, P. Bydžovský, D. Gazda, F. Křížek, J. Mareš, J. Óbertová

Preface

The 14th International Conference on Hypernuclear and Strange Particle Physics, HYP2022, took place in Prague, Czech Republic from June 27 to July 1, 2022. It was the next conference in a series that started in Heidelberg (1982), followed then by conferences at Brookhaven (1985), Padova (1988), Shimoda (1991), Vancouver (1994), Brookhaven (1997), Torino (2000), JLab (2003), Mainz (2006), Tokai (2009), Barcelona (2012), Sendai (2015), and Norfolk/Portsmouth (2018).

The HYP conference brings together theoreticians and experimentalists working in hadron and nuclear physics with strangeness. Over the years, the conference has been always open to cover fresh and challenging problems, expanding its frontiers to emerging directions in nuclear and particle physics. The scope of the HYP conference has been extended to related fields such as low-energy aspects of QCD, hot and dense matter, as well as hadronic and nuclear systems with heavier flavors.

The HYP2022 conference was attended by 160 participants; more than 2/3 of them took part in person, the rest participated in the conference online. The scientific program comprised invited talks in the plenary and parallel sessions, and contributed papers in the parallel and poster sessions. Topics discussed at the conference included production, structure and decay of hypernuclei, multistrange systems, strangeness and hypernuclei in elementary reactions and nuclear collisions, interactions of mesons and baryons with strangeness, strangeness in hadron structure, strange mesons in nuclei, strange exotic atoms, strangeness in astrophysics and in extreme forms of matter, heavy flavor systems, and future experiments and facilities.
The Local Organizing Committee thanks all the participants for their invaluable contribution to the conference program and form.

Editors of the HYP2022 proceedings
October 2022

HYP 2022 PRAGUE

14th International Conference on Hypernuclear and Strange Particle Physics

June 27–July 1, 2022
Prague, Czech Republic

Conference Topics

- Production, structure and decay of hypernuclei
- Multistrange systems
- Strange mesons in nuclei
- Strange exotic atoms
- Heavy flavor systems
- Strangeness and hypernuclei in elementary reactions and nuclear collisions
- Strangeness in astrophysics and in extreme forms of matter
- Interactions of mesons and baryons with strangeness
- Strangeness in hadron structure
- Future experiments and facilities

Hosts

Nuclear Physics Institute
Czech Academy of Sciences

Faculty of Nuclear Sciences and Physical Engineering
Czech Technical University in Prague

Local Organizing Committee

J. Bielčíková, P. Bydžovský, A. Cieplý, D. Gazda, F. Křížek, J. Mareš, D. Skoupil, P. Tlustý *(NPI)*
J. Bielčík, J. Óbertová, B. Tomášik *(FNSPE)*

International Advisory Committee

E. Botta *(Torino)*, T. Bressani *(Torino)*, H. Caines *(Yale)*, C. Curceanu *(Frascati)*, L. Fabbietti *(Munich)*, A. Feliciello *(Torino)*, A. Gal *(Jerusalem)*, B. F. Gibson *(Los Alamos)*, J. Haidenbauer *(Juelich)*, E. Hiyama *(Tohoku)*, T. Kishimoto *(Osaka)*, H. Lenske *(Giessen)*, Yu-Gang Ma *(Shanghai)*, J. Mareš *(Řež)*, D. J. Millener *(BNL)*, T. Motoba *(Osaka)*, T. Nagae *(Kyoto)*, S. N. Nakamura *(Tohoku)*, H. Noumi *(RCNP)*, M. Oka *(Tokyo)*, E. Oset *(Valencia)*, A. Parreño *(Barcelona)*, J. Pochodzalla *(Mainz)*, À. Ramos *(Barcelona)*, R. Schumacher *(CMU)*, H. Tamura *(Tohoku)*, L. Tang *(Hampton)*, I. Vidaña *(Catania)*, T. Yamazaki *(Tokyo)*, Shi-Lin Zhu *(Beijing)*, Bing-Song Zou *(Beijing)*

http://rafael.ujf.cas.cz/hyp2022

STRONG 2020
CTU
BNL–CZ
Research Infrastructure CERN-CZ
FAIR

edp sciences

Web of Conferences

Statement of Peer review

In submitting conference proceedings to *Web of Conferences*, the editors of the proceedings certify to the Publisher that

1. They adhere to its **Policy on Publishing Integrity** in order to safeguard good scientific practice in publishing.
2. All articles have been subjected to peer review administered by the proceedings editors.
3. Reviews have been conducted by expert referees, who have been requested to provide unbiased and constructive comments aimed, whenever possible, at improving the work.
4. Proceedings editors have taken all reasonable steps to ensure the quality of the materials they publish and their decision to accept or reject a paper for publication has been based only on the merits of the work and the relevance to the journal.

Title, date and place of the conference

HYP2022 - 14th International Conference on Hypernuclei and Strange Particle Physics
June 27 - July 1, 2022
Prague, Czech Republic

Proceedings editor(s):

Jana Bielčíková, Petr Bydžovský, Daniel Gazda, Filip Křížek, Jiří Mareš, Jaroslava Óbertová

Date and editor's signature

27 Oct. 2022,

High accuracy spectroscopy of 3- and 4-body Λ hypernuclei at Jefferson Lab

T. Gogami[1,*], P. Achenbach[2], T. Akiyama[3], D. Androic[4], A. Asaturyan[5], E. Brash[6], M. H. Bukhari[7], A. Camsonne[8], S. Covrig Dusa[8], K. Ebata[1], M. A. Elaasar[9], Y. Fujii[10], T. Fujiwara[3], M. Furic[4], F. Garibaldi[11,12], P. Gueye[13], D. W. Higinbotham[8], T. Ishige[3], K. Itabashi[14], M. Kaneta[3], R. Kino[3], N. Lashley[15], P. Markowitz[16], D. Meekins[8], M. Mizuno[3], H. G. Mkrtchyan[5], A. H. Mkrtchyan[5], S. Nagafusa[1], S. Nagano[3], S. Nagao[3], S. N. Nakamura[14], Y. R. Nakamura[3], G. Niculescu[17], I. Niculescu[17], K. Okuyama[3], B. Pandey[18], J. Pochodzalla[2], J. Reinhold[16], V. M. Rodriguez[19], C. Samanta[18], B. Sawatzky[8], M. H. Shabestari[20], A. Shahinyan[5], S. Sirca[21], K. N. Suzuki[1], K. Tachibana[3], L. Tang[8,15], Y. Toyama[22], K. Tsutsumi[1], K. Uehara[3], E. Umezaki[1], G. M. Urciuoli[11], D. Watanabe[3], and S. A. Wood[8] for the JLab Hypernuclear Collaboration

[1]Department of Physics, Graduate School of Science, Kyoto University, Kyoto, Kyoto 606-8502, Japan
[2]Institute for Nuclear Physics, Johannes Gutenberg-University, D-55099 Mainz, Germany
[3]Department of Physics, Graduate School of Science, Tohoku University, Sendai, Miyagi 980-8578, Japan
[4]Department of Physics, University of Zagreb, HR-10000 Zagreb, Croatia
[5]A.I. Alikhanyan National Science Laboratory, Yerevan 0036, Armenia
[6]Department of Physics, Computer Science & Engineering, Christopher Newport University, Newport News, VA, USA 23606
[7]Jazan University, Jazan 45142, Saudi Arabia
[8]Thomas Jefferson National Accelerator Facility (JLab), Newport News, VA 23606, USA
[9]Department of Physics, Southern University at New Orleans,New Orleans, LA 70126, USA
[10]Physics Section, Tohoku Medical and Pharmaceutical University, Sendai, Miyagi 981-8558, Japan
[11]INFN, Sezione di Roma, 00185 Rome, Italy
[12]Istituto Superiore di Sanità, 00161 Rome, Italy
[13]Facility for Rare Isotope Beams, Michigan State University, MI 48824, USA
[14]Department of Physics, Graduate School of Science, the University of Tokyo, Japan
[15]Department of Physics, Hampton University, Hampton, VA 23668, USA
[16]Department of Physics, Florida International University, Miami, FL 33199, USA
[17]Department of Physics and Astronomy, James Madison University, Harrisonburg, VA 22807, USA
[18]Department of Physics & Astronomy, Virginia Military Institute, Lexington, Virginia 24450, USA
[19]Division de Ciencias y Tecnología, Universidad Ana G. Mendez, Recinto de Cupey, San Juan 00926, Puerto Rico
[20]Department of Physics, University of West Florida, FL 32514, USA
[21]Faculty of Mathematics and Physics, University of Ljubljana, 1000 Ljubljana, Slovenia
[22]Center for Muon Science and Technology, Chubu University, Kasugai-shi, Aichi 487-8501, Japan

Abstract. JLab E12-19-002 Experiment is planned to measure the Λ-binding energies of ${}^{3}_{\Lambda}$H [$J^{\pi} = 1/2^{+}$ or $3/2^{+}(T = 0)$] and ${}^{4}_{\Lambda}$H (1^{+}) at JLab Hall C. The expected accuracy for the binding-energy measurement is $|\Delta B_{\Lambda}^{\rm total}| \simeq 70$ keV. The accurate spectroscopy for these light hypernuclei would shed light on the

*e-mail: gogami@jlab.org

puzzle of the small binding energy and short lifetime of ${}^{3}_{\Lambda}$H, and the charge-symmetry breaking in the ΛN interaction. We aim to perform the experiment in 2025.

1 Introduction

Experimental studies for light Λ hypernuclei have been playing an important role to investigate the ΛN interaction. Particularly hypertriton (${}^{3}_{\Lambda}$H) which is the lightest bound system ever confirmed, and the 4-body iso-doublet ($T = 1/2$) hypernuclei (${}^{4}_{\Lambda}$H and ${}^{4}_{\Lambda}$He) are of great importance.

The ground-state ($J^{\pi} = 1/2^{+}$) binding energy of ${}^{3}_{\Lambda}$H was found to be $B_{\Lambda} = 0.13 \pm 0.05^{\mathrm{stat.}}$ MeV in an emulsion experiment [1] whereas the STAR collaboration recently reported a larger value $B_{\Lambda} = 0.41 \pm 0.12^{\mathrm{stat.}} \pm 0.11^{\mathrm{sys.}}$ MeV [2]. There is an argument that the small binding energy from the emulsion experiment does not allow for the short lifetime that was reported by recent heavy-ion beam experiments. On the other hand, the binding energy as large as the report by the STAR collaboration may allow the predicted lifetime to be consistent with the measured short lifetime [3]. Here, we should say that the experimental data for both binding energy and lifetime are not determined well enough, and need to confirm them by more accurate experiments. There are many experimental attempts which are on-going or are being prepared to shed light on the puzzle of the hypertriton's binding energy and lifetime at the present day [4–8]. In addition, the first excited state ($3/2^{+}$) of ${}^{3}_{\Lambda}$H has not been observed, and its existence is open question [9–11]. A difference of the ground-state (0^{+}) binding energy between ${}^{4}_{\Lambda}$H and ${}^{4}_{\Lambda}$He is a firm evidence of the charge symmetry breaking (CSB) in the ΛN interaction [12, 13]. The energy levels for the first excited states (1^{+}) of these hypernuclei also give us fundamental information about CSB. A recent γ-ray measurement by Germanium detector at J-PARC updated the energy of ${}^{4}_{\Lambda}$He(1^{+}) [14], and it turned out that the binding energy is consistent with that of existing data for its mirror state ${}^{4}_{\Lambda}$H(1^{+}). Similarly, the data of ${}^{4}_{\Lambda}$H(1^{+}), which is a weighted average of three of γ-ray measurements by NaI counters [15–17], are awaited to be improved and/or confirmed by modern experimental techniques.

The context shown above motivated us to launch an experimental project (JLab E12-19-002 Experiment) in which the Λ-binding energies of ${}^{3}_{\Lambda}$H [$1/2^{+}$ or $3/2^{+}$ ($T = 0$)] and ${}^{4}_{\Lambda}$H (1^{+}) are determined with the accuracy of $|\Delta B^{\mathrm{total}}_{\Lambda}| \simeq 70$ keV by using the $(e, e'K^{+})$ reaction at Jefferson Lab (JLab) [18, 19]. The present experiment and other experimental attempts such as ${}^{3}_{\Lambda}$H ($1/2^{+}$, $T = 0$) measurement by decay-pion spectroscopy at MAMI [20], ${}^{4}_{\Lambda}$H (1^{+}) measurement by γ-ray spectroscopy at the J-PARC K1.1 beam line (J-PARC E63 Experiment) [21] etc. are complementary because they use totally different experimental techniques which may have different origins of systematic errors on results. The experiment E12-19-002 is being prepared aiming to be performed in 2025.

2 What to consider for the experiment

E12-19-002 was originally proposed to be carried out at JLab Hall A [22] to investigate the nuclear masses of hyperhydrogens, ${}^{3}_{\Lambda}$H and ${}^{4}_{\Lambda}$H [18, 19]. However, it turned out the experiment could be performed earlier in Hall C. We considered the experimental feasibility in Hall C in order to seek the earliest occasion of beam time because tackling the physics problems shown above is urgent. However, in the case of Hall C, given a required momentum resolution and acceptance, High resolution Election Spectrometer (HES) [23] of which the maximum central momentum is 0.844 GeV/c is the first option to use for the e' measurement. The limit of the maximum momentum of HES does not allow us to use the beam energy as high as the case of Hall A. The lowered beam energy causes an increase of background electrons from the Bremsstrahlung process occurred in an experimental target, possibly leading to worse S/N. In addition, the background rate due to the Bremsstrahlung process is proportional to about the square of proton number of target material ($Z^2_{\text{targ.}}$) [24]. The effect of the Bremsstrahlung background is not negligible even for our light mass targets, which are ^{3,4}He gases, because the gaseous targets need to be contained by metal cells which have an enough strength against inner pressures. However, there is an important merit in the Hall-C option. The missing-mass resolution gets better due to the lower energies for both the incident and scattered electrons.

We examined the background rate and S/N for considerable experimental conditions, based on particle rates from the previous hypernuclear experiment [25–28]. As a result, it was found that the experiment at Hall C is feasible, although a fine tuning for the experimental design is ongoing to find an optimal condition that maximizes physics output. This article shows the expected result for a certain experimental condition at JLab Hall C.

3 Expected result

3.1 Assumption of the experimental condition

E12-19-002 performs missing-mass spectroscopy with $(e, e'K^+)$ reaction. The scattered electron (e') and K^+ are detected by HES and High resolution Kaon Spectrometer (HKS) [29, 30], respectively. The HES is assumed to be a vertical bending spectrometer whereas the modification plan from the original horizontal bending to the vertical one is now being examined. The introduction of the vertical bending spectrometer is vital for the present experiment because we need a reaction vertex information for an analysis of the gas targets. The spectrometers HES and HKS will be combined with a new Pair of Charge-Separation dipole magnets (PCS) which was already constructed and transferred to JLab from Japan in 2022. The PCS makes the detectable scattering angle more forward to increase the hypernuclear yield. The continuous electron beam at E_e = 2.24 GeV with the rastered beam size of 2×2 mm^2 is impinged on the target. The beam current is assumed to be 20 μA. The helium-3,4 gases that have the areal densities of 192 and 262 mg/cm^2, respectively, are contained in cylindrical cells with the diameter of 200 mm. It is noted that the areal density and the cell diameter for a hydrogen gas, which is used for the energy calibration, are 54 mg/cm^2 and 220 mm, respectively. The cell is made of aluminum alloy Al-7075 with the wall thickness of 0.3 mm. The central momenta are set to 0.74 and 1.2 GeV/c and the scattering angles with respect the incident beam are $\theta_{ee'} = 8°$ and $\theta_{eK} = 10.2°$ for HES (e') and HKS (K^+), respectively. These conditions determine the square of four-momentum transfer taken with the negative sign ($Q^2 = -q^2$) and the K^+-scattering angle with respect to the virtual photon in the laboratory frame to be $Q^2 \sim 0.04$ GeV^2 and $\theta_{\gamma^* K} \sim 6°$, respectively. Assumed beam times for the helium-3 and 4 targets are 480 and 96 hours, respectively, which exclude calibration runs.

3.2 Expected spectra for the $^{3}_{\Lambda}$H and $^{4}_{\Lambda}$H productions

The expected number of events for the $^{3}_{\Lambda}$H and $^{4}_{\Lambda}$H is about 200 for each. The yield estimation is based on the following assumptions. The differential cross sections, which are 5 and 20 nb/sr for the ^{3,4}He$(\gamma^*, K^+)^{3,4}_{\Lambda}$H reactions, are taken from the previous measurement [31, 32]. The assumed detection efficiency (detector, analysis, data acquisition) and the K^+-survival ratio against its decay are 0.75 and 0.26, respectively. The gas density is expected

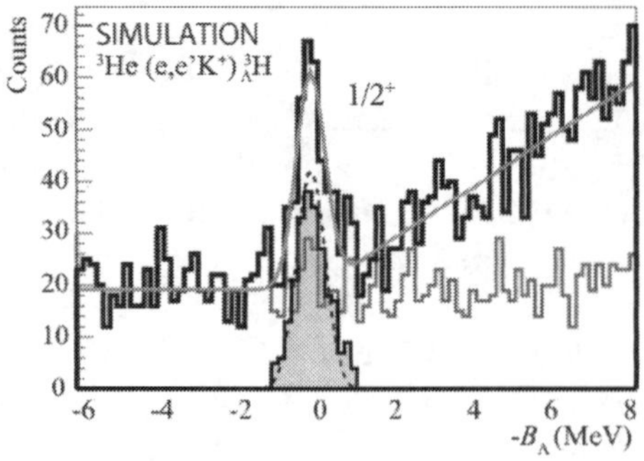

Figure 1. Expected spectrum for the $^{3}_{\Lambda}$H production in JLab E12-19-002.

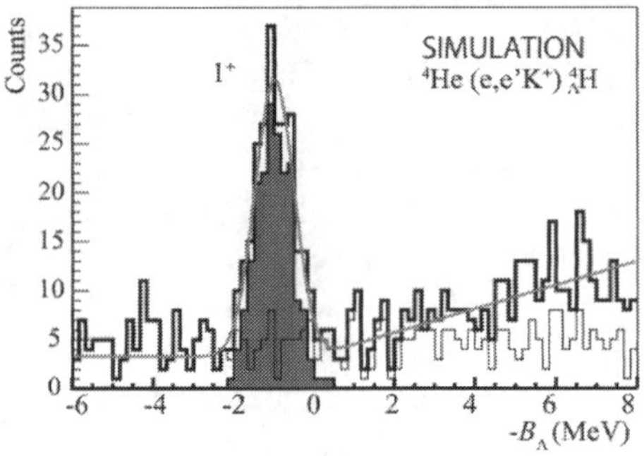

Figure 2. Expected spectrum for the $^{4}_{\Lambda}$H production in JLab E12-19-002.

be reduced due to a local temperature rise by the beam exposure, and the gas-density reduction is conservatively assumed to be 0.5. In the analysis, an event selection by the production vertex along the beam direction (v_z) is applied to eliminate a contamination from the target cell as was done in the previous hypernuclear experiment [33–35] with a gaseous tritium target [36]. The v_z resolution of HES is assumed to be $\Delta v_z = 2$ cm in FWHM. The fraction of remained signals by the v_z cut is assumed to be 0.8 which corresponds to the suppression of the event contamination from the target cell by 98%.

The missing-mass resolution is estimated to be about 1 MeV/c^2 in FWHM considering an effect of the finite volumes of gas targets. It is worth noting that the resolution could get better to about 0.6 MeV/c^2 (FWHM) in the case of solid targets which are much thinner; they are mostly as much as the order of 0.1 mm. Measurements of other hypernuclei, $^{40,48}_{\Lambda}$K [37] and $^{208}_{\Lambda}$Tl [38], which use solid targets with the areal density of about 100 mg/cm^2, are planned to be performed with the same experimental setup as that of the present experiment.

The expected spectrum for the ^{3}He$(e, e'K^+)^3_\Lambda$H is shown in Fig. 1. Only a peak for the ground state $(1/2^+, T = 0)$ is assumed to exist in the simulated $^{3}_{\Lambda}$H spectrum. However, the first excited state $(3/2^+, T = 0)$, which has not been observed, may have a larger cross section than that of the ground state by a factor of eight [39, 40]. Therefore, instead of the ground-state measurement, the binding energy for the $3/2^+$ state may be able to be determined for the first time if the state exists. In addition, the $1/2^+(T = 1)$ state may also be observed at about 2-MeV higher energy from the $1/2^+(T = 0)$ state if its production-cross section is reasonably large and decay width is small enough. Theoretical predictions of the production-cross sections as well as the energy levels are desired. Particularly relative cross-section strengths for the states would help a lot for the spectrum analysis. Figure. 2 shows the expected spectrum for the ^{4}He$(e, e'K^+)^4_\Lambda$H reaction. The cross section for the 1^+ is predicted to be dominant [41], and thus the 1^+ state would be a prominent peak as shown in the figure. The statistical error for the peak fitting is $|\Delta B^{\rm stat.}_\Lambda| \leq 40$ keV for both $^{3}_{\Lambda}$H and $^{4}_{\Lambda}$H measurements. The systematic error on the binding-energy measurement is evaluated to be $|\Delta B^{\rm sys.}_\Lambda| \simeq 60$ keV [42]. Therefore, the total accuracy that takes into account both statistical and systematic errors is $|\Delta B^{\rm total}_\Lambda| \simeq 70$ keV when we take a square root of their quadratic sum.

4 Summary

JLab E12-19-002 Experiment measures the Λ-binding energies of $^{3}_{\Lambda}$H [$J^\pi = 1/2^+$ or $3/2^+(T = 0)$] and $^{4}_{\Lambda}$H (1^+) at JLab Hall C. The total accuracy for the binding energy measurement is expected to reach $|\Delta B^{\rm total}_\Lambda| \simeq 70$ keV. The accurate spectroscopy for the light hypernuclei would shed light on the puzzle of the binding energy and lifetime of hypertriton, and the ΛN CSB. We have been preparing for the experiment aiming to perform it in 2025.

Acknowledgments

This work was partially supported by the Grant-in-Aid for Scientific Research on Innovative Areas "Toward new frontiers Encounter and synergy of state-of-the-art astronomical detectors and exotic quantum beams." This work was supported by JSPS KAKENHI grants nos. 18H05459, 18H05457, 18H01219, and 17H01121. This work was also supported by SPIRITS 2020 of Kyoto University, and the Graduate Program on Physics for the Universe, Tohoku University (GP-PU).

References

[1] M. Jurič et al., Nucl. Phys. B **52**, 1–30 (1973)
[2] The STAR Collaboration, Nat. Phys. **16**, 409–412 (2020)
[3] A. Pérez-Obiol, D. Gazda, E. Friedman, A. Gal, Phys Lett. B **811**, 135916 (2020)
[4] S. Nagao et al., Letter of Intent to ELPH.
[5] M. Agnello et al., Nucl. Phys. A **954**, 176–198 (2016)
[6] Y. Ma et al., Proposal to J-PARC, P73 (2018)
[7] A. Feliciello et al., Proposal to J-PARC, P74 (2019)
[8] T. R. Saito et al., Nat. Rev. Phys. **3**, 803–813 (2021)
[9] M. Schäfer et al., Phys. Lett. B **808**, 135614 (2020)
[10] M. Schäfer et al., Phys. Rev. C **103**, 025204 (2021)
[11] M. Schäfer et al., Phys. Rev. C **105**, 015202 (2022)
[12] A. Esser et al., Phys. Rev. Lett. **114**, 232501 (2015)
[13] F. Schulz et al., Nucl. Phys. A **954**, 149 (2016)
[14] T. O. Yamamoto et al., Phys. Rev. Lett. **115**, 222501 (2015)
[15] M. Bedjidian et al., Phys. Lett. B **62**, 4, 467–470 (1976)
[16] M. Bedjidian et al., Phys. Lett. B **83**, 2, 252–256 (1979)
[17] A. Kawachi, Ph.D. thesis, University of Tokyo (1997)
[18] T. Gogami et al., Letter of Intent to JLab, LOI12-18-003 (2018);
T. Gogami et al., Proposals to JLab, E12-19-002 (2019, 2020, and 2021)
[19] T. Gogami et al., AIP Conf. Proc. **2319**, 080019 (2021)
[20] P. Achenbach et al., Hadron Spectroscopy and Structure, 713–717 (2020)
[21] H. Tamura et al., Proposal to J-PARC, P63 (2016)
[22] J. Alcorn et al., Nucl. Instrum. Methods, Phys. Res. Sect. A **21**, 294 (2004)
[23] T. Gogami et al., Nucl. Instrum. Methods, Phys. Res. Sect. A **900**, 69 (2018)
[24] Y. Tsai, Rev. Mod. Phys. **46**, 4 (1974)
[25] L. Tang et al., Phys. Rev. C **90**, 034320 (2014)
[26] T. Gogami et al., Phys. Rev. C **93**, 034314 (2016)
[27] T. Gogami et al., Phys. Rev. C **94**, 021302(R) (2016)
[28] T. Gogami et al., Phys. Rev. C **103**, L041301 (2021)
[29] T. Gogami et al., Nucl. Instrum. Methods, Phys. Res. Sect. A **729**, 816–824 (2013)
[30] Y. Fujii et al., Nucl. Instrum. Methods, Phys. Res. Sect. A **795**, 351–363 (2015)
[31] F. Dohrmann et al., Phys. Rev. Lett. **93**, 242501 (2004)
[32] F. Dohrmann et al., Phys. Rev. C **76**, 054004 (2007)

[33] B. Pandey, L. Tang et al., Phys. Rev. C **105**, L051001 (2022)
[34] K. N. Suzuki, T. Gogami et al., Prog. Theor. Exp. Phys. **2022**, 013D01 (2022)
[35] K. Itabashi et al., Few-Body Systems **63**, 16 (2022)
[36] S. N. Santiesteban et al., Nucl. Instrum. Methods, Phys. Res. Sect. A **940**, 351 (2019)
[37] S. N. Nakamura et al., Proposal to JLab, E12-15-008 (2015)
[38] F. Garibaldi et al., Proposal to JLab, E12-20-013 (2020)
[39] T. Mart et al., Nucl. Phys. A **640**, 235-258 (1998)
[40] T. Mart and B.I.S. van der Ventel, Phys. Rev. C **78**, 014004 (2008)
[41] T. Motoba, JPS Conf. Proc. **17**, 011003 (2017)
[42] T. Toyoda, Master's thesis, Kyoto University (in Japanese) (2021)

Lifetime of the hypertriton

Daniel Gazda[1,*], *Axel* Pérez-Obiol[2], *Avraham* Gal[3], and *Eliahu* Friedman[3]

[1]Nuclear Physics Institute, 25068 Řež, Czech Republic
[2]Barcelona Supercomputing Center, 08034 Barcelona, Spain
[3]Racah Institute of Physics, The Hebrew University, Jerusalem 91904, Israel

Abstract. Conflicting values of the hypertriton lifetime $\tau({}^3_\Lambda\mathrm{H})$ were derived in relativistic heavy ion (RHI) collision experiments over the last decade. A very recent ALICE Collaboration measurement is the only experiment where the reported $\tau({}^3_\Lambda\mathrm{H})$ comes sufficiently close to the free-Λ lifetime τ_Λ, as expected naively for a very weakly bound Λ in ${}^3_\Lambda\mathrm{H}$. We revisited theoretically this ${}^3_\Lambda\mathrm{H}$ lifetime puzzle [1], using ${}^3_\Lambda\mathrm{H}$ and ${}^3\mathrm{He}$ wave functions computed within the ab initio no-core shell model employing interactions derived from chiral effective field theory to calculate the two-body decay rate $\Gamma({}^3_\Lambda\mathrm{H} \to {}^3\mathrm{He} + \pi^-)$. We found significant but opposing contributions arising from ΣNN admixtures in ${}^3_\Lambda\mathrm{H}$ and from $\pi^- - {}^3\mathrm{He}$ final-state interaction. To derive $\tau({}^3_\Lambda\mathrm{H})$, we evaluated the inclusive π^- decay rate $\Gamma_{\pi^-}({}^3_\Lambda\mathrm{H})$ by using the measured branching ratio $\Gamma({}^3_\Lambda\mathrm{H} \to {}^3\mathrm{He} + \pi^-)/\Gamma_{\pi^-}({}^3_\Lambda\mathrm{H})$ and added the π^0 contributions through the $\Delta I = \frac{1}{2}$ rule. The resulting $\tau({}^3_\Lambda\mathrm{H})$ varies strongly with the rather poorly known Λ separation energy $E_{\mathrm{sep}}({}^3_\Lambda\mathrm{H})$ and it is thus possible to associate each one of the distinct RHI $\tau({}^3_\Lambda\mathrm{H})$ measurements with its own underlying value of $E_{\mathrm{sep}}({}^3_\Lambda\mathrm{H})$.

1 Introduction

Hypertriton ${}^3_\Lambda\mathrm{H}$ is the lightest bound hypernucleus with spin-parity $J^\pi = \frac{1}{2}^+$ and isospin $I = 0$. It can be approximated as a bound state of a Λ hyperon and a deuteron (${}^2\mathrm{H}$) with tiny Λ separation energy of $E_{\mathrm{sep}}({}^3_\Lambda\mathrm{H}) = 0.165(44)\,\mathrm{MeV}$ [2]. The lifetime of such a loosely bound system is expected to be comparable to the lifetime of the free Λ, $\tau_\Lambda = 263(2)\,\mathrm{ps}$ [3], which is almost completely (99.7%) governed by its nonleptonic $\Lambda \to N+\pi$ weak decay mode. Yet, the world average of measured ${}^3_\Lambda\mathrm{H}$ lifetime, $\tau({}^3_\Lambda\mathrm{H}) = 223^{+12}_{-11}\,\mathrm{ps}$ [2], is by $\sim$ 20% shorter than τ_Λ. This so-called 'hypertriton lifetime puzzle' has been strengthened recently by the ALICE Collaboration's $\tau({}^3_\Lambda\mathrm{H})$ value [4] which is consistent with τ_Λ and in tension with previous STAR [5] and HypHI [6] Collaborations measurements. The latest STAR Collaboration's measurement [7] does not seem to be conclusive since their reported $\tau({}^3_\Lambda\mathrm{H})$ value is consistent within experimental uncertainties with τ_Λ but its central value is 20% shorter. The measured ${}^3_\Lambda\mathrm{H}$ lifetimes from recent RHI experiments together with latest microscopic calculations are summarized in table 1.

*e-mail: gazda@ujf.cas.cz

Table 1. ${}^{3}_{\Lambda}$H lifetime values (in ps) from recent RHI experiments and theoretical calculations.

Experiment / Theory	$\tau({}^{3}_{\Lambda}\mathrm{H})$	Reference
Exp.	$183^{+42}_{-32} \pm 37$	HypHI [6]
Exp.	$142^{+24}_{-21} \pm 29$	STAR [5]
Exp.	$242^{+34}_{-38} \pm 17$	ALICE [8]
Exp.	$221 \pm 15 \pm 19$	STAR [7]
Exp.	$253 \pm 11 \pm 6$	ALICE [4]
Th.	256	Kamada et al. [9]
Th.	213 ± 5	Gal, Garcilazo [10]
Th.	see table 2	Pérez-Obiol et al. [1]
Th.	$\approx \tau_\Lambda$	Hildenbrand, Hammer [11]

2 Method

2.1 Hypertriton decay

The main channels contributing to the ${}^{3}_{\Lambda}$H decay are the π^- and π^0 mesonic modes

$$ {}^{3}_{\Lambda}\mathrm{H} \to {}^{3}\mathrm{He}+\pi^-,\ {}^{2}\mathrm{H}+p+\pi^-,\ p+p+n+\pi^-, $$
$$ {}^{3}_{\Lambda}\mathrm{H} \to {}^{3}\mathrm{H}+\pi^0,\ {}^{2}\mathrm{H}+n+\pi^0,\ n+n+p+\pi^0 $$

due to weak-interaction $\Lambda \to \pi + N$ and possibly $\Sigma \to \pi + N$ transitions, accompanied by the rare non-mesonic modes ${}^{3}_{\Lambda}\mathrm{H} \to {}^{2}\mathrm{H}+n$, $n+n+p$ due to $\Lambda N \to NN$. In this work, we calculate microscopically the two-body π^- decay rate $\Gamma({}^{3}_{\Lambda}\mathrm{H}\to{}^{3}\mathrm{He}+\pi^-)$ and evaluate the inclusive π^- decay rate Γ_{π^-} by employing the world average experimental branching ratio $R_3 = \Gamma({}^{3}_{\Lambda}\mathrm{H}\to{}^{3}\mathrm{He}+\pi^-)/\Gamma_{\pi^-}({}^{3}_{\Lambda}\mathrm{H}) = 0.35 \pm 0.04$ [12]. To account for the contribution of the π^0 channels, we employ the empirical isospin $\Delta I = 1/2$ rule, which relates the π^- and π^0 rates by $\Gamma_{\pi^-} = 2\,\Gamma_{\pi^0}$. Finally, the non-mesonic $\Lambda N \to NN$ and true-absorption $\pi NN \to NN$ contributions to the decay rate, are incorporated through a reduction of the ${}^{3}_{\Lambda}$H lifetime $\tau_\pi({}^{3}_{\Lambda}\mathrm{H}) = 1/\Gamma_\pi$ by 1.5% and 0.8%, respectively [13–16]. Here, $\Gamma_\pi = 3/2 \cdot \Gamma({}^{3}_{\Lambda}\mathrm{H}\to{}^{3}\mathrm{He}+\pi^-)/R_3$ is the inclusive ${}^{3}_{\Lambda}$H pionic decay rate obtained using the relations above.

2.2 Two-body ${}^{3}_{\Lambda}\mathrm{H} \to {}^{3}\mathrm{He} + \pi^-$ decay rate

We follow reference [9] and write the two-body π^- decay rate of ${}^{3}_{\Lambda}$H as [1]

$$ \Gamma^{{}^3\mathrm{He}} = \frac{3}{4\pi}\frac{M_{{}^3\mathrm{He}}\, q_\pi}{M_{{}^3\mathrm{He}} + \sqrt{m_\pi^2 + q_\pi^2}} \sum_{m_{{}^3_\Lambda\mathrm{H}}} \sum_{m_{{}^3\mathrm{He}}} \left| \langle \Psi_{{}^3\mathrm{He}} \phi_\pi | \hat{O} | \Psi_{{}^3_\Lambda\mathrm{H}} \rangle \right|^2, \tag{1} $$

where $M_{{}^3_\Lambda\mathrm{H}} = 2991\,\mathrm{MeV}$ and $M_{{}^3\mathrm{He}} = 2809\,\mathrm{MeV}$ are the ${}^{3}_{\Lambda}$H and ^{3}He masses, and $q_\pi = (2M_{{}^3\mathrm{He}})^{\frac{1}{2}}[M_{{}^3_\Lambda\mathrm{H}} - (m_\pi^2 + 2M_{{}^3\mathrm{He}}M_{{}^3_\Lambda\mathrm{H}} - M^2_{{}^3\mathrm{He}})^{\frac{1}{2}}]^{\frac{1}{2}} = 114.4\,\mathrm{MeV}$ is the pion momentum, with $m_\pi = 138.04\,\mathrm{MeV}$ the pion mass. The summations run over spin projections $m_{{}^3\mathrm{He}}$ and $m_{{}^3_\Lambda\mathrm{H}}$ of the initial ${}^{3}_{\Lambda}$H and final ^{3}He wave functions Ψ discussed in Sect. 2.3. The ϕ_π is the π^- wave function discussed in Sect. 2.4. The operator $\hat{O}$ due to the weak-interaction $\Lambda, \Sigma \to N+\pi$ transitions

$$ \frac{\hat{O}}{i G_F m_\pi^2} = \sqrt{2}\left(\mathcal{A}_\Lambda + \frac{\mathcal{B}_\Lambda}{2\overline{M}_{\Lambda N}}\vec{\sigma}\cdot\vec{q}\right)\hat{P}_{t_{12}=0} + \frac{1}{\sqrt{2}}\mathcal{A}_{\Sigma^-}\hat{P}_{t_{12}=1} \tag{2} $$

contains contributions from $\Lambda \to \pi^- + p$, together with $\Sigma^- \to n+\pi^-$ and $\Sigma^0 \to p+\pi^-$ due to ΣNN admixtures in the ${}^3_\Lambda$H wave function. The $\vec{\sigma}$ and $\vec{\tau}$ are the spin and isospin Pauli matrices, $G_F m_\pi^2 = 2.21 \times 10^{-7}$, $\vec{q}$ is the pion on-shell momentum, and $\overline{M}_{\Lambda N} = 1027.3$ MeV is the average $\Lambda - N$ mass. The parity-violating (PV) $\mathcal{A}_\Lambda = 1.024$ and parity-conserving (PC) $\mathcal{B}_\Lambda = -9.431$ amplitudes were extracted from the free Λ lifetime $\tau_\Lambda = 263(2)$ ps [3], and the PC/PV decay rates ratio [17]. In equation (2), we neglect the PC Σ amplitudes [18], fix the PV $\Sigma^- \to n+\pi^-$ amplitude $\mathcal{A}_{\Sigma^-} = 1.364$ [18] by $\tau_{\Sigma^-} = 147.9$ ps, and use the χEFT Lagrangian to relate the $\Sigma^0 \to p+\pi^-$ amplitude by $\mathcal{A}_{\Sigma^0} = \frac{1}{\sqrt{2}}\mathcal{A}_{\Sigma^-}$. The projection operators $\hat{P}_{t_{12}}$ select ${}^3_\Lambda$H and ^{3}He wave function components with a specific isospin t_{12} of the NN subcluster. For more details and origin of the numerical factors in equations (1) and (2) see reference [1].

2.3 Nuclear and hypernuclear wave functions

The initial- and final-state ${}^3_\Lambda$H and ^{3}He wave functions in equation (1) are computed within the ab initio no-core shell model (NCSM) approach [19, 20], where nuclei and hypernuclei are described as systems of A nonrelativistic particles interacting through realistic nucleon–nucleon (NN), 3-nucleon (NNN) and hyperon-nucleon (YN) interactions. In this work we employed a version of NCSM formulated in translationally-invariant relative Jacobi-coordinate harmonic oscillator (HO) basis which is suitable for dealing with few-body systems [21, 22]. The many-body wave function is cast as an expansion in a complete set of basis states

$$|\Psi_E^{J^\pi T}\rangle = \sum_{N=0}^{N_{\max}} \sum_\lambda c_{N\lambda}^{J^\pi T} |N\lambda JT\rangle, \quad (3)$$

where the HO states $|N\lambda JT\rangle$ are characterized by the HO frequency $\hbar\omega$ and the expansion is truncated by the maximum number $N_{\max}$ of HO excitations above the lowest configuration allowed by Pauli principle. In equation (3), N is the total number of HO excitations of all particles and $J^\pi T$ are the total angular momentum, parity and isospin. The quantum number λ labels all additional quantum numbers and the sum over N is restricted by parity to an even or odd sequence. The energy eigenstates are obtained by solving the Schrödinger equation

$$\hat{H}\,|\Psi_E^{J^\pi T}\rangle = E(J^\pi T)\,|\Psi_E^{J^\pi T}\rangle \quad (4)$$

with the intrinsic Hamiltonian

$$H = \sum_{i=1}^{A} \frac{\vec{p}_i^{\,2}}{2m_i} + \sum_{1\le i<j}^{A-1} V_{NN,ij} + \sum_{1\le i<j<k}^{A-1} V_{NNN,ijk} + \sum_{i=1}^{A-1} V_{YN,Ai} + \Delta M - H_{\rm CM}. \quad (5)$$

Here, the masses m_i and single-particle momenta $\vec{p}_i$ for $i \le A-1$ correspond to the nucleons and for $i = A$ to hyperons. The $H_{\rm CM}$ is the free center-of-mass (CM) Hamiltonian and the mass term $\Delta M = \sum_{i<A} m_i - M_0$ with M_0 the reference mass of a hypernuclear system containing only nucleons and a Λ hyperon is introduced to account for the mass difference of coupled Λ- and Σ-hypernuclear states.

For the nuclear $V_{NN}+V_{NNN}$ interactions in equation (5) we employ the NNLO$_{\rm sim}$ potentials [23] that are based on chiral effective field theory (χEFT) up to next-to-next-to-leading order (NNLO). The NNLO$_{\rm sim}$ is a family of 42 different interactions where each potential is associated with one of seven different regulator cutoffs $\Lambda_{NN} = 450, 475, \ldots, 575, 600$ MeV and six different maximum scattering energies in the laboratory system $T_{\rm Lab}^{\max} = 125, 158, \ldots, 257, 290$ MeV at which the experimental NN scattering cross sections data base used to constrain the respective interaction was truncated. For the purpose of

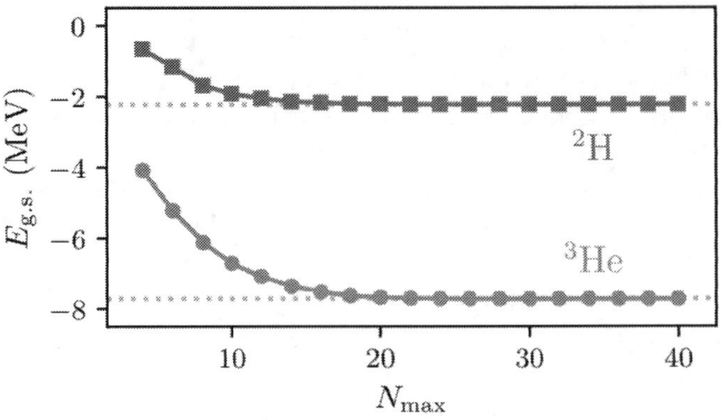

Figure 1. Dependence of ^{2}H and ^{3}He ground-state energies $E_{g.s.}$ on the NCSM model-space truncation N_{max} for fixed value of the HO frequency $\hbar\omega$ = 14 MeV, calculated using the NNLO$_{sim}$(Λ_{NN}= 500 MeV, T_{Lab}^{max}= 290 MeV) interaction. Experimental values are indicated by the gray dotted lines.

this work precise wave functions of ^{3}He and the 'core nucleus' ^{2}H are required. Since certain low-energy properties of ^{2}H and ^{3}He were included in the pool of fit data, their energies are accurately described for all NNLO$_{sim}$ interactions, $E_{^2\text{H},^3\text{He}} = -2.224^{(+0)}_{(-1)}, -7.717^{(+17)}_{(-21)}$ MeV [23]. For NNLO$_{sim}$, the NCSM ground-state (g.s.) energies of ^{2}H and ^{3}He exhibit a good convergence with the NCSM model-space truncation N_{max} as demonstrated in figure 1 for $\hbar\omega$ = 14 MeV using the NNLO$_{sim}$(Λ_{NN}= 500 MeV, T_{Lab}^{max}= 290 MeV) *NN*+*NNN* interaction. The g.s. energies are converged within few keV already at $N_{max} \approx 30$ for a wide range of HO frequencies.

For the *YN* sector (V_{YN} in equation (5)) we use the leading order (LO) coupled-channel Bonn–Jülich SU(3)-based χEFT model [24]. The potential consists of pseudoscalar π, K, η meson exchanges and baryon–baryon contact interaction terms. The interaction is regularized by a smooth momentum cutoff Λ_{YN} ranging from 550 to 700 MeV. Unfortunately, calculations of $^3_\Lambda$H g.s. energy exhibit a slower convergence with the NCSM model-space truncation N_{max}. This can be attributed to the very small binding energy of $^3_\Lambda$H and, consequently, the long tail of the $^3_\Lambda$H wave function in coordinate space. Nevertheless, well-converged results for the $^3_\Lambda$H g.s. energy can be obtained for $N_{max} \approx 70$ [25]. However, as shown in Sect. 3, it is not possible to achieve full convergence for the transition operator matrix element in equation (1) even in the largest feasible NCSM model spaces and it is necessary to extrapolate the finite-space results into infinite model space. We employ the infrared (IR) extrapolation scheme developed for nuclear NCSM in [26] and generalized recently to hypernuclear NCSM in [27]. In this scheme, the truncation of the HO basis in terms of N_{max} and $\hbar\omega$ is translated into the associated infrared (L_{IR}) and ultraviolet (UV) (Λ_{UV}) scales and IR correction formulae can be systematically derived for observables to extrapolate to infinite model space, $L_{IR} \to \infty$. The LO correction for energies and the expected magnitude of subleading corrections σ_{IR} [28] are

$$E^{UV}(L_{IR}) = E^{UV}_\infty + a_0^{UV}\, e^{-2k^{UV}_\infty L_{IR}}, \quad \sigma_{IR} \propto \frac{e^{-2k^{UV}_\infty L_{IR}}}{k^{UV}_\infty L_{IR}}, \tag{6}$$

where $(E^{UV}_\infty, a_0^{UV}, k^{UV}_\infty)$ are parameters determined from fit to the NCSM-calculated energies with weights proportional to the inverse of σ_{IR}. Note that this procedure slightly differs and should improve the one employed in [1]. Additional ultraviolet (UV) corrections to equation (6)

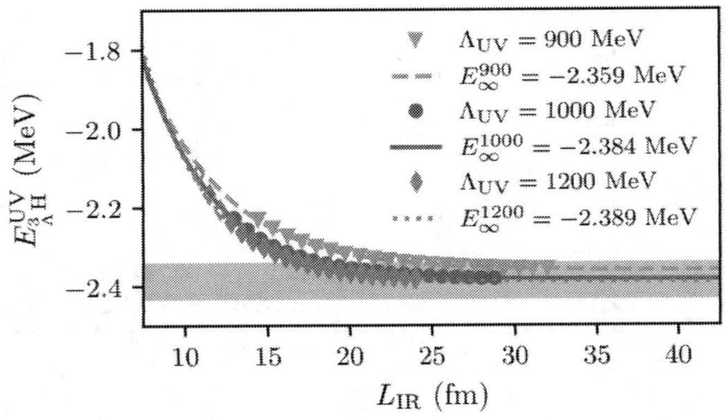

Figure 2. The $^3_\Lambda$H ground-state energies $E^{\rm UV}_{^3_\Lambda{\rm H}}$ as a function of the IR length $L_{\rm IR}$ (up to $N_{\rm max} = 68$) for the NNLO$_{\rm sim}$(Λ_{NN}= 500 MeV, $T^{\rm max}_{\rm Lab}$= 290 MeV) and LO YN(Λ_{YN}= 600 MeV) interactions, together with their extrapolated values for several fixed values of the UV cutoff, $\Lambda_{\rm UV}$ = 900, 1000, 1200 MeV. Only the points marked by filled symbols, corresponding to particle-stable $^3_\Lambda$H configurations, are included in the fits. Experimental value [2] is indicated by the gray shaded area.

can be substantial unless $\Lambda_{\rm UV} \gg \Lambda_{NN}, \Lambda_{YN}$. A large-enough $\Lambda_{\rm UV}$ scale can be identified by performing calculations at a fixed $\Lambda_{\rm UV}$—by choosing appropriate ($N_{\rm max}$, $\hbar\omega$) model-space parameters—and monitoring the UV dependence [28]. We find that $1000 \lesssim \Lambda_{\rm UV} \lesssim 1200$ MeV is sufficient to achieve UV-convergence and to perform IR extrapolations. This is shown in figure 2 where the $^3_\Lambda$H g.s. energy $E^{\rm UV}_{^3_\Lambda{\rm H}}$ is shown as a function of the IR length $L_{\rm IR}$ for several fixed values of the UV scale $900 \leq \Lambda_{\rm UV} \leq 1200$ MeV. The NCSM calculations were performed with model space truncation up to $N_{\rm max} = 68$ using the NNLO$_{\rm sim}$(Λ_{NN}= 500 MeV, $T^{\rm max}_{\rm Lab}$= 290 MeV) NN and LO YN(Λ_{YN}= 600 MeV) interactions. The extrapolated g.s. energies $E^{\rm UV}_\infty$ show only a small dependence on the $\Lambda_{\rm UV}$ scale of the HO basis once $\Lambda_{\rm UV} \gtrsim 1000$ MeV. E.g., for $\Lambda_{\rm UV}$ = 1000(1200) MeV, which translates into $\hbar\omega$ = 7.299(10.510) MeV for $N_{\rm max}$ = 68, the finite-space result is $E^{1000(1200)}_{^3_\Lambda{\rm H}} = -2.3807(-2.3814)$ MeV, while the extrapolated infinite-space result estimated using equation (6) is $E^{1000(1200)}_\infty = -2.384(-2.389)$ MeV.

2.4 Pion wave function

The pion wave function in equation (1) was generated from a standard optical potential constrained by data of pionic atoms across the periodic table [29, 30]. In order to extrapolate from the near-threshold region relevant for pionic atoms to q_π = 114.4 MeV in the $\pi^- - {}^3$He CM system, the potential parameters were adjusted using the πN [31], as well as π–nucleus elastic scattering amplitudes [32, 33].

3 Results

In figure 3, we show the 2-body $^3_\Lambda$H π^- decay rate $\Gamma(^3_\Lambda{\rm H}\to{}^3{\rm He}+\pi^-)$ as a function of the HO IR length scale $L_{\rm IR}$ (open and filled symbols), together with their extrapolations (solid lines), for several values of the HO basis UV cutoff $\Lambda_{\rm UV}$. Marked by the solid symbols are the points included in the fit which correspond to particle-stable $^3_\Lambda$H configurations with $E(^3_\Lambda{\rm H}) < E(^2{\rm H})$. The IR extrapolations were performed by adapting the relations in (6)

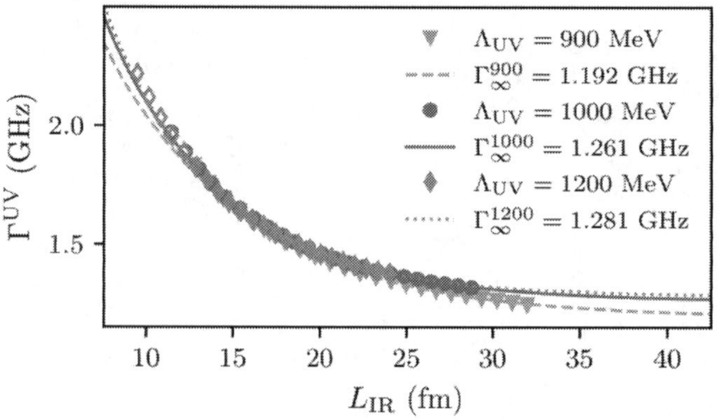

Figure 3. The two-body decay rates $\Gamma(^3_\Lambda\mathrm{H}\to{}^3\mathrm{He}+\pi^-)$ as a function of the IR length L_{IR} (up to $N_{\max}$ = 68) for the NNLO$_{\mathrm{sim}}$(Λ_{NN}= 500 MeV, $T^{\max}_{\mathrm{Lab}}$= 290 MeV) and LO YN(Λ_{YN}= 600 MeV) interactions, together with their extrapolated values for several fixed values of the UV cutoff, Λ_{UV} = 900, 1000, 1200 MeV. The calculated rates include contributions of pion DW as well as ΣNN channels. Marked by filled symbols are the points included in the fits which correspond to particle-stable $^3_\Lambda$H configurations. The ^{3}He wave function was held fixed, calculated using $N_{\max}$ = 36 and $\hbar\omega$ = 14 MeV NCSM model space parameters corresponding to a well-converged configuration.

as $\Gamma^{\mathrm{UV}}(L_{\mathrm{IR}}) = \Gamma^{\mathrm{UV}}_\infty + a^{\mathrm{UV}}_0 \exp(-2k^{\mathrm{UV}}_\infty L_{\mathrm{IR}})$ with $(\Gamma^{\mathrm{UV}}_\infty, a^{\mathrm{UV}}_0, k^{\mathrm{UV}}_\infty)$ the fit parameters. This is motivated by expanding the rate in powers of $[E_{\mathrm{sep}}(^3_\Lambda\mathrm{H})]^{\frac{1}{2}}$, see below. The extrapolated values $\Gamma^{\mathrm{UV}}_\infty$ show only a weak dependence on the UV scale for $\Lambda_{\mathrm{UV}} \gtrsim$ 1000 MeV. The rates in figure 3 are calculated using the NNLO$_{\mathrm{sim}}$(Λ_{NN}= 500 MeV, $T^{\max}_{\mathrm{Lab}}$= 290 MeV) and LO YN(Λ_{YN}= 600 MeV) interactions and include the effects of π^- DW and $\Sigma \to N+\pi$ transitions due to the ΣNN component of the $^3_\Lambda$H wave function. Replacing the π^- PW by $\pi^- - {}^3$He DW in equation (1) leads to increase of $\Gamma(^3_\Lambda\mathrm{H}\to{}^3\mathrm{He}+\pi^-)$ by $\approx$ 15%. On the contrary, including the contributions of $\Sigma \to N+\pi$ transitions was found to decrease $\Gamma(^3_\Lambda\mathrm{H}\to{}^3\mathrm{He}+\pi^-)$ by $\approx$ 10%.

The dependence of the extrapolated Λ separation energies E^{UV}_∞ (figure 2) and 2-body decay rates $\Gamma^{\mathrm{UV}}_\infty$ (figure 3) on the UV scale of the HO basis Λ_{UV} can be exploited to study the relationship of $E_{\mathrm{sep}}(^3_\Lambda\mathrm{H})$ and $\Gamma(^3_\Lambda\mathrm{H}\to{}^3\mathrm{He}+\pi^-)$. While the UV convergence for $\Lambda_{\mathrm{UV}} \lesssim$ 1000 MeV was not fully achieved, the calculations still give meaningful, sufficiently IR-converged results. The missing UV corrections to equation (6) depend only on short-range details of the employed interactions truncated by Λ_{UV}. In table 2, we show the extrapolated Λ separation energies, two-body π^- decay rates, and lifetime values for several HO basis UV scales, $800 \le \Lambda_{\mathrm{UV}} \le 1400$ MeV. The $\tau(^3_\Lambda\mathrm{H})$ values were deduced from $\Gamma(^3_\Lambda\mathrm{H}\to{}^3\mathrm{He}+\pi^-)$ using the procedure detailed in Sect. 2.1, tacitly assuming that the branching ratio R_3 is independent of $E_{\mathrm{sep}}(^3_\Lambda\mathrm{H})$. The Λ separation energy rapidly decreases with reducing the UV scale from Λ_{UV} = 1200 MeV to Λ_{UV} = 800 MeV which lowers the two-body π^- rate and enhances $\tau(^3_\Lambda\mathrm{H})$ due to the vanishing overlap between the wave functions of the loosely bound $^3_\Lambda$H and ^{3}He. The small increase in $E^{\mathrm{UV}}_{\mathrm{sep}}$ for Λ_{UV} = 1400 MeV indicates that higher-order IR corrections to equation (6) become relevant for $\Lambda_{\mathrm{UV}} \gtrsim$ 1400 MeV. The last column of table 2 contains estimated values of $\Gamma(^3_\Lambda\mathrm{H}\to{}^3\mathrm{He}+\pi^-)$ and $\tau(^3_\Lambda\mathrm{H})$ corresponding to the STAR Collaboration's reported value of $E_{\mathrm{sep}}(^3_\Lambda\mathrm{H})$ = 0.41(12)(11) MeV [34]. The estimate was obtained by expanding $\Gamma(^3_\Lambda\mathrm{H}\to{}^3\mathrm{He}+\pi^-)$ in powers of $[E_{\mathrm{sep}}(^3_\Lambda\mathrm{H})]^{\frac{1}{2}}$ as $a\,[E_{\mathrm{sep}}(^3_\Lambda\mathrm{H})]^{\frac{1}{2}} + b\,E_{\mathrm{sep}}(^3_\Lambda\mathrm{H})$ and fixing the two expansion parameters (a, b) by fitting to the $E^{\mathrm{UV}}_{\mathrm{sep}}$ and $\Gamma^{\mathrm{UV}}_\infty$ from table 2 for

Table 2. Extrapolated ${}^3_\Lambda$H Λ separation energies, two-body π^- decay rates, and lifetime values for several HO basis UV scales. The $E^{\rm UV}_{\rm sep} = E^{\rm UV}_{\infty} - E^{\rm UV}_{^2{\rm H}}$ is the extrapolated Λ separation energy calculated using the deuteron energy at the corresponding $\Lambda_{\rm UV}$ scale. The uncertainty in $\tau({}^3_\Lambda{\rm H})$ comes from the experimental uncertainty in R_3. The last column shows extrapolations to the STAR Collaboration [34] reported value of $E_{\rm sep}({}^3_\Lambda{\rm H})$, see text for details.

$\Lambda_{\rm UV}$ (MeV)	800	900	1000	1200	1400	–
$E^{\rm UV}_{\rm sep}$ (keV)	69	135	160	165	162	410
$\Gamma^{\rm UV}_{\infty}$ (GHz)	0.943	1.192	1.261	1.281	1.293	1.548
$\tau({}^3_\Lambda{\rm H})$ (ps)	234 ± 27	190 ± 22	180 ± 21	178 ± 20	176 ± 20	147 ± 17

$800 \le \Lambda_{\rm UV} \le 1200$ MeV. Given the strong dependence of $\tau({}^3_\Lambda{\rm H})$ on $E_{\rm sep}({}^3_\Lambda{\rm H})$ and considering the large experimental uncertainty of $E_{\rm sep}({}^3_\Lambda{\rm H})$, none of the recent RHI reported $\tau({}^3_\Lambda{\rm H})$ values listed in table 1 can be excluded but, rather can be correlated with its own underlying value of $E_{\rm sep}({}^3_\Lambda{\rm H})$.

Aside from the large measurement uncertainties, there are also considerable theoretical uncertainties associated with $E_{\rm sep}({}^3_\Lambda{\rm H})$ and $\tau({}^3_\Lambda{\rm H})$. Note that the values listed in table 2 are calculated using a particular hypernuclear Hamiltonian. Recently, we explored the limits of theoretical precision of $E_{\rm sep}({}^3_\Lambda{\rm H})$ [25, 27] and $\Gamma({}^3_\Lambda{\rm H}\to{}^3{\rm He}+\pi^-)$ due to model uncertainties in *YN* and *NN*+*NNN* interactions. For this purpose we employed 4 versions of the Bonn–Jülich LO *YN* interaction regularized at $\Lambda_{YN} = 550, 600, 650, 700$ MeV, together with the whole family of 42 NNLO$_{\rm sim}$ nuclear interactions (see Sect. 2.3). Fixing $\Lambda_{\rm UV} = 1200$ MeV for IR extrapolations, the combined spread of Λ separation energies for all combinations of $(\Lambda_{NN}, T^{\rm max}_{\rm Lab}, \Lambda_{YN})$ is found to be $90 \lesssim E_{\rm sep}({}^3_\Lambda{\rm H}) \lesssim 190$ keV, while the spread in calculated two-body π^- rates $1.0 \lesssim \Gamma({}^3_\Lambda{\rm H}\to{}^3{\rm He}+\pi^-) \lesssim 1.4$ GHz, which implies $160 \lesssim \tau({}^3_\Lambda{\rm H}) \lesssim 220$ ps for a fixed value of $R_3 = 0.35$. Full details of this study will be presented in our future work.

4 Summary

We performed a new microscopic calculation of the hypertriton π^- two-body decay rate $\Gamma({}^3_\Lambda{\rm H}\to{}^3{\rm He}+\pi^-)$ employing ${}^3_\Lambda$H and ^{3}He three-body wave functions generated by ab initio hypernuclear no-core shell model (Y-NCSM) using realistic *YN* and *NN*+*NNN* interactions derived from χEFT. Using the $\Delta I = \frac{1}{2}$ rule and the measured branching ratio R_3 to include the remaining π^0 and 3- plus 4-body ${}^3_\Lambda$H decay channels, we deduced the ${}^3_\Lambda$H lifetime $\tau({}^3_\Lambda{\rm H})$. The main findings of this study are: (i) Replacing pionic plane wave (PW) by realistic π^-–^{3}He distorted wave (DW) enhances $\Gamma({}^3_\Lambda{\rm H}\to{}^3{\rm He}+\pi^-)$ by ≈ 15%. (ii) The ΣNN admixtures in ${}^3_\Lambda$H reduce the purely ΛNN decay rate by ≈ 10% due to interference effects, despite their ≲ 0.5% contribution to the norm of the ${}^3_\Lambda$H wave function. (iii) The lifetime $\tau({}^3_\Lambda{\rm H})$ varies strongly with the rather poorly known $E_{\rm sep}({}^3_\Lambda{\rm H})$. It is then possible to associate each of the distinct RHI measured $\tau({}^3_\Lambda{\rm H})$ values with its own underlying value of $E_{\rm sep}({}^3_\Lambda{\rm H})$. Remarkably, the most recent ALICE Collaboration's [4] $E_{\rm sep}({}^3_\Lambda{\rm H}) = 72$ keV central value is almost the same as our lowest $E^{\Lambda_{\rm UV}=800}_{\rm sep} = 69$ keV and our lifetime $\tau^{\Lambda_{\rm UV}=800}({}^3_\Lambda{\rm H}) = 234 \pm 27$ ps is then consistent with their reported $\tau({}^3_\Lambda{\rm H}) = 253 \pm 11 \pm 6$ ps. Only future experiments expected at MAMI, JLab, J-PARC, and CERN will hopefully pin down $E_{\rm sep}({}^3_\Lambda{\rm H})$ with a better precision than 50 keV and lead to a resolution of the ${}^3_\Lambda$H lifetime puzzle.

Acknowledgments

We are grateful to Petr Navrátil for helpful advice on extending the nuclear NCSM codes to hypernuclei; to Johann Haidenbauer and Andreas Nogga for providing us with the input LO

Bonn–Jülich *YN* potential; and to Andreas Ekström for providing us the $\mathrm{NNLO_{sim}}$ *NN*+*NNN* interactions used in the present work. The work of D.G. was supported by the Czech Science Foundation GAČR grants 19-19640S, 22-14497S, and by the European Union's Horizon 2020 research and innovation programme under grant agreement No 824093. Some of the computational resources were supplied by the project "e-Infrastruktura CZ" (e-INFRA CZ LM2018140) supported by the Ministry of Education, Youth and Sports of the Czech Republic and IT4Innovations at Czech National Supercomputing Center under project number OPEN-24-21 1892.

References

[1] A. Pérez-Obiol, D. Gazda, E. Friedman, A. Gal, Phys. Lett. B **811**, 135916 (2020)
[2] P. Eckert et al., *Systematic treatment of hypernuclear data and application to the hypertriton*, in *19th International Conference on Hadron Spectroscopy and Structure* (2022)
[3] P.A. Zyla et al. (Particle Data Group), PTEP **2020**, 083C01 (2020)
[4] ALICE Collaboration (2022), arXiv:2209.07360 [nucl-ex]
[5] L. Adamczyk et al. (STAR), Phys. Rev. C **97**, 054909 (2018),
[6] C. Rappold et al., Nucl. Phys. A **913**, 170 (2013)
[7] M. Abdallah et al. (STAR), Phys. Rev. Lett. **128**, 202301 (2022)
[8] S. Acharya et al. (ALICE), Phys. Lett. B **797**, 134905 (2019)
[9] H. Kamada et al., Phys. Rev. C **57**, 1595 (1998)
[10] A. Gal, H. Garcilazo, Phys. Lett. B **791**, 48 (2019)
[11] F. Hildenbrand, H.W. Hammer, Phys. Rev. C **102**, 064002 (2020)
[12] G. Keyes, J. Sacton, J.H. Wickens, M.M. Block, Nucl. Phys. B **67**, 269 (1973)
[13] M. Rayet, R.H. Dalitz, Nuovo Cim. A **46**, 786 (1966)
[14] J. Golak et al., Nucl. Phys. A **631**, 740C (1998)
[15] J. Golak et al., Phys. Rev. C **55**, 2196 (1997), [Erratum: **56**, 2892 (1997)]
[16] A. Pérez-Obiol, D.R. Entem, A. Nogga, J. Phys. Conf. Ser. **1024**, 012033 (2018)
[17] M. Ablikim et al. (BESIII), Nature **606**, 64 (2022)
[18] J.F. Donoghue, E. Golowich, B.R. Holstein, *Dynamics of the standard model*, Vol. 2 (CUP, 2014)
[19] B.R. Barrett, P. Navrátil, J.P. Vary, Prog. Part. Nucl. Phys. **69**, 131 (2013)
[20] P. Navrátil, S. Quaglioni, I. Stetcu, B.R. Barrett, J. Phys. **G36**, 083101 (2009)
[21] D. Gazda, J. Mareš, P. Navrátil, R. Roth, R. Wirth, Few-Body Syst. **55**, 857 (2014)
[22] R. Wirth, D. Gazda, P. Navrátil, R. Roth, Phys. Rev. **C97**, 064315 (2018)
[23] B.D. Carlsson et al., Phys. Rev. **X6**, 011019 (2016)
[24] H. Polinder, J. Haidenbauer, U.G. Meissner, Nucl. Phys. **A779**, 244 (2006)
[25] T.Y. Htun, D. Gazda, C. Forssén, Y. Yan, Few Body Syst. **62**, 94 (2021)
[26] K.A. Wendt, C. Forssén, T. Papenbrock, D. Sääf, Phys. Rev. C **91**, 061301(R) (2015)
[27] D. Gazda, T.Y. Htun, C. Forssén (2022), arXiv:2208.02176 [nucl-th]
[28] C. Forssén et al., Phys. Rev. C **97**, 034328 (2018)
[29] E. Friedman, A. Gal, Phys. Lett. B **792**, 340 (2019),
[30] E. Friedman, A. Gal, Acta Phys. Polon. B **51**, 45 (2020)
[31] R.A. Arndt et al., Phys. Rev. C **74**, 045205 (2006)
[32] E. Friedman et al., Phys. Rev. Lett. **93**, 122302 (2004)
[33] E. Friedman et al., Phys. Rev. C **72**, 034609 (2005)
[34] J. Adam et al. (STAR), Nature Phys. **16**, 409 (2020)

Comparison of ${}^{3}_{\Lambda}$H/${}^{4}_{\Lambda}$H production cross-section via (K^{-}, π^{0}) reaction at J-PARC

Takaya Akaishi[1,*], *Hidemitsu* Asano[2], *Xurong* Chen[3], *Alberto* Clozza[4], *Catalina* Curceanu[4], *Raffaele* Del Grande[4], *Carlo* Guaraldo[4], *Chengdong* Han[3], *Tadashi* Hashimoto[5], *Mihai* Iliescu[4], *Kentaro* Inoue[6], *Shigeru* Ishimoto[7], *Kenta* Itahashi[2], *Masahiko* Iwasaki[2], *Yue* Ma[2], *Rie* Murayama[2], *Marco* Miliucci[4], *Hiroyuki* Noumi[6,7], *Hiroyuki* Ohnishi[8], *Shinji* Okada[9], *Haruhiko* Outa[2], *Kristian* Piscicchia[4,10], *Yuta* Sada[8], *Atsushi* Sakaguchi[14], *Fuminori* Sakuma[2], *Masaharu* Sato[7], *Alessandro* Scordo[4], *Diana* Sirghi[4,11], *Florin* Sirghi[4,11], *Kotaro* Shirotori[6], *Shoji* Suzuki[7], *Kiyoshi* Tanida[5], *Taiga* Toda[1], *Megumu* Tokuda[1], *Masato* Tsuruta[8], *Takumi* Yamaga[2], *Xiaohua* Yuan[3], *Hongqiang* Zhang[12], *Pengming* Zhang[13], and *Yapeng* Zhang[3]

[1]Department of Physics, Osaka University, 560-0043, Toyonaka, Japan
[2]RIKEN Cluster for Pioneering Research, RIKEN, 351-0198, Wako, Japan
[3]Institute of Modern Physics, 730000, Gansu, China
[4]Laboratori Nazionali di Frascati dell' INFN, I-00044, Frascati, Italy
[5]ASRC, Japan Atomic Energy Agency, 319-1195, Tokai, Japan
[6]Research Center for Nuclear Physics (RCNP), Osaka University, 567-0047, Suita, Japan
[7]High Energy Accelerator Research Organization (KEK), 305-0801, Tsukuba, Japan
[8]Research Center for Electron Photon science (ELPH), Tohoku University, 982-0826, Sendai, Japan
[9]Engineering Science Laboratory, Chubu University, 487-8501, Kasugai, Japan
[10]CENTRO FERMI - Museo Storico della Fisica e Centro Studi e Ricerche "Enrico Fermi", 00184 Rome, Italy
[11]Horia Hulubei National Institute of Physics and Nuclear Engineering (IFIN-HH), Magurele, Romania
[12]School of Nuclear Science and Technology, Lanzhou University, 730000, Gansu, China
[13]School of Physics and Astronomy, Sun Yat-sen University, Zhuhai 519082, China
[14]Center for the Study of Higher Education and Global Admissions(CHEGA), Osaka University, 560-0043, Toyonaka, Japan

Abstract.
Recent heavy-ion collision experiments reported a surprisingly short lifetime for the hypertriton, which has been recognized as the hypertriton lifetime puzzle. Our J-PARC E73 experiment contributes to solve this puzzle with an independent experimental method by employing ${}^{3}\mathrm{He}(K^{-}, \pi^{0})\,{}^{3}_{\Lambda}\mathrm{H}$ reaction. In this contribution, we will demonstrate our capability to provide ${}^{3}_{\Lambda}$H binding energy information by deriving the production cross section ratio, $\sigma_{{}^{3}_{\Lambda}\mathrm{H}}/\sigma_{{}^{4}_{\Lambda}\mathrm{H}}$. The production cross section data for ${}^{3}_{\Lambda}$H and ${}^{4}_{\Lambda}$H are already available as the pilot run of E73 experiment and data analysis is in progress.

*e-mail: akaishi@ne.phys.sci.oska-u.ac.jp

1 Introduction

The hypertriton($^3_\Lambda$H) is the lightest hypernucleus consisting of a proton, a neutron, and a Λ hyperon. Its properties, such as lifetime and binding energy, provide important benchmarks for hypernuclear physics. The $^3_\Lambda$H binding energy was measured to be 0.13 ± 0.05 MeV[1] by the old emulsion experiment so that $^3_\Lambda$H has been considered to be a loosely bound system of Λ hyperon and deuteron for a long time. As a very loosely bound system, the $^3_\Lambda$H is expected to possess a similar lifetime as the free Λ hyperon (τ = 263 ps). However, there is inconsistency between the lifetime of $^3_\Lambda$H in heavy-ion collision experiments [2–4], which has been called the "hypertriton lifetime puzzle". Measured $^3_\Lambda$H lifetime is close to the free Λ hyperon with the reporting of very recent heavy-ion collision experiments[5, 6]. In contrast, a value of the binding energy of $^3_\Lambda$H has not yet been determined. Recently, a value of the binding energy of $^3_\Lambda$H of 0.41 ± 0.12 MeV was reported by the STAR Collaboration[7], which was three times larger than the previous known value from emulsion data. On the other hand, preliminary value presented by the ALICE Collaboration confirms that the binding energy is small, 0.05 ± 0.06 MeV[8]. Therefore, new data by using other methods on both lifetime and binding energy are needed.

2 J-PARC E73 experiment

The J-PARC E73 experiment aims to pin down the "hypertriton liftme puzzle"[9]. The lifetime will be obtained from direct measurement in the time domain so that the result is complementary to that from the heavy-ion collision experiments, in which the lifetime is extracted from an indirect measurement for the decay length. The E73 experiment employs the (K^-, π^0) reaction to populate $^3_\Lambda$H and $^4_\Lambda$H as:

$$K^- + {}^{3,4}\mathrm{He} \rightarrow {}^{3,4}_{\Lambda}\mathrm{H} + \pi^0, \quad {}^{3,4}_{\Lambda}\mathrm{H} \rightarrow {}^{3,4}\mathrm{He} + \pi^-$$

This reaction is a novel production method to convert a proton into Λ hyperon. It is tagged by detecting gamma-ray from π^0 meson with a forward calorimeter [10]. The hypernucleus produced in this strangeness exchange reaction is mainly in the ground state. The decayed π^- from the hypernucleus is measured by a cylindrical detector system (CDS) covering the target developed for an experiment to search for kaonic nuclei[11]. The CDS is composed of a solenoid magnet, a drift chamber and timing counters. In the (K^-, π^0) reaction at p_{K^-} = 1.0 GeV/c, recent theoretical work reported that the information on $^3_\Lambda$H binding energy can be extracted from the production cross-section ratio $^3_\Lambda$H/$^4_\Lambda$H[12]. We can obtain production data with the same experimental setup, just by swapping targets.

3 Preliminary result

We have measured the $^3_\Lambda$H and $^4_\Lambda$H production as pilot experiments. The total beam times were 3 days at 51 kW beam power and 4.5 days at 60 kW beam power for the ^{3}He and ^{4}He targets, respectively. Since the hypernucleus decay at rest in our method, it can be identified by detecting the mono-energic π^- from two-body mesonic weak decay. As shown in the fig.1, hypernuclear signals from two-body mesonic weak decay can be clearly seen. This is part of full data sample. The main background in this reaction is the quasi-free Λ and Σ in-flight decays. The background evaluation is underway using Geant4, taking into account Fermi motion and detector resolution. The number of hypernuclear signals is obtained by subtracting the simulated background from the data. We will compare the measured ratio to theoretical calculation and discuss the magnitude of the binding energy of $^3_\Lambda$H.

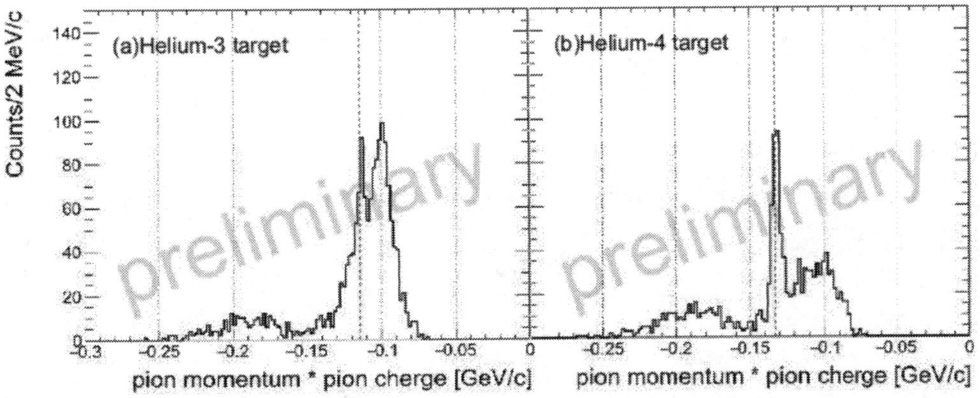

Figure 1. (a)The π^- momentum distribution in helium-3 target. Mono-energic π^- from two-body mesonic weak decay signal is at 114 MeV/c. (b)The π^- momentum distribution in helium-4 target. Mono-energic π^- from two-body mesonic weak decay signal is at 133 MeV/c.

4 Summary and outlook

The E73 experiment has obtained data with ^{3}He and ^{4}He targets to produce $^3_\Lambda$H and $^4_\Lambda$H, respectively. We presented the preliminary result of π^- momentum distribution in the (K^-, π^0) reaction. Mono-momentum π^- events have been clearly observed which come from two body decays of $^3_\Lambda$H and $^4_\Lambda$H. Finalization of the evaluation of the number of hypernuclear signals and systematic error are in progress. We will provide data on binding energy of $^3_\Lambda$H by using the ratio of the production cross-sections $^3_\Lambda$H and $^4_\Lambda$H.

References

[1] D.H. Davis, Nucl. Phys. A 754, 3c-13c (2005)
[2] C. Rappold *et al.* (HypHI Collaboration), Nucl. Phys. A 913, 170-184 (2013)
[3] J. Adam *et al.* (ALICE Collaboration), Phys. Lett. B 754, 360-372 (2016)
[4] L. Adamczyk *et al.* (STAR Collaboration), Phys. Rev. C 97, 054909 (2018)
[5] S.Acharya *et al.* (ALICE Collaboration), Phys. Lett, B 797, 134905 (2019)
[6] M. S. Abdallah *et al.* (STAR Collaboration), Phys. Rev. Lett. 128, 202301 (2022)
[7] J. Adam *et al.* (STAR Collaboration), Nat. Phys. 16, 409 (2020)
[8] P.Fecchio (ALICE Collaboration), Presented at the 19th International Conference on Strangeness in Quark Matter (SQM2021), 2021, https://indi.to/3MvSH
[9] Y. Ma *et al.*, HADRON2021, Rev. Mex. Fis. Suppl. 3, 3, 0308120 (2022)
[10] T. Akaishi *et al.*, PoS PANIC2021, 214 (2022)
[11] K. Agari *et al.*, Prog. Theo. Exp. Phys., 02B011 (2012)
[12] T. Harada, Y. Hirabayashi, Nucl. Phys. A 1015, 122301 (2021)

Single- & double-strangeness hypernuclei up to $A = 8$ within χEFT

Hoai Le[1,*]

[1]IAS-4, IKP-3 and JHCP, Forschungszentrum Jülich, D-52428 Jülich, Germany

Abstract. We investigate S=-1 and -2 hypernuclei with A = 4 – 8 employing the Jacobi-NCSM approach and in combination with baryon-baryon interactions derived within the frame work of chiral effective field theory. The employed interactions are transformed using the similarity renormalization group (SRG) so that the low- and high-momentum states are decoupled, and, thereby, convergence of the binding energies with respect to model space can be significantly speeded up. Such an evolution is however only approximately unitary when the so-called SRG induced higher-body forces are omitted. We first explore the impact of the SRG evolution on the Λ separation energies B_Λ in A = 3 – 5 hypernuclei when only SRG-evolved two-body and when both two- and three-body forces are included. For the latter scenario, we thoroughly study predictions of the two almost phase-equivalent NLO13 and NLO19 YN potentials for A = 4 – 7 hypernuclei. We further explore CSB splittings in the A = 7, 8 multiplets employing the two NLO YN potentials that include also the leading CSB potential in the ΛN channel, whose strength has been fitted to the presently established CSB in A = 4. Finally, we report on our recent study for Ξ hypernuclei based on the ΞN interaction at NLO.

1 Introduction

Experimental information on hyperon-nucleon (YN) and hyperon-hyperon (YY) scattering data is indispensable input for developing realistic YN and YY interactions [1–6]. However, despite significant progress over the last several decades, scattering data in the strangeness S=-1,-2 sectors remain extremely scarce and the situation is not expected to be substantially improved in the near future. Yet, other important source of information is hypernucler spectroscopy [7], in which not only ground-state energies but also energy level splittings in single- and double-strangness systems can be determined with high accuracy. This together with theoretical studies of hypernuclei, especially with microscopic approaches are essential tools to explore the underlying baryon-baryon (BB) interactions and thereby to improve interactions models.

In this study, we investigate single-Λ and Ξ hypernuclear systems with A=4 - 8 based on the "ab initio" Jacobi no-core shell model (NCSM) [8, 9] and in combination with BB interactions derived within chiral effective field theory (EFT). In this approach, potentials are established in terms of an expansion in powers of low momenta in accordance with an appropriate power counting scheme. The long-ranged parts of the interactions are fixed from chiral symmetry, whereas unresolved short-distance dynamics are parameterized in terms of

*e-mail: h.le@fz-juelich.de

the so-called low energy constants (LECs), whose strengths should be determined via a fit to experiments. In the non-strangeness sector, thanks to the wealth of NN scattering data, two- and three-body interactions have been consistently derived up to high order, e.g. N^4LO [10] and N^2LO [11], respectively, see also Ref. [12] for an overview on the current status of the chiral NN and 3N interactions. In the S=-1,-2 sectors, the situation is however less satisfactory due to, as already said, the lack of the appropriate scattering data. Despite of that, YN and YY potentials up to NLO have been derived [2–5] and recently extended to N^2LO [6]. Especially, at NLO, by pursuing different strategies for fixing LECs, two realizations of the YN interaction, NLO13 [2] and NLO19 [3], which yield similarly good description of all the available YN scattering data, could be established. However, as discussed in [3], NLO19 is characterized by a somewhat weaker $\Lambda N - \Sigma N$ transition potential, and as a result, the two potentials yield slightly different predictions for ${}^4_\Lambda$He and for nuclear matter. Also, in [9] we have explored effects of NLO13 and NLO19 on light hypernuclei with A=4 - 7, but only qualitatively. Below we revisit this topic and take a closer look at the separation energies of these systems predicted by the two potentials, and thereby, discuss the possible effect of chiral YNN forces. In addition, we explore charge symmetry breaking (CSB) effects in the A=7, 8 multiplets, employing the NLO13 and NLO19 interactions and with the inclusion of the leading CSB potential derived in Ref. [16]. The latter involves additional two LECs that have been determined via a fit to the separation energies difference ΔB_Λ in the ground and excited states of ${}^4_\Lambda$He and ${}^4_\Lambda$H, with $\Delta B_\Lambda(0^+) = 233 \pm 92$ keV and $\Delta B_\Lambda(1^+) = -83 \pm 94$ [17], respectively. Finally, we discuss our recent work [13], where the possible existence of Ξ hypernuclei up to A=7 for the NLO ΞN potential [5] has been explored.

2 SRG evolution for YN NLO and impact on B_Λ in A=3-5 hypernuclei

BB interactions generally contain short-range and tensor correlations that couple low- and high-momentum states, which make calculations expanded in harmonic oscillator (HO) bases such as the NCSM converge slowly. A very large number of HO functions is required in order to incorporate such strong correlations and therefore well-converged results for systems with $A \geq 5$ are practically computationally inaccessible. So, in order soften the interactions we will continuously evolve the Hamiltonian using the Similarity Renormalization Group (SRG). The latter can be written in term of a flow equation for potentials, which, in three-body YNN space, has following form

$$\frac{dV_s}{ds} = [[T_{\rm rel}, V_s], H_s]; \quad H_s = T_{\rm rel} + V_s + \Delta M = T_{\rm rel} + V^{\rm NN}_{12,s} + V^{\rm YN}_{32,s} + V^{\rm YN}_{31,s} + V^{\rm YNN}_s + \Delta M. \quad (1)$$

Here $T_{\rm rel}$ is the intrinsic kinetic energy, ΔM is the rest mass difference arising due to particle conversions such as $\Lambda N - \Sigma N$, and s is a flow parameter that varies from 0 to ∞ with $s = 0$ meaning no evolution or bare interactions. For characterising SRG-evolved interactions, we will however use the more intuitive λ parameter, $\lambda = (4\mu_N^2/s)^{1/4}$ with μ_N being nucleon reduced mass, that provides a measure of the width of the potentials in momentum space [18]. We solve the flow equation Eq. (1) in momentum space and follow the same approach as done in [18, 19] that explicitly separates the flow equations for two-body NN and YN, and three-body YNN interactions by employing the relation $\frac{dV_s}{ds} = \frac{dV^{\rm NN}_{12,s}}{ds} + \frac{dV^{\rm YN}_{32,s}}{ds} + \frac{dV^{\rm YN}_{31,s}}{ds} + \frac{dV^{\rm YNN}_s}{ds}$. In two-body NN and YN space, the flow equations for $V^{\rm NN}$ and $V^{\rm YN}$ can be written as [9]

$$\frac{dV^{\rm NN}_{12}}{ds} = [[t^{\rm NN}_{12}, V^{\rm NN}_{12}], t^{\rm NN}_{12} + V^{\rm NN}_{12}]$$
$$\frac{dV^{\rm YN}_{32}}{ds} = [[t^{\rm YN}_{32}, V^{\rm YN}_{32}], t^{\rm YN}_{32} + V^{\rm YN}_{32} + \Delta M]; \qquad \frac{dV^{\rm YN}_{31}}{ds} = [[t^{\rm YN}_{31}, V^{\rm YN}_{31}], t^{\rm YN}_{31} + V^{\rm YN}_{31} + \Delta M], \quad (2)$$

hence, the evolution equation for $V^{\rm YNN}$ reads,

$$\frac{dV^{\rm YNN}}{ds} = [[t_{12}^{\rm NN}, V_{12}^{\rm NN}], V_{32}^{\rm YN} + V_{31}^{\rm YN} + V^{\rm YNN}] + [[t_{32}^{\rm YN}, V_{32}^{\rm YN}], V_{12}^{\rm NN} + V_{31}^{\rm YN} + V^{\rm YNN}] \tag{3}$$
$$+[[t_{31}^{\rm YN}, V_{31}^{\rm YN}], V_{32}^{\rm YN} + V_{12}^{\rm NN} + V^{\rm YNN}] + [[T_{\rm rel}, V^{\rm YNN}], T_{\rm rel} + V_{12}^{\rm NN} + V_{32}^{\rm YN} + V_{31}^{\rm YN} + V^{\rm YNN} + \Delta M]$$

Note that now the Kronecker delta functions that appear when embedding two-body interactions in a three-body YNN basis, see the right hand side of Eq. (3), can be evaluated analytically [18, 19]. By projecting Eqs.(2,3) onto the corresponding two- and three-body partial-decomposed momentum-space bases, we obtain a system of coupled ordinary differential equations which will be solved simultaneously employing the ARKsover from SUNDIAL. In the current work, we do not include chiral YNN forces in the evolution, however, it should be clear from Eq. (3) that the SRG evolution will eventually induce three-body YNN interactions even if the initial (bare) potential $V_{s=0}^{\rm YNN}$ is set to zero. Therefore, omitting those induced higher-body forces, the SRG evolution will only be approximately unitary. Indeed, as it can clearly be seen in Fig. 1, the binding energies (square blue symbols) of the A = 3 − 5 hypernuclei, computed for SRG-evolved NN and YN interactions, exhibit a strong dependence on the λ flow parameter. With the inclusion of the induced YNN forces, we practically reproduce the hypertrition binding energy $E(^3_\Lambda\mathrm{H})$ (dashed green line) that was obtained when solving the Faddeev equation for the bare NN and YN potentials, see panel (a) in Fig. 1. Note the large estimated errors for the NCSM results because of the extremely small separation energy of $^3_\Lambda$H. In panel (b), we clearly demonstrate that $E(^3_\Lambda\mathrm{H})$ is indeed independent of the SRG parameter by increasing the strength of the YN interaction, say, by a factor of 1.5, so that the $^3_\Lambda$H hypernucleus is more tightly bound. Similarly, with both SRG-induced YNN and 3N forces included, the SRG-dependence of B_Λ in the A = 4, 5 systems is significantly removed, as clearly shown in panels (c-d). This is an important indication of likely negligibly small contributions of SRG-induced higher-body forces to the separation energies of these systems. However, for a more quantitative estimate of those contributions, well-converged separation energies computed for a wide range of flow parameter are required.

3 Effect of NLO YN interactions on B_Λ in A = 4 − 7 hypernuclei

In our work [9], using only SRG-evolved NN and YN forces, we have qualitatively studied predictions of the two practically phase equivalent NLO13 and NLO19 YN interactions for the separation energies in the A = 4 − 7 systems. Overall, the NLO19 potential, that is characterized by a weaker ΛN − ΣN transition potential [3], yields substantially larger B_Λ in all systems considered except for $^4_\Lambda\mathrm{H}/^4_\Lambda\mathrm{He}(0^+)$, see Fig. 10 in Ref.[9]. As also discussed in that work, such a large discrepancy between the NLO13 and NLO19 predictions could be attributed to possible contributions of chiral YNN forces, which according to the underlying power counting will appear at N^{2}LO [6, 20]. Now, taking into account both chiral and SRG-induced 3N interactions as well as SRG-induced YNN forces, we can study the actual separation energies predicted by the two YN potentials in more detail. These results together with the experimental B_Λ for the A = 4 − 7 hypernuclei are summarized in Table. 1. One sees that NLO13 and NLO19 yield very similar separation energies for the $^4_\Lambda\mathrm{H}(0^+)$ state, with B_Λ = 1.551 ± 0.007 MeV and 1.514 ± 0.007 MeV, respectively, that are by roughly 0.6 MeV smaller than the empirical value of B_Λ = 2.16 ± 0.08 MeV [21]. Likewise, the separation energies for $^4_\Lambda\mathrm{H}(1^+)$ and the ground states in $^5_\Lambda$He and $^7_\Lambda$Li computed for NLO13 are 0.823 ± 0.003 MeV, 2.22 ± 0.06 MeV and 5.28 ± 0.68, respectively. In comparison with the empirical B_Λ of 1.07 ± 0.08 MeV, 3.12 ± 0.02 MeV and 5.58 ± 0.03 MeV [21], respectively, the NLO13 potential clearly underestimates these systems. In contrast, the predictions

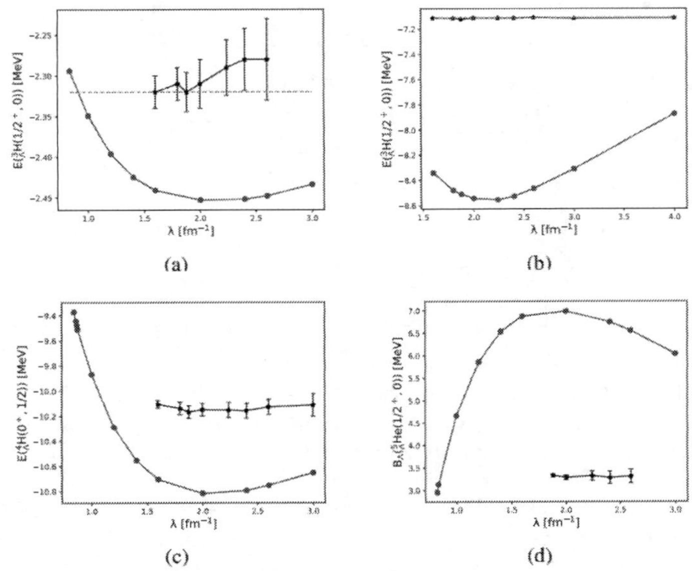

Figure 1. Dependence of the binding (separation) energies in A = 3 – 5 hypernuclei on λ flow parameter. The results (blue circles, panels (a, c,d)) are computed using the SMS N^{4}LO$^+$450 [10] NN potential SRG-evolved to λ_{NN} = 1.6 fm^{-1} and the NLO13(500) YN interaction (with CSB included) evolved to a wide range of flow parameter. In panel (b) the strength of the original NLO13 potential is amplified by a factor of 1.5 so that $^3_\Lambda$H is more tightly bound. Black triangles are obtained with inclusion of SRG-induced YNN (panels (a-b)) and of the 3N forces [11] (panels (c-d)). Dashed green lines are binding energies computed by solving the Faddeev equation for the bare NN and YN interactions.

	$^4_\Lambda$H		$^5_\Lambda$He	$^7_\Lambda$Li
	0$^+$	1$^+$	1/2$^+$	(1/2$^+$,0)
NLO13(500)	1.551 ± 0.007	0.823 ± 0.003	2.22 ± 0.06	5.28 ± 0.68
NLO19(500)	1.514 ± 0.007	1.27 ± 0.009	3.32 ± 0.03	6.04 ± 0.30
Exp [21, 22]	2.16 ± 0.08	1.07 ± 0.08	3.12 ± 0.02	5.85 ± 0.13(10) 5.58 ± 0.03

Table 1: B_Λ for the A = 4 – 7 hypernuclei computed for the YN potentials NLO13(500) and NLO19(500) including the CSB interaction. The SMS N^{4}LO$^+$(450) and N^{2}LO(450) are used to describe the 2N and 3N interactions, respectively. All the employed potentials are evolved to a SRG parameter of λ = 1.88 fm^{-1}. SRG-induced YNN forces are explicitly taken into account. All energies are given in MeV.

of NLO19 for these three states are quite close to experiment, only slightly above. For example, the computed separation energies for $^4_\Lambda$H(1$^+$) and $^5_\Lambda$He, $B_\Lambda(^4_\Lambda\text{H}, 1^+)) = 1.27 \pm 0.009$ and $B_\Lambda(^5_\Lambda\text{He}) = 3.32 \pm 0.03$, exceed the experiment by less than 0.2 MeV. Interestingly, for the ground state of $^7_\Lambda$Li we obtained the value of B_Λ = 6.04 ± 0.30 MeV, that is somewhat larger than the emulsion separation energy, B_Λ = 5.58 ± 0.03 [21], but it is in perfect agreement with the one deduced from counter experiments, with B_Λ = 5.85 ± 0.13(10) [22]. In conclusion, the discrepancies between B_Λ predicted by NLO13 and NLO19, and between the computed B_Λ and experiment could be an indication of the importance of YNN forces for properly describing light hypernuclei within the χ EFT approach.

4 CSB in $A = 7, 8$ isospin multiplets

Being able to accurately compute hypernuclear wave functions for hypernuclei up to the p-shell, we can now employ those wave functions to explore CSB effects in the A = 7, 8 multiplets. Here, we employ the NLO13(500) and NLO19(500) potentials both without and with CSB components [16] as YN interactions, while the SMS N^{4}LO$^+$(450) [10] and N^{2}LO(450)

		ΔT	$\Delta V_{\rm NN}$	$\Delta V_{\rm YN}$			ΔB_Λ
				$^1S_0$	$^3S_1$	total	
${}^7_\Lambda$Be-${}^7_\Lambda$Li*	NLO13	7	-24	-1	0	0	-17
	NLO13-CSB	8	-24	-49	26	-24	-40
	NLO19	6	-40	-1	0	0	-34
	NLO19-CSB	6	-41	-43	42	9	-35
	Hiyama [25]		-70			200	150
	Gal [24]	3	-70			50	-17
	experiment [17]						-100 ± 90
${}^8_\Lambda$Be-${}^8_\Lambda$Li	NLO13	12	8	-2	0	-4	16
	NLO13-CSB	12	7	100	56	159	178
	NLO19	7	-11	-1	0	-2	-6
	NLO19-CSB	6	-11	62	79	147	143
	Hiyama [25]		40				160
	Gal [24]	11	-81			119	49
	experiment [26]						40 ± 60

Table 2: Contributions to CSB in the A = 7 and 8 isospin multiplets, based on the YN potentials NLO13(500) and NLO19(500) (including 3N and SRG-induced YNN forces). The results are for the original potentials (i.e. without CSB force) and for the scenario CSB1 [16]. The estimated uncertainties for ΔB_Λ in A = 7, 8 multiplets are 30 and 50 keV, respectively (see the text). All energies are in keV.

[11] are used for NN and 3N interactions, respectively. All the employed potentials are evolved to a SRG parameter of λ = 1.88 fm^{-1}. In addition, both YNN and 3N forces that are generated during the evolution are explicitly included in all calculations. Let us further remark that, as one can also see in Table. 1, due to the limited accessible model space, B_Λ for systems with A $\geq$ 7 are afflicted with rather large uncertainties, which even exceed the experimentally found CSB splittings in those systems. Therefore, it will be more meaningful to extract the CSB results for the A = 7, 8 multiplets perturbatively based on the ${}^7_\Lambda$Li*(T = 1) and ${}^8_\Lambda$Li(T = 1/2) hypernuclear wave functions and the ^{6}Li*(T = 1) and ^{7}Li(T = 1/2) nuclear wave functions, respectively. Note that this perturbative treatment has been also applied to study CSB effects in the A = 4 systems in Ref. [16]. The employed wave functions are computed for the largest computationally accessible model space, namely $\mathcal{N}_{\rm max}$ = 10 (9) for A = 7 (8), and for the optimal HO frequency $\omega_{\rm opt}$ = 16 MeV that yields the lowest binding energies. For estimating the uncertainty, we will consider only two-body interactions and perform calculations using the model space $\mathcal{N} = \mathcal{N}_{\rm max}$, $\mathcal{N}_{\rm max} + 2$, and at the HO frequencies $\omega = \omega_{\rm opt} \pm 2$ MeV. The variation of the CSB results compared to the values obtained at $\mathcal{N}_{\rm max}$ and $\omega_{\rm opt}$ will be assigned as the uncertainty of our CSB estimations which amount to 30 (50) keV for A = 7 (8). Note that the latter is of the same magnitude as that for the uncertainties of the CSB results deduced from experiment, therefore, a direct comparison of our CSB predictions with experiment is possible.

Details of the computed CSB splittings for the mirror hypernuclei (${}^7_\Lambda$Be-${}^7_\Lambda$Li*) and (${}^8_\Lambda$Be-${}^8_\Lambda$Li) are summarized in Table. 2, where individual contributions of the kinetic energy ΔT, of the NN interaction $\Delta V_{\rm NN}$ and of the YN potential $\Delta V_{\rm YN}$ to the total CSB are also provided. Note that the 3N and SRG-induced YNN forces contributions are insignificantly small, they are therefore omitted from the table. Furthermore, in order to understand the relation between the CSB effect of the employed YN potentials and the $\Delta V_{\rm YN}$ for larger systems, which as shown in [16] contributes dominantly to the CSB splitting in the A = 4 doublets, we also estimate the $^1S_0$ and $^3S_1$ partial-wave contributions to $\Delta V_{\rm YN}$ separately.

One clearly sees that, despite the substantial discrepancies in their predicted B_Λ, the two YN potentials yield comparable CSB results in the A = 7 mutiplets. For both cases, with and without the CSB interactions included, the overall CSB effects are small and consistent with the experiment both in magnitude and in sign. It is further observed that the $^1S_0$ and $^3S_1$ states contribute with opposite signs to the total $\Delta V_{\rm YN}$, like that for the 1^+ states in the A = 4 systems [23], which in turns leads to a small total CSB in the A = 7 isotriplet. Surprisingly, our prediction for ${}^7_\Lambda$Be-${}^7_\Lambda$Li* for the original NLO13 potential (i.e. without CSB components), with ΔB_Λ(NLO13) = -17 keV, is identical to the value estimated by Gal [24], for which a simple shell-model approach in combination with effective potentials were employed. However, the individual contributions ΔT, $\Delta V_{\rm NN}$, and $\Delta V_{\rm YN}$ differ substantially. For example, the original NLO13 yields a vanishing $\Delta V_{\rm YN}$, whereas in Gal's calculation this contribution amounts to 50 keV. That value is also of opposite sign and larger than our estimated $\Delta V_{\rm YN}$ for the actual CSB potential. In addition, our NCSM estimate for $\Delta V_{\rm NN}$ (that also includes the Coulomb effect) is visibly smaller than the corresponding one used in Gal's calculation. Note that the latter is not computed consistently but rather taken from a cluster-model study of Hiyama *et al.* [25].

CSB results for ${}^{8}_{\Lambda}$Be-${}^{8}_{\Lambda}$Li are listed at the lower end of Table 2. Without the CSB components, the two NLO13 and NLO19 potentials predict a negligibly small CSB for ${}^{8}_{\Lambda}$Be-${}^{8}_{\Lambda}$Li, with $\Delta B_{\Lambda} = 16 \pm 50$ keV and -6 ± 50 keV, respectively, which are well in line with the separation energies difference of 40 ± 60 keV [26] deduced from emulsion data. A similarly small ΔB_{Λ} was also predicted by Gal in [24]. However, in contrast to the rather small $\Delta V_{\rm NN}$ contribution in our calculation, e.g. $\Delta V_{\rm NN} = -11$ keV for NLO19, Gal assigned a significantly larger value to $\Delta V_{\rm NN}$, with $\Delta V_{\rm NN} = -81$ keV that was taken from a shell model calculation by Millener [24]. Clearly, the individual contributions to ΔB_{Λ} are rather model-dependent, hence, it is very important that they are computed consistently. Interestingly, with the CSB potentials included, both NLO13 and NLO19 yield rather sizable CSB results, $\Delta B_{\Lambda} = 177 \pm 50$ keV and 143 ± 50 keV, respectively. In this case, the ${}^{1}S_{0}$ and ${}^{3}S_{1}$ partial-wave contributions are also large, and importantly, of the same sign, adding up to a pronounced total CSB. We found that this exactly resembles the situation for the 0^{+} states of the A = 4 mirror hypernuclei [23]. One could therefore conclude that a fairly large splitting in the A = $4(0^{+})$ states, as presently suggested by experiments [17], implies automatically a likewise significant CSB splitting in ${}^{8}_{\Lambda}$Be–${}^{8}_{\Lambda}$Li. Interestingly, the predictions of NLO13 and NLO19 with CSB interaction are comparable to the value of $\Delta B_{\Lambda} = 160$ keV obtained in a three-body cluster calculation by Hiyama *et al.* [25]. However, it should be stressed that the phenomenological CSB YN interaction employed in that work was fitted to an outdated CSB splittings for the A = 4 doublets, namely $\Delta B_{\Lambda}(0^{+}) = 350 \pm 60$ keV and $\Delta B_{\Lambda}(1^{+}) = 240 \pm 60$ keV. Also it was estimated there that the Coulomb interaction contributes about -80 keV to ΔB_{Λ}, which is substantially larger than our estimate of roughly 10 keV.

5 Results for $A = 4 - 7$ Ξ hypernuclei

In this section we discuss the results for $A = 4 - 7$ Ξ hypernuclei based on our recent study in Ref. [13]. Here all the calculations have been computed using the SMS N^{4}LO$^+$(450) NN [10] and NLO(500) ΞN [5] interactions. The original ΞN potential includes coupling to all other baryon-baryon channels in the $S = -2$ sector such as $\Lambda\Lambda$, $\Lambda\Sigma$ and $\Sigma\Sigma$. However, in practice, the coupling $\Lambda\Lambda - \Xi$N was omitted here and its contribution is effectively incorporated by appropriately re-adjusting the strength of the $V_{\Xi N-\Xi N}$ potential [13]. This facilitates a proper convergence of the energy calculations to the lowest lying Ξ state. Furthermore, the employed NN and ΞN potentials are evolved to a SRG flow parameter of $\lambda_{\rm NN} = \lambda_{\Xi \rm N} = 1.6$ fm^{-1}. For simplicity, the electromagnetic NN interaction [27] as well as the Coulomb point-like interaction between Ξ^{-} and a proton are only included after the evolution. And, SRG-induced higher-body forces are not included in that study. As a result, we observe a moderate dependence of the Ξ separation energies B_{Ξ} on the $\lambda_{\Xi \rm N}$ parameter [13]. However, the overall SRG dependence of B_{Ξ} is relatively small as compared to the actual separation energies, and therefore, it should not affect our conclusion on the existence of the found bound state [13].

The computed separation energies B_{Ξ} of $A = 4 - 7$ Ξ hypernuclei together with their perturbatively estimated decay width Γ are provided in Table 3 . It is estimated that the $\Xi^{-}p$ Coulomb interaction contributes only moderately to the total binding energies of the NNNΞ, ${}^{5}_{\Xi}$H and ${}^{7}_{\Xi}$H systems, about 0.2, 0.6, and 0.4 MeV, respectively. Thus, the binding of these systems should be predominantly due to the strong ΞN interaction. Overall, we

	B_Ξ [MeV]	Γ [MeV]
${}^{4}_{\Xi}\mathrm{H}(1^+,0)$	0.48 ± 0.01	0.74
${}^{4}_{\Xi}\mathrm{n}(0^+,1)$	0.71 ± 0.08	0.2
${}^{4}_{\Xi}\mathrm{n}(1^+,1)$	0.64 ± 0.11	0.01
${}^{4}_{\Xi}\mathrm{H}(0^+,0)$	-	-
${}^{5}_{\Xi}\mathrm{H}(\frac{1}{2}^+,\frac{1}{2})$	2.16 ± 0.10	0.19
${}^{7}_{\Xi}\mathrm{H}(\frac{1}{2}^+,\frac{3}{2})$	3.50 ± 0.39	0.2

Table 3: Ξ separation energies B_Ξ and estimated decay widths Γ for $A = 4-7$ Ξ hypernuclei. All calculations are based on the NLO(500) YY-ΞN interaction and the SMS N^{4}LO+(450) NN potential. Both interactions are SRG-evolved to a flow parameter of $\lambda_{NN} = \lambda_{\Xi N} = 1.6$ fm^{-1}. The values of B_Ξ in NNNΞ, ${}^{5}_{\Xi}$H and ${}^{7}_{\Xi}$H are measured with respect to the binding energies of the core nuclei ^{3}H, ^{4}He and ^{6}He, respectively.

found three weakly bound states $(1^+,0)$, $(0^+,1)$ and $(1^+,1)$ in NNNΞ, with quite similar B_Ξ's but substantially different decay widths. Interestingly, our result for the NNN$\Xi(1^+,0)$ state, $B_\Xi = 0.48 \pm 0.01$ MeV, is close to the value of 0.36 ± 0.16 MeV obtained with the HAL QCD potential [28]. The latter however does not support the binding of the two $(0^+,1)$ and $(1^+,1)$ states. In addition, there are substantial differences between our separation energies B_Ξ(NNNΞ) and the results for the ESC08c potential [29] also reported in [28]. Note that this version of the Nijmegen potential predicts an unrealistically large strength for the ΞN interaction in the $^{33}S_1$ state that even results in a bound ΞN state.

Our computed separation energy and decay width for ${}^{5}_{\Xi}$H are $B_\Xi = 2.16 \pm 0.1$ MeV and $\Gamma = 0.19$ MeV. Surprisingly, these values agree roughly with the estimations by Myint and Akaishi [30] of 1.7 MeV and 0.2 MeV, respectively. However, in contrast to our finding that ${}^{5}_{\Xi}$H is bound primarily due to the strong ΞN interaction, the authors in [30] estimated that most of the 1.7 MeV binding energy comes from the ^{4}He-Ξ^- Coulomb interaction. A quite similar separation energy, $B_\Xi({}^{5}_{\Xi}\mathrm{H}) = 2.0$ MeV, is also reported in the work by Friedman and Gal, where an optical potential is employed [31]. But, as discussed in [13], such an agreement could be more or less accidental.

For the ${}^{7}_{\Xi}$H $(\frac{1}{2}^+,\frac{3}{2})$ system, the chiral ΞN potential yields a separation energy of $B_\Xi = 3.50 \pm 0.39$ MeV and a narrow decay width, with $\Gamma = 0.2$ MeV. The computed B_Ξ is only slightly larger than the value of 3.15 MeV reported by Fujioka *et al.* [32] when the HAL QCD potentials is employed in combination with a four-body ($\alpha nn\Xi$) cluster model [33]. With an older $S = -2$ potential from the Nijmegen group, the authors however obtained somewhat smaller binding energy, namely $B_\Xi = 1.8$ MeV [33]. Let us finally remark that in Ref. [13] we have also explored the relation between properties of the employed chiral ΞN interaction and the binding of the A = 4 − 7 Ξ systems. We shown that the ΞN attraction in the $^{33}S_1$ state is essential for the binding of all the systems considered, which contributes to more than 50% of the ΞN potential expectation value $\langle V_{\Xi N}\rangle$.

6 Conclusion

In this work, we throughly studied single- and double-strangeness hypernuclei with A = 4 − 8 based on the Jacobi NCSM and in combination with the NN (3N), YN and ΞN interactions consistently derived within the chiral EFT approach. In order to speed up the convergence of the NCSM with respect to model space, all the employed two-body and 3N potentials are transformed using an SRG evolution. We explicitly showed that, with the inclusion of the SRG-induced three-body forces (3N and YNN), the strong dependence of B_Λ on SRG flow parameter is removed for ${}^{3}_{\Lambda}$H and significantly reduced for the A = 4, 5 systems. After that, we studied the predictions of the two practically phase-equivalent NLO13(500) and NLO19(500) YN interactions for the separation energies in the A = 4 − 7 hypernuclei. Both potentials underestimate the ground state of ${}^{4}_{\Lambda}\mathrm{H}(0^+)$. The NLO19 potential yields B_Λ for

${}^{4}_{\Lambda}$H(1^+) and the ground states in ${}^{5}_{\Lambda}$He and ${}^{7}_{\Lambda}$Li that are comparable with the experiments (only slightly above), whereas NLO13 substantially underestimates these systems. We also investigated CSB splittings in the A = 7, 8 multiplets using the two NLO YN potentials that also include the leading CSB interaction in the ΛN channel derived in Ref. [16]. Overall, the predicted CSB for the A = 7 systems are small and agree with experiment while the results for the A = 8 doublet are somewhat larger than the available empirical value deduced based on emulsion data. Finally, we discussed our recents results for Ξ hypernuclei with A = 4 − 7 based on the NLO(500) ΞN potential. We found three loosely bound states (1^+, 0), (0^+, 1) and (1^+, 1) for the NNNΞ system. The ${}^{5}_{\Xi}$H and ${}^{7}_{\Xi}$H ($\frac{1}{2}^+$, $\frac{3}{2}$) hypernuclei are more tightly bound. The decay widths of these Ξ bound states are expected to be very small which could allow them to be observed in experiment.

Acknowledgements: I am very grateful to Ulf-G. Meißner, Johann Haidenbauer and Andreas Nogga for the fruitful collaboration, for proofreading and providing valuable suggestions to improve the manuscript. Numerical calculations have been performed on JURECA and JURECA booster of the JSC, Jülich, Germany.

References

[1] Th. A. Rijken, V. G. J. Stoks, and Y. Yamamoto, Phys. Rev. C **59**, 21 (1999)
[2] J. Haidenbauer *et al.* , Nucl. Phys. A **915**, 24 (2013)
[3] J. Haidenbauer, U.-G. Meißner, and A. Nogga, Eur. Phys. J. A **56**, 91 (2020)
[4] J. Haidenbauer, U.-G. Meißner , and S. Petschauer, Nucl. Phys. **A954**, 273 (2015)
[5] J. Haidenbauer and U.-G. Meißner, Eur. Phys. J. A **55**, 23 (2019)
[6] J. Haidenbauer, Contribution to these proceedings
[7] O. Hashimoto, H. Tamura, Prog. Part. Nucl. Phys. **57** (2005,2006)
[8] S. Liebig, U.-G. Meißner, and A. Nogga, Eur. Phys. J. A **52**, 103 (2016)
[9] H. Le J. Haidenbauer, U.-G. Meißner, and A. Nogga, Eur. Phys. J. A **56**, 301 (2020)
[10] P. Reinert, H. Krebs, and E. Epelbaum, Eur. Phys. J. A **54**, 86 (2018)
[11] P. Maris *et al.* , Phys. Rev. C **103**, 054001 (2021)
[12] E. Epelbaum, Contribution to these proceedings
[13] H. Le, J. Haidenbauer, U.-G. Meißner, and A. Nogga, Eur. Phys. J. A **57**, 339 (2021)
[14] H. Le, J. Haidenbauer, U.-G. Meißner, and A. Nogga, Phys. Lett. B **801**, 135189 (2020)
[15] H. Le, J. Haidenbauer, U.-G. Meißner, and A. Nogga, Eur. Phys. J. A **57**, 217 (2021)
[16] J. Haidenbauer and U.-G. Meißner, and A. Nogga, Few Body Syst. **62(4)**:105, 2021
[17] E. Botta, AIP Conf. Proc., **2130(1)** 030003, (2019)
[18] S. K. Bogner, R. J. Furnstahl, and R. J. Perry, Phys. Rev. C **75**, 061001 (2007)
[19] K. Hebeler Phys. Rev. C **85**, 061001 (2012)
[20] S. Petschauer, N. Kaiser, J. Haidenbauer, and U.-G. Meißner, Phys. Rev. C (2016)
[21] M. Juric *et al.* Nucl. Phys. B **52**, 1 -30 (1973)
[22] M. Agnello *et al.* Phys. Lett. B **698**, 219-225 (2012)
[23] H. Le, J. Haidenbauer, U.-G. Meißner, and A. Nogga (in preparation)
[24] A. Gal, Phys. Lett. B **744** 352 (2015)
[25] E. Hiyama *et al.* , Phys. Rev. C **80** 054321 (2009)
[26] E. Botta, T. Bressani, and A. Feliciello, Nucl. Phys. A **960**, 165 (2010)
[27] R. B. Wiringa, V. G. J. Stoks, and R. Schiavilla, Phys. Rev. C **51** 38 (1995)
[28] E. Hiyama *et al.* , Phys. Rev. Lett. **124** 092501 (2020)
[29] M. M. Nagels and Th. A. Rijken and Y. Yamamoto, arXiv: 1504.02634 (2015)
[30] K. S. Myint and Y. Akaishi, Prog. Theor. Phys. Suppl. **117** 251 (1994)
[31] E. Friedman and A. Gal, Phys. Let B **820** 136555 (2021)
[32] H. Fujioka *et al.* , Few Body Syst. **62** 47 (2021)
[33] E. Hiyama *et al.* Phys. Rev. C **78** 054316 (2008)

Baryonic effective field theory for light hypernuclei

Nir Barnea[1,*], *Betzalel* Bazak[1], *Lorenzo* Contessi[2], *Avraham* Gal[1], *Jiří* Mareš[3], and *Martin* Schäfer[1]

[1]The Racah Institute of Physics,The Hebrew University, Jerusalem 9190401, Israel
[2]Université Paris-Saclay, CNRS-IN2P3, IJCLab, 91405 Orsay, France
[3]Nuclear Physics Institute of the Czech Academy of Sciences, 25068 Řež, Czech Republic

Abstract. Light hypernuclei containing one or two Λ baryons are the subject of an ongoing experimental campaign aiming to study the spectrum of these systems, as well as the 2 and 3-body interaction between Λ hyperons and nucleons. Here we shortly review the theoretical study of these systems within the framework of baryonic effective field theory.

1 Introduction

Baryonic effective field theory (BEFT) describing the dynamics of hypernuclei composed of protons (p), neutrons (n), and Λ-particles as the only degrees of freedom, has proven itself over the last few years as an useful tool to study bound and resonance states in light single- and double-Λ, hypernuclear systems. Constrained to reproduce the available low energy data, BEFT solves the longstanding overbinding problem of the $^{5}_{\Lambda}$He hypernucleus [1, 2], and predicts the existence of bound double-Λ hypernuclei still under debate. Its application to study the continuum spectrum of hypernuclear trios reveals the existence of a virtual state in the Λ-neutron-proton (Λnp) $J^\pi = \frac{3}{2}^+$ channel, leading to cross-section enhancement near the Λ-deuteron threshold. For the Λnn $J^\pi = \frac{1}{2}^+$ channel it predicts a resonance state, depending, however, on the value of the Λ-nucleon scattering length. Recently, BEFT was also applied to study the $^{4}_{\Lambda}$H-$^{4}_{\Lambda}$He charge symmetry breaking, yielding an estimate for the $\Lambda - \Sigma^0$ admixture amplitude $\mathcal{A}_{I=1} \approx 1.5\%$ in agreement with the value deduced by Dalitz and von-Hippel from the baryon octet mass [3].

In this article we shortly review the foundation of BEFT, the determination of its low energy constants (LECs) using the available experimental data, and its application to the study of light single- and double-Λ hypernuclear bound and resonance states, as presented in [1–6].

2 Baryonic EFT

BEFT is a nuclear effective field theory (EFT), see *e.g.* [7], having baryons as the only degrees of freedom. As such, BEFT provides a low energy realization of QCD, the fundamental theory of the strong interactions. The details of the QCD dynamics are encoded in the BEFT interaction strengths, the LECs. Being a low energy theory, the physical observables

*e-mail: nir@phys.huji.ac.il

in BEFT are calculated as expansions in Q/M_h, with Q being the typical momentum scale of the physics we want to describe, and M_h the scale where the EFT breaks down.

Here we shall limit our attention to the three lightest baryons - namely the neutron, the proton and the Λ-particle - and discuss the appropriate BEFT. Denoting by $N = (p, n)$ the nucleon field and by Λ the Λ-particle field, the BEFT Lagrangian takes the form

$$\mathcal{L} = N^\dagger(i\partial_0 + \frac{\nabla^2}{2M_N})N + \Lambda^\dagger(i\partial_0 + \frac{\nabla^2}{2M_\Lambda})\Lambda + \mathcal{L}_{2B} + \mathcal{L}_{3B} + \ldots, \quad (1)$$

with M_N the nucleon mass, M_Λ the Λ mass, and $\mathcal{L}_{2B}$, $\mathcal{L}_{3B}$ the two and three-baryon interaction Lagrangians. Naively, the various terms appearing in the Lagrangian are ordered by powers of (Q/M_h). However this naive ordering is modified to ensure that the EFT is renormalizable in the sense that at each order in the expansion the sensitivity to unaccounted short-range physics is small, of the order $O(Q/M_h)$.

Ignoring spin and flavor, the 2-body Lagrangian $\mathcal{L}_{2B}$ is nothing but a low energy expansion of the 2-body interaction, equivalent to the effective range expansion [7]. In momentum space, the resulting s-wave $2B$ potential can be written as

$$\begin{aligned} V_{\mathrm{N^nLO}} =& C_0 + C_2\left(p^2 + p'^2\right) \\ &+ C_4\left(p^4 + p'^4\right) + C_{22}p^2p'^2 + \cdots, \end{aligned} \quad (2)$$

where $\boldsymbol{p}$ is the incoming relative momentum and $\boldsymbol{p}'$ the outgoing momentum. To avoid the UV divergences embedded in these contact terms the interaction must be regularized.

Behind its apparent simplicity BEFT exhibits some non-trivial features, such as (a) The Thomas collapse, compelling the promotion of a 3-body contact term to leading-order (LO) [8], (b) The appearance of a 4-body force at next-to-leading-order (NLO) [9], and (c) The Wigner bound which limits the possible values of the effective range and forces a perturbative treatment of all but the LO terms [10, 11]. These features have led to the power counting scheme depicted in Fig. 1 [7]. The accuracy of the BEFT expansion depends on the ratio between the physical scale Q and the breakup scale M_h. For bound and resonance states we can take Q to be the binding momentum. For non-strange nuclei the breakup scale of BEFT is the pion mass m_π. Thus, considering the binding energy of light nuclei

$$\left(\frac{Q}{M_h}\right) = \frac{\sqrt{2B_N M_N}}{m_\pi} \approx 0.3 - 0.8$$

when the large range reflects the strong variation in the single nucleun binding energy B_N between the deutron and ^{4}He. For ΛN interaction the one-pion exchange diagram is forbidden, and therefore we may assume the breaking scale to be $M_h \approx 2m_\pi$. This leads to the expansion parameter

$$\left(\frac{Q}{M_h}\right) = \frac{\sqrt{2B_\Lambda M_\Lambda}}{2m_\pi} \approx 0.05 - 0.3\,.$$

In view of these expansion parameters, we expect that already al leading order (LO) BEFT would give a very reasonable predictions for the Λ particle binding in the nucleus. It should be noted however, that in practice non-strange BEFT seems to work better than indicated by the large value of the nuclear expansion parameter. It predicts the binding energy of ^{4}He with an accuracy of $\approx 10\%$.

3 The low energy constants

The BEFT of single- and double-Λ hypernuclei contains, at leading order, 10 LECs, see Fig. 2. In the non-strange sector (strangeness = 0) there are 3 LECs - 2 two-body LECs

 https://doi.org/10.1051/epjconf/202227101005

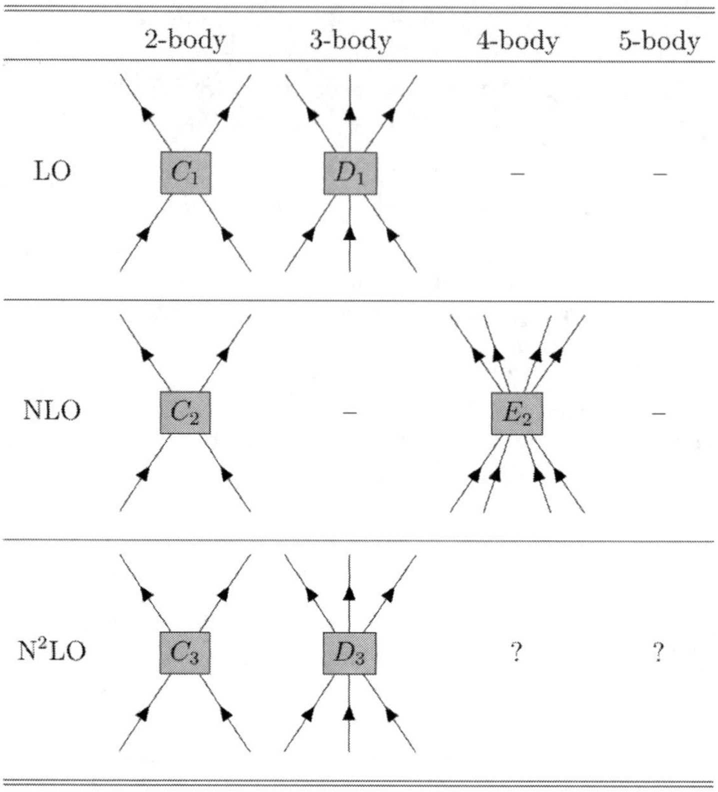

Figure 1. Diagrammatic representation of the multi baryon interaction in BEFT. The order n of each term is indicated by N^{n}LO. The letters in each diagram stand for the corresponding LECs.

which can be fitted either to the NN scattering lengths or to the np isospin triplet $I = 1$ scattering length and to the deutron binding energy, and 1 three-body term which can be fitted to the triton binding energy.

In the single-Λ sector (strangeness =-1) there are 5 LECs - 2 two-body LECs which can be associated with the spin-singlet $S = 0$ and spin-triplet $S = 1$ ΛN scatterering lengths, and 3 three-body LECs. The problem here is the lack of experimental data. As there is no available low energy ΛN scattering data, the ΛN LECs can be fitted to theoreticaly extracted ΛN scattering lengths rather than directly to experimental data, see *e.g.* [2]. In the three-body sector the situation is more severe as there is only one ΛNN bound state, no scattering data, and we have 3 LECs. The solution suggested in [2] was to fit 2 LECs to the 2 available $^4_\Lambda$H states.

In the double-Λ sector (strangeness =-2) there are 2 LECs - 1 two-body LEC which can be associated with the spin-singlet $S = 0$ $\Lambda\Lambda$ scatterering length, and one three-body $\Lambda\Lambda N$ LEC. Here the two-body LEC can be fitted again to theoretical estimates of the $\Lambda\Lambda$ scattering length. We note that the avaialable theoretical value is consistent with the values extracted from heavy ion femtoscopy experiments and from LQCD calculations [12]. The three-body $\Lambda\Lambda N$ LEC can be fitted to the $^{\ 6}_{\Lambda\Lambda}$He Nagara event [5].

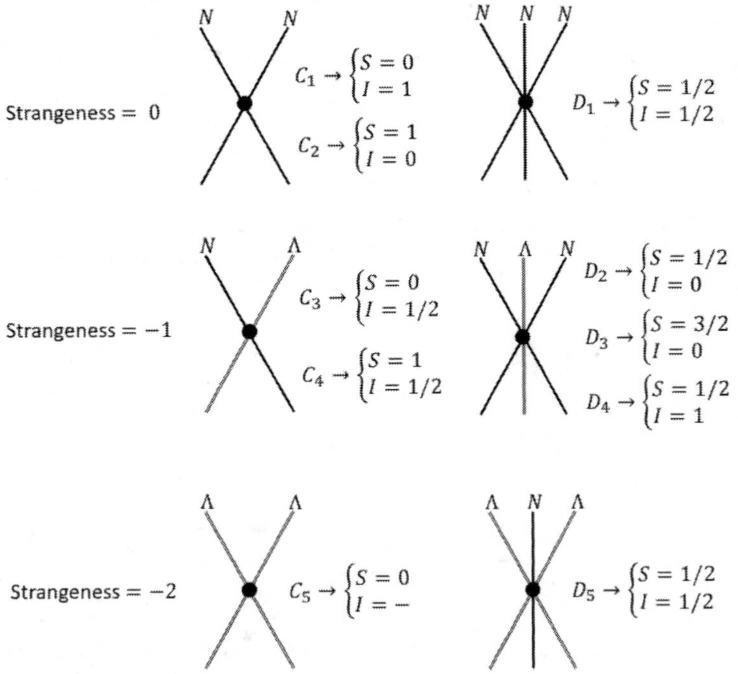

Figure 2. Diagrammatic representation of 2,3-baryon intearactions apearing at LO in single- and double-Λ BEFT. Here C_i stand for the two-body LECs, and D_j for the three-body LECs. S is the spin, and I the isospin. Figure taken from [5].

Inspecting Fig. 1 and Fig. 2 we can conclude that, at NLO, BEFT of single and double Λ hypernuclei contains 5 two-body LECs, each corresponding to the same quantum numbers as the LO two-body terms, no new three-body terms, and 5 four-body LECs. The first four-body LEC is associated with the nuclear - zero strangeness - sector, the next 2 four-body LECs are associated with the $S = 0, 1$ ${}^4_\Lambda$H states, and the last 2 are associated with the $S = 0, 1$ ${}^4_{\Lambda\Lambda}$H states. The lack of the needed experimental 3,4-body data hinders the development of BEFT at NLO.

4 Light single- and double-Λ hypernuclear systems

Table 1 summarizes the available low energy data used to fit the Λ LECs used for BEFT calculations in [1, 2, 5, 6]. The table also presents the main predictions obtained in these works regarding the properties of the light single- and double-Λ hypernuclei bound and resonance states.

It is worth noting that constrained to reproduce the available low energy data, BEFT reproduce within error bars the ${}^5_\Lambda$He hypernucleus [1, 2], and predicts the existence of a bound ${}^5_{\Lambda\Lambda}$H system. In the application of BEFT to study the continuum spectrum of hypernuclear trios, it was found that the excited $J^\pi = \frac{3}{2}^+$ hypertriton state ${}^3_\Lambda\mathrm{H}^{J=3/2}$ is a virtual state. For the

Table 1. Compilation of the experimental data reproduced or predicted by the BEFT. Here, a_J is the scattering length with J being the channel's total angular momentum, E is the state's energy, B_Λ the Λ particle separation energy and $B_{\Lambda\Lambda}$ the double Λ separation energy. All energies are given in MeV, scattering lengths are given in fm. Observables used to fit the LECs are denoted by (LEC).

System	observable	BEFT	Exp./Theo. extraction
ΛN	a_0 (LEC)	-2.35 ± 0.55	-2.35 ± 0.55 [13–16]
	a_1 (LEC)	-1.60 ± 0.1	-1.60 ± 0.1 [13–16]
$\Lambda\Lambda$	a_0 (LEC)	-1.2 ± 0.7	-1.2 ± 0.7 [17–19]
${}^3_\Lambda$H	B_Λ (LEC)	0.13	0.13± 0.05 [20]
Λnn	Re(E)	0.0±0.2 [6]	?
	Im(E)	-0.75± 0.15 [6]	?
${}^3_\Lambda$H$^{J=3/2}$	E (virtual)	-0.012 ± 0.08 [6]	?
Λ-deuteron	$a_{3/2}$	-16 ± 9 [6]	?
${}^4_\Lambda$H$_0$	B_Λ (LEC)	2.16	2.16± 0.08 [21]
${}^4_\Lambda$H$_1$	B_Λ (LEC)	1.07	1.07± 0.08 [22]
${}^5_\Lambda$He	B_Λ	3.01± 0.10 [1]	3.12± 0.02 [20]
${}^5_{\Lambda\Lambda}$H	B_Λ	$1.14^{+0.44}_{-0.26}$ [5]	?
${}^6_{\Lambda\Lambda}$He	$B_{\Lambda\Lambda}$ (LEC)	6.91	6.91 ± 0.17 [23]

Λnn $J^\pi = \frac{1}{2}^+$ state it was found that the pole is very close to the real axis, and that its exact position depends on the value of the ΛN scattering lengths. Therefore at this point we cannot conclude weather it is a physical resonance state. We note that while at the moment there is no available experimental data regarding these continuum states, both systems are currently a subject of an ongoing experimental study.

Taking into account the Coulomb corrections the charge symmetry breaking (CSB) between the mirror nuclei ^{3}H and ^{3}He is about 70 keV, or about 1% of its binding energy. In comparison the CSB between the $S = 0$ and $S = 1$ states in the Λ hypernuclei ${}^4_\Lambda$H-${}^4_\Lambda$He is about 300 keV, i.e. $\Delta E_{CSB}/\Delta E_{01} \approx 22\%$. The most plausible source for this large CSB is the $\Lambda - \Sigma^0$ admixture as suggested by Gal [24], generalizing the work of Dalitz and von-Hippel [25]. Building upon this hypothesis, BEFT was applied to study the ${}^4_\Lambda$H- ${}^4_\Lambda$He CSB in [3]. Utilizing SU$_3$ flavor symmetry BEFT was used to estimate the $\Lambda - \Sigma^0$ admixture amplitude $\mathcal{A}_{I=1} \approx 1.5\%$. The obtained result was found to be in agreement with the value deduced by Dalitz and von-Hippel from the baryon octet mass [25].

5 Summary

In this article we presented a short review of BEFT and its application to the study of light Λ hypernuclei. At leading order BEFT seems to be working rather well, reproducing the binding energy of ${}^5_\Lambda$He, and the charge symmetry breaking in the mirror hypernuclei ${}^4_\Lambda$H-${}^4_\Lambda$He. Going beyond LO seems to be a difficult step in view of the appearance of 4-body terms in Lagrangian and the shortage of experimental data to fit the LECs.

Acknowledgment The work of MS and NB was supported by the Pazy Foundation and by the Israel Science Foundation grant 1086/21. Furthermore, the work of AG and NB is part of a project funded by the European Union's Horizon 2020 research and innovation programme under grant agreement No. 824093.

References

[1] L. Contessi, N. Barnea, A. Gal, Phys. Rev. Lett. **121**, 102502 (2018)
[2] L. Contessi, N. Barnea, A. Gal, AIP Conf. Proc. **2130**, 040012 (2019)
[3] M. Schäfer, N. Barnea, A. Gal Phys. Rev. C **106**, L031001 (2022)
[4] M. Schäfer, B. Bazak, N. Barnea, A. Gal, J. Mareš, Phys.Rev.C **105**, 015202 (2022)
[5] L. Contessi, M. Schäfer, N. Barnea, A. Gal, J. Mareš, Phys. Lett. B **797**, 134893 (2019)
[6] M. Schäfer, B. Bazak, N. Barnea, J. Mareš, Phys. Lett. B **808** 135614 (2020)
[7] H.-W. Hammer, S. König, U. van Kolck, Rev. Mod. Phys. **92**, 025004 (2020)
[8] P. F. Bedaque, H.-W. Hammer, U. van Kolck, Phys. Rev. Lett. **82**, 463 (1999)
[9] B. Bazak, J. Kirscher, S. König, M. P. Valderrama, N. Barnea, and U. van Kolck, Phys. Rev. Lett. **122**, 143001 (2019)
[10] D. R. Phillips and T. D. Cohen, Phys. Lett. B **390**, 7 (1997)
[11] D. R. Phillips, S. R. Beane, and T. D. Cohen, Ann. Phys. **263**, 255 (1998)
[12] Y. Kamiya et al., Phys. Rev. C **105**, 014915 (2022).
[13] G. Alexander, U. Karshon, A. Shapira, et al. Phys. Rev. **173**, 1452 (1968)
[14] T.A. Rijken, V.J.G Stoks, Y. Yamamoto, Phys. Rev. C **59**, 21 (1999)
[15] H. Polinder, J. Haidenbauer, U.-G. Meißner, Nucl. Phys. A **779**, 244 (2006)
[16] J. Haidenbauer, S. Petschauer, N. Kaiser, U.-G. Meißner, A. Nogga, W. Weise, Nucl. Phys. A **915**, 24 (2013)
[17] K. Morita, T. Furumoto, A. Ohnishi, Phys. Rev. C 91 (2015) 024916; A. Ohnishi, K. Morita, K. Miyahara, T. Hyodo, Nucl. Phys. A **954**, 294 (2016)
[18] L. Adamczyk, et al. (STAR Collaboration), Phys. Rev. Lett. **114**, 022301 (2015)
[19] S. Acharya, et al. (ATLAS Collaboration), Phys. Rev. C **99**, 024001 (2019)
[20] D. H. Davis, Nucl. Phys. A **754**, 3c (2005)
[21] A. Esser et al. (MAMI A1 Collaboration), Phys. Rev. Lett. **114** 232501 (2015) ; F. Schulz et al. (MAMI A1 Collaboration), Nucl. Phys. A **954**, 149 (2016)
[22] T.O. Yamamoto et al. (J-PARC E13 Collaboration), Phys. Rev. Lett. **115**, 222501 (2015)
[23] H. Takahashi, et al., Phys. Rev. Lett. **87**, 212502 (2001)
[24] A. Gal, Phys. Lett. B **744**, 352 (2015)
[25] R.H. Dalitz and F. von Hippel, Phys. Lett **10**, 153 (1964)

Commissioning of the hypertriton binding energy measurement at MAMI

Philipp Eckert[1,*], *Patrick* Achenbach[1,**], *Takeru* Akiyama[2], *Jinhui* Chen[4,5], *Michael O.* Distler[1], *Anselm* Esser[1], *Julian* Geratz[1], *Christian* Helmel[1], *Michael* Hoek[1], *Kosuke* Itabashi[2], *Masashi* Kaneta[2], *Ryoko* Kino[2], *Pascal* Klag[1], *Yu-Gang* Ma[4,5], *Harald* Merkel[1], *Masaya* Mizuno[2], *Ulrich* Müller[1], *Sho* Nagao[2], *Satoshi N.* Nakamura[6,2], *Yuki R.* Nakamura[2], *Kazuki* Okuyama[2], *Josef* Pochodzalla[1,3], *Björn Sören* Schlimme[1], *Concettina* Sfienti[1], *Tian-hao* Shao[1,4], *Marcell* Steinen[3], *Koga* Tachibana[2], *Michaela* Thiel[1], *Yuichi* Toyama[2], and *Keita* Uehara[2] for the A1 Collaboration

[1]Institute for Nuclear Physics, Johannes Gutenberg University Mainz, 55099 Mainz, Germany
[2]Graduate School of Science, Tohoku University, Sendai, Miyagi 980-8578, Japan
[3]Helmholtz Institute Mainz, GSI Helmholtzzentrum für Schwerionenforschung, Darmstadt, Johannes Gutenberg University Mainz, 55099 Mainz, Germany
[4]Key Laboratory of Nuclear Physics and Ion-beam Application (MOE), Institute of Modern Physics, Fudan University, Shanghai 200433, China
[5]Shanghai Research Center for Theoretical Nuclear Physics, NSFC and Fudan University, Shanghai 200438, China
[6]Department of Physics, Graduate School of Science, The University of Tokyo, Tokyo 113-0033, Japan

Abstract. A high-precision hypernuclear experiment has been commissioned at the Mainz Microtron (MAMI) to determine the hypertriton Λ binding energy via decay-pion spectroscopy. The method has been successfully pioneered with ${}^{4}_{\Lambda}$H studies in the last decade. The experiment makes use of a novel high luminosity lithium target with a length of 45 mm while being only 0.75 mm thick to keep momentum smearing of the decay pions low. The target-to-beam alignment as well as the observation of the deposited heat is achieved with a newly developed thermal imaging system. Together with a precise beam energy determination via the undulator light interference method a recalibration of the magnetic spectrometers will be done to obtain a statistical and systematic error of about 20 keV. The experiment started in the summer of 2022 and initial optimization studies for luminosity and data quality are presented.

1 Λ binding energies of hydrogen hypernuclei

The hypertriton, ${}^{3}_{\Lambda}$H, is the simplest known hypernucleus and therefore a benchmark nucleus for hypernuclear structure calculations. It plays a fundamental role in strangeness nuclear physics, comparable to the deuteron in non-strange nuclear physics. Its Λ binding energy is of high importance for understanding the Λ–N interaction and can be used to constrain state-of-the-art calculations which describe the internal structure of ${}^{3}_{\Lambda}$H .

*e-mail: eckert@uni-mainz.de
**Now at Thomas Jefferson National Accelerator Facility (JLab), Newport News, Virginia 23606, USA

For almost 50 years, the most precise binding energy value is given by $B_\Lambda = 130 \pm 50\,\mathrm{keV}$, averaged and compiled by Jurič et al. from emulsion experiments [1, 2]. Just recently, two new values became available, one by the STAR Collaboration [3], $B_\Lambda = 406 \pm 120\,(\mathrm{stat.}) \pm 110\,(\mathrm{syst.})\,\mathrm{keV}$, and one by the ALICE Collaboration [4], $B_\Lambda = 72 \pm 63\,(\mathrm{stat.}) \pm 36\,(\mathrm{syst.})\,\mathrm{keV}$.

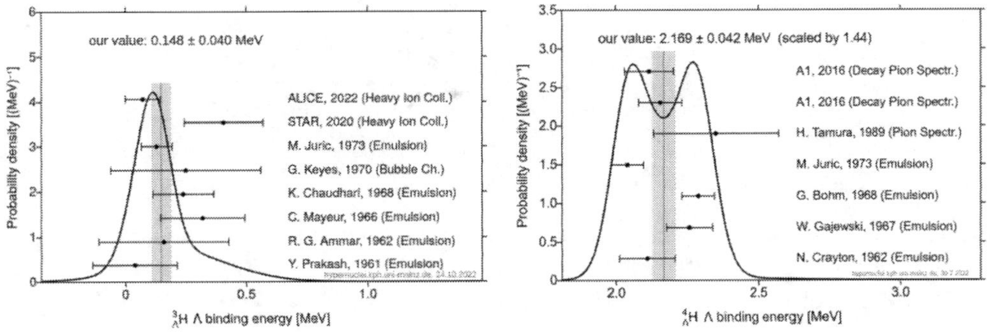

Figure 1. Ideograms of the world data sets on Λ binding energy measurements for ${}^{3}_{\Lambda}\mathrm{H}$ (left) and ${}^{4}_{\Lambda}\mathrm{H}$ (right) [5]. The error bars include statistical and systematic uncertainties. In case of ${}^{4}_{\Lambda}\mathrm{H}$, the data have been conservatively handled by computing a scaling factor $S = \sqrt{\chi^2/ndf}$ and multiplying it with the error. Our averages are $B_\Lambda({}^{3}_{\Lambda}\mathrm{H}) = 0.148 \pm 0.040\,\mathrm{MeV}$ and $B_\Lambda({}^{4}_{\Lambda}\mathrm{H}) = 2.169 \pm 0.042\,\mathrm{MeV}$.

Remarkably, the STAR value is about 8 times larger than the preliminary one from ALICE and they differ by two standard deviations. The STAR value also seems to be in tension with the emulsion value. This data situation along with other earlier measurements is visualized in an ideogram in the left panel of Fig. 1, obtained from the Chart of Hypernuclides [5]. An average value of $\overline{B_\Lambda} = 148 \pm 40\,\mathrm{keV}$ is computed with a relative error of still more than 25 %. It is to be noted, that in many cases the emulsion data values are missing a systematic error. So for these values, a systematic error of $\pm 40\,\mathrm{keV}$, given in Davis' reviews [6, 7], has been included in these averages. This is the so-far best estimate for the accuracy of these emulsion values.

In the right panel of Fig. 1 the ideogram for ${}^{4}_{\Lambda}\mathrm{H}$ is shown, including two values from previous decay-pion experiments by the A1 Collaboration [8, 9] and several emulsion values [1, 2]. Correlated errors in the A1 data were treated explicitly adopting the PDG procedure [10]. The resulting probability density distribution has a structure with two maxima and a $\chi^2/ndf = 12.5/6$ greater than 1, but still with a p value of 0.05, suggesting that the errors of one or more contributing measurements are underestimated. Not knowing which of the errors are underestimated, for averaging they are all multiplied by the same factor $\sqrt{\chi^2/ndf}$, again analogous to the PDG [11]. The resulting relative error of the average is $\sim 20\,\%$.

As the data situation for these light hypernuclei is clearly not satisfactory, a new high-precision experiment via decay-pion spectroscopy has been commissioned at MAMI with the goal to reach a 20 keV total error in the binding energy [12, 13]. Within the running experiment, both, ${}^{3}_{\Lambda}\mathrm{H}$ and ${}^{4}_{\Lambda}\mathrm{H}$, are expected to be observed.

2 Decay-pion spectroscopy at MAMI

2.1 Experimental technique

In general, hypernuclei can be produced from a target nucleus with a high-energy electron beam. The underlying process is the strangeness electroproduction, converting a proton to a

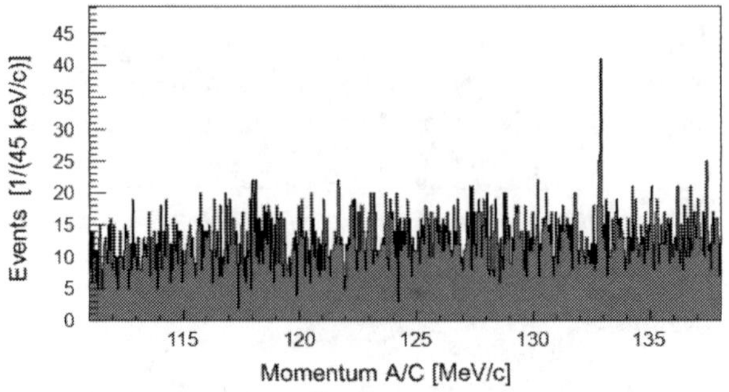

Figure 2. Momentum spectrum for strangeness tagged pions from the 2014 measurement at MAMI [14]. Mono-energetic decay-pions of ${}^{4}_{\Lambda}$H were observed at ~ 133 MeV/c. A signal from two-body decays of stopped ${}^{3}_{\Lambda}$H was not found at the expected momentum of ~ 114 MeV/c.

$\Lambda - K^+$ pair. If this Λ stays bound to the nucleus, a highly excited hypernucleus is formed. This will eventually de-excitate, possibly by the emission of one or more other nucleons, so that finally a smaller hypernuclear isotope in its ground state remains. Due to the comparably long Λ lifetime of $\tau_\Lambda = 263$ ps, the hypernucleus is expected to survive the production process as well as the loss of recoil momentum inside the target material, giving the opportunity to observe the decay of the hypernucleus at rest. In this experiment we are looking for the charged pionic two-body decay, which is mono-energetic. Hence, the mass of the hypernucleus $m_{\text{hyp}} = \sqrt{m^2_{\text{nucl}} + p^2_\pi} + \sqrt{m^2_\pi + p^2_\pi}$ can be determined by measuring the momentum of the decay-pions p_π with the magnetic spectrometers. The pion and nucleus masses are precisely known.

The probability for the formation of a hypernucleus is low compared to background events generated by other processes. In the experiment the Kaos spectrometer is used to detect the kaons in coincidence with the decay pions in the two spectrometers A and C. In that way "strangeness tagged" events can be identified. This method was successfully used at MAMI within the last decade [8, 14]. In that time, beryllium targets of thicknesses between 23 and 47 mg/cm^2 were used. Fig. 2 shows a representative momentum spectrum, where decay pions of ${}^{4}_{\Lambda}$H are clearly visible at around 133 MeV/c, resulting in a binding energy of $B_\Lambda = 2.157 \pm 0.005\,(\text{stat.}) \pm 0.077\,(\text{syst.})$ MeV. The systematic error was strongly dominating because the spectrometer calibration was limited by the available MAMI beam energy measurement with an accuracy of 160 keV. However, the beam spread and instability are known to be much smaller. Consequently, an improved beam energy measurement can reduce the error of the experiment. This will be made possible by the novel undulator light interference method with an accuracy of 18 keV [15].

2.2 Novel high-luminosity lithium target

Lithium is expected to have a higher ${}^{3}_{\Lambda}$H yield compared to beryllium, since it offers fewer fragmentation possibilities into other hypernuclei. Also, due to its low density of 0.534 g/cm^3, it allows for a completely new target geometry, shown in Fig. 3. Here, a length of 45 mm lithium will be traversed by the electron beam resulting in a total thickness of 2.5 g/cm^2. This is around 100 times thicker than the previous target to maximize the production rate of hypernuclei while minimizing the accidental background by radiation contamination, since the beam current can be reduced. At the same time, the lithium sheet

Figure 3. High-luminosity lithium target used for the hypertriton experiment. Left: The lithium sheet has a thickness of 2.5 g/cm^2 in beam direction. In the left part of the setup, pipes for the cooling liquid can be seen for the optimal removal of heat from the lithium sheet. Parts of the copper frame are painted with infrared absorbing black paint to prevent undesired reflections in the camera pictures. Right: The copper frame mounted inside the target chamber, additional targets on top. Below the frame, a rotary and a linear stepper motor are located.

is 750 μm thin perpendicular to the beam direction to minimize the amount of material to be traversed by the exiting decay pions.

Metallic lithium is challenging to handle due to its low melting point of 180°C and its high reactivity even with ambient air. To keep it below melting temperature during the experiment, it is mounted inside a water-cooled copper frame. Regarding the reactivity, lithium can be handled inside an argon atmosphere. However, the spatial alignment of the target inside the target chamber cannot be done with the common theodolite techniques, since the contact time with air would be too long. Therefore, the target holder is mounted onto a rotary and a linear motor to move the target into the correct position during the experiment. To observe the temperature of the lithium while being warmed up by the electron beam, a thermal camera surveillance was installed onto the target chamber. This serves the two purposes of measuring the target temperature as well as extracting information about the alignment from the heat distribution. Examples of this can be seen in Fig. 4. On the left, the target is rotated against the electron beam, so that it is only partially warmed up. On the right side, the full amount of lithium is traversed by the electrons. These pictures were taken with a beam intensity of 2 μA and the resulting temperature was around 28°C, equivalent to 23°C above the cooling water temperature.

Also the spectrometer rates give a good indication on how the target is aligned. In another study the linear as well as the rotary position of the target was varied with the step motors and the resulting rates were observed. By that, also the ideal target position can be found, see Fig. 5. In both cases, a clear peak can be seen.

3 Hypertriton experiment commissioning in 2022

With the commissioning of the complete spectrometer setup consisting of KAOS and the A1 spectrometers A and C, roughly 100 h decay pion spectroscopy data could already be taken. The used beam energy and current are 1508 MeV/c and 500 nA, respectively. The data taking rate was at around 60 Hz. Initially it was planned to use larger beam currents of up to 5 μA, but it was observed that this exceeds the maximum particle rate KAOS is capable of. One reason is that at around 1.6 μA the KAOS trigger reaches a maximal rate of 620 kHz, seen on the left panel of Fig. 6. For higher currents the rate even dropped due to overflow issues. The other

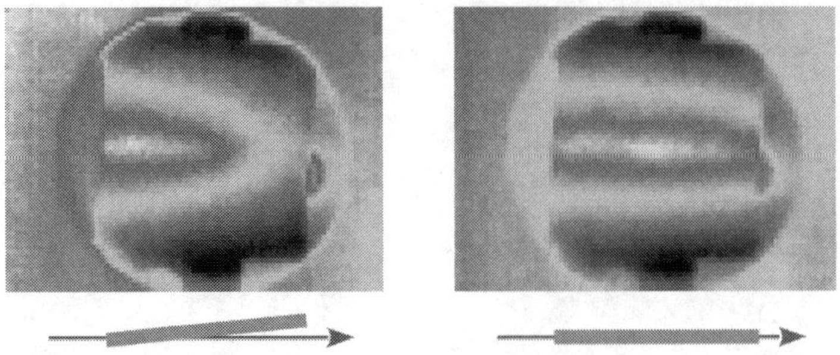

Figure 4. Thermal images of the lithium target during recent tests with a 1.5 GeV electron beam at MAMI. The orientation of the target with respect to the beam is schematically depicted at the bottom. Left: Target and beam misaligned, the lithium was only partially heated. Right: Correct alignment, the heat is equally distributed. In both cases a beam intensity of $2\,\mu$A was used and the maximum temperature of 28°C hardly exceeded room temperature. The red spot on the right side of the target results from a flange of the vacuum system in the background.

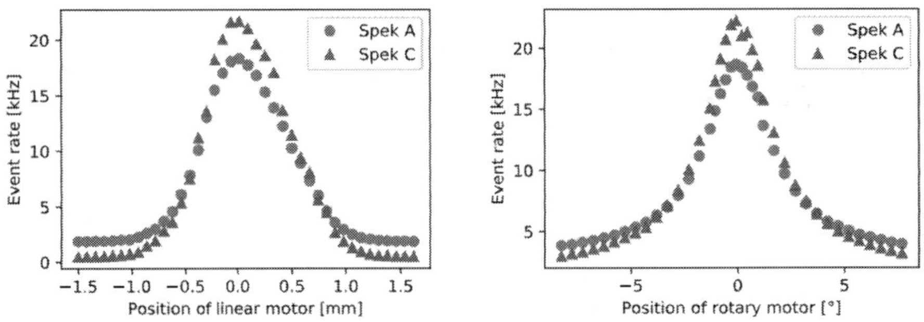

Figure 5. Spectrometer rates in relation to the target position. Left: the linear position of the target parallel to the beam was changed. Right: the rotary position was changed. In both cases, 855 MeV electrons with $3\,\mu$A beam current were used.

even more limiting reason is the quality of the recorded KAOS data, which decreases due to growing ambiguities in the reconstruction of multiple particle tracks inside the spectrometer. A measure for this is the amount of reconstructed true $\pi^+ - \pi^-$ coincidences per time between KAOS and either spectrometer A or C. As one can see in the right panel of Fig. 6, this rate initially increased linearly for low beam currents, but reached a maximum at around 600 nA. After that, the rate dropped rapidly, until at 1000 nA almost no coincidence events were reconstructed anymore. While 600 nA seemed to offer the highest rate compared to other currents, 500 nA yielded a 50% better signal to noise ratio. Therefore, it was decided to use 500 nA for the first data taking period.

Facing the second data taking period in September 2022, adjustments at the lead absorber inside KAOS will be made. Its purpose is to suppress the huge background by positrons, while still enabling kaons to enter the scintillators [16]. Right now, it is varying in thickness, being 100 mm for the low momentum range and 120 mm for the high range. Since especially at low momenta the event rate turned out to be the highest, the absorber thickness will be evened out to be 120 mm everywhere. By that, KAOS will be able to handle higher beam currents.

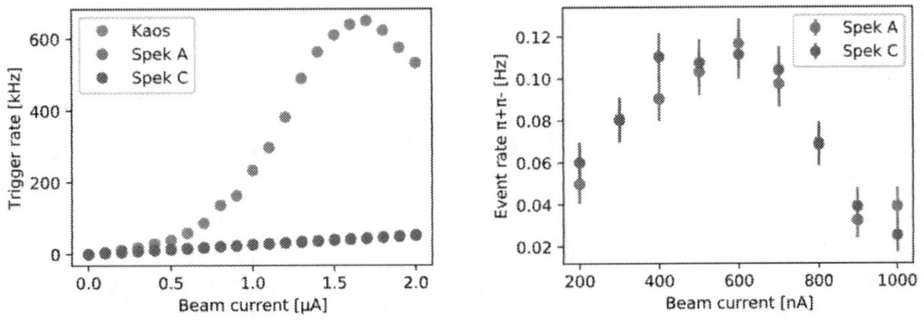

Figure 6. Rate study with the three spectrometers. Left: Trigger rate of the three spectrometers. While A and C are comparably low and continuously rising with the beam current, KAOS sees far more particles leading to a saturation of the trigger logic at around 1.6 μA before the rates drop even lower. Right: Rate of true particle coincidences of KAOS with A or C. Here, after a linear increase, a maximum at 600 nA can be seen. This is due to growing track reconstruction ambiguities with higher particle multiplicities in KAOS.

Acknowledgements

This project is supported by the Deutsche Forschungsgemeinschaft, Grant Number PO256/7−1 and the European Union's Horizon 2020 research and innovation programme No. 824093. The work is partially supported by JSPS KAKENHI Grant Numbers JP18H05459, 18H01219, 17H01121, 120002001, 15684005, 16GS0201 and GP-PU, Tohoku University. The presented data were in part collected within the framework of the PhD thesis of Philipp Eckert at the Johannes Gutenberg University Mainz.

References

[1] M. Jurić et al., Nucl. Phys. B **52**, 1 (1973)
[2] G. Bohm et al., Nucl. Phys. B **4**, 511 (1968)
[3] J. Adam et al. (STAR), Nat. Phys. **16**, 409 (2020)
[4] ALICE (2022), `2209.07360`
[5] P. Eckert, P. Achenbach et al., *Chart of hypernuclides — Hypernuclear structure and decay data* (2021), hypernuclei.kph.uni-mainz.de
[6] D.H. Davis, Nucl. Phys. A **547**, 369 (1992)
[7] D.H. Davis, Nucl. Phys. A **754**, 3c (2005)
[8] A. Esser et al. (A1), Phys. Rev. Lett. **114**, 232501 (2015)
[9] F. Schulz et al. (A1), Nucl. Phys. A **954**, 149 (2016)
[10] Particle Data Group, Prog. Theor. Exp. Phys. **2020**, 083C01 (2020)
[11] C.G. Wohl et al. (Particle Data Group), Rev. Mod. Phys. **56**, S1 (1984)
[12] P. Achenbach, S. Bleser, J. Pochodzalla, M. Steinen, PoS (Hadron2017) **207**, 1 (2018)
[13] P. Eckert, P. Achenbach, M. Aragonès Fontboté, T. Akiyama, M.O. Distler, A. Esser, J. Geratz, M. Hoek, K. Itabashi, M. Kaneta et al., PoS **PANIC2021**, 201 (2022)
[14] F. Schulz, Ph.D. thesis, Johannes Gutenberg-Universität, Mainz (2016)
[15] P. Klag et al., Nucl. Instrum. Methods Phys. Res. A **910**, 147 (2018)
[16] A. Esser, Ph.D. thesis, Johannes Gutenberg-Universität, Mainz (2013)

Studying ΛN interactions through the ^{208}Pb(e,e'K$^+$)$^{208}{}_\Lambda$Tl reaction

Franco Garibaldi[1,*], *Omar* Benhar[1], *Petr* Bydzovsky[2], *Toshiyuki* Gogami[3], *Silviu* Covrig[4], *Pete* E.C. Markowitz[5], *John* Millener[6], *Toshio* Motoba[7], *Sho* Nagao[8], *Satoshi* N. Nakamura[8,9], *Joerg* Reinhold[5], *Liguang* Tang[4,10], *Guido Maria* Urciuoli[1], and *Isaac* Vidana[11]

[1)]Istituto Nazionale di Fisica Nucleare, Sezione di Roma, P. A. Moro Rome, Italy
[2)]Nuclear Physics Institute, 250 68 Rez, Czech Republic
[3)]Department of Physics, Graduate School of Science, Kyoto University, Kyoto, Kyoto 606-8502 Japan
[4)]Thomas Jefferson National Accelerator Facility, Newport News, Virginia 23606, USA
[5)]Florida International University, Miami, Florida 33199, USA
[6)]Brookhaven National Laboratory, Upton, New York 11973, USA
[7)]Osaka University, Japan
[8)]Department of Physics, Graduate School of Science, Tohoku University, Sendai, Miyagi 980-8578, Japan
[9)]Department of Physics, Graduate School of Science, The University of Tokyo, Hongo, Tokyo 113-0033, Japan
[10)]Department of Physics, Hampton University, Hampton, Virginia, 23668, USA
[11)]Istituto Nazionale di Fisica Nucleare, Sezione di Catania, Catania, Italy

Abstract. The hyperon puzzle, the observation that the two-solar-mass neutron star existence is hardly explained by all models predicting the appearance of hyperons in the neutron star core, is currently one of the unsolved key issues in the physics of compact stars. To solve the hyperon puzzle an experimental program has been proposed by the Jefferson lab hypernuclear collaboration and approved by the Jefferson Lab (JLab) PAC. This program consists in the study of the reaction (e, e'K) on ^{40}Ca and ^{48}Ca nuclei, aimed at investigating the isospin dependence of hyperon dynamics, and in the study of the reaction (e, e'K) on ^{208}Pb nucleus, which is the subject of this paper and which will take advantage of the fact that ^{208}Pb properties largely reflects those of the uniform nuclear matter present in the interior of the neutron stars making it the ideal tool to investigate neutron star features. The proposal of the study of the reaction (e, e'K) on ^{208}Pb was approved by the Jlab PAC in 2020.

1 Introduction

An ambitious and challenging experimental program, aimed at obtaining high-resolution hypernuclear spectroscopy via the (e,e'K$^+$) reaction, was started at Jefferson Lab almost 20 years ago. The data, taken in both Hall A and Hall C using p-shell and medium-mass nuclear targets, have provided clear spectra with ~ 0.5-0.8-MeV energy resolution [1,2]. The process, whose feasibility has been established at JLab, is now widely recognized as a powerful tool to study hypernuclear spectroscopy, in addition to the (K$^-$, π$^-$) and (π$^+$, K$^+$) reactions. Electron- and hadron-induced reactions are in fact complementary to one another, being predominantly driven by spin-flip and non-spin-flip mechanisms, respectively. Furthermore, the (e,e'K$^+$) reaction allows for a much better energy resolution and produces mirror hypernuclei with respect to those produced with hadron probes.

The JLab hypernuclear collaboration proposed to the Jefferson Lab PAC43 a coherent series of studies of the (e,e'K$^+$) reaction, to be performed using targets spanning a wide range of mass. The purpose of this analysis was the investigation of the ΛN interactions in a variety of nuclear media. The PAC43 identified the study of the isospin dependence as the highest priority, and conditionally approved the $^{40}{}_\Lambda$K and $^{48}{}_\Lambda$K binding energy measurements. The JLab hypernuclear collaboration re-submitted to PAC 2016 the proposal which was approved.

The recent observation of two-solar-mass neutron stars, whose existence cannot be explained by all models predicting the appearance of hyperons in the neutron star core and which is known as "hyperon puzzle", strongly suggests that our understanding of nuclear interactions involving hyperons is not adequate for a fully quantitative description of neutron star matter. Unfortunately, the information provided by the available scattering data in the hyperon-nucleon (YN) sector is scarce, and does not allow the determination of the YN potential using the same scheme successfully employed in the nucleon-nucleon (NN) sector. Presently, hypernuclear spectroscopy appears to be the only practical way to study baryonic forces, and may play a significant role towards the solution of the hyperon puzzle.
In view of its important astrophysical implications, the extension of the (e,e'K$^+$) experimental program to a heavy nuclear target with large neutron excess, such as ^{208}Pb, is of primary importance. The study of the ^{208}Pb(e,e'K$^+$)$^{208}{}_\Lambda$Tl reaction which would provide information on the properties of a bound hyperon in an environment little affected by

1Corresponding author: franco.garibaldi7@gmail.com

surface and shell effects is complementary to the measurements on ^{40}Ca and ^{48}Ca targets, whose main purpose is the analysis of the isospin dependence of hyperon dynamics.

The measured charge density distribution of ^{208}Pb [3] clearly shows (see Fig. 1b) that the region of nearly constant density accounts for a very large fraction (~70 %) of the nuclear volume, thus suggesting that its properties largely reflect those of uniform nuclear matter in the nuclear star interior (see Fig. 1a). The validity of this conjecture has been long established by a comparison between the results of theoretical calculations and the data extracted from the ^{208}Pb(e,e´p)^{207}Tl cross sections measured at NIKHEF in the 1980s [4,5] (Fig. 1a).

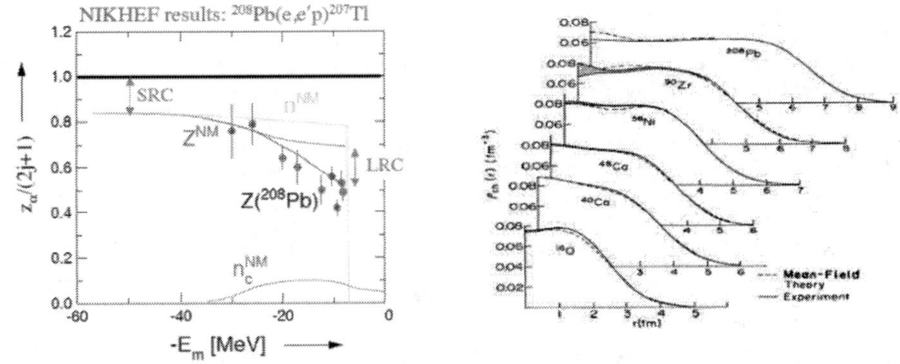

a b

Fig. 1a. Energy dependence of the spectroscopic factors (occupation probabilities *n* and hole-state strengths *Z*) extracted from the measured ^{208}Pb(e,e´p)^{207}Tl cross sections [4,5], compared to the theoretical results of Ref. [6]. The black and red solid lines, labelled Z(^{208}Pb) and Z^{NM}, correspond to ^{208}Pb and uniform nuclear matter, respectively. The effects of short- (SRC) and long-range-correlations (LRC), the latter arising from surface and shell effects, are indicated.
Fig. 1b. charge density distribution of ^{208}Pb compared to lighter nuclei.

As shown in Fig.1 the energy dependence of the spectroscopic factors, obtained from the analysis of the measured missing energy spectra, turns out to be in remarkably good agreement with the results reported in the pioneering work of Ref. [6]. Short-range correlations appear to be the most important mechanism leading to the observed quenching of the spectroscopic factor, while surface and shell effects only play an important role in the vicinity of the Fermi surface. The picture emerging from Fig. 1 suggests that deeply bound protons in the ^{208}Pb ground state behave as if they were in nuclear matter. Figure 2, provides a schematic representation of the (e,e´K$^+$) reaction, illustrating the connection with the corresponding (e,e´p) process. The left and right boxes highlight the amplitudes determined by nuclear and hypernuclear dynamics, described by the nucleon and hyperon spectral functions, respectively.

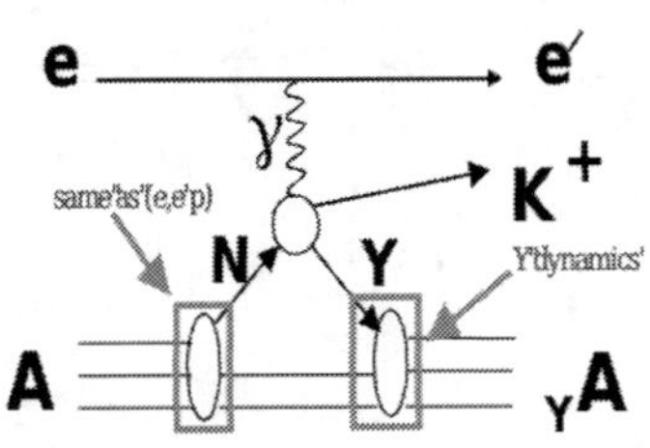

Fig. 2. Schematic representation of the (e,e´K$^+$) reaction. The left and right boxes highlight the amplitudes involved in the proton and Λ spectral functions, respectively.

The availability of the information obtained from the measured (e,e´p) cross sections is critical for the interpretation of (e,e´K$^+$) data. To see this, just consider that the hyperon binding energies are given by the difference between the missing energy measured in (e,e´K$^+$) and the proton binding energies obtained from (e,e´p) cross section. Hence, (e,e´p) data will provide the baseline needed to extract information on hyperon binding energies.

The results of theoretical calculations of the hyperon spectral function and the wealth of available ^{208}Pb(e,e´p) data, including binding energies and widths of the many proton states, will allow us to develop a fully realistic model of the ^{208}Pb(e,e´K$^+$) cross section, indispensable for the extraction of the information on hypernuclear dynamics.

2 Spectroscopic study of $^{208}{}_{\Lambda}$Tl trough the ^{208}Pb(e,e'K$^+$)$^{208}{}_{\Lambda}$Tl reaction

So far, the spectroscopy of heavy hypernuclei has been studied with pion and kaon beams which do not allow for adequate accuracy and precision. The main issue was the poor resolution, 1.6 MeV (FWHM). In addition, the spacing between Λ single-particle states is only 4 to 6 MeV. As a result, the existing data do not resolve the two series of states, introducing uncertainties in the theoretical analyses. The absolute binding energies of all (π^+, K^+) hypernuclear data are based on the 6 events of $^{12}{}_{\Lambda}C$ emulsion data, while in the case of (e,e'K$^+$) experiments the absolute energy scale can be calibrated exploiting the elementary p(e, e'K$^+$)Λ/Σ^0 processes. The recent and precise hypernuclear data on $^{12}{}_{\Lambda}B$ and $^{16}{}_{\Lambda}N$ collected at JLab, combined with a careful re-examination of the consistency between emulsion data, point to a possible half-MeV shift of the $^{12}{}_{\Lambda}C$ ground state energy, which would greatly impact on every hypernuclear binding energy measured using the reaction (π^+, K^+). Precise hypernuclear spectroscopy on medium to heavy hypernuclei with electron beams, which is only possible with the CEBAF electron beam, is critically important to obtain new information needed to constrain theoretical models of neutron star matter.

The (e,e'K$^+$) reaction enables the determination of binding energies with high precision because of the calibration provided by the elementary reaction on hydrogen and is currently the only method that can measure the absolute centroids of peak doublets, which naturally emerge in hypernuclear missing mass spectroscopy, with an unprecedented accuracy of <100 keV. It provides information on the cross section as well as on the binding energy.

Microscopic calculations of the Λ spectral function in a variety of nuclei, ranging from ^{5}He to ^{208}Pb, have been recently carried out [7]. The results of this analysis, based on a realistic hyperon-nucleon potential model constrained by the available scattering data, suggest that hyperon dynamics in ^{208}Pb may be somewhat different compared to lighter nuclei, as shown by a larger quenching of the spectroscopic factors.

The study of ^{208}Pb with the (e,e'K$^+$) reaction will provide a better binding energy resolution than that of the experiments performed so far with hadronic probes and thus a more detailed understanding of baryon behaviour deep inside of the nucleus, providing important information for studying the Λ single-particle nature under high nucleon density. ^{208}Pb is the ideal target to study hyperons in a medium closely resembling neutron star matter. This environment is best suited to investigate the effects of three body forces involving hyperons [8-11] which increase the stiffness of the nuclear matter equation of state, thus allowing for the existence of massive neutron stars compatible with the observational constraints.

In conclusion, even if the typical baryon density inside a neutron star is much higher than in a hypernucleus a precise knowledge of the ^{208}Pb level structure can, by constraining the hyperon-nucleon potential, contribute to more reliable predictions regarding the internal structure of neutron stars, and in particular their maximum mass.

3 The proposed experiment

The aim of the experiment is measuring the ^{208}Pb(e,e'K$^+$)$^{208}{}_{\Lambda}$Tl reaction. Fig. 3a shows the missing mass spectrum obtained by the ^{208}Pb(π^+,K$^+$)$^{208}{}_{\Lambda}$Pb reaction [12]. It shows a characteristic bump structure starting from the binding energies B_Λ around 25 MeV.

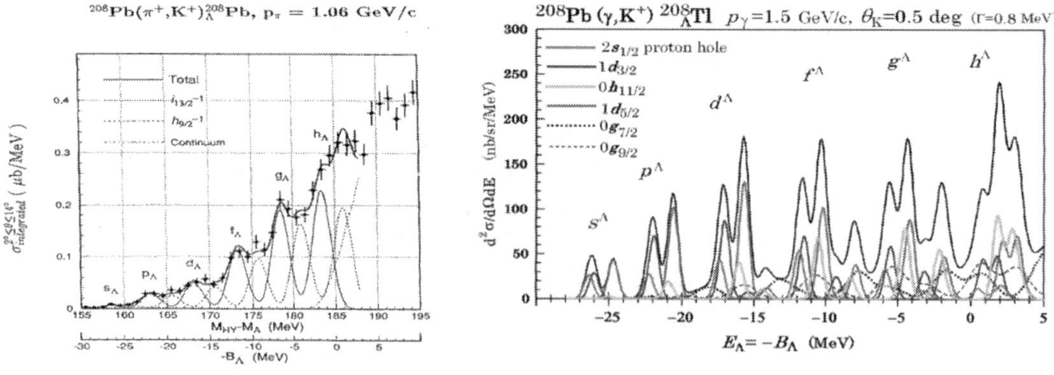

a b

Fig. 3a. Binding energy spectrum of $^{208}{}_{\Lambda}$Pb measured in E140 experiment. **Fig. 3b.** ^{207}Tl core nucleus level scheme (Motoba-Millener calculation)

The experiment will make it possible to determine with much better precision the binding energies thanks to the calibration achievable exploiting the elementary p(e,e'K$^+$)Λ/Σ^0 processes with hydrogen [1] and to test both the dynamical models and the computational techniques employed in the available theoretical approaches.

Spectroscopic data exist for few Λ hypernuclei obtained through the (e,e'K$^+$) reaction and few others would be available on ^{40}Ca and ^{48}Ca. Consequently it is extremely important to perform (e,e'K$^+$) experiment also on ^{208}Pb. The much better energy resolution of binding energy spectra of hypernuclei generated through the (e,e'K$^+$), a factor of ~ 3 better than in spectra of hypernuclei generated with π^+ or K^- as probes (see Fig. 3a and 3b), will make it possible the determination of much more precise Λ single-particle energies. It will be possible to "see" deep shells, in practice not visible with (π,K) reaction ("the observed small peaks *are assumed* to be the s_Λ states" [12]"). Fig. 3b shows the Excitation Energy Spectrum for the ^{208}Pb(e,e'K$^+$)$^{208}{}_\Lambda$Tl reaction, calculated by Millener and Motoba for the kinematics originally planned for the experiment in the Hall A of JLab [13] using the Saclay-Lyon elementary (SLA) amplitude [14]. The Λ is assumed to be weakly coupled to the proton-hole states of ^{207}Tl strongly populated in (e,e'p) or (d,^{3}He) reactions on ^{208}Pb [15]. The Λ single-particle energies were calculated from a Woods-Saxon well fitted to energies derived from the ^{208}Pb(π$^+$,K$^+$)$^{208}{}_\Lambda$Pb reaction. Just a few states cannot be resolved: states based on the closely-spaced p 2s1/2^{-1} and p 1d3/2^{-1} states, (blue bars and curves), the p 0h11/2^{-1} and p 1d5/2^{-1} states (red bars and curves). The successive red and blue peaks correspond to the population of the 0s, 0p, 0d, 0f, 0g, and 0h Λ orbits. The green lines correspond to the noded 1s, 1p, 1d/2s, and 1f Λ orbits. The remaining (wiggley) curves correspond to strength based on deeper and fragmented proton-hole strength. Once the Λ single-particle energies are known for Pb, many-body calculations using the Auxiliary Field Diffusion Monte Carlo (AFDMC) approach [8] can be used to try to determine the balance between the spin-independent components of the ΛN and ΛNN interactions required to fit Λ single-particle energies across the entire periodic table. The Λ binding energy extracted will provide most valuable information, needed to constrain the models of hyperon interaction in nuclear matter.

4 The target

A major concern in the design of the experiment was developing a Pb target that could operate at high beam currents without melting. We considered the setup used at NIKHEF for (e,e'p) experiment [5]. This would allow us to run safely with 10 μA of beam current and 100 mg/cm^2 of pure ^{208}Pb target. However, using the same setup but cooling with a cryogenic we will be allowed to use higher beam currents (25 μA) in order to increase the counting rate and hence to improve the "detectability" of the small peaks. Fig 4 shows that we would be able to run safely with a beam current of 25 μA provided that a raster area not smaller than 2 x 2 mm^2 is used [16]. The results were obtained with Computational Fluid Dynamics simulations, assuming that the Pb foil would be held in the target frame used in the experiment PREX-II.

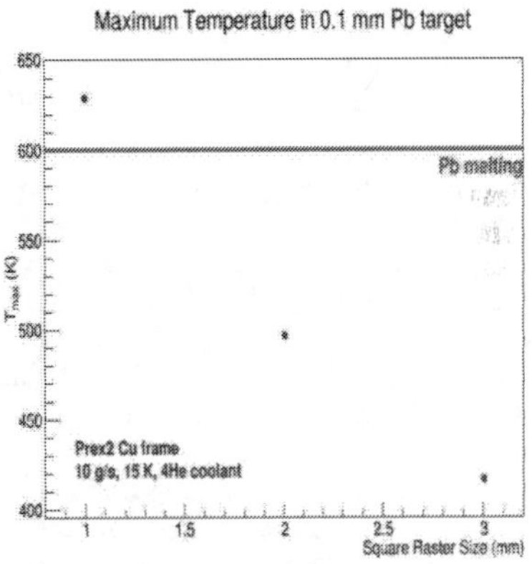

Fig. 4. Maximum temperature of the Pb target as function of the raster area for a beam current of 25 μA.

5 Particle identification

The identification of kaons detected in the hadron arm against a huge background of protons and pions is one of the major challenge of the experiment. To reduce the background level in produced spectra, a very efficient PID system is necessary for unambiguous kaon identification. In the electron arm, the gas Čerenkov counters [17] give pion rejection ratios up to 10^3. The dominant background (knock-on electrons) is reduced by 2 further orders of magnitude by the lead glass shower counters, giving a total pion rejection ratio of 10^5 [18]. The lead-glass shower counters and the gas Čerenkov are calibrated against each other.

The PID system in the hadron arm (HKS) is composed of: three planes of time-of-flight counters, two planes of water Čerenkov counters, and three planes of aerogel Čerenkov counters (see Fig. 5a). The power rejection capability of the HKS hadron identifier is 4.7 x 10^{-4} [19] (see Fig. 5b).

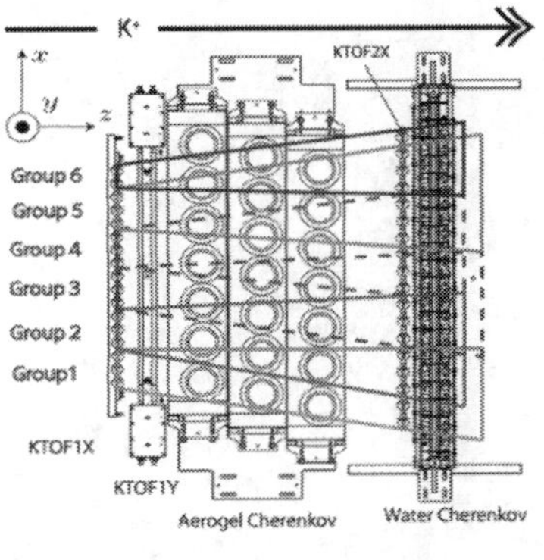

a

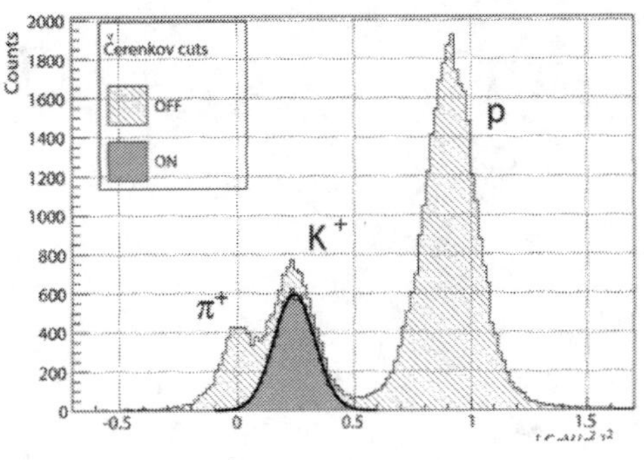

b

Fig. 5a. HKS PID (see text); **Fig. 5b.** HKS kaon power identification

Combining the HKS hadron identification detectors with the RICH would allow us to obtain the almost unprecedented separation power of 10^{12} [21].

Fig. 6 shows the performance of the Hall A RICH detector [20,21].

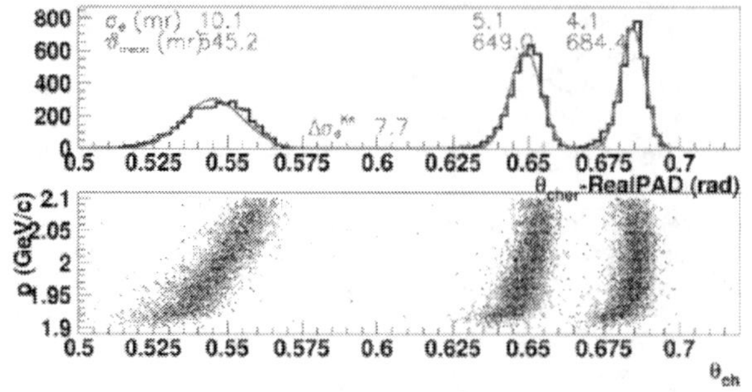

Fig. 6. Upgraded RICH simulated performance: Pion/Kaon Čerenkov angle distribution, with equal hadrons populations, at 2 GeV/c momentum, in the HRS spectrometer acceptance. The Montecarlo is tuned on Hall A hypernuclear experimental data.

6 Kinematics and counting rate

The experiment has been approved by the Jlab PAC to run in Hall A using the spectrometer HRS_L to detect scattered electrons and the spectrometer HKS to detect kaons. Nevertheless, due to unavailability of the Hall A for this experiment in the next 10 years we will be obliged to run in Hall C using the spectrometer HES to detect electrons and the spectrometer HKS to detect kaons. In Fig. 7 we show the new experimental setup. The new kinematics of the experiment is: E_{beam} = 2.34 GeV, $\Theta_{ee'} = 8^0$, $p_{e'}$ = 0.84 GeV/c, $\Theta_{e\gamma} = 4.5^0$, E_γ = 1.5 GeV, $\Theta_{K\gamma} = 8^0$, p_K = 1.2 GeV/c.

The counting rates, calculated in the new experimental setup in Hall C, at a current of 25 μA, an energy resolution of 800 keV and a target thickness of 100 mg/cm^2 are shown in Table 1. The counting rates and the Signal to Noise Ratio (SNR) were calculated by Montecarlo simulations as well as a comparison with the K^+ background detected during experiment E05-115 which used the same apparatus setup but a new pair of charge separation dipole magnets (PCS) instead of the splitter. The SNR refers to cross section calculations of the ^{208}Pb(e,e'K$^+$)$^{208}{}_\Lambda$Tl reaction only at a very preliminary stage.

Table 1. Counting rates

Target	Beam current (μA)	Target thickness (mg/cm^2)	Assumed cross section (nb/sr)	Expected yield (/hour)	Number of events	Requested Beam time (hours)	B.G. Rate (MeV/h)	S/N
^{208}Pb	25	100	80 (g.s.)	0.24	115	480	0.084	18.5

Fig. 7. The experimental setup in Hall C.

7 Summary and conclusions

The experiment on ^{208}Pb described in this paper is an essential part of the campaign of measurements the Jefferson Lab hypernuclear collaboration is doing. It allows us to extend the A (mass number) range of the hypernuclear binding energy spectroscopy to its extreme. The mass number dependence of the binding energy for each shell model orbital will be extended to A = 208. The study of the ^{208}Pb(e,e'K^{+})$^{208}{}_{\Lambda}$Tl reaction together with the other reactions proposed to Jefferson Lab PAC by the same collaboration will "complete" the systematic study of Λ hypernuclear bound states over a wide mass range. It will also deepen our understanding of the nature of the ΛN and ΛNN interactions in nuclear medium. Moreover, together with the spectroscopy studies already performed at JLab, and proposed to the JLab PAC (the approved experiment on ^{40}C and ^{48}Ca) may play a significant role towards the solution of the so-called hyperon puzzle.

8 References

1. F. Garibaldi et al., Phys. Rev. C **99**, 054309 (2019); arXiv:1807.09720
2. S.N. Nakamura, T. Gogami and L. Tang, JPS Conf. Proc., 011002 (2017)
3. B. Frois and C. N. Papanicolas, Ann. Rev. Nucl. Part. Sci. **37**, 133 (1987)
4. E. N. Quint et al., Phys. Rev. Lett. **56**, 186 (1986); E. N. Quint et al., ibid. **58**, 1727 (1987)
5. M. van Batenburg, PhD Thesis, University of Utrecht, 1996
6. O. Benhar, A. Fabrocini, and S. Fantoni, Phys. Rev. C **41**, R24 (1990)
7. I. Vidaña, Nucl. Phys. **958**, 468 (2017)
8. D. Lonardoni, F. Pederiva, and S. Gandolfi, Phys. Rev. C **89**, 014314 (2014)
9. D. Lonardoni, A. Lovato, S. Gandolfi, and F. Pederiva, Phys. Rev. Lett. **114**, 092301 (2015)
10. B. Sharma, Q.N. Usmani and A.R. Bodmer, arXiv:110 2.1542
11. Bobeldij et al. Phys. Rev. **73**, 2684 (1994)
12. O. Hashimoto and H. Tamura, Progress in Particle and Nuclear Physics **57**, 564 (2006)
13. J. Millener, personal communication
14. C. David, C. Fayard, G.-H. Lamot, B. Saghai, Phys. Rev. C 53, 2613 (1996); T. Mizutani, C. Fayard, G.-H. Lamot, B. Saghai, Phys. Rev. C **58**, 75 (1998)
15. M. Sotona, personal communication
16. S. Covrig, personal communication
17. M. Iodice et al., Nucl. Intrum Methods Phys. Res. A **411**, 223 (1998)
18. J. Alcorn et al., Nucl. Intrum. Methods Phys. Res. A **522,** 294 (2004)
19. T. Gogami et al, Nucl. Intrum Methods A **900**, 69 (2018)
20. F. Garibaldi et al., Nucl. Intrum. Methods Phys. Res. A **502**, 117 (2003)
21. E Cisbani, personal communication

Hypernuclear gamma-ray spectroscopy: summary and future prospect

Mifuyu Ukai[1,2,*]

[1]Institute of Particle and Nuclear Studies, KEK, Tsukuba, Ibaraki 305-0801, Japan
[2]Department of Physics, Tohoku University, Sendai, Miyagi 980-8578, Japan

Abstract. The present status and prospects of hypernuclear γ-ray spectroscopy are summarized. In particular, 4-body hypernuclear γ-ray spectroscopy, the recent result of ${}^4_\Lambda$He and the future plan of ${}^4_\Lambda$H and charge symmetry breaking in the ΛN interaction are presented. In addition, future plans to measure the Λ-spin-flip $B(M1)$ values of ${}^7_\Lambda(3/2^+ \rightarrow 1/2^+)$ and ${}^{12}_\Lambda$C$(2^- \rightarrow 1^-)$ transitions are introduced. They aim to study the g-factor of Λ in the nuclear medium.

1 Introduction

The γ-ray spectroscopy experiments of *s*-, *p*- and *sd*-shell hypernuclei have been carried out via the (K^-, π^-) and (π^+, K^+) reactions at KEK-PS, BNL-AGS and J-PARC since 1998. The γ-ray energies were measured using a series of germanium detector arrays, Hyperball, Hyperball-2, and Hyperball-J, which were developed particularly for high counting rate and high energy deposit environment for the hypernuclear experiment using secondary hadron beams.

Precise level structures of hypernuclei measured by the γ-ray spectroscopy give various information about the ΛN interaction. Accumulation of such data is important. In addition, the lifetime of the state which is deexcited by the Λ-spin-flip M1 transition contains information on the Λ g-factor at the nuclear density. However, such a lifetime measurement has not been successful yet.

2 Gamma-ray spectroscopy of light hypernuclei

2.1 4-body hypernuclear structure and charge symmetry breaking in the ΛN interaction

Charge symmetry is a basic concept in the *NN* interaction. It is expected to hold also in the ΛN interaction and Λ hypernuclei. Therefore the Λ binding energies (B_Λ) between a pair of mirror hypernuclei such as ${}^4_\Lambda$H and ${}^4_\Lambda$He should be identical under this symmetry. However, the old emulsion experiment data reported the charge symmetry breaking (CSB) in the ΛN interaction from a large difference between the B_Λ values of those ground 0^+ states to be $\Delta B_\Lambda = 350$ keV [1]. Theoretical efforts have been made since the 1960s, but have failed to give such a large ΔB_Λ value. Since the B_Λ difference for the excited 1^+ states may provide

*e-mail: mifukai@post.kek.jp

additional important information on the spin dependent CSB, an experiment to measure the ${}^{4}_{\Lambda}\mathrm{He}(1^+ \to 0^+)$ γ-ray energy was performed as J-PARC E13 at J-PARC K1.8 beam line. The γ-ray energy was measured using a germanium detector array, Hyperball-J.

Figure 1 shows the result of the J-PARC E13 experiment. (A) shows the missing mass spectrum measured in the ${}^{4}\mathrm{He}(K^-, \pi^-)$ reaction and (B) shows the γ-ray spectrum with event-by-event Doppler shift correction after gating the peak region in the mass spectrum. As shown in the γ-ray spectrum, a γ-ray peak energy corresponding to the ${}^{4}_{\Lambda}\mathrm{He}(1^+ \to 0^+)$ transition was determined to be 1.406 ± 0.002 ± 0.002 MeV [2].

Combined with the emulsion data of $B_\Lambda(0^+)$, the E13 result and ${}^{4}_{\Lambda}\mathrm{H}(1^+, 0^+)$ energy measured by NaI, level schemes of the A=4 mirror hypernuclei are obtained (Fig. 2). Comparing the ΔB_Λ values of the 1^+ and 0^+, the CSB effect is suggested to be strongly spin dependent.

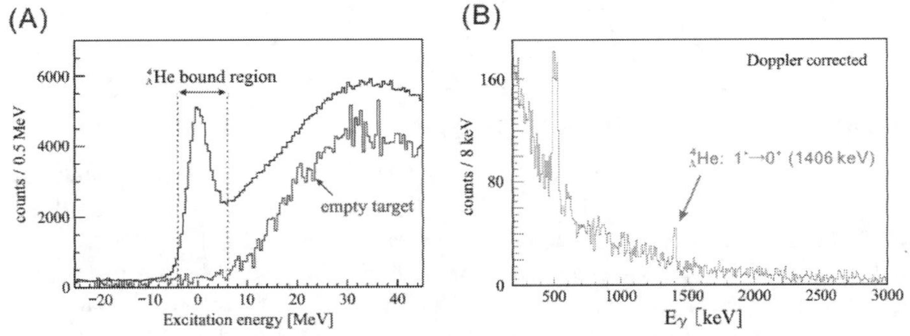

Figure 1. Result of the J-PARC E13 experiment for ${}^{4}_{\Lambda}\mathrm{He}$ γ-ray spectroscopy [2]. (A) Missing mass spectrum measured in the ${}^{4}\mathrm{He}(K^-, \pi^-)$ reaction. (B) γ-ray spectrum with event-by-event Doppler shift correction after gating the peak region in the mass spectrum.

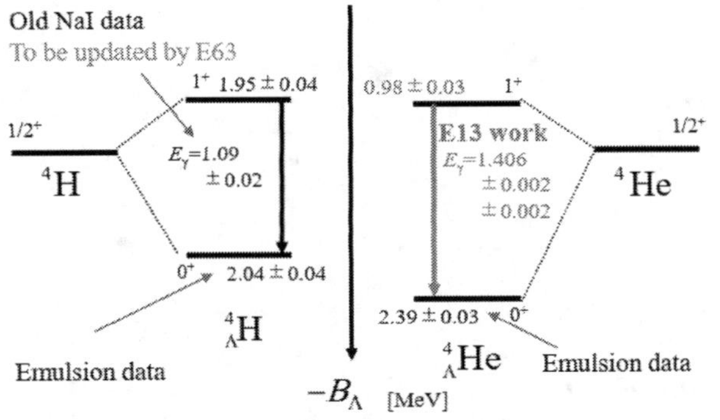

Figure 2. Level schemes of the A=4 mirror hypernuclei. The binding energies of both ground states are determined by old emulsion data [1]. The ${}^{4}_{\Lambda}\mathrm{H}(1^+ \to 0^+)$ transition energy is the average value of three independent experiments measured by NaI counters[3–5].

The ${}^{4}_{\Lambda}\mathrm{H}(1^+ \to 0^+)$ transition energy of 1.09 ± 0.02 MeV in Fig. 2 is the world average of the values reported by three old experiments using NaI detectors, 1.04 ± 0.04 MeV [3], 1.09 ± 0.03 MeV [4], and 1.114 ± 0.015 ± 0.015 MeV [5]. To measure this value precisely by Hyperball-J, a new experiment was proposed and accepted as a part of J-PARC E63[6]. Since ${}^{4}_{\Lambda}\mathrm{H}(1^+)$ cannot be produced by a non-charge exchange reactions. Therefore it was propose to produce ${}^{4}_{\Lambda}H$ as a secondary hypernucleus in the ${}^{3}\mathrm{He}$ emission decay of the ${}^{7}_{\Lambda}\mathrm{Li}$ highly excited unbound states, as shown in Fig. 3. This experiment will be perfomed at K1.1 beam line via the (K^-, π^-) reaction on the lithium target using a $p_K = 0.9$ GeV/c beam.

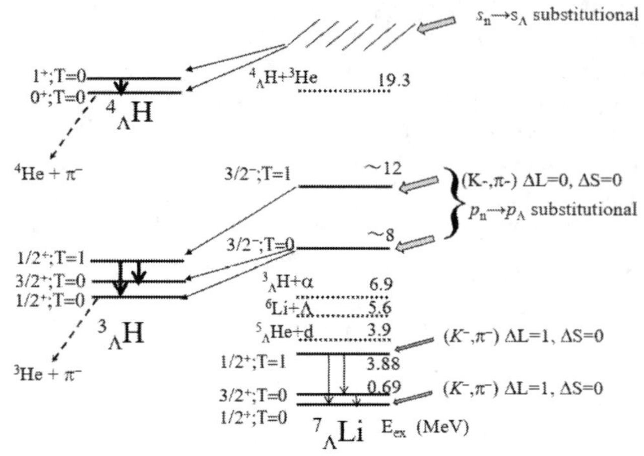

Figure 3. Level scheme of the $^7_\Lambda$Li and the particle emission decay from the highly excited state of the $^7_\Lambda$Li.

2.2 Search for the γ deexcited state of hypertriton

$^3_\Lambda$H is the lightest hypernucleus which has a bound state. The GSI data reported the $^3_\Lambda n$ is bound [7]. If so, excited state of $^3_\Lambda$H(1^+,T=1) is also below the $N + N + \Lambda$ threshold. Figure 4 shows the level schemes of the NN and $NN\Lambda$ systems in the case that $^3_\Lambda n$ is bound. Since the excited state of $^3_\Lambda$H(1^+,T=1) cannot decay to the d+Λ channel due to the isospin conservation, it decays to the T=0 ground state doublet by the γ transitions. On the other hand, the JLab data reported the negative result for the existence of the $^3_\Lambda n$ bound state [8]. Even if $^3_\Lambda n$ is not bound, the $^3_\Lambda$H(1^+,T=1) is expected to be very close to the $N + N + \Lambda$ threshold to have a finite γ transition probability.

Search for the γ rays from the $^3_\Lambda$H(1^+,T=1) state is to be performed as a by-product of the $^4_\Lambda$H($1^+ \rightarrow 0^+$) measurement (J-PARC E63). As shown in Fig. 3, the $^3_\Lambda$H(1^+,T=1) state can be also produced in the α emission decay from the $^7_\Lambda$Li highly excited unbound states.

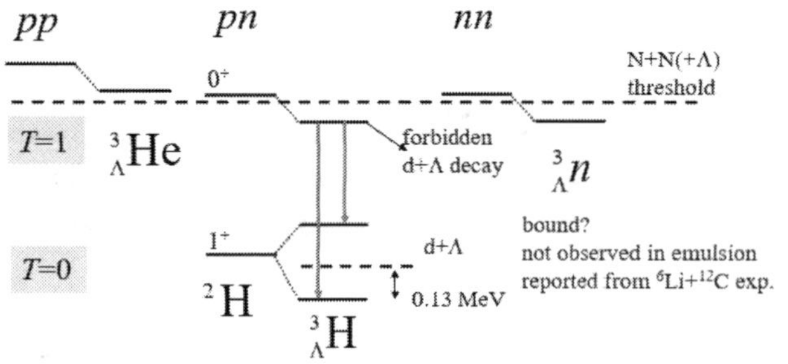

Figure 4. Level schemes of the NN and $NN\Lambda$ systems in case of the $^3_\Lambda n$ is bound as reported in the GSI data[7].

3 B(M1) measrurement of Λ spin-flip M1 transitions

3.1 g-factor of Λ in the nuclear medium

Magnetic moment of a Λ hyperon in the nuclear medium may be different from that in the free space. However the direct measurement of g-factor of hypernuclei is extremely difficult.

When the core nucleus has non-zero spin (J_c), the hypernuclear level splits into the spin-doublet state ($J_c \pm 1/2$) as shown in Fig. 5. The uppler level of the doublet decays to the lower level by the Λ-spin-flip M1 transition. In the weak coupling limit between Λ and the core nucleus, the reduced transition probability $B(M1)$ can be expressed as [9]

$$B(M1) = (2J_{up}+1)^{-1}| < \phi_{low}||\mu||\phi_{up} > |^2 = \frac{3}{8\pi}\frac{(2J_{low}+1)}{(2J_c+1)}(g_c - g_\Lambda)^2 = \frac{9}{16\pi}\frac{1}{E_\gamma^3}\frac{B.R.(M1)}{\tau_{up}}, \quad (1)$$

where g_c and g_Λ denote effective g-factors of the core nucleus and Λ, $\mathbf{J}_c$ and $\mathbf{J}_\Lambda$ denote their spins, respectively, $\mathbf{J} = \mathbf{J}_c + \mathbf{J}_\Lambda$ is the spin of the hypernucleus, τ_{up} denotes the lifetime of the upper state of the spin doublet and $B.R.(M1)$ denotes the branching ratio of the M1 transition.

Table 1 shows the experimentally determined Λ-spin-flip M1 transition energies of hypernuclear ground state doublets and the calculated M1 decay widths. Except for the ${}^{12}_{\Lambda}$C case, the branching ratios of the M1 transitions are almost 100% due to their much shorter lifetimes than the typical hypernuclear lifetime to be ~ 200 ps.

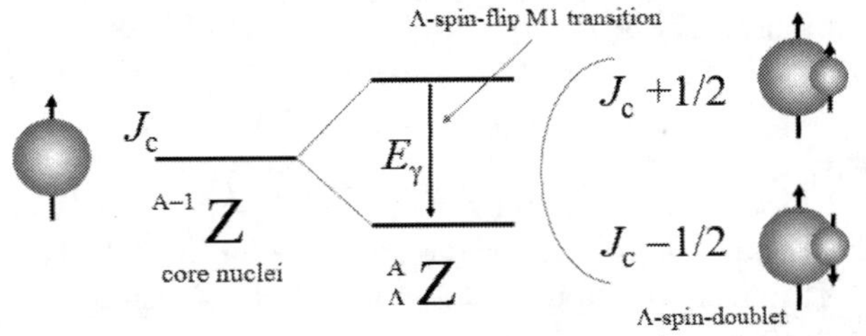

Figure 5. Schematic drawing of the hypernuclear spin-doublet states and the Λ-spin-flip M1 transition.

Table 1. Measured Λ-spin-flip M1 transitions between hypernuclear ground-state doublets and calculated M1 decay widths.

	g.s. doublet J_{up}, J_{low}	E_γ [keV]	Core g-factor [μ_N]	Calculated $1/\Gamma_{M1}$ [ps]
${}^{4}_{\Lambda}$He	$1^+, 0^+$	1406	−4.2552	0.1
${}^{7}_{\Lambda}$Li	$3/2^+, 1/2^+$	692	0.8220	0.5
${}^{11}_{\Lambda}$B	$7/2^+, 5/2^+$	262	0.6002	9
${}^{12}_{\Lambda}$C	$2^-, 1^-$	160	−0.643	440
${}^{19}_{\Lambda}$F	$3/2^+, 1/2^+$	316	0.849(Calc.)	6

3.2 B(M1) of the ${}^{7}_{\Lambda}$Li ground-state doublet by Doppler shift attenuation method

When the stopping time of the recoiling excited hypernucleus in the target material is of the same order as the lifetime of the γ-emitting excited state, the lifetime can be obtained by analyzing a partly Doppler-broadened γ-ray peak shape, which is called the Doppler Shift Attenuation Method (DSAM). To measure the lifetimes precisely by DSAM, matching between the lifetime and the stopping time is essential. In our simulation, the ${}^{7}_{\Lambda}$Li ($3/2^+ \rightarrow 1/2^+$) using a lithium oxide (Li_2O) crystal with a density of 2.01 g/cm^3 is found to be the best candidate for the lifetime measurement applying DSAM.

To measure the B(M1) of the ${}^{7}_{\Lambda}$Li ($3/2^+ \rightarrow 1/2^+$), a new experiment was proposed and accepted as a part of J-PARC E63 [6].

This experiment will be performed at K1,1 beam line using the (K^-, π) reaction on a Li_2O target. K^- beam momentum will be determined after taking a pilot data for 0.9 GeV/c and 1.1 GeV/c momenta to maximize the lifetime sensitivity.

3.3 B(M1) of the ${}^{12}_{\Lambda}$C gound-state doublet by gamma-weak coincidence method

When the lifetime of the upper level of the spin-doublet is of the same order as of the hypernuclear weak decay ($\sim$ 200 ps), the lifetime can be determined by measuring the timing of the particles such as π^- emitted by weak decays. The branching ratio of the M1 transition can be measured by the yield of the M1 γ ray. To apply this technique, ${}^{12}_{\Lambda}$C is the best candidate.

Figure 6 shows the level scheme of the ${}^{12}_{\Lambda}$C bound states. Excitation energies were measured in the Hyperball project [10]. The weak decay lifetime of the ${}^{12}_{\Lambda}$C was measured to be 212^{+7}_{-6} ps [11]. This measured lifetime value is slightly affected by the weak decay of 2^- state and feeding from the slow $2^- \rightarrow 1^-$ M1 transition. Expected lifetime and M1 branching ratio of the M1($2^- \rightarrow 1^-$) is calculated to be $\sim$140 ps and 0.32, respectively.

Figure 7 shows the conceptual design to measure the weak decay lifetime and γ rays of ${}^{12}_{\Lambda}$C. In this figure, to measure the timing of the weak decay particles, employment of the dipole magnet and time projection chamber of the type used in J-PARC E42 [12] is assumed. Comparing the lifetimes accompanying γ_1 or γ_2 and γ_3 events, the lifetime of the 2^- can be determined. Further more, the 2^- lifetime can be measured directly using a fast response scintillation counter as a γ-ray detector. For this purpose, $CeBr_3$ scintillator of 17 ps decay constant is to be developed. R&D of this counter and realistic simulation is now undergoing. This experiment is planned to be performed at K1.1 or K1.8BR beam line using a $p_K = 0.9$ GeV/c beam.

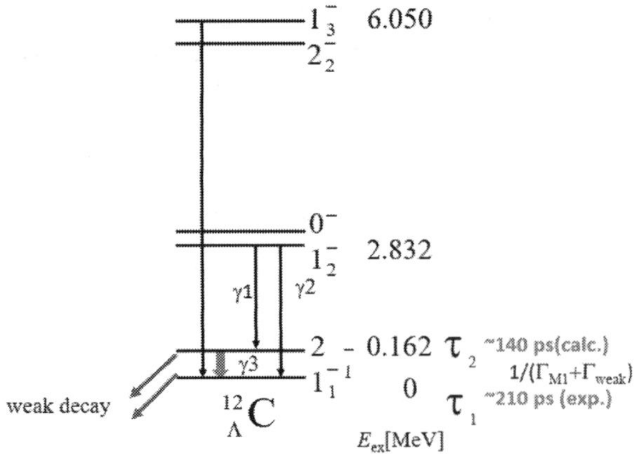

Figure 6. Level scheme of ${}^{12}_{\Lambda}$C bound states. The excitation energies are measured in the Hyperball project [10].

4 Summary

The γ-ray spectroscopy of ${}^{4}_{\Lambda}$He confirmed the existence of large CSB and its spin dependence in the ΛN interaction. Further study of CSB in hypernuclei, precise γ-ray spectroscopy of ${}^{4}_{\Lambda}$H

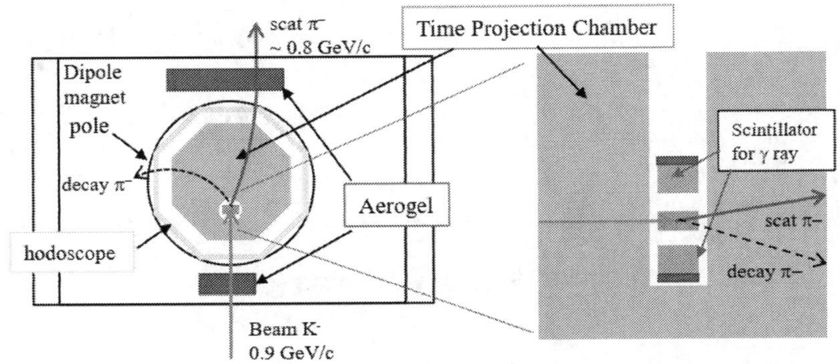

Figure 7. Conseptual design of the hypernuclear γ-weak coincidence experiment to measure the lifetime of the state in which the γ decay and the weak decay are competing. Dipole magnet and time projection chamber are assumed as the type of J-PARC E42 ones[12].

is to be performed via the (K^-, π^-) reaction on the lithium target. As the by-product, search for the ${}^{3}_{\Lambda}$H γ ray transition is also to be performed. In addition, measurement the Λ-spin-flip $B(M1)$ values of ${}^{7}_{\Lambda}(3/2^+ \to 1/2^+)$ and ${}^{12}_{\Lambda}$C$(2^- \to 1^-)$ transitions will be perfomed to study the g-factor of a Λ in the nuclear medium.

References

[1] M. Jurić *et al.*, Nucl. Phys. **B52**,1 (1973)
[2] T. O. Yamamoto et al., Phys. Rev. Lett. **115**, 222501 (2015)
[3] M. Bedjidian *et al.*, Phys. Lett. B **83**, 252 (1979)
[4] M. Bedjidian *et al.*, Phys. Lett. B **62**, 467 (1976)
[5] A. Kawachi, Docter thesis, University of Tokyo, 1997
[6] H. Tamura *et al.*, J-PARC E63 proposal, https://j-parc.jp/researcher/Hadron/en/pac_1601/pdf/P63_2016-2.pdf
[7] C. Rappold *et al.*, Phys. Rev. C **88**, 041001(R) (2013)
[8] K. N. Suzuki *et al.*, Prog. Theor. Exp. Phys. **2022**, 013D01 (2022)
[9] R. H. Dalitz and A. Gal, Ann. Phys. 16 (1978) 167; J. Phys. G **6**, 889 (1978)
[10] K. Hosomi *et al.*, Prog. Theor. Exp. Phys. **2015**, 081D01 (2015)
[11] S. Kameoka *et al.*, Nuclear Physics A **754**, 173c (2005)
[12] Y. Ichikawa *et al.*, JPS Conf. Proc. **33**, 011103 (2021)

Feasibility study for measurement of beta-decay rates of Λ hypernuclei

Kento Kamada[1*], *Manami* Fujita[2], and *Hirokazu* Tamura[1,2]

[1] *Department of Physics, Tohoku University, Sendai, 980-0845, Japan*
[2] *Advanced Science Research Center, Japan Atomic Energy Agency, Tokai, 319-1184, Japan*

Abstract. The baryon structure may be modified in nuclear medium. The quark meson coupling model predicts that the beta decay rate of a Λ hyperon in Λ hypernuclei decreases by 20% at maximum. We propose an experiment to precisely measure the beta-decay rate of $^{5}_{\Lambda}$He at J-PARC via the (π^+, K^+) reaction. A feasibility study via GEANT4 simulation has been done, showing that the background can be suppressed down to ~ 3.9% of the beta-decay electron yield, using a calorimeter covering the whole solid angle around the target. The measurement of the beta-decay of a Λ in a nucleus is found to be feasible.

1 Modification of the baryon structure in nuclear medium

Since a nucleon is a composite particle made of quarks and gluons, it is natural to consider that the properties and the structure of the nucleon may be modified in a nucleus via the interactions with other nucleons. The only clear evidence for the nucleon modification is the EMC effect [1]. The EMC effect indicates a change of the momentum distribution of quarks in the nucleon in nuclear matter, but the mechanism of the change has not been clarified very well. We thought of an idea that possible modification of baryon structure in a nucleus may be clearly detected by using a Λ hypernucleus. A promising probe for the baryon modification are weak decay properties of a Λ in a nucleus. It has been conjectured that the size of a nucleon (baryon) may be swelled since the spatial distribution of u and d quark wavefunctions in a baryon may be spread due to their interaction with the meson field in the nucleus. Figure 1 shows a schematic view of a Λ in the free space and in the nucleus. It is expected that the spatial distribution of u and d quarks is spread and that of s quarks is almost unchanged because the coupling of s quarks with the meson field is much smaller than that of u and d quarks. When the beta-decay of Λ takes place in the nucleus, the spatial overlap between the u quark wavefunction and the unchanged s quark wavefunction is expected to decrease. This structure change of Λ results in reduction of the axial charge g_A, and then the beta decay-rate of Λ, Γ_β in a nucleus. According to a calculation by the quark meson coupling model, g_A is predicted to be reduced by 10 % at maximum, which corresponds to a reduction of Γ_β by 20% at maximum in a nucleus [2]. The g_A/g_V value of a Λ in the free space was

* *Email:* kamada@lambda.phys.tohoku.ac.jp

measured to be $g_A/g_V = -0.718 \pm 0.015$ [3]. However, there is no experimental data on beta decay of Λ in a nucleus and the in-medium g_A value.

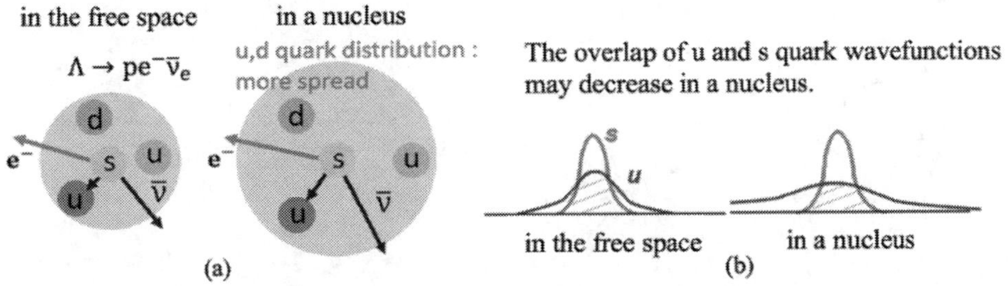

Fig. 1(a). Conceptual illustration for beta-decay of Λ in the free space (left) and in a nucleus (right).
Fig. 1(b). Illustration for wavefunctions of the initial state s quark and the final state u quark in the free space (left) and in a nucleus (right).

2 Experiment for measuring beta-decay rate of Λ

When measuring g_A reduction of Λ from the baryon modification via measurement of the beta-decay rate, other effects which cause the reduction of the beta-decay rate become a problem. In the beta-decay of ordinary nuclei, nuclear many-body effects as well as hadronic effects such as meson exchange current are known to reduce the beta-decay rate of Gamow Teller transition. Such reduction is ascribed to quenching of the effective g_A value in nuclei [3]. To measure the change of g_A from the baryon modification clearly, these quenching effects should be small. The quenching effects become larger in heavier nuclei, and they are small enough (<~5%) for the s-shell (A ≤ 4) nuclei [4,5]. Thus, we propose an experiment to measure the beta-decay rate of $^{5}_{\Lambda}$He. In the proposed experiment, $^{5}_{\Lambda}$He will be produced by the ^{6}Li$(\pi^+, K^+)^{6}_{\Lambda}$Li, $^{6}_{\Lambda}\text{Li} \rightarrow {}^{5}_{\Lambda}\text{He} + \text{p}$ reaction at J-PARC K1.1 beam line with a setup as shown in Fig. 2. Momenta of beam π^+'s and scattered K^+'s are measured by the K1.1 beam line spectrometer and the SKS spectrometer. After producing $^{5}_{\Lambda}$He hypernuclei, the beta-decay rate, Γ_β, will be obtained by measuring the branching ratio of the beta-decay, BR_β, and the lifetime of $^{5}_{\Lambda}$He, τ_Λ. Considering the available beam time at J-PARC K1.1 beam line, we expect that the BR_β can be measured within a 4% statistical error and τ_Λ can be measured within a 2% statistical error. Then we will be able to confirm the reduction of Γ_β (20% at maximum) with more than 3σ confidence level. The lifetime τ_Λ was previously measured with a 4% statistical error in the KEK E462 experiment [6]. In our proposed experiment, we will use a similar setup as the E462 and achieve a statistical error of 2% [7].

When measuring the BR_β, the beta-decay electron from $^{5}_{\Lambda}$He should be measured by suppressing the background from other decay modes of Λ. The expected background comes from mesonic weak decay, $\Lambda \rightarrow p\pi^-$ ($BR_{\pi^-} = 0.4$) and $\Lambda \rightarrow n\pi^0$ ($BR_{\pi^0} = 0.2$), and nonmesonic weak decay, Λn → nn and Λp → np (BR_{nm}= 0.4), where the *BR* values were measured for $^{5}_{\Lambda}$He [6]. The ^{6}Li target is surrounded by plastic counters, lucite Cerenkov counters and a BGO calorimeter, as shown in Fig. 3. The plastic counters distinguish the beta-decay electron from neutral particles such as γ and neutron. The lucite Cerenkov counters distinguish the beta-decay electron from charged particles such as $\pi^-$and proton. The BGO calorimeter separates the beta decay-electron from π^0 and π^- by using the number of clusters in the BGO.

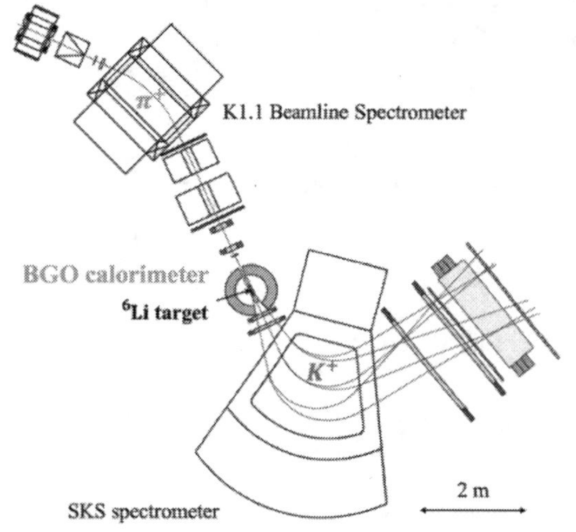

Fig. 2. Setup at the K1.1 beam line at J-PARC.

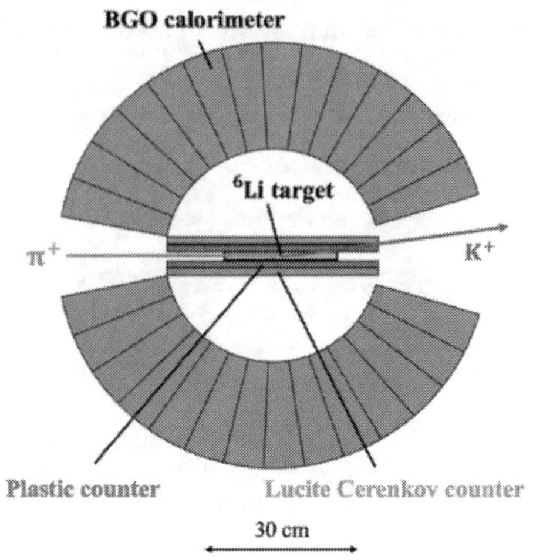

Fig. 3. Apparatus around the ^{6}Li target.

3 Background reduction study for measuring the branching ratio

In this section, we report simulation results for the background reduction study by the GEANT4 simulation code, and discuss feasibility of the beta decay measurement.

3.1 The cluster analysis via the BGO calorimeter

The background from π^0 and π^- from mesonic weak decay of Λ can be greatly suppressed via the cluster analysis in the BGO calorimeter. Figure 4 shows simulated distributions of the number of clusters in the BGO calorimeter. A cluster is identified when one or several adjacent segments in the BGO calorimeter have an energy deposit more than 1 MeV each. As shown in Fig. 4, a beta-decay electron produces one cluster. A π^0 mainly produces two clusters since π^0 mainly decays to 2γ, and one cluster events of π^0 are caused by the escape of one γ from the upstream or downstream hole of the BGO calorimeter. A π^-produces many clusters caused by many neutrons and nuclear γ-rays from interaction between π^- and nuclei in BGO. It was found that, by selecting one cluster events, 97% of π^0 and 92.8% of π^- events are rejected. Thus, the π^0 and π^- backgrounds are reduced down to $BR_{\pi^0} \times (1-0.97) = 0.006$ and $BR_{\pi^-} \times (1-0.928) = 0.028$, respectively.

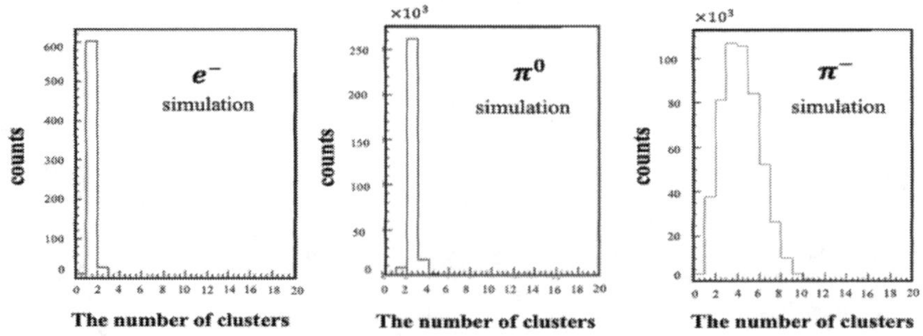

Fig. 4. The number of clusters in the BGO for the beta-decay electron, π^0 and π^- from $^5_\Lambda$He decay.

3.2 Analysis in the plastic counter and the lucite Cerenkov counter

Figure 5 shows a simulated distribution of the energy deposit per path length. The path length of particles in the plastic counter is determined by the vertex point in the target and the hit position at the BGO calorimeter. The distribution of the beta-decay electron has a minimum ionization peak at 0.2 MeV as shown in Fig. 5. The distribution for π^0 has two peaks corresponding to either of e^-or e^+, and to both $e^-$and e^+. The distribution for π^- has a peak at 0.5 MeV due to the pion velocity of $\beta \sim 0.6$. By selecting 0.12 − 0.25 MeV/mm in the energy deposit per path length and requiring a hit of the lucite Cerenkov counter, the π^0 and π^- backgrounds are reduced down to 0.006×0.12 (Plastic) $\times$ 0.1 (Lucite) = 7.2×10^{-5} and 0.0288×0.001 (Plastic) $\times$ 0.045 (Lucite) = 1.3×10^{-6} , respectively.

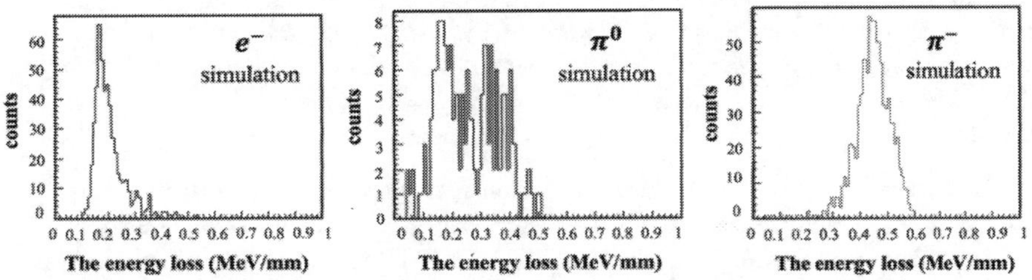

Fig. 5. The energy deposit per path length in the plastic counter for the beta-decay electron, π^0 and π^-.

3.3 Simulation results

Besides the analysis of each detector described above, the π^0 and π^- backgrounds can be further suppressed by analysis in photon veto counters to be installed at the exit of the downstream hole of the BGO calorimeter. Via all the analyses, the remaining π^0 and π^- backgrounds are reduced down to branching ratios of 7.2×10^{-5} and 1.3×10^{-6} , respectively. These values correspond to 3.6 % and 0.26% of the branching ratio of the beta-decay electron. The systematic error from the background is expected to be smaller than the required statistical error (4%) of the beta-decay measurement. Thus, the beta-decay measurement is found to be feasible.

4 Conclusion and future prospects

We found that the measurement of the beta-decay rate of Λ in the ${}^{5}_{\Lambda}$He hypernucleus is feasible. Most of the present results have been reported as a Letter-Of-Intent to J-PARC. In the near future, we will make more realistic simulation considering the performance of each detector and prepare a full proposal to J-PARC.

References

[1] D. F. Geesaman, K. Saito, A. W. Thomas, Ann. Rev. Nucl. Part. Sci. **45**, 337 (1995)
[2] M. Takahashi et al. (Particle Data Group), Phys. Rev. D **98**, 030001 (2018)
[3] P. A. M. Guichon, A.W. Thomas, Phys. Lett. B **773**, 332 (2017)
[4] F. C. Khanna, I. S. Towner, H. C. Lee, Nucl. Phys. A **305**, 349 (1978)
[5] W. T. Chou et al., Phys. Rev. C **47**, 163 (1993)
[6] S. Kameoka et al., Nucl. Phys. A **754**, 173c (2005)

Hypernuclear production spectroscopy with an extended shell model

Atsushi Umeya[1,*], *Toshio* Motoba[2,3], and *Kazunori* Itonaga[4,5]

[1]Liberal Arts and Sciences, Nippon Institute of Technology, 345-8501 Miyashiro, Saitama, Japan
[2]Research Center for Nuclear Physics, Osaka University, 567-0047 Ibaraki, Osaka, Japan
[3]Laboratory of Physics, Osaka Electro-Communication University, 572-8530 Neyagawa, Japan
[4]Faculty of Medicine, Miyazaki University, 889-1692 Miyazaki, Japan
[5]Faculty of Education, Gifu University, 501-1193 Gifu, Japan

Abstract. The $(e, e'K^+)$ reaction experiments with several nuclear targets performed recently at the Jefferson laboratory have provided high-resolution spectroscopic data. Among others, the ^{10}B $(e, e'K^+)$ $^{10}_{\Lambda}$Be reaction data are particularly interesting due to new appearance of extra sub-peaks which seem difficult to be explained within the conventional model with the p-shell nuclear natural-parity configurations. To describe this novel hypernuclear structure, we have extended the model space by introducing the new configurations which include unnatural-parity nuclear core-excited states. This configuration mixing successfully explains the sub-peaks in the ^{10}B $(e, e'K^+)$ reaction data. For adjacent hypernuclei with $A = 9$–11, we show the DWIA spectra of (K^-, π^-), (π^+, K^+) and (γ, K^+) reactions that are calculated within the extended model in a consistent way. We discuss the theoretical spectra of $^{9,10}_{\Lambda}$Be and $^{11}_{\Lambda}$B production reactions.

1 Introduction

The $(e, e'K^+)$ reaction experiments with several nuclear targets performed recently at the Jefferson laboratory (JLab) have provided high-resolution spectroscopic data [1–3]. They are very fruitful in disclosing hypernuclear structure details and also in understanding hyperon-nucleon fundamental interaction properties. Among others, the ^{10}B $(e, e'K^+)$ $^{10}_{\Lambda}$Be reaction data [2] are quite interesting, because they have shown extra sub-peaks which cannot be explained within the conventional model with the p-shell nuclear natural-parity configurations [4]. In order to describe this novel hypernuclear structure, we have extended the model space by introducing the new configurations which include unnatural-parity nuclear core-excited states [5, 6].

For adjacent hypernuclei with $A = 9$–11, we have calculated not only the energy levels of positive and negative parities but also the DWIA cross-sections of hypernuclear production reactions within the extended model space. We focus on the interesting behavior of Λ p-state. In $^{9}_{\Lambda}$Be, it is well known that the Λ p-state, p^{Λ}, splits into two orbital states, $p^{\Lambda}_{\perp}$ and $p^{\Lambda}_{//}$, which is due to the strong coupling between the Λ particle and the nuclear core deformation characterized by the $\alpha\alpha$ structure [8, 9]. We show the theoretical spectra of ^{9}Be (K^-, π^-) $^{9}_{\Lambda}$Be, ^{10}B $(e, e'K^+)$ $^{10}_{\Lambda}$Be and ^{11}B (K^-, π^-) $^{11}_{\Lambda}$B reactions, and discuss the characteristic peaks in relation to the behavior of the $p^{\Lambda}_{\perp}$ and $p^{\Lambda}_{//}$ states.

*e-mail: aumeya@nit.ac.jp

2 Model

In the conventional shell model applied so far, only configurations with natural-parity nuclear-core states and s^Λ state had been taken into account. In the extended model we have proposed recently, each hypernuclear state of $J^\pm$ is described by taking four types of configurations, $[J^+_{c(i)} \otimes s^\Lambda]_{J^+}$, $[J^+_{c(i)} \otimes p^\Lambda]_{J^-}$, $[J^-_{c(i)} \otimes s^\Lambda]_{J^-}$ and $[J^-_{c(i)} \otimes p^\Lambda]_{J^+}$ where $J^\pm_{c(i)}$ denotes all the possible spin-parity states of core nucleus, and s^Λ and p^Λ denote the Λ single-particle states. We take into account all the states of core nuclei in the $0\hbar\omega$ and $1\hbar\omega$ space, which are labeled with i. It is interesting to point out that the nuclear-core eigenstates with different parities are mixed by the Λ participation. For the ${}^{10}_{\Lambda}$Be hypernucleus, for example, natural- and unnatural-parity nuclear-core configurations, $[J^+_{c(i)} \otimes p^\Lambda]_{J^-}$ and $[J^-_{c(i)} \otimes s^\Lambda]_{J^-}$, can be mixed easily by the ΛN interaction at appropriate excitation energy. Furthermore, in order to allow the hypernuclear production through various configurations, we extend the model space for target nucleus up to $2p$-$2h$ $2\hbar\omega$ configurations in order to meet the consistency requirement with the four configurations shown above. We emphasize that, for the first time, this configuration-mixing calculation successfully explains the extra sub-peaks in the ^{10}B $(e, e'K^+)$ ${}^{10}_{\Lambda}$Be reaction experiments [5, 6]. For the $\gamma p \to \Lambda K^+$ amplitudes, we employ the SLA model [7].

3 Results

Figure 1 shows the theoretical and experimental energy levels of ${}^{10}_{\Lambda}$Be together with those of the ^{9}Be core nucleus. We obtain various states of ${}^{10}_{\Lambda}$Be with positive and negative parity by using the extended model. We note that, among the various states, typical states can be extracted through the different reaction characteristics pertinent to the (K^-, π^-), (π^+, K^+) and (γ, K^+) reactions. Through the (γ, K^+) reactions, $(1^-_{\rm g.s.}, 2^-_1)$ doublet states are extracted and these states correspond to the #1 peak in the JLab experiments [2]. Similarly, $(2^-_2, 3^-_1)$, $(3^-_2, 4^-_1)$ and $(3^-_4, 4^-_2)$ states are extracted, and these states correspond to the #2, #3 and #4 peaks, respectively, as indicated in the experimental spectrum [2] and also indicated in the space between columns 3 and 4 of Fig. 1. These four peaks have already been predicted in the DWIA calculation with the wave functions obtained in the smaller model space of the conventional shell model [4]. Here we particularly emphasize that the 2^+_3, 1^+_3, 3^+_2, 2^+_4, 2^+_5 and 3^+_3 states are newly obtained in the extended calculation, and in fact these new states well correspond to the new bump (indicated as #a in Fig. 1) at $E_x = 8.34$ MeV observed in the JLab experiment [2]. These non-normal parity states have not been obtained in the conventional shell model calculations exploited so far. The present configuration mixing successfully explains the extra sub-peaks confirmed for the first time in the ^{10}B $(e, e'K^+)$ ${}^{10}_{\Lambda}$Be reaction experiment [2].

In contrast to the four sharp peaks (#1–#4) in ^{10}B (γ, K^+) ${}^{10}_{\Lambda}$Be, the new bump (#a) at $E_x^{\rm exp} = 8.34$ MeV is found to have a large experimental peak width of 0.78 MeV (FWHM) [2]. In fact, the calculation confirms that the peak consists of a group of several states (2^+_3, 1^+_3, 3^+_2, 2^+_4, 2^+_5, 3^+_3) obtained at $E_x = 7.8$–9.0 MeV. This fact naturally explains the experimental width of the bump #a. In order to know the characteristics of these states, it is interesting to show the dominant components of the calculated wave functions for some examples. Both of the first two states among them can be described approximately with the following structures: $\Psi(2^+_3, 1^+_3) \sim 81\%[C(3/2^-_1)(p_{3/2}p_{1/2})^\Lambda] + 17\%[C(5/2^-_1)(p_{3/2}p_{1/2})^\Lambda]$, where $C(J)$ denotes the ^{9}Be states. On the other hand, the other four members are obtained to have mixed configuration of $[C(5/2^+_1)(s_{1/2})^\Lambda]$, $[C(3/2^-_1)(p_{3/2}p_{1/2})^\Lambda]$ and $[C(5/2^-_1)(p_{3/2}p_{1/2})^\Lambda]$. In other words, the different parity core-excited states are strongly admixed mediated by the Λ particle. The principal reason of such admixture is because the energy gain of the $p^\Lambda_{//}$ orbit under the influence of core deformation. The energy change of the $p^\Lambda_{//}$ orbit can be seen by comparing the different reactions as discussed below.

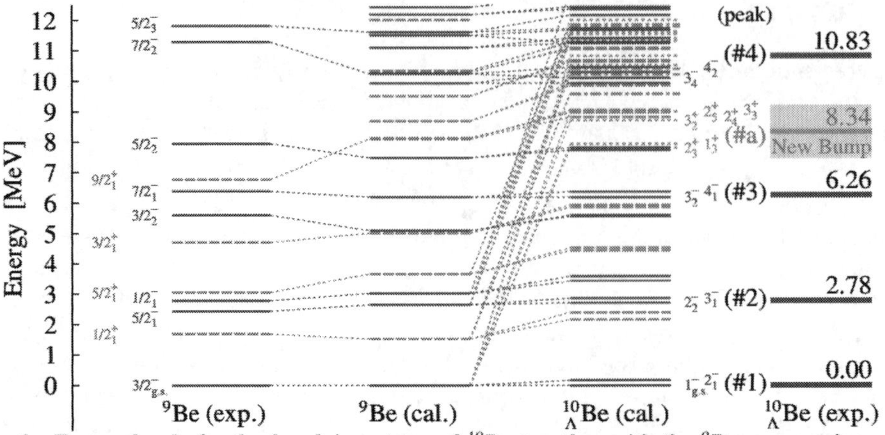

Figure 1. Energy levels for the low-lying states of ${}^{10}_{\Lambda}$Be together with the ^{9}Be core nucleus. The columns 2 and 3 show our results, and the columns 1 and 4 show the experimental data [2]. In the column 3, the spin-parities are labeled for dominant states produced by the (γ, K^+) reaction.

In order to investigate appearance of the new bump in the ${}^{10}_{\Lambda}$Be production, we calculate the DWIA cross-sections for three kinds of hypernuclear production reactions compared in Fig. 2. The comparison aims to make use of the different merits of these reactions such as the recoilless excitation, the spin-flip dominance and the high-spin selectivity. Figure 2 shows the DWIA spectra of ^{9}Be$(K^-, \pi^-)\,{}^{9}_{\Lambda}$Be, ^{10}B$(e, e'K^+)\,{}^{10}_{\Lambda}$Be and ^{11}B$(\pi^+, K^+)\,{}^{11}_{\Lambda}$B reactions. In ${}^{9}_{\Lambda}$Be, it is well known that the p^{Λ} state splits into two orbital states expressed by $p^{\Lambda}_{\perp}$ and $p^{\Lambda}_{//}$, which is due to the strong coupling with nuclear core deformation having the $\alpha\alpha$ structure [8, 9]. In Fig. 2 (a), the recoilless reaction leads to the exclusive population of the $p^{\Lambda}_{\perp}$ state as the strangeness analog state of the ground state of the target ^{9}Be$(3/2^-_{\rm g.s.})$. Therefore one easily knows the energy positions of $p^{\Lambda}_{\perp}$ and $p^{\Lambda}_{//}$ states. The calculated spectrum (a) is in good agreement with the cluster-model calculation [8]. The recoilless large peak obtained at $E_{\Lambda} = 5$ MeV ($E_x = 12$ MeV), and the small peak obtained at $E_{\Lambda} = 0$ MeV ($E_x = 7.2$ MeV) correspond to the $p^{\Lambda}_{\perp}$ and $p^{\Lambda}_{//}$ states, respectively. Note that in the calculation we use the Λ single-particle energies $\varepsilon_{\Lambda}(p) - \varepsilon_{\Lambda}(s) = 10.5$ MeV which is consistent with the 3^+ peak energy in ${}^{12}_{\Lambda}$B. As seen in Fig. 2 (b), we could obtain a group of strongly excited states to explain the experimental bump (#a) at the right position $E_x = 8.34$ MeV. This fact suggests that the energy position of the $p^{\Lambda}_{//}$ orbit stays around 8.4 MeV above the $\varepsilon_{\Lambda}(s)$ orbit, which energy difference is a slightly larger than that in ${}^{9}_{\Lambda}$Be but is still sufficiently low in energy. The situation is realized under the influence of the $\alpha\alpha$-like core deformation also in this adjacent hypernucleus ${}^{10}_{\Lambda}$Be, and hence causes large mixings between the $[C(J^+_c)s^{\Lambda}]$ and $[C(J^-_c)p^{\Lambda}_{//}]$ configurations as dicussed above. These parity-mixed states obtained at around $E_{\Lambda} = 0$ MeV ($E_x = 8.4$ MeV) can be strongly excited in the DWIA calculations because of the wave function admixtures and explain the extra broad bump (#a) observed in the JLab experiment. Although the $p^{\Lambda}_{\perp}$-dominant states are not distinguished in Fig. 2 (b), we note it is selectively excited in the recoilless ^{10}B$(K^-, \pi^-)\,{}^{10}_{\Lambda}$B reaction (not shown here).

On the other hand, as shown in Fig. 2 (c), the $5/2^-$ peak having the $p^{\Lambda}_{//}$ nature is obtained at $E_{\Lambda} = -1$ MeV ($E_x = 9.0$ MeV) in the (π^+, K^+) spectrum. The peak having the $p^{\Lambda}_{\perp}$ nature is not distinguished in the (π^+, K^+) spectrum, but can be found clearly at $E_x = 12$ MeV in the (K^-, π^-) spectrum. We remark that the energy of $p^{\Lambda}_{//}$ orbit with respect to the ground state increases as going from ${}^{9}_{\Lambda}$Be $\rightarrow$ ${}^{10}_{\Lambda}$Be $\rightarrow$ ${}^{11}_{\Lambda}$B. Correspondingly the energy splitting between $p^{\Lambda}_{\perp}$ and $p^{\Lambda}_{//}$ states becomes smaller with increasing mass number. Therefore we remark such a

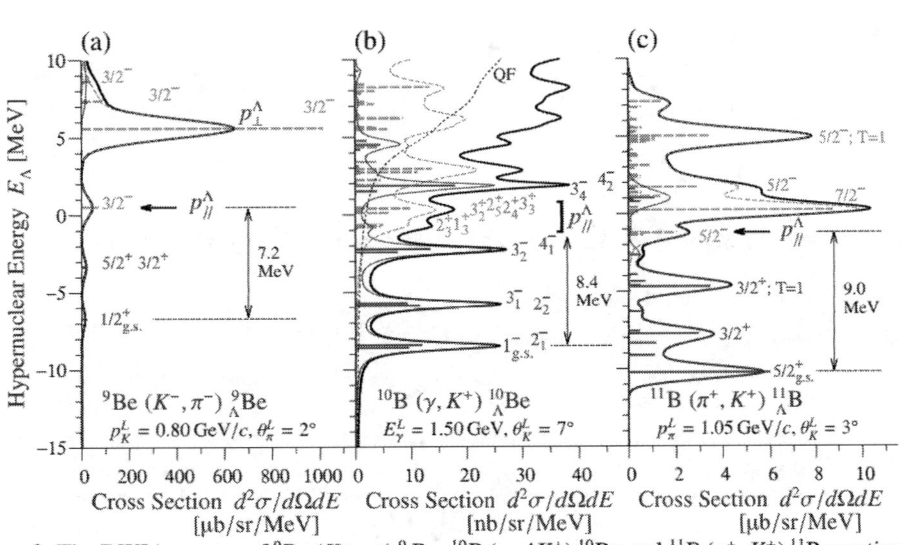

Figure 2. The DWIA spectra of $^{9}\mathrm{Be}\,(K^-,\pi^-)\,^{9}_{\Lambda}\mathrm{Be}$, $^{10}\mathrm{B}\,(e,e'K^+)\,^{10}_{\Lambda}\mathrm{Be}$ and $^{11}\mathrm{B}\,(\pi^+,K^+)\,^{11}_{\Lambda}\mathrm{B}$ reactions.

possibility that the hypernuclear spectra detects or reflects the change of nuclear core deformations of several adjacent species. Also we confirm that, in the spherical nuclear core such as ^{12}C, p^{Λ} state does not split into $p^{\Lambda}_{\perp}$ and $p^{\Lambda}_{//}$ states. The new-type wave function should appear in $^{9,10}_{\Lambda}$Be and $^{11}_{\Lambda}$B depending on the amount of core deformation, but not in spherical systems without appreciable deformation.

4 Concluding remarks and acknowledgements

For hypernuclei with $A = 9$–11, we have applied an extended shell model which takes both natural- and unnatural-parity core-excited states on the same footing. The new-type wave functions have been used to estimate the production cross sections in DWIA. In $^{9,10}_{\Lambda}$Be and $^{11}_{\Lambda}$B, the p^{Λ} state splits into $p^{\Lambda}_{\perp}$ and $p^{\Lambda}_{//}$ orbits, and then the lower $p^{\Lambda}_{//}$ orbit comes down in energy and couples easily with s^{Λ} state. In fact, the finding of peak #a in $^{10}\mathrm{B}\,(e,e'K^+)\,^{10}_{\Lambda}\mathrm{Be}$ reaction is a novel evidence for genuine hypernuclear wave function with parity-mixing realized in deformed hypernuclei. The present study will be useful for $(K^-,\pi^-\gamma)$ and $(\pi^+,K^+\gamma)$ reaction experiments with high-intensity and high-resolution that are being planned at J-PARC.

We are grateful to P. Bydžovský for providing us with an available form of the SLA amplitudes. This work was supported by JSPS KAKENHI Grant Number JP20K03950.

References

[1] T. Gogami et al., Phys. Rev. C **103**, L041301 (2021), and references therein
[2] T. Gogami et al., Phys. Rev. C **93**, 034314 (2016)
[3] G. M. Urciuoli et al., Phys. Rev. C **91**, 034308 (2015), and references therein
[4] T. Motoba, M. Sotona, and K. Itonaga, Prog. Theor. Phys. Suppl. **117**, 123 (1994)
[5] A. Umeya, T. Motoba, and K. Itonaga, JPS Conf. Proc. **26**, 023016 (2019)
[6] A. Umeya, T. Motoba, and K. Itonaga, J. Phys. Conf. Ser. **1643**, 012110 (2020)
[7] T. Mizutani, C. Fayard, G.-H. Lamot, and B. Shagai, Phys. Rev. C **58**, 75 (1998)
[8] T. Motoba, H. Bando, K. Ikeda, and T. Yamada, Prog. Theor. Phys. Suppl. **81**, 42 (1985)
[9] R. H. Dalitz and A. Gal, Phys. Rev. Lett. **36**, 362 (1976); Ann. Phys. **131**, 314 (1981)

Spectrum of few-body systems in a finite volume

Betzalel Bazak[1,*], *Martin* Schäfer[1], *Roee* Yaron[1], and *Nir* Barnea[1]

[1]The Racah Institute of Physics, The Hebrew University, Jerusalem 9190401, Israel

Abstract. Lattice quantum chromodynamics calculations of multi-baryon systems with physical quark masses could bridge the gap between nuclear physics and its fundamental theory. Such calculations are performed on a finite grid, and thus extrapolation of finite-volume numerical results to the physical-relevant free-space quantities is needed. A common approach for the extrapolation utilizes the two-body Lüscher formula and its generalization to larger systems. To understand better the accuracy of this approach we study here two model systems, bosonic and fermionic. The few-body systems are solved with the stochastic variational method in a periodic box, and the extrapolated values are compared to those calculated directly in free space.

1 Introduction

The underlying theory of the strong force is quantum chromodynamics (QCD), and thus it is of great interest to understand nuclear physics directly from QCD. *Ab initio* QCD predictions of few-nucleon systems can be achieved from numerical calculations on a finite lattice, known as lattice QCD (LQCD). Such calculations are performed with discrete spacing between lattice sites and in a finite volume with periodic boundary conditions. For infinitesimal spacing and infinite box size, LQCD provides an exact solution of QCD [1]. Since the box sizes in LQCD calculation are in the same length scale as the size of the wave function, an appropriate extrapolation method is required.

A popular method for extrapolating finite volume results to the physical free-space values utilizes the Lüscher formula [2], relevant for the two body case in the limit of short-range interaction and large boxes. A similar formula is valid also for few-body bound states, assuming that the asymptotic part of their wave function describes separation of the system into two subclusters, and treating each subcluster as a point-like particle [3].

The applicability of such an approach for light nuclei was studied recently in the framework of pionless effective field theory [4]. Here we expand this work by studying two other few-particle systems, namely spinless bosons interacting with local Gaussian interaction, as well as nucleons interacting with the Minnesota potential [5].

We perform few-body calculations in a periodic box and free space using a correlated Gaussians basis, which is optimized with the stochastic variational method (SVM) [6]. The resulting shifts in the binding energy induced by the finite volume are compared to the values predicted by the relevant formulae using parameters extracted directly from free-space calculations.

*e-mail: betzalel.bazak@mail.huji.ac.il

2 Models and methods

The leading correction to the binding energy of a two-body s-wave bound state ΔB_2, in the limit of large box size L, was derived by Lüscher [2],

$$\Delta B_2 = \frac{6\kappa_2|\mathcal{A}_2|^2}{\mu_2 L} e^{-\kappa_2 L}, \tag{1}$$

where μ_2 is the reduced mass, $\kappa_2 = \sqrt{2\mu_2 B_2^{\text{free}}}$ is the binding momentum, and $\mathcal{A}_2$ is the dimensionless two-body asymptotic normalization coefficient (ANC). Here we use $\hbar = 1$.

A similar approach was used to analyze N-body bound states, once the lowest s-wave threshold of the system corresponds to a breakup into two subclusters, assuming that each of these subclusters behaves as a point-like particle. In this case, the leading correction is [3],

$$\Delta B_N = C_N \frac{6\kappa_N|\mathcal{A}_N|^2}{\mu_N L} e^{-\kappa_N L}, \tag{2}$$

where μ_N is the reduced mass of the subclusters and $\kappa_N = \sqrt{2\mu_N(B_N^{\text{free}} - B_a^{\text{free}} - B_b^{\text{free}})}$ is the binding momentum calculated from the free-space N-body binding energy B_N^{free} and the binding energies B_a^{free}, B_b^{free} of subclusters a and b. Here, C_N stands for a combinatorial factor that counts the number of options to partition N particles into two subclusters and $\mathcal{A}_N$ denotes the ANC corresponding to the asymptotic part of the N-body wave function given by the inter-cluster coordinate. In principle, one should sum over corrections from all possible cluster configurations, however, the contribution of the lowest threshold is usually the dominant one.

Here, we study the finite-volume implications on two model cases. The first case is a system of a few spinless bosons interacting via local Gaussian potential,

$$V(r) = V_0 \exp(-r^2/R^2), \tag{3}$$

where the potential parameters, $V_0 = -5$ and $R = 1$, are taken from Ref. [3], and we work in the natural units where $m = 1$.

The second case represents a series of light nuclei with $A \leq 4$ nucleons interacting via the Minnesota nucleon-nucleon potential [5], with the exchange-mixture parameter $u = 1$. This spin-dependent interaction was fitted to reproduce nucleon-nucleon effective range parameters and gives reasonable binding energies for the deuteron, triton, and ^{4}He nuclei treated here. In this work, we neglect the Coulomb interaction.

Considering only the ground states of the s-wave nuclei ^{2}H, ^{3}H, and ^{4}He, we note that the lowest separation thresholds are those associated with breakup into $(N-1)$-particle subcluster and one nucleon. This is also the case for the bosonic system.

The binding energies and wave functions are solved numerically with the SVM. The N-body wave function is expanded in a correlated Gaussian basis, where the basis functions parameters are optimized to the system at hand in a stochastic way. Putting the system in a periodic box, the boundary conditions are enforced by including all possible shifts of the basis functions in an integer multiplication of the box size [7]. Our implementation of the SVM in a box is described in Ref. [8], while to enable fast and accurate calculations such as those presented here, a significant upgrade of the code was needed [4]. For more details see Refs. [4, 7, 8].

3 Results

3.1 Systems of few bosons

First, we show the energy shift due to the periodic box for the bosonic systems studied in Ref. [3]. The energy shifts for the $N = 2$, 3, and 4-body systems are shown as a function of the box sizes in figure 1. For large boxes, the results are in agreement with the Lüscher-like formulae, Eqs. (1) and (2). Both κ_N and $\mathcal{A}_N$ were extracted from free-space calculations, and therefore no free parameter was used. For small boxes, however, significant deviations exist (note the log scale), and the validity of the asymptotic formulae deteriorates.

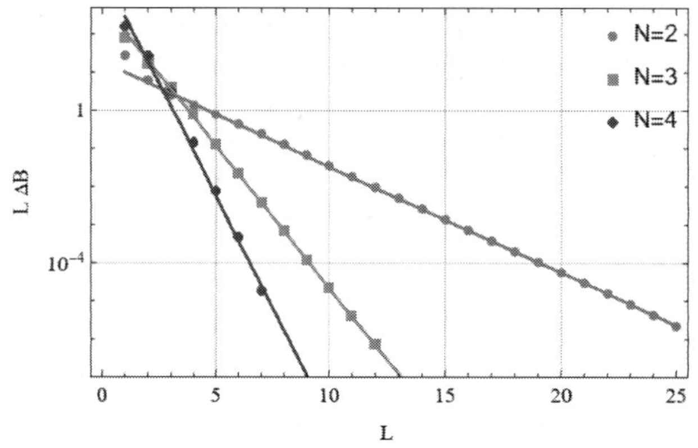

Figure 1. The energy shift ΔB_N due to the finite volume, multiplied by the box size L, as a function of L for the bosonic case, in natural units. Shown are results for dimer (red), trimer (green), and tetramer (blue). The symbols represent SVM results, while the curves are the the Lüscher-like formulae, Eqs. (1,2) using free-space $\mathcal{A}_N$ and κ_N values.

The binding energies and the ANCs for the bosonic systems are listed in table 1, and are compared to the results of Ref. [3]. Small differences between the values result from the fact that the kinetic energy was calculated in Ref. [3] with a finite difference scheme.

Table 1. Binding energies (in natural units) and dimensionless ANCs for the bosonic case, from Ref. [3] and as calculated here. ANCs are calculated by fitting the energy shifts in a box (fit), by the expectation value of a short-range operator (operator), and deduced directly from the free-space wave function (free).

N	BE		$\mathcal{A}_N$			
	Ref. [3]	This work	Ref. [3]		This work	
			fit	operator	fit	free
2	0.449	0.4061	1.634(3)	1.633(9)	1.58(1)	1.57682(1)
3	2.916	2.6263	3.92(5)	4.11(6)	3.8(1)	3.707(2)
4	–	7.5135	–	–	8(2)	6.99(1)

3.2 Light nuclei

Now we move to the second case, i.e. light nuclei with the Minnesota potential. The energy shift in a box is shown in figure 2. For large boxes, the SVM results fit nicely the Lüscher-like formulae, Eqs. (1,2) using free-space $\mathcal{A}_N$ and κ_N values. Also here we see significant differences for small boxes, showing the limitation of the asymptotic formulae.

The binding energies and the ANCs for light nuclei with the Minnesota nucleon-nucleon potential, as well as ANC values extracted from experiment results [9–11], are listed in table 2.

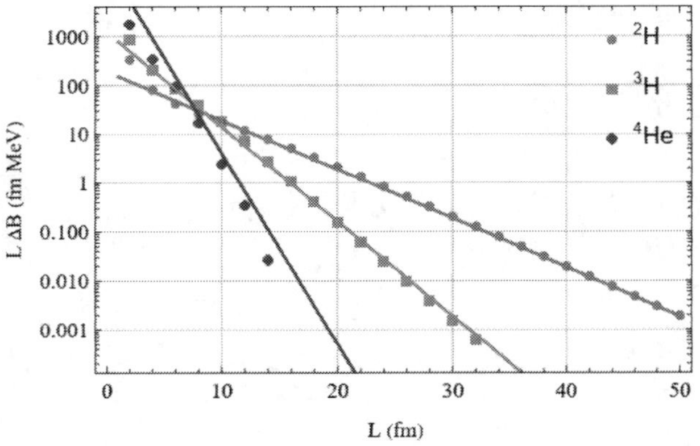

Figure 2. The energy shift ΔB_N due to the finite volume, multiplied by the box size L, as a function of L, for light nuclei using the Minnesota potential. Shown are results for deuteron (red), triton (green), and ^{4}He (blue). Symbols represent SVM results, and curves are the predictions of Lüscher-like formulae, Eq. (1,2) using free-space $\mathcal{A}_N$ and κ_N values.

Table 2. Binding energies (in MeV) and dimensionless ANCs of one nucleon separation for light nuclei. ANCs are calculated by fitting the energy shifts in a box (fit), deduced directly from the free-space wave function (free), and compared also to the values extracted from experiments (exp.).

sys.	BE (MeV)	$\mathcal{A}_N$ fit	free	exp.	Ref.
^{2}H	2.2023	1.3(1)	1.28703	1.2902(65)	[9]
^{3}H	8.3858	1.8(1)	1.9145(1)	2.19(2)	[10]
^{4}He	30.782	4(1)	5.265(3)	4.11-6.42	[11]

4 Conclusion

Here we studied the accuracy of the formulae for the binding energy shift in a periodic box, and their applicability to extract free-space parameters, like the ANC, from finite-volume calculations. We explored two cases: a few bosons interacting via local Gaussian potential, and light nuclei with the Minnesota nucleon-nucleon interaction. In both cases, these formulae indeed give the leading correction, but quite large boxes are needed for performing accurate extrapolation.

References

[1] K. G. Wilson, Phys. Rev. D **10**, 2445 (1974)
[2] M. Lüscher, Commun. Math. Phys. **104**, 177 (1986); Commun. Math. Phys. **105**, 153 (1986)
[3] S. König and D. Lee, Phys. Lett. B **779**, 9 (2018); S. König, Few-Body Syst. **61**, 20 (2020)

[4] R. Yaron, B. Bazak, M. Schäfer, and N. Barnea, Phys. Rev. D **106**, 014511 (2022)
[5] D. R. Thompson, M. Lemere, and Y. C.Tang, Nucl. Phys. A **286**, 53 (1977)
[6] K. Varga and Y. Suzuki, Phys. Rev. C **52**, 2885 (1995); Y. Suzuki and K. Varga, *Stochastic Variational Approach to Quantum-Mechanical Few-Body Problems* (Springer, Berlin, 1998)
[7] X. Yin and D. Blume, Phys. Rev. A **87**, 063609 (2013)
[8] M. Eliyahu, B. Bazak and N. Barnea, Phys. Rev. C **102**, 044003 (2020)
[9] I. Borbély et al., Phys. Lett. B **160**, 17 (1985)
[10] N. K. Timofeyuk, Phys. Rev. C **81**, 064306 (2010)
[11] L. D. Blokhintsev, I. Borbely, and E. I. Dolinskii, Sov. J. Part. Nuclei **8**, 485 (1977)

Mean field and beyond mean field calculations of hypernuclei for study of electroproduction

P. Veselý[1,*], *G. De Gregorio*[2,3], *F. Knapp*[4], *D. Petrellis*[1], *P. Bydžovský*[1], *D. Denisova*[1,4], *J. Pokorný*[1], *D. Skoupil*[1], and *N. Lo Iudice*[5]

[1]Nuclear Physics Institute, Czech Academy of Sciences, 250 68 Řež, Czech Republic
[2]Dipartimento di Matematica e Fisica, Università degli Studi della Campania "Luigi Vanvitelli", viale Abramo Lincoln 5, I-81 100 Caserta, Italy
[3]Istituto Nazionale di Fisica Nucleare, Complesso Universitario di Monte S. Angelo, Via Cintia, I-80 126, Italy
[4]Institute of Particle and Nuclear Physics, Faculty of Mathematics and Physics, Charles University, V Holešovičkách 2, 180 00 Prague, Czech Republic
[5]Dipartimento di Fisica, Università di Napoli Federico II, 80126 Napoli, Italy

Abstract. We present two methods, the Nucleon-Lambda Tamm Dancoff Approximation (NΛ TDA) and the Equation of Motion Phonon Method (EMPM) suitable for calculating hypernuclear energy spectra and structure. These methods are applicable for hypernuclei of wide range of masses with one Λ particle replacing one nucleon in an even-even nuclear cores. Using an effective Lambda-nucleon (ΛN) potential both methods were applied to calculate the energy spectrum of ${}^{12}_{\Lambda}$B and also one body density matrix elements (OBDME). The OBDME were applied to calculate the electroproduction cross section of ${}^{12}_{\Lambda}$B. We obtained better agreement with the experimental data by using EMPM than NΛ TDA. We plan to provide theoretical prediction (by applying the same methods and ΛN potentials) of the cross section in electroproduction of ${}^{40}_{\Lambda}$K and ${}^{48}_{\Lambda}$K.

1 Introduction

It is of great interest to study the energy spectra of Λ hypernuclei and explore to what extent the coupling to the nucleonic excitations of the nuclear core affects the properties of these systems. This goal can be reached by applying many-body methods which go beyond the mean-field approach as NΛ TDA [1]. The EMPM extended for hypernuclei is suitable for this purpose. Here we study the cross section for electroproduction of ${}^{12}_{\Lambda}$B by using both methods.

2 Extension of EMPM for Λ hypernuclei

The EMPM was introduced to describe the structure of ordinary nuclei composed of protons and neutrons [2–4]. The extension of EMPM for single Λ hypernuclei with odd-even nuclear cores is provided by coupling of the NΛ TDA phonons to (multi)phonon core excitations. For a detailed introduction into the hypernuclear extension of EMPM see [6].

*e-mail: p.vesely@ujf.cas.cz

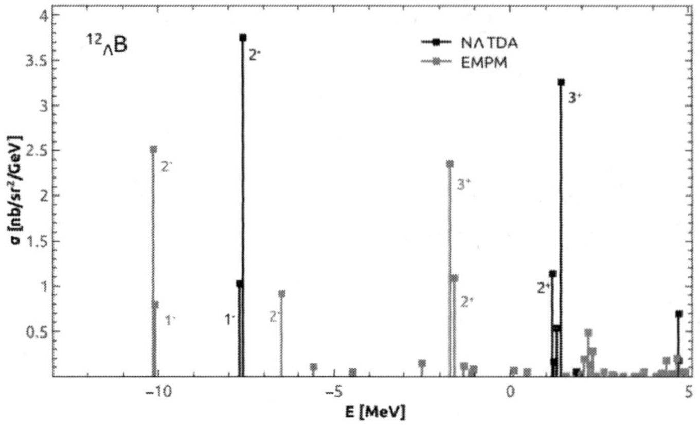

Figure 1. Cross sections of $^{12}_{\Lambda}$B electroproduction, calculated by the $N\Lambda$-TDA and the EMPM, depending on the binding energy E.

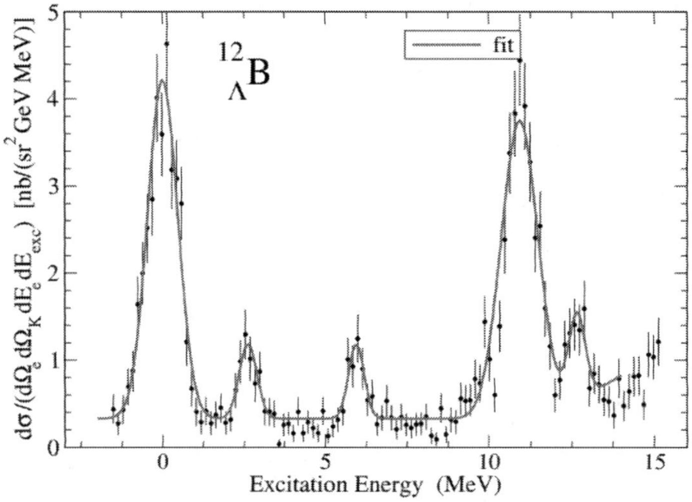

Figure 2. Experimental cross sections of $^{12}_{\Lambda}$B electroproduction (dots with error bars) and fit of these data (green line). Taken from [10].

3 Results

The NΛ TDA and EMPM were used to study properties and electroproduction cross sections of $^{12}_{\Lambda}$B.

For these calculations we adopted a Hamiltonian which consists of the intrinsic kinetic term, the nucleon potential $NNLO_{sat}$ [7] (which includes two-body NN and three-body NNN interactions), and the effecive ΛN potential derived as a Gaussian expansion of the G-matrix on the Nijmegen-F potential [8]. For the technical details of the calculations see [6].

The calculation of the cross section of electroproduction depends on the amplitude of the elementary process and the kinematics of the reaction. For the elementary amplitude we applied an isobar model using the parametrization BS3 [9], while the used kinematics is consistent with Hall-A settings [6, 10]. For a more detailed description of the theoretical formalism for the calculation of electroproduction cross section see [11].

The strength of the electroproduction cross section of $^{12}_{\Lambda}$B calculated within the EMPM is more fragmented with respect to the NΛ TDA method as is seen in figure 1. The fragmentation is the result of the coupling of NΛ TDA states to the excitations of the nuclear core. The fragmentation of the calculated strength makes the cross section more realistic with respect to the experimental one plotted in figure 2.

Table 1. Ground-state Λ separation energies B_Λ obtained by the NΛ TDA and EMPM in comparison with the experimental value [12].

	B_Λ [MeV]
NΛ TDA	7.70
EMPM	10.29
exper.	11.10 ± 0.50

The other effect of the coupling of NΛ TDA states to the excitations of the nuclear core is the downward shift of the energy spectrum of $^{12}_{\Lambda}$B. In table 1 we show the ground-state Λ separation energies obtained by the NΛ TDA and EMPM in comparison with the experimental value [12].

4 Conclusion and outlook

The cross sections of $^{12}_{\Lambda}$B electroproduction calculated with the beyond mean-field approach EMPM are more fragmented than the results obtained by the mean-field approach NΛ TDA. This effect is crucial for the description of the experimental data on this hypernucleus. Although the EMPM provides better qualitative agreement with the experimental data than the NΛ TDA, we plan to improve further our calculations by applying more recent ΛN interactions.

We plan to study the electroproduction cross sections of $^{40}_{\Lambda}$K and $^{48}_{\Lambda}$K within the NΛ TDA and EMPM methods in order to provide theoretical predictions.

5 Acknowledgment

We thank P. Navrátil for having provided the matrix elements of the $NNLO_{sat}$. This work was supported by the Czech Science Foundation GACR Grant No. 19-19640S and by the Charles University Research Center UNCE/SCI/013. F. Knapp and P.Veselý thank the INFN for financial support.

References

[1] P. Veselý, G. De Gregorio, J. Pokorný, Phys. Scr. **94**, 014006 (2019)
[2] D. Bianco et al., Phys. Rev. C **85**, 014313 (2012)
[3] F. Knapp et al., Phys. Rev. C **92**, 054315 (2015)
[4] G. De Gregorio, F. Knapp, N. Lo Iudice, P. Veselý, Phys. Rev. C **93**, 044314 (2016)
[5] G. De Gregorio, F. Knapp, N. Lo Iudice, P. Veselý, Phys. Rev. C **94**, 061301(R) (2016)
[6] D. Petrellis, P. Veselý, G. De Gregorio, F. Knapp, P. Bydžovský, D. Denisova, J. Pokorný, D. Skoupil, N. Lo Iudice, *Investigation of Λ Hypernuclei within the Equation of Motion Phonon Method*, in preparation
[7] A. Ekström et al., Phys. Rev. C **91**, 051301(R) (2015)
[8] Y. Yamamoto et al., Prog. Theor. Phys. Suppl. **117**, 361 (1994)
[9] D. Skoupil, P. Bydžovský, Phys. Rev. C **97**, 025202 (2018)
[10] F. Garibaldi et al., Phys. Rev. C **99**, 054309 (2019)
[11] P. Bydžovský, D. Denisova, D. Skoupil, P. Veselý, arXiv:2209.00881v1, Phys. Rev. C subm.
[12] T. Hiyoshi et al., Phys. Rev. Lett **90**, 232502 (2003)

Kinematical effects in electroproduction of hypernuclei

Daria Denisova[1,2,*]

[1]Nuclear Physics Institute, Czech Academy of Sciences, Řež, Czech Republic
[2]Institute of Particle and Nuclear Physics, Charles University, Prague, Czech Republic

Abstract. On the example of electroproduction of ${}^{12}_{\Lambda}$B we show effects in the cross section caused by various choices of the momentum of the initial proton. We have found that the effects are quite noticeable, mainly for small kaon angles. This extention of our model calculations beyond the frosen-proton approximation represents an important improvement of theoretical predictions.

1 Motivation

Studying spectra in electroproduction of hypernuclei provides important information about details of the effective ΛN interaction, mainly on its spin-dependent part. A reasonable analysis of experimental data, however, requires a good knowledge of the reaction mechanism. One of uncertainties in the DWIA calculations of the cross sections is the Fermi motion of the initial (target) proton. Here we will discuss effects due to various choices of the proton momentum in the optimal factorization approximation. Assuming non-zero values of the proton momentum improves on previous calculations [2] where the zero momentum (the frozen-proton approximation) was used.

2 Calculation of the cross section and the results

The cross sections for electroproduction of hypernuclei are calculated in the impulse approximation. Details on the calculation can be found in Ref. [1], here we mention only some facts concerning the extention of our previous model calculations presented in Ref. [2].

In the optimal factorization approximation the Fermi averaging integral is factorized by assuming the elementary amplitude at an effective (non zero) proton momentum. To this end, the two-component form of the elementary amplitude was extended also by including dependence on the proton momentum. The remaining parts of the electroproduction amplitude, the radial integrals and one-body density matrix elements (OBDME) were treated as in the previous calculations [2].

The kaon momentum is calculated from the energy conservation. In the elementary amplitude, we use the kaon momentum calculated from the energy conservation in the elementary vertex (2-body), the amplitude is therefore on-energy-shell, but the kaon momentum used in the radial integrals and in a kinematical factor at the cross section [1] is calculated from the overall (many-body) energy conservation. This hybrid scheme with two different kaon momenta was also used in the previous calculations [1].

*e-mail: denisova@ujf.cas.cz

Using a new form of the elementary amplitude with a non-zero proton momentum, however, allows to consider an optimum value of the proton momentum which fulfils the energy conservation both in the 2-body and many-body systems. Then the calculation proceeds with one common value of the kaon momentum and the elementary amplitude is on-energy-shell.

We consider the following values of the effective proton momentum (p_p):
a) $p_p = 0$, the frozen-proton approximation;
b) p_p equal to momentum transfer between photon and kaon and, therefore $p_\Lambda = 0$ (the frozen-Λ approximation);
c) p_p equal to the optimum value (the optimum on-shell approximation).

In the figure we show the angular dependence of the electroproduction cross sections for excited states 1^+, 2^+, and 3^+ of ${}^{12}_{\Lambda}$B. The results are for the three selected proton momenta given above calculated in PWIA with the elementary amplitude BS3 [3]. The nucleus-hypernucleus structure (OBDME) is described in the Shell-model calculations by John Millener as in Ref. [2]. Similar results in DWIA and more details on the analysis can be found in [1].

The calculations in the proton and Λ-frozen approximations are done with two different kaon momenta as discussed above, whereas in the optimum on-shell approximation one value of kaon momentum is used. The results differ more for very small kaon angles and the Λ-frozen result is mostly larger than the other results. One can conclude that the differences in PWIA due to the Fermi motion of the proton amount up to about 10–20% for the dominant (3^+) hypernucleus state. Noticeable is also the different ordering of the green (optimum) and red (frosen-proton) curves for the states 1^+ and 3^+ and for 2^+ which is mainly given by selective contributions from the longitudinal part of the electroproduction amplitude [1].

3 Summary and outlook

A general two-component form of the amplitude in electroproduction of kaons on the nucleon was constructed. This new form, which contains a dependence on the proton momentum, allows to extend our calculations beyond the frozen-proton approximation and show effects from a proton Fermi motion in the target nucleus. In the PWIA calculations, the effects amount to about 10–20% for the most strongly populated states and they are larger at small kaon angles. We suggest using the optimum on-shell approximation, which makes kinematics (the kaon momentum) more consistent.

We plan to continue our analysis by assuming the sensitivity of the cross sections to various descriptions of the nuclear and hypernuclear structures.

This work was supported by the Czech Science Foundation GACR Grants No. 19-19640S.

References

[1] P. Bydžovský, D. Denisova, D. Skoupil, P. Veselý, "Fermi motion effects in electroproduction of hypernuclei", Phys. Rev. C **106**, 044609 (2022).

[2] F. Garibaldi, *et al*, Phys. Rev. C **99**, 054309 (2010); P. Bydžovský, *et al*, AIP Conf. Proc. 2130, 020016 (2019).

[3] D. Skoupil, P. Bydžovský, Phys. Rev. C **97**, 025202 (2018).

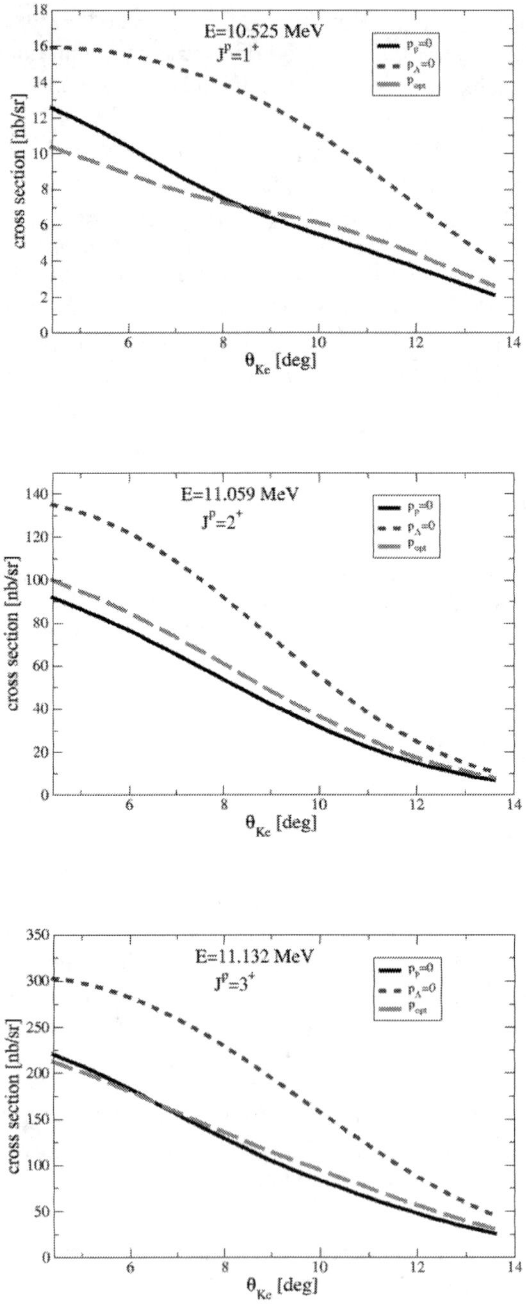

Figure 1. Angular dependence of the electroproduction cross sections induced by virtual photons for three excited states of ${}^{12}_{\Lambda}B$ with the energy E and spin-parity J^P.

The strangeness program at GlueX

Peter Pauli[1,*] , for the GlueX collaboration

[1]School of Physics and Astronomy, G12 8QQ, Glasgow, UK

Abstract. The GlueX experiment located at Jefferson Lab studies the spectrum of hadrons using photoproduction on a LH_2 target in a wide variety of final states. With its detector system capable of measuring neutral and charged final state particles over almost the full solid angle, and very good particle identification capabilities, GlueX can measure many different hadrons containing strangeness. A linearly polarized photon beam allows the measurement of polarization observables, which contain information about the production mechanisms involved in generating strange particles in photoproduction. In addition, GlueX can perform precise cross-section measurements, which help to study the spectrum of strange hadrons. In this presentation, the GlueX experiment is introduced, and recent progress of its strangeness program is discussed. We present recent results on Σ^0 beam asymmetries, $\Lambda(1520)$ spin-density matrix elements and ongoing studies of the $\Lambda(1405)$ lineshape. We also present our recent progress on measurements of $\Lambda\bar{\Lambda}$ and $\Xi^{(*)}$ photoproduction. Also, future prospects for strangeness measurements at GlueX are discussed.

1 Introduction

Although much progress has been made over the last years in studying the hadron spectrum of light and heavy quark states, the spectrum of hadrons containing strange quarks is only poorly understood. For example, at present only a fraction of the hyperon states predicted by theory are experimentally confirmed. One possibility to expand the available data is the study of strange hadrons in photoproduction. It provides a complementary production process to hadron beam experiments and utilizing linearly polarized photons enables the study of production mechanisms.

GlueX is a photoproduction experiment located in Jefferson Lab's Hall D. A linearly polarized photon beam is produced off a diamond radiator via the coherent bremsstrahlung technique. The diamond is oriented such that the largest fraction of polarization is located at photon energies between 8.2 GeV to 8.8 GeV. This fixed target experiment uses a liquid hydrogen target which is surrounded by the GlueX detector setup, consisting of multiple tracking detectors and calorimeters as well as particle identification detectors providing almost 4π acceptance. The main motivation behind the GlueX experiment is the search for exotic hybrid mesons. Since these are expected to decay into charged and neutral final state particles, GlueX was designed to have good energy and momentum resolution for both. A comprehensive description of the GlueX detector system and beamline can be found in Ref. [1]. The large acceptance paired with good resolutions for different particle types makes GlueX

*e-mail: Peter.Pauli@glasgow.ac.uk

an ideal detector system to study strange hadrons, which often decay into final states with charged and neutral particles. We will show that GlueX has an exciting strangeness program with high statistical precision for many strange hadron reactions.

GlueX's first campaign of data taking (GlueX-I) ran from 2017-18 and resulted in an integrated luminosity of about 440 pb^{-1} for 6 GeV < E_γ < 11.6 GeV. Afterwards GlueX was upgraded with a new DIRC detector to improve π/K separation and GlueX-II is taking data since 2020. Unless otherwise stated the data presented in this talk represents the full GlueX-I statistics.

2 $\Lambda\bar{\Lambda}$ production

GlueX is well suited to study baryon-anti-baryon photoproduction and some preliminary results for $p\bar{p}$ and $\Lambda\bar{\Lambda}$ have already been reported at previous conferences [2, 3]. These reactions are interesting as they allow to search for baryonium, a bound baryon-anti-baryon system which has so far not been observed. However, BES [4] and BESIII [5] reported interesting cross-section enhancements near threshold for $e^+e^- \to \gamma p\bar{p}$ and $e^+e^- \to \Lambda\bar{\Lambda}$, respectively. This motivated a closer examination of these two final states at GlueX, where these reactions can be measured with high statistical accuracy over a wide range of photon beam energies.

We focus here on $\Lambda\bar{\Lambda}$, which we measure in the reaction $\gamma p \to \Lambda\bar{\Lambda}p \to \{\pi^- p\}\{\pi^+\bar{p}\}p$. Little theoretical work has been done so far to describe the reaction, so a first focus is on the production mechanisms. Studying angular distributions of Λ (Fig. 1 left), $\bar{\Lambda}$ and p in the centre-of-mass frame reveals that the reaction process seems to be dominated by t-channel exchanges in which the $\Lambda\bar{\Lambda}$ pair is produced recoiling against the proton (Fig. 1 middle). However, there is also a sizeable contribution of events where $p\bar{\Lambda}$ are produced as pair recoiling via strangeness exchange against a Λ (Fig. 1 right). There is no indication of production of a $p\Lambda$ pair recoiling against $\bar{\Lambda}$. These two reaction mechanisms are intriguing as they

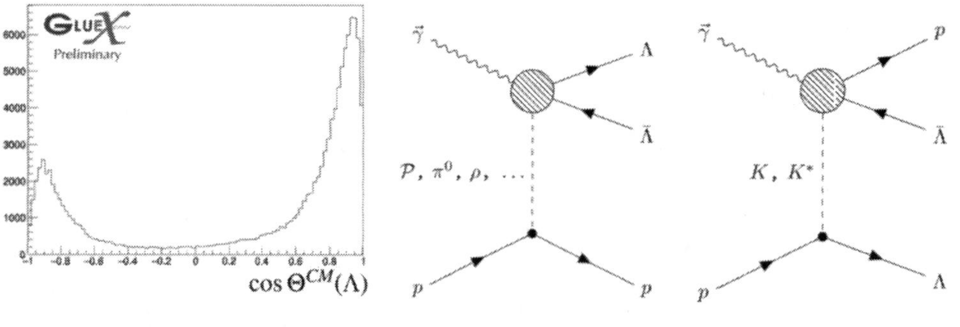

Figure 1. The left panel shows the $\cos\theta$ distribution of Λ in the center-of-mass frame for the reaction $\gamma p \to \Lambda\bar{\Lambda}p$. The forward and backward peaks can be explained by the two possible production mechanisms shown to the right of it. The middle panel shows the $\Lambda\bar{\Lambda}$ pair being produced at the top vertex recoiling against a proton. The right panel depicts a process in which strangeness is exchanged and a $p\bar{\Lambda}$ pair is produced at the top vertex recoiling against a Λ. All taken from Ref. [3].

could proceed through intermediate mesons. In order to study this further, the differential cross-sections $d\sigma/d\mathbf{IM}(\Lambda\bar{\Lambda})$ and $d\sigma/d\mathbf{IM}(p\bar{\Lambda})$ are measured for the respective systems. Preliminary results are shown in Fig. 2. The data is shown in blue and the grey area indicates the expected distribution for events produced uniformly in phase-space. Interesting enhancements at threshold are visible in both cases and studies are ongoing to investigate their origin.

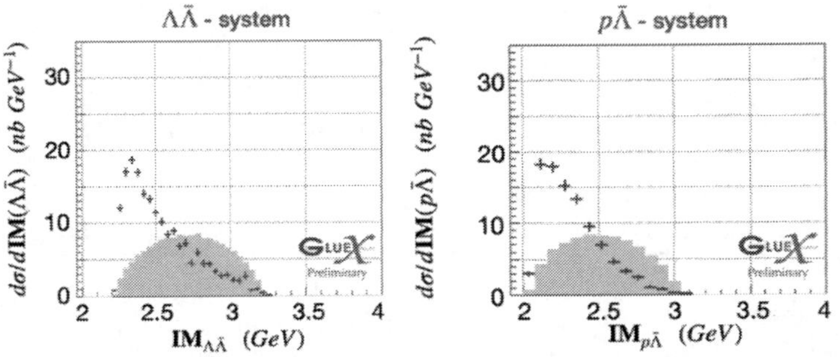

Figure 2. Preliminary differential cross-sections for the $\Lambda\bar{\Lambda}$ (left) and the $p\bar{\Lambda}$ (right) systems. The grey areas indicate the distribution expected for production uniform in phase-space.

Further measurements are planned for this reaction. The Λ, being self-analyzing due to its weak decay, offers the opportunity to measure the polarization of the produced baryon pair. In addition, the linearly polarized photon beam of GlueX can be used to study the beam asymmetry Σ, which contains information about the nature of the exchanged particle in the t-channel exchange (c.f. Sec. 5).

3 $\Lambda(1405)$ line shape measurements

The $\Lambda(1405)$ is arguably one of the most interesting hyperons. Although its existence is long established beyond any doubt, it is still controversial in its nature. Many current analyses support a two-pole structure. For a recent review of the $\Lambda(1405)$ see e.g. Ref. [6]. The Particle Data Group include a review of the pole structure of the $\Lambda(1405)$ (Chapter 83 in Ref. [7]) and recently added the $\Lambda(1380)$, the lower of the two poles, to its particle listings. One key measurement to help with the interpretation of the $\Lambda(1405)$ is the precise measurement of its line shape. Due to its good energy and momentum resolutions, GlueX is well suited for this task.

The reaction of interest is $\gamma p \to K^+\Lambda(1405) \to K^+\Sigma^0\pi^0$ with $\Sigma^0 \to \Lambda\gamma \to p\pi^-\gamma$ and $\pi^0 \to \gamma\gamma$. The $\Sigma^0\pi^0$ decay of the $\Lambda(1405)$ is ideally suited to study the resonance as it is free from $\Sigma(1385)$ background since it is isospin $I = 0$. The measurement is done in the photon energy range $E_\gamma = 6.5\,\text{GeV}$ to $11.6\,\text{GeV}$. In total, about 13,350 events where identified in the mass region of the $\Lambda(1405)$. The mass resolution for these events is less than $5\,\text{MeV}$. The data was split in three bins of momentum transfer $-t = -(p_{\text{beam}} - p_{K^+})^2$. The preliminary results are shown in Fig. 3. One can easily identify the $\Lambda(1405)$ and $\Lambda(1520)$ as two peaking structures. The $\Lambda(1405)$ clearly shows a non-Breit-Wigner shape with a sharp drop at the $\bar{K}N$ threshold. It seems that the line shape of the $\Lambda(1405)$ shows a slight t-dependence, which is consistent with a two-pole structure. Investigations into the extraction of the number and positions of poles are currently ongoing.

A comprehensive report on preliminary results by GlueX can be found elsewhere in these proceedings [8].

4 $\Lambda(1520)$ spin-density matrix elemements and cross-sections

Another excited Λ hyperon under investigation at GlueX is the $\Lambda(1520)$. This $J^P = \frac{3}{2}^-$ state is observed in its decay to K^-p. Therefore, the reaction of interest is $\gamma p \to K^+\Lambda(1520) \to K^+K^-p$. The $\Lambda(1520)$ is a very well-established state and a better understanding of its production process will be very helpful for other analyses at GlueX where strangeness exchange in the t-channel is important.

GlueX has published a paper on spin-density matrix elements (SDMEs) in this reaction [9]. SDMEs parametrize the angular decay distribution of the K^- in the $\Lambda(1520)$ rest

 https://doi.org/10.1051/epjconf/202227102001

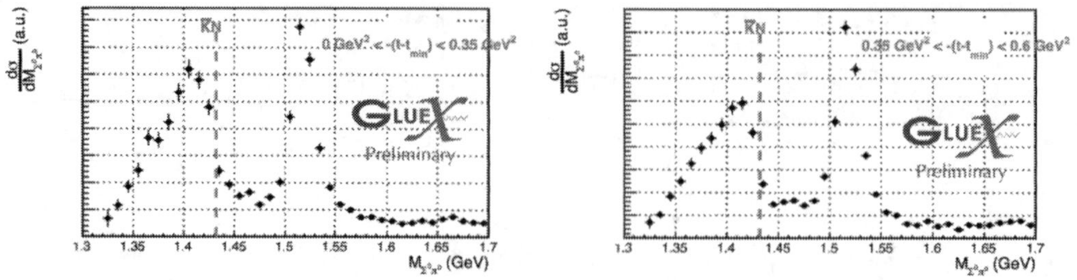

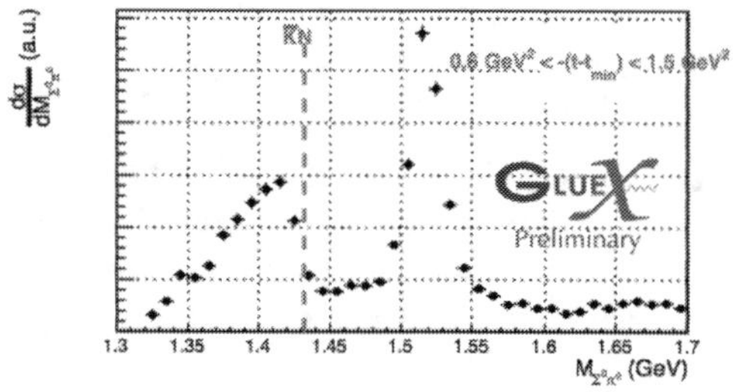

Figure 3. Preliminary results for the differential cross-section measurement of $\gamma p \rightarrow K^+\Sigma^0\pi^0$ in three bins of momentum transfer $-t = -(p_{\text{beam}} - p_{K^+})^2$. The $\Lambda(1405)$ and $\Lambda(1520)$ are clearly identifiable as two peaking structures. The $\Lambda(1405)$ shows a very much, non-Breit-Wigner shape with a sharp drop at the $\bar{K}N$ threshold.

frame. Studying them allows to draw conclusions about the production mechanisms involved in the reaction. More specifically, assuming t-channel exchange as the dominant production mode, one can study the naturality $\eta = P(-1)^J$ of the exchanged particle with quantum numbers J^P. Utilizing the linear polarization of our photon beam, we measured six polarized SDMEs for the first time, in addition we also extracted three independent unpolarized SDMEs. We found, that the production of $\Lambda(1520)$ between $E_\gamma = 8.2\,\text{GeV}$ to $8.8\,\text{GeV}$ is dominated by natural exchanges, i.e. vector or tensor meson exchange. In order to make more precise statements on the exact nature of the exchanged mesons, additional information is needed. For this reason, we are currently working on the extraction of the differential cross-section $\frac{d\sigma}{dt}$. We study 20% of the GlueX-I data and measure $\frac{d\sigma}{dt}$ in 18 bins in momentum transfer $-t = -(p_{\text{beam}} - p_{K^+})^2$ and seven bins in photon beam energy E_γ. Our preliminary results are shown in Fig. 4 in red and compared to the blue data points measured by an experiment at SLAC [10]. SLAC performed their measurements at $E_\gamma = 11\,\text{GeV}$ but in their paper they claim an expected dependency of $1/E_\gamma^2$ for the total cross-section. For this reason, and to compare to our results, we scaled the SLAC data points according to this factor to our energy bins and plotted them in each E_γ bin. We see good agreement between the SLAC data and our data. The agreement is better for higher photon beam energies and gets slightly worse towards lower E_γ, indicating that the scaling factor quoted by SLAC might not be accurate at low energies.

In order to measure total cross-sections, we fit a function with the form $c * t^a * e^{bt}$ to the differential cross-section results and integrate it to large $-t$. The preliminary total cross-section results are shown in Fig. 5 in red. The green data points come from CLAS [11] and the blue data point at 11 GeV was measured by SLAC [10]. In order to compare the results to our preliminary data, we fitted the CLAS and SLAC data with $c * E_\gamma^{2a-2}$ (black dashed curve), which we found to better describe the data than $c * E_\gamma^{-2}$, as suggested by SLAC. The preliminary GlueX results, which were not part of the fit, agree very well with the fit. We were also able to take some data at lower photon beam energies during our 2018 data taking campaign. This data which is yet to be analysed will cover most of the gap between the CLAS data and our preliminary results.

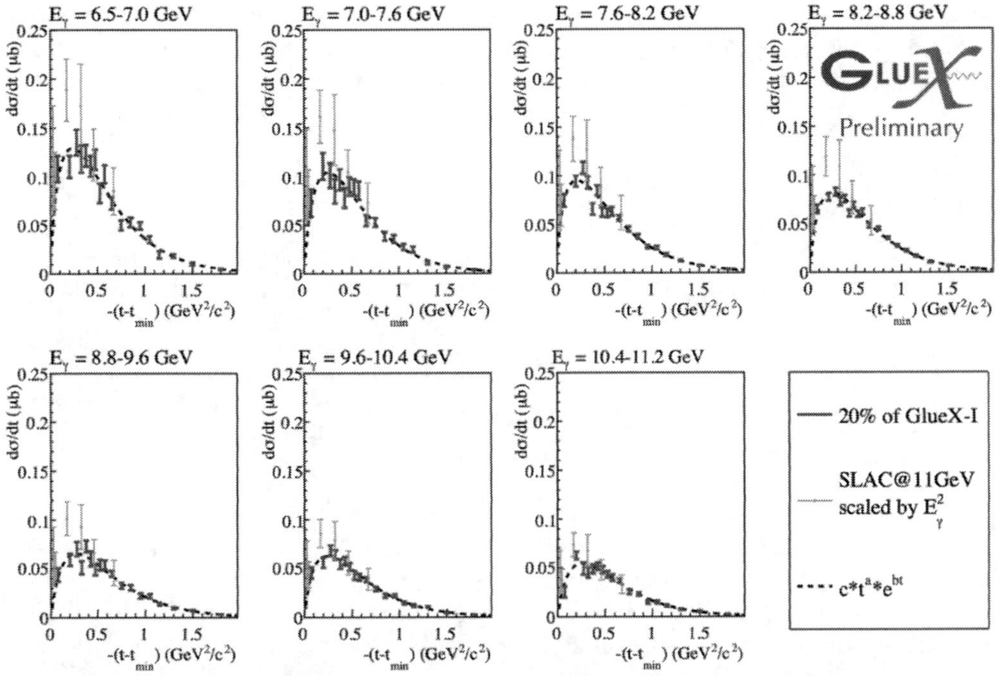

Figure 4. Preliminary differential cross-section measurements for $\Lambda(1520)$ phoptoproduction in seven bins of beam energy and 18 bins of momentum transfer $-t = -(p_{\text{beam}} - p_{K^+})^2$ are shown in red. The blue data points show a previous measurements performed at SLAC [10] at $E_\gamma = 11\,\text{GeV}$ and scaled down by E_γ^2. The black dashed line is the fit used to integrate the differential cross-section data to get the total cross-section.

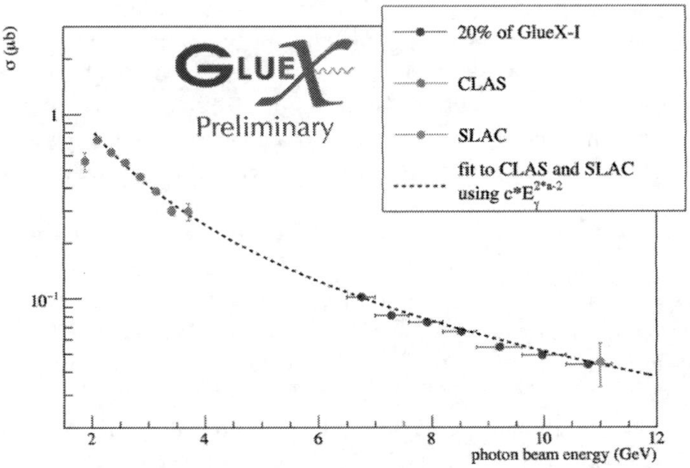

Figure 5. Preliminary total cross-section results are show in red. The green data at low energies comes from the CLAS experiment [11] the data point at 11 GeV is from SLAC [10]. The black dashed curve is a phenomenological fit of the form $c * E_\gamma^{2a-2}$ to the CLAS and SLAC data. Our preliminary data, which was not fitted, agrees well with the black curve.

5 Σ^0 hyperon beam asymmetry

In addition to the various Λ hyperons, GlueX also studied the ground state Σ^0. It is observed in the reaction $\gamma p \to K^+\Sigma^0$ with the subsequent decays $\Sigma^0 \to \Lambda\gamma \to p\pi^-\gamma$. In order to learn more about how the Σ^0 is produced, we have measured the beam asymmetry Σ, which is accessible thanks to GlueX's linearly polarized photon beam, in bins of momentum transfer

$-t = -(p_{\text{beam}} - p_{K^+})^2$. Σ can be measured as a counting rate asymmetry of Σ^0 events between two orthogonal settings of the polarization axis of the beam. The results are published in Ref. [12] and shown in Fig. 6. The black points show the GlueX results, which are consistent

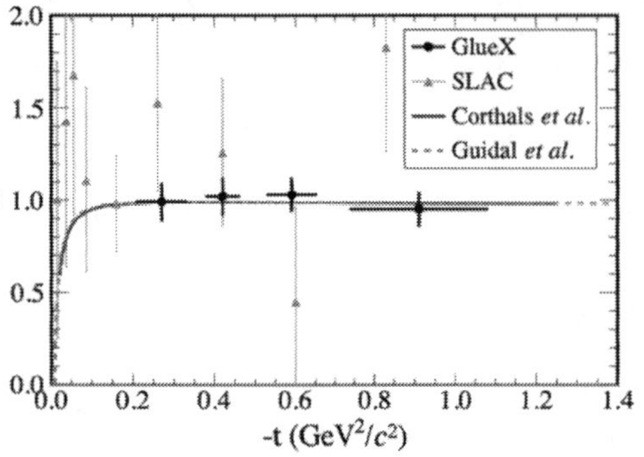

Figure 6. GlueX results for the beam asymmetry Σ in bins of momentum transfer $-t = -(p_{\text{beam}} - p_{K^+})^2$ for Σ^0 photoproduction based on about 20% of GlueX-I are shown in black. The grey triangles show previous measurements carried out at SLAC [13] at $E_\gamma = 16\,\text{GeV}$. The data is compared to two model calculations by Corthals *et al.* [14, 15] and Guidal *et al.* [16], both predicting a beam asymmetry of almost one over the measured t-range. Taken from Ref. [12].

with one over the whole measured range in $-t$. The grey triangles are previous measurements performed by SLAC [13] at $E_\gamma = 16\,\text{GeV}$. They have large uncertainties but generally agree with the GlueX data. Also shown are two model calculations by Corthals *et al.* [14, 15] and Guidal *et al.* [16]. Both predict a beam asymmetry close to one over most of the range in momentum transfer. This large beam asymmetry implies natural parity exchange, similar to what is also observed in the $\Lambda(1520)$ measurement.

6 Cascade production

GlueX can not only provide important data for single strangeness hyperons but also observes various doubly strange hyperons. This is especially interesting since there seems to be a large discrepancy between the number of expected and number of known states with strangeness $S = -2$. GlueX can provide important data to help identify new states in this sector.

Cascade baryons are expected to be produced slightly differently compared to Λ or Σ hyperons. Instead of being produced in a direct t-channel exchange, which can only generate one unit of strangeness, cascade baryon production has to proceed through a decay of a high mass hyperon with $S = -1$, which decays into a kaon and the cascade hyperon adding another unit of strangeness. Therefore, we expect to observe the reactions through $\gamma p \to K^+(\Lambda^*/\Sigma^*) \to K^+K^{+/0}\Xi^{-/0}$.

The ground state cascades decay predominantly into $\pi\Lambda$. Therefore, in the search for the charged ground state cascade the reaction of interest is $\gamma p \to K^+K^+\Xi^- \to K^+K^+\pi^-\Lambda \to K^+K^+\pi^-\pi^-p$. Figure 7 shows the preliminary invariant mass spectrum of the $\Lambda\pi^-$ system. One can clearly identify a strong and narrow signal peak at the mass of the $\Xi^-(1320)$. In order to observe the neutral ground state cascade $\Xi^0(1315)$ we study the reaction $\gamma p \to K^+K^0_s\Xi^0 \to K^+\{\pi^+\pi^-\}\pi^0\Lambda \to K^+\{\pi^+\pi^-\}\gamma\gamma\pi^-p$. This final state has a higher multiplicity and contains photons in the final state. Nevertheless, GlueX is able to identify the $\Xi(1315)$ in the $\pi^0\Lambda$ invariant mass spectrum as a peak on top of some background, as shown in Fig. 8.

The excited $\Xi^{*-}(1530)$ is known to decay into $\pi^0\Xi^-$. In order to search for it, the reaction $\gamma p \to K^+K^+\Xi^{*-} \to K^+K^+\pi^0\Xi^- \to K^+K^+\gamma\gamma\pi^-\Lambda \to K^+K^+\gamma\gamma\pi^-\pi^-p$ is studied. This is again a final state with both charged and neutral particles and high multiplicity. Nevertheless, studying about 50% of the GlueX-I data set reveals a peaking structure at the expected mass, as shown in Fig. 9.

The highest mass excited cascade we have identified so far is the $\Xi^{*-}(1820)$, which was seen for the first time in photoproduction by GlueX. We observe it in the reaction $\gamma p \to$

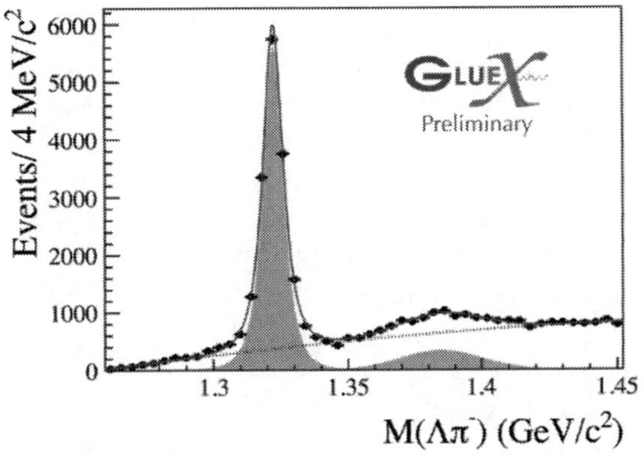

Figure 7. Preliminary invariant mass spectrum of the $\Lambda\pi^-$ subsystem in the reaction $\gamma p \to K^+K^+\Xi^- \to K^+K^+\pi^-\Lambda \to K^+K^+\pi^-\pi^-p$. The $\Xi^-(1320)$ can be easily identified as a narrow peak at the expected mass. The bump at higher masses is due to the $\Sigma(1385)$ with a miss-identified pion.

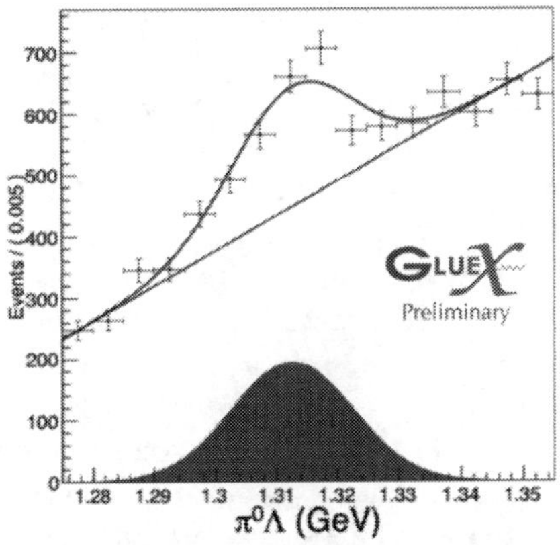

Figure 8. Preliminary invariant mass spectrum of the $\pi^0\Lambda$ subsystem in the reaction $\gamma p \to K^+K^0_s\Xi^0 \to K^+\{\pi^+\pi^-\}\pi^0\Lambda \to K^+\{\pi^+\pi^-\}\gamma\gamma\pi^-p$. The $\Xi^0(1315)$ can be identified as a peak at the expected mass on top of some moderate background.

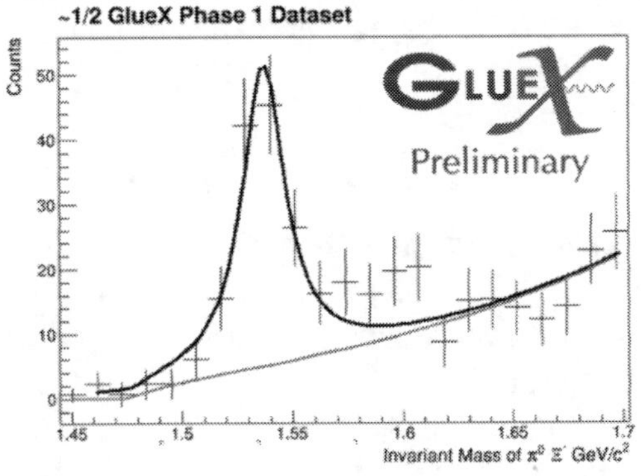

Figure 9. Preliminary invariant mass spectrum of the $\pi^0\Xi^-$ subsystem in the reaction $\gamma p \to K^+K^+\Xi^{*-} \to K^+K^+\pi^0\Xi^- \to K^+K^+\gamma\gamma\pi^-\Lambda \to K^+K^+\gamma\gamma\pi^-\pi^-p$. The $\Xi^{*-}(1530)$ can be clearly identified as a peak at the expected mass.

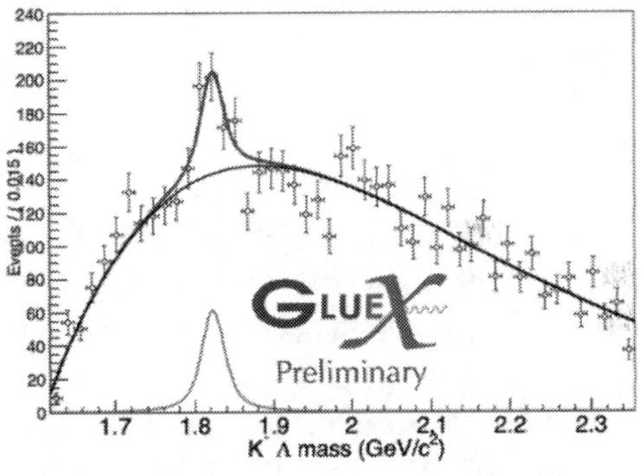

Figure 10. Preliminary invariant mass spectrum of the $K^-\Lambda$ subsystem in the reaction $\gamma p \to K^+K^+\Xi^{*-} \to K^+K^+K^-\Lambda \to K^+K^+K^-\pi^-p$ for $E_\gamma = 7.4\,\text{GeV}$ to $8.0\,\text{GeV}$. The $\Xi^{*-}(1820)$ can be identified as a peak on some background distribution.

$K^+K^+\Xi^{*-} \to K^+K^+K^-\Lambda \to K^+K^+K^-\pi^-p$. It can be identified as a peak in the preliminary $K^-\Lambda$ invariant mass spectrum as shown in Fig. 10. It is noteworthy that it is the only excited cascade hyperon that clearly shows up as peak in the broad range of invariant mass shown in the plot.

Despite experimental challenges like mixed final states and high multiplicities GlueX can clearly identify a range of excited and ground state cascade hyperons. Going forward we will work towards measurements of physics observables for these reactions.

7 Summary and outlook

The GlueX experiment at Jefferson Lab is an exceptional photoproduction facility which, thanks to its comprehensive detector system, is able to detect a wide range of charged and neutral final state particles. This allows GlueX to measure many strangeness related reactions. GlueX establishes the production mechanisms for a wide range of different hyperons and, thereby, paves the way for future partial wave analyses which rely on good knowledge of the involved reactions. In addition, GlueX can search for and identify many (excited) hyperons with strangeness $S = -2$. This is very encouraging for future searches for new strange particles.

In addition to the hyperon measurements reported here, GlueX also has a range of active strangeness analyses in the meson sector. We observe many excited kaons and other mesons decaying to final states containing strangeness. With the addition of the DIRC detector to our GlueX spectrometer setup, we expect improved kaon identification in our GlueX-II data. This will enable us to continue to make important contributions to strange hadron spectroscopy. The planned measurements comprise, but are not limited to, searches for strangeonia, the $Y(2175)$ and strange hybrid mesons. In addition, we will continue our program of searches for hyperons.

We would like to acknowledge the outstanding efforts of the staff of the Accelerator and the Physics Divisions at Jefferson Lab that made the experiment possible. This work was supported in part by the UK Science and Technology Facilities Council. This material is based upon work supported by the U.S. Department of Energy, Office of Science, Office of Nuclear Physics under contract DE-AC05-06OR23177. For a complete list of founding bodies and acknowledgements please consult gluex.org/thanks.

References

[1] S. Adhikari et al. (GlueX), Nucl. Instrum. Meth. A **987**, 164807 (2021)
[2] R. Schumacher, H. Li (GlueX), AIP Conf. Proc. **2130**, 020007 (2019)
[3] H. Li, R.A. Schumacher (GlueX), AIP Conf. Proc. **2249**, 030037 (2020)
[4] J.Z. Bai et al. (BES), Phys. Rev. Lett. **91**, 022001 (2003)
[5] M. Ablikim et al. (BESIII), Phys. Rev. D **97**, 032013 (2018)
[6] M. Mai, Eur. Phys. J. ST **230**, 1593 (2021)
[7] P.A. Zyla et al. (Particle Data Group), PTEP **2020**, 083C01 (2020)
[8] N. Wickramaarachchi (GlueX), elsewhere in these proceedings (2022)
[9] S. Adhikari et al. (GlueX), Phys. Rev. C **105**, 035201 (2022)
[10] A. Boyarski, R.E. Diebold, S.D. Ecklund, G.E. Fischer, Y. Murata, B. Richter, M. Sands, Phys. Lett. B **34**, 547 (1971)
[11] K. Moriya et al. (CLAS), Phys. Rev. C **88**, 045201 (2013), [Addendum: Phys.Rev.C 88, 049902 (2013)]
[12] S. Adhikari et al. (GlueX), Phys. Rev. C **101**, 065206 (2020)
[13] D.J. Quinn, J.P. Rutherford, M.A. Shupe, D.J. Sherden, R.H. Siemann, C.K. Sinclair, Phys. Rev. D **20**, 1553 (1979)
[14] T. Corthals, J. Ryckebusch, T.V. Cauteren, Phys. Rev. C **73**, 045207 (2006)
[15] T. Corthals, T.V. Cauteren, J. Ryckebusch, D.G. Ireland, Phys. Rev. C **75**, 045204 (2007)
[16] M. Guidal, J.M. Laget, M. Vanderhaeghen, Nuclear Physics A **627**, 645 (1997)

EPJ Web of Conferences **271**, 02002 (2022) https://doi.org/10.1051/epjconf/202227102002
HYP2022

Cross-section measurement of virtual photoproduction of iso-triplet three-body hypernucleus, Λnn

T. Gogami[1,*], K. N. Suzuki[1], B. Pandey[3], K. Itabashi[2], S. Nagao[2], K. Okuyama[2], S. N. Nakamura[2], L. Tang[3,4], D. Abrams[5], T. Akiyama[2], D. Androic[6], K. Aniol[7], C. Ayerbe Gayoso[8], J. Bane[9], S. Barcus[8], J. Barrow[9], V. Bellini[10], H. Bhatt[11], D. Bhetuwal[11], D. Biswas[3], A. Camsonne[4], J. Castellanos[12], J-P. Chen[4], J. Chen[8], S. Covrig[4], D. Chrisman[13,14], R. Cruz-Torres[15], R. Das[16], E. Fuchey[17], K. Gnanvo[5], F. Garibaldi[10,18], T. Gautam[3], J. Gomez[4], P. Gueye[3], T. J. Hague[19], O. Hansen[4], W. Henry[4], F. Hauenstein[20], D. W. Higinbotham[4], C. E. Hyde[20], M. Kaneta[2], C. Keppel[4], T. Kutz[16], N. Lashley-Colthirst[3], S. Li[21,22], H. Liu[23], J. Mammei[24], P. Markowitz[12], R. E. McClellan[4], F. Meddi[10,25], D. Meekins[4], R. Michaels[4], M. Mihovilovic[26,27,28], A. Moyer[29], D. Nguyen[15,30], M. Nycz[19], V. Owen[8], C. Palatchi[5], S. Park[16], T. Petkovic[6], S. Premathilake[5], P. E. Reimer[31], J. Reinhold[12], S. Riordan[31], V. Rodriguez[32], C. Samanta[33], S. N. Santiesteban[21], B. Sawatzky[4], S. Širca[26,27], K. Slifer[21], T. Su[19], Y. Tian[34], Y. Toyama[2], K. Uehara[2], G. M. Urciuoli[10], D. Votaw[13,14], J. Williamson[35], B. Wojtsekhowski[4], S. A. Wood[4], B. Yale[21], Z. Ye[31], J. Zhang[5], and X. Zheng[5]

[1]Graduate School of Science, Kyoto University, Kyoto 606-8502 Japan
[2]Graduate School of Science, Tohoku University, Sendai, Miyagi 980-8578, Japan
[3]Department of Physics, Hampton University, Virginia 23668, USA
[4]Thomas Jefferson National Accelerator Facility, Newport News, Virginia 23606, USA
[5]Department of Physics, University of Virginia, Charlottesville, Virginia 22904, USA
[6]Department of Physics & Department of Applied Physics, University of Zagreb, HR-10000 Zagreb,Croatia
[7]Physics and Astronomy Department, California State University, Los Angeles, California 90032, USA
[8]Department of Physics, The College of William and Mary, Virginia 23185, USA
[9]Department of Physics, University of Tennessee, Knoxville, Tennessee 37996, USA
[10]INFN, Sezione di Roma, 00185, Rome, Italy
[11]Department of Physics, Mississippi State University, Mississippi State, Mississippi 39762, USA
[12]Department of Physics, Florida International University, Miami, Florida 33199, USA
[13]Department of Physics and Astronomy, Michigan State University, East Lansing, Michigan 48824, USA
[14]National Superconducting Cyclotron Laboratory, Michigan State University, East Lansing, MI 48824, USA
[15]Department of Physics, Massachusetts Institute of Technology, Cambridge, Massachusetts 02139, USA
[16]Department of Physics, State University of New York, Stony Brook, New York 11794, USA
[17]Department of Physics, University of Connecticut, Storrs, Connecticut 06269, USA
[18]Istituto Superiore di Sanità, 00161, Rome, Italy
[19]Department of Physics, Kent State University, Kent, Ohio 44242, USA
[20]Department of Physics, Old Dominion University, Norfolk, Virginia 23529, USA
[21]Department of Physics, University of New Hampshire, Durham, New Hampshire 03824, USA
[22]Nuclear Science Division, Lawrence Berkeley National Laboratory, Berkeley, CA 94720, USA

*e-mail: gogami@jlab.org

[23]Department of Physics, Columbia University, New York, New York 10027, USA
[24]Department of Physics and Astronomy, University of Manitoba, Winnipeg, Manitoba R3T 2N2, Canada
[25]Sapienza University of Rome, I-00185, Rome, Italy
[26]Faculty of Mathematics and Physics, University of Ljubljana, 1000 Ljubljana, Slovenia
[27]Jožef Stefan Institute, Ljubljana, Slovenia
[28]Institut für Kernphysik, Johannes Gutenberg-Universität Mainz, DE-55128 Mainz, Germany
[29]Department of Physics, Christopher Newport University, Newport News, Virginia 23606, USA
[30]University of Education, Hue University, Hue City, Vietnam
[31]Physics Division, Argonne National Laboratory, Lemont, Illinois 60439, USA
[32]División de Ciencias y Tecnologia, Universidad Ana G. Méndez, Recinto de Cupey, San Juan 00926, Puerto Rico
[33]Department of Physics & Astronomy, Virginia Military Institute, Lexington, Virginia 24450, USA
[34]Department of Physics, Syracuse University, New York, New York 10016, USA
[35]School of Physics & Astronomy, University of Glasgow, Glasgow, G12 8QQ, Scotland, UK

Abstract. Missing-mass spectroscopy with the ^{3}H$(e, e'K^+)$ reaction was carried out at Jefferson Lab's (JLab) Hall A in Oct–Nov, 2018. The differential cross section for the ^{3}H$(\gamma^*, K^+)\Lambda nn$ was deduced at $\omega = E_e - E_{e'} = 2.102$ GeV and at the forward K^+-scattering angle ($0° \leq \theta_{\gamma^* K} \leq 5°$) in the laboratory frame. Given typical predicted energies and decay widths, which are $(B_\Lambda, \Gamma) = (-0.25, 0.8)$ and $(-0.55, 4.7)$ MeV, the cross sections were found to be $11.2 \pm 4.8(\text{stat.})^{+4.1}_{-2.1}(\text{sys.})$ and $18.1 \pm 6.8(\text{stat.})^{+4.2}_{-2.9}(\text{sys.})$ nb/sr, respectively. The obtained result would impose a constraint for interaction models particularly between Λ and neutron by comparing to theoretical calculations.

1 Introduction

Existing data of binding energies of Λ hypernuclei and scattering experiments between Λ and proton lead to a prediction that three-body hypernuclei of the iso-spin triplet state, such as Λpp and Λnn, do not exist [1]. However, the HypHI collaboration at GSI identified events that may be interpreted as the bound states of Λnn and even a two-body system of Λn in invariant-mass spectra [2]. It is noted that the experiment at GSI measured decayed particles from weak-decay processes of hypernuclei, and essentially had sensitivity to only the bound state. The bound state of Λnn is hard to be theoretically explained maintaining consistency of energy levels of other hypernuclei [3–9]. On the other hand, some theoretical calculations support the existence of resonant state [7–14]. Experiments which can also investigate the resonant state should help in order to pin down the bound Λnn problem. The experiment at Jefferson Lab (JLab) which is described in the present article (JLab E12-17-003 Experiment) has sensitivity to both bound and resonant states because it measures the reaction at the strangeness production (missing-mass method). The data analysis is divided into three as follows: (i) Λnn-peak search with a count-base spectrum [20], (ii) measurement of the cross-section of ^{3}H$(\gamma^*, K^+)\Lambda nn$ [21], and (iii) study of the final state interaction between Λ and n based on the shape analysis of quasi-free Λ spectrum [22]. The present article shows a part of result of the analysis (ii).

2 Λnn-search experiment by electron beam

We performed the experiment to investigate the Λnn state through the missing-mass spectroscopy with the reaction of ^{3}H$(e, e'K^+)$ at JLab's Experimental Hall A in 2018. The continuous electron beam at $E_e = 4.32$ GeV with the rastered beam size of 2×2 mm^2 was impinged on a gaseous tritium target [23]. The typical beam intensity on the target was 1.4×10^{14} electrons per second (22.5μA). The scattered electron (e') and K^+ were measured by two HRSs [24] which are existing spectrometers at JLab Hall A. The central momenta measured by the HRSs were set to be 2.218 and 1.823 GeV/c for the e' and K^+, respectively. The acceptance of the K^+ scattering angle with respect to the virtual photon was about $0° \leq \theta_{\gamma^* K} \leq 5°$ in the laboratory frame. The beam energy on the target was well controlled, and total of its energy spread and fluctuation ($\Delta E_e/E_e$) was down to less than a few of 10^{-4} in FWHM. Momentum vectors of the e' and K^+ at a reaction point were reconstructed by the HRSs, and they were used to obtain the missing-mass spectrum. The Λ binding energy B_Λ is calculated once the missing mass $M_{\Lambda nn}$ is obtained as follows, $B_\Lambda = 2M_n + M_\Lambda - M_{\Lambda nn}$ where $M_{n,\Lambda}$ are the masses of neutron and Λ [15].

The momentum vectors of e' and K^+ were obtained by a backward transfer matrix (BTM) for each spectrometer [16]. The matrix elements were optimized by using data with sieve slits (SSs), a hydrogen target. The SSs are 2.54-cm thick tungsten plates that have specific patterns of holes, and we took special data with the SS attached to a first quadrupole magnet of each spectrometer. The elements of BTMs for the scattering angle were tuned so as to reproduce the 2-D images of SSs' hole patterns. As for the elements of the momentum, Λ and Σ^0 productions from the hydrogen target were used as references because the masses of Λ and Σ^0 are well known. The obtained mass resolution of Λ is $\sigma_\Lambda = 1.3 \pm 0.1$ MeV/c^2 which is consistent with the expectation by a Monte Carlo (MC) simulation based on Geant4 [17–19]. The MC simulation determined the expected mass resolution for the Λnn production to be $\sigma^{\text{pred.}}_{\Lambda nn} = 1.5 \pm 0.2$ MeV/c^2 when its decay width is assumed to be $\Gamma = 0$ MeV. In addition, the systematic uncertainty on the binding-energy measurement was found to be $|\Delta B^{\text{sys.}}_\Lambda| = 0.4$ MeV for the present analysis. Refer to Ref. [21] for details.

3 Analysis and result

Figure 1 shows the differential cross section for the ^{3}H(γ^*, K^+) reaction as a function of $-B_\Lambda$. In the present analysis, a momentum cut that allowed the particle distribution in the MC simulation to be consistent with that in the experimental data was applied to minimize the systematic error on the differential cross section due to the acceptance correction. It is noted that the momentum cut reduced the number of events by about 10–20%.

The background distribution of the accidental coincidence between e' and K^+ was obtained by mixed event analysis (MEA) as shown in Fig. 1. The MEA takes random combinations of e' and K^+ from the data sample to estimate the distribution with a much smaller statistical uncertainty. The MEA was used and proven to work well in the past hypernuclear experiments with the ($e, e'K^+$) reaction [25–30]. On the other hand, the distribution of the quasi-free (QF) Λ production which rises at $-B_\Lambda = 0$ MeV is unknown. The QF-Λ distribution might be approximated as a linear function unless the considered range is not far from the threshold ($-B_\Lambda = 0$ MeV). Decent treatment of the final state interaction (FSI) between Λ and n (and nn) is necessary to be a closer estimate. However, no theory to estimate the QF-Λ distribution with FSI exists so far. Therefore, we assumed the QF-Λ distribution as the linear function in the present analysis.

Spectral fits were performed with the unbinned maximum likelihood (UBML). The fit function for the signal is a convolution of the experimental response and Breit-Wigner functions. The probability density function that we defined for the fit was composed of the signal function, linear function for the QF-Λ production, and the 4th-order polynomial function for the accidental coincidence background. Here, parameters for the 4th-order polynomial function were determined by using the accidental coincidence-background distribution obtained by MEA in advance of the UBML fit. Figure 2 shows results of the spectral fits assuming

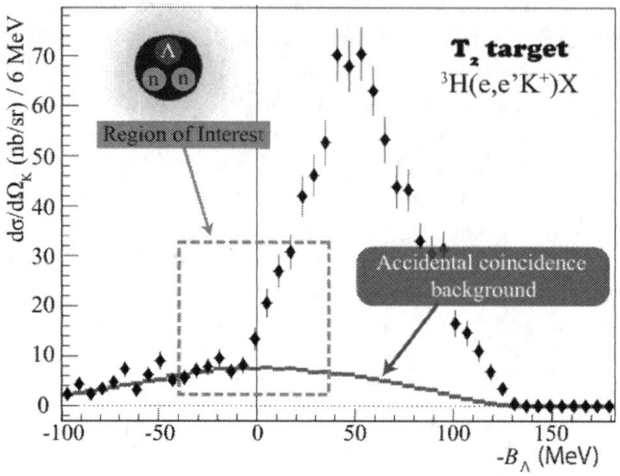

Figure 1. The differential cross section of the ^{3}H(γ^*, K^+) reaction as a function of $-B_\Lambda$. Distribution of the accidental coincidence between e' and K^+ was obtained by mixed event analysis (MEA).

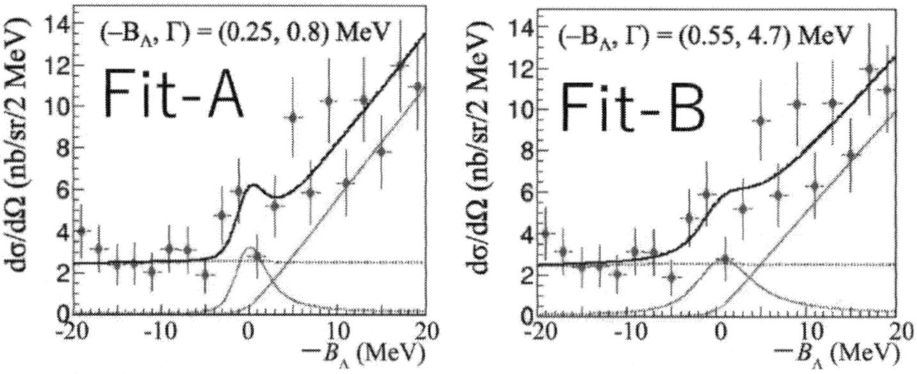

Figure 2. Results of the spectral fits by UBML assuming typical theoretical predictions of the energies and decay widths, $(-B_\Lambda, \Gamma)$ = (0.25, 0.8) [9] and (0.55, 4.7) MeV [10], which are labeled Fit-A and Fit-B, respectively.

typical theoretical predictions $(-B_\Lambda, \Gamma)$ = (0.25, 0.8) [9] and (0.55, 4.7) MeV [10]; the former (Fit-A) has a smaller decay width and the latter (Fit-B) has a larger one. It is noted that the Γ was taken into account in the Breit-Wigner function of the signal function.

The obtained cross sections for the signal were $11.2 \pm 4.8(\text{stat.})^{+4.1}_{-2.1}(\text{sys.})$ and $18.1 \pm 6.8(\text{stat.})^{+4.2}_{-2.9}(\text{sys.})$ nb/sr for Fit-A and Fit-B, respectively. There is surplus that could be attributed to the Λnn signal for both cases. However, the peak significance is less than three sigma which does not allow us to reach decisive conclusion about the Λnn state. Therefore, we also derived the upper limit of the differential cross section. A 2-D scan of the upper limit of the cross section was performed for $-20 \leq -B_\Lambda \leq 20$ MeV and $0 \leq \Gamma \leq 10$ MeV. The result of the cross-section upper limits is shown in Ref. [21].

4 Conclusion

We performed the experiment with the ^{3}H$(e, e'K^+)$ reaction at JLab Hall A to investigate the Λnn state. The present article focuses on the data analysis of the differential cross section for the ^{3}H$(\gamma^*, K^+)\Lambda nn$ reaction. Given typical predictions of the energies and decay widths, which are $(B_\Lambda, \Gamma) = (-0.25, 0.8)$ and $(-0.55, 4.7)$ MeV, the cross sections were obtained to be 11.2 ± 4.8(stat.)$^{+4.1}_{-2.1}$(sys.) and 18.1 ± 6.8(stat.)$^{+4.2}_{-2.9}$(sys.) nb/sr, respectively. The obtained result would impose a constraint for interaction models particularly between Λ and n. Theoretical calculations to be compared with the obtained result are awaited. In addition, the result has significant information that allows us to design a future experiment in which we would pin down the bound Λnn problem.

Acknowledgments

We thank the JLab staff of the Division of Physics, Division of Accelerator, and the Division of Engineering for providing support for conducting the experiment. We acknowledge the outstanding contribution of the Jefferson Lab target group for the design and safe handling of the tritium target for the present experiment. This work was supported by U.S. Department of Energy (DOE) grant DE-AC05-06OR23177 under which Jefferson Science Associates LLC operates the Thomas Jefferson National Accelerator Facility. The work of the Argonne National Laboratory group member is supported by DOE grant DE-AC02-06CH11357. The Kent State University contribution is supported under grant no. PHY-1714809 from the U.S. National Science Foundation. The hypernuclear program at JLab is supported by U.S. DOE grant DE-FG02-97ER41047. This work was partially supported by the Grant-in-Aid for Scientific Research on Innovative Areas "Toward new frontiers Encounter and synergy of state-of-the-art astronomical detectors and exotic quantum beams." This work was supported by JSPS KAKENHI grants nos. 18H05459, 18H05457, 18H01219, 17H01121, 19J22055, and 18H01220. This work was also supported by SPIRITS 2020 of Kyoto University, and the Graduate Program on Physics for the Universe, Tohoku University (GP-PU).

References

[1] B. W. Downs and R. H. Dalitz, Phys. Rev. **114**, 593 (1959)
[2] C. Rappold et al., Phys. Rev. C **88**, 041001 (2013)
[3] A. Gal et al., Phys. Lett. B **736**, 93 (2014)
[4] H. Garcilazo and A. Valcarce, Phys. Rev. C **89**, 057001 (2014)
[5] E. Hiyama, B. F. Gibson S. Ohnishi, and Th. A. Rijken, Phys. Rev. C **89**, 061302 (2014)
[6] K. Miyagawa, H. Kamada, W. Glöckle, and V. Stoks, Phys. Rev. C **92**, 054608 (2015)
[7] I. R. Afnan and B. F. Gibson, Phys. Rev. C **92**, 054608 (2015)
[8] I. Filikhin, V. Suslov, and B. Vlahovic, EPJ Web of Conf. 113, 08006 (2016)
[9] H. Kamada, K. Miyagawa, and M. Yamaguchi, EPJ Web Conf. **113**, 07004 (2016)
[10] V. B. Belyaev, S. A. Rakityansky, and W. Sandhas, Nucl. Phys. A **803**, 210–226 (2008)
[11] M. Schäfer et al., Phys. Lett. B **808**, 135614 (2020)
[12] M. Schäfer et al., Phys. Rev. C **103**, 025204 (2021)
[13] M. Schäfer et al., Phys. Rev. C **105**, 015202 (2022)
[14] T. Y. Htun and Y. Yan, Phys. Rev. C **105**, 064001 (2022)
[15] P. A. Zyla et al., Prog. Theor. Exp. Phys. **2020**, 083C01 (2020)
[16] T. Gogami et al., AIP Conf. Proc. **2319**, 080019 (2021)

[17] S. Agostinelli et al., Nucl. Instrm. Methods, Phys. Res. Sect. A **506**, 250 (2003)
[18] S. Allison et al., IEEE Trans. Nucl. Sci. **53**, 270 (2006)
[19] S. Allison et al., Nucl. Instrum. Methods, Phys. Res. Sect. A **835**, 186 (2016)
[20] B. Pandey, L. Tang et al., Phys. Rev. C **105**, L051001 (2022)
[21] K. N. Suzuki, T. Gogami et al., Prog. Theor. Exp. Phys. **2022**, 013D01 (2022)
[22] K. Itabashi et al., Few-Body Systems **63**, 16 (2022)
[23] S. N. Santiesteban et al., Nucl. Instrum. Methods, Phys. Res. Sect. A **940**, 351 (2019)
[24] J. Alcorn et al., Nucl. Instrum. Methods, Phys. Res. Sect. A **21**, 294 (2004)
[25] S. N. Nakamura et al., Phys. Rev. Lett. **110**, 012502 (2013)
[26] L. Tang et al., Phys. Rev. C **90**, 034320 (2014)
[27] T. Gogami et al., Phys. Rev. C **93**, 034314 (2016)
[28] T. Gogami et al., Phys. Rev. C **94**, 021302(R) (2016)
[29] T. Gogami et al., Nucl. Instrum. Methods, Phys. Res. Sect. A **900**, 69 (2018)
[30] T. Gogami et al., Phys. Rev. C **103**, L041301 (2021)

EPJ Web of Conferences **271**, 02003 (2022) https://doi.org/10.1051/epjconf/202227102003
HYP2022

Study of the Λ/Σ^0 electroproduction in the low-Q^2 region at JLab

K. Okuyama[1,*], K. Itabashi[1], S. Nagao[1], S. N. Nakamura[1,2], K. N. Suzuki[3], T. Gogami[3], B. Pandey[4], L. Tang[4,5], D. Abrams[6], T. Akiyama[1], D. Androic[7], K. Aniol[8], C. Ayerbe Gayoso[9], J. Bane[10], S. Barcus[9], J. Barrow[10], V. Bellini[11], H. Bhatt[12], D. Bhetuwal[12], D. Biswas[4], A. Camsonne[5], J. Castellanos[13], J-P. Chen[5], J. Chen[9], S. Covrig[5], D. Chrisman[14,15], R. Cruz-Torres[16], R. Das[17], E. Fuchey[18], K. Gnanvo[6], F. Garibaldi[11,19], T. Gautam[4], J. Gomez[5], P. Gueye[4], T. J. Hague[20], O. Hansen[5], W. Henry[5], F. Hauenstein[21], D. W. Higinbotham[5], C. E. Hyde[21], M. Kaneta[1], C. Keppel[5], T. Kutz[17], N. Lashley-Colthirst[4], S. Li[22,23], H. Liu[24], J. Mammei[25], P. Markowitz[13], R. E. McClellan[5], F. Meddi[11,26], D. Meekins[5], R. Michaels[5], M. Mihovilovic[27,28,29], A. Moyer[30], D. Nguyen[16,31], M. Nycz[20], V. Owen[9], C. Palatchi[6], S. Park[17], T. Petkovic[7], S. Premathilake[6], P. E. Reimer[32], J. Reinhold[13], S. Riordan[32], V. Rodriguez[33], C. Samanta[34], S. N. Santiesteban[22], B. Sawatzky[5], S. Širca[27,28], K. Slifer[22], T. Su[20], Y. Tian[35], Y. Toyama[1], K. Uehara[1], G. M. Urciuoli[11], D. Votaw[14,15], J. Williamson[36], B. Wojtsekhowski[5], S. A. Wood[5], B. Yale[22], Z. Ye[32], J. Zhang[6], and X. Zheng[6]

[1]Department of Physics, Graduate School of Science, Tohoku University, Sendai, Miyagi 980-8578 Japan
[2]Department of Physics, Graduate School of Science, The University of Tokyo, Hongo, Tokyo 113-0033 Japan
[3]Department of Physics, Graduate School of Science, Kyoto University, Kyoto, Kyoto 606-8502 Japan
[4]Department of Physics, Hampton University, Hampton, Virginia 23668, USA
[5]Thomas Jefferson National Accelerator Facility, Newport News, Virginia 23606, USA
[6]Department of Physics, University of Virginia, Charlottesville, Virginia 22904, USA
[7]Department of Physics & Department of Applied Physics, University of Zagreb, HR-10000 Zagreb, Croatia
[8]Physics and Astronomy Department, California State University, Los Angeles, California 90032, USA
[9]Department of Physics, The College of William and Mary, Virginia 23185, USA
[10]Department of Physics, University of Tennessee, Knoxville, Tennessee 37996, USA
[11]Istituto Nazionale di Fisica Nucleare, Sezione di Roma, 00185, Rome, Italy
[12]Department of Physics, Mississippi State University, Mississippi State, Mississippi 39762, USA
[13]Department of Physics, Florida International University, Miami, Florida 33199, USA
[14]Department of Physics and Astronomy, Michigan State University, East Lansing, Michigan 48824, USA
[15]National Superconducting Cyclotron Laboratory, Michigan State University, East Lansing, MI 48824, USA
[16]Department of Physics, Massachusetts Institute of Technology, Cambridge, Massachusetts 02139, USA
[17]Department of Physics, State University of New York, Stony Brook, New York 11794, USA
[18]Department of Physics, University of Connecticut, Storrs, Connecticut 06269, USA
[19]Instituto Superiore di Sanità, 00161, Rome, Italy
[20]Department of Physics, Kent State University, Kent, Ohio 44242 USA
[21]Department of Physics, Old Dominion University, Norfolk, Virginia 23529, USA

*e-mail: kazuki@lambda.phys.tohoku.ac.jp

 https://doi.org/10.1051/epjconf/202227102003

[22]Department of Physics, University of New Hampshire, Durham, New Hampshire 03824, USA
[23]Nuclear Science Division, Lawrence Berkeley National Laboratory, Berkeley, CA 94720, USA
[24]Department of Physics, Columbia University, New York, New York 10027, USA
[25]Department of Physics and Astronomy, University of Manitoba, Winnipeg, Manitoba R3T 2N2, Canada
[26]Sapienza University of Rome, I-00185, Rome, Italy
[27]Faculty of Mathematics and Physics, University of Ljubljana, 1000 Ljubljana, Slovenia
[28]Jožef Stefan Institute, Ljubljana, Slovenia
[29]Institut für Kernphysik, Johannes Gutenberg-Universität Mainz, DE-55128 Mainz, Germany
[30]Department of Physics, Christopher Newport University, Newport News, Virginia 23606, USA
[31]University of Education, Hue University, Hue City, Vietnam
[32]Physics Division, Argonne National Laboratory, Lemont, Illinois 60439, USA
[33]División de Ciencias y Tecnologia, Universidad Ana G. Méndez, Recinto de Cupey, San Juan 00926, Puerto Rico
[34]Department of Physics & Astronomy, Virginia Military Institute, Lexington, Virginia 24450, USA
[35]Department of Physics, Syracuse University, New York, New York 10016, USA
[36]School of Physics & Astronomy, University of Glasgow, Glasgow, G12 8QQ, Scotland, UK

Abstract. We performed an experiment using tritium and hydrogen cryogenic gas targets at Thomas Jefferson National Accelerator Facility (JLab) in 2018 (E12-17-003)[1, 2]. In this article, we discuss the Λ/Σ^0 hyperon electroproduction from hydrogen target. Elementary Λ/Σ^0 hyperon production processes are important not only for an absolute mass scale calibration in our experiment, but also for the study of the electroproduction mechanisms themselves. In this article, we reported the results of the differential cross section for the $p(e, e'K^+)\Lambda/\Sigma^0$ reaction at $Q^2 \sim 0.5$ (GeV/c)^2.

1 Introduction

We have started and proceeded Λ hypernuclear experiments at JLab to understand the hyperon-nucleon interaction from the binding energy measurements of hypernuclei. Our experimental method enables us to investigate not only Λ hyperon but also Σ^0 hyperon production processes simultaneously as described later on. Λ/Σ^0 photoproduction $p(\gamma, K^+)\Lambda/\Sigma^0$ and electroproduction $p(e, e'K^+)\Lambda/\Sigma^0$ have been studied over several decades. In both reactions, protons are converted to hyperons. The only difference between them is that the photon involved in the reaction is real or virtual.

Experimentally, hyperon photoproduction has been studied very well and an abundance of data has been reported by many experimental groups[3–8]. Meanwhile, hyperon electroproduction can be associated with hyperon photoproduction and studied in a complementary manner under the same theoretical framework. The contribution from the virtual photon can be extracted by factorizing the triple differential cross section as follows[9]:

$$\frac{d^3\sigma}{d\omega d\Omega_{e'} d\Omega_K^{c.m.}} = \Gamma \frac{d\sigma_{\gamma^*}}{d\Omega_K^{c.m.}}, \tag{1}$$

where Γ is a so-called virtual photon flux. Through its definition, $d\sigma_{\gamma^*}/d\Omega_K^{c.m.}$ is regarded as a differential cross section for the kaon-hyperon production from virtual photons. Four-momentum of a virtual photon is denoted as $q^\mu := (\omega, \boldsymbol{q}) = (E_e - E_{e'}, \boldsymbol{P}_e - \boldsymbol{P}_{e'})$.

In the case of hyperon electroproduction, data can be taken at forward angles along the virtual photon direction, which are regions of interest for the photoproduction because it is missing. Besides that, $Q^2 := -q^2$ dependency appears in the electroproduction and this is also interesting and promising. However, experimental data on electroproduction is less abundant than that on photoproduction. Thus, providing the results of the differential cross section for the hyperon electroproduction from experimental side is necessary.

2 Experiment

2.1 Missing mass method

We used 4.32-GeV electron beams provided by CEBAF (Continuous Electron Beam Accelerator Facility) at JLab and hydrogen gas as a fixed proton target. We measured momenta of the scattered final electrons (~ 2.10 GeV/c) and positively charged kaons (~ 1.82 GeV/c) associated with the neutral charged hyperon production in coincidence by using large magnetic spectrometers, HRS-L and HRS-R, respectively. Detailed descriptions of the experimental setup can be found in following papers[1, 10, 11]. Eventually, missing mass is calculated from the momentum-energy conservation for every coincidence event by

$$\text{Missing Mass} = \sqrt{E^2_{\text{miss}} - \boldsymbol{P}^2_{\text{miss}}} = \sqrt{\left(E_{\text{e}} - E_{\text{e}'} + M_{\text{p}} - E_{\text{K}}\right)^2 - (\boldsymbol{P}_{\text{e}} - \boldsymbol{P}_{\text{e}'} - \boldsymbol{P}_{\text{K}})^2}. \quad (2)$$

With proton target, peaks at masses of Λ and Σ^0 are observed in the spectrum as shown in figure 1. Backgrounds come from the Al$(e, e'K^+)X$ reaction because of the target cell made of Aluminum and the $p(e, e'\pi^+)X$ reaction because of the high rate of pions. Amount and distribution of these background source are estimated and indicated by a yellow line in figure 1 even though they are very small. Therefore, remaining events are considered to be truly the hyperon production events. The tail structure from each peak is considered as a radiative tail. Radiative tail is well-understood phenomenon in the electron scattering experiment because electrons easily emit radiations. In this case, it was mainly caused from the scattered electrons when they pass through the Aluminum target cell obliquely. In order to derive the differential cross sections, radiative tail also needs to be taken into account as a hyperon production events. Those estimations were performed by fitting the distribution in some ways, combined with adequate simulation. The fluctuations of the fitting results are included as systematic uncertainties.

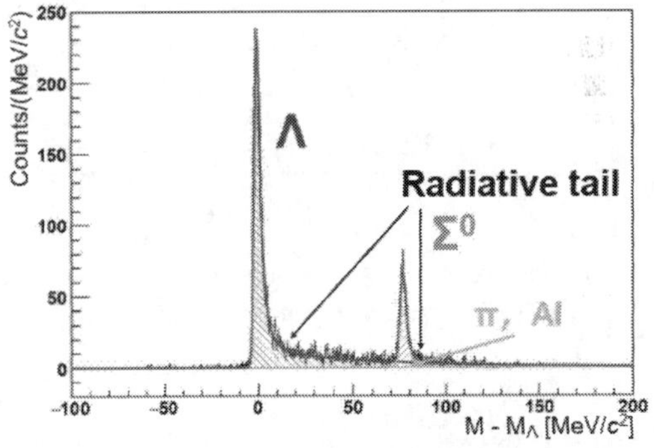

Figure 1. Missing mass spectrum obtained by the $p(e, e'K^+)X$ reaction. It was fitted to estimate number of hyperons including radiative tails (see text).

2.2 Differential cross section derivation

Differential cross sections are deduced using a following formula,

$$\overline{\left(\frac{d\sigma_{\gamma^* p \to K^+ \Lambda(\Sigma^0)}}{d\Omega_{K^+}}\right)}_{\text{HRS-R}} = \frac{1}{N_{\text{T}}} \cdot \frac{1}{N_{\gamma^*}} \cdot \frac{1}{\bar{\varepsilon}} \cdot \sum_{i=1}^{N_{\Lambda(\Sigma^0)}} \frac{1}{\varepsilon_i^{\text{DAQ}} \cdot \varepsilon_i^{\text{Decay}} \cdot \Delta\Omega_{\text{HRS-R},i}}, \quad (3)$$

where, N_{T} is the number of proton targets per unit area, N_{γ^*} is the number of virtual photons contributing to the production, and $\bar{\varepsilon}$ is the cut efficiency in the event selection. In particular, data acquisition efficiency $\varepsilon_i^{\text{DAQ}}$, K^+'s survival ratio without any decay $\varepsilon_i^{\text{Decay}}$, and solid angle of HRS-R $\Delta\Omega_{\text{HRS-R},i}(p_{\text{K}}, Z)$ as a function of momentum and reaction point are applied every i-th event. Estimated quantities are summarized in table 1.

Table 1. Estimated quantities with statistical errors for deriving the differential cross sections

	Λ	Σ^0
N_{T}	0.0375 ± 0.0013 [b^{-1}]	
$N_{\gamma^*}/10^{13}$	3.53 ± 0.01	4.95 ± 0.01
$\bar{\varepsilon}$	45.4 ± 1.2 [%]	44.3 ± 1.2 [%]
$N_{\mathrm{Hyperon}}/10^3$	1.36 ± 0.04	0.37 ± 0.02
$\varepsilon^{\mathrm{Decay}}$	~ 14 [%]	~ 14 [%]
$\varepsilon^{\mathrm{DAQ}}$	~ 96 [%]	~ 96 [%]
$\Delta\Omega^{\mathrm{Lab}}_{\mathrm{HRS\text{-}R}}$	~ 5.5 [msr]	

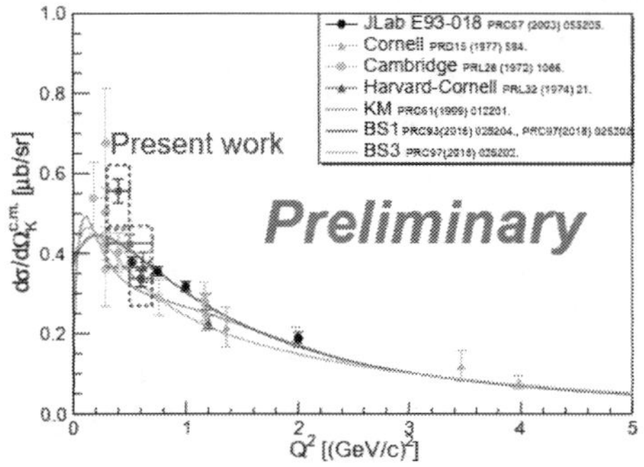

Figure 2. The Q^2 dependency of the differential cross sections for the Λ hyperon electroproduction (see text).

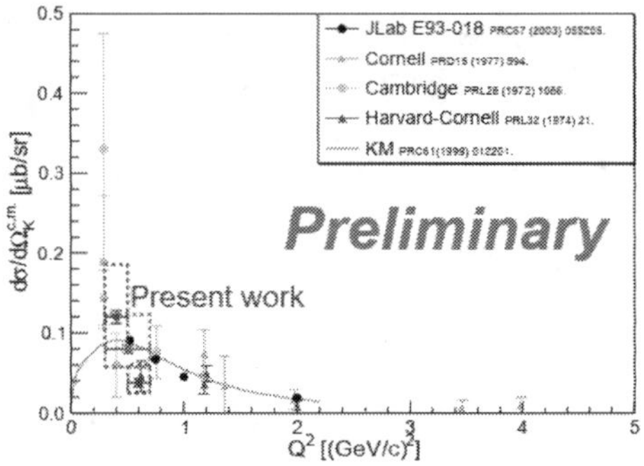

Figure 3. The Q^2 dependency of the differential cross sections for the Σ^0 hyperon electroproduction (see text).

3 Results

Using the dataset of our experiment ($W \sim 2.14$ GeV, $Q^2 \sim 0.5$ (GeV/c)^2, $\theta^{\mathrm{c.m.}}_{\gamma\mathrm{K}} \sim 8$ deg), we deduced the differential cross sections. In next subsections, we discuss these results comparing with other experimental data and theoretical calculations by showing the Q^2 dependency and angle dependency.

3.1 Q^2 dependency

The obtained results are shown with Q^2 dependency of the differential cross section in figure 2 and figure 3. Red points are the results using all data and blue points are the results with data divided into 2 sets based on Q^2 to see its dependency. Our results show the statistical error by the solid lines and the dotted boxes include systematic error. The other experimental results show only the statistical error[12–15]. Theoretical results are shown by curved lines[16–18]. As Q^2 decreases, our results tend to increase, which are similar to the other results, but with a higher slope.

3.2 Angle dependency

The obtained results are shown with angle dependency of the differential cross section in figure 4 and figure 5. Experimental data and theoretical calculations for the photoproductions are also shown without corrections[3–8, 16, 19, 20] . We deduced the differential cross section in low-Q^2 regions at forward angles which is missing data in the photoproductions. Our results tend to decrease as they approach to the forward angles in both reactions, but almost flat for Λ within statistical errors.

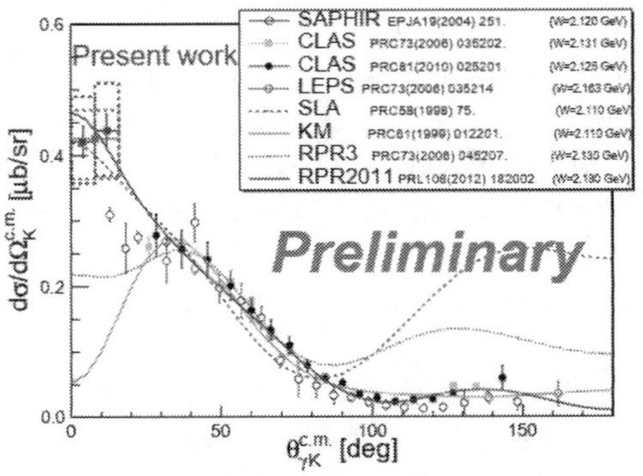

Figure 4. The $\theta_{\gamma K}^{c.m.}$ dependency of the differential cross sections for the Λ hyperon production. Our data is at $Q^2 \sim 0.5$ (GeV/c)^2, the others are at $Q^2 = 0$ (see text).

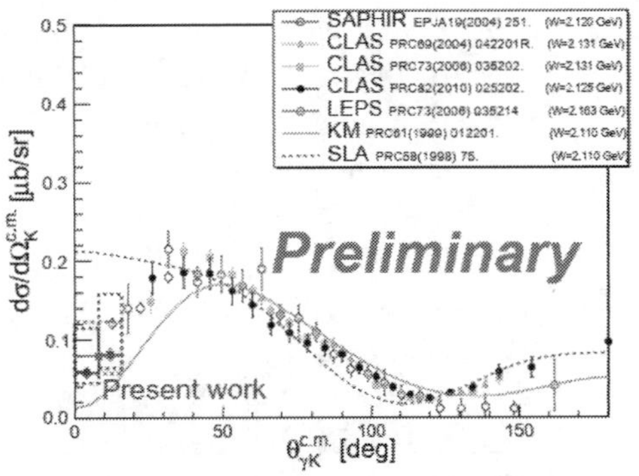

Figure 5. The $\theta_{\gamma K}^{c.m.}$ dependency of the differential cross sections for the Σ^0 hyperon production. Our data is at $Q^2 \sim 0.5$ (GeV/c)^2, the others are at $Q^2 = 0$ (see text).

4 Summary

We performed E12-17-003 experiment at JLab in 2018. This article described the results of the differential cross sections for the Λ/Σ^0 hyperon electroproduction at forward angles in low-Q^2 region. This is a region of interest where there was no data in the photoproduction so far. This work helps understanding hyperon photoproduction and electroproduction in the same theoretical framework.

5 Acknowledgement

We thank the JLab staff of the Division of Physics, Division of Accelerator, and the Division of Engineering for providing support for conducting the experiment. This work was supported by U.S. Department of Energy (DOE) grant DE-AC05-06OR23177 under which Jefferson Science Associates, LLC, operates the Thomas Jefferson National Accelerator Facility. The work of the Argonne National Laboratory group member is supported by DOE grant DE-AC02-06CH11357. The Kent State University contribution is supported under grant no. PHY-1714809 from the U.S. National Science Foundation. The hypernuclear program at JLab is supported by U.S. DOE grant DE-FG02-97ER41047. This work was partially supported by the Grant-in-Aid for Scientific Research on Innovative Areas "Toward new frontiers Encounter and synergy of state-of-the-art astronomical detectors and exotic quantum beams." This work was supported by JSPS KAKENHI grants nos. 18H05459, 18H05457, 18H01219, 17H01121, 19J22055, and 18H01220. This work was also supported by the Graduate Program on Physics for the Universe, Tohoku University (GP-PU), and SPIRITS 2020 of Kyoto University.

References

[1] K.N. Suzuki, T. Gogami et al., Prog. Theor. Exp. Phys. **2022**, 013D01 (2022)
[2] B. Pandey, L. Tang et al., Phys. Rev. C **105**, L051001 (2022)
[3] K.H. Glander, J. Barth et al., Eur. Phys. J. A **19**, 251 (2004)
[4] R. Bradford, R.A. Schumacher et al., Phys. Rev. C **73**, 035202 (2006)
[5] M.E. McCracken, M. Bellis et al., Phys. Rev. C **81**, 025201 (2010)
[6] M. Sumihama, J.K. Ahn et al., Phys. Rev. C **73**, 035214 (2006)

[7] J.W.C. McNabb, R.A. Schumacher et al., Phys. Rev. C **69**, 042201 (2004)
[8] B. Dey, C.A. Meyer et al., Phys. Rev. C **82**, 025202 (2010)
[9] E. Amaldi, S. Fubini et al., *Pion Electroproduction. Electroproduction at Low- Energy and Hadron Form-Factors*, Vol. 83 (Springer Tracts in Modern Physics, 1979)
[10] J. Alcorn, B. Anderson et al., Nucl. Instrum. Methods, Phys. Res. Sect. A **522**, 294 (2004)
[11] S. Santiesteban, S. Alsalmi et al., Nucl. Instrum. Methods, Phys. Res. Sect. A **940**, 351 (2019)
[12] R.M. Mohring, D. Abbott et al., Phys. Rev. C **67**, 055205 (2003)
[13] C.J. Bebek, C.N. Brown et al., Phys. Rev. D **15**, 594 (1977)
[14] C.N. Brown, C.R. Canizares et al., Phys. Rev. Lett. **28**, 1086 (1972)
[15] C.J. Bebek, C.N. Brown et al., Phys. Rev. Lett. **32**, 21 (1974)
[16] T. Mart, C. Bennhold, Phys. Rev. C **61**, 012201 (1999)
[17] D. Skoupil, P. Bydžovský, Phys. Rev. C **93**, 025204 (2016)
[18] D. Skoupil, P. Bydžovský, Phys. Rev. C **97**, 025202 (2018)
[19] T. Corthals, J. Ryckebusch et al., Phys. Rev. C **73**, 045207 (2006)
[20] L. De Cruz, T. Vrancx et al., Phys. Rev. Lett. **108**, 182002 (2012)

https://doi.org/10.1051/epjconf/202227102004

Possible ΣNN resonances in the ^{3}H(e,e′K$^+$) reaction

Benjamin F. Gibson[1,*] and *Iraj* R. Afnan[2,**]

[1]Theoretical Division, Los Alamos National Laboratory, Los Alamos, New Mexico 87545, USA
[2]College of Science and Engineering, Flinders University, Bedford Park, South Australia 5042, Australia

Abstract. In a recent JLab ^{3}H(e,e′K$^+$) experiment a structure near the ΣNN threshold was observed, which was interpreted to be a T = 1 resonance. Using separable NN and $\Lambda N{-}\Sigma N$ potentials in the Faddeev equations, we demonstrate that it is possible that both T = 0 and T = 1 resonances may have been excited. Moreover, their pole positions, in our model calculation, are sufficiently close to one another that it is unlikely that these resonances in the ΣNN spectrum can be resolved.

1 Introduction

In a recent JLab experiment, in which the ^{3}H(e,e′K$^+$)Λnn reaction was exploited to investigate the existence of a threshold $^3_\Lambda$n resonance that would place constraints upon the Λn scattering length, a structure was observed in the spectrum near the ΣNN threshold that was interpreted to be a ΣNN resonance [1, 2]. Because the electromagnetic operator does not conserve isospin, such a ΣNN resonance could have isospin T = 0, 1, or 2. Garcilazo argued in 1987, on the basis of rank-one separable potentials, that no T = 2 ΣNN bound state or resonance should exist [3]. This agreed with the 1982 analysis of Dover and Gal [4]. Stadler *et al.* later demonstrated that there was little possibility of a T = 2 bound state or narrow resonance based upon the Jülich one-boson exchange potential [5]. This confirmed Garcilazo's result and demonstrated the usefulness of separable potential calculations regarding low-energy properties of three-body systems. However, continuum Faddeev-type calculations were needed to address the existence of T = 0 and T = 1 resonant states.

2 Past T = 0 and T = 1 Faddeev-type resonance results

In 1993 Afnan *et al.* found, while exploring Λd elastic scattering using a separable potential model for the NN interaction and the $\Lambda N - \Sigma N$ coupled channel interaction, that a near-threshold T = 0 ΣNN resonance could exist [6]. (These calculations were performed using a code initially developed for $^3_\Lambda$H bound-state calculations [7].) The position of the poles, on the second Riemann sheet of the complex energy plane, were determined by examining the eigenvalues of the kernel of the Faddeev equations using contour rotation methodology. Four different $\Lambda N - \Sigma N$ potential models were explored; here we consider only the TGE-*B* model

*e-mail: bfgibson@lanl.gov
**e-mail: iraj.afnan@icloud.com

[8] result. The T = 3/2 ΣN interaction was constructed to fit the Nijmegen 1989 effective range parameters [9].

The pole position for this T = 0 resonance is

$$75.5 - 8.34i \text{ MeV},$$

which lies near the ΣNN model threshold at 77 MeV. [The thresholds due to the different masses of the (Σ^-, Σ^0, Σ^+) triplet were not distinguished in this model calculation.]

A decade later Garcilazo *et al.* utilized a separable potential approximation to a chiral constituent quark model of the hyperon-nucleon interaction to explore Λd and Σd scattering [10]. They concluded that the T = 0 and T = 1 spin-1/2 channels of the ΣNN system were the only channels exhibiting attraction. By examining the Fredholm determinant for the ΣNN system, they concluded that only the T = 1 channel was sufficiently attractive to support a near-threhsold resonance. Presumably, had they explored a larger energy range for the T = 0 channel, they would have found a resonance in that channel, also.

3 Additional ΣNN resonance information

In 1992 Barakat *et al.* reported a null result in a ^{3}He(K$^-$,π^+) in-flight K$^-$ experimental search at BNL (Brookhaven National Laboratory) for a $\Sigma^- np$ resonance [11]. This was motivated by the ^{4}He(K$^-$,π^-) at-rest experiment performed with K$^-$ stopping in a He bubble chamber [12] and later in-flight ^{4}He(K$^-$,$\pi^\pm$) experiments [13]. That is, a $^4_\Sigma$He state was observed by Hayano *et al.*

In 2014 Harada *et al.* performed a distorted wave impulse approximation calculation that agreed with the BNL ^{3}He(K$^-$,π^+) result of no resonance [14]. However, their model results indicated that one should see a T = 1 resonance in the conjugate ^{3}He(K$^-$,π^-) reaction.

Thus, various theoretical calculations have provided indications that T = 0 and T = 1 resonances could be seen in the ΣNN spectrum. In terms of charge states, the $\Sigma^- nn$ system, having the lowest threshold, was suggested to be the most likely candidate in the recently reported ^{3}H(e,e′K$^+$) reaction [1].

4 Recent Faddeev results

As noted above, an s-wave separable potential ΣNN T = 0 pole was found in Ref.[6]. The pole position is

$$75.5 - 8.34i \text{ MeV},$$

which lies below the ΣNN model threshold of 77 MeV. In the Garcilazo *et al.* separable potential approximation to the chiral quark model, the T = 0 channel was found to be attractive but not sufficiently so as to support a pole near the ΣNN threshold. This difference with our calculation provides an indication of the model dependence in the system.

We have now extended our Faddeev separable potential calculation to search for the T = 1 pole position in our model. We find that pole position to be

$$76.9 - 6.44i \text{ MeV},$$

which lies closer to the ΣNN threshold than does our T = 0 pole.

The Garcilazo *et al.* model can be considered to be more sophisticated than our s-wave separable potential model in that it includes a tensor force in both the NN and YN interactions. We, therefore, extended our model to include an NN tensor force. The T = 0 pole is essentially unaffected, as one would anticipate. The T = 1 pole moved slightly closer to the ΣNN threshold. We note that in our calculation we also allowed for $L = 2$ for the spectator particle when two-body tensor interactions are included. In Ref.[10] only $L = 0$ for the spectator particle was included.

Thus, in our TGE-B model calculation we find both a T = 0 and a T = 1 pole in the ΣNN spectrum, poles that are rather close to one another in position and have similar widths. Our results agree qualitatively with those of Ref.[10] and Ref.[14] for the T = 1 resonance.

5 Summary

The primary conclusions from our model calculations are (*i*) that one may see both a T = 0 resonance as well as a T = 1 resonance in the ^{3}H(e,e$'$K$^+$)ΣNN spectrum near the ΣNN threshold and (*ii*) that the two resonances in our model reside too close to one another to be easily resolved experimentally.

6 Acknowledgement

The work of BFG was performed under the auspices of the National Nuclear Security Administration of the U.S. Department of Energy at Los Alamos National Laboratory under Contract No. DE-AC52-06NA25396.

References

[1] B. Pandy *et al.*, Phys. Rev. C **105**, L051001 (2022)
[2] K. N. Suzuki *et al.*, Prog. of Theor. and Exper. Physics **2022**, 013D01 (2022)
[3] H. Garcilazo, J. Phys. G **13**, L65 (1987)
[4] C. B. Dover and A. Gal, Phys. Lett. B **110**, 443 (1982)
[5] A. Stadler and B. F. Gibson, Phys. Rev. C **50**, 512 (1994)
[6] I. R. Afnan and B. F. Gibson, Phys. Rev. C **47**, 1000 (1993)
[7] I. R. Afnan and B. F. Gibson, Phys. Rev. C **41**, 2787 (1990)
[8] G. Toker, A. Gal, and J. M. Eisenberg, Nucl. Phys. **A 362**, 405 (1981)
[9] P. M. M. Maessen, Th. A. Rijken, and J. J. de Swart, Phys. Rev. C, **40**, 2226 (1989)
[10] H. Garcilazo, T. Fernandez-Carames, and A. Valcarce, Phys. Rev. C **75**, 034002 (2007)
[11] M. Barakat and E. V. Hungerford, Nucl. Phys. **A 547**, 157c (1992)
[12] R. Hayano *et al.*, Nuovo Cimento **102**, 437 (1989)
[13] R. S. Hayano *et al.*, Phys. Lett. B **231**, 355 (1989)
[14] T. Harada and Y. Hirabayshi, Phys. Rev. C **89**, 054603 (2014)

The impact of final-state interaction on the $\pi\Sigma$ photoproduction in the $\Lambda(1405)$ region

Aleš Cieplý[1,*] and *Peter C.* Bruns[1,**]

[1]Nuclear Physics Institute of the Czech Academy of Sciences, 250 68 Řež, Czechia

Abstract. It was recently demonstrated [1] that the meson-baryon rescattering in the final state has a major impact on the magnitude and structure of the $\pi\Sigma$ mass distributions observed in the $\gamma p \to K^+\pi\Sigma$ photoproduction. We discuss briefly several aspects of this work emphasizing the model dependence of the meson-baryon amplitudes, used to represent the final-state interaction, and the role of the adopted photoproduction mechanism.

1 Introduction

The modern theoretical approaches to meson-baryon (MB) interactions are based on effective chiral Lagrangians. When such theory is applied to the coupled channel $\pi\Sigma - \bar{K}N$ system the enigmatic $\Lambda(1405)$ resonance emerges as a two-pole state (on the complex energy plane), generated dynamically as a result of sufficiently strong inter-channel couplings. The theoretical predictions tend to agree on the position of the narrower pole located around 1425 MeV and often interpreted as a $\bar{K}N$ molecular state, while the position of the second pole is not determined so well, lying at significantly lower energies and much further from the real axis. Since the model parameters are typically fixed by fitting K^-p reactions and threshold data, it is not so surprising that the theoretical predictions are quite varied at energies below the $\bar{K}N$ threshold. In this respect, the *two-meson photoproduction* reaction $\gamma p \to K^+\pi\Sigma$ enables a scan in the invariant mass of the MB system down to its production threshold and the calculated mass spectra also reveal a strong model dependence [1]. The determined line-shapes of the $\pi\Sigma$ mass distributions cover the energies relevant for the $\Lambda(1405)$ formation and put additional constraints on the parameters of the strangeness $S = -1$ meson-baryon system.

The corresponding experiment was performed by the CLAS collaboration several years ago, taking very precise, high-resolution data [2]. These data have already made a significant impact on the field, see e.g. Refs. [3–6] where various photoproduction models were tested and the model predictions compared with the data. In this contribution we follow the recent work [1] applying the photoproduction model presented in [7]. While doing so we demonstrate the sensitivity of the calculated $\pi\Sigma$ mass spectra to ambiguities inherent in extending the $\pi\Sigma - \bar{K}N$ coupled-channel models to energies below the $\bar{K}N$ threshold.

2 Formalism

The MB coupled channel models mentioned in the previous section provide us with partial wave scattering amplitudes $f^{c',c}_{\ell\pm}(M_{\pi\Sigma})$ that describe the transitions from channel c to channel

*e-mail: cieply@ujf.cas.cz
**e-mail: bruns@ujf.cas.cz

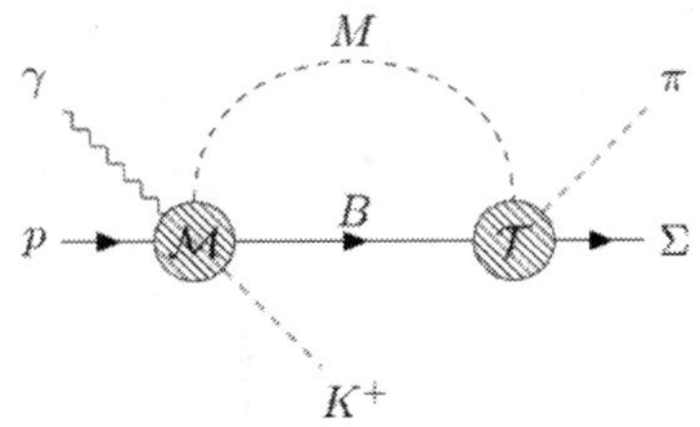

Figure 1. Illustration of the final-state interaction (FSI) of the meson-baryon pair. $\mathcal{M}$ represents the amplitude for the $\gamma p \to K^+ MB$ without FSI, and $\mathcal{T}$ is the meson-baryon scattering amplitude, which can be decomposed into partial waves $f_{\ell\pm}$.

c' ($c, c' = \pi\Lambda$, $\pi\Sigma$, $\bar{K}N$, $\eta\Lambda$, $\eta\Sigma$, $K\Xi$) at a given two-body c.m. energy (invariant MB mass) $M_{\pi\Sigma}$ and for total angular momentum $\ell \pm \frac{1}{2}$ and orbital angular momentum $\ell = 0, 1, 2, \ldots$. These amplitudes are designed to be consistent with ChPT up to a fixed order of the low-energy expansion and fulfill the requirement of coupled-channel unitarity above the lowest reaction threshold. Considering the photoproduction reaction $\gamma p \to K^+ MB$ as illustrated in Fig. 1, the s-wave scattering amplitudes f_{0+} can be implemented to describe the final-state interaction of the produced MB pair. It was shown in Refs. [1, 7] that one can obtain four independent photoproduction amplitudes $\mathcal{A}^{i(\text{tree})}_{0+}(s, M_{\pi\Sigma})$[1] that combine the structure functions contributing to the tree-level photoproduction amplitude $\mathcal{M}$ which is projected on the $\ell = 0$ state of the $\pi\Sigma$ pair. Since the outgoing K^+ can be treated effectively as a spectator particle in the unitarization procedure, the FSI is then restricted to the $S = -1$ meson-baryon subspace, the sector relevant for the formation of the $\Lambda(1405)$.

The unitarized photoproduction amplitudes for $\gamma p \to K^+ MB$ are taken as the coupled-channel vector

$$[\mathcal{A}^{i}_{0+}(s, M_{\pi\Sigma})] = [\mathcal{A}^{i(\text{tree})}_{0+}(s, M_{\pi\Sigma})] + [f_{0+}(M_{\pi\Sigma})]\,[8\pi M_{\pi\Sigma} G(M_{\pi\Sigma})]\,[\mathcal{A}^{i(\text{tree})}_{0+}(s, M_{\pi\Sigma})]\,, \qquad (1)$$

where $G(M_{\pi\Sigma})$ is a diagonal matrix with entries given by suitably regularized MB loop integrals and the square brackets mark the vector or matrix character of the involved quantities in the MB channel-space. The $MB = \pi\Sigma$ entries of the vector $\mathcal{A}^{i}_{0+}$ can then be used to obtain the required s-wave cross sections. We note that the tree-level amplitudes $\mathcal{A}^{i(\text{tree})}_{0+}$ are constructed from 16 $\mathcal{M}$-structures representing the Weinberg-Tomozawa, Born, and anomalous graphs adhering strictly to constraints arising from unitarity, gauge invariance and chiral perturbation theory, see Refs. [1, 7] for details. The whole formalism does not contain any free parameters and refrains from including additional graphs based e.g. on vector meson exchanges or on triangular singularities that were considered by other authors [5, 6].

We find it illustrative to compare our formalism with the early phenomenological approaches adopted in [3, 4]. There, the FSI was implemented in the photoproduction amplitudes as

$$[\mathcal{A}_{0+}(s, M_{\pi\Sigma})] = [f_{0+}(M_{\pi\Sigma})]\,[8\pi M_{\pi\Sigma} G(M_{\pi\Sigma})]\,[C(s)]\,, \qquad (2)$$

where the channel-space vector $C(s)$ was introduced to simulate the reaction mechanism. A direct comparison of the two equations shows that the phenomenological approach based on Eq. (2) corresponds to setting the tree graphs in the first term on the r.h.s. of Eq. (1) to zero and at the same time replaces them by energy-dependent constants in the second term. It is obvious that such an ad-hoc treatment cannot reflect fully the complexity of the photoproduction process and the resulting MB amplitudes may not be quite reliable. However, by fitting

[1] In fact, the $\mathcal{A}$ amplitudes depend also on t_K which we do not show explicitly to keep the notation concise.

the constants $C^c(s)$ (at each energy $W = \sqrt{s}$ separately) the authors of [3, 4] were able to describe quite well the CLAS data [2] on the $\pi\Sigma$ mass spectra.

3 Results and discussion

The formalism described in the previous section and based on Eq. (1) was used in [1] to calculate the $\pi\Sigma$ mass distributions observed in the CLAS experiment [2]. In Fig. 2 we show the results obtained in [1] at two sample energies, $W = 2.0$ (left figures) and 2.4 GeV (right figures) considering four different $\pi\Sigma - \bar{K}N$ models that provide the final state rescattering amplitudes f_{0+}. The models include the recent version of the Prague amplitudes [8], tagged as P model, and three versions of the Bonn model amplitudes: B2, B4 [4], and BW [9]. First of all, we note that the P model seems to work reasonably well for the $\pi^0\Sigma^0$ and $\pi^+\Sigma^-$ cross sections at the c.m. energy $W = 2.0$ GeV. Unfortunately, one cannot say the same about the $\pi^-\Sigma^+$ mass spectrum at the same energy, but we should bear in mind that all theoretical predictions shown in Fig. 2 were made without any adjustment of the MB rescattering amplitudes, and without introducing any free parameter in the formalism. The varied predictions provided by the selected models reflect the ambiguities when extending their application below the $\bar{K}N$ threshold, to a completely different sector of experimental data and to energies much lower

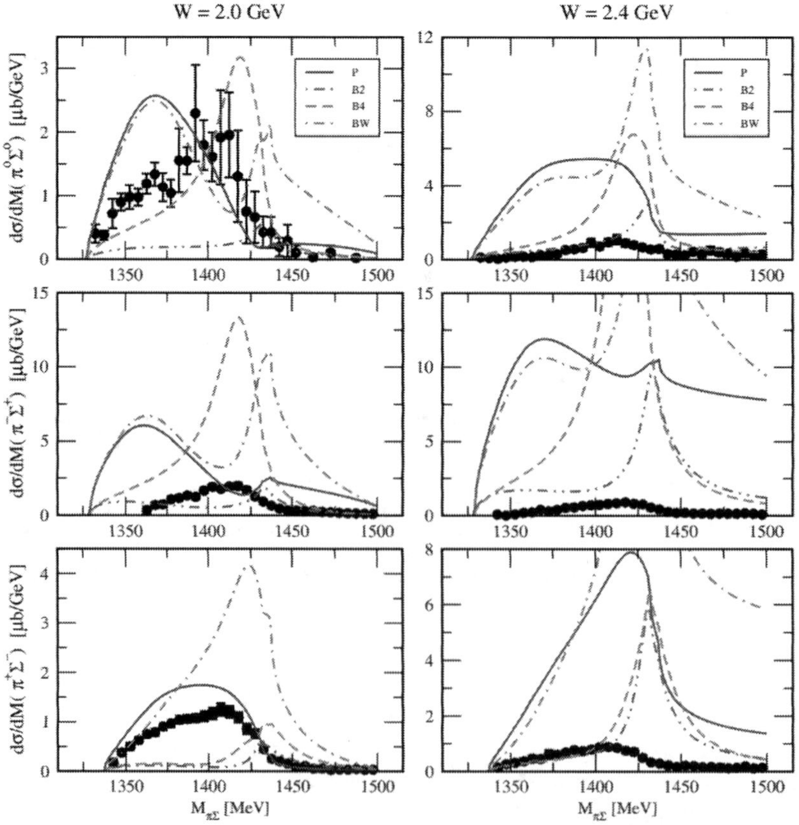

Figure 2. Comparison of $\pi\Sigma$ mass distributions calculated at two c.m. energies, $W = 2.0$ and 2.4 GeV, while employing MB amplitudes generated by four different coupled-channel models. The experimental data are taken from [2].

than those of the $K^- p$ reaction data the f_{0+} model parameters were fitted to. The Fig. 2 also shows that the Bonn model amplitudes do not reproduce the CLAS data when employed in the current formalism based on Eq. (1), despite providing adequate description of the same data when the phenomenological $C(s)$ factors defining the formalism based on Eq. (2) were adjusted to fit the photoproduction data in Refs. [4, 9]. Apparently, the calculated $\pi\Sigma$ mass distributions depend strongly on both, the adopted photoproduction scheme as well as on the MB model used to describe the FSI, so both aspects are intertwined and should be considered in the same framework.

Taking aside the partial success of the P model at $W = 2.0$ GeV, all other calculated spectra provide much higher cross sections than those observed experimentally. It was demonstrated in [1] that this problem can be addressed by tuning down the mass-scales used to regularize the MB loop functions that connect the tree-level amplitudes $\mathcal{M}$ with the FSI amplitudes $\mathcal{T}$ in Fig. 1. In addition, since the formalism adopted to calculate the tree-level amplitudes $\mathcal{A}_{0+}^{i(\text{tree})}(s, M_{\pi\Sigma})$ is based on LO ChPT, we anticipate that at $W = 2.4$ GeV the kaon momentum is probably too large for the LO ChPT photo-kernel making the theoretical predictions less reliable. It is also understood that the energy dependence can be augmented and a better agreement with experimental data reached by including contributions which go beyond the standard ChPT approach, see Refs. [5, 6] for examples of such model extensions.

4 Summary

We have reviewed the recent implementation [1] of the MB FSI into the two-meson photoproduction amplitude for the $\gamma p \to K^+\pi\Sigma$ reaction. Although all employed $\pi\Sigma - \bar{K}N$ coupled channel models provide practically equivalent description of the experimental data for $K^- p$ reactions, the results differ significantly when the models are used to calculate the $\pi\Sigma$ mass spectra in the photoproduction process. This reflects the ambiguities in the MB scattering amplitudes when extended below the $\bar{K}N$ threshold [10], and also demonstrates the crucial role played by the FSI in the reaction. Although the calculated $\pi\Sigma$ mass distributions (in general) fail to reproduce the data reported by the CLAS experiment [2] it is understood that a better agreement can be reached when including the photoproduction data in future fits that would narrow the parameter space of the MB models.

Acknowledgement: The authors thank M. Mai for providing them with the f_{0+} transition amplitudes generated by the $\bar{K}N$ Bonn models and for his personal contribution to the completion of Ref. [1]. The present work was also supported by the Grant Agency of the Czech Republic under Grant No. 19-19640S.

References

[1] P. C. Bruns, A. Cieplý, M. Mai, Phys. Rev. **D 106**, 074017 (2022)
[2] K. Moriya et al. [CLAS], Phys. Rev. **C 87**, 035206 (2013)
[3] L. Roca, E. Oset, Phys. Rev. **C 88**, 055206 (2013)
[4] M. Mai, U.-G. Meißner, Eur. Phys. J. **A 51**, 30 (2015)
[5] S. X. Nakamura, D. Jido, PTEP **2014**, 023D01 (2014)
[6] E. Wang, J.-J. Xie, W.-H. Liang, F.-K. Guo, E. Oset, Phys. Rev. **C 95**, 015205 (2017)
[7] P. C. Bruns, arXiv:2012.11298 [nucl-th] (2020)
[8] P. C. Bruns, A. Cieplý, Nucl. Phys. **A 1019**, 122378 (2022)
[9] D. Sadasivan, M. Mai, M. Döring, Phys. Lett. **B 789**, 329–335 (2019)
[10] A. Cieplý, M. Mai, U.-G. Meißner, J. Smejkal, Nucl. Phys. **A 954**, 17-40 (2016)

Study of Λn FSI with Λ quasi-free productions on the ^{3}H$(e, e'K^+)X$ reaction at JLab

K. Itabashi[1*], *K.N.* Suzuki[2], *B.* Pandey[3], *K.* Okuyama[4], *T.* Gogami[2], *S.* Nagao[4], *S.N.* Nakamura[1], *L.* Tang[3,5], *D.* Abrams[6], *T.* Akiyama[4], *D.* Androic[7], *K.* Aniol[8], *C.* Ayerbe Gayoso[9], *J.* Bane[10], *S.* Barcus[9], *J.* Barrow[10], *V.* Bellini[11], *H.* Bhatt[12], *D.* Bhetuwal[12], *D.* Biswas[3], *A.* Camsonne[5], *J.* Castellanos[13], *J-P.* Chen[5], *J.* Chen[9], *S.* Covrig[5], *D.* Chrisman[14,15], *R.* Cruz-Torres[16], *R.* Das[17], *E.* Fuchey[18], *K.* Gnanvo[6], *F.* Garibaldi[11,19], *T.* Gautam[3], *J.* Gomez[5], *P.* Gueye[3], *T.J.* Hague[20], *O.* Hansen[5], *W.* Henry[5], *F.* Hauenstein[21], *D.W.* Higinbotham[5], *C.E.* Hyde[21], *M.* Kaneta[4], *C.* Keppel[5], *T.* Kutz[17], *N.* Lashley-Colthirst[3], *S.* Li[22,23], *H.* Liu[24], *J.* Mammei[25], *P.* Markowitz[13], *R.E.* McClellan[5], *F.* Meddi[11,26], *D.* Meekins[5], *R.* Michaels[5], *M.* Mihovilovic[27,28,29], *A.* Moyer[30], *D.* Nguyen[16,31], *M.* Nycz[20], *V.* Owen[9], *C.* Palatchi[6], *S.* Park[17], *T.* Petkovic[7], *S.* Premathilake[6], *P.E.* Reimer[32], *J.* Reinhold[13], *S.* Riordan[32], *V.* Rodriguez[33], *C.* Samanta[34], *S.N.* Santiesteban[22], *B.* Sawatzky[5], *S.* Širca[27,28], *K.* Slifer[22], *T.* Su[20], *Y.* Tian[35], *Y.* Toyama[4], *K.* Uehara[4], *G.M.* Urciuoli[11], *D.* Votaw[14,15], *J.* Williamson[36], *B.* Wojtsekhowski[5], *S.A.* Wood[5], *B.* Yale[22], *Z.* Ye[32], *J.* Zhang[6], and *X.* Zheng[6]

[1]Department of Physics, Graduate School of Science, the University of Tokyo, Hongo, Tokyo 113-0033 Japan
[2]Department of Physics, Kyoto University, Kyoto, 606-8502, Japan
[3]Department of Physics, Hampton University, Hampton, Virginia 23668, USA
[4]Department of Physics, Graduate School of Science, Tohoku University, Sendai, 980-8578 Japan
[5]Thomas Jefferson National Accelerator Facility, Newport News, Virginia 23606, USA
[6]Department of Physics, University of Virginia, Charlottesville, Virginia 22904, USA
[7]Department of Physics & Department of Applied Physics, University of Zagreb, HR-10000 Zagreb, Croatia
[8]Physics and Astronomy Department, California State University, Los Angeles, California 90032, USA
[9]Department of Physics, The College of William and Mary, Virginia 23185, USA
[10]Department of Physics, University of Tennessee, Knoxville, Tennessee 37996, USA
[11]Istituto Nazionale di Fisica Nucleare, Sezione di Roma 00185, Rome, Italy
[12]Department of Physics, Mississippi State University, Mississippi State, Mississippi 39762, USA
[13]Department of Physics, Florida International University, Miami, Florida 33199, USA
[14]Department of Physics and Astronomy, Michigan State University, East Lansing, Michigan 48824, USA
[15]National Superconducting Cyclotron Laboratory, Michigan State University, East Lansing, MI 48824, USA
[16]Department of Physics, Massachusetts Institute of Technology, Cambridge, Massachusetts 02139, USA
[17]Department of Physics, State University of New York, Stony Brook, New York 11794, USA
[18]Department of Physics, University of Connecticut, Storrs, Connecticut 06269, USA
[19]Instituto Superiore di Sanitá, 00161, Rome, Italy
[20]Department of Physics, Kent State University, Kent, Ohio 44242, USA
[21]Department of Physics, Old Dominion University, Norfolk, Virginia 23529, USA
[22]Department of Physics, University of New Hampshire, Durham, New Hampshire 03824, USA
[23]Nuclear Science Division, Lawrence Berkeley National Laboratory, Berkeley, CA 94720, USA
[24]Department of Physics, Columbia University, New York, New York 10027, USA
[25]Department of Physics and Astronomy, University of Manitoba, Winnipeg, Manitoba R3T 2N2, Canada

*e-mail: kosuke.itabashi@kek.jp

[26]Sapienza University of Rome, I-00185, Rome, Italy
[27]Faculty of Mathematics and Physics, University of Ljubljana, 1000 Ljubljana, Slovenia
[28]Jožef Stefan Institute, Ljubljana, Slovenia
[29]Institut für Kernphysik, Johanness Gutenberg-Universität Mainz, DE-55128 Mainz, Germany
[30]Department of Physics, Christopher Newport University, Newport News, Virginia 23606, USA
[31]University of Education, Hue University, Hue City, Vietnam
[32]Physics Division, Argonne National Laboratory, Lemont, Illinois 60439, USA
[33]División de Ciencias y Tecnologia, Universidad Ana G. Méndez, Recinto de Cupey, San Juan 00926, Puerto Rico
[34]Department of Physics & Astronomy, Virginia military Institute, Lexington, Virginia 24450, USA
[35]Department of Physics, Syracuse University, New York, New York 10016, USA
[36]School of Physics & Astronomy, University of Glasgow, Glasgow, G12 8QQ, Scotland, UK

Abstract. An $nn\Lambda$ is a neutral baryon system with no charge. The study of the pure Λ-neutron system such as $nn\Lambda$ gives us information on the Λn interaction. The $nn\Lambda$ search experiment (E12-17-003) was performed at JLab Hall A in 2018. In this article, the Λn FSI was investigated by a shape analysis of the ^{3}H$(e,e'K^+)X$ missing mass spectrum, and a preliminary result for the Λn FSI study is given.

1 Introduction

In the nucleon-nucleon (NN) interaction, realistic nuclear potentials have been constructed based on rich NN scattering data. On the other hand, there are relatively large uncertainties due to limited ΛN scattering data in case of the ΛN interaction. In addition, since there is no Λn scattering data, the Λn interaction have been established from the limited Λp scattering data assuming charge symmetry (CS). However, it has been experimentally observed that charge symmetry is broken between A=4 mirror Λ hypernuclei ($^4_\Lambda$H, $^4_\Lambda$He) [1]. Therefore, it is important to derive the Λn interaction experimentally. One of the major experimental investigation methods for the ΛN interaction is the study of the ΛN final state interaction (FSI). The FSI is the reaction between the recoil Λ and a nucleon in a nucleus, and it is known to make an enhanced structure in the missing mass spectrum [2, 3]. Therefore, the spectroscopic study of a pure Λ-neutron system such as $nn\Lambda$ is expected to give us information on the Λn interaction. The ^{3}H$(e,e'K^+)X$ missing mass spectrum was obtained by using two HRS spectrometers and a tritium target (^{3}H) which is a radioactive material in 2018 at Jefferson Lab (JLab) Hall A [4, 5]. In this study, the Λn FSI interaction was investigated by analyzing the ^{3}H$(e,e'K^+)X$ missing mass spectrum obtained by this experiment (E12-17-003).

2 ^{3}H$(e,e'K^+)X$ missing mass spectrum

In this experiment, a cryogenic tritium gas target (40 K) with a thickness of 84.8 mg/cm^2 was irradiated with an electron with an energy (E_e) of 4.3 GeV, and measured momenta of scattered electrons ($p_{e'}$ = 2.2 GeV/c) and K^+ mesons (p_K = 1.8 GeV/c) by two high resolution spectrometers (HRS). The missing mass (M_X) was calculated with the momentum vectors ($\vec{p}_{e'}$, $\vec{p}_K$) and energies ($E_{e'}$, E_K) as follows :

$$M_X = \sqrt{(E_e + m_T - E_{e'} - E_K)^2 - (\vec{p}_e - \vec{p}_{e'} - \vec{p}_K)^2}. \tag{1}$$

where m_T and $\vec{p}_e$ are the mass of tritium and momentum vector of the electron beam, respectively. As a function of the measured the ^{3}H$(e,e'K^+)X$ missing mass, the missing mass spectrum is shown by black dots with error bars in Fig. 1. This vertical axis represents the differential cross section for the ^{3}H$(e, e'K^+)X$ reaction. Any parameter values such as momentum acceptances ($d\Omega_{e'}$, $d\Omega_K$) used to calculate the cross section are explained in Ref. [6].

2.1 Monte Carlo Simulation (SIMC)

The solid black line in Fig. 1 shows the Λ quasi-free (Λ-QF) distribution calculated by the Monte Carlo Simulation (SIMC); SIMC is a JLab standard Monte Carlo simulation code, which takes into account the effects such as the proton Fermi momentum, kaon decay. Comparing the SIMC spectrum with the missing mass spectrum, the region over 60 MeV is good agreement. However, there is some enhancement around $nn\Lambda$ mass threshold ($-B_\Lambda \sim 0$ MeV) and 20 MeV regions. Around the $nn\Lambda$ mass threshold where a $nn\Lambda$ peak is expected to exist, there are excess events which cannot reproduced by SIMC.

Though the statics is not enough to conclude anything about the existence of the $nn\Lambda$ system [7], the excess events around 20 MeV are expected to be produced by the Λn FSI. In the past Λ-hypernuclear experiment in the ^{3}He$(e, e'K^+)X$ reaction at JLab, a similar enhanced structure was observed within a range of $-B_\Lambda > 0$ MeV. Moreover, this enhanced structure was successfully reproduced by considering the ΛN FSI effect [2]. Therefore, in this study, the Λn FSI effect was investigated by analyzing the structure around 20 MeV.

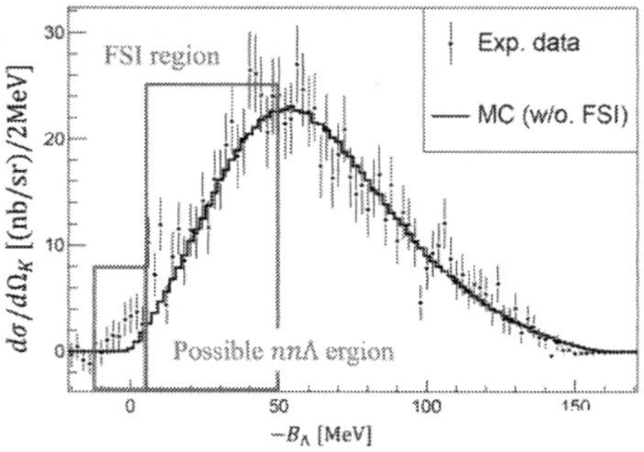

Figure 1. ^{3}H$(e, e'K^+)X$ missing mass spectrum. The horizontal and vertical axes indicate the binding energy of the Λ (MeV) and the differential cross section of the missing mass (nb/sr/2 MeV), respectively. The black points with bar shows the experimental data. The black solid line is the simulation result (SIMC) of the Λ-QF distribution without any final state interaction effects.

3 Calculation of the Λn final state interaction (FSI)

The cross section for Λ-QF productions including FSI is written as follows :

$$\left(\frac{d\sigma}{d\Omega}\right)_{\mathrm{FSI}} = \mathrm{I}(\vec{k}_{\Lambda n})\left(\frac{d\sigma}{d\Omega}\right)_{\mathrm{w/oFSI}}, \quad (2)$$

where $\mathrm{I}(\vec{k}_{\Lambda n})$ is the influence factor depending on a relative momentum ($\vec{k}_{\Lambda n}$) between a neutron in tritium and a recoil Λ. In the two-body ($\Lambda - n$) scattering model, the influence factor can be written by using the Jost function (J) as $\mathrm{I}(\vec{k}_{\Lambda n}) = 1/|J(\vec{k}_{\Lambda n})|^2$ [8]. Moreover, in the effective range approximation ($k_{\Lambda n}\cot\delta = -1/a + r/2k^2_{\Lambda n}$), the Jost function is written as :

$$J(\vec{k}_{\Lambda n}) = \frac{k_{\Lambda n} - i\beta}{k_{\Lambda n} - i\alpha}, \quad (3)$$

$$\frac{1}{2}r(\alpha - \beta) = 1, \quad \frac{r}{2}\alpha\beta = -\frac{1}{a}, \quad (4)$$

where a and r are a scattering length and an effective range, respectively. Figure 2 shows the calculation results of the influence factor with various Λn potential models. The weighting ratio of the spin singlet ($^1S_0$) and triplet ($^3S_1$) factors is one to three. From Eq. (2)-(4), the cross section of the Λ-QF productions including Λn FSI can be obtained.

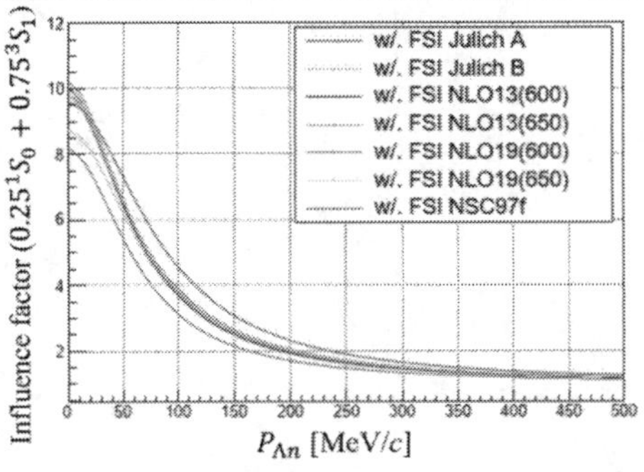

Figure 2. The influence factor depending on a relative Λn momentum in each theoretical model. The influence factor is composed of four spin states, a singlet (I_s) and triplet (I_t) states with a ratio of one to three.

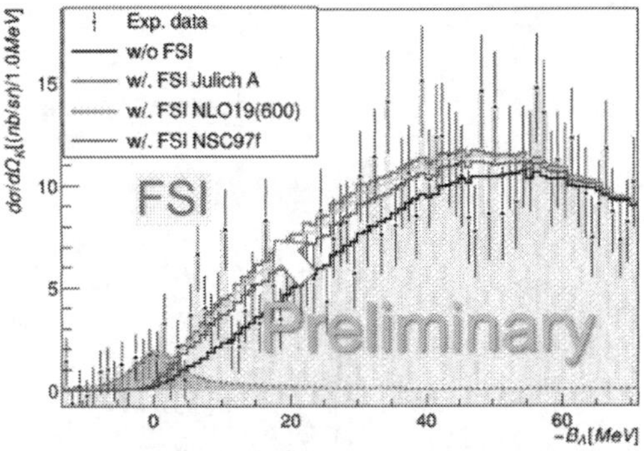

Figure 3. The missing mass spectra with Λn FSI. The black and colored solid lines show the SIMC spectrum without the Λn FSI and with the Λn FSI in each potential model.

4 Results

4.1 Λn potential model dependence

The Λn FSI effect was estimated by the χ^2 fitting of the $^3\mathrm{H}(e, e'K^+)X$ missing mass spectrum. The structure around $nn\Lambda$ mass threshold ($-B_\Lambda \sim 0$ MeV) exists, which could not be reproduced by the SIMC spectrum. Therefore, this structure was described as background by using the Breit-Wigner function (f_{WB}) because it was successful to reproduced well by using this function in Ref.[6]. The fitting function of the $^3\mathrm{H}(e, e'K^+)X$ missing mass spectrum was defined as :

$$\left(\frac{d\sigma}{d\Omega}\right)_{\mathrm{FSI}} = w_{\mathrm{FSI}} \cdot \mathrm{I}(\vec{k}_{\Lambda n}) \left(\frac{d\sigma}{d\Omega}\right)_{\mathrm{SIMC}} + w_{WB} \cdot f_{\mathrm{WB}} , \tag{5}$$

where w_{FSI} and w_{WB} are scaling factors for the SIMC spectrum and Breit-Wigner function, respectively. These weighting factors were scaled by the χ^2-fitting with the missing mass spectrum within a range from 0 to 60 MeV. The fitting results of the SIMC spectrum by each Λn FSI model are shown in Fig. 3.

4.2 Search for the best Λn FSI parameters

The scattering length (a) and effective range (r) are parameters that determine the characteristics of the Λn potential. Since these parameters (a, r) were used for calculation of the Λn FSI in Eq.(3)-(4), they can be evaluated by the χ^2-fitting of the experimental spectrum. Figure 4 shows the χ^2-distribution depending on average Λn scattering length ($\bar{a} = (a_s + 3a_t)/4$) and effective range ($\bar{r} = (r_s + 3r_t)/4$). As a result, χ^2 values had the minimum value at $(\bar{a}, \bar{r}) = (-2.6, 5.0)$ fm. Each colored marker in Fig. 4 shows a value of (a, r) in the Λn potential model. Moreover, regions hatched in rad show the experimental results of the singlet and triplet values for Λp FSI [3]. Since the potential values of Λn FSI ($-2.6, 5.0$ fm) took values between these of Λp FSI of the spin singlet and triplet, the potential values at $(-2.6, 5.0)$ fm does not conflict with a result of the Λp FSI. On the other hand, the black solid, dashed and dashed-dot lines in Fig. 4 show the contour lines added one, two and three from the chi-square minimum ($\chi^2_{\mathrm{min}} = 59$). Especially, the contour line at $\chi^2_{\mathrm{min}} + 1$ represents the statistics error, so when the scattering length (a) is -2.6 fm, the effective range (r) is successfully limited to be $3.8 < r < 6.3$ fm (preliminary).

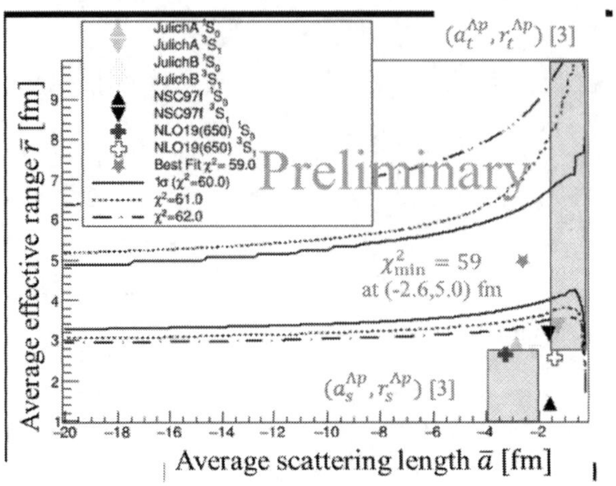

Figure 4. The χ^2 distribution depending on an average scattering length ($\bar{a}$) and effective range ($\bar{r}$). The average values are assumed to be same value of spin singlet and triplet. The χ^2 has minimum value at $(-2.6, 5.0)$ fm (star marker in pink). The colored markers show the potential value in each Λn potential model. The hatched boxes in red show a experimental result of the Λp FSI in the singlet and triplet, respectively [3]. The black lines show the contour lines at $\chi^2 = 60,\ 61,\ 62$.

5 Summary

The search for the $nn\Lambda$ experiment (E12-17-003) was performed at JLab Hall A in 2018. By analyzing the ^{3}H$(e, e'K^+)X$ missing mass spectrum by the χ^2-fitting, the Λn FSI was investigated in this study. As a result, the effective range (r) was successfully given a limit for a certain scattering length (a) from Fig. 4 (preliminary).

Acknowledgements We would like to thank Dr. Hidenbauer for helpful discussions and provide us the theoretical calculation results. And this work was supported in part by Aid for Graduate Program on Physics for the Universe. Also, the work was partially supported by JSPS KAKENHI Grants Nos. JP18H05459, 18H01219, 17H01121, 120002001, 15684005, 16GS0201, GP-PU Tohoku Univ., US Department of Energy Contracts Nos. DE-AC05-84ER40150, DEFG02-00ER41110, DE-AC02-98CH10886, DE-AC05-06OR23177, DE-FG02-99ER41065, DE-FG02-97ER41047, DE-AC02-06CH11357, and US NSF Contracts No. 013815 and 0758095.

References

[1] T.O. Yamamoto *et al.*, Phys. Rev. Lett. **115**, 222501 (2015)
[2] F. Dohrmann *et al.*, Phys. Rev. C **76**, 054004 (2007)
[3] F. Hinterberger *et al.*, Eur. Phys. J. A **21**, 313-321 (2004)
[4] S.N. Santiesteban *et al.*, Nucl. Inst. And Meth. A 940, 351-358 (2019)
[5] J. Alcorn *et al.*, Nucl. Inst. And Meth. A 522, 294-346 (2004)
[6] K.N. Suzuki *et al.*, Prog. Theor. Exp. Phys. **2022** 013D01 (2022)
[7] B. Pandey *et al.*, Phys. Rev. C **105**, L051001 (2022)
[8] K. Willner *et al.*, Phys. Rev. A **74**, 052715 (2006)

The BGOOD experiment at ELSA

Exotic structures in the strange quark sector?

Thomas Jude[1,*], *Stefan* Alef[1], *Reinhard* Beck[2], *Alessandro* Braghieri[3], *Philip* Cole[4], *Rachele* Di Salvo[5], *Daniel* Elsner[1], *Alessia* Fantini[5,6], *Oliver* Freyermuth[1], *Frank* Fromberger[1], *Francesco* Ghio[7,8], *Johannes* Groß[1], *Katrin* Kohl[1], *Paolo* Levi Sandri[9], *Guiseppe* Mandaglio[10,11], *Paolo* Pedroni[3], *Mariia* Romaniuk[1,12], *Georg* Scheluchin[1], and *Hartmut* Schmieden[1]

[1]Rheinische Friedrich-Wilhelms-Universität Bonn, Physikalisches Institut, Bonn, Germany
[2]Rheinische Friedrich-Wilhelms-Universität Bonn, HISKP, Bonn, Germany
[3]INFN sezione di Pavia, Pavia, Italy
[4]Lamar University, Department of Physics, Beaumont, Texas, USA
[5]INFN Roma "Tor Vergata", Rome, Italy
[6]Università di Roma "Tor Vergata", Dipartimento di Fisica, Rome, Italy
[7]INFN sezione di Roma La Sapienza, Rome, Italy
[8]Istituto Superiore di Sanità, Rome, Italy
[9]INFN - Laboratori Nazionali di Frascati, Frascati, Italy
[10]INFN sezione Catania, Catania, Italy
[11]Università degli Studi di Messina, Dipartimento MIFT, Messina, Italy
[12]Institute for Nuclear Research of NASU, Kyiv, Ukraine

Abstract. The BGOOD photoproduction experiment accesses forward meson angles and low momentum exchange kinematics in the *uds* sector, which may be sensitive to molecular-like hadronic structure. Recent highlights are summarised in these proceedings.

The recent discoveries of the pentaquark, P_C, states [1, 2] and XYZ mesons in the charmed quark sector [3, 4] have initiated a new epoch in hadron physics, where the existence of exotic multi-quark states beyond the conventional three and two quark systems has been realised. Such states could manifest as colour-singlet objects, or evolve from meson-baryon and meson-meson interactions, creating molecular like systems and re-scattering effects near production thresholds. Equivalent structures may be evidenced in the light, *uds* sector. This includes, for example the $\Lambda(1405)$ which is now considered the archetypal molecular state in the strangeness sector [5]. A cusp-like structure was also recently observed in $K^0\Sigma^+$ photoproduction at the $K^*\Sigma$ threshold [6] which was suggested to derive from a vector meson-baryon dynamically generated state, the $N^*(2030)$ [7]. This was considered as the strangeness counterpart to the P_C states, which were predicted by an equivalent model [8].

To experimentally study any potentially loosely bound molecular hadronic structure, access to a low momentum exchange and forward meson production region is mandatory. The BGOOD experiment [9] at the ELSA electron accelerator facility [10] is ideally suited for such photoproduction measurements. A 3 GeV electron beam impinges upon a thin radiator to produce an energy tagged bremsstrahlung photon beam which is subsequently incident

*e-mail: jude@physik.uni-bonn.de

 https://doi.org/10.1051/epjconf/202227102007

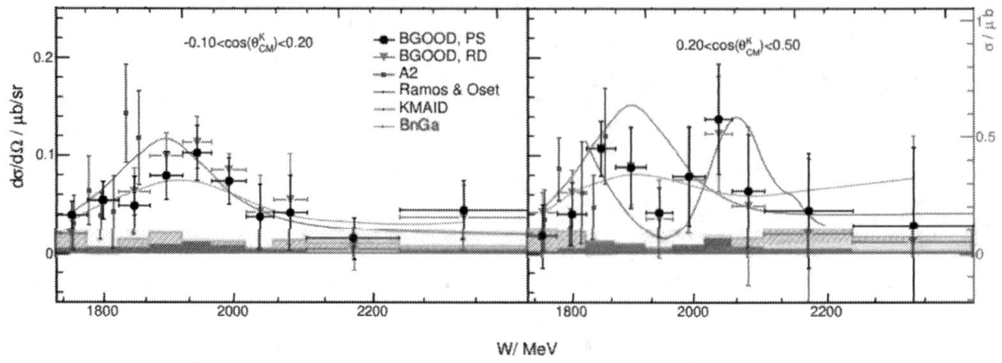

Figure 1. Differential cross section of $\gamma n \to K^0\Sigma^0$ as a function of W for two intervals in $\cos\theta^K_{\mathrm{CM}}$ and two different fitting methods (red triangles, and black circles). Data from the A2 Collaboration and calculations from KMAID and BnGa are labelled in the legend. The predicted total cross section from Ramos and Oset [7] is included in the angular bin $0.2 < \cos\theta^K_{\mathrm{CM}} < 0.5$ (red solid line) at an arbitrary scale denoted by the red axis on the right. Figure (adapted) and references in Ref. [11].

upon the target at the centre of the BGOOD experiment. BGOOD is comprised of two main parts, a central calorimeter region ideal for the reconstruction of neutral mesons via their decays, and a forward spectrometer used for charged particle identification and momentum reconstruction at forward angles.

BGOOD has an extensive strangeness photoproduction physics programme, recent of results of which are highlighted below. Due to space constraints, previous data or calculations that are not cited above are referenced in the corresponding BGOOD publication.

Measurement of the $\gamma n \to K^0\Sigma^0$ differential cross section over the K^* threshold [11]

The calculation of Ramos and Oset, which suggested the dynamically generated $N^*(2030)$ amplified the cusp in the $K^0\Sigma^+$ channel, predicted constructive interference in $K^0\Sigma^0$ photoproduction resulting in a peak [7]. Observing this experimentally would therefore be direct evidence of such a molecular state in the *uds* sector.

Using a deuterium target, the reaction $\gamma n(p) \to K^0\Sigma^0$ was identified via the decays $K^0\pi^0\pi^0$ and $\Sigma^0 \to \gamma(\Lambda \to p\pi^-)$. After numerous selection criteria, the $\pi^0\pi^0$ invariant mass was fitted to separate signal and background events from other channels, and reactions off the proton were subtracted by an equivalent analysis using a hydrogen target.

Two intervals in the cosine centre-of-mass polar angle of the kaon, $\cos\theta^K_{\mathrm{CM}}$ are shown in fig. 1. The data are in reasonable agreement with the previous data from the A2 collaboration, calculations from the KaonMAID isobar model and the Bonn Gatchina partial wave analysis (BnGa). In the more forward interval, a peak is observed which is consistent with the predicted peak from the model of Ramos and Oset. Further data has now been taken to improve the statistical precision and enable a firm interpretation.

Photoproduction of $K^+\Lambda(1405) \to K^+\pi^0\Sigma^0$ extending to forward angles and low momentum transfer [12]

The BGOOD setup enables a clean identification of the decay $\Lambda(1405) \to \pi^0\Sigma^0$ over a $\cos\theta^K_{\mathrm{CM}}$ range enabling access to previously unattainable low momentum transfer regions. The reaction channel was identified by identifying all final state particles and using a kinematic fit. Subsequent fitting to two-dimensional $\Sigma^0\pi^0$ and $\gamma\Lambda$ invariant mass spectra enabled a clean separation of signal to background.

Fig. 2 shows the cross section versus photon beam energy, E_γ, integrated over all $\cos\theta^K_{\mathrm{CM}}$, with good agreement to previous CLAS data however much improved E_γ resolution. The purple line is a calculation by Wang et al., of a triangle singularity being driven by the $N^*(2030)$ resonance. The excellent agreement supports the $N^*(2030)$ as being a molecular $K^*\Sigma$ system.

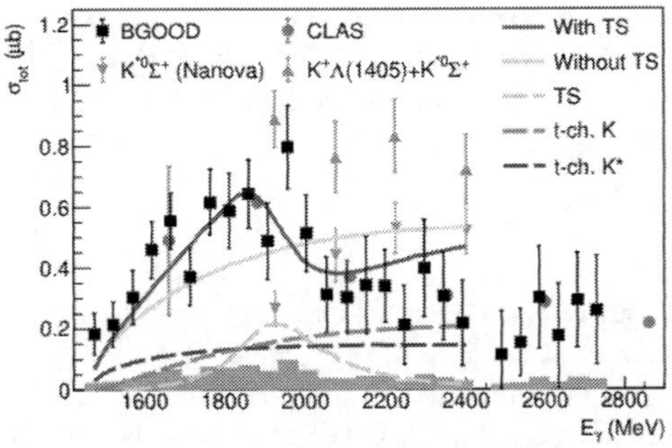

Figure 2. Integrated $\gamma p \to K^+\Lambda(1405)$ cross section. The purple and cyan line is the model of Wang *et al.* with and without the triangle singularity, the $K^{*0}\Sigma^+$ data from CBELSA/TAPS are the grey triangles and the sum of the $K^{*0}\Sigma^+$ and the BGOOD $K^+\Lambda(1405)$ data are the magenta triangles. Red circles are CLAS data. Figure and references in Ref. [12].

$K^+\Lambda$ and $K^+\Sigma^0$ photoproduction at forward angles and low momentum transfer [13, 14]

Forward $K^+\Lambda$ photoproduction provides low-t data which is an important constraint on the electroproduction of hypernuclei, where the Q^2 is required to be small. It is also particularly sensitive to intermediate high spin N^*. Fig. 3 shows the differential cross section data for $\cos\theta^K_{\mathrm{CM}} > 0.9$. The statistical precision is markedly improved and enables a discrimination between previous conflicting datasets.

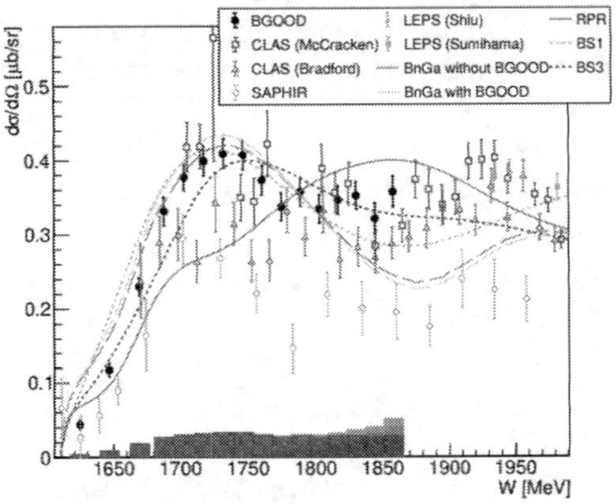

Figure 3. $\gamma p \to K^+\Lambda$ differential cross section for $\cos\theta^K_{\mathrm{CM}} > 0.90$ (black filled circles). The systematic uncertainties on the abscissa are in three components, where the grey bars are the total. Previous data (only including statistical errors) is indicated in the legend. The CLAS data are at the more backward angle of $0.85 < \cos\theta^K_{\mathrm{CM}} < 0.95$. The RPR, BS1, and BS3 are the Regge plus resonant model and isobar models BS1 and BS3 of Skoupil and Bydžovský. The Bonn-Gatchina PWA solutions are shown with and without the inclusion of the new data. Figure and references in Ref. [13].

The $K^+\Sigma^0$ differential cross section for $\cos\theta^K_{\mathrm{CM}} > 0.9$ is shown in fig. 4. Similarly for $K^+\Lambda$, the statistical precision of the data is able to discriminate between differences in previous datasets. A cusp-like structure is resolved close to the pK^+K^- threshold at $W \sim 1900$ MeV. Fig. 5 shows the data extrapolated to minimum momentum transfer, t_{min} and $\cos\theta^K_{\mathrm{CM}} = 1$, where the cross section drops by 75 %. No firm interpretations have been made, but it is interesting to note the proximity of multiple thresholds and predicted bound states immediately at this centre-of-mass energy. The extent of this cusp-like structure changes quickly over $\cos\theta^K_{\mathrm{CM}}$ at forward angles, demonstrating the importance of accessing these forward kinematics.

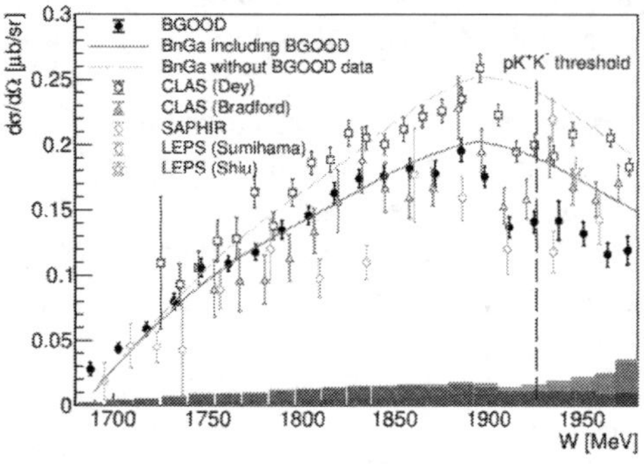

Figure 4. $\gamma p \to K^+\Sigma^0$ differential cross section for $\cos\theta^K_{\mathrm{CM}} > 0.90$ (black circles). The systematic uncertainties on the abscissa are in three components, where the grey bars are the total. Previous data (only including statistical errors) is indicated in the legend. The CLAS data are at the more backward angle of $0.85 < \cos\theta^K_{CM} < 0.95$. The Bonn-Gatchina PWA solutions with and without the inclusion of the new data are the solid magenta and dashed cyan lines respectively. The pK^+K^- threshold is indicated by the dashed black line. Figure and references in Ref. [14].

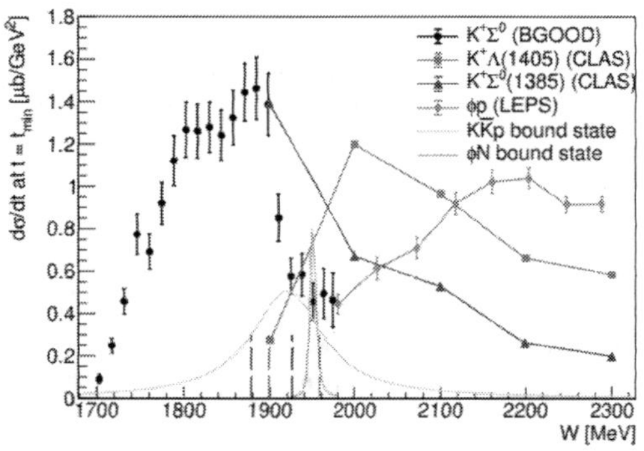

Figure 5. $\gamma p \to K^+\Sigma^0$ differential cross section, $d\sigma/dt$ extrapolated to $t_{\min}$ versus W (filled black circles). Previous data of other final statesis indicated in the legend. The vertical dashed lines indicate the respective thresholds, with the addition of the K^+K^-p threshold indicated by the black dashed line. Predictions of $K\bar{K}N$ and ϕN bound states are shown. Figure and references in Ref. [14].

This work is supported by the Deutsche Forschungsgemeinschaft Project Numbers 388979758 and 405882627, the Third Scientific Committee of the INFN and has received funding from the European Union's Horizon 2020 research and innovation programme under grant agreement STRONG–2020 No. 824093.

[1] R. Aaij *et al.* (LHCb), Phys. Rev. Lett. **122**, 222001 (2019)
[2] R. Aaij *et al.* (LHCb), Phys. Rev. Lett. **115**, 072001 (2015)
[3] H.-X. Chen *et al.*, Phys. Rept. **639**, 1 (2016)
[4] Y.-R. Liu *et al.*, Prog. Part. Nucl. Phys. **107**, 237 (2019)
[5] F. K. Guo et al., Rev. Mod. Phys. **90** 015004 (2018) and references therein
[6] R. Ewald *et al.*, Phys. Lett. **B 713**, 180 (2012)
[7] A. Ramos and E. Oset, Phys. Lett. **B 727**, 287 (2013)
[8] J.-J. Wu, R. Molina, E. Oset, and B.S. Zhou, Phys. Rev. Lett. **105**, 232001 (2010)
[9] S. Alef, et al. (BGOOD), Eur. Phys. J. **A 56** 104 (2020)
[10] W. Hillert, Eur. Phys. J. **A 28** 139 (2006)
[11] K. Kohl, T.C. Jude, et al. (BGOOD), arXiv:2108.13319 (2021)
[12] G. Scheluchin, T.C. Jude, et. al (BGOOD), Phys. Lett. B**B 833** 137375 (2022)
[13] S. Alef, et. al. (BGOOD), Eur. Phys. J. **A 57** 80 (2021)
[14] T.C. Jude, et. al. (BGOOD), Phys. Lett. **B 820** 136559 (2021)

Λ polarization measurement of the $\pi^- p \to K^0\Lambda$ reaction in J-PARC E40 experiment

Tamao Sakao[1,*], *Ryotaro* Honda[2], *Koji* Miwa[1,2], *Yoshiyuki* Nakada[3], *Takuya* Nanamura[4,5], *Toshiyuki* Takahashi[2], *Mifuyu* Ukai[1,2], *Takeshi* Yamamoto[5]
, and J-PARC E40 Collaboration

[1]Department of Physics, Tohoku University, Sendai, Miyagi 980-8578, Japan.
[2]High Energy Accelerator Research Organization (KEK), Tokai, Ibaraki 319-1106, Japan.
[3]Department of Physics, Osaka University, Toyonaka, Osaka 560-0043, Japan.
[4]Department of Physics, Kyoto University, Kyoto 606-8502, Japan.
[5]Japan Atomic Energy Agency (JAEA), Tokai, Ibaraki 319-1195, Japan.

Abstract. We performed the J-PARC E40 experiment to measure the Σp scattering cross sections from 2018 to 2020. Together with the $\pi^- p \to K^+\Sigma^-$ data, the $\pi^- p \to K^0\Lambda$ data were accumulated as a byproduct. The analysis confirmed that Λ could be identified with an S/N ratio of $\sim$ 2.67. The polarization of Λ (P_Λ) was preliminarily derived as 1.009 ± 0.049 for the K^0 angular range of $0.7 < \cos\theta_{K^0,\,CM} < 0.8$. It is more accurate than the past data [1]. The high polarization enables us to measure not only the differential cross section but also spin observables of the Λp scattering in the future J-PARC experiment.

1 Introduction

The nucleon-nucleon (NN) interaction is the foundation of nuclear physics. The nuclear force has an attractive potential in the long-range region, which is well described by the one-boson exchange (OBE) model. In contrast, it has a repulsive core in the short-range region. Although the origin of the core is still unknown, it is expected that the quark-gluon dynamics would dominate because the two nucleons overlap in this range. That is why most of us are engaging in hypernuclear physics to expand the nuclear force study by adding strange quarks.

Representative baryon-baryon (BB) interation models are, for example, Nijmegen model [2], quark cluster model [3], and chiral EFT [4] [5]. To update the theories based on experimental results, precise two-body hyperon-nucleon (YN) data are necessary to make the YN interaction models more realistic. Furthermore, accurate binding energy and level scheme of hypernuclei are also important to compare with hypernuclear calculations based on the YN interaction.

Historically, many hypernuclear experiments in KEK, Jefferson Lab, J-PARC, etc., have accumulated many hypernuclear data. In addition, precise two-body YN data should be taken by YN scattering and hypertriton experiments as the fundamental information. The realistic YN interactions would be established by feeding back two-body YN data into theories. Such YN interactions will derive from many-body interactions in the hypernuclear data.

*e-mail: sakao@lambda.phys.tohoku.ac.jp

The J-PARC E40 (Σp scattering) experiment recently pioneered the YN scattering technique in J-PARC [6]. We now plan the next-generation Λp scattering experiment (J-PARC E86) using the same procedure as the E40 experiment. The J-PARC E86 aims to measure many scattering observables such as the differential cross section $(d\sigma/d\Omega)_{\Lambda p}$, analyzing power ($A_y$), and depolarization ($D_y^y$).

2 Λ polarization measurement for the J-PARC E86

2.1 Λp scattering observables and Λ polarization

In the momentum range of 0.4 – 0.8 GeV/c, The J-PARC E86 will measure the Λp scattering observables, such as the differential cross section ($(d\sigma/d\Omega)_{\Lambda p}$), analyzing power ($A_y$), and depolarization ($D_y^y$).

The Λ beam should be polarized in this measurement. Besides, we need to measure the Λ beam polarization (P_Λ) for deriving A_y and D_y^y. It is desirable to confirm the high polarization of Λ produced in the $\pi^- p \to K^0\Lambda$ reaction and to establish the analysis method in advance. Therefore, we analyzed the $\pi^- p \to K^0\Lambda$ data accumulated as a byproduct in the J-PARC E40. Although the past experiment reported P_Λ in the $\pi^- p \to K^0\Lambda$ reaction is $\sim$ 100% for the forward angular range of K^0 [1], its statistic is not enough. Hence, our re-measurement of P_Λ would be valuable.

2.2 Analysis of the reaction $\pi^- p \to K^0\Lambda$ in the J-PARC E40 byproduct data

The J-PARC E40 measured the Σp scattering and pioneered the new hyperon scattering experimental technique. It was performed at the K1.8 beamline of the J-PARC Hadron Experimental Facility. The detector setup is shown in Figure 1. In its byproduct data of the (π^-, K^0) reaction, K^0 was reconstructed from π^+ and π^-. They are detected by the KURAMA spectrometer, and CATCH system [7], respectively. Λ was tagged by calculating the missing mass of the $\pi^- p \to K^0 X$ reaction. The detail of detector configuration and K^0 reconstruction is described in Ref. [9]. The J-PARC E40 byproduct data includes enough Λ beam yield for the P_Λ analysis of 2.80×10^4 with the S/N ratio of ~ 2.67. These values were calculated by requiring the $\Lambda \to p\pi^-$ decay detection with the CATCH system.

2.3 Λ polarization calculation

As we mentioned, the $\Lambda \to p\pi^-$ decay was identified with the CATCH system. Due to the parity violation in the $\Lambda \to p\pi^-$ decay, the emission angle of decay proton (θ_p) distributes asymmetrically. By measuring the θ_p distribution in the rest frame of Λ, we obtain P_Λ from the following equation,

$$\frac{1}{N_0}\frac{dN}{d\cos\theta_p} = \frac{1}{2}(1 + \alpha P_\Lambda \cos\theta_p). \tag{1}$$

Here, N_0 is the yield of the decay proton, and α is the asymmetry parameter (which is recently updated as $0.750 \pm 0.009 \pm 0.004$ [10]). The raw θ_p distribution was corrected by the acceptance of the CATCH system, estimated by a Geant4-based Monte Carlo simulation. Figure 2 shows the θ_p distribution after the acceptance correction in the K^0 angular range of $0.7 < \cos\theta_{K^0,\,CM} < 0.8$. Due to parity violation in the $\Lambda \to p\pi^-$ decay, a clear asymmetry called "Up-Down asymmetry" can be seen.

We calculated the P_Λ for the K^0 angular range of $0.6 < \cos\theta_{K^0,\,CM} < 1.0$ by fitting the $\cos\theta_{K^0,\,CM}$ distribution with Eq. (1). The comparison between the present values of P_Λ

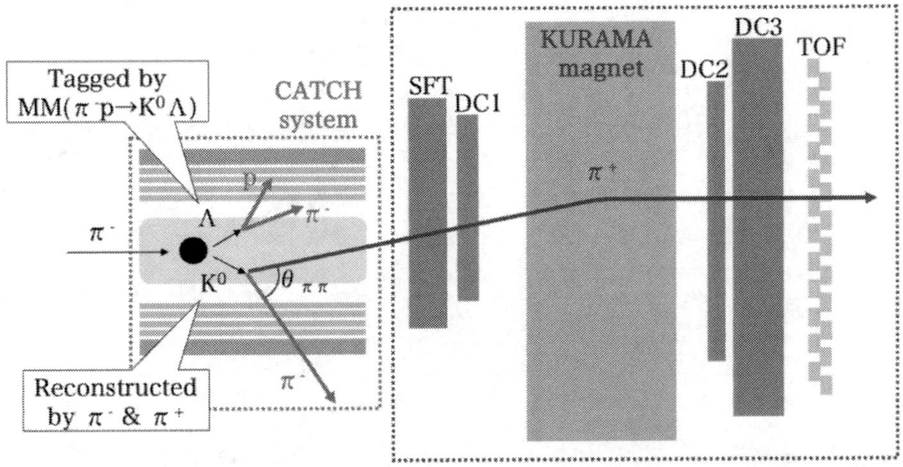

Figure 1. The schematic of K^0 reconstruction and Λ identification in the J-PARC E40 byproduct data of the (π^-, K^0) reaction. The cylindrical detector cluster (CATCH) is represented by pink parts, and the forward magnetic spectrometer (KURAMA), blue ones. The detail of the analysis technique is described in Ref. [9].

and the past experiment is shown in Figure 3. Here, the past data are replotted with the updated asymmetry parameter. The higher statistics of the Λ production in the J-PARC E40 allowed us to get smaller error bars. This result indicates that the Λ produced in the reaction $\pi^- p \to K^0\Lambda$ has high polarization.

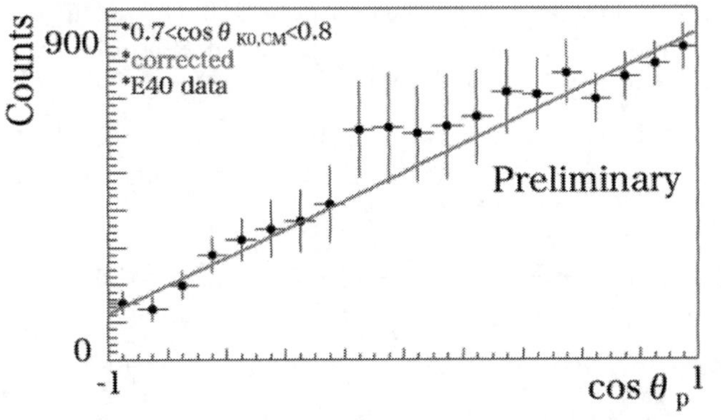

Figure 2. The distribution of the emission angle of decay proton (θ_p) in the K^0 angular range of $0.7 < \cos\theta_{K^0,\,CM} < 0.8$. We measured this in the rest frame of Λ. The Λ polarization (P_Λ) is calculated by fitting this distribution with Eq. (1).

3 Conclusion

To establish the realistic *YN* interactions, we need feedback on two-body *YN* data into theories. Such realistic *YN* interactions are necessary to derive many-body interactions in the

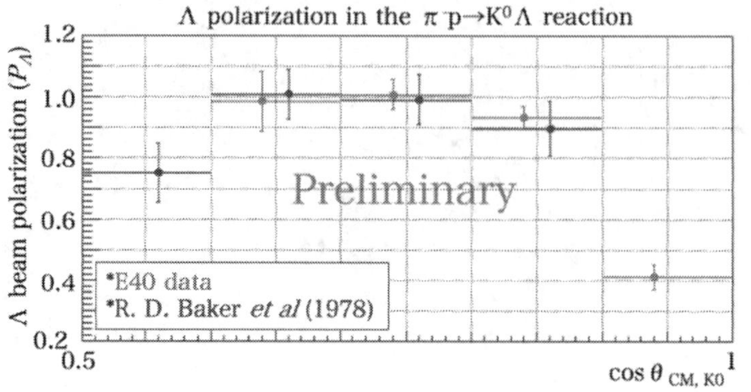

Figure 3. Λ beam polarization (P_Λ) in the reaction $\pi^- p \rightarrow K^0\Lambda$. The present value (the J-PARC E40 byproduct data) is marked by red points, and the past data [1], by blue points.

hypernuclear data. J-PARC E40 (Σp scattering) recently pioneered the hyperon scattering experimental technique for the first time. Following this success, we plan the next-generation Λp scattering experiment, J-PARC E86. It aims to measure the spin observables and the differential cross section of the Λp scattering precisely. In the spin observables measurement, we need the highly polarized Λ beam. Although the past experiment reported that the Λ polarization (P_Λ) was ~ 100% for the forward angular region of K^0[1], its accuracy is not enough due to the low statistics. Therefore, we decided to re-measure P_Λ.

In the P_Λ analysis, we used the (π^-, K^0) reaction data accumulated as a byproduct in the J-PARC E40. We have identified the Λ beam yield of 2.80×10^4 with the S/N ratio of ~ 2.67. By measuring the θ_p distribution in the rest frame of Λ, P_Λ was derived from its slope. Now, P_Λ was preliminarily derived as 1.009 ± 0.049 for the K^0 angular range of $0.7 < \cos\theta_{K^0,\,CM} < 0.8$. Using such a highly-polarized Λ beam, Λp spin observables measurement is possible in J-PARC E86.

References

[1] R. D. Baker *et al.*, Nucl. Phys. B **141**, 29 (1978)
[2] M. M. Nagels, T. A. Rijken, and Y. Yamamoto, Phys. Rev. C **99**, 044003 (2019)
[3] Y. Fujiwara, Y. Suzuki, C. Nakamoto, Nucl. Phys. **58**, 439 (2007)
[4] J. Haidenbauer, S. Petschauer, N. Kaiser, U.-G. Meißner, A. Nogga, and W. Weise, Nucl. Phys. A **915**, 24 (2013)
[5] J. Haidenbauer, U.-G. Meißner, and A. Nogga, Eur. Phys. J. A **56**, 91 (2020)
[6] K. Miwa *et al.*, Phys. Rev. C **104**, 045204 (2021)
[7] Y. Akazawa *et al.*, NIM. Phys. A **1029**, 166430 (2022)
[8] T. Sakao, Master Thesis of Graduate school Tohoku Univ., (2021)
[9] T. Sakao *et al.*, JPS Conf. Proc. **33**, 011133 (2021)
[10] M. Ablikim *et al.*, Nature Physics **15**, 631 (2019)

$K^+\Sigma^-$ photoproduction within an isobar model

Dalibor Skoupil[1,*], *Petr* Bydžovský[1], *Aleš* Cieplý[1], *Dimitrios* Petrellis[1], and *Nicholas* Zachariou[2]

[1]Nuclear Physics Institute, Czech Academy of Sciences, Řež, Czech Republic
[2]University of York, York, YO10 5DD, United Kingdom

Abstract. We used an isobar model to study the $K^+\Sigma^-$ photoproduction reaction on a neutron target with focus on the resonance region. In order to achieve a reasonable agreement with the data, we included spin-3/2 and spin-5/2 nucleon resonances in a consistent formalism together with a Δ resonance and two kaon resonances on top of the Born terms. The free parameters of the model were adjusted to the data from the CLAS and LEPS Collaborations on differential cross sections and photon beam asymmetry. The cornerstone of this analysis was an upgrade of the fitting method. Previously, we used only the plain χ^2 minimization, which could not prevent us from overfitting the data. We, therefore, introduced a regularization method, the least absolute selection shrinkage operator (LASSO), which, together with information criteria, restricts the number of nonzero parameters and prevents us from overfitting the data. In our analysis, we arrived at two models, fit M, whose parameters were fitted with the MINUIT code only, and fit L, where we used the more advanced LASSO. Both models describe the data in a similar way and we observe only slight differences in the $d\sigma/d\Omega$ data description at very forward angles where the fit M is flat whereas the fit L produces two broad peaks, and in the photon beam asymmetries above 2 GeV at backward kaon angles where the fit M produces a bump. Surprisingly, no hyperon resonances are needed for the correct data description in these models. On the other hand, the $N(1720)3/2^+$ nucleon resonance was found to be very important in both models.

1 Introduction and methodology

The investigation of kaon-hyperon photo- and electroproduction from nucleons provides us with much needed information on the spectrum of baryon resonances and interactions in the systems of hyperons and nucleons. We can also gather information on the existence and properties of the so-called missing resonances, i.e. states which were predicted in the quark models but have not been seen in the π production or πN scattering processes. These states may have escaped experimental confirmation due to their stronger decay coupling to strangeness channels rather than to the more well-known pion final states.

Much of the work in the kaon-hyperon production is focused on the $K^+\Lambda$ channel which is most abundant with experimental data. For the time being, the database for the channels using neutron targets is very limited, with available measurements of the differential cross

*e-mail: skoupil@ujf.cas.cz

section only in the $K^+\Sigma^-$ and $K^0\Lambda$ channels. So far, only two measurements of the photon-beam asymmetry Σ were performed: the one by the LEPS Collaboration [1] with a limited kinematical coverage and the other by the CLAS Collaboration [2] which covers a wide range of kinematics. It was this recent analysis which brought our attention to the $K^+\Sigma^-$ channel.

The isobar model we use in our study is based on effective Lagrangians in the tree-level approximation, i.e. we do not take into account any higher-order corrections (such as final-state interaction or rescattering). The nonresonant part of the amplitude includes the Born terms and exchanges of resonances in the t channel (K^* and K_1 kaon resonances) and u channel (Σ^* hyperon resonances). The resonant part is then modelled by s-channel exchanges of nucleon and Δ resonances whose masses range from the threshold of the process up to approximately 2 GeV. Within the isobar model, we can investigate the kaon-hyperon photoproduction in the energy range from the threshold of 1.609 GeV up to around 2.5 GeV.

Since the kaon photoproduction occurs in the so-called third nucleon resonance region, where there are many nucleon resonances which can contribute and none of them is dominant in this process, we consider around 20 resonances with masses up to around 2 GeV. In the studied energy region, contributions of high-spin resonances play a crucial role. The formalism for baryon fields with a higher spin is problematic because of the presence of non-physical degrees of freedom, which connect to the lower-spin modes of the Rarita-Schwinger field. When this field is off its mass shell, the non physical components of the field can participate in the interaction (we call such interaction inconsistent). Following Refs. [3, 4], we can get rid of these non physical terms in the amplitude: the corresponding vertices are transversal to the momentum of the exchanged particle, which provides only the contributions from the highest-spin components of the propagator (for details, see [5]).

In order to account for the inner structure of hadrons, which is basically neglected in our hadrodynamical approach, and also to fine-tune the behaviour of high-spin resonances, whose contributions can grow strongly due to their high-power momentum dependence, we introduce hadronic form factors. The cutoff parameter of these form factors, which is fitted to experimental data, influences the effectiveness of the hadronic form factor in taming the contributions of high-spin resonances.

The set of considered nucleon resonances was motivated predominantly by analyses which we did in the past on the $K^+\Lambda$ channel [5, 6] and another analysis of the $K\Sigma$ channel [7]. Besides the N^*'s included in our studies of $K^+\Lambda$ channel, we had to consider also the Δ and Σ resonances in the s and u channels, respectively. The free model parameters, which are the couplings of individual resonances and cut-off parameters for the hadronic form factors, were adjusted to experimental data and the quality of fit was checked by comparing its prediction with the data. Altogether, we used 674 data points to fit the free parameters of the model. Currently available data in the $K^+\Sigma^-$ channel are only on the differential cross section, photon-beam asymmetry Σ, and beam-target asymmetry E. We did not use the beam-target asymmetry data in the fits, as their errors are much larger than the errors of the other two data sets on other observables.

At the end of the fitting procedure, we selected variants with the smallest values of χ^2/n.d.f. and acceptable values of the parameters. Firstly, we used the plain χ^2 minimization with help of MINUIT library [8] to fit the free parameters of the model. The resulting best fit was dubbed "fit M" (or fit MINUIT). Minimizing χ^2, however, is a good measure to determine underfitting but it may lead to overfitting the data. In the energy region we are concerned with, there are many resonant states with widths at the order of hundreds of MeV, which effectively overlap each other. It is thus of great importance to decrease the number of states which we consider in our analysis. In order to do so, we introduce a penalty term to the standard χ^2 definition, which then restricts the number of nonzero parameters. For more details on the fitting method, see Ref. [9]. With this so-called LASSO technique we arrived at a more economical solution which we called "fit L" (or fit LASSO).

2 Discussion

As a result of the fitting procedure, there are two different models. The fit M includes 14 resonances and 25 parameters: two kaon resonances, several nucleon resonances, one Δ resonance and no hyperon resonance, which comes as a surprise since the hyperon resonances contribute significantly in the $K^+\Lambda$ channel. Among the set of included N^*'s in the fit M, the most important is the contribution of the $N(1720)3/2^+$ state. Once we omit this state, we underestimate the differential cross sections by a half. A noteworthy effect of this nucleon state was observed also in the $K^0\Sigma^+$ channel [10]. We also point out significant contributions of the $N(1895)1/2^-$ state and the $\Delta(1900)1/2^-$ resonance. All of these states are among the states which were identified as the most important ones in a multi-channel analysis [11].

The fit L includes only 9 resonances and 17 parameters: there is only one kaon resonance (K^*) and no Δ and hyperon resonances. This result further corroborates the result of the fit M that for a reasonable description of the current $K^+\Sigma^-$ data the hyperon resonances are not needed. In the fit L, the most important contributions come from the K^*, $N(1720)3/2^+$, and $N(2060)5/2^+$. The role of the kaon resonance lies in capturing the experimental data mainly at forward angles. Without this resonance the cross section drops substantially in this region. The $N(1720)3/2^+$ state creates the first peak in the cross sections and in this region dominates the data description. On the other hand, the second peak in the cross sections is created mainly by the $N(2060)5/2^+$ state. Apart from these states, no other resonance has substantial effect on the cross-section description. In other words, the importance of the other states is more in their mutual interference.

In Fig. 1, we show differential cross sections in dependence on the photon laboratory energy E_γ^{lab} for ten values of $\cos\theta_K^{\mathrm{c.m.}}$. We compare our two results, fits M and L, with the experimental data from the CLAS [2, 12] and LEPS [1] Collaborations. We see that the models are in a satisfactory agreement with the data in all angular bins shown. The fit M diverges at forward kaon angles beyond 2.5 GeV, whereas the fit L tends to underestimate these data. The structures produced by the fit L at forward angles are also notable: While the fit M produces a cross section which is almost flat, the fit L creates two rather broad peaks. The latter behaviour is supported by experiment.

In Fig. 2, we show the photon-beam asymmetry Σ as a function of the cosine of the kaon center-of-mass angle $\theta_K^{\mathrm{c.m.}}$. In the threshold region, the two fits produce almost the same beam asymmetry which is large and positive at the central kaon angles and decreases gradually at backward angles; at forward angles the drop is rather sharp. Both models are in a good agreement with experimental data. The only difference is at energies beyond 2 GeV around $\cos\theta_K^{\mathrm{c.m.}} = -0.5$ where the fit L creates a dip while the fit M shows a peak. The former behaviour is more in agreement with experiment as data indicate a dip rather than a peak in this region.

3 Conclusion

In this work, we used the isobar model to study the $K^+\Sigma^-$ photoproduction on a neutron target. In the isobar model, we use the tree-level Feynman diagrams with exchanges of particles in their ground and excited states. The main feature of the present analysis is an improvement of the fitting method by adding a penalty function to the definition of χ^2. We arrived at two models that give us good agreement with data. There are only slight differences in the data

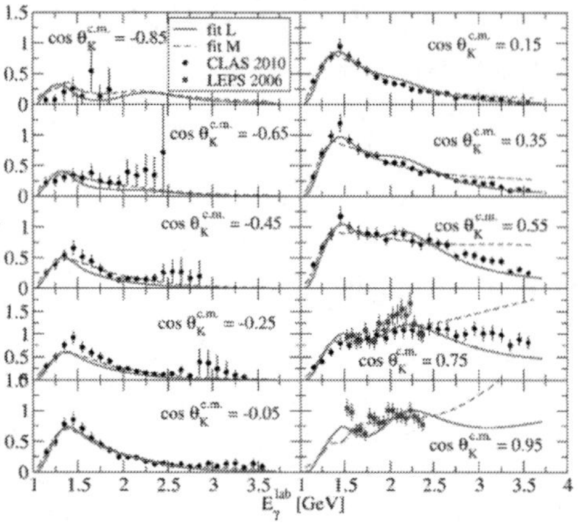

Figure 1. The description of the differential cross section as a function of the incident photon energy E_γ^{lab} by fits M (dashed line) and L (solid line). Data are from the CLAS [2, 12] (indicated by circles) and LEPS [1] (squares) Collaborations.

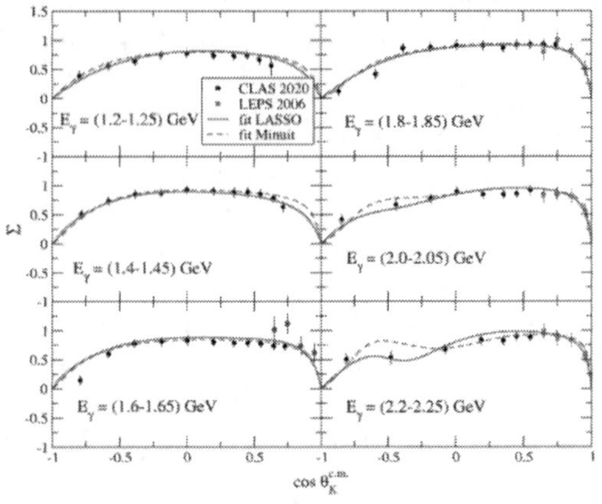

Figure 2. The description of the photon-beam asymmetry for six energy bins in dependence on the kaon center-of-mass angle $\theta_K^{\text{c.m.}}$. Notation is the same as in Fig. 1.

description by both models, the most important one being the description of differential cross sections at very forward angles where the fit L produces two broad peaks while the fit M is flat.

Our plans for the future include using the LASSO method to perform a multichannel analysis of the Σ photoproduction channels. In this way, we hope we will be able to arrive at the optimal set of resonances which contribute to the process.

This work was supported by the Czech Science Foundation GACR Grants No. 19-19640S and No. 19-14048S.

References

[1] H. Kohri *et al.* (LEPS Collaboration), Phys. Rev. Lett. 97, 082003 (2006)
[2] N. Zachariou *et al.* (CLAS Collaboration), Phys. Lett. B 827 (2022), 136985
[3] V. Pascalutsa, Phys. Rev. D **58**, 096002 (1998)
[4] T. Vrancx, L. De Cruz, J. Ryckebusch, and P. Vancraeyveld, Phys. Rev. C **84**, 045201 (2011)
[5] D. Skoupil, P. Bydžovský, Phys. Rev. C **93**, 025204 (2016)
[6] D. Skoupil, P. Bydžovský, Phys. Rev. C **97**, 025202 (2018)
[7] J. C. David, C. Fayard, G. H. Lamot, and B. Saghai, Phys. Rev. C **53**, 2613 (1996)
[8] F. James and M. Roos, MINUIT, CERN Report No. D506, 1981 (unpublished)
[9] P. Bydžovský, A. Cieplý, D. Petrellis, D. Skoupil, and N. Zachariou, Phys. Rev. C **104**, 065202 (2021)
[10] T. Mart, Phys. Rev. C **62**, 038201 (2000)
[11] T. Mart and M. J. Kholili, J. Phys. G: Nucl. Part. Phys. **46**, 105112 (2019)
[12] S. A. Pereira *et al.* (CLAS Collaboration), Phys. Lett. B **688**, 289 (2010)

Application of model selection criteria for $K^+\Sigma^-$ photoproduction within an isobar approach

Dimitrios Petrellis[1,*], *Petr* Bydžovský[1], *Aleš* Cieplý[1], *Dalibor* Skoupil[1], and *Nicholas* Zachariou[2]

[1]Nuclear Physics Institute, Czech Academy of Sciences, Řež, Czech Republic
[2]University of York, York YO10 5DD, United Kingdom

Abstract. We study kaon photoproduction off a neutron target using an isobar model. Recently acquired polarization data from the CLAS collaboration are incorporated in our analysis, while a novel fitting procedure is employed that allows us to reduce the number of parameters used with a minimal cost on the information content.

1 Introduction

Reactions such as $\gamma n \rightarrow K^+\Sigma^-$, which is the focus of the current study, play an important role in our understanding of the baryon spectrum. However, due to the plethora of nucleon resonances in the energy region of interest (W $\approx 1.6 - 2.6$ GeV) we are faced not only with a large number of parameters to use for fitting, but most importantly, a huge number of possible combinations of resonances to include in our model. It should be stressed that by "model selection" we refer to the selection of an optimal subset out of a group of candidate resonances. The techniques presented in section 3 have been used recently in similar contexts [1–3] and are well-established statistical learning tools [4]. The results of the current investigation have been reported in [5] and the data used are from [6] and [7].

2 The isobar approach

Isobar models, due to their phenomenological nature, rely on a number of approximations [8, 9]: the fundamental fields are hadrons and their interactions are based on effective Lagrangians, while reaction amplitudes are constructed as sums of lowest-order Feynman diagrams. Higher-order effects are implicitly taken into account in the *effective* couplings, which are obtained by fits to experimental data. At this level of approximation, there are three types of diagrams, depending on the nature of the exchanged particle (nucleon, kaon or hyperon), each corresponding to the **s**, **t** and **u** Mandelstam variables. Another distinction, based on whether the exchanged particle is in its ground, or in a resonant state (N^*, K^*, Y^*), results in further classifying these diagrams as being of the Born or non-Born types, respectively. Of these six types of interactions, only the exchange of N^* or Δ^* (the **s**, non-Born diagram) produces resonant structures in the cross-section, with the rest of the diagrams contributing only to the "background".

*e-mail: petrellis@ujf.cas.cz

3 Regularized least-squares fitting and information criteria

The question of optimal model complexity is central in statistical modeling. Most commonly, it refers to the optimal number of parameters needed by a certain model to describe a set of data. Using more parameters than necessary may lead to overfitting the data, while using fewer will lead to underfitting. Overfitting is caused by fitting the noise in the sample, thus resulting in a model of very low predictive power. A common way to deal with this problem is through regularization, where the χ^2 error function, defined as

$$\chi^2 = \sum_{i=1}^{N} \frac{[d_i - p_i(w_1, \ldots, w_k)]^2}{(\sigma_{d_i}^{stat})^2}, \tag{1}$$

is modified by the addition of a penalty term $P(\lambda)$ containing the k parameters $\{w_1, w_2 \ldots w_k\}$, with respect to which the χ^2 function is to be minimized. We thus have a penalized χ_T^2, which reads

$$\chi_T^2 = \chi^2 + P(\lambda), \quad \text{with} \quad P(\lambda) = \lambda \sum_{j=1}^{k} |w_j|^q \tag{2}$$

The effect of the $P(\lambda)$ term in the error function is to prevent the parameters from taking large values that would allow them to overfit the data. The magnitude of the λ regularization parameter determines the extent of parameter shrinkage. The case of $q = 1$ is known as Least Absolute Shrinkage and Selection Operator (LASSO) regularization [4] and results in some of the parameters to be driven to zero, with the value of λ determining the degree of sparsity of the resulting model.

At this point, it is the use of information criteria (IC), namely the Akaike (AIC) [10] and Bayesian (BIC) [11], that decide on the optimal value of λ, as the one that minimizes the corresponding expressions: AIC $= 2k + \chi_T^2$ and BIC $= k\log(N) + \chi_T^2$, where k is the number of parameters of the model and N is the size of the data set. The result of this process can be seen in Fig. 1. As can be observed, all criteria give similar results that differ only by a scale factor.

4 Numerical results and discussion

In the present work we conducted two fits: one, denoted as "fit M", where we performed ordinary χ^2 minimization (using the Minuit package) on a set of resonances derived from previous studies [9, 12] and a second one denoted as "fit L", where we applied LASSO regularization in conjunction with the aforementioned information criteria that led to a sparser model.

The free parameters of our model, apart from the coupling constants, comprise cut-off values for hadron form factors, as well. Moreover, resonances with spin-1/2 contribute one parameter and resonances with spins-3/2 or 5/2 contribute two parameters each. Masses and widths of resonances were taken from [13]. Table 1, shows the reduction in the number of resonances and parameters as a result of our fitting procedure.

The photon beam asymmetry $\mathbf{\Sigma}$ is defined as $\mathbf{\Sigma} = (d\sigma^{\perp} - d\sigma^{\parallel})/(d\sigma^{\perp} + d\sigma^{\parallel})$, where $d\sigma^{\parallel}$ denotes the differential cross section when the incident photon beam is linearly polarized in the x-direction and $d\sigma^{\perp}$ the differential cross section when the photon beam is polarized in the y-direction.

Figures 2 and 3 (taken from [5]) show the results of the two fittings M and L as they compare to data on the $\mathbf{\Sigma}$ asymmetry. Another useful result that can be drawn from these

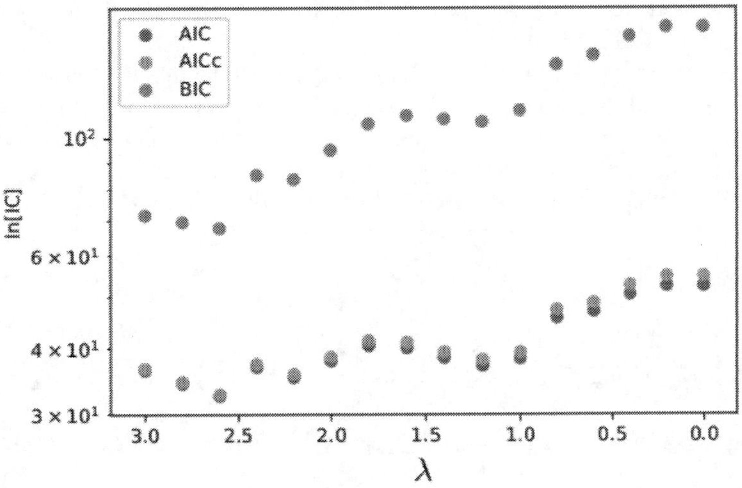

Figure 1. The Akaike (blue), a modified version known as 'corrected Akaike' (orange), and Bayesian (green) information criteria for various values of the regularization parameter λ. Please note the logarithmic scale of the vertical axis. (Taken from [5]).

Table 1. Summary of the features of the two fits.

	M: Minuit	L: LASSO + IC
no. of resonances	14	9
no. of parameters	25	17
χ^2 / n.d.f.	2.4	3.2

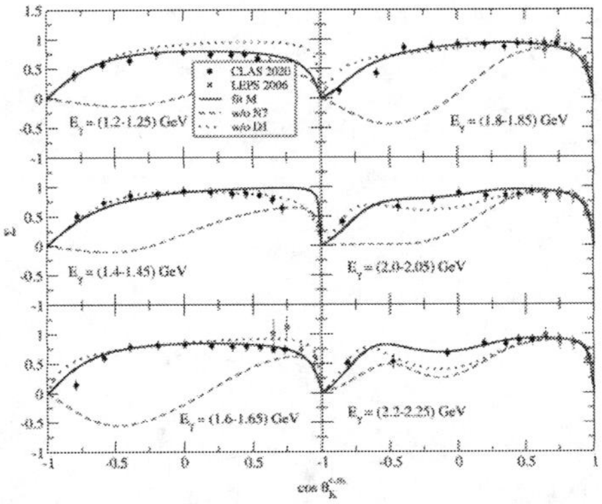

Figure 2. Photon beam asymmetry data compared to the full "fit M" results (solid line). Same fit, with the $N(1720)$ $3/2^+$ (dashed line) and $\Delta(1900)$ $1/2^-$ (dotted line) resonances omitted. The data are from CLAS [6] and LEPS [7] experiments.

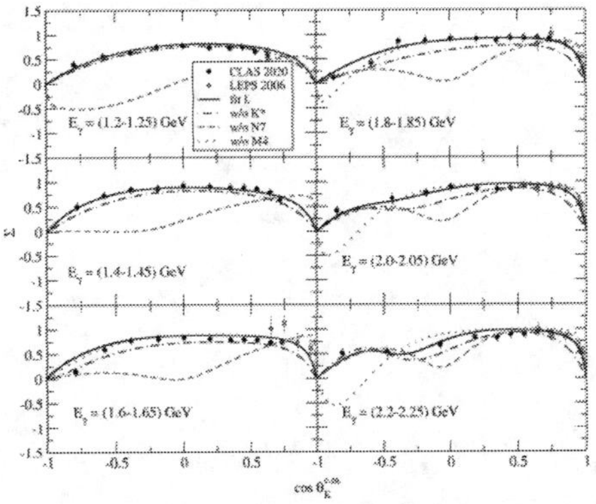

Figure 3. Photon beam asymmetry data as in Fig. 2, compared to "fit L" results (solid line). Same fit, with the $K^*(892)$ (dash-dotted line), $N(1720)$ $3/2^+$ (dashed line) and $N(2060)$ $5/2^-$ (dotted line) resonances omitted.

 https://doi.org/10.1051/epjconf/202227102010

figures is the importance of the $N(1720)$ $3/2^+$ resonance, given the disagreement with the data when it is not included in the fits.

In conclusion, the application of LASSO regularization in combination with information criteria provides a tool for automatic selection of parameters based on their information content. This can be especially useful in our case, where the number of possible combinations of resonances is prohibitively large in the region of interest.

This work was supported by the Czech Science Foundation GACR Grant No. 19-19640S.

References

[1] B. Guegan, J. Hardin, J. Stevens, and M. Williams, JINST **10**, P09002 (2015)
[2] J. Landay, M. Döring, C. Fernández-Ramírez, B. Hu and R. Molina, Phys. Rev. C **95**, 015203 (2017)
[3] J. Landay, M. Mai, M. Döring, H. Haberzettl and K. Nakayama, Phys. Rev. D **99**, 016001 (2019)
[4] T. Hastie, R. Tibshirani, and J. Friedman *The Elements of Statistical Learning: Data Mining, Inference, and Prediction* (Springer Series in Statistics, New York, 2009)
[5] P. Bydžovský, A. Cieplý, D. Petrellis, D. Skoupil and N. Zachariou, Phys. Rev. C **104**, 065202 (2021)
[6] N. Zachariou et al., Phys. Lett. B **827**, 136985 (2022)
[7] H. Kohri *et al.* (LEPS Collaboration), Phys. Rev. Lett. **97**, 082003 (2006)
[8] D. Skoupil and P. Bydžovský, Phys. Rev. C **97**, 025202 (2018)
[9] D. Skoupil and P. Bydžovský, Phys. Rev. C **93**, 025204 (2016)
[10] H. Akaike, IEEE Transactions on Automatic Control, **19** (6), 716 (1974)
[11] G. Schwarz, Ann. Stat. **6**(2), 461 (1978)
[12] L. De Cruz *Bayesian model selection for electromagnetic kaon production in the Regge-plus-resonance framework* (PhD Thesis, Ghent University, 2012)
[13] P. Zyla *et al.*, Prog. Theor. Exp. Phys. 2020, 083C01 (2020)

Exploring the neutron skin by hyperon–antihyperon production in antiproton–nucleus interactions

Martin Christiansen[1], *Falk* Schupp[1], *Patrick* Achenbach[1,2], *Michael* Bölting[1], *Josef* Pochodzalla[1,2,*], and *Marcell* Steinen[1]

[1]Helmholtz Institute Mainz, GSI Helmholtzzentrum für Schwerionenforschung, Darmstadt, Johannes Gutenberg University, 55099 Mainz, Germany
[2]Institute for Nuclear Physics, Johannes Gutenberg University, 55099 Mainz, Germany

Abstract. In this work we propose a new method to measure the evolution of the neutron skin thickness between different isotopes. We consider antiproton–nucleus interactions close to the production threshold of $\Lambda\overline{\Lambda}$ and $\Sigma^-\overline{\Lambda}$ pairs. At low energies, $\Lambda\overline{\Lambda}$ pairs are produced in $\overline{p}$ + p collisions, while $\Sigma^-\overline{\Lambda}$ pairs can only be produced in $\overline{p}$ + n interactions. Measuring these cross sections provides information on the neutron skin thickness.

1 Introduction

Protons and neutrons are the building blocks of conventional atomic nuclei. Their distribution in nuclei are related to bulk properties of nuclear matter, which are often expressed in terms of a nuclear equation of state (EoS). Indeed, the isospin dependence of the EoS correlates strongly with the distribution of neutrons in nuclei [1].

The neutron skin thickness, Δr_{np}, is defined as the difference between the root-mean-squared (rms) point radii of the neutron and proton density distributions. The focus of their studies often lies on neutron rich doubly magic nuclei such as ^{40}Ca, ^{48}Ca, and ^{208}Pb. However, it is the evolution of the neutron skin and proton distributions along isotope chains that provide important information for our understanding of the nuclear structure over the whole nuclear chart.

In this work we propose a new method to measure the difference of the neutron skin thickness between two different isotopes of a given element. We consider antiproton–nucleus interactions close to the thresholds of $\Lambda\overline{\Lambda}$ and $\Sigma^-\overline{\Lambda}$ pair production. At low energies, $\Lambda\overline{\Lambda}$ pairs are produced in $\overline{p}$+p collisions, while $\Sigma^-\overline{\Lambda}$ pairs can only be produced in $\overline{p}$+n interactions. Measuring the probabilities $p_{\Lambda\overline{\Lambda}}$ and $p_{\Sigma^-\overline{\Lambda}}$ for the two processes for a reference isotope (I) and a second isotope (II), allows to determine the double ratio

$$DR = \frac{p^{II}_{\Sigma^-\overline{\Lambda}} \big/ p^{II}_{\Lambda\overline{\Lambda}}}{p^{I}_{\Sigma^-\overline{\Lambda}} \big/ p^{I}_{\Lambda\overline{\Lambda}}}. \tag{1}$$

Within a simple geometrical picture we show that this ratio is strongly related to the difference of the neutron skin thicknesses of the two considered isotopes (I) and (II). In a second step, we explore this ratio with the Gießen Boltzmann–Uehling–Uhlenbeck transport model for a xenon isotope chain.

*e-mail: pochodza@uni-mainz.de

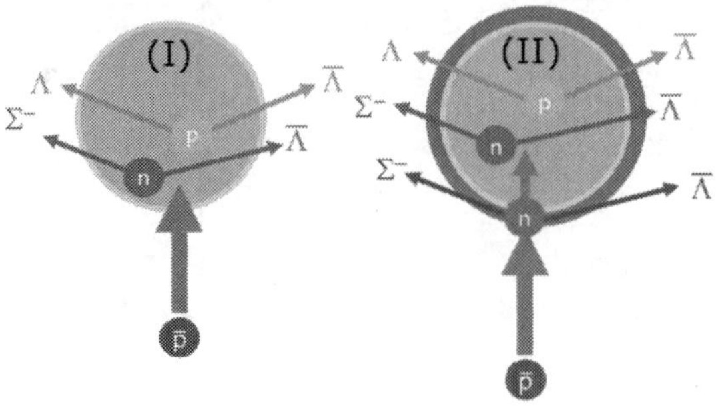

Figure 1. Illustration of $\Lambda\overline{\Lambda}$ and $\Sigma^-\overline{\Lambda}$ pair production in antiproton–nucleus interactions close to their thresholds. At low energies, $\Lambda\overline{\Lambda}$ pairs are produced in $\overline{\mathrm{p}}$+p collisions, while $\Sigma^-\overline{\Lambda}$ pairs can only be produced in $\overline{\mathrm{p}}$+n interactions. We consider two isotopes (I) and (II) which differ by an additional outer neutron layer with thickness Δ_n, indicated in pink.

2 Motivation of the method

Because of the strong absorption of antiprotons in nuclei the production of hyperon–antihyperon pairs happens in the nuclear periphery. For simplicity we consider a nearly central antiproton–nucleus collision (see left part of Fig. 1). Close to threshold, the probability $p^I_{\Lambda\overline{\Lambda}}$ to produce a $\Lambda\overline{\Lambda}$ pair within the reference nucleus (I) can be written as

$$p^I_{\Lambda\overline{\Lambda}} = \kappa_{\Lambda\overline{\Lambda}} \cdot \frac{\rho(p)}{\rho(p)+\rho(n)} \cdot \frac{\sigma_{\Lambda\overline{\Lambda}}}{\sigma_{\text{tot}}}. \tag{2}$$

Here σ_{tot} denotes the total $\overline{\mathrm{p}} + A$ cross section and $\sigma_{\Lambda\overline{\Lambda}}$ is the elementary $\overline{\mathrm{p}} + \mathrm{p} \to \Lambda\overline{\Lambda}$ cross section. $\rho(p)$ and $\rho(n)$ denote the effective densities in the periphery of the target nucleus of protons and neutrons, respectively. The factor $\kappa_{\Lambda\overline{\Lambda}}$ describes the loss of $\Lambda\overline{\Lambda}$ pairs by the absorption of the produced Λ or $\overline{\Lambda}$ after their production. Similarly, the production probability of $\Sigma^-\overline{\Lambda}$ pairs can be approximated by

$$p^I_{\Sigma^-\overline{\Lambda}} = \kappa_{\Sigma^-\overline{\Lambda}} \cdot \frac{\rho(n)}{\rho(p)+\rho(n)} \cdot \frac{\sigma_{\Sigma^-\overline{\Lambda}}}{\sigma_{\text{tot}}}. \tag{3}$$

Because of the large annihilation cross section of antibaryons in nuclei, both, $\kappa_{\Lambda\overline{\Lambda}}$ and $\kappa_{\Sigma^-\overline{\Lambda}}$ are dominated by the $\overline{\Lambda}$ absorption. Therefore, we assume $\kappa_{\Lambda\overline{\Lambda}} \approx \kappa_{\Sigma^-\overline{\Lambda}} \equiv \kappa_I$.

We now turn to a isotope (II) with a larger neutron number and hence a more extended neutron distribution. For simplicity, we assume that the proton distribution remains identical to the one of nucleus (I) and that the neutron distribution is only extended by an additional halo Δ_n at the surface, see pink area in the right part of Fig. 1. The production of $\Lambda\overline{\Lambda}$ pairs is reduced by the absorption probability p_{abs} of the incident antiprotons within this additional neutron skin Δ_n:

$$p^{II}_{\Lambda\overline{\Lambda}} = (1 - p_{\text{abs}}) \cdot \kappa_{II} \cdot \frac{\rho(p)}{\rho(p)+\rho(n)} \cdot \frac{\sigma_{\Lambda\overline{\Lambda}}}{\sigma_{\text{tot}}}. \tag{4}$$

The absorption probability p_{abs} can be expressed in terms of the total $\overline{\mathrm{p}}$ + n reaction cross section $\sigma_{\overline{\mathrm{p}}\mathrm{n}}$ [2] and the integrated skin density $\int_{\Delta_n} \rho_n \, dr_n$ of the additional neutron skin of nucleus (II) with respect to the reference nucleus (I):

$$1 - p_{\text{abs}} \approx \exp\Big\{ - \sigma_{\overline{\mathrm{p}}\mathrm{n}} \cdot \int_{\Delta_n} \rho_n \mathrm{d}r_n \Big\} \tag{5}$$

Like the incoming antiprotons, the produced $\overline{\Lambda}$ are also absorbed in the additional neutron layer. We use the factorisation ansatz $\kappa_{II} = \kappa_I \cdot \kappa_n$. With this simplification we can express Eq. (4) as

$$p^{II}_{\Lambda\overline{\Lambda}} = \kappa_n \cdot (1 - p_{abs}) \cdot p^{I}_{\Lambda\overline{\Lambda}}. \quad (6)$$

The production of $\Sigma^-\overline{\Lambda}$ pairs gains an additional component from the additional neutron skin. On the other hand, the contribution from the inner part of the nucleus (II) is reduced by the loss of antiprotons in the additional neutron layer:

$$p^{II}_{\Sigma^-\overline{\Lambda}} = \kappa_{II} \cdot p_{\text{abs}} \cdot \frac{\sigma_{\Sigma^-\overline{\Lambda}}}{\sigma_{\text{tot}}} + (1 - p_{\text{abs}}) \cdot \kappa_{II} \frac{\rho(n)}{\rho(p) + \rho(n)} \cdot \frac{\sigma_{\Sigma^-\overline{\Lambda}}}{\sigma_{\text{tot}}}. \quad (7)$$

Here, we assume that the loss of outgoing Σ^- and/or $\overline{\Lambda}$ for pairs produced in the additional neutron skin is the same as for pairs produced within the core nucleus. We thus obtain for the double ratio

$$DR = \frac{p^{II}_{\Sigma^-\overline{\Lambda}} / p^{II}_{\Lambda\overline{\Lambda}}}{p^{I}_{\Sigma^-\overline{\Lambda}} / p^{I}_{\Lambda\overline{\Lambda}}} = \frac{p_{\text{abs}} \cdot \frac{\rho(p)+\rho(n)}{\rho(n)} + \left(1 - p_{\text{abs}}\right)}{1 - p_{\text{abs}}} \quad (8)$$

With the assumption $\rho(p) = \rho(n)$, we finally find for the double ratio the simple expression

$$DR = \frac{1 + p_{\text{abs}}}{1 - p_{\text{abs}}} \quad (9)$$

To evaluate Eq. (9), we use neutron skin thicknesses which are predicted during the initialization stage of the GiBUU simulations described in the next section (cf. Fig. 2). In the following we will compare the predictions of Eq. (9) with the result of microscopic GiBUU simulations of $\overline{\text{p}} + A$ interactions at 2.4 GeV/c.

3 Transport calculations for a Xe isotope chain

Of course, this extremely simplified geometrical picture has several deficiencies. A more realistic description of the hyperon pair production can be achieved by microscopic transport calculations. As such a model, GiBUU describes many features of $\overline{\text{p}}$–nucleus interactions in the FAIR energy range [5] .

In the following we employ the GiBUU model to study $\overline{\text{p}}$ + Xe for a series of isotopes in the mass range from $A = 129$ to $A = 136$. These isotopes have a sizable abundance and can in principle be used for experimental studies. The simulations were performed with an incident antiproton momentum of 2.4 GeV/c. For each isotope about 109 million events were generated, requiring a computational time of approximately four days on the MOGON2 high performance computing cluster at the University of Mainz. During the initialization of these simulations, the proton and neutron distributions of the target nuclei were generated by a self consistent relativistic mean-field model.

The filled red squares and blue points in Fig. 2 show the rms radii of the proton and neutron distributions as a function of the Xe mass number. In these calculations not only the neutron radius but, unlike in the simple geometrical picture presented above also the proton radius is rising with increasing neutron number. This increasing charge radius is qualitatively consistent with experimental data, see open squares in Fig. 2 [3, 4], though the experimental slope with the mass number is only about half as large.

The filled red circles in Fig. 3 show the double ratio DR predicted by the GiBUU simulations. For comparison, the blue squares in this figure show the results of Eq. (9), using the same proton and neutron radii which are used as starting point for the GiBUU simulations.

 https://doi.org/10.1051/epjconf/202227102011

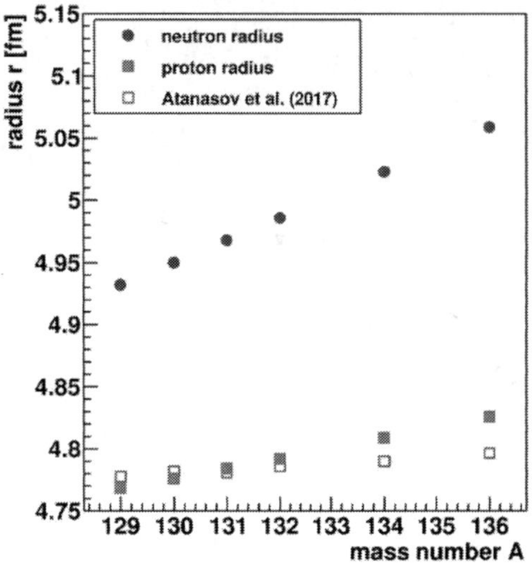

Figure 2. Proton (filled red squares) and neutron (blue points) rms radii of the initial distributions for stable or long-lived xenon isotopes used by the GiBUU simulations. The open red squares show experimental values for the proton radii [3, 4].

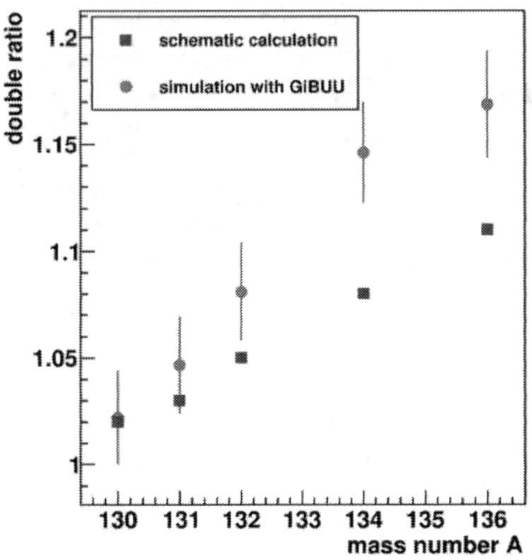

Figure 3. Red points show the double ratio predicted by the full GiBUU simulations for isotopes pairs of ^{A}Xe and ^{129}Xe. For comparison, the blue squares are the results of Eq. (9), adopting the same proton and neutron radii as shown in Fig. 2.

While the absolute agreement is less relevant, we find a remarkable correlation between the double ratio evaluated with our simple analytical expression and the double ratio predicted by the complete GiBUU simulations. This correlation shows that the double ratio may be a good measure for the neutron skin difference. In future we will apply this method to the case of ^{40}Ca and ^{48}Ca.

Acknowledgements

We thank Horst Lenske and Theo Gaitanos for valuable comments. This project has received funding from the European Union's Horizon 2020 research and innovation programme under grant agreement No 824093. The presented data were in part collected within the framework of the bachelor theses of Martin Christiansen and the PhD thesis of Falk Schupp at the Johannes Gutenberg University Mainz.

References

[1] J. Piekarewicz, F.J. Fattoyev, Physics Today **72**, 30 (2019)

[2] M. Tanabashi, K. Hagiwara, K. Hikasa, K. Nakamura, Y. Sumino, F. Takahashi, J. Tanaka, K. Agashe, G. Aielli, C. Amsler et al. (Particle Data Group), Phys. Rev. D **98**, 030001 (2018)

[3] I. Angeli, K. Marinova, Atomic Data and Nuclear Data Tables **99**, 69 (2013)

[4] D. Atanasov, D. Beck, K. Blaum, C. Borgmann, R.B. Cakirli, T. Eronen, S. George, F. Herfurth, A. Herlert, M. Kowalska et al., Journal of Physics G: Nuclear and Particle Physics **44**, 044004 (2017)

[5] O. Buss, T. Gaitanos, K. Gallmeister, H. van Hees, M. Kaskulov, O. Lalakulich, A. Larionov, T. Leitner, J. Weil, U. Mosel, Physics Reports **512**, 1 (2012)

High resolution spectroscopy of the "ΣN cusp" by using the $d(K^-, \pi^-)$ reaction

Yudai Ichikawa[1,*], *Jungkeun* Ahn[2], *Sungwook* Choi[2], *Manami* Fujita[1], *Toshiyuki* Gogami[3], *Shoichi* Hasegawa[1], *Takeshi* Harada[1,3], *Shuhei* Hayakawa[4], *Shigeru* Ishimoto[5], *Woose-ung* Jung[2], *Byungmin* Kang[2], *Shinhyung* Kim[1], *Koji* Miwa[4], *Tomofumi* Nagae[3], *Takuya* Nanamura[1,3], *Hiroyuki* Sako[1], *Susumu* Sato[1], *Shoji* Suzuki[5], *Hirokazu* Tamura[1,4], *Kiyoshi* Tanida[1], *Mifuyu* Ukai[1], *Takeshi* Yamamoto[1], and *Seongbae* Yang[3]

[1]ASRC, Japan Atomic Energy Agency, Ibaraki 319-1195, Japan
[2]Department of Physics, Korea University, Seoul 02841, Republic of Korea
[3]Department of Physics, Kyoto University, Kyoto 606-8502, Japan
[4]Department of Physics, Tohoku University, Sendai 980-8578, Japan
[5]High Energy Accelerator Research Organization (KEK), Tsukuba, 305-0801, Japan

Abstract. We present a new proposal, J-PARC E90, to measure a missing-mass spectrum near the ΣN threshold for the $d(K^-, \pi^-)$ reactions at 1.4 GeV/c. While many previous experiments support apparent enhancement near the ΣN threshold, the dynamical origin of this so-called "ΣN cusp" remains yet unsolved. The enhancement suggests either a cusp structure or a weakly bound state. One of the keys to making it clear is improving the missing-mass resolution and statistics. Our new experiment can achieve the missing-mass resolution of 0.4 MeV in σ using the K1.8 beam line and S-2S spectrometers at J-PARC. Furthermore, we can suppress quasi-free background processes with the time projection chamber (HypTPC), which operated nicely for the H-dibaryon search experiment (J-PARC E42). The J-PARC E90 aims to extract the scattering length of the ΣN system with isospin T = 1/2 and spin-triplet channels.

1 Introduction

An enhancement near the ΣN threshold (~2.13 GeV/c^2) was clearly observed in the $K^- d \to \pi^- \Lambda p$ reaction at rest more than 50 years ago [1]. This enhancement is called as "ΣN cusp", while whether the "ΣN cusp" is cusp (inelastic virtual state) or an unstable bound state has not been confirmed yet. The "ΣN cusp" was measured by various experiments using $K^- d \to \pi^- \Lambda p$, $\pi^+ d \to K^+ \Lambda p$, and $pp \to K^+ \Lambda p$ reactions [2–8]. Recently, the "ΣN cusp" has also been observed by the Λ-p femtoscopy [9]. At J-PARC, E27 collaboration reported a clear enhancement due to the "ΣN cusp" in the inclusive missing-mass spectrum of the $d(\pi^+, K^+)$ reaction [10]. There exists a lot of experimental observations; however, the dynamical origin of the "ΣN cusp" remains unclear as yet. One of the reasons is the insufficient resolution and statistics of the past experiments.

*e-mail: yudai@post.j-parc.jp

2 "ΣN cusp" for the $K^- d \to \pi^- \Lambda p$ reaction

Theoretical work for the "ΣN cusp" was performed by Dalitz and Deloff [11, 12]. The detail is also described in the J-PARC E90 proposal [13]. In the framework of Ref. [11], the "ΣN cusp" is generated through the following two-step processes,

$$K^- + d \to \pi^- + (\Sigma N)^+ \to \pi^- + \Lambda + p. \tag{1}$$

The reaction amplitude for the net process shown in Eq. (1) can be expressed by three factors as,

$$T(\bar{K}d \to \pi\Lambda N) \sim T(\bar{K}N \to \pi\Sigma)F_d(Q_\Sigma, k_\Sigma)T(\Sigma N \to \Lambda N). \tag{2}$$

The first factor, $T(\bar{K}N \to \pi\Sigma)$, is the T-matrix of the elementary process. The second one is the deuteron factor reflecting deuteron properties. The third factor, the T-matrix of $\Sigma N \to \Lambda p$ process, is an important term to describe the interaction of an intermediate Σ particle with the second nucleon of the deuteron target. $T(\Sigma N \to \Lambda N)$ can be expressed in terms of the ΣN ($T = 1/2$, ${}^3S_1$) scattering length $A_0 = (a + ib)$. Because the measured final state is Λp, the total isospin of ΣN system should be $T = 1/2$. Moreover, the amplitude of the elementary reaction $\bar{K}N \to \pi Y$ has no spin-flip component for the (K^-, π^-) reaction at 0°. Therefore, the YN states also are necessary to be in spin-triplet by reflecting the target-deuteron spin configuration, as ${}^3S_1$.

The third factor of Eq. (2), $T(\Sigma N \to \Lambda N)$, can be expressed as,

$$T(\Sigma N \to \Lambda N) \sim T^t_S(\Sigma N \to \Lambda N) = \frac{\beta^t_{\Sigma\Lambda}}{1 - ik_\Sigma A_0}, \tag{3}$$

where $\beta^t_{\Sigma\Lambda}$ is the $\Sigma N \to \Lambda N$ element of the reaction matrix and b is proportional to $(\beta^t_{\Sigma\Lambda})^2$ (ΛN phase space). The k_Σ denotes the relative momentum of ΣN system in the final ΛN at rest frame. Note that it is clear that the reaction amplitude has a pole at $k_\Sigma = -i/A_0$.

Above the threshold, the reaction rate related to $T(\Sigma N \to \Lambda N)$ term can be expressed as,

$$R^t_S \sim (k_\Sigma \sigma^t_S(\Sigma N \to \Lambda N)) = \frac{4\pi b}{(1 + k_\Sigma b)^2 + (k_\Sigma a)^2}. \tag{4}$$

Below the threshold, k_Σ is replaced by $+i|k_\Sigma|$ due to the analytic continuation, and the corresponding reaction rate is given by

$$R^t_S \sim \frac{4\pi b}{(1 + |k_\Sigma|a)^2 + k^2_\Sigma b^2}. \tag{5}$$

3 J-PARC E90 experiment

3.1 Experimental setup

The E90 experiment will be performed at the K1.8 beamline by using the S-2S spectrometer [14] and HypTPC [15]. Figure 1 shows the schematic view of the proposed experiment. The beam K^- will be measured by using the existing K1.8 beam-line spectrometer [16]. The design value of the momentum resolution is $\delta p/p = 3.3 \times 10^{-4}$ (FWHM) with the position accuracy of 0.2 mm in rms.

The scattered π^- in the (K^-, π^-) reaction are momentum analyzed by using the S-2S spectrometer [14]. The momentum resolution of S-2S will be $\delta p/p = 6.0 \times 10^{-4}$ in FWHM. Then, the total mass resolution 0.4 MeV in σ will be achieved, which is two times better than the past experiment (HIRES at COSY [7]).

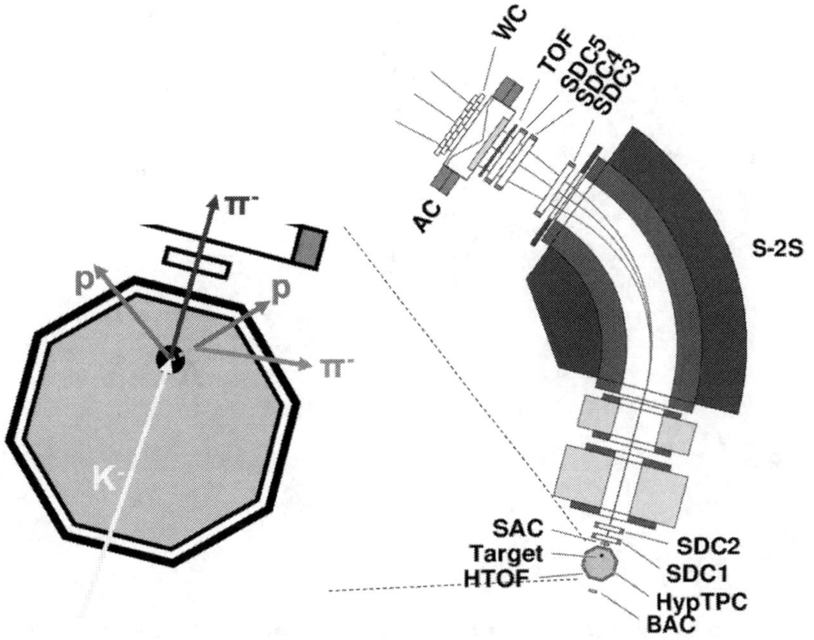

Figure 1. The schematic view of the J-PARC E90 experiment.

In this experiment, we plan to install the Time Projection Chamber (HypTPC) to suppress the quasi-free background and improve the signal-to-noise ratio by requiring the charged multiplicity, $Mt = 3$. The detailed strategy to suppress the background by using HypTPC is described in J-PARC E90 proposal [13]. The liquid deuterium target will be installed inside the TPC volume to achieve the large acceptance. We plan to use the same liquid target system as J-PARC E45 experiment [17], which concerns the baryon spectroscopy using $p(\pi^{\pm}, 2\pi)$ reaction.

3.2 Expected result

The simulated missing-mass distributions are presented in Fig. 2. Here, 3 choices of the scattering length are shown to demonstrate different shapes, namely, the theoretical values of Jülich potential–A [18], ((a), shallow bound), Nijmegen model (D ((b), deeply bound) and F ((c), sharp cusp)) [19]. Note that there is no reason to select these three models by considering theoretical justification. The points with error bars are the smeared spectra with the experimental resolution of ΔM = 0, 0.4, 1, and 2 MeV in σ shown by the different colors. In these plots, 1.4×10^4 events, corresponding to the statistics of the E90 experiment, are generated and the statistical errors are shown.

The mass resolutions of the past bubble-chamber experiments using in-flight $d(K^-, \pi^-)$ reaction were 2 MeV ("Braun" [2]) and 3 MeV ("Eastwood" [3]). As shown in Fig. 2, the original shapes are significantly distorted in the case of ΔM = 2 MeV. The high mass-resolution is really important for the "ΣN cusp". Moreover, the statistics of these past experiments were poor as 603 events ("Braun") and 217 events ("Eastwood"). By improving statics to 1.4×10^4 events and mass resolution to 0.4 MeV, we can deduce the scattering length with the statistical error $\lesssim 0.3$ fm.

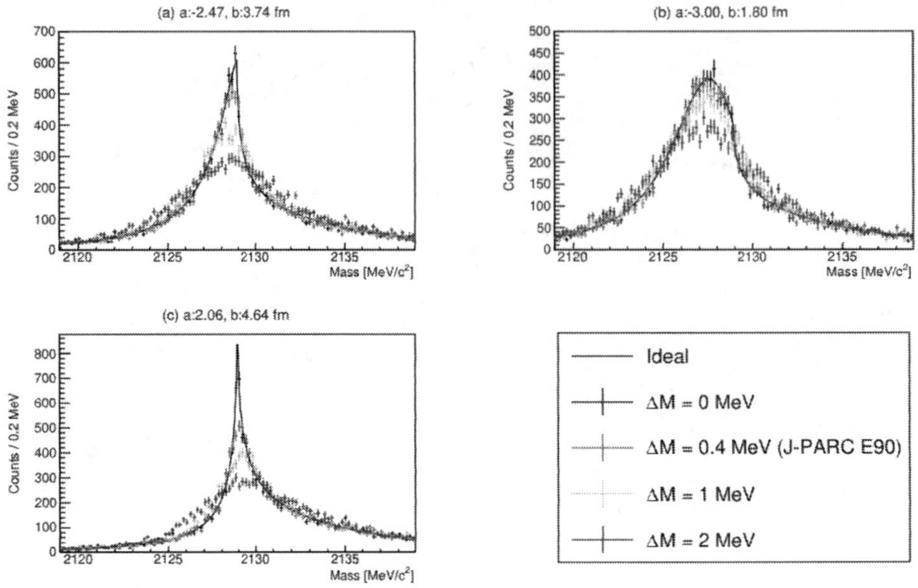

Figure 2. The simulated mass distributions for three choices of the scattering-length, which are shown in the title of the plots. The solid black-lines show the original distribution given by Eqs. (2). The points with error bars show the smeared spectra with the experimental resolution of $\Delta M = 0, 0.4, 1$, and 2 MeV in σ and the statistic errors with 1.4×10^4 events.

4 Summary

We proposed the high resolution missing-mass spectroscopy of the "ΣN cusp" by using the S-2S spectrometer and HypTPC. The mass resolution is a quite important key to investigate the nature of "ΣN cusp". We can achieve 0.4 MeV resolution in σ, which is twice better than the past COSY HIRES experiment [7]. Moreover, the in-flight $d(K^-, \pi^-)$ reaction is unique to derive $^3S_1$ spin contribution with low background. By fitting the measured mass spectra, we are able to deduce the $(T = 1/2, S = 1)$ ΣN scattering length with good accuracy, statistical error of $\lesssim 0.3$ fm.

References

[1] O.I. Dahl *et al.*, Phys. Rev. Lett., **6**, 3 (1961)
[2] O. Braun *et al.*, Nucl. Phys. **B124**, 45 (1977)
[3] D. Eastwoood *et al.*, Phys. Rev. D **3**, 2603 (1971)
[4] C. Pigot *et al.*, Nucl. Phys. **B249**, 172 (1985)
[5] T.H. Tan, Phys. Rev. Lett. **23**, 395 (1969)
[6] R. Siebert *et al.*, Nucl. Phys. **A567**, 819 (1994)
[7] A. Budzanowski *et al.*, Phys. Lett. B **692**, 10 (2010)
[8] S. Abd El–Samad *et al.*, Eur. Phys. J. A **49**, 41 (2013)
[9] ALICE Collaboration, Phys. Lett. B **833**, 10 (2022)
[10] Y. Ichikawa *et al.*, Prog. Theor. Exp. Phys. **2014**, 101D03 (2014)
[11] R.H. Dalitz, Nucl. Phys. **A354**, 101, (1981)
[12] R.H. Dalitz and A. Deloff, Czech. J. Phys. **B**, 32, (1982)

[13] Y. Ichikawa *et al.*, *High resolution spectroscopy of the "ΣN cusp" by using the* $d(K^-, \pi^-)$ *reaction*, J-PARC E90 proposal,
https://j-parc.jp/researcher/Hadron/en/pac_2201/pdf/P90_2022-03.pdf

[14] T. Nagae *et al.*, *Proposal for the next E05 run with the S-2S spectrometer*,
http://j-parc.jp/researcher/Hadron/en/pac_1801/pdf/P70_2018-10.pdf

[15] S.H. Kim *et al.*, Nucl. Instrum. Meth. A **940**, 359 (2019)

[16] T. Takahashi (the Hadron Beamline Group and the K1.8 Experimental Group), Nucl. Phys. A **835**, 88 (2010)

[17] Technical Design Report of the J-PARC E45 experiment

[18] J. Haidenbauer and Ulf-G. Meißner, Phys. Rev. C **72**, 044005 (2005)

[19] K. Miyagawa and H. Yamamura, Phys. Rev. C **60**, 024003 (1999)

Strangeness production in the NA61/SHINE experiment at the CERN SPS energy range

Yuliia Balkova[1,*] for NA61/SHINE collaboration

[1]University of Silesia, 12 Bankowa str., 40-007 Katowice, Poland

Abstract. NA61/SHINE is a multi-purpose fixed-target experiment located at the H2 beamline of the CERN North Area. One of the main goals of the experiment is to study the phase transition and search for the critical point of the strongly interacting matter. Strangeness production is a long-known valuable probe for understanding particle production in high-energy physics due to the absence of strange valence quarks in the initial collision state.
This talk will present the results of strangeness production in p+p, Be+Be, and Ar+Sc collisions in the SPS energy range ($\sqrt{s_{NN}}$ = 5.1 – 17.3 GeV) measured by NA61/SHINE. The talk will emphasise the importance of the results for discussion of the onset of deconfinement and onset of fireball. The obtained results will be compared to available world data and selected theoretical models.

1 Introduction

NA61/SHINE is a fixed-target experiment, which is located at the H2 beamline of the CERN North Area and takes its beams from the CERN SPS [1]. The NA61/SHINE detector layout is presented in Figure 1.

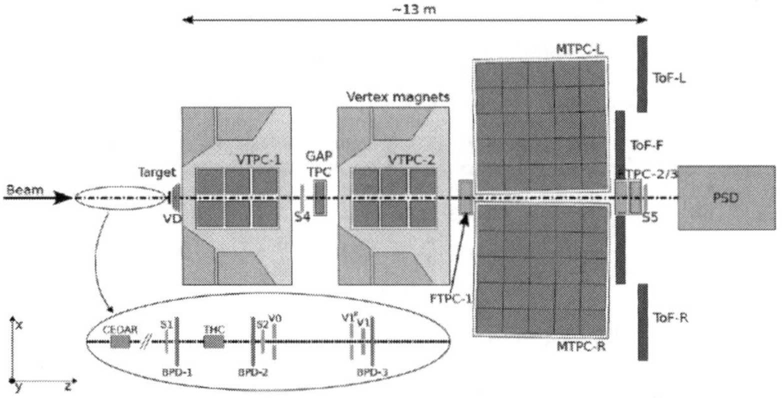

Figure 1. Schematic view of the NA61/SHINE detector system.

*e-mail: yuliia.balkova@cern.ch

It is a large acceptance hadron spectrometer equipped with a set of Time Projection Chambers and Time-of-Flight detectors, which allows for charged particle momentum measurements and identification. The high-resolution forward calorimeter, the Projectile Spectator Detector (PSD), measures the energy of spectators, which is related to the centrality of the collision. Thanks to modular construction, PSD modules can be used to optimize centrality selection for different energies and systems.

One of the main scientific goals of the experiment is to study the phase transition and search for the critical point of the strongly interacting matter. In order to study the phase diagram of the strongly interacting matter, the NA61/SHINE experiment performed the two-dimensional scan in collision energy and system size.

2 Results on strangeness production

Strangeness production is a long-known valuable probe for understanding particle production in high-energy physics due to the absence of strange valence quarks in the initial collision state [2].

When compiling the results for different system sizes and collision energies, intriguing results ought to be seen. Figure 2 presents the energy dependence of the inverse slope parameter T of transverse mass m_T spectra of charged kaons. A plateau structure is clearly visible and was first observed in heavy-ion collisions [3]. It was interpreted by the Statistical Model of Early Stage [4] as a result of the coexistence of two phases of strongly interacting matter: hadron gas and quark-gluon plasma. The results for different system sizes – from the lightest proton-proton to the heaviest Pb+Pb – show qualitatively similar behaviour for the inverse slope parameter, though the magnitude of T naturally increases with the system size.

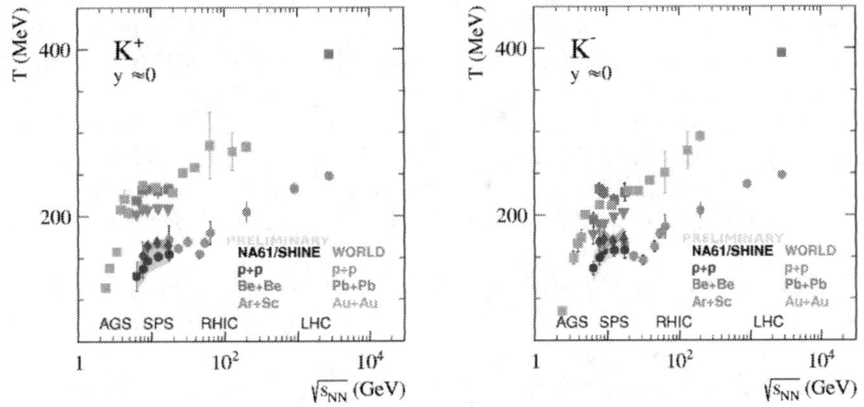

Figure 2. Inverse slope parameter T of K^+ (*left*) and K^- (*right*) m_T spectra as a function of collision energy for p+p, Be+Be, Ar+Sc and Pb+Pb/Au+Au collisions.

Figure 3 shows the multiplicity ratio of positively charged kaons and pions at mid-rapidity and in 4π acceptance. A rapid increase and following plateau were observed for large systems [5] and interpreted as a sign of the occurring phase transition. While the results from small systems, such as $p+p$ and Be+Be, indicate no such structure, the results from the intermediate system, Ar+Sc, are quite surprising as quantitatively, they lay a lot closer to heavier systems, nevertheless, without an indication of the maximum.

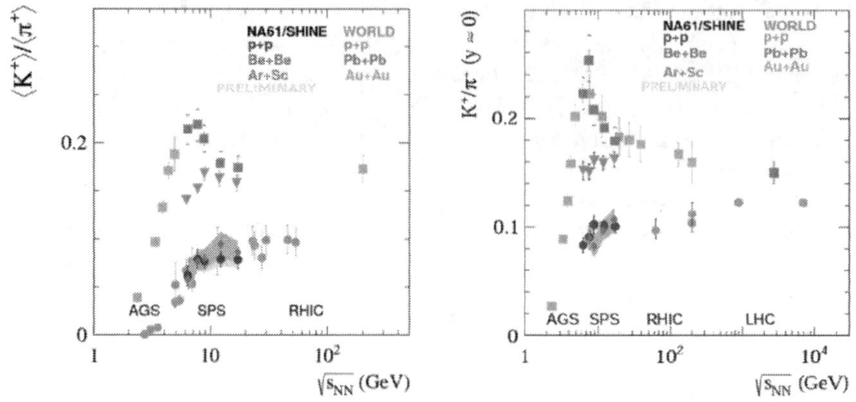

Figure 3. $\langle K^+\rangle$ / $\langle\pi^+\rangle$ ratio in full 4π phase space (*left*) and $\langle K^+\rangle$ / $\langle\pi^+\rangle$ ratio at mid-rapidity (*right*) as a function of collision energy for p+p, Be+Be, Ar+Sc and Pb+Pb/Au+Au collisions.

The first two panels of Fig. 4 present the system size dependence of $\langle K^+\rangle$ / $\langle\pi^+\rangle$ ratio at mid-rapidity, and the last panel shows the system size dependence of the inverse slope parameter T of positively charged kaons. Data points are obtained by NA61/SHINE for $p + p$ [6], Be+Be [7] and Ar+Sc (preliminary results), and by NA49 for Pb+Pb [5] at 150A GeV/c beam momentum. An increase in both observables is visible when going from small systems to intermediate and large, which can be interpreted as a signature of the beginning of the formation of large clusters of strongly interacting matter, i.e. onset of fireball. The results are then compared with various dynamical and statistical models. The predictions of dynamical models without phase transition – EPOS [8], UrQMD [11] and SMASH [10] – agree with the results from small systems, failing to describe the results from larger systems. In contrast, PHSD [9], the model with phase transition, well describes the results from larger systems. Both tested statistical models – SMES [12], with the phase transition, and HRG [11], without one, tend to overestimate the data, especially for small systems. Clearly, none of the tested models can describe the data well for the full number of wounded nucleons $\langle W\rangle$ range.

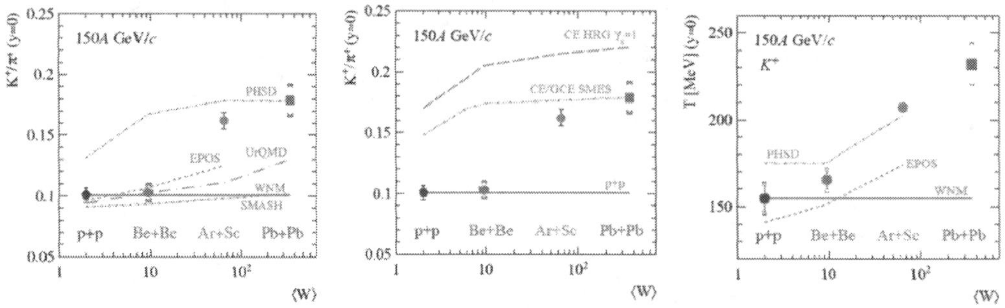

Figure 4. The system size dependence of $\langle K^+\rangle$ / $\langle\pi^+\rangle$ ratio at mid-rapidity compared to dynamical (*left*) and statistical (*middle*) models; the system size dependence of inverse slope parameter T of K^+ m_T spectra compared to dynamical models (*right*).

3 Conclusions

This contribution concentrates on the compilation of selected results on strangeness production in various collision systems at the CERN SPS energy range. As was shown, presented results are of high importance when discussing the onset of deconfinement and the onset of fireball. However, continuous theoretical input is crucial to allow models to describe existing data.

Acknowledgements

This work was supported by the funds granted by the National Science Center, Poland Grant No. 2019/34/H/ST2/00585, by the Helmholtz Institute Mainz and the European Union's Horizon 2020 research and innovation programme under grant agreement No. 824093 and by the Polish Minister of Education and Science, contract No. 2021/WK/10.

References

[1] NA61/SHINE Collaboration, JINST **9**, P06005 (2014)
[2] J. Rafelski and B. Muller, Phys. Rev. Lett. **48**, 1066 (1982) [Erratum: Phys. Rev. Lett. **56**, 2334 (1986)]
[3] NA49 Collaboration, Phys. Rev. C **77**,024903 (2008)
[4] M. Gazdzicki and M. Gorenstein, Acta Phys.Polon. **B30**, 2705 (1999)
[5] NA49 Collaboration, Phys. Rev. C **66**, 054902 (2002)
[6] NA61/SHINE Collaboration, Eur. Phys. J. C **77**, 671 (2017)
[7] NA61/SHINE Collaboration, Eur. Phys. J. C **81**, 73 (2021)
[8] K. Werner, Nucl. Phys. Proc. Suppl. **175–176**, 81–87 (2008)
[9] V. Kireyeu et al., Eur. Phys. J. A **56**, 223 (2020)
[10] J. Mohs et al., J. Phys. G **47**, 065101 (2020)
[11] A. Motornenko et al., Phys. Rev. C **99**, 034909 (2019)
[12] R.V. Poberezhnyuk at el., Acta Phys. Polon. B **46**, 1991 (2015)

X ray spectroscopy on Ξ^- atoms (J-PARC E03, E07 and future)

Takeshi. O. Yamamoto[1,*], *M.* Fujita[1], *T.* Gogami[2], *T. K.* Harada[2,1], *S. H.* Hayakawa[3], *K.* Hosomi[1], *Y.* Ichikawa[1], *Y.* Ishikawa[3], *K.* Kamada[3], *H.* Kanauchi[3], *T.* Koike[3], *K.* Miwa[3], *T.* Nagae[2], *F.* Oura[3], *T.* Takahashi[4], *H.* Tamura[3,1], *K.* Tanida[1], *M.* Ukai[4,3], and E03/E07/E96 Collaborations

[1]Advanced Science Research Center (ASRC), Japan Atomic Energy Agency (JAEA), Tokai, Ibaraki 319-1195, Japan
[2]Department of Physics, Kyoto University, Kyoto 606-8502, Japan
[3]Department of Physics, Tohoku University, Sendai, Miyagi 980-8578, Japan
[4]Institute of Particle and Nuclear Studies (IPNS), High Energy Accelerator Research Organization (KEK), Tsukuba, Ibaraki 305-0801, Japan

Abstract. X-ray spectroscopy of hadronic atoms is a powerful method to study strong interaction between hadrons and nuclei. At J-PARC, we have conducted two experiments, J-PARC E07 and E03, for hadronic atoms with a doubly strange hyperon, Ξ^-, aiming at the world-first detection of their X-rays. The first measurement is performed as a byproduct of J-PARC E07 experiment with the hybrid emulsion technique. The second one, J-PARC E03, is a dedicated experiment for detection of Ξ^- Fe atom X rays. The preliminary results and the present status of E07 and E03 are shown in this article. Future prospects of Ξ^--atomic X-ray spectroscopy are also discussed. A new measurement has been proposed for detecting Ξ^- C atom X rays, where a novel Ξ^- tracking method will be applied to realize an improved signal to noise ratio.

1 Introduction

X-ray spectroscopy of hadronic atoms gives us various information on strong interaction between hadrons and nuclei. This method has been successfully applied for negatively charged hadrons, such as π^-, K^-, $\overline{p}$, and Σ^-. The negatively charged hadrons can approach the nuclear surface, leading to energy shifts of atomic states by attractive/repulsive strong interaction as well as their energy widths by nuclear absorption. These energy shifts and widths, which provide information on the real and imaginary parts of the hadron-nucleus optical potential, can be determined by measuring X rays from the transition between atomic states. In the S (strangeness) $= -1$ sector, more than twenty data points are available for Σ^- atoms for a wide mass range of nuclei, giving constraints for the Σ-nuclear optical potential [1]. In the $S = -2$ sector, strong mixing between ΞN and $\Lambda\Lambda$ is expected because their mass difference is as small as 28 MeV, being much smaller than the case of the $S = 0$ and 1 sectors. Such baryon-baryon coupling effects may play a dominant role, especially in the $S = -2$ systems.

*e-mail: tyamamo@post.j-parc.jp

Furthermore, understanding of the $\Xi^- N$ interaction is essential to clarify the role of Ξ^- in neutron stars. However, little is known experimentally on the $S = -2$ systems.

Ξ^--atomic X-ray data for light nuclei may have another impact on emulsion data for $S = -2$ hypernuclei. For example, the "NAGARA" event, which gives the binding energy of double Λs ($B_{\Lambda\Lambda}[^6_{\Lambda\Lambda}\text{He}]$) of 6.91±0.16 MeV [3], was analyzed based on the binding energy of Ξ^- C atom (B_{Ξ}) of 0.13 MeV assuming that the Ξ^- absorption occurred in the $3D$ Ξ^--atomic state. However, the atomic state before the absorption may be different if the absorption strength differs from the expectation. The absorption strength, and thus the absorption state, should be determined by Ξ^--atomic X-ray spectroscopy. Therefore, Ξ^--atomic X-ray data for nuclei contained in the emulsion (carbon, nitrogen, and oxygen) help the emulsion analysis.

In this situation, experimental data of Ξ^- atomic X rays have been eagerly awaited [2], but no experiment has been performed yet. Aiming at the world-first detection of the X rays, we have conducted two experiments, J-PARC E07(2016-2017) and J-PARC E03(2020-2021). Preliminary reports from these experiments are given in sections 2 and 3. Future prospects of Ξ^--atomic X-ray spectroscopy are also discussed in section 4.

2 First measurement in J-PARC E07

Our first measurement of Ξ^--atomic X rays has been performed as a byproduct of the J-PARC E07 experiment, a search for double Λ hypernuclei and Ξ hypernuclei with the hybrid emulsion technique [4]. The data taking was done in 2016-2017. In this experiment, hypernuclei were produced via Ξ^- capture by nuclei in emulsion stack. Ξ^-s were produced via the (K^-, K^+) reaction and injected into the emulsion. Stopped Ξ^-s form Ξ^- atoms before the hypernuclear production reaction. The tracks of Ξ^- were measured by silicon strip detectors and used in the microscope analysis for Ξ^- track following in the emulsion. By applying this method, called the hybrid emulsion technique, we found Ξ^- stop events efficiently. A germanium (Ge) detector array, Hyperball-X [5], was installed near the emulsion stack for X-ray measurement. When Ξ^-s stop in emulsion, they form Ξ^- atoms with nuclei contained in the emulsion, C, O, N, O, Br and Ag. It is noted that yields of Ξ^- atoms with Ag and Br are expected to be large because Ξ^- is mostly captured by heavy nuclei. In addition, some of Ξ^-s may also stop inside the Ξ^- production target, a 3.2 g/cm^2-thick diamond, and form Ξ^- C atoms. Ξ^- stop identification is essential for the X-ray measurement because more than 90% of the produced Ξ^- will decay before stopping, leading to a huge number of background events. In this measurement, we applied two methods to observe Ξ^--atomic X rays: (1) combined analysis with Ξ^- stop identification from emulsion image and (2) coincidence with only Ξ^- production without emulsion image.

In the combined analysis, a good signal to noise ratio for the X-ray spectrum is expected because Ξ^- stop events can be unambiguously identified using emulsion image. However, the X-ray yield rate is limited by a low Ξ^- survival ratio due to decay before reaching the emulsion, the small abundance of each nucleus contained in the emulsion and a not-optimized configuration of the X-ray detector array. X rays from Ξ^- atoms on Br and Ag are expected to be detected with this method. The expected X-ray transitions to the "last" Ξ^--atomic states, from which the Ξ^- is absorbed, are $(n, l) = (7, 6) \rightarrow (6, 5)$ for Br (315 keV) and $(8, 7) \rightarrow (7, 6)$ for Ag (370 keV). In the present analysis, X-ray peaks are not observed. It is because the expected X-ray yield is as small as a few counts due to insufficient emulsion analysis efficiency. Further progress in the emulsion analysis is necessary.

In the analysis without emulsion, we tried to select events in which Ξ^-s do not reach silicon strip detectors placed just downstream of the diamond target for the X-ray measurement of Ξ^- C atoms. The signal to noise ratio of the X-ray spectrum is lower than the emulsion combined analysis, while the ratio is partly improved by an additional analysis selecting Ξ^-s

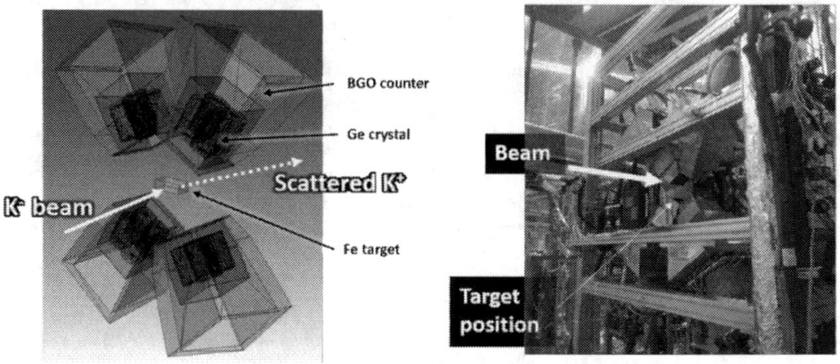

Figure 1. A schematic view and a photo of the detector setup around the experimental target in J-PARC E03. A Ge detector array Hyperball-X' consists of four Ge + BGO units covering the upper and lower directions from the target.

with low recoil momenta. The expected X-ray yield is also low as in the case of the combined analysis. The X-ray transition to the "last" Ξ^--atomic state is expected to be $3D \to 2P$ in the case of weak absorption suggested by a theoretical study with the lattice QCD potential [6], of which the corresponding X-ray energy is 154 keV. A preliminary result shows no clear peak structure in the X-ray spectrum. The upper limit for the branching ratio of the $3D \to 2P$ transition was obtained to be ~40%/ Ξ^- stop [7], which is close to a value suggested by a theoretical case study [6]. Recently, we made a plan for a new measurement for Ξ^- C atoms with improved sensitivity. Details are shown in section 4.

3 Second measurement in J-PARC E03

Our second measurement of Ξ^--atomic X rays was performed as a dedicated X-ray spectroscopy experiment (J-PARC E03) [8]. We selected iron as a target because (1) iron is dense enough (7.9 g/cm^3) with a higher Ξ^- stopping probability, and (2) the expected absorption strength corresponding to the energy width of 4 keV is suitable for the measurement. Since the J-PARC beam intensity is currently insufficient to take the requested full statistics data, we decided to take 10% of the full statistics data as the 1st-phase run. Even with the 1st-phase statistics, a peak of the X-ray transition to the "last" Ξ^- Fe atomic state, the $(6,5) \to (5,4)$ transition (~286 keV), is expected to be observed if the energy width is smaller than 1 keV. Furthermore, the upper X-ray transition of $(7,6) \to (6,5)$ (172 keV) will be observed without peak broadening. The ratio between the yields of the upper and the lower X-ray transitions gives information on the absorption strength. In this measurement, Ξ^-s were produced by the (K^-, K^+) reaction and stopped in the target. Ξ^- production was tagged with the missing mass method by reconstructing momenta of the beam K^-s and the scattered K^+s with the K1.8 beam line spectrometer and the KURAMA spectrometer, respectively. For X-ray detection, a Ge detector array, Hyperball-X', was installed near the target as shown in Fig. 1. Hyperball-X' consists of four Ge + BGO units covering the upper and lower directions from the target to avoid self-absorption of X rays inside the horizontally wide iron target. "Clover-type" Ge detectors were mounted to the array to realize both a large solid angle in total and a low counting rate for each Ge crystal. To optimize the Ge detector readout with respect to a low beam intensity of ~250 kHz, a conventional type of the shaping amplifiers, ORTEC 671, were used for Hyperball-X', while a high-rate type of the shaping amplifiers should be used in the full

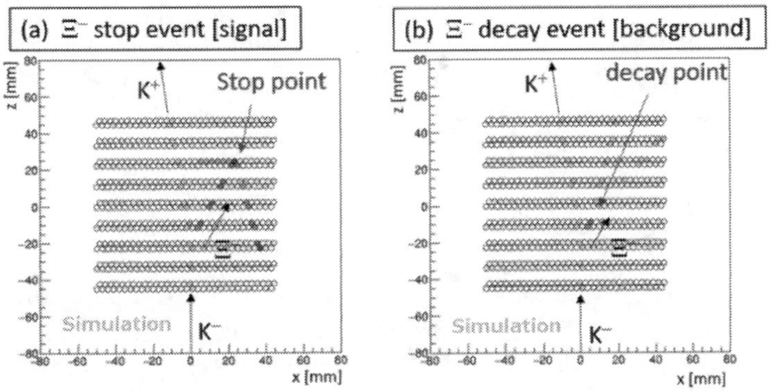

Figure 2. Simulated event displays of AFT for a Ξ^- stop event (a) and a Ξ^- decay event (b) in the J-PARC E96.

statistics run in the future. The typical energy resolution was ~2.3 keV (FWHM) for 307 keV. Data taking for the 1st-phase was done in Apr. 2021 with the total irradiated K^- of 9.5×10^{10}. Analysis has started for selecting the (K^-, K^+) reaction and detecting X rays. We confirmed that Ξ^- production events can be clearly tagged by the missing mass method through the momentum reconstruction with the magnetic spectrometers. The performances of Hyperball-X' were also evaluated for resolutions, efficiencies and background suppression. In the E03 as well as the E07 experiments, a precise in-beam energy calibration was successfully carried out through the whole beam time with a trigger generated by LSO scintillators with natural activity of ^{176}Lu [5]. Further analysis for event selections and calibrations is necessary to obtain the final X-ray spectrum. The result will be reported in the near future.

4 Future measurement in J-PARC E96

We are planning future measurements of Ξ^--atomic X rays simultaneously with high resolution Ξ hypernuclear spectroscopy experiments using the S-2S spectrometer [9]. Systematic measurements with a carbon target (J-PARC E70 [10]) and a lithium target (J-PARC E75 [11]) will be performed shortly. By using the Ξ^- hyperons produced in these experiments with the S-2S spectrometer, we will be able to take Ξ^--atomic X-ray data in parallel. The first experiment with the S-2S spectrometer (J-PARC E70) will start in 2023 with a carbon target. We will retry the X-ray measurement for Ξ^- C atoms in this experiment. Active Fiber Target (AFT) will be used as a carbon target. AFT consists of ϕ3 mm scintillating fibers, of which hit information will be used for correcting energy loss of K^-s and K^+s in the target. For the X-ray measurement, the hit information of the AFT fibers allows us to identify Ξ^- stop events and gives a good signal to noise ratio, although a low density of the target will limit the Ξ^- stopping probability. Event selection can be made by finding tracks of Ξ^-s produced via the (K^-, K^+) reaction followed by a vertex of the Ξ^- absorption on carbon nuclei. We can reject dominant background sources such as (1) Ξ^-s passing through and exiting the target and (2) Ξ^-s decaying in-flight. Figure 2 shows simulated event displays of a Ξ^- stop event and a Ξ^- decay event. The background rejection ratio is expected to be ~95%. This event selection will improve the signal to noise ratio for X-ray measurement. We estimated the sensitivity for Ξ^- C atomic X rays to be better than the previous measurement in J-PARC E07 by a factor of ~3. Detection of X rays are expected for both upper $4F \rightarrow 3D$ and lower

$3D \rightarrow 2P$ transitions of Ξ^- C atoms. We submitted a new proposal (J-PARC E96 [12]) for this X-ray measurement of Ξ^- C atoms and started preparing the detector system. The S-2S magnets were installed in the J-PARC K1.8 beam line. Detector installation for the spectrometer is in progress. Data-taking will start in 2023 after installing the Ge detector array, Hyperball-X', after modification of geometrical configuration.

5 Summary

X-ray spectroscopy of hadronic atoms is a powerful tool to investigate strong interaction between hadrons and nuclei. We are aiming at the world-first measurement of X rays from Ξ^- atoms in J-PARC experiments. We have conducted two experiments for Ξ^- C atoms (J-PARC E07) and Ξ^- Fe atoms (J-PARC E03). The sensitivity will be improved in the newly proposed experiment (J-PARC E96) for Ξ^- C atoms again by introducing the AFT system. This measurement will be performed in 2023. Furthermore, we plan to take Ξ^--atomic X-ray data for a wide mass range of nuclei in future measurements in order to understand the ΞN interaction.

References

[1] E. Friedman, A. Gal, the International School of Physics Enrico Fermi (2007)
[2] C. J. Batty, E. Friedman, and A. Gal, Phys. Rev. **C59**, 295 (1999)
[3] H. Takahashi *et al.*, Phys. Rev. Lett. **87**, 212502 (2001)
[4] K. Imai, K. Nakazawa, H. Tamura, *Systematic Study of Double Strangeness System with an Emulsion-Counter Hybrid Method*, Proposal to J-PARC, E07 (2006)
[5] M. Fujita *et al.*, Nucl. Instrum. Methods A, in press
[6] T. Koike, private communication
[7] M. Fujita, Doctoral thesis, Tohoku University (2019)
[8] K. Tanida *et al.*, *Measurement of X Rays from Ξ^- Atom*, Proposal to J-PARC, E03 (2006)
[9] T. Nagae *et al.*, AIP Conf. Proc. **2130 1**, 020015 (2019)
[10] T. Nagae *et al.*, *Proposal for the next E05 run with the S-2S spectrometer*, Proposal to J-PARC, E70 (2018)
[11] H. Fujioka *et al.*, *Decay Pion Spectroscopy of ${}^{5}_{\Lambda\Lambda}H$ Produced by Ξ-hypernuclear Decay*, Proposal to J-PARC, E75 (2019)
[12] T.O. Yamamoto *et al.*, *Measurement of X rays from Ξ^- C atom with an active fiber target system*, Proposal to J-PARC, E96 (2022)

Ξ-nuclear constraints from Ξ^- emulsion capture events

Eliahu Friedman[1,*] and *Avraham* Gal[1,**]

[1] *Racah Institute of Physics, The Hebrew University, Jerusalem 9190401, Israel*

Abstract. All five KEK and J-PARC two-body $\Xi^- + {}^{A}Z \rightarrow {}^{A'}_{\Lambda}Z' + {}^{A''}_{\Lambda}Z''$ capture events in light emulsion nuclei, including KISO and IBUKI in ^{14}N, are consistent with Coulomb-assisted $1p_{\Xi^-}$ nuclear states. The underlying Ξ-nuclear potential is strongly attractive, with nuclear-matter depth V_{Ξ} larger than 20 MeV. The recent ^{14}N capture events KINKA and IRRAWADDY assigned by J-PARC E07 to $1s_{\Xi^-}$ nuclear states, and implying considerably shallower V_{Ξ}, have also another interpretation as $1p_{\Xi^0}$ nuclear states.

1 Introduction: Ξ^- capture events

The nuclear interactions of Ξ hyperons are poorly known [1, 2]. Because of the large momentum transfer in the standard (K^-, K^+) production reaction, induced by the two-body $K^-p \rightarrow K^+\Xi^-$ strangeness exchange reaction, the Ξ^- hyperons are produced dominantly in the quasi-free continuum region, with less than 1% expected to form Ξ^--nuclear bound states that decay by a $\Xi^-p \rightarrow \Lambda\Lambda$ strong-interaction capture reaction. In fact, no Ξ^- or $\Lambda\Lambda$ hypernuclear bound states have ever been observed unambiguously in such experiments, e.g., KEK-PS E224 [3] and BNL-AGS E885 [4], both on ^{12}C, and BNL-AGS E906 [5] on ^{9}Be. A potential depth of $V_{\Xi} = 17 \pm 6$ MeV was deduced recently from the quasi-free Ξ^- spectrum taken in the ^{9}Be(K^-, K^+) reaction [6].

Here we focus on Ξ nuclear constraints derived by observing Ξ^- capture events in exposures of light-emulsion CNO nuclei to the (K^-, K^+) reaction. A small fraction of the produced high-energy Ξ^- hyperons slows down in the emulsion, undergoing an Auger process to form high-n atomic states, and cascades down radiatively. Strong-interaction capture takes over atomic radiative cascade in a 3D atomic orbit bound by 126, 175, 231 keV in C, N, O, respectively, affected to less than 1 keV by the strong interaction [7]. Interestingly, all two-body Ξ^- capture events $\Xi^- + {}^{A}Z \rightarrow {}^{A'}_{\Lambda}Z' + {}^{A''}_{\Lambda}Z''$ to twin single-Λ hypernuclei reported in KEK and J-PARC light-emulsion K^- exposure experiments [8–12] are consistent with Ξ^- capture from a lower orbit: a Coulomb-assisted $1p_{\Xi^-}$ nuclear state bound by $\sim$1 MeV. These capture events are listed in Table 1. Expecting the final two Λ hyperons in $\Xi^-p \rightarrow \Lambda\Lambda$ capture to form in a spin $S = 0$, $1s^2_{\Lambda}$ configuration, the initial Ξ^- hyperon and the proton on which it is captured must satisfy $l_{\Xi^-} = l_p$ [13], which for p-shell nuclear targets favors the choice $l_{\Xi^-} = 1$. Not listed in the table are multi-body capture events that require for their interpretation undetected capture products, usually neutrons, on top of a pair of single-Λ hypernuclei. Several of these new J-PARC E07 events [12], like KINKA and IRRAWADDY, imply Ξ^- capture from $1s_{\Xi^-}$ nuclear states, with estimated capture rates of order 1% of capture rates from the $1p_{\Xi^-}$ nuclear states considered here [13, 14]. We discuss these states below.

*Eliahu.Friedman@mail.huji.ac.il
**avragal@savion.huji.ac.il

Table 1. Two-body Ξ^- capture emulsion events from KEK and J-PARC experiments.

Experiment	Event	^{A}Z	$^{A'}_{\Lambda}Z' + ^{A''}_{\Lambda}Z''$	B_{Ξ^-} (MeV)
KEK E176 [8]	10-09-06	^{12}C	$^{4}_{\Lambda}$H+$^{9}_{\Lambda}$Be	0.82±0.17
KEK E176 [8]	13-11-14	^{12}C	$^{4}_{\Lambda}$H+$^{9}_{\Lambda}$Be*	0.82±0.14
KEK E176 [8]	14-03-35	^{14}N	$^{3}_{\Lambda}$H+$^{12}_{\Lambda}$B	1.18±0.22
KEK E373 [10]	KISO	^{14}N	$^{5}_{\Lambda}$He+$^{10}_{\Lambda}$Be*	1.03±0.18
J-PARC E07 [11]	IBUKI	^{14}N	$^{5}_{\Lambda}$He+$^{10}_{\Lambda}$Be	1.27±0.21

2 Ξ nuclear optical potential

Ξ^- atomic and nuclear bound states are calculated using a standard $t\rho$ optical-potential form [15]

$$V_{\rm opt}(r) = -\frac{2\pi}{\mu}(1+\frac{A-1}{A}\frac{\mu}{m_N})\,[b_0\,\rho(r) + b_1\,\rho_{\rm exc}(r)], \quad (1)$$

where μ is the Ξ^--nucleus reduced mass and the complex strength parameters b_0 and b_1 are effective, generally density dependent ΞN isoscalar and isovector c.m. scattering amplitudes respectively. The density $\rho = \rho_n + \rho_p$ is a nuclear density distribution normalized to the number of nucleons A and $\rho_{\rm exc} = \rho_n - \rho_p$ is a neutron-excess density with $\rho_n = (N/Z)\rho_p$, implying that $\rho_{\rm exc} = 0$ for the $N = Z$ emulsion nuclei ^{12}C and ^{14}N considered here. A finite-size Coulomb potential V_c, including vacuum-polarization terms is added. For densities we used mostly harmonic-oscillator (HO) densities [16] where the r.m.s. radius of ρ_p was set equal to that of the nuclear charge density [17]. Folding reasonably chosen ΞN interaction ranges other than corresponding to the proton charge radius, or using Modified Harmonic Oscillator densities, or replacing HO densities by realistic three-parameter Fermi density distributions, made little difference: all the calculated binding energies changed by a small fraction, about 0.03 MeV, of the uncertainty imposed by the ±0.15 MeV experimental uncertainty of the 0.82 MeV $1p_{\Xi^-}$ binding energy in ^{12}C listed in Table 1. This holds also for adding a $\rho_{\rm exc} \neq 0$ term induced by considering realistic differences of neutron and proton r.m.s. radii.

Accepting the binding energy interval $B^{1p}_{\Xi^-}$=0.82±0.15 MeV for the two KEK E176 events listed in Table 1, a Ξ-nuclear potential strength of ${\rm Re}\,b_0 = 0.32 \pm 0.01$ fm follows for a fixed value ${\rm Im}\,b_0 =$ 0.01 fm. The sensitivity to variations of ${\rm Im}\,b_0$ is minimal: choosing ${\rm Im}\,b_0 = 0.04$ fm [7] instead of 0.01 fm increases ${\rm Re}\,b_0$ by 0.01 fm to 0.33 ± 0.01 fm. The value ${\rm Re}\,b_0 = 0.32 \pm 0.01$ fm implies in the limit $A \to \infty$ and $\rho(r) \to \rho_0$=0.17 fm^{-3} a depth value $V_\Xi = 24.3 \pm 0.8$ MeV in nuclear matter, compatible with that derived from AGS-E906 in Ref. [6] and in agreement with the range of values 21–24 MeV extracted from old emulsion events [18].

So far we have discussed a density independent t-matrix element b_0 in $V_{\rm opt}$, Eq. (1), to fit the Ξ^- capture events in ^{12}C from Table 1. To explore how robust the deduced Ξ potential-depth value $V_\Xi = 24.3 \pm 0.8$ MeV is, we introduce the next to leading-order density dependence of $V_{\rm opt}$, replacing ${\rm Re}\,b_0$ in Eq. (1) by

$${\rm Re}\,b_0(\rho) = \frac{{\rm Re}\,b_0}{1+\frac{3k_F}{2\pi}{\rm Re}\,b_0^{\rm lab}}, \qquad k_F = (3\pi^2\rho/2)^{\frac{1}{3}}, \quad (2)$$

where k_F is the Fermi momentum corresponding to nuclear density ρ and $b_0^{\rm lab} = (1+\frac{m_{\Xi^-}}{m_N})b_0$ is the lab transformed form of the c.m. scattering amplitude b_0. Eq. (2) accounts for Pauli exclusion correlations in ΞN in-medium multiple scatterings [19, 20]. Shorter-range correlations, disregarded here, were shown in Ref. [21] to contribute less than ~30% of the long-range Pauli correlation term. Applying Eq. (2) in the present context, $B^{1p}_{\Xi^-}(^{12}{\rm C})$=0.82 MeV is refitted by ${\rm Re}\,b_0$=0.527 fm. The nuclear-matter Ξ-nuclear potential depth V_Ξ decreases from 24.3±0.8 to 21.9±0.7 MeV, a decrease of merely 10%.

3 $1p_{\Xi^-}$ states in ^{14}N

Applying Eqs. (1,2) to ^{14}N, with $\mathrm{Re}\, b_0$ fitted to $B^{1p}_{\Xi^-}(^{12}\mathrm{C})$=0.82±0.15 MeV, results in $B^{1p}_{\Xi^-}(^{14}\mathrm{N})$=1.96±0.26 MeV, Pauli correlations included. This binding energy is considerably higher than the value B_{Ξ^-}=1.15±0.20 MeV obtained from the three events assigned in Table 1 to Ξ^- capture in ^{14}N. To resolve this apparent dicrepancy, we note that the calculated $B^{1p}_{\Xi^-}(^{14}\mathrm{N})$ corresponds to a $(2\Lambda+1)$-average of binding energies for a triplet of states $\Lambda^\pi = (0^-, 1^-, 2^-)$ obtained by coupling a $1p_{\Xi^-}$ state to $J^\pi(^{14}\mathrm{N_{g.s.}})=1^+$, as shown in Fig. 1.

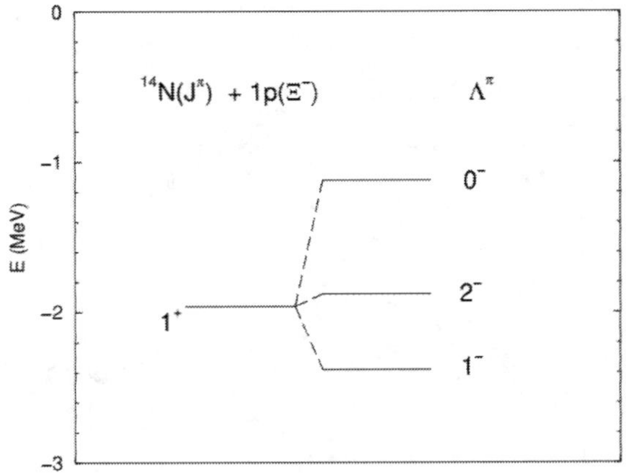

Figure 1. Energies (in MeV) of $\Lambda^\pi = (0^-, 1^-, 2^-)$ triplet of $^{14}\mathrm{N_{g.s.}} + 1p_{\Xi^-}$ states, split by a $Q_N \cdot Q_\Xi$ residual interaction (3). The $(2\Lambda+1)$-averaged energy −1.96 MeV was calculated using the same Pauli-corrected optical potential parameter b_0 that yields a $^{12}\mathrm{C_{g.s.}} + 1p_{\Xi^-}$ state at −0.82 MeV, corresponding to the Ξ^- capture events in ^{12}C listed in Table 1. Figure updating Fig. 2 in Ref. [15].

The energy splittings marked in Fig. 1 follow from a shell-model quadrupole-quadrupole spin independent residual interaction $\mathcal{V}_{\Xi N}$,

$$\mathcal{V}_{\Xi N} = F^{(2)}_{\Xi N} Q_N \cdot Q_\Xi, \quad Q_B = \sqrt{\frac{4\pi}{5}} Y_2(\hat{r}_B), \tag{3}$$

where $F^{(2)}$ is the corresponding Slater integral. A representative value of $F^{(2)}_{\Xi N} = -3$ MeV is used here, smaller than the value $F^{(2)}_{\Lambda N} = -3.7$ MeV established empirically for p-shell Λ hypernuclei [22], in accordance with a ΞN strong interaction somewhat weaker than the ΛN strong interaction. A single $^3D_1$ $^{14}\mathrm{N_{g.s.}}$ component providing a good approximation to the full intermediate-coupling g.s. wavefunction [23] was assumed in the present evaluation.

Fig. 1 shows a triplet of $^{14}\mathrm{N_{g.s.}} + 1p_{\Xi^-}$ levels, spread over more than 1 MeV. The least bound of these states, $\Lambda^\pi = 0^-$, is shifted upward by 0.84 MeV from the $(2\Lambda+1)$ averaged position at -1.96 ± 0.26 MeV to $E(0^-) = -1.12 \pm 0.26$ MeV. This is consistent with the averaged position $\bar{E} = -1.15 \pm 0.20$ MeV of the three Ξ^- $^{14}\mathrm{N_{g.s.}}$ capture events listed in Table 1. The $\Lambda^\pi = 0^-$ state assumes spin-parity $J^\pi = \frac{1}{2}^-$ when Pauli-spin $s_{\Xi^-} = \frac{1}{2}$ is introduced, but its position is unaffected by spin dependent ΞN residual interactions in leading order. We are not aware of any good reason why capture has not been seen from the other two states with $\Lambda^\pi = 1^-, 2^-$. This may change when more events are collected at the next stage of the ongoing J-PARC E07 emulsion experiment.

4 $1s_{\Xi^-}$ states in ^{14}N?

In addition to the Ξ^-_{1p}–^{14}N capture events listed as KISO and IBUKI in Table 1, the J-PARC light-nuclei emulsion experiment E07 reported also two other events KINKA and IRRAWADDY, assigned as Ξ^-_{1s}–^{14}N states, see Fig. 2. We note that $2P \to 1S$ radiative decay rates are of order 1% of $3D \to 2P$ radiative decay rates [13, 14] suggesting that Ξ^- capture from a nuclear Ξ^-_{1s}–^{14}N state is suppressed to this order relative to capture from a nuclear Ξ^-_{1p}–^{14}N state. Assigning a Ξ^-_{1s}–^{14}N bound state to IRRAWADDY, and by default also KINKA which–given its large uncertainty–is not inconsistent with IRRAWADDY, is therefore questionable.

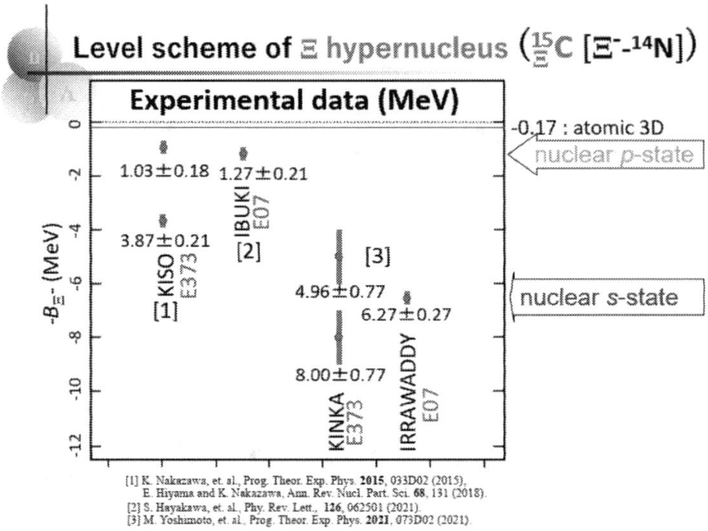

Figure 2. Ξ^-_{1s} and Ξ^-_{1p} nuclear states in ^{14}N, assigned in KEK-E373 and J-PARC E07 emulsion experiments by interpreting Ξ^- capture events that lead to observed twin-Λ hypernuclear decays. Figure provided by Dr. K. Nakazawa, based on recent results from Refs. [10–12].

It has been conjectured by us recently [24] that IRRAWADDY is a near-threshold Ξ^0_{1p}–^{14}C bound state that has nothing to do with a Ξ^-_{1s}–^{14}N bound state suggested by E07. The mixing induced by the ΞN strong interaction between a Ξ^-_{1p}–^{14}N bound state identified with IBUKI and a Ξ^0_{1p}–^{14}C bound state lying about 5 MeV below IBUKU, within the J-PARC E07 experimental uncertainty of IRRAWADDY, is sufficiently strong to make the $E1$ radiative deexcitation of the Ξ^-_{3D}–^{14}N atomic state populate equally well *both* Ξ^-_{1p}–^{14}N and Ξ^0_{1p}–^{14}C nuclear states.

5 Discussion

Two Ξ-nuclear scenarios are listed in Table 2. In the first one, two KEK-E176 ^{12}C events [8], with $B^{1p}_{\Xi^-} = 0.82 \pm 0.15$ MeV, serve as input for setting up the strength of the Ξ-nuclear optical potential. Other Ξ^- binding energies are then predicted, as listed in the first row of Table 2. We note that a value $V_\Xi \gtrsim 20$ MeV implies a substantially stronger in-medium ΞN attraction than reported by some recent free-space model evaluations (HAL-QCD [25], EFT@NLO [26, 27] and RMF [28]), all of which satisfy $V_\Xi \lesssim 10$ MeV. A notable exception is provided by versions ESC16*(A,B) of the latest

Table 2. Ξ^-–^{12}C and Ξ^-–^{14}N binding energies in $1s$ and $1p$ states, $B^{1s}_{\Xi^-}$ and $B^{1p}_{\Xi^-}$, plus Ξ nuclear potential depths $V_\Xi(\rho_0)$ at nuclear-matter density $\rho_0 = 0.17$ fm^{-3} calculated using a density-dependent optical potential Eqs. (1,2) with Re b_0 fitted to binding energies underlined for each input choice. All entries are in MeV.

Input	$B^{1s}_{\Xi^-}(^{12}\mathrm{C})$	$B^{1p}_{\Xi^-}(^{12}\mathrm{C})$	$B^{1s}_{\Xi^-}(^{14}\mathrm{N})$	$B^{1p}_{\Xi^-}(^{14}\mathrm{N})$	$V_\Xi(\rho_0)$
KEK E176 [8]	9.82	<u>0.82</u>	11.78	1.96	21.9
J-PARC E07 [12]	4.94	0.31	<u>6.27</u>	0.50	13.8

Nijmegen extended-soft-core ΞN interaction model [29], in which values of V_Ξ higher than 20 MeV are derived. However, these large values are reduced substantially by ΞNN three-body contributions within the same ESC16* model.

Choosing instead the J-PARC E07 ^{14}N IRRAWADDY event, with $B^{1s}_{\Xi^-} = 6.27 \pm 0.27$ MeV [12] as input, gives rise to different predictions as listed in the second row of the table. The difference between the two sets of predictions is striking, particularly for the Ξ^-_{1s} binding energies. This large difference is reflected also in the Ξ-nuclear potential depths at nuclear matter density, $V_\Xi(\rho_0)$, listed in the last column of the table. Equally interesting is the difference between the two sets with regard to Ξ^-_{1p} bound states. In particular, the Ξ^-_{1p}–^{12}C binding energy constrained by $B^{1s}_{\Xi^-}(^{14}\mathrm{N})=6.27\pm0.27$ MeV comes out in the second row exceedingly small, substantially disagreeing with that determined from the two KEK E176 capture events [8] underlined in the first row. As for the calculated Ξ^-_{1p}–^{14}N binding energies, since $J(^{14}\mathrm{N}) \neq 0$, see Fig. 1, these cannot be compared *directly* with IBUKI's binding energy of 1.27±0.21 MeV from Fig. 2.

Recent Skyrme-Hartree-Fock (SHF) calculations [30] presented global fits to comprehensive Ξ-nuclear data, including $B^{1s}_{\Xi^-}$=8.00 MeV for KINKA and $B^{1p}_{\Xi^-}$=1.13 MeV for the mean of KISO and IBUKI. Apart from small nonlocal potential terms and effective mass corrections, the SHF Ξ-nuclear mean-field potential $V_\Xi(\rho_N)$ consists of two terms: $V^{(2)}_\Xi(\rho_N) \propto \rho_N$ & $V^{(3)}_\Xi(\rho_N) \propto \rho_N^2$. Large-scale SHF fits of the corresponding Ξ potential depths, and their sum, are listed in the first row of Table 3. Listed in the lower rows are Ξ potential depths obtained in the optical potential methodology [31] when fitting just the two Ξ^- states in ^{14}N as used in the SHF calculations [30]. Good agreement is observed between the two methods, with $V_\Xi(\rho_0) \approx 14$ MeV, similar to the depth value listed in the second row of Table 2 using IRRAWADDY alone. We note that the Ξ^-_{1p} state in ^{12}C (see Table 2) comes out unbound in the SHF calculations unless ^{12}C is made artificially deformed [30].

Table 3. Ξ-nuclear potential depths (in MeV) from a large-scale SHF fit [30] and from our $V_{\rm opt}$ two-parameter fits to just $B^{1s}_{\Xi^-}$=8.00 MeV (KINKA) and $B^{1p}_{\Xi^-}$=1.13 MeV (KISO,IBUKI) in ^{14}N. 'Pauli' refers to Eq. (2).

Method	Pauli	$V^{(2)}_\Xi(\rho_0)$	$V^{(3)}_\Xi(\rho_0)$	$V_\Xi(\rho_0)$
SHF	No	34.1	−20.4	13.7
$V_{\rm opt}$	No	27.5	−12.6	14.9
$V_{\rm opt}$	Yes	24.6	−11.0	13.6

The solution proposed here to the difficulty of interpreting IRRAWADDY as a Ξ^-_{1s} bound state in ^{14}N is by pointing out that it could correspond to a Ξ^0_{1p}–^{14}C bound state, something that cannot occur kinematically in the other light-emulsion nuclei ^{12}C and ^{16}O. Given that in this nuclear mass range capture rates from $1s_{\Xi^-}$ states are estimated to be two orders of magnitude below capture rates from $1p_{\Xi^-}$ states [13, 14], this Ξ^0_{1p}–^{14}C new assignment addresses satisfactorily the capture rate hierarchy.

Acknowledgments

One of us (A.G.) thanks Jiří Mareš and other members of the HYP2022 organizing team for their generous hospitality during the Conference. This work is part of a project funded by the EU Horizon 2020 Research & Innovation Programme under grant agreement 824093.

References

[1] A. Gal, E.V. Hungerford, D.J. Millener, Rev. Mod. Phys. **88**, 035004 (2016)
[2] E. Hiyama, K. Nakazawa, Annu. Rev. Nucl. Part. Sci. **68**, 131 (2018)
[3] T. Fukuda *et al.* (E224 Collaboration), Phys. Rev. C **58**, 1306 (1998)
[4] P. Khaustov *et al.* (AGS E885 Collaboration), Phys. Rev. C **61** 054603 (2000)
[5] J.K. Ahn *et al.* (AGS E906 Collaboration), Phys. Rev. Lett. **87**, 132504 (2001)
[6] T. Harada, Y. Hirabayashi, Phys. Rev. C **103**, 024605 (2021)
[7] C.J. Batty, E. Friedman, A. Gal. Phys. Rev. C **59**, 295 (1999)
[8] S. Aoki *et al.* (KEK E176 Collaboration), Nucl. Phys. A **828**, 191 (2009), and earlier E176 publications cited therein
[9] J.K. Ahn *et al.* (KEK-PS E373 Collaboration), Phys. Rev. C **88**, 014003 (2013)
[10] H. Nakazawa *et al.* (KEK PS-E373 Collaboration), Prog. Theor. Exp. Phys. **2015**, 033D02 (2015)
[11] S.H. Hayakawa *et al.* (J-PARC E07 Collaboration), Phys. Rev. Lett. **126**, 062501 (2021)
[12] M. Yoshimoto *et al.* (J-PARC E07 Collaboration), Prog. Theor. Exp. Phys. **2021**, 073D02 (2021)
[13] D. Zhu, C.B. Dover, A. Gal, M. May, Phys. Rev. Lett. **67**, 2268 (1991)
[14] T. Koike, JPS Conf. Proc. **17**, 033011 (2017)
[15] E. Friedman, A. Gal. Phys. Lett. B **820**, 136555 (2021)
[16] L.R.B. Elton, *Nuclear Sizes*, (Oxford Univ. Press, Oxford, 1961)
[17] I. Angeli, K.P. Marinova, At. Data Nucl. Data Tables **99**, 69 (2013)
[18] C.B. Dover, A. Gal. Ann. Phys. (NY) **146**, 309 (1983).
[19] C.B. Dover, J. Hüfner, R.H. Lemmer, Ann. Phys. (NY) **66**, 248 (1971)
[20] T. Waas, M. Rho, W. Weise, Nucl. Phys. A **617**, 449 (1997)
[21] W. Weise, R. Härtle, Nucl. Phys. A **804**, 173 (2008)
[22] R.H. Dalitz, A, Gal, Ann. Phys. (NY) **131**, 314 (1981)
[23] D.J. Millener, in *Topics in Strangeness Nuclear Physics*, Edited by P. Bydžovský, A. Gal, and J. Mareš, Lecture Notes in Physics **724**, 31-79 (Springer, Heidelberg, 2007)
[24] E. Friedman, A. Gal, *Has J-PARC E07 observed a Ξ^-_{1s} nuclear state?*, accepted to Phys. Lett. B, arXiv:2209.01606
[25] T. Inoue (for HAL QCD Collaboration), AIP Conf. Proc. **2130**, 020002 (2019)
[26] J. Haidenbauer, U.-G. Meißner, Eur. Phys. J. A **55**, 23 (2019)
[27] M. Kohno, Phys. Rev. C **100**, 024313 (2019)
[28] T. Gaitanos, A. Chorozidou, Nucl. Phys. A **1008**, 122153 (2021)
[29] M.M. Nagels, Th.A. Rijken, Y. Yamamoto, Phys. Rev. C **102**, 054003 (2020); see in particular Tables XXI, XXIV for models ESC16*(A,B) respectively
[30] J. Guo, X.-R. Zhou, H.-J. Schulze, Phys. Rev. C **104**, L061307 (2021)
[31] E. Friedman, A. Gal, *Constraints from Λ hypernuclei on the ΛNN content of $V_\Lambda(\rho_0)$*, arXiv:2204.02264

EPJ Web of Conferences **271**, 03003 (2022) https://doi.org/10.1051/epjconf/202227103003
HYP2022

A Skyrme force for Ξ^- hypernuclei

Hans-Josef Schulze[1], *Jing* Guo[2], and *Xian-Rong* Zhou[2]

[1]INFN Sezione di Catania, Universitá di Catania, Via Santa Sofia 64, I-95123 Catania, Italy
[2]Department of Physics, East China Normal University, Shanghai 200241, China

Abstract. Experimental data for the cascade hypernuclei ${}^{15}_{\Xi}$C, ${}^{12}_{\Xi}$Be, and ${}^{13}_{\Xi}$B are analyzed within a Skyrme-Hartree-Fock theoretical approach and optimal Skyrme parameters are determined. The important role of deformation for ${}^{13}_{\Xi}$B is pointed out. Predictions for ${}^{7}_{\Xi}$H and ${}^{10}_{\Xi}$Li are made.

1 Introduction

While the properties of Λ hypernuclei and the ΛN interaction are fairly well known [1–3], only a few data regarding the heavier Ξ hyperons are available that are furthermore often hampered by ambiguities in their identification and interpretation [4–6]. Recently, new data on Ξ^- hypernuclei have become available at the J-PARC and KEK experimental facilities [7]. In this work we try to determine an optimal set of ΞN interaction parameters within the Skyrme-Hartree-Fock (SHF) theoretical approach, capable to describe consistently most data [8].

We first briefly review the available data for the Ξ hypernuclei ${}^{15}_{\Xi}$C (^{14}N+Ξ^-), ${}^{13}_{\Xi}$B (^{12}C+Ξ^-), and ${}^{12}_{\Xi}$Be (^{11}B+Ξ^-), which are mostly nuclear emulsion events occurring after a $K^-p \to K^+\Xi^-$ reaction. Regarding ${}^{15}_{\Xi}$C, analysis of the KEK-E373-T2 (KISO) event [6, 9] provided clear evidence that this is a (Coulomb-assisted) strongly-bound hypernucleus, produced in the reaction $\Xi^-+{}^{14}\mathrm{N} \to {}^{15}_{\Xi}\mathrm{C} \to {}^{10}_{\Lambda}\mathrm{Be} + {}^{5}_{\Lambda}\mathrm{He}$. It is interpreted [6, 10–12] as production of a Ξ^- $1p$ state with a removal energy $B_\Xi = 1.03 \pm 0.18$ MeV, leaving ${}^{10}_{\Lambda}$Be in an excited state after its decay. Compatible results were obtained by the J-PARC-E07-T006 (IBUKI) event, $B_\Xi = 1.27 \pm 0.21$ MeV [13], and also KEK-E176 #14-03-35, $B_\Xi = 1.18 \pm 0.22$ MeV [14]. The combined analysis of KISO and IBUKI yields $B_\Xi = 1.13 \pm 0.14$ MeV [13].

Recently larger B_Ξ values interpreted as possible Ξ^- $1s$ bound states [11] have been extracted for the J-PARC-E07-T010 (IRRAWADDY), $B_\Xi = 6.27 \pm 0.27$ MeV, and the KEK-E373-T3 (KINKA), $B_\Xi = 8.00 \pm 0.77$ MeV or $B_\Xi = 4.96 \pm 0.77$ MeV, events. As the error bars are non overlapping, at most one of these values may correspond to the ground state energy. We demonstrate in the following that the KINKA (8.00 MeV) result yields the best consistency with the other data.

Regarding the ${}^{12}_{\Xi}$Be hypernucleus, in the BNL-E885 counter experiment [15] the cross section for the ${}^{12}\mathrm{C}(K^-, K^+){}^{12}_{\Xi}\mathrm{Be}$ reaction was interpreted by assuming a Ξ^--nucleus Woods-Saxon (WS) potential with a depth of about 14 MeV. The result is consistent with the estimate $V_{\mathrm{WS}} < 20$ MeV of the preceding KEK-E224 [16], and $V_{\mathrm{WS}} = 17 \pm 6$ MeV in ${}^{9}_{\Xi}$He (^{8}Li+Ξ^-) of BNL-E906 [17], while the sequel J-PARC-E05 experiment on ${}^{12}_{\Xi}$Be remained so far inconclusive, reporting a preliminary possible one-peak interpretation with $B_\Xi \approx 6.3$ MeV [18].

Finally, there are several older KEK-E176 emulsion data [14, 19] that can be understood as possible formation of ${}^{13}_{\Xi}$B, but all of them are ambiguous. In particular, the events KEK-E176 #10-9-6, $B_\Xi = 0.82 \pm 0.17$ MeV, and KEK-E176 #13-11-14, $B_\Xi = 0.82 \pm 0.14$ MeV, might both be interpreted as ${}^{13}_{\Xi}$B ($1p$) states [19–21].

2 Formalism

In the present work, we study Ξ hypernuclei with $A = 12, 13$, and 15 and focus on the Ξ^- $1s$ and $1p$ states and their hyperon separation energy,

$$B_\Xi \equiv E([n, p, -]) - E([n, p, \Xi^-]) = E[^{A-1}(Z+1)] - E[^A_\Xi Z] \,. \tag{1}$$

We employ a model based on the self-consistent SHF method [22, 23], first extended to the theoretical description of Λ hypernuclei in Ref. [24], and now used for Ξ^- hyperons here. The fundamental SHF local energy-density functional of hypernuclear matter is written as

$$\varepsilon_{\text{SHF}} = \varepsilon_N + \varepsilon_Y \,, \tag{2}$$

and depends on the one-body densities ρ_q, kinetic densities τ_q, and spin-orbit currents $\boldsymbol{J}_q$,

$$\left[\rho_q,\ \tau_q,\ \boldsymbol{J}_q\right] = \sum_{i=1}^{N_q} n_q^i \left[|\phi_q^i|^2,\ |\boldsymbol{\nabla}\phi_q^i|^2,\ \phi_q^{i\,*}(\boldsymbol{\nabla}\phi_q^i \times \boldsymbol{\sigma})/i\right], \tag{3}$$

where ϕ_q^i $(i = 1, N_q)$ are the self-consistently calculated single-particle (s.p.) wave functions of the N_q occupied states for the species $q = n, p, Y$ in a hypernucleus.

The functional ε_N is the usual nucleonic part [22, 23] and a possible standard parametrization for the hyperonic part is [10, 20, 24–26]

$$\begin{aligned}\varepsilon_Y = \frac{\tau_Y}{2m_Y} &+ a_0\rho_Y\rho_N + a_3\rho_Y\rho_N^\alpha + a_1(\rho_Y\tau_N + \rho_N\tau_Y) \\ &- a_2(\rho_Y\Delta\rho_N + \rho_N\Delta\rho_Y)/2 - a_4(\rho_Y\boldsymbol{\nabla}\cdot\boldsymbol{J}_N + \rho_N\boldsymbol{\nabla}\cdot\boldsymbol{J}_Y)\,,\end{aligned} \tag{4}$$

from which one obtains the corresponding hyperonic SHF mean fields

$$V_Y = a_0\rho_N + a_1\tau_N - a_2\Delta\rho_N - a_4\boldsymbol{\nabla}\cdot\boldsymbol{J}_N + a_3\rho_N^\alpha \,, \tag{5}$$

$$V_N^{(Y)} = a_0\rho_Y + a_1\tau_Y - a_2\Delta\rho_Y - a_4\boldsymbol{\nabla}\cdot\boldsymbol{J}_Y + a_3\alpha\rho_Y\rho_N^{\alpha-1} \,, \tag{6}$$

and a hyperon effective mass

$$\frac{1}{2m_Y^*} = \frac{1}{2m_Y} + a_1\rho_N \,, \tag{7}$$

which appear in the SHF Schrödinger equation

$$\left[\boldsymbol{\nabla}\cdot\frac{1}{2m_q^*(\boldsymbol{r})}\boldsymbol{\nabla} - V_q(\boldsymbol{r}) - e_q V_C(\boldsymbol{r}) + i\boldsymbol{W}_q(\boldsymbol{r})\cdot(\boldsymbol{\nabla}\times\boldsymbol{\sigma})\right]\phi_q^i(\boldsymbol{r}) = e_q^i\,\phi_q^i(\boldsymbol{r}) \,, \tag{8}$$

where V_C is the Coulomb field and $\boldsymbol{W}_N$ the nucleonic spin-orbit mean-field [23]. The 'three-body' parameter α is kept here to its standard value of 2, but also an alternative value of $7/6$, used in several NN Skyrme forces, is evaluated.

Contrary to Λ hypernuclei, the Coulomb interaction is very important for Ξ^- hypernuclei discussed here and provides a substantial part of the Ξ^- binding. An approximate c.m. correction is applied as usual [22, 23, 27] by replacing the bare masses: $1/m_q \to 1/m_q - 1/M$, where $M = (N_n + N_p)m_N + N_Y m_Y$ is the total mass of the (hyper)nucleus. Solving the Schrödinger equation provides the wave functions $\phi_q^i(\boldsymbol{r})$ and the s.p. energies $-e_q^i$ for the different s.p. levels i and species q. We use in this work the standard nucleonic Skyrme force SLy4 [28], but the results for hyperonic observables hardly depend on that choice [29].

It should be noted that the Ξ^- hypernuclei decay into double-Λ hypernuclei by the ΞN–$\Lambda\Lambda$ coupling [30]. Therefore, the ΞN interaction should have an imaginary part to represent

Table 1. The removal energies B_Ξ (in MeV) for the Ξ^- $1s$ and $1p$ states of ${}^{13}_{\Xi}$B, ${}^{12}_{\Xi}$Be, ${}^{10}_{\Xi}$Li, and ${}^{7}_{\Xi}$H, obtained with ΞN Skyrme forces of different parameters α, a_0[MeV fm^3], a_2[MeV fm^5], a_3[MeV fm$^{3\alpha}$] and $a_1 = a_4 = 0$. With these parameters, the Ξ^- $1s$ and $1p$ removal energies of ${}^{15}_{\Xi}$C are fixed to 8.00 and 1.13 MeV, respectively. ${}^{13}_{\Xi}$B values in brackets are for spherical calculations.

α	a_2	a_0	a_3	${}^{13}_{\Xi_s}$B	${}^{13}_{\Xi_p}$B	${}^{12}_{\Xi_s}$Be	${}^{10}_{\Xi_s}$Li	${}^{7}_{\Xi_s}$H
	0	-200.3	704.8	6.62 (6.71)	0.86 (-0.12)	5.83	4.42	1.59
	10	-198.7	641.3	6.49 (6.65)	0.75 (-0.11)	5.72	4.23	1.31
2	20	-196.9	576.8	6.37 (6.59)	0.65 (-0.09)	5.61	4.04	1.05
	30	-194.7	509.3	6.25 (6.53)	0.56 (-0.08)	5.51	3.86	0.80
	40	-192.5	443.0	6.14 (6.46)	0.47 (-0.07)	5.40	3.67	0.56
	0	-498.6	551.3	6.44 (6.84)	0.84 (-0.07)	5.87	4.37	1.52
	10	-470.8	503.0	6.32 (6.76)	0.73 (-0.06)	5.76	4.17	1.25
7/6	20	-439.4	449.3	6.21 (6.69)	0.63 (-0.06)	5.64	3.98	1.00
	30	-409.0	397.3	6.10 (6.61)	0.54 (-0.05)	5.53	3.79	0.76
	40	-378.9	345.5	6.02 (6.54)	0.47 (-0.04)	5.42	3.62	0.54

the decay width. However, since we have so far no useful experimental information on this coupling by Refs. [9, 11, 13–16], here the imaginary part is omitted.

While the core nuclei ^{14}N and ^{11}B are (nearly) spherical, ^{12}C is an axially-deformed oblate nucleus [20, 31–36]. In our approach, the deformed SHF Schrödinger equation is solved in cylindrical coordinates (r, z) within the axially-deformed harmonic-oscillator basis [22, 23]. The geometric quadrupole deformation parameter of the nuclear core is expressed as

$$\beta_2 \equiv \sqrt{\frac{\pi}{5}} \frac{\langle 2z^2 - r^2 \rangle}{\langle r^2 + z^2 \rangle} . \tag{9}$$

As there are currently not enough data to determine uniquely the five ΞN interaction parameters a_i, we proceed as follows. The effective-mass parameter a_1 is kept zero, since recent Brueckner-Hartree-Fock calculations [37] indicate a rather flat Ξ^- s.p. spectrum and thus m^*_Ξ/m_Ξ close to unity. Also the spin-orbit parameter is disregarded, $a_4 = 0$, as the inclusion of Ξ^- spin-orbit splitting is clearly premature. For the same reason we do not introduce further parameters for the isospin dependence of the interaction at this stage.

The surface-energy parameter a_2 is essential to determine the *shape* of the Ξ^- mean field V_Ξ in the hypernucleus, Eq. (5), which is important for comparison with the WS mean field that was used in the experimental analysis of BNL-E885 for ${}^{12}_{\Xi}$Be, Refs. [9, 15],

$$V_{\mathrm{WS}}(r) = \frac{-V_0}{1 + \exp[(r - 2.52\,\mathrm{fm})/0.65\,\mathrm{fm}]} \tag{10}$$

with $V_0 = 12, 14, 16, 18, 20$ MeV and $V_0 \approx 14$ MeV identified as the preferred value. Motivated by the equivalent parameter value ($a_2^{\Lambda N} \approx 20\,\mathrm{MeV\,fm}^5$) of the recently derived SLL4 ΛN Skyrme force [26], we use several trial values of this parameter, $a_2 = 0, 10, \ldots, 40\,\mathrm{MeV\,fm}^5$.

For fixed $a_1 = a_4 = 0$ and chosen a_2, the remaining volume parameters a_0 and a_3 are then determined by fitting the removal energies $B_\Xi = 8.00$ and 1.13 MeV for the Ξ^- $1s$ and $1p$ states in ${}^{15}_{\Xi}$C, respectively, as claimed for the KINKA and KISO+IBUKI events. The resulting three-parameter force is termed SLX3.

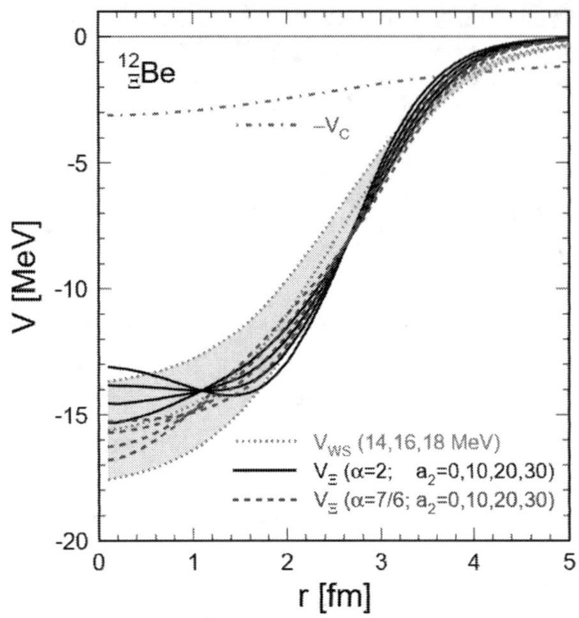

Figure 1. Local Ξ^- mean field V_Ξ and Coulomb field $-V_C$ in $^{12}_{\Xi}$Be, obtained with the SHF models with various forces in Table 1. Increasing values of a_2 produce deeper well depths $V_\Xi(0)$. The Woods-Saxon mean field deduced in [15] is shown for comparison.

3 Results and discussion

Table 1 lists the parameter values $a_{0,2,3}$ obtained, together with the Ξ^- $1s$ and $1p$ removal energies of $^{13}_{\Xi}$B, $^{12}_{\Xi}$Be, $^{10}_{\Xi}$Li, and $^{7}_{\Xi}$H that are predicted. Strongly-bound $1p$ states with positive removal energies are only found for deformed $^{13}_{\Xi}$B nuclei, but not in spherical approximation (numbers in brackets). $^{10}_{\Xi}$Li and $^{7}_{\Xi}$H are bound for all versions of the ΞN force, although these nuclei are probably too light systems to be calculated reliably within the SHF approach. One obtains reasonable values for the parameters a_0 and a_3. For comparison, in the SLL4 ΛN Skyrme force [25, 26], the equivalent optimal parameters are $a_0^{\Lambda N} \approx 320\,\mathrm{MeV\,fm}^3$ and $a_3^{\Lambda N} \approx 700\,\mathrm{MeV\,fm}^6$. For increasing surface parameter a_2, more Ξ^- binding is provided by the associated terms in Eqs. (4,5), and therefore both a_0 and a_3 decrease in magnitude.

In Fig. 1, the different SHF mean field potentials in the $^{12}_{\Xi}$Be hypernucleus are plotted, including the local Coulomb field V_C, the strong mean field V_Ξ, Eq. (5), and for comparison the WS mean field Eq. (10) with $V_0 = 14, 16, 18$ MeV, as used in the analysis of BNL-E885 [9, 15]. One observes that the SHF results cover the range between the 14 and 18 MeV curves. We consider this as a very good agreement with the analysis of E885. It can clearly be seen that the zero-momentum value $V_\Xi(0)$ alone is not a useful unique indicator of the interaction strength, as the shape of the mean field controlled by the parameter a_2 plays an essential role. A value of about $a_2 \approx 20 - 30$ MeV fm^5 gives the closest correspondence to a WS shape, whereas $a_2 = 0$ generates a flat or well shape in the core region.

The hypernuclei $^{15}_{\Xi}$C and $^{12}_{\Xi}$Be and their core nuclei discussed so far are spherical nuclei. The case of $^{13}_{\Xi}$B is more delicate, as the core nucleus ^{12}C is axially deformed in the SHF approach [20, 31–36]. Its deformation derived from the proton quadrupole moment Q_p is rather large [38],

$$\beta \equiv \frac{\sqrt{5\pi}}{3}\frac{Q_p}{ZR_0^2} \approx -0.58 \pm 0.03 \tag{11}$$

with $R_0 \equiv 1.2\,A^{1/3}$ fm. In the SHF approach the size of nuclear deformation can be controlled by adjusting the spin-orbit parameter of the NN Skyrme force [20, 31–36], and we follow this procedure to reproduce the proper β value. As discussed in detail in [20], the oblate deformation favors the binding of a Ξ^- $1p$ orbital because of the improved geometrical overlap of wavefunction and embedding potential, such that this state becomes more bound than in spherical approximation. The results of Table 1 for ${}^{13}_{\Xi}$B are obtained in this way and demonstrate that the proposed interpretation of the KEK-E176 events as ${}^{13}_{\Xi p}$B states might be solely due to the fact that ^{12}C is a strongly deformed nucleus. Excellent agreement with the experimental value $B_\Xi \approx 0.82$ MeV is obtained with small values of a_2 and both choices of α.

4 Summary

We have tried to fit all current data for the cascade hypernuclei ${}^{15}_{\Xi}$C, ${}^{12}_{\Xi}$Be, and ${}^{13}_{\Xi}$B within a global SHF approach for the ΞN interaction. Our main conclusions are:

1) Of all proposed $B_\Xi({}^{15}_{\Xi s}\text{C})$ values, the KINKA (8.00 MeV) interpretation seems to be the one most compatible with both the ${}^{12}_{\Xi}$Be and ${}^{13}_{\Xi}$B data, although slightly lower values down to about 7 MeV might also be possible and consistent with ${}^{12}_{\Xi}$Be, but would not allow interpretation of the KEK-176 events as ${}^{13}_{\Xi p}$B states.

2) Combining KINKA (${}^{15}_{\Xi s}$C) and KISO+IBUKI (${}^{15}_{\Xi p}$C) data, the predicted Ξ^- mean field in ${}^{12}_{\Xi}$Be is very similar to the best-choice WS mean field for BNL-E885 with a depth of about 14–16 MeV [15], and also compatible with KEK-E224 [16] and BNL-E906 [17].

3) ${}^{13}_{\Xi}$B is a delicate case, as in the current model this hypernucleus and its core nucleus ^{12}C are strongly axially deformed, and the claimed $B_\Xi \approx 0.8$ MeV values of some KEK-E176 emulsion data might be explained as a consequence of this deformation, which energetically favors the extended Ξ^- $1p$ orbit. Treating these nuclei as undeformed does not produce a sufficiently bound Ξ^- $1p$ state (which is actually another possible interpretation of the data).

Furthermore, ${}^{7}_{\Xi}$H is predicted as a bound hypernucleus. The proposed Skyrme force SLX3 with moderate values of the parameter a_2 is thus able to fit satisfactorily all current data, although due to their limited accuracies and ambiguities a firm statement cannot yet be made. Hopefully a large number of emulsion events obtained in KEK-E373 and J-PARC-E07 will be analyzed soon for confrontation. Future improvements of the approach include constraining better the SHF ΞN interaction parameters, going beyond the mean-field treatment, and including the imaginary parts due to the ΞN–$\Lambda\Lambda$ decay.

Acknowledgments

This contribution was supported by the National Science Foundation of China under contract No. 11775081, the Natural Science Foundation of Shanghai under contract No. 17ZR1408900, the Helmholtz Institute Mainz, and the European Union's Horizon 2020 research and innovation programme under grant agreement No. 824093.

References

[1] O. Hashimoto and H. Tamura, Prog. Part. Nucl. Phys. **57**, 564 (2006)
[2] A. Feliciello and T. Nagae, Rep. Prog. Phys. **78**, 096301 (2015)
[3] A. Gal, E. V. Hungerford, and D. J. Millener, Rev. Mod. Phys. **88**, 035004 (2016)
[4] B. Dover and A. Gal, Annals of Physics **146**, 309 (1983)
[5] Y. Yamamoto, T. Motoba, T. Fukuda, M. Takahashi, and K. Ikeda, Prog. Theor. Phys. Suppl. **117**, 281 (1994)
[6] E. Hiyama and K. Nakazawa, Ann. Rev. Nucl. Part. Sci. **68**, 131 (2018)
[7] H. Ohnishi, F. Sakuma, and T. Takahashi, Prog. Part. Nucl. Phys. **113**, 103773 (2020)
[8] J. Guo, X.-R. Zhou, and H.-J. Schulze, Phys. Rev. **C104**, L061307 (2021)

[9] K. Nakazawa et al., Prog. Theor. Exp. Phys. 033D02 (2015)
[10] T. T. Sun, E. Hiyama, H. Sagawa, H.-J. Schulze, and J. Meng, Phys. Rev. **C94**, 064319 (2016)
[11] M. Yoshimoto et al., PTEP 073D02 (2021)
[12] Y. Tanimura, H. Sagawa, T.-T. Sun, and E. Hiyama, Phys. Rev. **C105**, 044324 (2022)
[13] S. H. Hayakawa et al. (J-PARC E07 Collaboration), Phys. Rev. Lett. **126**, 062501 (2021)
[14] S. Aoki et al. (KEK E176 Collaboration), Prog. Theor. Phys. **89**, 493 (1993); Phys. Lett. **B355**, 45 (1995); Nucl. Phys. **A828**, 191 (2009)
[15] P. Khaustov et al. (AGS E885 Collaboration), Phys. Rev. **C61**, 054603 (2000)
[16] T. Fukuda et al. (E224 Collaboration), Phys. Rev. **C58**, 1306 (1998)
[17] T. Tamagawa et al. (BNL-E906 Collaboration), Nucl. Phys. **A691**, 234c (2001); T. Harada and Y. Hirabayashi, Phys. Rev. **C103**, 024605 (2021)
[18] T. Nagae et al. (J-PARC E05 collaboration), AIP Conf. Proc. **2130**, 020015 (2019)
[19] Y. Yamamoto and T. Motoba, Nucl. Phys. **A585**, 371c (1995); Y. Yamamoto, Few Body Syst. Suppl. **9**, 145 (1995); M. Yamaguchi, K. Tominaga, Y. Yamamoto, and T. Ueda, Prog. Theor. Phys. **105**, 627 (2001)
[20] Y. Jin, X.-R. Zhou, Y.-Y. Cheng, and H.-J. Schulze, Eur. Phys. J. **A56**, 135 (2020)
[21] E. Friedman and A. Gal, Phys. Lett. **B820**, 136555 (2021)
[22] D. Vautherin and D. M. Brink, Phys. Rev. **C5**, 626 (1972); D. Vautherin, Phys. Rev. **C7**, 296 (1973)
[23] M. Bender, P.-H. Heenen, and P.-G. Reinhard, Rev. Mod. Phys. **75**, 121 (2003); J. R. Stone and P.-G. Reinhard, Prog. Part. Nucl. Phys. **58**, 587 (2007); J. Erler, P. Klüpfel, and P.-G. Reinhard, J. Phys. **G38**, 033101 (2011)
[24] M. Rayet, Ann. Phys. (N.Y.) **102**, 226 (1976); Nucl. Phys. **A367**, 381 (1981)
[25] H.-J. Schulze and E. Hiyama, Phys. Rev. **C90**, 047301 (2014)
[26] H.-J. Schulze, AIP Conf. Proc. **2130**, 020009 (2019)
[27] M. Bender, K. Rutz, P.-G. Reinhard, and J. A. Maruhn, Eur. Phys. J. **A7**, 467 (2000)
[28] E. Chabanat, P. Bonche, P. Haensel, J. Meyer, and R. Schaeffer, Nucl. Phys. **A627**, 710 (1997); **A635**, 231 (1998); **A643**, 441 (1998)
[29] Y. Zhang, H. Sagawa, and E. Hiyama, Phys. Rev. **C103**, 034321 (2021)
[30] Y. Yamamoto, M. Wakai, T. Motoba, and T. Fukuda, Nucl. Phys. **A625**, 107 (1997)
[31] T. Suzuki, H. Sagawa, and K. Hagino, Phys. Rev. **C68**, 014317 (2003)
[32] H. Sagawa, X.-R. Zhou, X.-Z. Zhang, and T. Suzuki, Phys. Rev **C70**, 054316 (2004)
[33] H.-J. Schulze, M. Thi Win, K. Hagino, and H. Sagawa, Prog. Theor. Phys. **123**, 569 (2010)
[34] J.-W. Cui, X.-R. Zhou, and H.-J. Schulze, Phys. Rev **C91**, 054306 (2015)
[35] J.-W. Cui, X.-R. Zhou, L.-X. Guo, and H.-J. Schulze, Phys. Rev. **C95**, 024323 (2017)
[36] J.-W. Cui and X.-R. Zhou, Prog. Theor. Exp. Phys. 093D04 (2017)
[37] H.-J. Schulze and T. Rijken, Phys. Rev. **C88**, 024322 (2013); Th.A. Rijken and H.-J. Schulze, Eur. Phys. J. **A52**, 21 (2016)
[38] N. J. Stone, Atomic Data and Nuclear Data Tables **90**, 75 (2005); IAEA Nuclear Data Services, Reference Input Parameter Library, http://www-nds.iaea.org/RIPL-2/

Production of light Ξ-hypernuclei, ${}^{7}_{\Xi}$H

Hiroyuki Fujioka[1,*] for the J-PARC E75 Collaboration

[1]Department of Physics, Tokyo Institute of Technology, Tokyo, 152-8551 Japan

Abstract. A light Ξ hypernucleus, ${}^{7}_{\Xi}$H, is expected to be bound. Furthermore, it is pointed out that ${}^{7}_{\Xi}$H will decay into ${}^{\;5}_{\Lambda\Lambda}$H with a large branching ratio. We plan to study the ^{7}Li(K^-, K^+) reaction to investigate the ${}^{7}_{\Xi}$H system and to evaluate the experimental feasibility to perform decay pion spectroscopy of ${}^{\;5}_{\Lambda\Lambda}$H in future.

1 Introduction

Experimental studies of double strange hypernuclei — Ξ hypernuclei and double-Λ hypernuclei — are one of important subjects in J-PARC. In particular, with the advent of a high-resolution magnetic spectrometer S-2S, high-resolution spectroscopy of ${}^{12}_{\Xi}$Be, which can be produced with the ^{12}C(K^-, K^+) reaction, will be realized as the J-PARC E70 experiment in the near future [1, 2]. As detailed properties of spin- and isospin-dependent Ξ-nucleon interaction are not well-known at present, the structure of light Ξ hypernuclei, to be explored with the S-2S spectrometer, is considered to be important in revealing the nature of Ξ-nucleon interaction [3].

We plan to investigate ${}^{7}_{\Xi}$H produced by the ^{7}Li(K^-, K^+) reaction [4]. The (K^-, K^+) reaction on nuclear targets other than ^{12}C has not been attempted, whereas the aforementioned ${}^{12}_{\Xi}$Be had been studied in the BNL-AGS E885 experiment [5] and is to be investigated with a much better resolution in the forthcoming J-PARC E70 experiment. The physics motivation to investigate experimentally the ${}^{7}_{\Xi}$H system, is twohold. The first is to gain some insight into the Ξ-nucleus and Ξ-nucleon interactions through the formation spectrum in the reaction, as is the case with single-Λ hypernuclei. The second is the relevance to double-Λ hypernuclei; formation of ${}^{\;5}_{\Lambda\Lambda}$H from ${}^{7}_{\Xi}$H, owing to the $\Xi N \to \Lambda\Lambda$ conversion, is plausible [6].

A proposal of an experiment, being afterward referred to as the P75 experiment, to determine the mass of ${}^{\;5}_{\Lambda\Lambda}$H by means of decay pion spectroscopy, was submitted to J-PARC in 2018 [7]. As the first phase of the P75 experiment, we will measure the formation spectrum of the ^{7}Li$(K^-, K^+){}^{7}_{\Xi}$H reaction and derive the formation cross section of ${}^{7}_{\Xi}$H. The proposal of the phase-1 experiment, submitted in 2019, can be found in Ref. [4].

2 Very light Ξ hypernuclei — ${}^{7}_{\Xi}$H

The level structure is expected to be simple, because the ^{6}He core is an even-even nucleus. The existence of a bound state of ${}^{7}_{\Xi}$H is supported by various theoretical calculations [8–10]. Hiyama *et al.* [8] calculated the binding energy and width of ${}^{7}_{\Xi}$H with a four-body cluster

*e-mail: fujioka@phys.titech.ac.jp

model, in which three types of ΞN interaction models, ESC04d, ND and HAL QCD, were adopted. Whereas the ESC04d interaction results in a moderately large decay width, the decay width could be narrower than 1 MeV in case of the other two interactions. In particular, the decay width for the HAL QCD interaction could be of the order of 10 keV [9], because of a small coupling between ΞN and $\Lambda\Lambda$ predicted by the HAL QCD calculation [11]. A calculation by H. Le *et al.* with the chiral effective field theory led to the decay width of 0.2 MeV [10]. On the contrary, the binding energy is less model-dependent; it would lie around 2–3 MeV below the $\Xi^- + {}^6$He threshold. Moreover, ${}^7_\Xi$H is expected to be close to the onset of Ξ binding. While the existence of bound states of $NNN\Xi$ systems are suggested by recent theoretical calculations [10, 12], production in the ${}^4\mathrm{He}(K^-, K^{0/+})$ reactions are challenging from an experimental point of view.

For these reasons, ^{7}Li will be one of the most suited target candidates for populating Ξ hypernuclei after a high-resolution spectroscopy of ${}^{12}_\Xi$Be with a ^{12}C target is performed. According to a calculation on the ${}^7\mathrm{Li}(K^-, K^+)$ reaction [13, 14], the formation cross section is expected to be several tens of nb/sr, which is more or less similar to that for ${}^{12}_\Xi$Be derived in the BNL-AGS E885 experiment [5].

3 J-PARC E75 phase-1 experiment

The experiment will be performed at the K1.8 beamline at J-PARC, similar to the E70 experiment [1, 2]. The S-2S spectrometer will be utilized for missing-mass spectroscopy of the ${}^7\mathrm{Li}(K^-, K^+)$ reaction. The missing-mass resolution is expected to be 3.5 MeV (FWHM) if a lithium target with the mass thickness of 10 g/cm^2 is used. According to formation cross sections with various inputs for the ΞN interaction [13, 14], the yield in the bound region below the $\Xi^- + {}^6$He threshold is estimated 56–104 for one-week beamtime [4].

4 Toward production of double-Λ hypernuclei — ${}^5_{\Lambda\Lambda}$H

Another noteworthy feature of ${}^7_\Xi$H is that kinematically allowed decay modes are limited: (i) ${}^5_{\Lambda\Lambda}\mathrm{H} + 2n$, (ii) ${}^4_\Lambda\mathrm{H} + \Lambda + 2n$, (iii) ${}^4_\Lambda\mathrm{H}^* + \Lambda + 2n$, and (iv) ${}^3\mathrm{H} + 2\Lambda + 2n$ [6]. It is because the released energy of $\Xi^- p \to \Lambda\Lambda$ is largely canceled by the proton separation energy of ^{6}He, which is the core of ${}^7_\Xi$H.

If the branching ratio into the decay mode (i) is sufficiently large, ${}^5_{\Lambda\Lambda}$H can be exclusively produced as a decay particle of ${}^7_\Xi$H. Theoretically, it is calculated to be ~ 90% [6] or 40–60% with the remaining part being the decay modes (ii) and (iii) [15]. This feature is in contrast to the Ξ^- capture in nuclear emulsion, resulting in production of double-Λ hypernuclei with mass numbers ranging from 5 to 18.

The P75 experiment [7] will attempt to produce ${}^5_{\Lambda\Lambda}$H via production and decay of ${}^7_\Xi$H. Not only for confirmation of ${}^5_{\Lambda\Lambda}$H production but also for determination of its mass, a decay pion from two-body decay, ${}^5_{\Lambda\Lambda}\mathrm{H} \to {}^5_\Lambda\mathrm{He} + \pi^-$, which should have a discrete momentum, will be momentum analyzed. The technique to determine the mass of the parent hypernucleus is called decay pion spectroscopy, and is established for single-Λ hypernuclei, especially ${}^4_\Lambda$H decaying into ${}^4\mathrm{He} + \pi^-$ [16]. For this purpose, a cylindrical detector system, equipped with a time projection chamber and timing counters etc., is being developed.

5 Conclusion and Outlook

The E75 phase-1 experiment which will be performed in series with the E70 experiment will provide the first experimental data on ${}^7_\Xi$H. It would provide further information on Ξ-nucleus and Ξ-nucleon interactions. Moreover, it will be the first step toward exclusive production of ${}^{5}_{\Lambda\Lambda}$H. If the P75 experiment is realized, it may open a new era of double-Λ hypernuclear physics; the properties of ${}^{5}_{\Lambda\Lambda}$H, such as the lifetime and the weak decay modes, may be assessed in a counter experiment. By using a different nuclear target, e.g. ^{9}Be and ^{10}B, other light double-Λ hypernuclei may be produced in a similar way.

This work was supported by a Grant-in-Aid for Scientific Research on Innovative Areas (MEXT KAKENHI Grant Number JP21H00118) and a Research Grant in the Natural Sciences, the Mitsubishi Foundation.

References

[1] T. Gogami, in these proceedings.
[2] T.K. Harada, in these proceedings.
[3] E. Hiyama, in these proceedings.
[4] H. Fujioka *et al.*, Proposal for J-PARC 50 GeV Proton Synchrotron, *Phase-1 of the P75 experiment: Measurement of the formation cross section of* ${}^{7}_{\Xi}$H *in the* ^{7}Li(K^-, K^+) *reaction*
https://j-parc.jp/researcher/Hadron/en/pac_2001/pdf/P75_2020-02.pdf
[5] P. Khaustov *et al.*, Phys. Rev. C **61**, 054603 (2000).
[6] I. Kumagai-Fuse and Y. Akaishi, Phys. Rev. C **54**, R24 (1996).
[7] H. Fujioka *et al.*, Proposal for J-PARC 50 GeV Proton Synchrotron, *Decay Pion Spectroscopy of* ${}^{5}_{\Lambda\Lambda}$H *Produced by* Ξ*-hypernuclear Decay* ^{7}Li(K^-, K^+) *reaction*
https://j-parc.jp/researcher/Hadron/en/pac_1901/pdf/P75_2019-09.pdf
[8] E. Hiyama, Y. Yamamoto, T. Motoba, Th. A. Rijken, and M. Kamimura, Phys. Rev. C **78**, 054316 (2008).
[9] E. Hiyama, private communications (2019).
[10] H. Le *et al.*, Eur. Phys. J. A **57**, 339 (2021).
[11] K. Sasaki *et al.*, Nucl. Phys. A **998**, 121737 (2020).
[12] E. Hiyama *et al.*, Phys. Rev. Lett. **124**, 092501 (2020).
[13] T. Koike and E. Hiyama, Few-Body Syst. **54**, 1275 (2013).
[14] T. Koike and E. Hiyama, private communications (2020).
[15] A. Ohnishi *et al.*, Prog. Theor. Exp. Phys. **2020**, 063D01 (2020).
[16] F. Schulz *et al.*, Nucl. Phys. A **954**, 149 (2016).

EPJ Web of Conferences **271**, 03005 (2022) https://doi.org/10.1051/epjconf/202227103005
HYP2022

Results of the Ξ^- atomic X-ray measurement in J-PARC E07

Manami Fujita[1,*], *Y.*Ishikawa[2], *M.*Ukai[2,3], *H.*Kanauchi[2], *T.*Koike[2], *H.*Tamura[1,2], *K.*Hosomi[1], *T.O.*Yamamoto[1], *H.*Ekawa[4], *S.H.*Hayakawa[2], *K.*Nakazawa[5,6], *J.*Yoshida[2], *M.*Yoshimoto[5], *A.*Kasagi[4,6], *N.*Nishimura[5], and *K.*Hayashi[5]

[1]Advanced Science Research Center, Japan Atomic Energy Agency, Tokai 319-1195, Japan
[2]Department of Physics, Tohoku University, Sendai 980-8578, Japan
[3]Institute of Particle and Nuclear Studies, High Energy Accelerator Research Organization (KEK), Tsukuba 305-0801, Japan
[4]High Energy Nuclear Physics Laboratory, RIKEN, Wako 351-0198, Japan
[5]Faculty of Education, Gifu University, Gifu 501-1193, Japan
[6]Graduate School of Engineering, Gifu University, Gifu 501-1193, Japan

Abstract. Ξ^- atomic X-ray spectroscopy is one of the most useful methods for investigation of the Ξ-nucleus strong interaction. A serious problem in the measurement is the significant background coming from in-flight Ξ^- decay. For the first Ξ^- atomic X-ray spectroscopy experiment, a novel method of identifying stopped Ξ^- events using nuclear emulsion was developed to reject background photons from in-flight Ξ^- decay. We succeeded in reducing the background to 1/170 by this method employing coincidence measurements using the nuclear emulsion and X-ray detectors.

1 Introduction

The interaction between a Ξ hyperon and nucleons/nuclei is currently little understood due to the extremely limited experimental data. The Ξ hypernuclear missing mass spectroscopy has been studied [1, 2], but no clear Ξ hypernuclear peak was observed due to the low resolution. These data only suggest that the depth of the Ξ-nucleus potential is 10–20 MeV. On the other hand, the nuclear emulsion experiments have been conducted as well [3, 4], but no clear Ξ hypernuclear peak was observed due to the limited resolution. S (strangeness) $= -2$ systems of double Λ hypernuclei and twin Λ hypernuclei were observed. The KISO event identified as a Ξ-^{14}N bound system revealed that the Ξ-nucleus interaction is attractive [5].

In addition to the above two experimental methods, we have established a new approach, Ξ^- atomic X-ray spectroscopy. Since the Ξ^- hyperon has a negative charge, it forms a Ξ^- atom via the Coulomb interaction with the nucleus. The captured Ξ^- hyperon cascades down to the inner orbits by emitting X-rays or Auger electrons. Finally, the Ξ^- is absorbed by the nucleus via the strong interaction of $\Xi^- p \to \Lambda\Lambda$ in the " last " orbit where the strong interaction is dominant. Affected by the strong interaction between the Ξ^- and the nucleus, the energy levels of the Ξ^- atom shift and/or broaden compared to those calculated with the Coulomb interaction alone. The values of the energy shifts, ΔE, and the widths, $\Gamma_{\rm abs}$, are determined from the Ξ atomic wave function and the potential. When the strong-interaction optical potential is denoted by $U = V + iW$, the measurement of the shifts and widths give

*e-mail: fujitam@post.j-parc.cjp

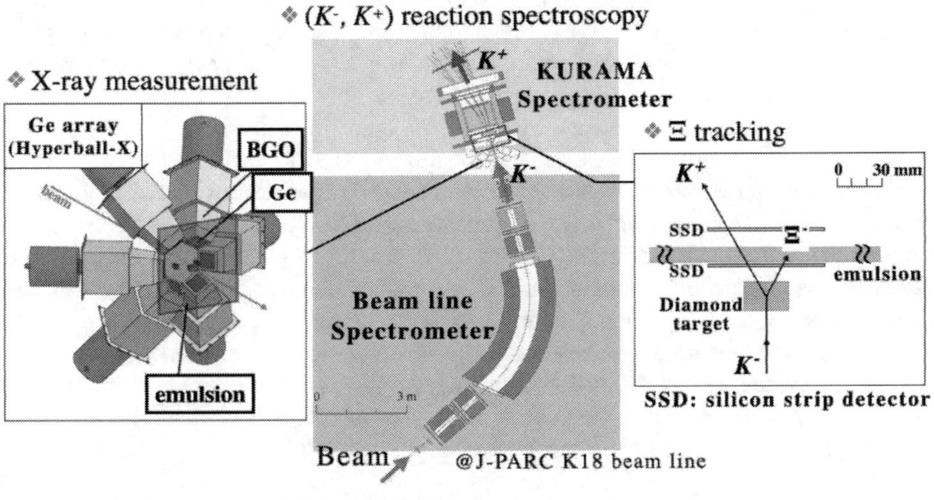

Figure 1. Experimental setup of the J-PARC E07 experiment.

information on the real (V) and imaginary (W) parts of the potential, respectively. In addition, the branching ratio of the last transition is useful to investigate the W potential.

2 Experiment

Since the lifetime of Ξ^- hyperons is as short as ~164 ps, most of them decay, before they stop and form Ξ^- atoms in matter. A serious difficulty in the measurement is that the high energy photons associated with the in-flight Ξ^- decay make a significant background in the X-ray spectra. In addition to selecting Ξ^- production events via the $p(K^+, K^+)\Xi^-$ reaction, it is necessary to identify the stopped Ξ^- events. We developed a novel method employing nuclear emulsion images for the Br and Ag Ξ^- atomic X-rays and carried out a measurement as the J-PARC E07 experiment alongside the experiment searching for $S = -2$ hypernuclei in nuclear emulsion.

The last transitions for Ξ^- Br and Ag atoms are predicted to be the Br($7I \rightarrow 6H$) and Ag($8J \rightarrow 7I$) transitions, and their X-ray energies without shifts are 316 keV and 370 keV, respectively. According to the theoretical calculations assuming the $t\rho$ potential, the shift and width values for the last transitions of the Ξ^- Ag atom are predicted as 0.28 keV and 0.15 keV, respectively [6, 7].

As shown in Fig. 1, a triple coincidence measurement was performed using three detectors: the two magnetic spectrometers for identifying the Ξ^- production reaction, the nuclear emulsion as a Ξ^- tracking detector, and the germanium (Ge) detector array, named Hyperball-X [8], for X-ray measurement. The silicon strip detectors (SSD) were installed upstream of the emulsion for Ξ^- track analysis, and the point where the Ξ^- hyperon reached the emulsion surface was predicted. Then, the emulsion image analysis was performed with a microscope, and the Ξ^- track was identified and followed to the end where the Ξ^- was stopped and absorbed or decayed in flight.

3 Analysis

Firstly, Ξ^- production events were identified by analysis of the $p(K^-, K^+)\Xi^-$ reaction; K^- and K^+ mesons were analyzed by the two magnetic spectrometers, and the events having the vertex of K^- and K^+ tracks within the reaction target were selected. In addition, the consistency between the Ξ^- track detected by SSD and the reconstructed missing momentum of the (K^-, K^+) reaction was checked. As a result, the number of the selected stopped Ξ^- events corresponds to 80% of the expected yield in simulation.

Secondly, the stopped Ξ^- events were identified in the emulsion. A total of 35438 Ξ^- tracks in all 118 emulsion modules were analyzed. The stopped Ξ^- hyperons were observed as one of the two types of images shown in Fig. 2. For the first case (I), classified as a "σ-stop" event, one or more charged fragments/particles are observed at the Ξ^- hyperon stopping point. For the second case (II), classified as a "ρ-stop" event, no charged particle tracks were observed at the stopping point. This happened when only neutral particles, including n, Λ, and π^0, were produced in the Ξ^- absorption process, or when heavy fragments were produced but their tracks were too short to be observed. The numbers of the observed σ- and ρ-stop events were 2208 and 9961, respectively. Assuming a ratio of σ- and ρ-stop events of 2:1 [9], the number of Ξ^- events in the ρ-stop events is estimated to be $\sim$ 1100, and the others are considered to be proton contaminations. Auger electrons associated with the de-excitation of Ξ^- atoms can be observed. To reject proton contamination, only ρ-stop events with Auger electrons were selected.

Finally, coincidence analysis of selected stopped Ξ^- events and X-ray data was performed. The energies of the Ge detectors were calibrated by a newly-developed in-beam energy calibration method [8].

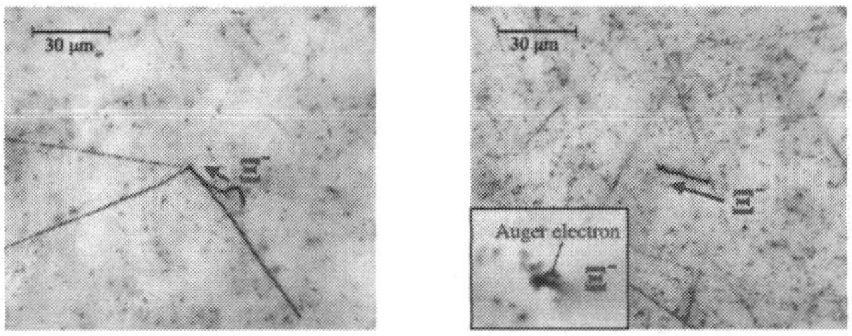

(I) σ-stop **(II) ρ-stop**

Figure 2. Emulsion images for (I) a σ-stop event and (II) a ρ-stop event

4 Results and discussion

The X-ray spectrum for the Ξ^- production events identified by the reaction analysis was obtained as shown in Fig. 3 (a). Several γ rays produced by the scattered beam or secondary particles/photons hitting the Ge detectors or surrounding materials were observed. Assuming that the signal is negligibly small compared to the background, the spectrum (a) was fitted with a linear function, and the background level was found to be 63.5 counts/keV at 370 keV. The spectrum for the σ-stop events and "ρ-stop with Auger electron" events was obtained as shown in Fig. 3 (b). For a peak search, the spectrum (b) was fitted by a Gaussian function plus a constant background. The spectrum (b) from 200 to 400 keV is shown at the top panel of Fig. 4, and the obtained peak significance values are plotted at the bottom. No evident peak with significance larger than 3σ was observed. By fitting spectrum (b) with a constant function, assuming that the signal is negligible, we found that the background level was 0.34 counts/keV in the region of 300–400 keV. The background level was successfully suppressed to 1/170 by employing the selection using emulsion images. The expected X-ray yield for the Ag(8J $\rightarrow$ 7I) transition is 2.2 counts, and the signal to noise ratio, $S/\sqrt{N}$, was estimated

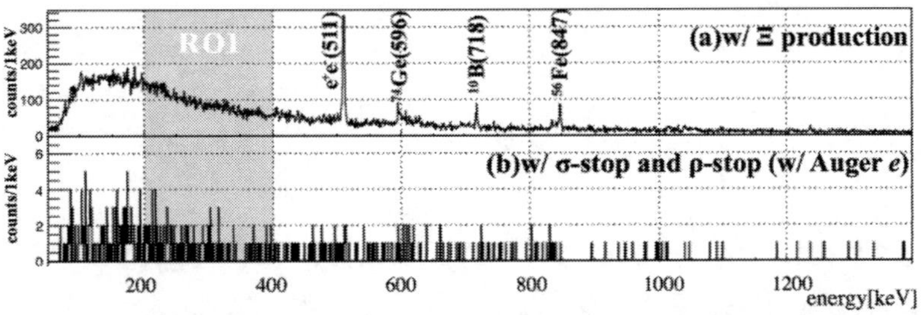

Figure 3. X-ray spectra (a) for the Ξ^- production events, and (b) for the σ-stop events and the "ρ-stop with Auger electron" events.

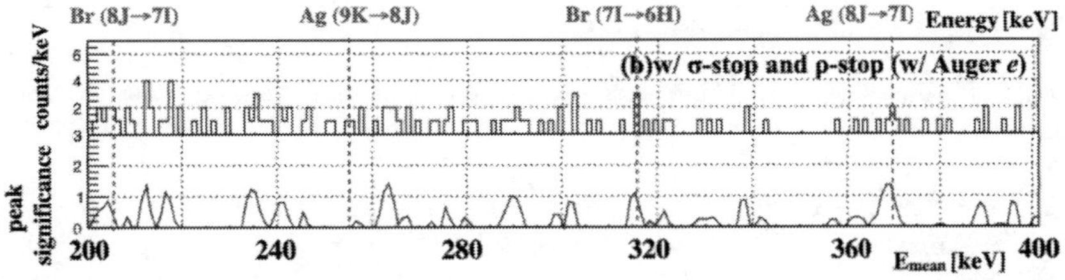

Figure 4. X-ray spectrum (top) and peak significance (bottom) for the spectrum (b) for the σ-stop events and the "ρ-stop with Auger electron" events. The dotted lines show the X-ray energies from Ξ^- Br and Ag atoms without shifts. The E_{mean} is the mean value of the Gaussian function.

to be 1.7. Thus, the Ξ^- Ag atomic X-ray peaks with a 3.8σ significance would be observed if five times more data are taken.

5 Summary

The first Ξ^- atomic X-ray spectroscopy experiment (J-PARC E07) was conducted at J-PARC K1.8 beam line to investigate Ξ-nucleus interaction by measuring the shift and width of the last transitions. To reject the significant background due to the in-flight Ξ^- decay, a novel method of identifying stopped Ξ^- events by a coincidence of magnetic spectrometers, nuclear emulsion, and X-ray detectors was employed. The X-rays of the last transitions of Ξ^- Br and Ξ^- Ag atoms were measured by the Ge detector array. Although no evident peak was observed, the background level was drastically reduced by employing the triple coincidence measurement. With five times higher statistics, successful identification of the Ξ^- atomic X-ray peaks would be possible in the future.

References

[1] T. Fukuda *et al.* (E224 Collaboration), Phys. Rev. C **58**, 1306 (1998)
[2] P. Khaustov *et al.* (The AGS E885 Collaboration), Phys. Rev. C **61**, 054603 (2000)
[3] KEK E176 Collaboration, S. Aoki *et al.*, Nucl. Phys. A **828**, 191 (2009)
[4] J.K. Ahn *et al.* (E373 (KEK-PS) Collaboration), Phys. Rev. C **88**, 014003 (2013)
[5] K. Nakazawa *et al.*, Prog. Theor. Exp. Phys. **2015**, 033D02 (2015)
[6] C. J. Batty *et al.*, Phys. Rev. C **59**, 295 (1999)
[7] E. Friedman and A. Gal, private communications (2001)
[8] M. Fujita *et al.*, Nucl. Instr. and Meth. A**1042**, 167439 (2022)
[9] A. M. M. Theint *et al.*, Prog. Theor. Exp. Phys. **2019**, 021D01 (2019)

High resolution spectroscopy of Ξ hypernuclei with active fiber target

Takeshi K. Harada[1, 2*], *K.* Ebata[1, 2], *T.* Gogami[1], *S. H.* Hayakawa[3], *R.* Honda[4], *Y.* Ichikawa[2], *K.* Miwa[3, 4], *T.* Nagae[1], *T.* Nanamura[1, 2], *T.* Takahashi[4], *M.* Ukai[4] and *T. O.* Yamamoto[2], for the J-PARC E70 collaboration

[1]Department of Physics, Graduate School of Science, Kyoto University, Kyoto 606-8502, Japan
[2]Advanced Science Research Center (ASRC), Japan Atomic Energy Agency (JAEA), Tokai, Ibaraki 319-1195, Japan
[3]Graduate School of Science, Tohoku University, Sendai, Miyagi 982-0826, Japan
[4]Institute of Particle and Nuclear Studies (IPNS), High Energy Accelerator Research Organization (KEK), Tsukuba, Ibaraki 305-0801, Japan

Abstract. We are planning to carry out a high-resolution spectroscopy of Ξ hypernucleus at the J-PARC K1.8 beamline, which provides a high-intensity K^- beam (J-PARC E70 experiment). The high-resolution spectroscopy aims to be realized by introducing a new magnetic spectrometer S-2S and an active fiber target AFT. In this article, the role of the AFT in this experiment and its development status are described.

1 Introduction

Recently, a few Ξ-hypernuclear events have been reported from emulsion experiments [1–4], and they allow us to study the baryon-baryon interaction for the strangeness $(S) = -2$ systems. As a result, the Ξ-nucleus interaction is found to be attractive. The ΞN interaction is studied by various theoretical models such as the mean-field theory [5], based on the observed Ξ hypernuclei from the emulsion experiments. However, there is still a large statistical uncertainty, and further experimental information is desired. A Ξ-hypernuclear production was investigated by missing-mass spectroscopy at KEK [6], BNL [7] and J-PARC [8] in the past. In these experiments, signals were observed in the bound region of the Ξ hypernuclei, but the peak structures were not clarified due to insufficient missing-mass resolution. Therefore, more precise spectroscopic experiments with higher signal sensitivity are needed.

2 J-PARC E70 experiment

We are going to perform a high-resolution spectroscopy of Ξ hypernucleus in a missing-mass method via the (K^-, K^+) reaction (J-PARC E70 experiment). The experiment will be

* e-mail : harada.takeshi.24e@st.kyoto-u.ac.jp

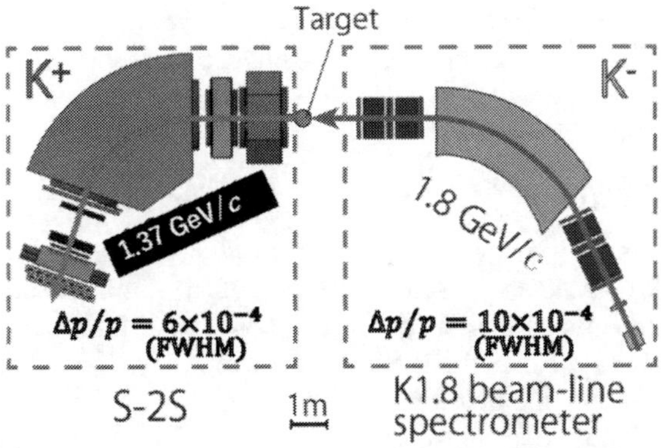

Fig. 1. Experimental setup of the J-PARC E70 experiment.

carried out at the J-PARC K1.8 beamline, which provides a high-intensity K^- beam. The experimental setup is shown in Fig. 1. The momentum of the incident K^- and scattered K^+ particles will be analyzed using the K1.8 beamline spectrometer ($\Delta p/p = 10 \times 10^{-4}$ (FWHM)) and the S-2S spectrometer ($\Delta p/p = 6 \times 10^{-4}$ (FWHM)), respectively. S-2S is a new magnetic spectrometer. An active fiber target (AFT) will be used as an experimental target with a thickness of about 9 g/cm^2. The expected yield and missing-mass resolution are about 100 counts and 2 MeV/c^2 (FWHM), respectively. The resolution of 2 MeV/c^2 (FWHM) is achievable by a correction of the energy losses of particles by AFT as shown in the next section.

3 Active fiber target

K^- and K^+ which are related to the Ξ-hypernuclear production lose their energies in AFT. The energy straggling due to the energy loss deteriorates the missing-mass resolution. AFT directly measures the energy losses event by event. The energy-loss information obtained by AFT is then used to correct the particles' momenta event by event, leading to a large suppression of the missing-mass deterioration. Therefore, AFT is a key detector to achieve the high resolution of 2 MeV/c^2 (FWHM), maintaining the significant statistics from such a thick target (9 g/cm^2) in J-PARC E70.

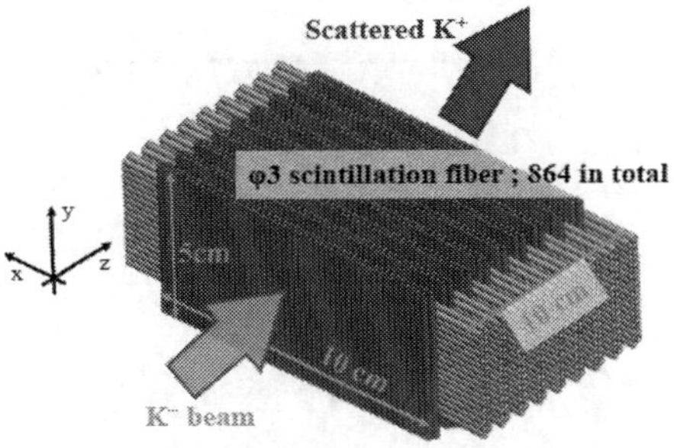

Fig. 2. Schematic drawing of AFT in J-PARC E70. AFT directly measures the energy losses of K^- and K^+ which are related to the Ξ-hypernuclear production event by event.

AFT is composed of about 900 scintillation fibers (Saint-Gobain BCF-10SC, ϕ3 mm), and the carbon nuclei contained in the scintillation fibers are used as the target nuclei to produce $^{12}_{\Xi}$Be hypernuclei. Fig. 2 shows a conceptual drawing of the AFT. The direction of the incident K^- beam is the z-axis and the fibers are arranged orthogonally to it (XX'YY'). The dimension is $5^H \times 10^W \times 10^T$ cm^3, designed to cover the beam area. The energy loss of the charged particles in AFT is converted into a scintillation light. The scintillation light is detected by Multi-Pixel Photon Counters (MPPC, Hamamatsu S13360-3075PE) attached to both ends of each fiber. Therefore, the total number of channels is up to 1800. Electrical signals induced by the MPPC are readout by VME-EASIROC modules. AFT enables us to directly measure the energy loss of $K^{\pm}$ particles in the target event by event because the yield of scintillation light is proportional to the energy loss. The energy loss measured by AFT is used as a correction when the missing mass is calculated in analyses. As a result, the target missing-mass resolution of 2 MeV/c^2 (FWHM) is expected to be achieved.

3.1 Evaluation test of energy resolution at RCNP

We performed an evaluation test of energy resolution at RCNP, Osaka University in 2017 using a detector (AFT-proto1) which consisted of a scintillation fiber, MPPCs, and EASIROC [9]. Proton beams at the energies of E = 64.6 and 295 MeV were irradiated onto the single fiber. The energy loss in the fiber was measured by a high-resolution spectrometer, Grand Raiden. Fig. 3 shows the obtained energy resolution of the detector (AFT-proto1) as a function of the energy loss measured by Grand Raiden. In J-PARC E70, a required energy resolution is $\sigma(E)/E \leq 10\%$ for the energy loss of 0.6 MeV which is a typical energy loss in a fiber. It was found that the resolution is less than or about 10% for the energy loss of 0.6 MeV as shown in Fig. 3. Fig. 4 shows simulated binding-energy spectra with and without the AFT correction based on the result of the above beam test. The energy resolution gets better from 4.1 to 2 MeV in FWHM thanks to the correction by AFT.

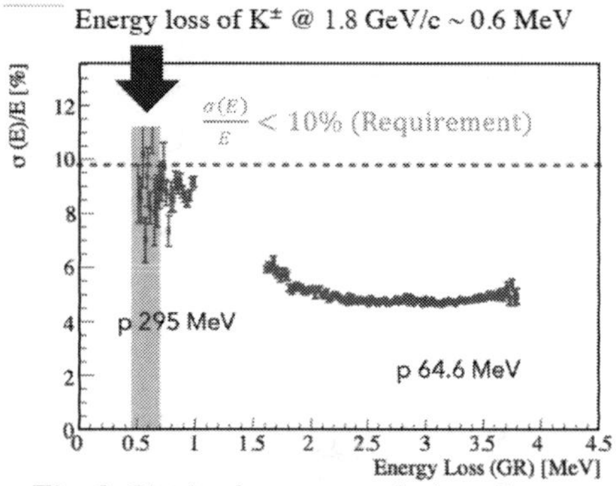

Fig. 3. Obtained energy resolution of detector (AFT-proto1) as a function of the energy loss which was measured by Grand Raiden at RCNP.

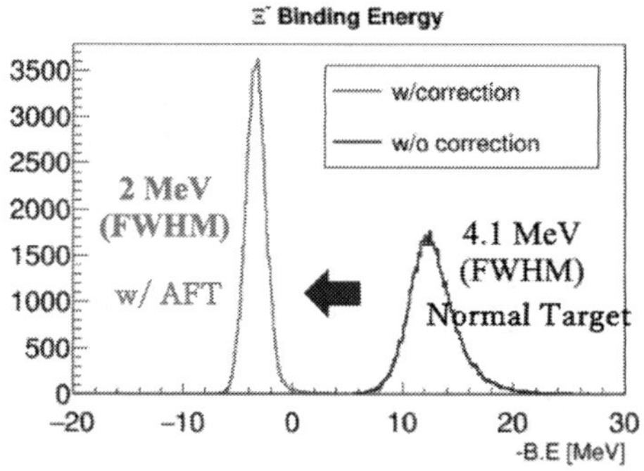

Fig. 4. Simulated spectra for the Ξ-binding energy for the ^{12}C$(K^-, K^+)^{12}_{\Xi}$Be reaction. The energy resolution is improved to 2 MeV from 4.1 MeV (FWHM) by the correction using AFT.

3.2 Multi-channel readout test at ELPH

A prototype (AFT-proto2) with a total of about 600 channels, which is 1/3 of the actual AFT, was built (Fig. 5). AFT-proto2 was tested to establish a multi-channel readout system at ELPH, Tohoku University in 2019 [10]. In the beam test, a positron beam at the momentum of 800 MeV/c was impinged on the prototype. AFT-proto2 was sandwiched by fiber trackers (SST) that had a better position resolution than the prototype. The efficiency of AFT-proto2 was analyzed by using a tracking information by the two SSTs. As a result, it was found that the efficiency is more than 95%, which meets our requirement. In addition, the system of multi-channel readout was found to work well.

Fig. 5. A photograph of AFT-proto2 tested at ELPH, Tohoku University in 2019.

The successful result allowed us to develop AFT which has three times more channels than AFT-proto2.

3.3 Simulation of pileup effect

Electrical signals output from the MPPC are input to the EASIROC chip in VME-EASIROC for shaping and amplification. The time constant for the signal shaping and amplification is 2 μs. A resolution of ADC, which corresponds to the energy loss information in fibers, could be deteriorated due to pileups of the signals. Particularly, the fibers at the center of AFT may suffer from high rates because the particle density is higher at the beam center. We performed a simulation, in which typical waveforms were taken into account, to evaluate the missing-mass deterioration due to the signal pileup. As a result, the missing-mass resolution may become 3 MeV/c^2 in FWHM which is worse than our goal resolution. Therefore, we also simulated a case of TOT (Time Over Threshold) instead of ADC for the energy-loss analysis. TOT should have a less effect for the signal pileup because it has a short time constant. It was found the TOT analysis allows us to achieve 2.4 MeV/c^2 resolution (FWHM) in a resulting missing-mass spectrum. The simulation result is in a scope of our goal resolution.

4 Summary

We are planning to perform a high-resolution spectroscopy of Ξ hypernucleus at the J-PARC K1.8 beamline with a high intensity K^- beam (J-PARC E70). The missing mass of the Ξ hypernucleus is reconstructed by using the K1.8 beamline spectrometer and the new spectrometer, S-2S. In J-PARC E70, AFT is a key detector to achieve the high resolution of about 2 MeV (FWHM), maintaining a reasonable yield of 100 counts from a thick target material. Prototypes of AFT were tested by using proton and positron beams at RCNP and ELPH, respectively. The successful beam tests for the prototypes allowed us to develop a full scale AFT. AFT is now being constructed and is going to be installed by the end of 2022. A simulation study in which AFT is used for the missing-mass reconstruction was performed. As a result, it is found that a missing-mass resolution of 2.4 MeV/c^2 (FWHM) is reachable by using AFT. The AFT simulations are ongoing to find better ways of analyses.

References

[1] K. Nakazawa et al., Prog, Theor, Exp. Phys. **2015**, 33D02 (2015)
[2] H. Ekawa *et al.*, Prog. Theor. Exp. Phys. **2019**, 021D02 (2019)
[3] S. H. Hayakawa *et al.*, Phys. Rev. Lett. **126**, 062501 (2021)
[4] M. Yoshimoto *et al.*, Prog. Theor. Exp. Phys. **2021**, 073D02 (2021)
[5] Y. Tanimura et al Phys. Rev. C. **105**, 044324 (2022)
[6] T. Fukuda et al., Phys. Rev. C 58, 2 (1998)
[7] P. Khaustov et al., Phys. Rev. C 61, 054603 (2000)
[8] T. Nagae et al., AIP Conference Prceedings 2130, 020015 (2019)
[9] A. Koshikawa, Master's Thesis, Kyoto University, Kyoto, 2018 (in Japanese)
[10] T. K. Harada, Master's Thesis, Kyoto University, Kyoto, 2020 (in Japanese)

A spectroscopic study of $^{10}_{\Xi}$Li hypernucleus via the ^{10}B(K$^-$, K$^+$)X reaction

Atsushi Tokiyasu[1,*] and $S-2S$ collaboration

[1]Research center for ELectron PHoton science (ELPH), Tohoku University, Sendai 982-0826, Japan.

Abstract. We are planning a $^{10}_{\Xi}$Li production experiment using the $^{10}B(K^-,K^+)$X reaction at the J-PARC, K1.8 beamline. We perform spectroscopy of missing masses using the $S-2S$ spectrometer, which can achieve high resolution. Since the cross section of Ξ-hypernuclei is very small, we would like to use a thick target. To ensure the missing mass resolution, we propose a semi-inclusive method using a tracking device to detect the decay particles from Ξ-hypernuclei.

1 Introduction

The Ξ-nucleon interaction is the last piece of the nuclear force study that has been extended to the strange quark, and the Ξ-hypernuclear spectroscopy experiment will provide us with rich information. We are planning a series of Ξ-hypernuclear spectroscopy experiments using (K^-, K^+) reactions at the Japan Proton Accelerator Re-search Complex (J-PARC), K1.8 beamline.

As the first experiment, a $^{12}_{\Xi}$Be spectroscopy experiment using a CH_2 active target is planned. In order to further investigate the Ξ-nucleon interaction precisely, it is necessary to systematically measure the binding energies of various Ξ-hypernuclei using different targets. We are particularly interested in spectroscopic experiments on $^{10}_{\Xi}$Li. $^{10}_{\Xi}$Li consists of two α particles, one neutron, and one Ξ particle, and it is very unique because information about the interaction between spin-dependent Ξ-neutron can be obtained. Recently, Hiyama *et al.* pointed out that the phase shift S-wave Ξ-neutron ionteraction is sensitive to the theoretical models [1]. In the ESC model, triplet term is strongly attractive, while in the HAL-QCD, triplet term is weakly attractive. This difference might play an important role for the binding energy or level ordering of $^{10}_{\Xi}$Li. Hiyama *et al.* also calclated the binding energy and conversion energy of $^{10}_{\Xi}$Li [2], but all calculated values of the conversion width vary from model to model, but all calculated results predict the existence of the bound state, and the binding energy is large enough to observe peak structures in the experiment using $S-2S$ spectrometer.

To produce $^{10}_{\Xi}$Li, it is necessary to use the ^{10}B(K^-,K^+)X reaction with a boron-10 target. The momentum of K^- is analyzed with K1.8 beamline spectrometer. The resolution $\Delta p/p$ is 3.3 × 10$^-$4 in FWHM. Scattered K^+ is analyzed through $S-2S$ spectrometer with the resolution of 6.0×10^{-4} in FWHM. By using $S-2S$ spectrometer, we can perform the missing mass spectroscopy with a very good mass resolution.

*e-mail: tokiyasu@lns.tohoku.ac.jp

The resolution of the missing mass can be written as:

$$\Delta M = \sqrt{(\frac{\partial M}{\partial p_{K^-}})^2 \Delta p_{K^-}^2 + (\frac{\partial M}{\partial p_{K^+}})^2 \Delta p_{K^+}^2 + (\frac{\partial M}{\partial \theta})^2 \Delta \theta^2 + \Delta E_{target}^2} \quad (1)$$

Here, M is the missing mass, $p_{K^-}(p_{K^+})$ is the momentum of $K^-(K^+)$, and θ scattering angle. This equation can be divided into two parts. The first three terms are derived from the performance of the spectrometer, which has a value of approximately 1.8 MeV in FWHM for $S-2S$. To achieve a resolution of less than 3 MeV in FWHM in total, the last term, which originates from energy fluctuations inside the target, must be less than 2.5 MeV in FWHM.

Figure 1 shows the energy loss of K^+ mesons inside the boron target for different thickness. If the target thickness of 40 mm is selected, the energy fluctuation is estimated to be approximetely 2 MeV in FWHM. To use a target of this thickness, the mean energy loss must be corrected by the Bethe-Bloch formula. To calulate the mean energy loss, it is necessary to reconstruct the vertices of the reaction with a resolution of about 1 mm.

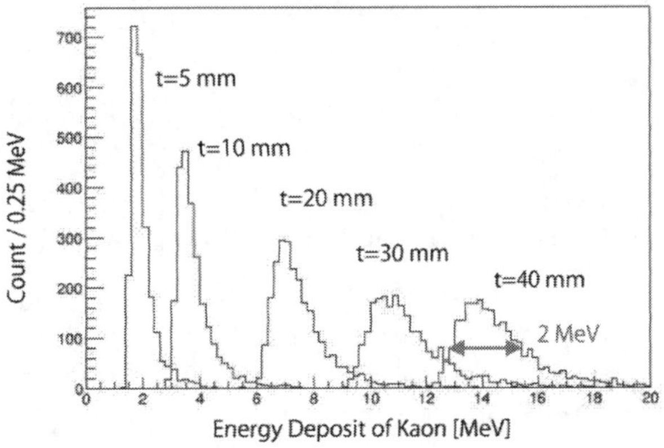

Figure 1. Energy loss of K^+ inside the targer for differnt thickness

2 Experimental setup

Figure 2 shows a conceptual setup near the target system. From the right side of the figure, K^- beam is injected to the segmented boron target. A $^{10}_{\Xi}$Li , the Ξ-hypernucleus is produced inside the target assosiated with the scattered K^+ going to the left side. The boron target is segmented and tracking devices are placed around it. Single-sided Silicon Detectors(SSDs) or Double-sided Silicon Detectors (DSSDs) are used for the tracking devices, and scintillation counters are used to identify particle species. Pions from the decay of hypernuclei are detected by this system and used to reconstruct the vertex position, and the segment of the target where the event occurred is identified. Then, the caluclated energy loss by Bethe-Bloch formula are used to correct the momentum for the K^- and K^+. The vertex resolution of $K^{-/+}$ and pion is estimated to be less than 1 mm.

The acceptance of this system is evaluated via a simple assumption that Ξ-N pair inside $^{10}_{\Xi}$Li is converted to two Λ's, which decay into neutron and π^0 or proton and π^-. The Q-value of this decay is 28 MeV. Here, the momentum of Ξ-N is calculated from the ratio of the mass of Ξ-N to the total mass. In this assumption, vertex reconstruction efficiency is estimated to be approximately 30%. However, in real situations, $^{10}_{\Xi}$Li has many decay modes such as twin hypernuclei, single hypernuclei and free Λ, etc. A detailed feasibility study of this method is currently underway using a more realistic assumption.

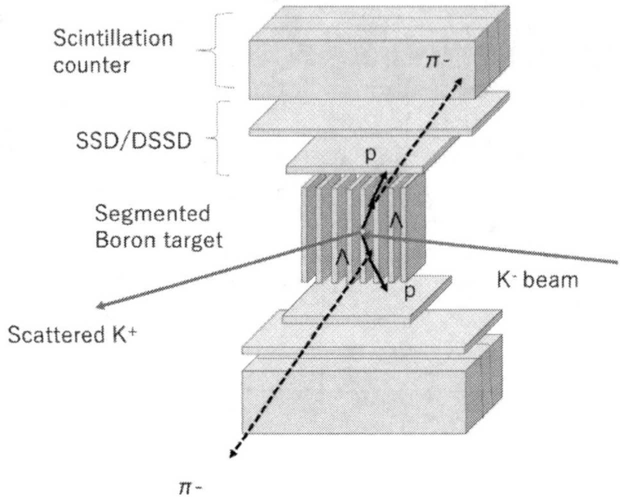

Figure 2. Conceptual experimental setup near the target system. The target is segmented to identify the reaction position. SSDs or DSSDs are installed to detect the decay particle. The scintillation counters are used to the identification of the particles.

3 Summary

We are planning to conduct a $^{10}_{\Xi}$Li production experiments using boron targets at J-PARC, K1.8 beamline. This Ξ-hypernucleus is very important for obtaining a spin-dependent interaction between Ξ and neutron. A missing mass spectroscopy will be performed by using $S-2S$ spectrometer. In order to achieve high yield and high resolution at the same time, we will prepare the target as thick as possible, and in addition, we will prepare a tracking system around the target. A resolution of 3.5 MeV missing mass can be achieved with a 40 mm thick target. Under simple assumptions, we expect to achieve about 30% acceptability with this system, but a more detailed validation check is needed.

References

[1] E. Hiyama *et al.*, Phys. Rev. C **78**, 054316 (2008)
[2] E. Hiyama *et al.*, Phys. Rev. Lett. **124**, 092501 (2020)

EPJ Web of Conferences **271**, 03008 (2022) https://doi.org/10.1051/epjconf/202227103008
HYP2022

Preparation status of missing-mass spectroscopy for Ξ hypernuclei with S-2S magnetic spectrometer

Kengo Ebata[1,2*], *H.* Fujioka[3], *M.* Fujita[2], *T.* Gogami[1], *T. K.* Harada[1,2], *S. H.* Hayakawa[4], *R.* Honda[5], *Y.* Ichikawa[2], *K.* Kamada[4], *T.* Kobori[3], *K.* Miwa[4], *T.* Nagae[1], *T.* Nanamura[1,2], *R.* Negishi[3], *F.* Oura[4], *T.* Sakao[4], *C.* Son[4], *T.* Takahashi[5], *H.* Takahashi[5], *H.* Tamura[2,4], *A. O.* Tokiyasu[6], *M.* Ukai[4,5], and *T. O.* Yamamoto[2], for the J-PARC E70 Collaboration.

[1]Department of Physics, Graduate School of Science, Kyoto University, Kyoto, Kyoto 606-8502, Japan
[2]Advanced Science Research Center, Japan Atomic Energy Agency (JAEA), Tokai, Ibaraki 319-1195, Japan
[3]Department of Physics, Tokyo Institute of Technology, Meguro, Tokyo 152-8551, Japan
[4]Department of Physics, Graduate School of Science, Tohoku University, Sendai, Miyagi 980-8578, Japan
[5]Institute of Particle and Nuclear Studies, High Energy Accelerator Research Organization (KEK), Tsukuba, Ibaraki 305-0801, Japan
[6]Research Center for Electron Photon Science (ELPH), Tohoku University, Sendai, Miyagi 982-0826, Japan

Abstract. J-PARC E70 experiment measures the missing-mass of Ξ hypernuclei ($^{12}_{\Xi}Be$) in Hadron Experimental Facility at J-PARC. We aim to reach the best missing-mass resolution of 2 MeV/c^2 in FWHM with a new magnetic spectrometer S-2S. The high-resolution spectroscopy of Ξ hypernuclei will play an important role to understand the unknown ΞN interaction. The experiment will start at the beginning of 2023. This article presents the preparation status.

1 Introduction

For the strangeness $S=-1$system, the ΛN interaction has been studied through spectroscopic investigations of Λ hypernuclei at experimental facilities such as KEK, J-PARC, BNL, and JLab. On the other hand, for the strangeness $S=-2$system, the experimental data are scarce. A few events of ΛΛ and ΞN hypernuclei were reported in emulsion experiments [1, 2]. However, the statistics were limited to determine the binding energy and decay width clearly. The data of Ξ hypernuclei which have higher statics and higher precision are awaited to understand the ΞN interaction. The missing-mass spectroscopy for Ξ hypernuclei was done at KEK [3] and BNL [4] in the past. However, they could not observe clear peak structures in the bound region. To further investigate the

* e-mail: ebata.kengo.84r@st.kyoto-u.ac.jp

ΞN interaction including spin-dependent forces, peak structures in the missing-mass spectra need to be resolved by better precision experiments. Our goal is to investigate Ξ hypernuclei with the world's highest mass-resolution of 2 MeV/c^2 (FWHM) to study the unknown ΞN interaction [5]. The first target to study is ${}^{12}_{\Xi}Be$ via ^{12}C (K$^-$, K$^+$) reaction A new magnetic spectrometer S-2S was designed to realize such a high-resolution spectroscopy of Ξ hypernuclei. Now, S-2S is being installed in the K1.8 beamline at Hadron Experimental Facility of the J-PARC. We aim to make the experimental apparatus ready by the end of 2022, and to start commissioning with the beam at the beginning of 2023. The present article shows the preparation status of the J-PARC E70 experiment [6].

2 Methods and detectors

In J-PARC E70, the high-intensity and purity K$^-$ beam is transported to a ^{12}C target. The K$^-$

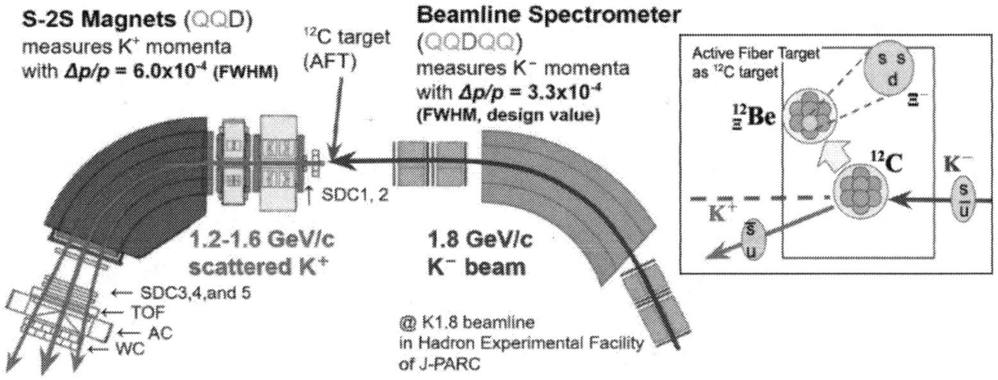

beam which has momenta of 1.8 GeV/c is used for production of Ξ hypernuclei (${}^{12}_{\Xi}Be$) via

Figure 1. Schematics of the experimental setup for J-PARC E70 and the ^{12}C(K$^-$, K$^+$)${}^{12}_{\Xi}Be$ reaction.

(K$^-$, K$^+$) reaction. The missing-mass is reconstructed by using momenta of K$^-$ beam and scattered K$^+$. These momenta are obtained by beam-optics analyses of the K1.8 beamline spectrometer and S-2S (Fig. 1). The designed momentum resolutions are $\Delta p/p$ = 3.3×10^{-4} and 6.0×10^{-4} in FWHM, respectively. An information about positions and angles for the momentum analysis is measured by a tracking analysis with 5 drift chambers (SDC1–5). Particles passing through S-2S are not only K$^+$s, but also 100–1000 times more π^+s and
protons as background. For K$^+$s identification, an aerogel-Cherenkov detector (AC) and a water-Cherenkov detector (WC) are installed. AC and WC suppress π^+s and protons, respectively. Time-of-flight information to identify K$^+$ is obtained by plastic-scintillation detectors (BH2 and TOF). In addition, one of important detectors for E70 is active fiber target (AFT). AFT is not only used as the ^{12}C target, but also measures energy loss of particle in AFT to prevent a deterioration of resolution due to the energy straggling. Table1 shows the specifications of detectors for E70.

3 Preparation status

The preparation is in progress for a commissioning beam time in January 2023.

3.1 S-2S magnets

S-2S magnets were newly designed and constructed. The magnetic fields for typical positions were measured in 2016 [7]. In January 2022, S-2S magnets were installed at the K1.8 beam line. S-2S magnets were excited several times to check the field stability.

3.2 Active fiber target (AFT)

An operation test for a single fiber was performed in RCNP in 2017 [8]. A test for multi-channelized fibers was performed in ELPH, in 2019 [9]. The design of the frame is currently in progress, and AFT will be installed in October 2022.

3.3 Drift chambers

SDC1, 3–5 were used in the former experiment at K1.8 beamline. On the other hand, SDC2 is a new drift chamber. A commissioning of SDC1–5 with a radiation source was performed. SDC3–5 were installed in May 2022, and a cabling of their signals was completed in September. SDC1 and 2 will be installed in September 2022.

3.4 PID counters

TOF and WC [10, 11] were newly designed. On the other hand, AC was used for the experiment a long time ago. The condition of silica aerogel tiles as radiator was checked, and they were reloaded in April 2022. TOF and AC were installed in July 2022. WC will be installed in September 2022.

4 Summary

The missing-mass spectroscopy of Ξ hypernucleus ($^{12}_{\Xi}Be$) is planned to be carried out in 2023 at the K1.8 beam line in J-PARC. We designed a new magnetic spectrometer S-2S to realize the highest resolution of 2 MeV/c^2 (FWHM) in a resulting mass spectrum. The installation for S-2S is in progress, and S-2S is going to be ready for the experiment by the end of 2022. The commissioning with the beam is going to start at the beginning of 2023.

References

1. H. Takahashi et al., Phys. Rev. Lett. **87**, 212502 (2001)
2. K. Nakazawa et al., Prog. Theor. Exp. Phys. **2015** 033D02 (2015)
3. T. Fukuda et al., Phys. Rev. C **58,** 1306 (1998)
4. P. Khaustov et al., Phys. Rev. C **61,** 054603 (2000)
5. T. Motoba, S. Sugimoto, Nuclear Physics A **835,** 223–230 (2010)
6. T. Nagae, Proposal for the next E05 run with the S-2S spectrometer (2018)
7. T. Nanamura, Master's thesis, Kyoto University (in Japanese) (2016)
8. A. Koshikawa, Master's thesis, Kyoto University (in Japanese) (2017)
9. T. K. Harada, Master's thesis, Kyoto University (in Japanese) (2019)
10. K. Takenaka, Master's thesis, Kyoto University (in Japanese) (2014)
11. T. Gogami et al., Nucl. Instrum. and Methods Phys. Rev. A **817,** 70-84 (2016)

Recent progress and future prospects of hyperon nucleon scattering experiment

Koji Miwa[1,2,*], *Takuya* Nanamura[3,4,], *Tamao* Sakao[1,], *J. K.* Ahn[5,], *Y.* Akazawa[2,], *T.* Aramaki[1,], *S.* Ashikaga[3,], *S.* Callier[6,], *N.* Chiga[1,], *N.* Chiga[1,], *S. W.* Choi[3,], *H.* Ekawa[7,], *P.* Evtoukhovitch[8,], *N.* Fujioka[1,], *M.* Fujita[4,], *T.* Gogami[3,], *T.* Harada[3,4,], *S.* Hasegawa[4,], *S. H.* Hayakawa[1,], *R.* Honda[2,], *S.* Hoshino[9,], *K.* Hosomi[4,], *M.* Ichikawa[3,14,], *Y.* Ichikawa[4,], *M.* Ieiri[2,], *M.* Ikedai[1,], *K.* Imai[4,], *Y.* Ishikawa[1,], *S.* Ishimoto[2,], *W. S.* Jung[5,], *S.* Kajikawa[1,], *H.* Kanauchi[1,], *H.* Kanda[10,], *T.* Kitaoka[1,], *B. M.* Kang[5,], *H.* Kawai[11,], *S. H.* Kim[5,], *K.* Kobayashi[9,], *T.* Koike[1,], *K.* Matsuda[1,], *Y.* Matsumoto[1,], *S.* Nagao[1,**], *R.* Nagatomi[9,], *Y.* Nakada[9,], *M.* Nakagawa[7,], *I.* Nakamura[2,], *M.* Naruki[3,], *S.* Ozawa[1,], *L.* Raux[6,], *T. G.* Rogers[1,], *A.* Sakaguchi[9,], *H.* Sako[4,], *S.* Sato[4,], *T.* Shiozaki[1,], *K.* Shirotori[10,], *K. N.* Suzuki[3,], *S.* Suzuki[1,], *M.* Tabata[11,], *C. d. L.* Taille[6,], *H.* Takahashi[2,], *T.* Takahashi[2,], *T. N.* Takahashi[15,], *H.* Tamura[1,4,], *M.* Tanaka[2,], *K.* Tanida[4,], *Z.* Tsamalaidze[8,12,], *M.* Ukai[2,1,], *H.* Umetsu[1,], *S.* Wada[1,], *T. O.* Yamamoto[4,], *J.* Yoshida[1,***], and *K.* Yoshimura[13,]

[1]1. Graduate School of Science, Tohoku University
[2]2. Institute of Particle and Nuclear Studies (IPNS), High Energy Accelerator Research Organization (KEK)
[3]3. Department of Physics, Kyoto University
[4]4. Advanced Science Research Center (ASRC), Japan Atomic Energy Agency (JAEA)
[5]5. Department of Physics, Korea University
[6]6. OMEGA Ecole Polytechnique-CNRS/IN2P3
[7]7. High Energy Nuclear Physics Laboratory, RIKEN
[8]8 Joint Institute for Nuclear Research (JINR), Dubna
[9]9 Department of Physics, Osaka University
[10]10 Research Center for Nuclear Physics (RCNP), Osaka University
[11]11 Department of Physics, Chiba University
[12]12 Georgian Technical University (GTU)
[13]13 Department of Physics, Okayama University
[14]14 Meson Science Laboratory, Cluster for Pioneering Research, RIKEN
[15]15 Nishina Center for Accelerator-based Science, RIKEN

Abstract. A new hyperon-proton scattering experiment, dubbed J-PARC E40, was performed to measure differential cross sections of the Σ^+p, Σ^-p elastic scatterings and the $\Sigma^-p \to \Lambda n$ scattering by identifying a lot of Σ particles in the momentum ranging from 0.4 to 0.8 GeV/c produced by the $\pi^\pm p \to K^+\Sigma^\pm$ reactions. We successfully measured the differential cross sections of these three channels with a drastically improved accuracy with a fine angular step. These new data will become important experimental constraints to improve the theories of the two-body baryon-baryon interactions. Following this success, we proposed a new experiment to measure the differential cross sections and

*e-mail: miwa9@lambda.phys.tohoku.ac.jp
**Present institution : Graduate Shook of Science, University of Tokyo
***Present institution : International Center for Synchrotron Radiation Innovation Smart, Tohoku University

spin observables by using a highly polarized Λ beam for providing quantitative information on the ΛN interaction. The results of three Σp channels and future prospects of the Λp scattering experiment are described.

1 Introduction

The realistic Nucleon-Nucleon (NN) potentials such as CD Bonn [1] and AV18 [2] NN potentials have been already established. At the stages of updating these theories, a lot of scattering observables of NN scattering were essential inputs. Nowadays, the realistic NN interactions have been making a solid base for nuclear studies. Thanks to these realistic NN interactions, three-nucleon force was confirmed from proton-deuteron scatterings [3]. This three-nucleon force is known as the necessary interaction to reproduce the nuclear binding energies and the equation of state of neutron stars.

The situation is completely different in the Baryon-Baryon (BB) interactions. Theoretical models are developed based on the limited hyperon-nucleon scattering data and the flavor SU(3) symmetry. The Nijmegen model is a representative boson-exchange model [4]. Quark cluster model (QCM) treats the degrees of freedom of the consituent quark [5, 6]. Recently chiral effective field theory, which is widely used in normal nuclear physics, has been extended to the hyperon-nucleon sectors [7, 8]. However, the experimental data for constraining these theories are extremely scarce and not updated from the 1970s. Good quality two-body scattering data are necessary to feed back to and to test the theories including Lattice QCD BB interaction.

In order to change the experimental situation, we started new hyperon-nucleon scattering experiments at J-PARC. We proposed a new hyperon-proton scattering experiment where the scattering events were identified by only the kinematical consistency check without imaging methods. We finally succeeded in measuring the differential cross sections of the $\Sigma^- p$ [9] and $\Sigma^+ p$ [10] elastic scatterings and $\Sigma^- p \to \Lambda n$ reaction [11] with much better accuracy (E40). Then, we proposed a Λp scattering experiment using a polarized Λ beam as a new project at J-PARC (E86) [12]. In this proceedings, We would like to summarize the results of the Σp scattering experiment and mention the future prospects of the Λp scattering experiment.

2 Measurement of the differential cross sections of the Σp reactions at J-PARC E40

Our first step toward precise hyperon-nucleon scattering experiments is the differential cross section measurements of Σp scatterings. One of the biggest motivations in this experiment is to verify the quark Pauli repulsion. In the quark picture, the $\Sigma^+ p$ (spin 1) system contains an up-quark pair having the same spin and color with a high probability. The Pauli effect in the quark level could make the $\Sigma^+ p$ potential quite repulsive. This was firstly predicted by Oka and Yazaki in the quark cluster model [5] and this prediction was confirmed by the Lattice QCD simulation [13, 14]. Another important motivation is to put strong constraints on the BB interaction theories by providing the systematic $d\sigma/d\Omega$ measurements with much better accuracy than past measurements. The theoretical predictions for the Σp differential cross sections are different depending on theories. Therefore, we planned the systematic measurements of the $\Sigma^+ p$ and $\Sigma^- p$ elastic scatterings and the $\Sigma^- p \to \Lambda n$ reaction.

Here we shortly summarize the experimental method. The detailed information can be found in [9]. We detected two successive two-body reactions in a liquid hydrogen (LH_2) target: the first one was the Σ production by the (π, K^+) reaction and the second one was the Σp scattering. The Σ production was identified using the K1.8 beam-line and KURAMA spectrometers and the momentum of Σ particle was tagged using the missing momentum. The interaction of this momentum tagged Σ particle was identified by CATCH detector [15], which comprised a cylindrical fiber tracker and BGO calorimeter.

The Σ particles running in the LH_2 target were regarded as Σ beam. To tag these Σ particles, we detected K^+ with the KURAMA spectrometer. Then, we reconstruct Σ from the missing mass. Each momentum was able to be reconstructed from the missing momentum. The typical momentum range was from 0.45 GeV/c to 0.8 GeV/c. These Σ particles sometimes reacted with a proton in LH_2 target and the recoil proton and decay particles from Σ decay were detected by CATCH. The particle identification for proton and π was performed using the dE-E information between CFT and BGO as shown in FIG. 8 in [9].

In order to identify the $\Sigma^- p$ scattering, the kinematical consistency between the energy and scattering angle was checked. We defined two values, that is, $E_{measured}$ as the measured kinetic energy for the recoil proton and E_{calc} as the calculated energy from the scattering angle based on the $\Sigma^- p$ elastic scattering kinematics. We checked the difference between these two values, that is, $\Delta E = E_{measured} - E_{calc}$. The $\Sigma^- p$ scattering events made a peak at $\Delta E = 0$ as shown in FIG. 15 (c) in [9]. If we changed the kinematical assumption, for example, $\Sigma^- p \to \Lambda n$ reaction, we could also identify the $\Sigma^- p \to \Lambda n$ event from this kinematical consistency check as shown in FIG. 1 (d) in [11].

Finally, we have obtained these differential cross sections for $\Sigma^- p$ elastic scattering and $\Sigma^- p \to \Lambda n$ inelastic scattering as shown in FIG. 17 in [9] and FIG.3 in [11], respectively. The black points are our experimental data and the accuracy has been improved to 10% level for a narrow angular step of $d\cos(\theta) = 0.1$. Now, we can identify the clear forward peaking angular dependence in the $\Sigma^- p$ elastic scattering and moderate forward peaking dependence in the $\Sigma^- p \to \Lambda n$ scattering. The fss2 [6] and chial EFT [7, 8] show a reasonable angular dependence for both channels. On the other hand, Nijmegen ESC models [16, 17] clearly underestimate the forward angle.

Next, we summarize the results obtained for the $\Sigma^+ p$ channel. As we mentioned, this channel is very interesting in relation to the quark Pauli effect. The more repulsive the $^3S_1$ potential is, the larger the cross section is predicted like fss2. However, we have found that the value is much smaller than the fss2 prediction and the E289 results as shown in Fig. 24 in [10]. The size parameter in fss2, which determines the strength of the quark Pauli repulsion, must be smaller than the present value to explain the present experimental data. In Chiral EFT, the differential cross section gets larger in higher momentum. This momentum dependence does not match the present data. On the other hand, Nijmegen models are rather consistent. However, we should keep in mind that the NSC96f predicts the attractive potential for this channel and this does not agree with the present understanding for the ΣN interaction.

Furthermore, we performed a phase-shift analysis for the derived differential cross sections to determine the strength of the quark Pauli repulsion quantitatively. The detailed description can be found in Section 6.2 in [10]. An important point in this analysis is that the ΣN (I=3/2) channel is simply represented by 27-plet and 10-plet. Here, the channels represented by 27-plet, such as the $^1S_0$ channel in the $\Sigma^+ p$ system, are less uncertain because 27-plet is well- estimated from the NN ($I = 1$) interaction based on the flavor SU(3) symmetry. In the phase shift analysis, we set the phase shifts of $^3S_1$ ($\delta_{^3S_1}$) and $^1P_1$ ($\delta_{^1P_1}$) as the fitting parameters, because these two phase shifts are theoretically the most uncertain. Other phase shifts up to D wave were fixed at some values from the Nijmegen model or pp scattering based on the explanation for the 27-plet. Finally, we obtained the momentum dependence of the $^3S_1$ phase shift as shown in Fig. 28 (a) in [10]. The absolute values of $\delta_{^3S_1}$ range from 20 to 35 degrees in the momentum range of 0.44 - 0.80 GeV/c. Because the sign is expected to be negative from the Σ production spectra in nuclei [18], we have concluded that the quark Pauli repulsion is moderately repulsive.

In E40 experiment, the systematic $d\sigma/d\Omega$ measurements of the Σp interaction were performed. Because all channels are related to each other within the framework of flavor SU(3) symmetry, these data must impose strong constraints on the theories of two-body BB interactions. We expect that "realistic" BB interaction will be constructed in the near future with a collaboration between theories and experiments.

3 Future prospects of the Λp scattering (E86)

Next, we would like to move to future prospects toward Λp scattering. The detailed study of the ΣN interaction can only be performed by the scattering experiment, because Σ is merely bound by nuclei. In contrast, the situation is different for the ΛN interaction. Reliable two-body ΛN interaction is a key to deepen the Λ hypernuclear physics. Indeed, such reliable two-body ΛN interaction is indispensable for deriving the ΛNN three-body interaction from Λ hypernuclear binding energies [17, 19]. This is crucially important to solve the so-called hyperon puzzle in neutron stars [20]. Nowadays, new two-body ΛN data are coming from femtoscopies in heavy ion collision experiments [21], new Λp cross section data from Jlab CLAS [22]. We are also planning a new project at J-PARC to measure the Λp scattering with a polarized Λ beam [12]. Here, we describe a feasibility study using the E40 data at first. Then the expected results in the new experiment are also briefly summarized.

3.1 Feasibility study with E40 data

Important experimental issue is how to identify the Λ production via the $\pi^- p \to K^0\Lambda$ reaction. Actually, the (π^-, K^0) spectroscopy method was not established for a long time due to the difficulty of the K^0 detection. In such a situation, we have newly established a new K^0 detection method where the forward emitted π^+ and the transversely emitted π^- from the K^0 decay are measured by the forward KURAMA spectrometer and CATCH surrounding the target, respectively. The detailed analysis for the Λ identification is written in [23]. Then, Λ can be identified in the missing mass spectrum of the $\pi^- p \to K^0 X$ reaction as shown in Fig. 8 in [23].

Based on the past measurement [24], the Λ spin is expected to be highly polarized with respect to the Λ production plane. We also measured the Λ polarization by measuring the angular dependence of protons from the Λ decay with CATCH. In the coordinate system at rest for Λ, the angular distribution of the emitted proton is represented by the following equation,

$$\frac{1}{N_0}\frac{dN}{d\cos\theta_p} = \frac{1}{2}(1 + \alpha P_\Lambda \cos\theta_p), \tag{1}$$

where P_Λ denotes the Λ polarization and α denotes asymmetry parameter. This asymmetry parameter has been measured with a good precision by BES3 collaboration [25] as

$$\alpha = 0.750 \pm 0.009 \pm 0.004. \tag{2}$$

For the polarization measurement, this α value was used. θ_p is the angle of the proton direction from the polarization axis (normal vector of the Λ production plane) in the coordinate system at rest for Λ. The Λ polarizations were obtained by fitting the proton's angular distribution after the CATCH's acceptance correction with the equation (1). We obtained approximately 100% polarization, which was consistent with the past measurements [24]. The details are described in the Sakao's proceedings of HYP2022. Therefore, it is a big advantage that the highly polarized Λ beam can be used for the Λp scattering experiment.

3.2 Next Λp scattering experiment at J-PARC

This high spin polarization of Λ enables us to measure not only the differential cross section but also spin observables such as analyzing power from left-right asymmetry of Λp scattering and depolarization from the up-down asymmetry of protons from scattered Λ's decay. We have proposed a new Λp scattering experiment using the spin polarized Λ beam at the J-PARC K1.1 beam line [12]. This Λp scattering experiment is one of the flagship experiments at the K1.1 beam line of the Hadron experimental facility extension project [26].

Figure 1 shows the sensitivities of analyzing power (A_y) which were obtained from the left-right asymmetry for the Λp scattering for 5×10^7 Λ beam. The expected statistical error is about 10% for the angular step of $d\cos\theta = 0.2$. In this simulation, $A_y = 0$ is assumed to check the experimental validity easily. Figure 2 shows the simulated results for the depolarization (D_y^y) measurement for 10^8 Λ beam. For this measurement, the polarization of the scattered Λ has to be measured from the up-down asymmetry of the decay proton from Λ with respect to the scattering plane. For the spin observables, there is no experimental constraint so far. Therefore, the theoretical prediction is completely different, especially around the ΣN threshold energy [27]. In the middle momentum range, 10% level accuracy can be achieved for both measurements. We believe that these new scattering data become an important constraint to determine spin-dependent ΛN interaction.

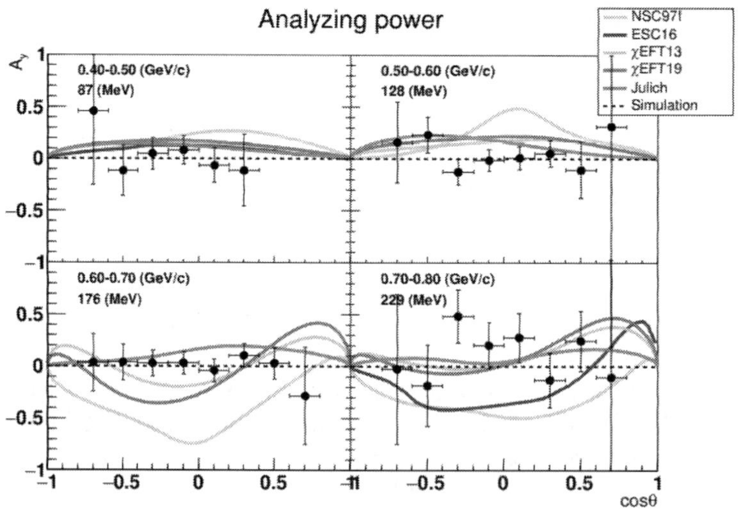

Figure 1. Simulated results for A_y measurement of the Λp scattering in the momentum range from 0.4 to 0.8 GeV/c with 0.1 GeV/c momentum interval for 5×10^7 Λ beam. Two types of Nijmegen models (NSC97f and ESC16 [17]) and Jülich model [28] are presented as the typical example of the boson exchange picture. The two results by chiral EFT (chiral EFT13 and 19) [7, 8] are calculated with the different sets of the LEC parameters both of which reasonably reproduce the YN scattering cross section.

4 Summary

Experimental research on the BB interactions has made great progress in recent years. Femtoscopy has played an important role for studying hadron-hadron interactions between various hadrons produced in high-energy heavy-ion collision experiments. New and accurate

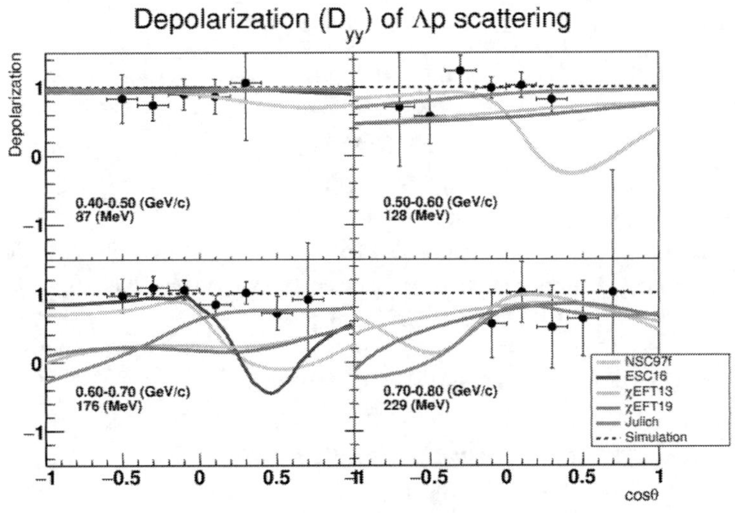

Figure 2. Expected D_y^y accuracy for 10^8 Λ beam. In this simulation, no spin transfer ($D_y^y = 0$) and no induced polarization ($P = 0$) in the Λp scattering are assumed for simplicity. The beam polarization of $P_\Lambda = 1$ is also assumed based on the past measurement. Theoretical calculations are also shown together. Due to the limitation of the statistics, errors in the high momentum region (0.7-0.8 GeV/c) are still large. However, in the middle momentum region (0.5 to 0.7 GeV/c) model difference between the chiral EFT and Nijmegen models can be clearly separated.

hyperon-nucleon scattering data also have become available at J-PARC and Jlab. Especially, we, J-PARC E40 collaboration, succeeded in systematically providing the differential cross sections of the $\Sigma^+ p$, $\Sigma^- p$ and $\Sigma^- p \rightarrow \Lambda n$ channels. The accuracy improved to 10% level for a narrow angular step of $d\cos\theta = 0.1$. Any theories cannot reproduce these data consistently. We believe that these data are essential for improving the theories to more realistic ones. In addition, we derived the $^3S_1$ phase shift in the $\Sigma^+ p$ channel for the first time by performing a phase-shift analysis for the obtained differential cross section. This phase shift value is directly related to the strength of the quark Pauli repulsion in this channel. We have concluded that the interaction in this channel is moderately repulsive from the obtained phase shift values.

We plan new Λp scattering experiment, J-PARC E86, as a future project. By utilizing a highly polarized Λ beam, we can measure not only the differential cross sections but also spin observables such as analyzing power and depolarization. We believe that these new scattering data become an important constraint to determine the spin-dependent ΛN interaction.

5 Acknowledgements

We would like to thank the staff of the J-PARC accelerator and the Hadron Experimental Facility for their support for providing the beam during the beam time. We would like to express thanks to Y. Fujiwara for the theoretical support from an early period of the experimental design and thank T. A. Rijken and J. Haidenbauer for their theoretical calculations. We also thank KEKCC and SINET4. This work was supported by JSPS KAKENHI Grant Number 23684011, 15H00838, 15H05442, 15H02079 and 18H03693. This work was also supported by Grants-in-Aid Number 24105003 and 18H05403 for Scientific Research from the Ministry of Education, Culture, Science and Technology (MEXT) Japan.

References

[1] R. Machleidt, Phys. Rev. C **63**, 024001 (2001)
[2] R. Wiringa, V. Stoks, R. Schiavilla, Phys. Rev. C **51**, 38 (1995)
[3] K. Sekiguchi et al., Phys. Rev. C **65**, 034003 (2002)
[4] T.A. Rijken, V.G.J. Stoks, Y. Yamamoto, Phys. Rev. C **59**, 21 (1999)
[5] M. Oka, K. Shimizu, K. Yazaki, Nucl. Phys. A **464**, 700 (1987)
[6] Y. Fujiwara, Y. Suzuki, C. Nakamoto, Prog. Part. Nucl. Phys. **58**, 439 (2007)
[7] J. Haidenbauer, S. Petschauer, N. Kaiser, U.G. Meißner, A. Nogga, W. Weise, Nucl. Phys. A **915**, 24 (2013)
[8] J. Haidenbauer, U.G. Meißner, A. Nogga, Eur. Phys. J. A **56**, 91 (2020)
[9] K. Miwa et al., Phys. Rev. C **104**, 045204 (2021)
[10] T. Nanamura et al., Prog. Theor. Exp. Phys. **2022**, 093D01 (2022)
[11] K. Miwa et al., Phys. Rev. Lett. **128**, 072501 (2022)
[12] K. Miwa et al., J-PARC E86 proposal, http://j-parc.jp/researcher/Hadron/en/pac_2107/pdf/P86_2021-14.pdf (2021)
[13] T. Inoue et al., Nucl. Phys. A **881**, 28 (2012)
[14] H. Nemura et al., EPJ Web of Conf. **175**, 05030 (2018)
[15] Y. Akazawa, N. Chiga, N. Fujioka, S. Hayakawa, R. Honda, M. Ikeda, K. Matsuda, K. Miwa, Y. Nakada, T. Nanamura et al., Nucl. Instrum. Meth. A **1029**, 166430 (2022)
[16] T.A. Rijken, M.M. Nagels, Y. Yamamoto, Prog. Theor. Phys. **185**, 14 (2010)
[17] M.M. Nagels, T.A. Rijken, Y. Yamamoto, Phys. Rev. C **99**, 044003 (2019)
[18] P. Saha et al., Phys. Rev. C **70**, 044613 (2004)
[19] Y. Yamamoto, T. Furumoto, N. Yasutake, T.A. Rijken, Phys. Rev. C. **90**, 045805 (2014)
[20] P. Demorest et al., Nature **467**, 1081 (2010)
[21] ALICE Collaboration, arXiv:2104.04427 [nucl-ex] (2021)
[22] J. Rowley et al., arXiv:2108.03134 (2021)
[23] T. Sakao et al., JPS Conf. Proc. **33**, 011133 (2021)
[24] R.D. Baker et al., Nucl. Phys. B **141**, 29 (1978)
[25] M. Ablikim et al., Nature Physics **15**, 631 (2019)
[26] K. Aoki et al., arXiv:2110.04462 [nucl-ex] (2021)
[27] J. Haidenbauer, U.G. Meißner, Chin. Phys. C **45**, 094104 (2021)
[28] J. Haidenbauer, U.G. Meißner, Phys. Rev. C **72**, 044005 (2005)

EPJ Web of Conferences **271**, 04002 (2022) https://doi.org/10.1051/epjconf/202227104002
HYP2022

Results of analysis of $\Sigma^+ p$ scattering events in J-PARC E40 experiment: differential cross sections and phase shifts of $^3S_1$ and $^1P_1$ states

Takuya Nanamura[1,2,*], *Koji* Miwa[3,4], *Shuhei H.* Hayakawa[3], *Ryotaro* Honda[4], *Yoshiyuki* Nakada[5], *Toshiyuki* Takahashi[4], *Mifuyu* Ukai[3,4], *Takeshi O.* Yamamoto[2], and J-PARC E40 collaboration

[1]Department of Physics, Kyoto University, Kyoto 606-8502, Japan
[2]Advanced Science Research Center (ASRC), Japan Atomic Energy Agency (JAEA), Tokai, Ibaraki 319-1195, Japan
[3]Department of Physics, Tohoku University, Sendai 980-8578, Japan
[4]Institute of Particle and Nuclear Studies (IPNS), High Energy Accelerator Research Organization (KEK), Tsukuba 305-0801, Japan
[5]Department of Physics, Osaka University, Toyonaka 560-0043, Japan

Abstract. We performed a novel $\Sigma^+ p$ scattering experiment at the J-PARC Hadron Experimental Facility. Approximately 2400 $\Sigma^+ p$ scattering events were identified by a kinematical consistency check for the recoil proton. The differential cross sections of the $\Sigma^+ p$ elastic scattering were derived with better precision than in previous experiments. By exploiting high-quality differential cross section data and the simple representation of $\Sigma^+ p$ interaction in the SU(3) flavor symmetry, we performed a phase-shift analysis on hyperon-nucleon scattering data for the first time. The absolute value of the phase shift of the $^3S_1$ channel, where a large repulsive force was predicted due to the Pauli exclusive effect between quarks, was evaluated. These results indicate that the interaction of the $^3S_1$ channel in the $\Sigma^+ p$ channel is moderately repulsive, as the Nijmegen extended-soft-core models predicted.

1 Introduction

By investigating interactions between octet baryons (*BB* interactions), the role of quarks and gluons in the short-range baryon-baryon interaction can be examined. The *BB* interactions are labeled by six irreducible multiplets as $8 \otimes 8 = 27 \oplus 8_s \oplus 1 \oplus 10^* \oplus 10 \oplus 8_a$. The nuclear force is represented by the 27-plet and 10*-plet and these multiplets are well-known. The other four multiplets are expected to have different features from the nuclear force. Especially, S-wave interactions of the 8_s-plet and 10-plet are predicted to have a quite large repulsive force due to Pauli exclusion principle at a quark level (quark Pauli effect) [1]. Because the $\Sigma N(I = 3/2)$ channel is simply represented by the 27-plet and 10-plet depending on spin and orbital angular momentum states, the $\Sigma^+ p$ interaction is one of the best channels for studying the repulsive nature of the 10-plet. In particular, the $^3S_1$ state in the $\Sigma^+ p$ system is expected to have a large repulsive force due to the quark Pauli effect. This expectation is supported by

*e-mail: nanamura.phys@gmail.com

the experimental information on Σ-nucleus interactions: the spin-isospin-averaged repulsive Σ potential [2, 3] and the isospin dependence in $A = 4$ system [4]. All of recent theoretical models, such as the ESC models [5, 6], fss2 [7], the chiral EFT (χEFT) models [8, 9], and lattice QCD calculation [10], predict repulsive ${}^3S_1$ interactions in the $\Sigma^+ p$ system. However, the strength of the repulsion, that is, the phase shift value of the ${}^3S_1$ channel, is different from each other and it should be determined from the $\Sigma^+ p$ scattering experiment.

Hyperon-nucleon scattering experiments have been difficult due to the short lifetime of hyperons. Regarding the $\Sigma^+ p$ scattering for the intermediate energy, two experiments to measure the differential cross section were performed at KEK [11, 12]. In these experiments, however, no conclusions could be reached owing to the insufficient precision stemming from low statistics. We, J-PARC E40 collaboration, have recently succeeded in the systematic measurements of $\Sigma^\pm p$ scatterings with high statistics. $\Sigma^- p$ results have already reported in Ref. [13, 14]. In this paper, we report the $\Sigma^+ p$ results, the differential cross sections and phase shifts. Details of analysis are described in Ref. [15].

2 Experiment and analysis

2.1 Experimental concept

The experiment was performed at the K1.8 beam line [16] in the J-PARC Hadron Experimental Facility with π^+ beams of 1.41 GeV/c. A conceptual drawing of the experiment is shown in Fig. 1. Σ^+ particles were produced in a liquid hydrogen (LH_2) target via the $\pi^+ p \to K^+ \Sigma^+$ reaction. Then $\Sigma^+ p$ scattering can occur during the Σ^+ flight in the LH_2 target. In order to identify $\Sigma^+ p$ scattering events, the incident Σ^+' s momentum, the recoil proton's kinetic energy, and recoil angle should be determined. The momentum of each Σ^+ particle can be reconstructed as the missing momentum of the π^+ beam and scattered K^+, analyzed using two magnetic spectrometers. Charged particles involved in the produced Σ^+, such as the recoil proton from $\Sigma^+ p$ scattering and a decay proton from $\Sigma^+ \to p\pi^0$ decay, were detected using the CATCH detector system [17] surrounding the LH_2 target.

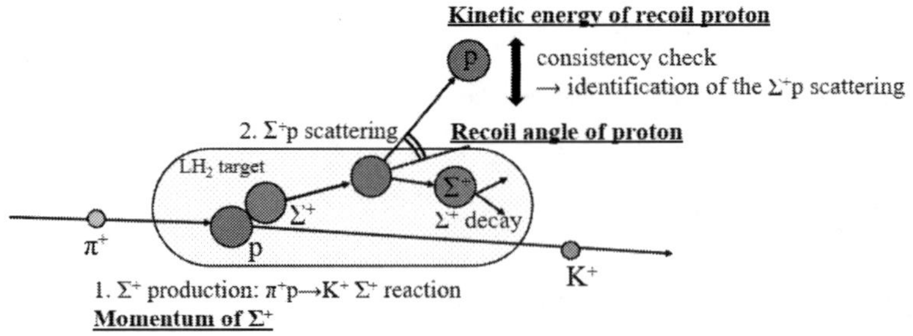

Figure 1. Experimental concept of the $\Sigma^+ p$ scattering experiment. The LH_2 target was used as the target for both Σ^+ production and $\Sigma^+ p$ scattering. $\Sigma^+ p$ scattering events can be kinematically identified by measuring the incident Σ^+'s momentum, the recoil proton's kinetic energy, and recoil angle.

2.2 Identification of $\Sigma^+ p$ scattering events

To identify $\Sigma^+ p$ scattering events, we define the kinematic consistency for the recoil proton as $\Delta E = E_{\text{meas}} - E_{\text{calc}}$, where E_{meas} means the recoil proton's kinetic energy measured using

CATCH and E_{calc} represents the calculated recoil proton's kinetic energy from the incident Σ^+'s momentum and recoil angle. For $\Sigma^+ p$ scattering signal events, ΔE values distribute around 0. Therefore, $\Sigma^+ p$ scattering events could be identified as a peak on backgrounds in the ΔE distribution. Contribution of background reactions, such as the pp scattering following the $\Sigma^+ \to p\pi^0$ decay and the accidental coincidences were estimated using a Monte Carlo simulation. In total, we identified approximately 2400 $\Sigma^+ p$ scattering events, which is 80 times larger than those achieved in past experiments.

3 Results and discussion

3.1 Differential cross sections

In addition to the numbers of $\Sigma^+ p$ scattering events, we evaluated the total flight length of incident Σ^+s in the LH_2 target and efficiency for the $\Sigma^+ p$ scattering events, including the detection and analysis efficiency, by Monte Carlo simulations. Then, we could calculate differential cross sections. Derived differential cross sections were shown in Fig. 2. The data quality in the present experiment has been improved significantly, with errors of typically less than 20% and a fine angular step ($\Delta \cos\theta = 0.1$). The systematic errors, shown in Fig. 2 as error boxes, came from the background estimation and ambiguity of detection efficiency for low momentum protons. Compared to theoretical calculations, our data is smaller than most theoretical predictions. Especially, FSS and fss2, based on Quark Cluster Model are much larger than the present data. In terms of the differential cross sections, NSC97f[18] agrees well with our data for $p_\Sigma > 0.55$ GeV/c. However, it predicts an attractive $\Sigma^+ p$ interaction, which does not agree with our understanding of the ΣN interaction as mentioned in Sect. 1.

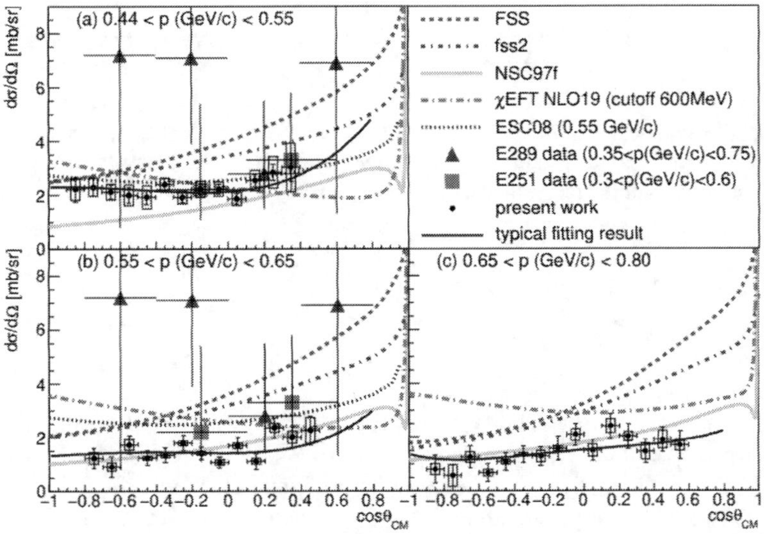

Figure 2. Derived differential cross sections of the $\Sigma^+ p$ scattering for the three momentum regions. The error bars and boxes show the statistical and systematic uncertainties, respectively. The red boxes and blue triangles shows the data of past experiments, KEK E251 and KEK E289. The navy blue lines represents the typical fitting results in the phase shift analysis, described in 3.2. The other lines shows the theoretical calculations.

3.2 Phase shift analysis

Extracting the contribution of the $^3S_1$ states is important to study the repulsive nature of Σ^+p system due to the quark Pauli effect. For this purpose, we performed a phase-shift analysis. This is the first attempt for hyperon-nucleon scattering data. We represented the differential cross section as a function of eleven phase shifts up to D waves. Nine phase shifts except for the $^3S_1$ and $^1P_1$ states are certain to some extent and can be fixed to reasonable values. Therefore, phase shifts of the remaining two states can be obtained by fitting. The typical fitting results are shown in Fig. 2 and χ^2/ndf are around 1. Obtained phase shifts are shown in Fig. 3. The absolute values of $\delta_{^3S_1}$ are determined within 4 degrees of error. Note that both signs of $\delta_{^3S_1}$ are numerically possible, which can be roughly understood from the fact that differential cross sections would be proportional to $(\sin\delta_0/k)^2$ in the S-wave limit. If the sign is assumed to be negative, the momentum dependence is consistent with ESC16, suggesting that the repulsive force is moderate. In contrast, the obtained $\delta_{^1P_1}$ values deviate considerably in the range of $-5° < \delta_{^1P_1} < 25°$ depending on the set of fixed parameters. Although the results of $\delta_{^1P_1}$ are ambiguous, they may support the predictions of fss2, ESC, and NSC97f, in which the interaction of the $^1P_1$ state in the Σ^+p system is weakly attractive.

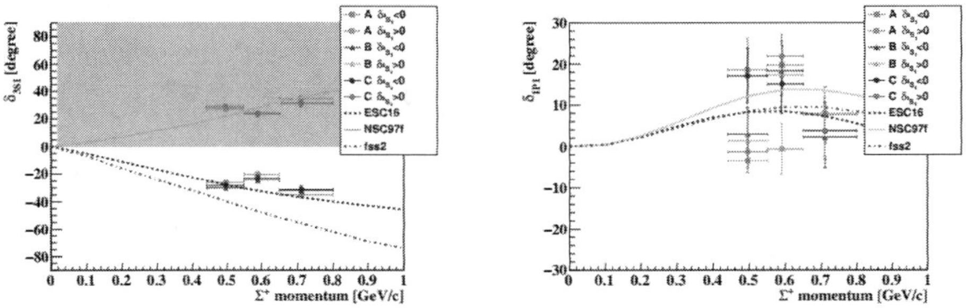

Figure 3. Obtained phase shifts $\delta_{^3S_1}$ and $\delta_{^1P_1}$ as a function of the incident Σ^+ momentum. A, B, and C in the legend correspond to the examined sets of fixed parameters. The black dashed, green solid, and blue dotted lines represent the theoretical calculations of ESC16, NSC97f, and fss2, respectively.

4 Summary

The Σ^+p channel is one of the best channels for investigating 10-plet of the BB interaction, which is closely related to the quark Pauli effect. To verify the strong repulsive force due to the quark Pauli effect in the Σ^+p channel, we performed a novel high-statistics Σ^+p scattering experiment at J-PARC. Approximately 2400 Σ^+p scattering events were identified for the incident Σ^+ momentum range 0.44-0.80 GeV/c. The differential cross sections with less than 20 % of errors and a fine angular step were derived. Owing to the precise data points and simple representation of the Σ^+p system, we could derive the phase shifts of the $^3S_1$ and $^1P_1$ states for the first time by performing a phase-shift analysis. Especially, the absolute values of $\delta_{^3S_1}$ range from 20° to 35° in the present momentum range. Comparing these results to theoretical calculations, the Nijmegen ESC models are consistent with present data, which suggests the $^3S_1$ interaction in the Σ^+p system is moderately repulsive. Because theoretical models has been built on little data, our data will be an important input for improving them.

References

[1] M. Oka, K. Shimizu, K. Yazaki, Nucl. Phys. A **464**, 700 (1987)
[2] H. Noumi et al., Phys. Rev. Lett. **89**, 072301 (2002)
[3] P.K. Saha et al., Phys. Rev. C **70**, 044613 (2004)
[4] T. Nagae et al., Phys. Rev. Lett. **80**, 1605 (1998)
[5] T.A. Rijken, M.M. Nagels, Y. Yamamoto, Prog. Theor. Phys. **185**, 14 (2010)
[6] M.M. Nagels, T.A. Rijken, Y. Yamamoto, Phys. Rev. C **99**, 044003 (2019)
[7] Y. Fujiwara, Y. Suzuki, C. Nakamoto, Prog. Part. Nucl. Phys. **58**, 439 (2007)
[8] J. Haidenbauer, S. Petschauer, N. Kaiser, U.G. Meiner, A. Nogga, W. Weise, Nucl. Phys. A **915**, 24 (2013)
[9] J. Haidenbauer, U.G. Meiner, A. Nogga, Eur. Phys. J. A **56**, 91 (2020)
[10] T. Inoue et al., AIP Conf. Proc. **2130**, 020002 (2019)
[11] J.K. Ahn et al., Nucl. Phys. A **648**, 263 (1999)
[12] J.K. Ahn et al., Nucl. Phys. A **761**, 41 (2005)
[13] K. Miwa et al., Phys. Rev. C **104**, 045204 (2021)
[14] K. Miwa et al., Phys. Rev. Lett. **128**, 072501 (2022)
[15] T. Nanamura, K. Miwa et al., PTEP **2022**, 093D01 (2022)
[16] K. Agari et al., PTEP **2012**, 02B009 (2012)
[17] Y. Akazawa et al., Nucl. Inst. Meth. A **1029**, 166430 (2022)
[18] T.A. Rijken, V.G.J. Stoks, Y. Yamamoto, Phys. Rev. C **59**, 21 (1999)

Status of the hyperon-nucleon interaction in chiral effective field theory

Johann Haidenbauer[1,*] and *Ulf-G.* Meißner[2,1,**]

[1]Institute for Advanced Simulation, Institut für Kernphysik and Jülich Center for Hadron Physics, Forschungszentrum Jülich, D-52425 Jülich, Germany
[2]Helmholtz-Institut für Strahlen- und Kernphysik and Bethe Center for Theoretical Physics, Universität Bonn, D-53115 Bonn, Germany

Abstract. The Jülich-Bonn group aims at an extensive study of the baryon-baryon (BB) interaction involving strange baryons (Λ, Σ, Ξ) within SU(3) chiral effective field theory. An overview of achievements and new developments over the past few years is provided. The topics covered are: 1) Derivation of the leading charge-symmetry breaking (CSB) interaction for the ΛN system and its application in a study of CSB effects in A=4 Λ-hypernuclei. 2) Updated results for the ΞN interaction at NLO and predictions for $\Xi^- p$ correlation functions. 3) Extension of the ΛN-ΣN interaction to next-to-next-to-leading order.

1 Introduction

The Jülich-Bonn group explores the baryon-baryon (BB) interaction involving hyperons within SU(3) chiral effective field theory (EFT). In this approach a potential is established via an expansion in terms of small momenta, subject to an appropriate power counting, so that the results can be improved systematically by going to higher orders, while at the same time theoretical uncertainties can be estimated [1, 2]. Furthermore, two- and three-baryon forces can be constructed in a consistent way. The resulting interaction potentials can be readily employed in standard two- and few-body calculations. They consist of contributions from an increasing number of pseudoscalar-meson exchanges, determined by the underlying chiral symmetry, and of contact terms which encode the unresolved short-distance dynamics and whose strengths are parameterized by a priori unknown low-energy constants (LECs).

The studies, performed so far up to next-to-leading order (NLO) in the chiral expansion, have shown that for the strangeness $S = -1$ (ΛN, ΣN) [3, 4] and $S = -2$ ($\Lambda\Lambda$, ΞN) [5, 6] sectors a consistent and satisfactory description of the available scattering data and other experimental constraints can be achieved within the assumption of (broken) SU(3) flavor symmetry. Applications of the resulting potentials in bound-state calculations for light hypernuclei [4, 7–10] led to results close to the empirical values. In addition, the exploration of neutron-star properties with the strangeness $S = -1$ interaction indicate the potential to resolve the so-called hyperon puzzle [11], when combined with consistently derived ΛNN and ΣNN three-body forces [12, 13]. Finally, the BB potentials have been successfully tested

*e-mail: j.haidenbauer@fz-juelich.de
**e-mail: meissner@hiskp.uni-bonn.de

in the analysis of two-particle momentum correlation functions involving strange baryons [14–16].

In this contribution we present some highlights from recent investigations. In Sect. 2 preliminary results of an extension of the ΛN-ΣN interaction up to next-to-next-to leading order (N^{2}LO) [17] are reported. In Sect. 3 a calculation of charge-symmetry breaking (CSB) of the separation energies of the A=4 Λ-hypernuclei ${}^4_\Lambda$He and ${}^4_\Lambda$H is reviewed [18]. Finally, in Sect. 4 selected results involving the ΞN interaction [6] are summarized. Specifically, predictions for the $\Xi^- p$ momentum correlation function are provided [15].

2 ΛN-ΣN interaction at next-to-next-to-leading order

While the description of the NN interaction within chiral EFT has been already pushed up to the fifth order [19, 20], corresponding applications of that framework to the YN interaction are lagging far behind. Here, NLO is presently the state of the art [3–6]. That status is primarily a consequence of the unsatisfactory situation with regard to the data base where practically only cross sections are available and primarily for energies near the thresholds. In particular, differential observables that would allow to fix the LECs in P- and/or higher partial waves, which arise in the chiral expansion when going to higher order, are rather scarce and of low statistics. Only within the last few years the overall circumstances became more promising, thanks to experiments performed by the E40 Collaboration at the J-PARC facility. That collaboration has already published differential cross sections for the $\Sigma^+ p$ and $\Sigma^- p$ channels for momenta from 440 MeV/c to 850 MeV/c [21–23] and corresponding measurements for Λp are in the stage of preparation, including possibly even spin-dependent observables.

This development was one of the reasons to extend our study of the ΛN-ΣN interaction to the next order. However, there are also several theoretical aspects which make an extension to N^{2}LO rather interesting. One of them is that in the Weinberg counting three-baryon forces emerge at this order. Calculations of the four-body systems ${}^4_\Lambda$H and ${}^4_\Lambda$He for the NLO13 [3] and NLO19 [4] potentials based on the Faddeev-Yakubovsky equations indicate that the experimental separation energies are underestimated [4]. Thus, there is obviously a need for including ΛNN and possibly also ΣNN three-body forces. Another appealing factor is (in view of the mentioned scarcity of data) that no new LECs appear at this order. At the same time pertinent results for NN scattering indicate that there is some improvement in the energy dependence of the S-waves and, specifically, in several P-waves once the contributions involving the sub-leading πN vertices that enter at N^{2}LO are taken into account. Certainly, no major improvement can be expected with regard to the residual regulator dependence [24]. In general, a substantial reduction of regulator artifacts can be only achieved by going to high order where then the larger number of LECs allows one to absorb the regulator dependence more efficiently. Since our calculation is at low order it is advantageous to keep such artifacts as small as possible. For that purpose a novel regularization scheme proposed and applied in Ref. [19] seems to be rather promising. Here a local regulator is applied to the pion-exchange contributions and only the contact terms, being non-local by themselves, are regularized with a non-local function. (In earlier works the latter has been applied to the whole potential [3, 4, 25].) Accordingly, the resulting interactions were called "Semilocal momentum-space regularized (SMS) chiral NN potentials" [19]. A local regulator for pion-exchange contributions leads to a reduction of the distortion in the long-range part of the interaction and, thereby, facilitates a more rapid convergence with increasing chiral order. Of course, this effect cannot be directly quantified in case of ΛN and ΣN because of the lack of corresponding empirical information. Nonetheless, given that we want to compare with the new J-PARC data at momenta around 500 MeV/c a reduction of regulator artifacts is definitely desirable.

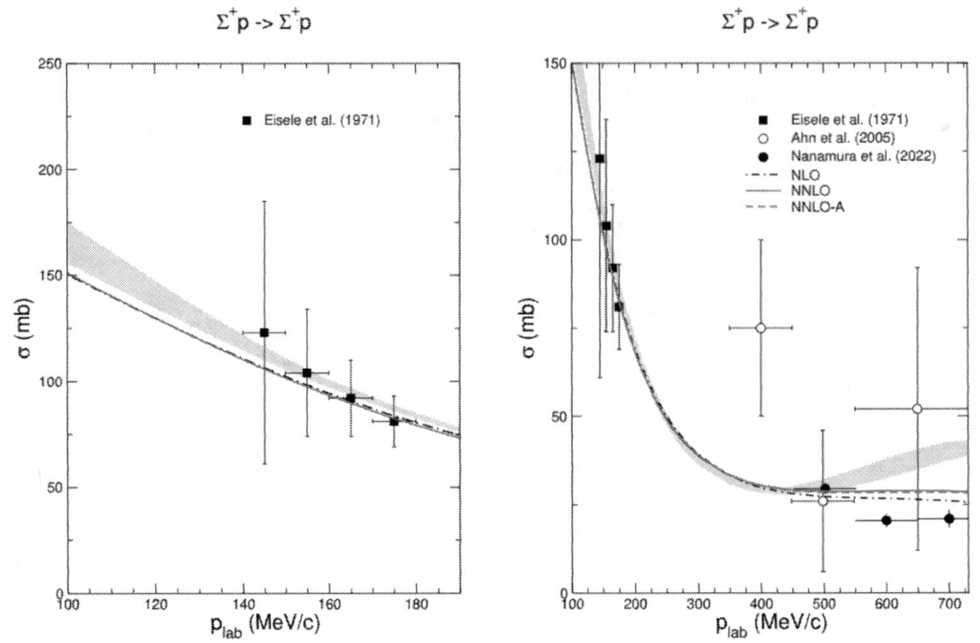

Figure 1. Cross section for $\Sigma^+ p$ scattering as a function of p_{lab}. Results are shown for the SMS NLO (dash-dotted) and N^{2}LO (solid) *YN* potentials. The dashed line corresponds to an alternative fit at N^{2}LO, see text. The cyan band is the result for NLO19 [4]. Data are from the E40 Collaboration [23] and from Refs. [28, 29].

The extension of our *YN* interaction up to N^{2}LO builds on the SMS scheme proposed in Ref. [19]. Details of the formalism will be reported elsewhere [17]. However, for illustration we show how the non-local exponential regulator employed in our *YN* potentials NLO13 [3] and NLO19 [4] for the meson-exchange part is replaced by a local regularization,

$$V_P^{\text{non-local}} \propto \frac{e^{-\frac{p'^4+p^4}{\Lambda^4}}}{\vec{q}^2 + M_P^2} \quad \rightarrow \quad V_P^{\text{local}} \propto \frac{e^{-\frac{\vec{q}^2+M_P^2}{\Lambda^2}}}{\vec{q}^2 + M_P^2}\,. \tag{1}$$

Here P stands for π, K, or η, $\vec{p}$ and $\vec{p}\,'$ are the incoming and outgoing center-of-mass momenta of the baryons, $\vec{q} = \vec{p}\,' - \vec{p}$ is the momentum transfer, and Λ is the cutoff parameter.

In the *NN* case, where only pion exchanges are taken into account, cutoff values in the range $\Lambda = 350 - 550$ MeV were considered where $\Lambda = 450$ MeV yielded the best result [19]. The choice of the cutoff mass for the *YN* interaction is more delicate, because we want to keep the underlying approximate SU(3) flavor symmetry as well as the explicit SU(3) breaking in the long-range part of the potential due to the mass splitting between the pseudoscalar mesons π, K, and η. Since the kaon mass is around 495 MeV it seems advisable to use cutoff masses that are at least 500 MeV, in order to capture the pertinent physics. On the other hand, large values, say above 650 MeV, are questionable because, like in *NN*, we want to avoid spurious bound states. These considerations suggest that two-meson exchange contributions involving a K and/or η (πK, KK, etc.) should and can no longer be included explicitly but have to be absorbed into the contact terms. Thus, contrary to our earlier work [3, 4] we allow and expect some SU(3) breaking between the LECs in the ΛN and ΣN systems.

In the following we present preliminary results for the cutoff value of 550 MeV. We focus on the ΣN channels where new data from J-PARC have become available. We start

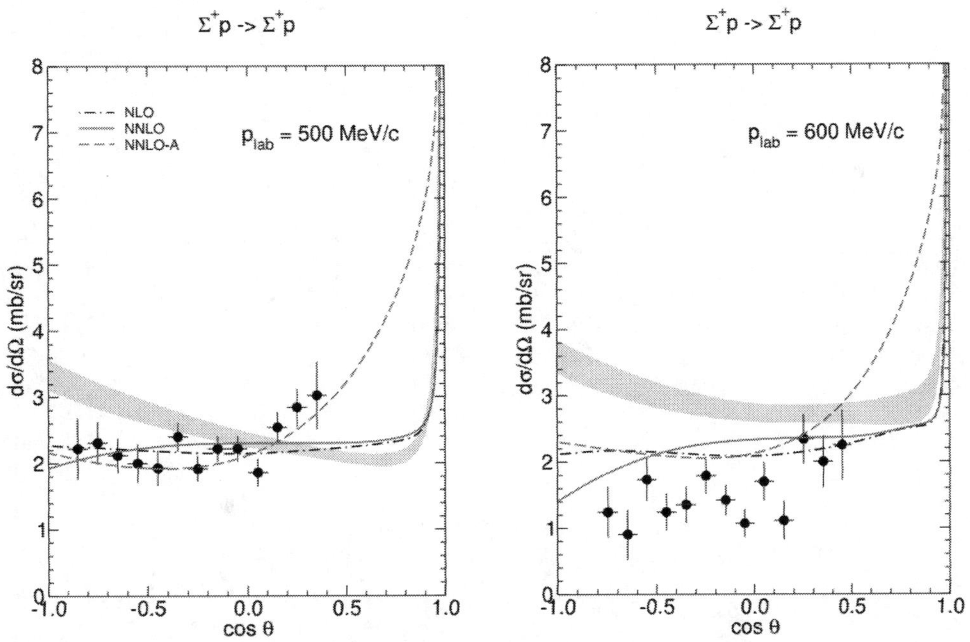

Figure 2. Differential cross section for $\Sigma^+ p$. Same description of curves as in Fig. 1. Data are from Nanamura et al. [23] (momentum regions 440-550 MeV/c and 550-650 MeV/c).

the discussion with $\Sigma^+ p$ scattering which is of particular interest for theory. Since the total isospin is $I = 3/2$ there is no coupling to the ΛN channel which simplifies the dynamics. Moreover, SU(3) symmetry provides strong constraints on several amplitudes. Specifically, space-spin antisymmetric partial waves ($^1S_0$, $^3P_{0,1,2}$, ...) belong all to the {27} irrep. of SU(3) symmetry [3, 4] and, thus, the corresponding interactions would be identical to those in the NN system provided that SU(3) symmetry is exactly fulfilled. While we know that there is a sizable SU(3) breaking in case of the $^1S_0$ partial wave [26], the amplitudes in the P- and higher partial waves could be much closer to those found for NN scattering.

In fitting to the YN data we proceed as before [3, 4], i.e. we consider only the set of 36 data for Λp, $\Sigma^- p$ and $\Sigma^+ p$ scattering at low energies for determining the LECs in the S waves. SU(3) symmetry is imposed for the contact terms at the initial stage but eventually relaxed for the LO LECs, in line with the power counting where SU(3) breaking terms arise from mass insertions in the chiral Lagrangian at the NLO level [27]. Anyway, as said we do expect some SU(3) breaking in the contacts terms in view of the fact that two-meson exchange contributions from πK, $\pi\eta$, etc. are not explicitly included. The achieved χ^2 is comparable to the one found for our NLO interactions [3, 4], and typically around 16 for the 36 data points. The pertinent results for $\Sigma^+ p$ are presented in Fig. 1 (left) and compared with data and with the results obtained from the NLO19 potential. The latter are shown as bands, representing the cutoff dependence [4].

Once the S-wave LECs are fixed, a fit to the differential cross sections reported by the E40 Collaboration is performed, starting with the $\Sigma^+ p$ data for the reasons discussed above. Interestingly, in the NLO case taking over the LECs in the $^3P_{0,1,2}$ partial waves from the corresponding NN potential by Reinert et al. [19], in accordance with SU(3) symmetry, (and assuming the LEC in the $^1P_1$ to be zero) yields already a good description of the E40 data taken in the region 440-550 MeV/c, cf. Fig. 2 (left). For the N^{2}LO interaction all P-wave LECs are fitted to the data. Actually, here we explore two scenarios, one where the resulting

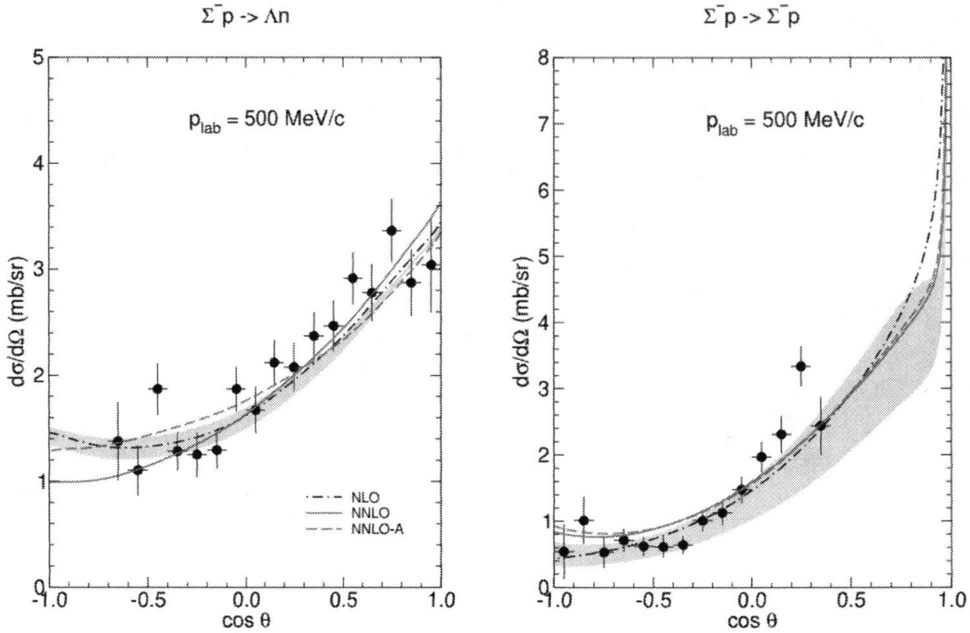

Figure 3. Differential cross sections for $\Sigma^- p \to \Lambda n$ and $\Sigma^- p \to \Sigma^- p$. Same description of curves as in Fig. 1. Data are from the E40 Collaboration [21, 22] (momentum region 470-550 MeV/c).

angular distribution is similar to that obtained for NLO and one which produces an overall more pronounced angular dependence. The latter is clearly preferred by the available data in that momentum range. Unfortunately, the data for the next momentum region, 550-650 MeV/c, suggest an overall somewhat different angular dependence, see Fig. 2 (right), so that conclusions are difficult to draw at present. In any case the predictions by NLO19 are clearly at odds with the data.

The integrated $\Sigma^+ p$ cross section over a larger energy range is shown in Fig. 1 (right). Note that the common angular averaging is performed here, cf. e.g. Eq. (24) in Ref. [3], for the E40 data (and accordingly for the theory results) in order to compensate for the incomplete angular coverage. Again the NLO19 potential does not reproduce the trend of the data. We believe that the rise of the cross section for larger p_{lab} could be an artifact of the employed regulator. Anyway, p_{lab} = 600 MeV/c corresponds to a laboratory energy of $T_{lab} \approx$ 150 MeV so that we are certainly in a region where NLO and possibly even N^{2}LO cannot be expected to be quantitatively reliable. Moreover, one should keep in mind that the $\Lambda N \pi$ channel opens around that energy.

Results for the differential cross sections in the reactions $\Sigma^- p \to \Sigma^- p$ and $\Sigma^- p \to \Lambda n$ are presented in Fig. 3. In this case there is good agreement with the E40 data for all considered YN potentials.

3 Charge symmetry breaking in A=4 Λ-hypernuclei

In Ref. [18] we have studied effects from CSB in the YN interaction based on the potentials NLO13 and NLO19. Specifically, we have utilized the experimentally known difference of the Λ separation energies in the mirror nuclei ${}^4_\Lambda$He and ${}^4_\Lambda$H to constrain the Λ-neutron interaction. The CSB part of the ΛN potential at NLØ (in the notation of [25]) is given by

Table 1. Λp and Λn scattering lengths (in fm) in the singlet and triplet S waves for the NLO19 potential, for cutoffs of 500-650 MeV. In addition, the resulting level splittings for the $A = 4$ mirror nuclei ${}^4_\Lambda$H and ${}^4_\Lambda$He (in keV) are listed.

	$a_s^{\Lambda p}$	$a_s^{\Lambda n}$	$a_t^{\Lambda p}$	$a_t^{\Lambda n}$	$\Delta E(0^+)$	$\Delta E(1^+)$
NLO19(500)	-2.649	-3.202	-1.580	-1.467	249	-75
NLO19(550)	-2.640	-3.205	-1.524	-1.407	252	-72
NLO19(600)	-2.632	-3.227	-1.473	-1.362	243	-67
NLO19(650)	-2.620	-3.225	-1.464	-1.365	250	-69

[18]

$$
\begin{aligned}
V^{CSB}_{\Lambda N\to\Lambda N} &= \Bigg[- f^{(\Lambda-\Sigma^0)}_{\Lambda\Lambda\pi} f_{NN\pi} \frac{(\sigma_1\cdot\mathbf{q})(\sigma_2\cdot\mathbf{q})}{\mathbf{q}^2+M^2_{\pi^0}} \\
&- f^{(\eta-\pi^0)}_{\Lambda\Lambda\pi} f_{NN\pi} (\sigma_1\cdot\mathbf{q})(\sigma_2\cdot\mathbf{q}) \left(\frac{1}{\mathbf{q}^2+M^2_{\pi^0}} - \frac{1}{\mathbf{q}^2+M^2_{\eta}} \right) \\
&+ \frac{1}{4}(1-\sigma_1\cdot\sigma_2)\, C^{CSB}_{^1S_0} + \frac{1}{4}(3+\sigma_1\cdot\sigma_2)\, C^{CSB}_{^3S_1} \Bigg] \tau_N\,. \qquad (2)
\end{aligned}
$$

The CSB contributions arise from a non-zero $\Lambda\Lambda\pi$ coupling constant which is estimated from $\Lambda - \Sigma^0$ ($f^{(\Lambda-\Sigma^0)}_{\Lambda\Lambda\pi}$) and $\eta - \pi^0$ ($f^{(\eta-\pi^0)}_{\Lambda\Lambda\pi}$) mixing [30], respectively, and from two contact terms, $C^{CSB}_{^1S_0}$ and $C^{CSB}_{^3S_1}$, that represent short-ranged CSB forces. In addition, there is a small contribution due to the mass difference between $K^\pm$ and K^0 [18]. $\tau_p = 1$ and $\tau_n = -1$.

In order to fix the LECs $C^{CSB}_{^1S_0}$ and $C^{CSB}_{^3S_1}$ in Eq. (2) the observed CSB splittings for the $A = 4$ hypernuclei, defined in the usual way in terms of the separation energies,

$$
\Delta E(0^+) = E^{0^+}_\Lambda({}^4_\Lambda\mathrm{He}) - E^{0^+}_\Lambda({}^4_\Lambda\mathrm{H}), \qquad \Delta E(1^+) = E^{1^+}_\Lambda({}^4_\Lambda\mathrm{He}) - E^{1^+}_\Lambda({}^4_\Lambda\mathrm{H})\,, \qquad (3)
$$

is considered. We aim at a reproduction of the present experimental situation, established by the recent measurements of the ${}^4_\Lambda$H 0^+ state in Mainz [31] and the one of the ${}^4_\Lambda$He 1^+-0^+ splitting at J-PARC [32], which implies $\Delta E(0^+) = 233 \pm 92$ keV and $\Delta E(1^+) = -83 \pm 94$ keV.

Our results for NLO19 are summarized in Table 1. One can see that the reproduction of the splittings $\Delta E(0^+)$ and $\Delta E(1^+)$ requires a sizable difference between the strength of the Λp and Λn interactions in the $^1S_0$ state. The modifications in the $^3S_1$ partial wave are much smaller and the effect goes also in the opposite direction, i.e. while for $^1S_0$ the Λp interaction is found to be noticeably less attractive than Λn, in case of $^3S_1$ it is slightly more attractive. In terms of the pertinent scattering lengths we predict for $\Delta a^{CSB} = a_{\Lambda p} - a_{\Lambda n}$ a value of 0.62 ± 0.08 fm for the $^1S_0$ partial wave and -0.10 ± 0.02 fm for $^3S_1$. An investigation of CSB in $A = 7, 8$ Λ-hypernuclei is in preparation [38].

4 Aspects of the ΞN interaction

Chiral potentials up to NLO for the BB interaction in the strangeness $S = -2$ system ΞN have been already established by us in 2016 [5]. Thereby constraints from the $\Lambda\Lambda$ scattering length in the $^1S_0$ state together with experimental upper bounds on the cross sections for ΞN scattering and for the transition $\Xi N \to \Lambda\Lambda$ have been exploited. This allowed us to fix the additional LECs that arise in the $\{1\}$ irreducible representation of SU(3) [5]. Furthermore, the consideration of those empirical constraints necessitated to add SU(3) symmetry breaking

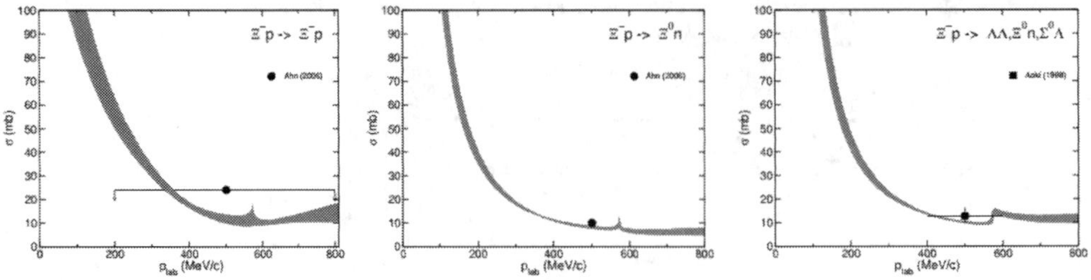

Figure 4. Results for the ΞN cross sections of our NLO potential [6]. Data are from Refs. [36, 37].

contact terms in other irreps ({27}, {10}, {10*}, {8_s}, {8_a}), with regard to those determined from the ΛN and ΣN data. This is anyway expected and fully in line with the power counting of SU(3) chiral EFT, as already mentioned in Sect. 2. In 2019 a modified version has been suggested [6] which is more attractive in the ${}^3S_1$ partial wave with isospin $I = 1$. That potential yields a moderately attractive (in-medium) Ξ-nuclear interaction [6, 33] and supports the existence of bound Ξ-hypernuclei [9], in accord with experimental evidence [34, 35]. The interactions in the $(I = 0, 1)$ ${}^1S_0$ partial waves are the same in the two versions.

Results for the ΞN cross sections based on the ΞN potential from 2019 are presented in Fig. 4. An interesting and independent test for the ΞN interaction is provided by two-particle momentum correlations. For $\Xi^- p$, correlation functions have been measured recently by the ALICE Collaboration in p-Pb collisions at 5.02 TeV [39] and in pp collisions at 13 TeV [40]. There are also new and still preliminary results from Au+Au collisions at 200 GeV by the STAR Collaboration [41]. In Fig. 5 we present predictions for $C(k)$ for the $S = -2$ interaction from 2019. The bands reflect the residual cutoff dependence [6]. Details on the evaluation of such correlation functions can be found, e.g., in Refs. [42, 43]. Clearly, the correlation functions, evaluated for the source radii R taken from the corresponding pp fits by ALICE [44] (1.43 fm for 5.02 TeV and 1.18 fm for 13 TeV), agree nicely with the measurements. For a more thorough discussion on the choice of R and of the other parameters that enter into the calculation, see [15, 43].

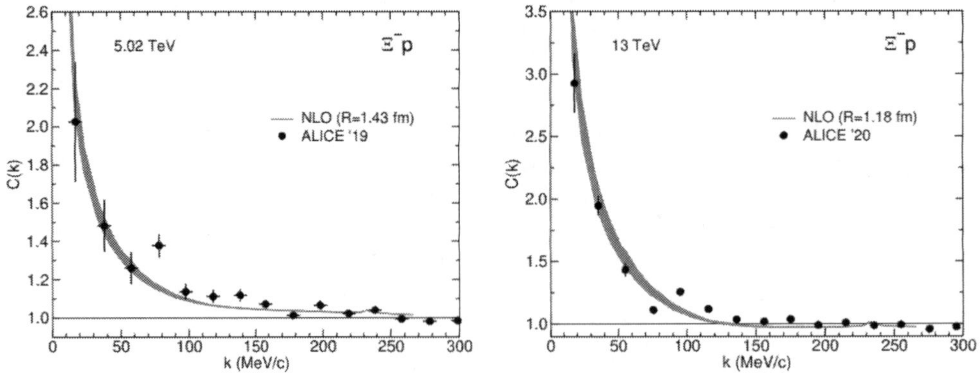

Figure 5. Predictions for the $\Xi^- p$ two-particle momentum correlation function $C(k)$ of our NLO potential [6]. Data are from the ALICE Collaboration [39, 40].

Acknowledgements: Work supported by the European Research Council (ERC) under the European Union's Horizon 2020 research and innovation programme (grant no. 101018170, EXOTIC), and by the DFG and the NSFC through funds provided to the Sino-German CRC 110 "Symmetries and the Emergence of Structure in QCD" (DFG grant. no. TRR 110).

References

[1] E. Epelbaum, H.-W. Hammer and U.-G. Meißner, Rev. Mod. Phys. **81**, 1773 (2009)
[2] R. Machleidt and D. R. Entem, Phys. Rept. **503**, 1 (2011)
[3] J. Haidenbauer, S. Petschauer, N. Kaiser, U.-G. Meißner, A. Nogga and W. Weise, Nucl. Phys. A **915**, 24 (2013)
[4] J. Haidenbauer, U.-G. Meißner and A. Nogga, Eur. Phys. J. A **56**, 91 (2020)
[5] J. Haidenbauer, U.-G. Meißner and S. Petschauer, Nucl. Phys. A **954**, 273 (2016)
[6] J. Haidenbauer and U.-G. Meißner, Eur. Phys. J. A **55**, 23 (2019)
[7] H. Le, J. Haidenbauer, U.-G. Meißner and A. Nogga, Eur. Phys. J. A **56**, 301 (2020)
[8] H. Le, J. Haidenbauer, U.-G. Meißner and A. Nogga, Eur. Phys. J. A **57**, 217 (2021)
[9] H. Le, J. Haidenbauer, U.-G. Meißner and A. Nogga, Eur. Phys. J. A **57**, 339 (2021)
[10] Hoai Le, Contribution to these proceedings, arXiv:2210.02860
[11] D. Chatterjee and I. Vidaña, Eur. Phys. J. A **52**, 29 (2016)
[12] J. Haidenbauer, U.-G. Meißner, N. Kaiser and W. Weise, Eur. Phys. J. A **53**, 121 (2017)
[13] D. Gerstung, N. Kaiser and W. Weise, Eur. Phys. J. A **56**, 175 (2020)
[14] S. Acharya *et al.* [ALICE], Phys. Lett. B **833**, 137272 (2022)
[15] J. Haidenbauer and U.-G. Meißner, [arXiv:2201.08238 [nucl-th]]
[16] ALICE, doi:10.1016/j.physletb.2022.137223 [arXiv:2204.10258 [nucl-ex]]
[17] J. Haidenbauer et al., in preparation
[18] J. Haidenbauer, U.-G. Meißner and A. Nogga, Few Body Syst. **62**, 105 (2021)
[19] P. Reinert, H. Krebs and E. Epelbaum, Eur. Phys. J. A **54**, 86 (2018)
[20] D. R. Entem, R. Machleidt and Y. Nosyk, Phys. Rev. C **96**, 024004 (2017)
[21] K. Miwa *et al.* [J-PARC E40], Phys. Rev. C **104**, 045204 (2021)
[22] K. Miwa *et al.* [J-PARC E40], Phys. Rev. Lett. **128**, 072501 (2022)
[23] T. Nanamura *et al.* [J-PARC E40], Prog. Theor. Exp. Phys. **2022**, 093D01 (2022)
[24] E. Epelbaum, H. Krebs and U.-G. Meißner, Eur. Phys. J. A **51**, 53 (2015)
[25] E. Epelbaum, W. Glöckle and U.-G. Meißner, Nucl. Phys. A **747**, 362 (2005)
[26] J. Haidenbauer, U.-G. Meißner and S. Petschauer, Eur. Phys. J. A **51**, 17 (2015)
[27] S. Petschauer and N. Kaiser, Nucl. Phys. A **916**, 1 (2013)
[28] F. Eisele, H. Filthuth, W. Föhlisch, V. Hepp and G. Zech, Phys. Lett. **37B**, 204 (1971)
[29] J. K. Ahn *et al.* [KEK-PS E289], Nucl. Phys. A **761**, 41 (2005)
[30] R. H. Dalitz and F. Von Hippel, Phys. Lett. **10**, 153 (1964)
[31] F. Schulz *et al.* [A1 Collaboration], Nucl. Phys. A **954**, 149 (2016)
[32] T. O. Yamamoto *et al.*, Phys. Rev. Lett. **115**, 222501 (2015)
[33] M. Kohno, Phys. Rev. C **100**, 024313 (2019)
[34] K. Nakazawa *et al.*, Prog. Theor. Exp. Phys. **2015**, 033D02 (2015)
[35] M. Yoshimoto *et al.*, PTEP **2021**, 7 (2021)
[36] J. K. Ahn *et al.*, Phys. Lett. B **633**, 214 (2006)
[37] S. Aoki *et al.*, Nucl. Phys. A **644**, 365 (1998)
[38] Hoai Le, J. Haidenbauer, U.-G. Meißner and A. Nogga, arXiv:2210.03387
[39] S. Acharya *et al.* [ALICE], Phys. Rev. Lett. **123**, 112002 (2019)
[40] S. Acharya *et al.* [ALICE], Nature **588**, 232 (2020)
[41] M. Isshiki [STAR], EPJ Web Conf. **259**, 11015 (2022)
[42] J. Haidenbauer, Nucl. Phys. A **981**, 1 (2019)
[43] Y. Kamiya *et al.*, Phys. Rev. C **105**, 014915 (2022)
[44] S. Acharya *et al.* [ALICE], Phys. Lett. B **797**, 134822 (2019)

Study on hadron-hadron interaction with femtoscopic technique

Yuki Kamiya[1,2,*]

[1]Helmholtz Institut für Strahlen- und Kernphysik and Bethe Center for Theoretical Physics, Universität Bonn, D-53115 Bonn, Germany
[2]RIKEN Interdisciplinary Theoretical and Mathematical Science Program (iTHEMS), Wako 351-0198, Japan

Abstract. We study the method to study the exotic hadrons using the femtoscopic correlation function in the high-energy collisions. The formula for the correlation function extended to the coupled-channel case is employed. We see that how the coupled-channel effect works for the correlation function from the different hadron emitting sources with the case of the $K^- p$ correlation function. The method is applied to the prediction for the $DD^*/D\bar{D}^*$ correlation function for the study of the $T_{cc}/X(3872)$ state.

1 Introduction

Recently, the new technique so called femtoscopy, for the hadoron-hadron interaction is beginning to be used to study hadron-hadron interaction. With this technique, we utilize the two-hadron intensity correlation in the high-energy nuclear collisions which is generated by the effect of quantum statistics and the final state interaction. This technique is originally used to investigate the detail of the hadron emitting source using the hadron pairs whose strong interaction is well known. For the study of hadron interaction, once the effective source size has been determined, the pairwise interaction of various hadron pairs can be studied by the measurement of the correlation function with the high statistics. This technique is quite useful in particular for the channel whose interaction is difficult to study through the scattering experiments, such as short-lived particles and heavy particles.

It is known that the correlation function is sensitive to the low-energy hadron interaction. Among various hadron pairs, the interesting systems to study in femtoscopy are systems related to the exotic hadrons. This is because, according to the Weinberg's weak-binding relation [1, 2], it is important to know the low-energy hadron interaction in detail in order to distinguish the nature of the near-threshold exotic states. As the XYZ states and T_{cc} state are representative, the many exotic hadron states in the heavy sector are found in the vicinity of the two-hadron threshold energies. Thus, by applying the femtoscopic study to the heavy hadron pairs, the nature of various exotic states can be investigated.

In this proceedings, we first review the recent femtoscopic study on the $\bar{K}N$ interaction. Next we discuss how we can apply the femtoscopic technique to the study of T_{cc} and $X(3872)$ states, whose nature is less known. The original contents on the study of $\bar{K}N$ (DD^* and $D\bar{D}^*$) interaction are found in Ref. [3] (Ref. [4]).

[*]e-mail: kamiya@hiskp.uni-bonn.de

2 Formalism

To calculate the correlation functions $C(q)$ with the coupled-channel effects, we employ the Koonin-Pratt-Lednicky-Lyuboshitz-Lyuboshitz formula (KPLLL) formula [3, 5, 6] given by

$$C(q) = \int d^3r \sum_{i=1}^{2} \omega_i S_i(r) |\Psi_i^{(-)}(q; \mathbf{r})|^2 , \tag{1}$$

where $\Psi^{(-)}$ is the wave function in the i-th channel written as a function of the relative coordinate r. This wave function must satisfy the outgoing boundary condition where the outgoing wave in the measured channel (channel 1) needs to be normalized. Because we consider the small momentum region, the modification of the s-wave component of the wave function is taken into account. Furthermore, $S_i(r)$ and ω_i are the normalized hadron emitting source function and its weight in i-th channel. This weight ω_i representing the ratio of the pair production yield in the i-th channel with respect to the measured channel must be determined in order to compare the theoretical models and experimental data. Due to the normalization of $C(q) = 1$ for the non-interacting case, ω_1 must be equal to unity. In general, this ratio can be written as [7]

$$\frac{\omega_j}{\omega_i} = \frac{\alpha_j}{\alpha_i} \frac{N(j_1)N(j_2)}{N(i_1)N(i_2)}, \tag{2}$$

where i_1, i_2, j_1 and j_2 are the labels of particles in each channel and α_i is the ratio of the number of particle pairs assigned to channel i among $N(i_1) \times N(i_2)$ with the number of particle i_k emitted from the hadron source $N(i_k)$. The factor α_i can be determined from the property of the channels, e.g. the spin degree and the ratio of particles. For $N(i_k)$, it is difficult to determine these values from experimental observables as we cannot directly observe the source function, however, it is reasonable to estimate it using the statistical model considering that it well reproduces hadron yields in the high-energy nuclear collisions. In this proceedings, we focus on the theoretical feature of the correlation function and we use the common Gaussian source function $S_i(r) = S_R(r) = \exp(-r^2/R^2)$ with the source size R and $\omega_i = 1$ for simplicity.

3 K^-p correlation function

First, we discuss the K^-p correlation function with the chiral SU(3) based effective potential. It is known that the detailed determination of the low-energy $\bar{K}N$ interaction and its coupling to the decay channels are very important for the study of nature of the $\Lambda(1405)$ state lying below the $\bar{K}N$ threshold energy. The scattering amplitude derived from the chiral SU(3) dynamics [8, 9] has been fitted by using an effective local $\bar{K}N$-$\pi\Sigma$-$\pi\Lambda$ coupled channels potential [10]. Solving the Schrödinger equation with the K^-p outgoing boundary condition, we obtain the squared wave functions for the KPLLL formula. The obtained wave function at $q = 50$ MeV/c and $q = 100$ MeV/c is shown in Fig. 1. We find that the wave functions in channels other than K^-p take significant values at small distances, while they are much smaller than that in K^-p at large distances. This is because of the K^-p outgoing boundary condition. From these wave functions, we can guess that contribution of the wave function components of these channels to K^-p correlation function is large for small size sources, but the K^-p wave function would be enough for large sources. Note that the coupled-channel effect contributes to the correlation function through the coupled-channel wave function $\Psi_i^{(-)}$ as well as $\Psi_{K\text{-}p}^{(-)}$. This is because the $\Psi_{K\text{-}p}$ itself is affected by the channel coupling to the other channels.

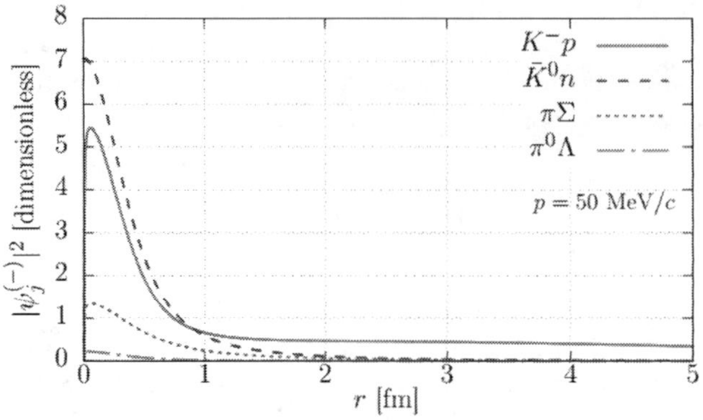

Figure 1. The square of each component of the coupled-channel wave function with $q = 50$ MeV (thin lines) and 100 MeV (thick lines). The solid, dashed, dotted, and dash-dotted line denote the K^-p, $\bar{K}^0n$, $\pi\Sigma$, and $\pi^0\Lambda$ component. For the $\pi\Sigma$ component, we show the sum of three components of $\pi^-\Sigma^+$, $\pi^0\Sigma^0$, and $\pi^+\Sigma^-$.

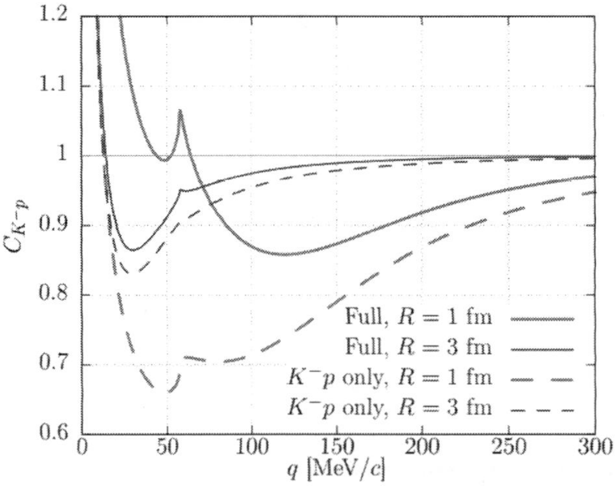

Figure 2. K^-p correlation function with $R = 1$ fm (thick lines) and $R = 3$ fm (thin lines). The long-dashed line denotes the result with K^-p component only. The solid lines show the full calculation result in which the contributions from all coupled-channel components are included.

We shall now calculate the K^-p correlation function using the coupled-channel wave function and the KPLLL formula. In Fig. 2, the results calculated with all the channel components of coupled-channel wave function (full calculation) and that calculated only with the K^-p component are shown for source sizes of $R = 1$ fm and 3 fm. For both source radii R, we see the strong enhancement due to the Coulomb attraction at small momenta and the cusp structure at the $\bar{K}^0n$ threshold at $q \simeq 58$ MeV/c. For the $R = 1$ fm case, the effect of the coupled-channel source, which is shown by the difference between the full calculation and the result with the K^-p component only, enhances the $C(q)$ in the low-momentum region and makes the cusp structure prominent. On the other hand, in the case of $R = 3$ fm, the coupled-channel source effect is moderate. This source size dependence of the coupled-channel source effect is due to the wave function components other than K^-p taking significant values only at small distances as shown in Fig. 1. Thus, for the femtoscopy of the coupled-channel system, the correlation data from the small hadron sources is useful for the detailed study of the coupled-channel interaction while that from the large hadron sources is useful to directly see the interaction of the measured channel (K^-p).

4 DD^* and $D\bar{D}^*$ correlation function

Next we discuss the DD^* and $D\bar{D}^*$ correlation function for the study of T_{cc} and $X(3872)$. We assume that T_{cc} ($X(3872)$) state lies just below the D^0D^{*+} ($\{D^0\bar{D}^{*0}\} = (D^0\bar{D}^{*0} + \bar{D}^0D^{*0})/\sqrt{2}(C = +)$) threshold. This channel couples to the D^+D^{*0} (D^+D^{*-}) whose threshold energy is higher than that of D^0D^{*+} ($D^0\bar{D}^{*0}$). Since these channels lie close to each other, we take into account their coupled-channel effect. On the other hand, because the other channels that couple to these states are far from the energy region of interest, we do not include the coupled-channel source effect of these channels, however, the effect of the decay to the lower channels are partly included by using the complex optical potential for DD^* and $D\bar{D}^*$ interaction. We also assume that the $I = 0$ interaction, which is considered to generate the T_{cc} and $X(3872)$ states, gives the dominant contribution and we neglect the $I = 1$ interaction for simplicity.[1]

To calculate the wave function for the KPLLL formula, we use the Gaussian potential $V_{I=0}(r) = V_0 \exp(-m_\pi^2 r^2)$ with the pion mass m_π for DD^* and $D\bar{D}^*$ interaction. The complex interaction strength V_0 is determined so as to reproduce the empirically obtained scattering lengths. For the DD^* system, we use the scattering length $a_0^{D^0D^{*+}} = -7.16 + i1.85$ fm, which is determined in the experimental analysis in Ref. [11].[2] For the $D\bar{D}^*$ system, $a_0^{D^0\bar{D}^{*0}(C=+)} = -4.23 + i3.95$, which is determined by the eigenenergy $E_h = -0.04 - i0.60$ MeV in PDG [12] measured from the $D^0\bar{D}^{*0}$ threshold, is employed. We obtained $V_0 = -36.569 - i1.243$ MeV for the DD^* system and $V_0 = -43.265 - i6.091$ MeV for the $D\bar{D}^*$ system.

The D^0D^{*+} and $D^0\bar{D}^{*0}$ correlation functions are calculated with the determined potential as shown in Fig. 3. We see that both of the correlation functions show the strong source size dependence which is typical to the system with the near-threshold eigenstates. The cusp structure is found at the eigenmomentum corresponding to D^+D^{*0} (D^+D^{*-}) threshold energy in $C_{D^0D^{*+}}(q)$ ($C_{D^0\bar{D}^{*0}}(q)$) while it is moderate compared to that found in the K^-p correlation function. On the other hand, we omit the results of correlation function of the higher channels (D^+D^{*0} and D^+D^{*-}) because these channels do not show such characteristic behavior. This is because the eigenenegies of T_{cc} and $X(3872)$ are much smaller than that of the deviation of threshold energies of isospin partners so that the only the wave functions of lower channels reflect the coupling to these states.

5 Summary

In this proceedings, we have discussed how the exotic states can be studied with the momentum correlation function obtained from the high-energy collision experiments. With the result of the K^-p correlation function, we find that the coupled-channel source effect gives significant additional enhancement on the K^-p correlation function especially for the small source case. On the other hand, the correlation function from the large source is not affected by this coupled-channel effect so much. Thus, with the careful comparison of the femtoscopic data

[1] The discussion with the different assumptions on the $X(3872)$ and $D\bar{D}^*$ interaction can be found in Ref. [4].

[2] Here we use the the high-energy physics convention for the scattering length where the positive (negative) real value corresponds to the weakly attractive (repulsive or strongly attractive) interaction.

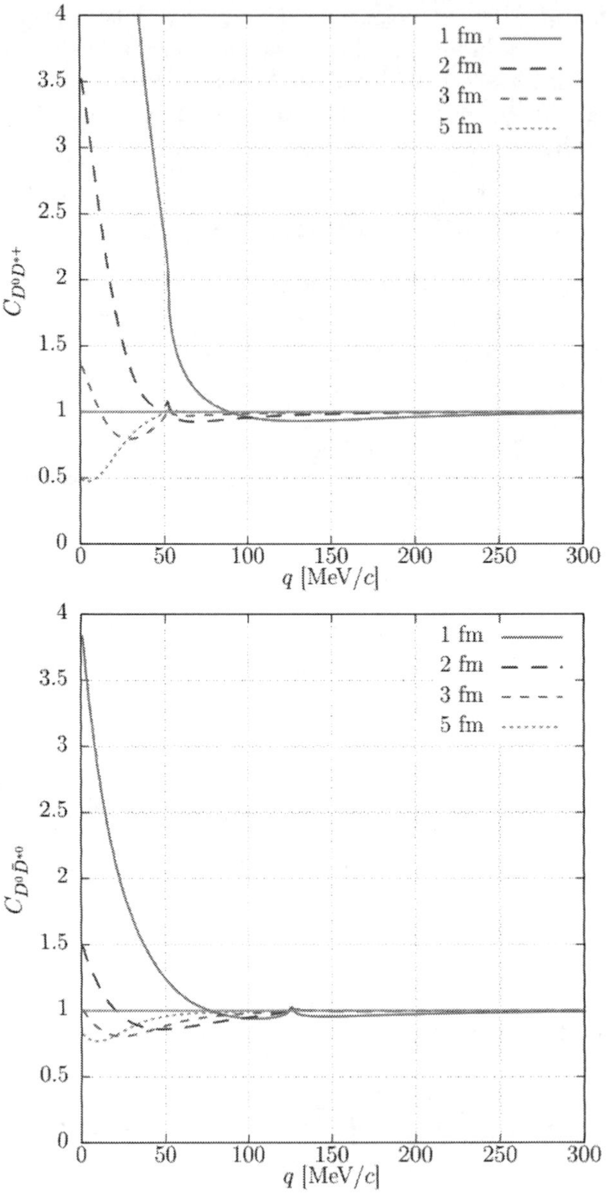

Figure 3. The D^0D^{*+} (upper panel) and $D^0\bar{D}^{*0}$ (lower panel) correlation function for several source sizes.

from the different source sizes and the theoretical models, the detailed $\bar{K}N$ interaction in the low-energy region can be determined.

In Ref. [4], it is found that the SU(3) chiral model works well to describe the ALICE K^-p correlation data from high-energy pp collisions [13] with the reasonable femtoscopic parameters. On the other hand, in the latest detailed analysis using both of the ALICE pp collisions and pPb collisions data [14], it is found that the theoretical chiral SU(3) model with the carefully determined source function cannot fully reproduce the data with the different source sizes. This deviation indicates that there is a possibility to further constrain the $\bar{K}N$ interaction to elucidate the nature of $\Lambda(1405)$.

Finally, we applied the KPLLL formula to the DD^* and $D\bar{D}^*$ correlation functions to elucidate the nature of the T_{cc} and $X(3872)$ states. We have found that both of the D^0D^{*+} and $D^0\bar{D}^{*0}$ correlation functions show the strong source size dependence due to the near-threshold states. According to Ref. [15], with the ALICE 3 upgrade with the large acceptance and the high luminosity, the DD^* and $D\bar{D}^*$ correlation data from both different colliding systems (pp and PbPb) can be measured with the great resolution, which is enough to see the characteristic behavior. Thus, by investigating the source size dependence of the DD^* and $D\bar{D}^*$ correlation function in the future experiments, the origin of the T_{cc} and $X(3872)$ can be distinguished.

Acknowledgements

This work is supported by the Deutsche Forschungsgemeinschaft (DFG) and the National Natural Science Foundation of China (NSFC) through the funds provided to the Sino-German Collaborative Research Center "Symmetries and the Emergence of Structure in QCD" (NSFC Grant No. 12070131001, DFG Project-ID 196253076 – TRR 110).

References

[1] S. Weinberg, Phys. Rev. **137**, B672 (1965)
[2] Y. Kamiya and T. Hyodo, PTEP **2017**, 023D02 (2017)
[3] Y. Kamiya, T. Hyodo, K. Morita, A. Ohnishi, and W. Weise, Phys. Rev. Lett. **124**, 132501 (2020)
[4] Y. Kamiya, T. Hyodo, and A. Ohnishi, Eur. Phys. J. A **58**, 131 (2022)
[5] S. E. Koonin, Phys. Lett. **70B**, 43 (1977)
[6] R. Lednicky, V. V. Lyuboshitz, and V. L. Lyuboshitz, Phys. Atomic Nuclei **61**, 2950 (1998)
[7] Y. Kamiya *et al.*, Phys. Rev. C **105**, 014915 (2022)
[8] Y. Ikeda, T. Hyodo, and W. Weise, Phys. Lett. **B706**, 63 (2011)
[9] Y. Ikeda, T. Hyodo, and W. Weise, Nucl. Phys. **A881**, 98 (2012)
[10] K. Miyahara, T. Hyodo, and W. Weise, Phys. Rev. **C98**, 025201 (2018)
[11] R. Aaij *et al.* (LHCb Collaboration), Nature Commun. **13** no.1, 3351 (2022)
[12] R. L. Workman (Particle Data Group), PTEP **2022**, 083C01(2022)
[13] S. Acharya *et al.* (ALICE Collaboration), Phys. Rev. Lett. **124**, 092301 (2020)
[14] ALICE Collaboration, arXiv:2205.15176 [nucl-ex] (2022)
[15] ALICE collaboration, CERN-LHCC-2022-009 (2022)

SU(3)$_f$ constraints on hypernuclear energy density functionals

Horst Lenske[1,*] and *Madhumita* Dhar[2,**]

[1]Institut für Theoretische Physik, Justus-Liebig-Universität Gießen, D-35392 Gießen, Germany
[2]Cooch Behar Government Engineering College, Cooch Behar, India

Abstract. A covariant hypernuclear energy density functional (EDF) is derived from in–medium nucleon–meson vertex functionals, assuring the proper description of nuclear mean–field dynamics. The fundamental SU(3) coupling constants for the mean–field relevant vector ($m = V$) and scalar ($m = S$) interactions as functionals of the total baryon density ρ_B are determined. Scalar and vector potentials and the resulting hyperon mean–fields in asymmetric nuclear matter are constructed and discussed, addressing also effects from 3–body interactions. Λ–Σ^0 mixing in asymmetric nuclear through the coupling to the background isovector mean–field is addressed.

1 Introduction

With the focus on mean–field dynamics and the description of bulk properties like single particle separation and total binding energies, energy density functionals (EDF) have been developed for hypernuclear and neutron star studies [1–5], the latter often with special focus on the still unsolved so–called hyperon puzzle, limiting the maximal masses of neutron stars. Covariant Lagrangian approaches and non–relativistic Skyrme-type EDFs were constructed in parallel. In order to overcome the scarce data base, Skyrme EDFs were derived from G–matrix interactions [6, 7], still being used and recently applied in investigations of light Cascade–nuclei [8]. First attempts to implement SU(3) symmetry into hypernuclear physics by using χEFT were made already more than a decade ago, see e.g. [9].

The standard approach to microscopic SU(3)–based in–medium dynamics follows closely the rather successful strategies developed over the last decades for pure nucleon dynamics, namely first fixing free space two–body interactions which in a second step are inserted into an in–medium scheme like Brueckner G–matrix theory, eventually extended by 3–body forces, e.g. [10, 11]. The resulting in–medium scattering amplitudes are used to construct mean–field self–energies given by static potentials, depending on momentum if non–localities due to anti–symmetrization are treated explicitly. That is also the strategy adopted in Density Dependent Relativistic Hadron (DDRH) theory, developed some time ago at Giessen (see [12–14] and further references therein) and used here as the gateway for SU(3) mean–field dynamics.

The DDRH scheme is based on covariant Dirac–Brueckner–Hartree–Fock (DBHF) G–matrix calculations. The DBHF self–energies are projected to an effective Hartree theory.

*e-mail: horst.lenske@physik.uni-giessen.de
**e-mail: madhumita.dhar007@gmail.com

For practical calculations, effective density–dependent NN–meson coupling functionals are derived such that the DBHF self–energies are reproduced. The functionals, incorporating naturally anti–symmetrization contributions, serve to recast the DBHF results into a covariant EDF. The interaction part is given by the mean–field producing isoscalar (σ, ω) and isovector $(\delta/a_0(980), \rho)$ meson fields. Applying the variational rules of covariant density functional theory (DFT), Dirac equations for baryons and (classical) Klein–Gordan equations for the static meson fields have to be solved. The approach and results are summarized in Fig. 1 where the vertex equation is illustrated graphically and the four NN–meson functionals are shown together with results for nuclear binding energies of stable nuclei all over the mass table. Results for infinite nuclear matter are found in Tab.1. The DDRH vertices were used, in fact, also in investigations of single–Λ nuclei [15, 16] by using a simple scaling approach, widely used at that time. A look into the original research paper [15] will be quite instructive for understanding of the uncertainties of hypernuclear parameters at that time, which in fact are still waiting for experimental support. Also in Fig. 1 these early results for Λ–separation energies are compared to data and corresponding results by a G–matrix folding approach using the non–relativistic Nijmegen–interaction [17].

In the following, we take a fresh view on the in–medium physics of octet baryons. Although the use of SU(3) relations is a standard tool in octet physics and beyond, the new aspect of the approach presented below is to exploit those relations not on the Born–level but on the level of the already fully resummed diagrams of nucleon in–medium interactions. The four NN–vertex functionals, available from DDRH theory, are sufficient to derive the mean–field producing in–medium interactions for the full set of octet baryons.

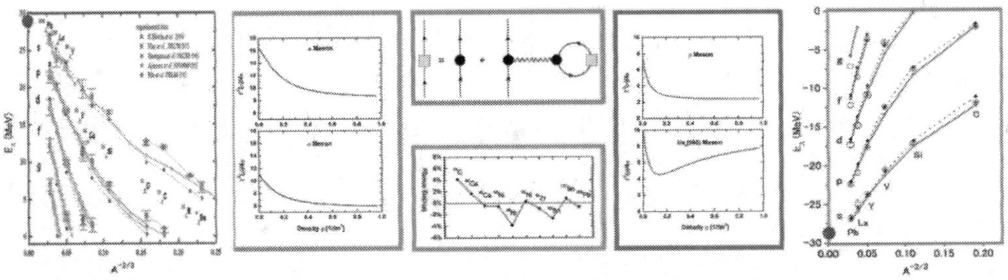

Figure 1. DDRH vertices derived from DBHF self–energies. The diagram in the upper center represents the dressed vertices (boxes) as given by the bare NN–meson couplings (filled circles). Baryon and meson propagators are indicated by straight and wavy lines, respectively. The resulting isoscalar and isovector coupling functionals are shown at the left (ω, σ) and the right $(\rho, \delta/a_0(980))$, respectively. Applications to stable nuclei over the mass chart are illustrated in the lower center by displaying the relative deviations of calculated and experimental binding energies. Also shown are mean–field results for single–Λ separation energies: on the far left covariant DDRH results and on the very right non–relativistic results obtained by a folding approach using the G–matrix from the ESC-08 interaction [17] are shown. For further details see Ref. [13].

2 In–medium SU(3) density functional theory

2.1 Formal aspects of SU(3) mean–field theory

The purpose of this section is to give an overview of the construction of a Lagrangian density describing octet baryons and their interactions with the meson nonets. Interactions are de-

Model	ρ_{sat}/fm^{-3}	E/A/MeV	K_∞/MeV	E_{sym}/MeV	L_{sym}/MeV	K_{sym}/MeV
2-body	0.180	-15.603	281.945	31.154	88.627	201.399
3-body	0.160	-16.000	283.136	32.000	90.000	133.502

Table 1. Properties of the DDRH equation of state of infinite symmetric nuclear matter, obtained with 2–body interactions only (first row) and by adding 3–body interactions (second row).

fined by the SU(3) rules introduced by de Swart [18]. Denoting the eight Gell–Mann SU(3) matrices by λ^a, the baryons are represented as usual by the 3–by–3 matrix $\mathcal{B} = \frac{1}{\sqrt{2}}\sum_{a=1}^{8}\lambda^a B^a$, see e.g. [13]. The octet meson matrices are defined accordingly, e.g. $\mathcal{V}_8 = \frac{1}{\sqrt{2}}\sum_{a=1}^{8}\lambda^a V^a$ for the vector mesons with $J^\pi = 1^-$. Singlet mesons are included by the 3–by–3 unity matrix λ^0, e.g. $\mathcal{V}_1 = \lambda^0 V^0$ where in the vector case $V^0 = \phi_1$ is given by the (unmixed) singlet vector meson ϕ_1. Scalar ($J^\pi = 0^+$) and pseudo–scalar ($J^\pi = 0^-$) mesons are described accordingly.

We obtain the Lagrangian density $\mathcal{L} = \mathcal{L}_B + \mathcal{L}_M + \mathcal{L}_{BBM}$. The baryon and meson Lagrangians $\mathcal{L}_{B,M}$ are superpositions of standard Dirac– and Klein–Gordan–type Lagrangians for the free motion of fermions and bosons, respectively. SU(3) symmetry breaking is included by using physical hadron masses. For finite systems, electromagnetic interactions are supplemented.

Of central interest for the theory is the baryon–meson interaction Lagrangian given by the sum of Lagrangians describing the three classes of interactions, $\mathcal{L}_{BBM} = \sum_{m=P,S,V}\mathcal{L}_{int}^{(m)}$. The interaction Lagrangians are of the same formal structure, e.g. for the vector case.:

$$\mathcal{L}_{int}^{(V)} = -\sqrt{2}\left(g_D^{(V)}\left[\overline{\mathcal{B}}\mathcal{B}\mathcal{V}_8\right]_D + g_F^{(V)}\left[\overline{\mathcal{B}}\mathcal{B}\mathcal{V}_8\right]_F\right) - g_S^{(V)}\frac{1}{\sqrt{3}}\left[\overline{\mathcal{B}}\mathcal{B}\mathcal{V}_1\right]_S \tag{1}$$

$$\left[\overline{\mathcal{B}}\mathcal{B}\mathcal{V}_8\right]_D = tr\left(\left\{\overline{\mathcal{B}},\mathcal{B}\right\}\mathcal{V}_8\right); \quad \left[\overline{\mathcal{B}}\mathcal{B}\mathcal{V}_8\right]_F = tr\left(\left[\overline{\mathcal{B}},\mathcal{B}\right]\mathcal{V}_8\right); \quad \left[\overline{\mathcal{B}}\mathcal{B}\mathcal{V}_1\right]_S = tr\left(\overline{\mathcal{B}}\mathcal{B}\right)tr\left(\mathcal{V}_1\right). \tag{2}$$

where the D and F couplings are given by the baryon anti–commutator and the baryon commutator, respectively:

2.2 SU(3) mean–field theory

The guiding principles are that firstly by the fit to scattering data NN–interactions fully inherit SU(3) symmetry, but e.g. by the use of physical masses incorporate also symmetry breaking effects, and secondly that both free space and in-medium Bethe-Salpeter equations conserve the fundamental symmetries. Thirdly, we note in addition that this is also true for the vertex equations by which the coupling functionals are derived from the DBHF G–matrix interactions. Thus, we formulate the *SU(3) DFT–Program*: For each of the nonet interaction channels, pseudo–scalar (P), scalar (S), and vector (V) meson exchange, only three physical couplings are sufficient to fix the set of fundamental SU(3) couplings $\{g_D^{(m)}, g_F^{(m)}, g_S^{(m)}\}_{m=P,S,V}$. If those are known, the full set of vertex functionals for the remaining baryon–meson couplings are determined. In the forthcoming, only the mean–field relevant scalar and vector parts will be discussed.

The full treatment of SU(3) baryon octet/meson nonet dynamics requires to consider octet–singlet meson mixing. For simplicity, but not as limitation by principle, we assume throughout ideal mixing, i.e. the charged–neutral octet mesons (η, σ, ω) do not contain $s\bar{s}$–components while their physical singlet partners $(\eta'(960), \sigma' \sim f_0(980), \phi(1020))$ are pure $s\bar{s}$ configurations. These constraints are fulfilled by the ideal mixing angle $\tan\theta = 1/\sqrt{2}$

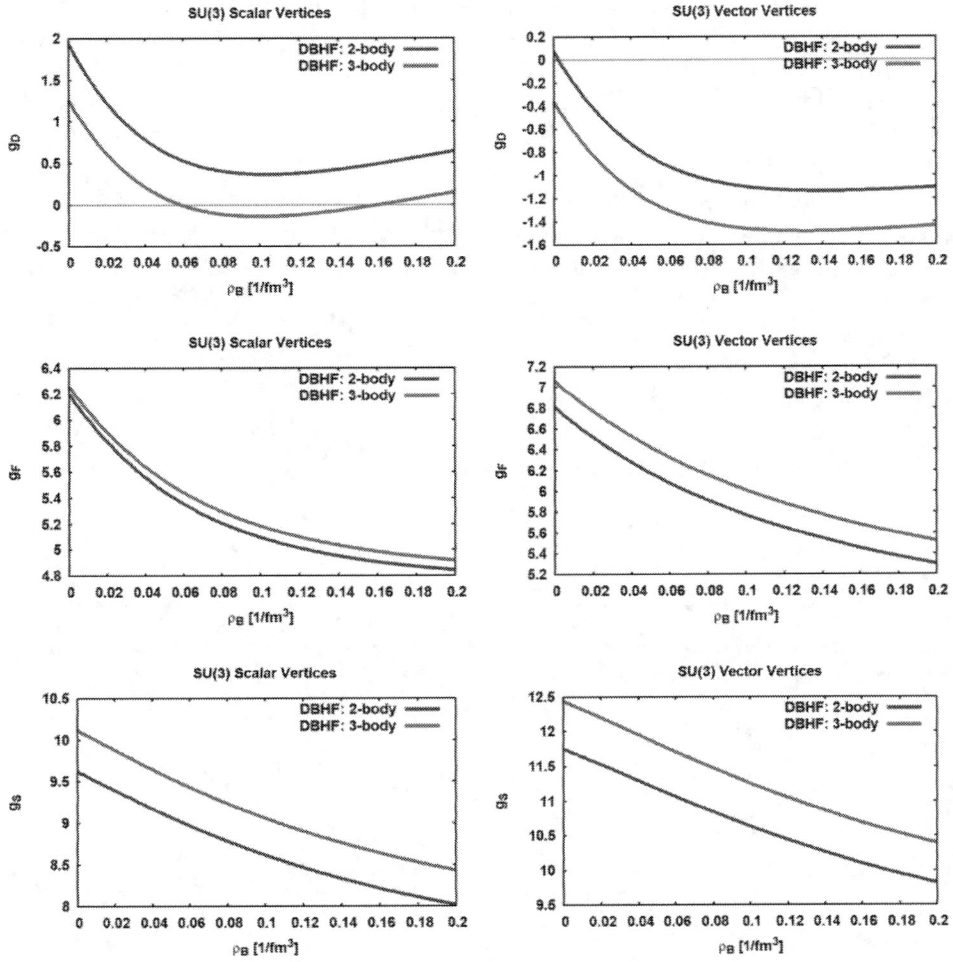

Figure 2. Elementary SU(3) in–medium scalar (left) and vector (right) vertices. In each column from top to bottom g_D, g_F and g_S are displayed as functions of the baryon number density in symmetric nuclear matter. As indicated, results are shown without (blue) and with three–body (red) interactions, the latter are of Fujita–Miyazawa type [19] and act only among nucleons.

($\theta \sim 35.26°$) which is common for all three interaction sectors. For example, denoting the vector meson mixing angle by θ_V the vector SU(3) couplings are explicitly:

$$\begin{aligned} g_D^{(V)} &= \frac{1}{2}\left[3g_{NN\rho} - \sqrt{3}\left(g_{NN\omega}\sin\theta_V - g_{NN\phi}\cos\theta_V\right)\right] \\ g_F^{(V)} &= \frac{1}{2}\left[g_{NN\rho} + \sqrt{3}\left(g_{NN\omega}\sin\theta_V - g_{NN\phi}\cos\theta_V\right)\right] \\ g_S^{(V)} &= \sqrt{2}\left[g_{NN\omega}\cos\theta_V + g_{NN\phi}\sin\theta_V\right], \end{aligned} \tag{3}$$

which – together with the corresponding relation for the scalar (and pseudo–scalar) couplings – are the key relations for the whole approach. Compatible with ideal mixing is the hypothesis that nucleons do not couple to the (physical) singlet mesons. Taking that view, we fix $g_{NN\eta'}(\rho_B) = g_{NNf_0}(\rho_B) = g_{NN\phi}(\rho_B) \equiv 0$. By means of Eq.(3) and the two sets of scalar and vector DDRH Hartree functionals, the prerequisites are at hand for the derivation of the full set of scalar and vector SU(3) in–medium couplings $g^{(S,V)}_{D,F,S}(\rho_B)$. [1] Thus, a SU(3)–based EDF can be constructed. However, numerically the values of the deduced SU(3) density–dependent coupling functionals will depend on the in–medium NN–interaction model.

In Fig. 2 the fundamental SU(3) coupling functionals for scalar and vector interactions are shown as a function of the baryon number density. The chosen scheme leads to a suppression of the D-couplings $g^{(S,V)}_D(\rho_B)$ such that mean–field dynamics are dominated by the octet F-couplings $g^{(S,V)}_F(\rho_B)$ and the singlet couplings $g^{(S,V)}_S(\rho_B)$. Similar results are found for the empirical linear or the Gell–Mann–Okuba mass relation (see also Ref. [20]).

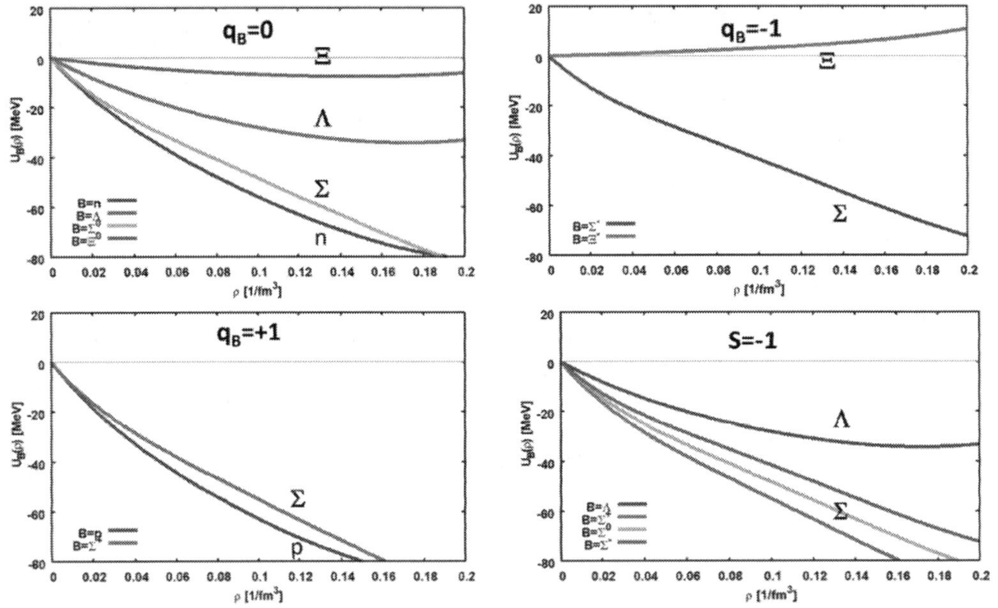

Figure 3. Baryon covariant mean–fields in infinite asymmetric matter are shown where $U_B(\rho) = U^{(S)}_B(\rho) + U^{(V)}_B(\rho)$ is the sum of scalar and vector potentials, including both isoscalar and isovector interactions. Results for $Z/A = 0.4$ are displayed, realized e.g. in ^{10}Be and approximately in ^{48}Ca, ^{124}Sn, and ^{208}Pb. Results are shown for baryons of equal charge (left and right top) and compared for the $S = -1$ multiplet of hypercharge $Y = 0$ consisting of the Σ iso–triplet and the Λ iso–singlet.

2.3 Mean–fields of octet baryons

The standard variational Euler–Lagrange rules serve to obtain the equations of motion, finally to be solved numerically. As seen in Fig. 1 and Tab. 1 the covariant DDRH–EDF leads already in calculations with only 2–body interactions to rather satisfactory results both for finite nuclei and infinite nuclear matter, much better than found in corresponding non–relativistic BHF studies. The reason is that the underlying DBHF calculations include important field–theoretical aspects like the coupling to the negative–energy Fermi sea, acting like a 3–body interaction (the so–called polarization– or Z–diagrams), causing e.g. the difference between

[1] Lorentz–invariance of the theory demands $\rho_B \equiv \sqrt{j_{B\mu} j^\mu_B}$ with the total baryon 4–current $j_{B\mu} = \sum_{b=N,Y} \overline{\Psi}_b \gamma_\mu \Psi_b$ which in the nuclear rest frame simplifies to the standard expression for the total baryon number density.

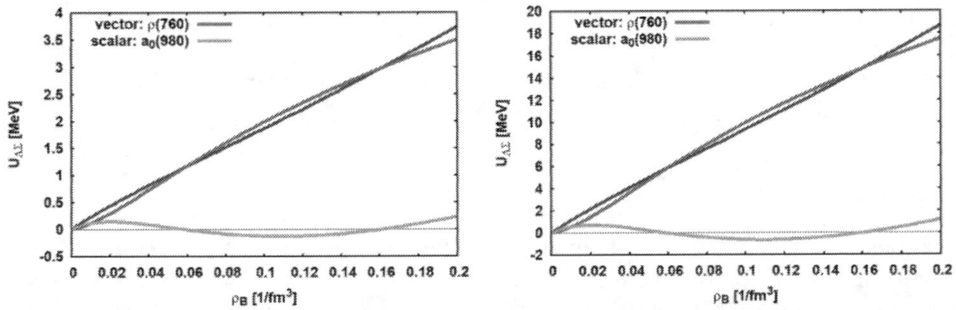

Figure 4. Density dependence of the Λ–Σ^0 in–medium mixing potential in nuclear matter with Z/A = 0.4 and in pure neutron matter, $Z/A = 0$. The scalar (green) and vector (red) components are shown together with the full mixing potential (blue) as defined in Eq.(4) .

Dirac scalar and vector densities. What is missing, however, are the Fujita–Miyazawa 3–body contributions involving excitation of nucleon resonances, studied in the past in much detail e.g. by the Urbana group [19]. Here, these were simulated by effective density dependent 2–body interactions [10] where the strengths were adjusted to reproduce the saturation properties of the AV18+UIX–results of Akmal et al. [21]. Overall modifications of scalar and vector NN–vertices on a level of about 5% or less are obtained which as seen in Tab. 1 are sufficient to adjust the equation of state closer to the empirical realm. However, as seen in Fig. 2, the nucleonic 3–body corrections have a pronounced influence on the SU(3) couplings, lowering $g_D^{(m)}$ considerably, compensated in part by increasing $g_{F,S}^{(m)}$.

Rather than considering separately the Dirac scalar and vector fields $U_B^{(S,V)}$, more reasonable quantities are their sums $U_B = U_B^{(S)}+U_B^{(V)}$ which are the leading order non–relativistic reductions. With additional correction terms [22], they become finally Schroedinger–equivalent potentials. The potentials U_B are displayed in Fig. 3 for infinite asymmetric nuclear matter with charge–mass ratio $Z/A = 0.4$, as encountered in ^{10}Be and approximately in nuclei up to ^{208}Pb. Remarkably, the Λ–potential agrees almost perfectly well with the potential derived much earlier in Ref. [15] by a χ^2–fit to single-Λ separation energies, see Fig. 1.

3 Λ–Σ^0 mixing by the isovector mean–field

In closing the paper we point to the interesting fact that the approach naturally predicts Λ–Σ^0 mixing in asymmetric matter by the isovector mean–field. The mixing potential depends on the nuclear isovector scalar and vector densities, but may be written to a good approximation as

$$U_{\Lambda\Sigma}(\rho_B) \approx \left(1 - 2\frac{Z}{A}\right)\left(U_N^{(S)}(\rho_B)\frac{g_{\Lambda\Sigma}^{(\delta)}}{g_N^{(\delta)}} + U_{NN}^{(V)}(\rho_B)\frac{g_{\Lambda\Sigma}^{(\rho)}}{g_{NN}^{(\rho)}}\right) \tag{4}$$

where we have extracted the proton–neutron asymmetry factor from the scalar and vector isovector background potentials $U_{NN}^{(S,V)}$ and $g_{\Lambda\Sigma}^{(m)} = \sqrt{\frac{2}{3}}g_D^{(S,V)}$ are the (density–dependent) coupling constants. As seen in Fig. 4 the form factor of the potential is almost independent of asymmetry, while the strength evolves with Z/A closely to the scaling law, Eq.(4). Hence, while in asymmetric matter as found in the interior of stable nuclei Λ and Σ^0 hyperons conserve essentially their individual characters, in extremely proton–poor matter as in neutron stars they become strongly mixed effective baryons. Moreover, the mixing affects also the lifetimes: The in–medium Λ–like states will have a shorter lifetime than the pure state because of the admixed Σ^0-component and, vice versa, the lifetime of the Σ^0–like states will increase. We emphasize that this kind of mixing is a genuine many–body effect, different from the well known mixing on the level of 2–body matrix elements.

4 Summary and outlook

Well constrained hypernuclear energy density functionals are highly demanded for exploratory investigations of medium– and heavy–mass hypernuclei, allowing also safe extrapolations to neutron stars. For that goal, it was pointed out that well studied nuclear EDFs together with the general rules of SU(3) physics provide in principle a promising and appropriate entry point. Here, we have presented the formal aspects and derived vertex functionals relevant for hypernuclear mean–field dynamics. The covariant approach was used to derive the fundamental SU(3) scalar and vector vertices as functionals of the invariant baryon number density. Additional three–body interactions, contributing at a level of a few percent to the nuclear equation of state, led, however, to enhanced effects in the SU(3) vertices. As an illustrative application, for the complete set of octet baryons covariant mean–field potentials in infinite nuclear matter were derived including isoscalar and – is applicable – isovector potentials. Further applications to physical systems from hypernuclei of medium and heavy mass to neutrons stars are in preparation.

References

[1] N.K. Glendenning, S.A. Moszkowski, Phys. Rev. Lett. **67**, 2414 (1991)
[2] N.K. Glendenning, D. Von-Eiff, M. Haft, H. Lenske, M.K. Weigel, Phys. Rev. C **48**, 889 (1993)
[3] K. Maslov, E.E. Kolomeitsev, D.N. Voskresensky, Phys. Lett. B **748**, 369 (2015)
[4] Y. Lim, C.H. Hyun, K. Kwak, C.H. Lee, Int. J. Mod. Phys. E **24**, 1550100 (2015)
[5] A. Kanakis-Pegios et al., Symmetry **13**, 183 (2021)
[6] D.E. Lanskoy, Y. Yamamoto, Phys. Rev. C **55**, 2330 (1997)
[7] H.J. Schulze, E. Hiyama, Phys. Rev. C **90**, 047301 (2014)
[8] J. Guo, X.R. Zhou, H.J. Schulze, Phys. Rev. C **104**, L061307 (2021)
[9] P. Finelli, N. Kaiser, D. Vretenar, W. Weise, Nucl. Phys. A **831**, 163 (2009)
[10] V. Soma, P. Bozek, Phys. Rev. C **78**, 054003 (2008)
[11] J.W. Holt, M. Kawaguchi, N. Kaiser, Front. in Phys. **8**, 100 (2020)
[12] H. Lenske, Lect. Notes Phys. **641**, 147 (2004)
[13] H. Lenske, M. Dhar, T. Gaitanos, X. Cao, Prog. Part. Nucl. Phys. **98**, 119 (2018)
[14] H. Lenske, M. Dhar, Lect. Notes Phys. **948**, 161 (2018)
[15] C.M. Keil, F. Hofmann, H. Lenske, Phys. Rev. C **61**, 064309 (2000)
[16] C. Keil, H. Lenske, Phys. Rev. C **66**, 054307 (2002)
[17] Y. Yamamoto, T. Furumoto, N. Yasutake, T.A. Rijken, Phys. Rev. C **90**, 045805 (2014)
[18] J.J. de Swart, Rev. Mod. Phys. **35**, 916 (1963), [Erratum: Rev. Mod. Phys. **37**,326 (1965)]
[19] S.C. Pieper, AIP Conf. Proc. **1011**, 143 (2008)
[20] C. Adamuscin et al., PoS **EPS-HEP2017**, 729 (2018)
[21] A. Akmal, V.R. Pandharipande, D.G. Ravenhall, Phys. Rev. C **58**, 1804 (1998)
[22] G.G. Adamian, N.V. Antonenko, H. Lenske, L.A. Malov, S.G. Zhou, Eur. Phys. J. A **57**, 89 (2021)

Exploration of in-medium hyperon-nucleon interactions

Madhumita Dhar[1,*] and *Horst* Lenske[2,**]

[1]Cooch Behar Government Engineering College, Cooch Behar, West Bengal, India 736170
[2]Institut für Theoretische Physik, Justus-Liebig-Universität Gießen, D-35392 Gießen, Germany

Abstract. The study focuses on exploring the changes in the hyperon-baryon interaction at various nuclear densities. This approach starts by building a vacuum hyperon-nucleon interaction model based on Boson -Exchange maintaining SU(3) flavor symmetry. Bethe-Goldstone equation is then explored to investigate the medium properties over the bare interaction. A detailed investigation of the density dependence revealed clear changes in the low energy parameters with the variation of the medium density shown for different strangeness channels.

1 Introduction

Nuclear physics has gone through an exciting period after the discovery of strange particles introducing certain un-explained phenomenon like the 'hyperon puzzle', charge symmetry breaking, hypertriton lifetime puzzle, possibility of $NN\Lambda$ resonant state and so on [1]. All these requires knowledge about hyperon-baryon interaction with a special focus on the medium properties.

With this motivation, this study focuses on the understanding of the hyperon-nucleon interaction in presence of medium that was developed some time ago at Giessen [2]. For that purpose, a microscopic approach is used by first considering the bare interaction and then gradually adding the medium effect to have a thorough understanding of the subject. The major goal of this work is to study the density effects qualitatively first so that with improved experimental inputs, the results can be quantified to a great extent.

Here in this paper, first the methodology for studying vacuum interaction is discussed followed by in-medium effective interaction. The results for each will be discussed.

2 Methodology

In this work, the meson exchange approach is used to develop a hyperon-nucleon (YN) interaction, applicable to the full baryon and meson octets, for the latter including also the meson singlets. In the past, that strategy has been used extensively, see e.g. [4, 5], although large uncertainties remain in view of the scarcity of hyperon-nucleon scattering data. In our approach, the main focus is on in–medium interactions, aiming finally at using hypernuclear data as independent additional constraints on interactions among octet baryons.

*e-mail: madhumita.dhar@cgec.org.in
**e-mail: horst.lenske@physik.uni-giessen.de

 https://doi.org/10.1051/epjconf/202227105004

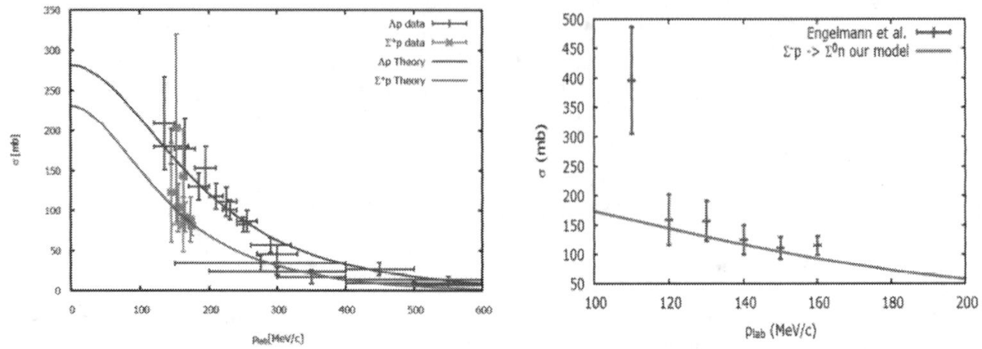

Figure 1. Calculated cross section (theory) for Σ^+p (left), Λp (left) and $\Sigma^-p \Rightarrow \Sigma^0 n$ (right) channels plotted in comparison to the bubble chamber data [7].

2.1 Vacuum hyperon-baryon interaction

Being mainly focused on studying the density effects of the interaction, a revived version of the One- Boson -Exchange potential (OBEP) was developed following SU(3) flavor symmetry to compensate the experimental data set scarcity from theory [3]. The interaction Lagrangian is defined by a superposition of pseudo-scalar (P), scalar (S), and vector (V) SU(3) scalars of the form for each meson type (denoted by x)

$$\mathcal{L}^x_{int} = -g^x_8\alpha^x_8 Tr([B,\bar{B}]\,\phi^x) + g^x_8(1-\alpha^x_8)Tr(\{\bar{B},B\}\,\phi^x) + g^x_1 Tr(B\bar{B}\phi^x) \quad (1)$$

with x= P, S, V, octet coupling constants $g_8, \alpha_8 = \frac{F}{F+D}$ and the singlet coupling g_1, $SU(3)_f$ invariant baryon and meson matrices B, ϕ. The mixing of octet and singlet mesons is taken into account by mixing angles θ.

In order to overcome the uncertainties by the lack of YN-data, here the baryon-meson coupling constants are derived from the elementary g and α couplings by using strictly the fundamental SU(3)-flavor relations [3]. Hence, the three sets of $(g_8, \alpha_8, g_1)_{P,S,V}$ are factors remaining as free parameters. The mixing angles were chosen as the ideal mixing values being found not to be crucial as determining factor if changed from ideal case in affecting the interaction. For g^P_8, the value fixed by pion-nucleon scattering is used. Owing to these, octet strengths ($g^{S,V}_8$) for scalar and vector meson exchange, singlet strengths ($g^{P,S,V}_1$) and $\alpha^{P,S,V}_8$ for each three of the meson nonets are fitted in this model.

With this construction and including dipole vertex form factors then the parameters were fitted with the available data set by solving a set of coupled 3-D Lippmann Schwinger equations for the octet T-matrix (Eq. 2)

$$T(\mathbf{q}',\mathbf{q}) = V(\mathbf{q}',\mathbf{q}) + \int d^3k\, V(\mathbf{q}',\mathbf{k})\, G(\mathbf{k})\, T(\mathbf{k},\mathbf{q}) \quad (2)$$

serving to determine phase shifts, cross sections, and other observabels like low energy parameters with (q, q',k) denoting initial, final and intermediate relative momenta. The OBEP and Green's function are denoted by V and G, respectively.

2.2 Vacuum interaction results

For the discussed energy sector of hyperon interaction, low energy behavior is dominant. Therefore, the constructed potential is first fitted to available set of Σ^+p and Λp cross section data with $^1S_0$ partial wave only for preliminary studies. The obtained fit parameter results with a χ^2 fit of 6.68 is shown in Figure 1 (left) and the obtained parameter set is given in Table 1 (left). As an application, the obtained parameter set were then used to evaluate $\Sigma^-p \Rightarrow \Sigma^0 n$ cross section as shown in Figure [1] (right) to produce a satisfactory output remembering

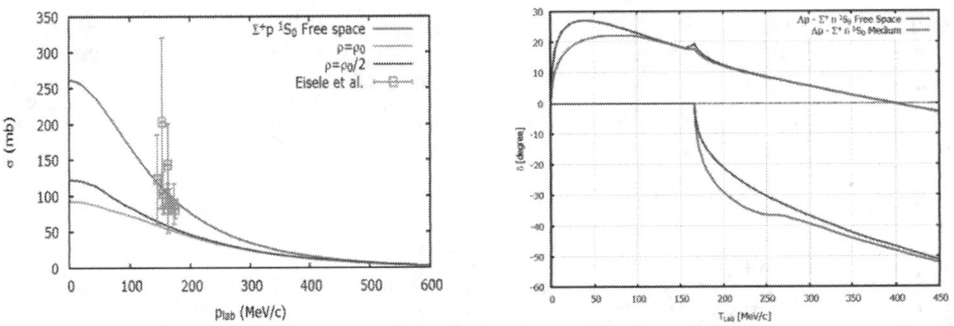

Figure 2. Theoretical integrated $\Sigma^+ p$ cross-section for nuclear saturation density and half saturation density as a function of laboratory momenta (left) showing significant medium effect on channel opening 'cusp' as can be seen in the phase shift plot (right).

the large error bars of the scattering data. For the present study, only S=-1, -2 results were calculated for few channels as a starting point. Thus, having a satisfactory vacuum interaction model, the next step is to apply the medium effect as discussed in next section.

2.3 Medium effect

To investigate the medium effect, infinite nuclear matter stands as rich laboratory for strong interaction that are crucial for hypernuclei to dense matter to heavy-ion experiments to name a few. In this work, medium effects are incorporated in terms of the Pauli Projector $Q_F = \Theta(k_1 - k_{F_1})\Theta(k_1 - k_{F_2})$ where k_F represents nucleon Fermi momentum. $Q_F = 1$ in free space. The Bethe-Goldstone equation in momentum space is then given by

$$T(\mathbf{q}', \mathbf{q}) = V(\mathbf{q}', \mathbf{q}) + \int d^3k \, V(\mathbf{q}', \mathbf{k}) \, \mathbf{Q_F}(\mathbf{k}, \mathbf{k_F}) \, G(\mathbf{k}) \, T(\mathbf{k}, \mathbf{q}) \tag{3}$$

2.4 In-medium results

When the effect of medium was studied for cross sections, a clear weakening of strength is found as shown in Figure 2 (left). A significant effect of dense medium is seen as suppressing the channel mixing as prominent from the weakening in the sharpness of the 'cusp'- a signature of channel mixing as shown in Figure 2 (right) phase shift plot. In order to get direct information for this kind of low energy scattering interaction, effective range parameters are suitable to gain more insights of the interaction that is obtained by using a series expansion of $q \cot \delta$ in powers of q that for S-wave takes the following form as shown in Eq. 4.. The low energy behavior stands as a convenient measure for the baryon sector as the core information is limited here.

$$\lim_{q \to 0} q \cot \delta \approx -\frac{1}{a_s} + \frac{1}{2} r_e q^2, \quad a_s = \frac{m}{2\pi\hbar^2} \int d^3r \, V(r) \tag{4}$$

The scattering length (a_s) and effective range (r_e) provide information about the nature of interaction. It is important to keep in mind here that for this sector, there are not precise quantitative results yet, hence a qualitative measure can stand as a theoretical prediction.

Table 1. Fitted parameter values of the revived SU(3) OBEP hyperon model (left), Calculated scattering lengths and effective range in free space and at nuclear saturation density (right).

	g_8	g_1	α_8	θ
P	3.795	0.1913	0.355	-23
S	1.2274	3.5434	0.9605	37.05
V	1.1566	3.4431	1.0	35.26

Channel	a_s^{free}	a_s^{sat}	r_e^{free}	r_e^{sat}
$\Lambda n \to \Lambda n$	-1.50	-0.76	2.34	2.01
$\Sigma^+ p$	-1.44	-0.86	5.18	5.34
$\Sigma^+ \Sigma^+$	-5.75	-1.23	1.94	2.78

To explore the density effect on the hyperon interaction, nuclear density was varied and the effect on scattering length and effective range was studied for S=-1 and S=-2 channels. The results obtained are reported in Figure 3. Here it shows that the scattering length saturates at saturation density that means at high densities. The calculated values are given in Table 1 (right). The obtained results are in line with the existing predictions as reported by other groups (Extended Soft Core (ESC) [4], Juelich [5], chiral effective field theory (χEFT) [6] models).

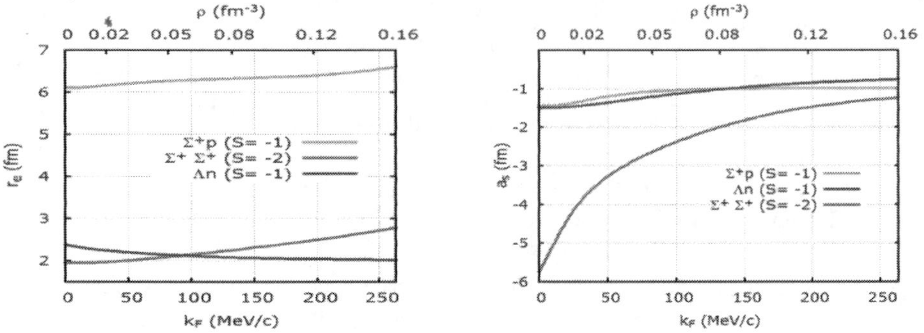

Figure 3. Variation of effective range (r_e) (left) and scattering length (a_s) (right) with nuclear density (top) and Fermi momentum (bottom) revealing channel dependent behavior.

3 Summary and outlook

The work reported a revived version of the one-boson exchange potential for hyperon-nucleon interaction primarily aiming at in-medium studies of interactions. As a preliminary step, the vacuum interaction showed good agreement for S=-1 scattering data. With the vacuum input, Bethe-Goldstone formalism then successfully describes the dense medium behavior of YN interaction. The investigation also revealed channel dependent behavior deserving further exploration.

As a future step, the approach can be extended for higher strangeness channels as well. The G-matrix formalism can be extended for various in-medium studies of hypernuclei as well as to study hypernuclear structure properties in terms of Hartree-Fock formalism. As a working qualitative model for hyperon, future promising scattering data availability will be helpful in quantifying the model parameters and thus to have more inputs on the dense matter behavior as well. Inclusion of three–body forces and comparison to the interactions derived independently by the DBHF energy density functional approach in [8] are in preparation.

References

[1] A. Gal, E. V. Hungerford, and D. J. Millener, Rev. Mod. Phys. **88**, 035004 (2016)
[2] H. Lenske, M. Dhar, T. Gaitanos, X. Cao, Prog. Part. Nucl. Phys. **98**, 119-206 (2018); H. Lenske, M. Dhar, Lect. Notes Phys. **948**, 161-253 (2018)
[3] J.J. de Swart, Rev. Mod. Phys. **35**, 916 (1963)
[4] T. A. Rijken, V. G. J. Stoks, and Y. Yamamoto, Phys. Rev. C **59**, 21 (1999); T. A. Rijken and H. -F. Schulze, Eur. Phys. J. A **52**, 21 – 31 (2016)
[5] A. Reuber, K. Holinde and J. Speth, Nucl. Phys. A **570**, 543 (1994) ; J. Haidenbauer, U.-G. Meißner, Phys. Rev. C **72**, 044005 (2005)
[6] J. Haidenbauer, U.-G. Meißner and S. Petschauer, Eur. Phys. J. A **51**, 17 (2015); J. Haidenbauer, U.-G. Meißner, S. Petschauer, Nucl. Phys. A **954**, 273-293 (2016)
[7] F. Eisele, H. Filthuth, W. Foehlisch, V. Hepp, G. Zech, Phys. Lett. **37B**, 204 (1971); H.G. Dosch, R. Engelmann, H. Filthuth, V. Hepp, E. Kluge, Phys. Lett. **21**, 236 (1966); H. Filthuth, W. Lughofer, Phys. Rev. **173**, 1452 (1968); B. Sechi-Zorn, B. Kehoe, J. Twitty, R.A. Burnstein, Phys. Rev. **175**, 1735 (1968)
[8] H. Lenske and M. Dhar, *$SU(3)_f$ Constraints on Hypernuclear Energy Density Functionals*, the HYP22 proceedings

$\Lambda\Lambda$ pairing effects in multi-Λ hyperisotopes within a Skyrme-Hartree-Fock model

Jing Guo[1] and *Xian-Rong* Zhou[1,*]

[1]Department of Physics, East China Normal University, Shanghai 200241, China

Abstract.
The $\Lambda\Lambda$ pairing effects are investigated in multi-Λ hyperisotopes with the Skyrme-Hartree-Fock approach. The shell closure of hyperon occurrs at $-S$ = 2, 8, 18, 20, 34, 58, 68, as well as 70 in spherical hypernuclei, which are attributed from the Λ hyperon potential being close to the Woods-Saxon potential. The more possible Λ hyperon magic numbers $-S$ = 4, 6, 10, 14, 26, 30, and 32 in deformed hypernuclei are found based on the analysis of the single-particle energy levels, and are all sensitive to the quadrupole deformation parameter β_2.

1 Introduction

Since 1953, when the first hypernucleus was discovered in a stack of photographic emulsions exposed to cosmic rays [1], some hypernuclei have been synthesized [2]. Most of the produced hypernuclei are single-Λ hypernuclei, and only few light double-Λ hypernuclei are known.

To make predictions for the multi-Λ hypernuclei theoretically, reliable $\Lambda\Lambda$ interactions are necessary. However, the strength of the force is still one of the open questions for multistrange systems. A large variety of phenomena were predicted for multi-Λ hypernuclei during the 1980s and the early 1990s [3–5], yet these studies assumed very attractive $\Lambda\Lambda$ interactions. In 2017, an empirical correction for the Nijmegen interactions NSC89, NSC97a, NSC97f in the $\Lambda\Lambda$ channel has been proposed and fitted to the bond energies of ${}^{6}_{\Lambda\Lambda}$He [6].

Furthermore, the pairing force between hyperons may modify the understanding of finite hypernuclei. There are several theoretical works about $\Lambda\Lambda$ pairing effects for spherical multi-Λ hyperisotopes. In the Hartree-Fock-Bogoliubov (HFB) model [7], the strength of $\Lambda\Lambda$ pairing was calibrated to match the maximal value for the prediction of the $\Lambda\Lambda$ pairing gap in uniform matter [8], so that an upper bound for the prediction of the Λ pairing gap and its effects in hypernuclei were provided. Soon after, with a relativistic-Hartree-Bogoliubov (RHB) approach based on the quark model, the strength of the pairing interaction was set as 4/9 of that for nucleons [9]. The strength of $\Lambda\Lambda$ pairing is therefore still an open question.

The present work aims to extend the method of determining the $\Lambda\Lambda$ pairing strength proposed in Ref. [9] to the non-relativistic SHF model. Furthermore, investigate the pairing effects of multi-Λ Mg, Fe, and Zr hypernuclei in detail.

2 FORMALISM

*e-mail: xrzhou@phy.ecnu.edu.cn

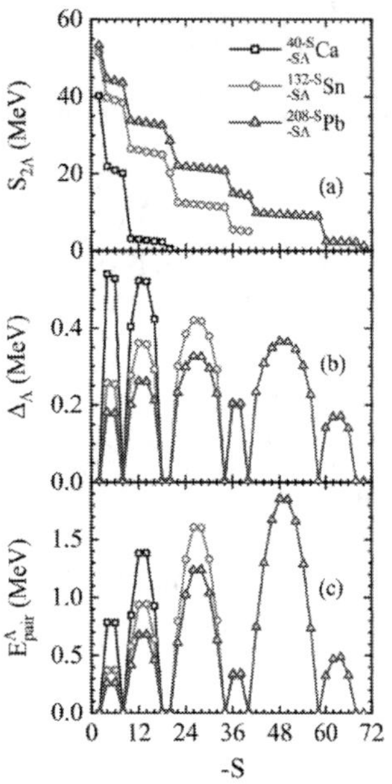

Figure 1. The 2Λ separation energies $S_{2\Lambda}$ (a), pairing gaps Δ_Λ (b), and pairing energies E^Λ_{pair} (c) for Ca, Sn, and Pb hyperisotopic chains.

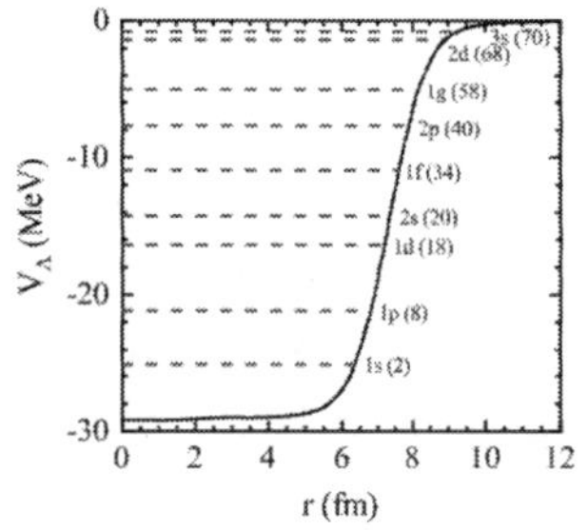

Figure 2. The Λ hyperon potential V_Λ and s.p. levels for $^{278}_{70\Lambda}$Pb.

The deformed SHF model proposed in Ref. [10] is used to investigate $\Lambda\Lambda$ pairing effects in multi-Λ hypernuclei. For the mean-field part, we use the same parameters as HFB [7], i.e., the Skyrme force SLy5 for the *NN* part [11], the Nijmegen interaction NSC97f for the $N\Lambda$ part, and the empirical prescription EmpC for the $\Lambda\Lambda$ interaction part [6].

The pairing interaction is taken as a δ-type force [12]:

$$V_q(\boldsymbol{r}_1, \boldsymbol{r}_2) = V_0^{(q)}\delta(\boldsymbol{r}_1 - \boldsymbol{r}_2), \quad q = p, n, \Lambda, \quad (1)$$

with the pairing strength $V_0 = 323$ MeV fm^{-3} for nucleons as in Ref. [7]. For the Λ hyperon, as in Ref. [9], it is set as 4/9 of that for nucleons, i.e., $V_0 = 144$ MeV fm^{-3}, which is smaller overall than that presented in Ref. [7]. A smooth energy cutoff is included in the BCS approximation [12].

3 RESULTS

3.1 Spherical hypernuclei

In Fig. 1, the 2Λ separation energies, pairing gaps Δ_Λ, and pairing energies E^Λ_{pair} [12] for $^{40-S}_{-S\Lambda}$Ca, $^{132-S}_{-S\Lambda}$Sn, and $^{208-S}_{-S\Lambda}$Pb are presented. As indicated by the panels in Fig. 1 (a), the $S_{2\Lambda}$ of all nuclei are positive and decrease with the hyperon number $-S$ increasing. There are obvious drop at points of $-S$=2, 8, 18, 20, 34, 40, 58, 68, and 70, which indicates the magic number of hyperisotopes.

Fig. 1 (b) shows that the average pairing gaps Δ_Λ evolve as the shape of arches with zero values at the location of hyperon magic number identified by the drop points of $S_{2\Lambda}$. Due to absence of relevant Λ spin-orbit forces, the individual shells are fairly large, and thus pairing correlations for hyperisotopes located close to their center are strong. The $\Lambda\Lambda$ pairing gap of Ca, Sn, and Pb isotopes reaches about 0.55, 0.40, and 0.35 MeV, respectively in half-filled-shell hypernuclei. Futhurmore, as shown in Fig. 1 (c), the same shell structure is also found by the pairing energies E^Λ_{pair}.

With the HFB approach, only the pairing energies were reported and the maximum value was about 3 MeV for Sn isotopes, which is originated from the larger pairing strength. In the RHB calculations, the structures of subshell closure are not exactly the same for different kinds of hyperisotopes.

To explore the reason of different magic numbers in detail, we plot the potentials V_Λ and single-particle levels of hyperons for $^{278}_{70\Lambda}$Pb ($-S = 70$) in Fig. 2. Due to lacking of spin-orbit term, the magic numbers of hyperons are not the same as the nucleons as we all know (2, 8, 20, 28, 50, 82, etc.). The V_Λ potential is of Woods-Saxon type, which results in the corresponding shell structure. A similar shell structure can be seen in Ref. [7] because of the same mean-field potential of hyperons. The V_Λ obtained in the RHB calculation [9] is similar to the shape of a harmonic oscillator potential and leads to a bit different shell closure structure.

3.2 Deformed hypernuclei

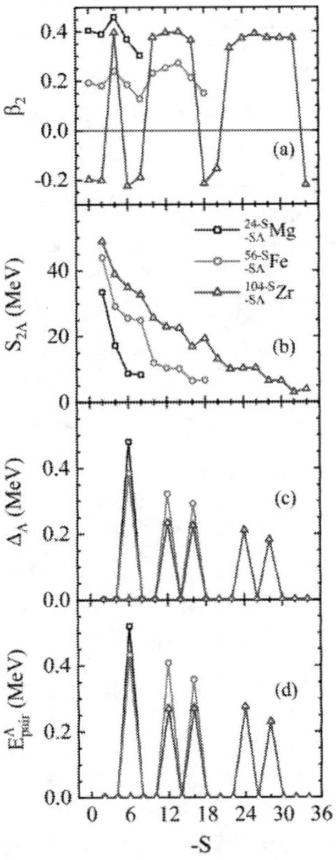

Figure 3. The quadrupole deformations β_2 (a), the 2Λ separation energies $S_{2\Lambda}$ (b), the pairing gaps Δ_Λ (c), and the pairing energies $E^\Lambda_{\rm pair}$ (d) for Mg, Fe, and Zr hyperisotopes.

In Fig. 3, the deformations (a), 2Λ separation energies $S_{2\Lambda}$ (b), average $\Lambda\Lambda$ pairing gaps Δ_Λ (c), and $\Lambda\Lambda$ pairing energies E^Λ_{pair} (d) for $^{24-S}_{-S\Lambda}$Mg, $^{56-S}_{-S\Lambda}$Fe, and $^{104-S}_{-S\Lambda}$Zr are presented. As indicated by Fig. 3 (a), normal nuclei and hypernuclei for all isotopes are clearly deformed. In Fig. 3 (b), there is an obvious drop at the point of $-S=4$ for $^{24-S}_{-S\Lambda}$Mg isotopes, which is different with other isotopes. For the average $\Lambda\Lambda$ pairing gaps Δ_Λ , and $\Lambda\Lambda$ pairing energies E^Λ_{pair}, all deformed hyperisotopes present zero values at $-S$=2, 4, 8, 10, 14, 18, 20, 26, 30, 32, 34, wheras the values for $^{110}_{6\Lambda}$Zr are also zero.

To further explore the above features, we show in Fig. 4 (a) the hyperon s.p. energies as functions of deformation β_2 for $^{130}_{26\Lambda}$Zr. One observes that there are large gaps at $-S = 2$, 4, 8, 10, 14, 18, 20, 22, 26, 30, 32, and 34 on the prolate side, which explains the behavior of $S_{2\Lambda}$, Δ_Λ, and $E^\Lambda_{\rm pair}$ shown in Fig. 3. The sudden drop of $S_{2\Lambda}$ in $^{24-S}_{-S\Lambda}$Mg at $-S = 4$ is attributed to the large gap between the [110] and [101] orbits at $\beta_2 \approx 0.4$ (the actual deformation of $^{28}_{4\Lambda}$Mg and $^{30}_{6\Lambda}$Mg). For prolate and oblate hypernuclei, the [101 1/2] and [101 3/2] orbits are degenerate, which explains why the pairing of hyperons vanishes for the oblate $^{110}_{6\Lambda}$Zr, but not for the prolate $^{30}_{6\Lambda}$Mg and $^{62}_{6\Lambda}$Fe.

As shown in panels (b) and (c) of Fig. 4, the gap drops fast from its undeformed value ≈ 0.4 MeV to vanish at $\beta_2 \approx 0.2$, remaining zero at the actual deformation $\beta_2 \approx 0.4$. This is caused by the splitting of the undeformed $1f$ level into a completely occupied [310 + 321] level that becomes well separated from the neighboring empty [301 + 312] level. Once the separation of these levels becomes much larger than the (spherical) pairing gap, the gap vanishes. For oblate deformation, $\Delta_\Lambda(\beta_2)$ exhibits a nonmonotonic behavior due to the crossing of the [310 + 321] and [301 + 312] levels, and vanishes only at $\beta_2 \approx -0.4$. The dependence of the pairing gap on β_2 is also illustrated by the corresponding occupations of the various involved $1f$ sublevels shown in panel (c). In other words, the pairing effect depends on the s.p. level density around

https://doi.org/10.1051/epjconf/202227105005

the Fermi surface. In general, however, deformation always leads to a weakening or complete suppression of pairing.

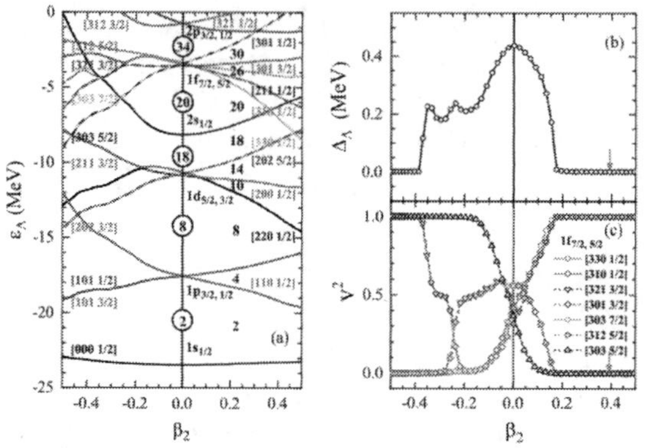

Figure 4. Several quantities of the hypernucleus $^{130}_{26\Lambda}$Zr as a function of deformation β_2: (a) The Λ s.p. energies. The Fermi surface is indicated by the dotted black curve. The quantum numbers for the spherical orbits are labeled around $\beta_2 = 0$, and the deformed quantum numbers $[Nn_z m_l\ \Omega]$ along the curves. Also indicated are the magic numbers in the spherical and prolate cases. (b) The Λ pairing gap. (c) The Λ occupation probability v^2 of the $1f$ sublevels. The actual deformation $\beta_2 \approx 0.40$ of this hypernucleus is also indicated.

4 Conclusion

The $\Lambda\Lambda$ pairing effects in spherical and deformed multi-Λ hypernuclei are explored with the SHF approach. For the spherical hyperisotopes, due to a Woods-Saxon-like Λ hyperon potential, the occurrences of magic numbers are at $-S = 2, 8, 18, 20, 34, 58, 68$, and 70 which can be evidenced by the sudden drop of 2Λ separation energies, as well as by the vanishing average pairing gaps and pairing energies. For the deformed hyperisotopes, more possible hyperon magic numbers $-S = 4, 6, 10, 14, 22, 26, 30$, and 32, corresponding to vanishing pairing, appear. All possible hyperon magic numbers of the deformed hyperisotopes are sensitive to β_2 based on the analysis of the s.p. energy levels.

References

[1] M. Danysz and J. Pniewski, The London, Edinburgh, and Dublin Philosophical Magazine and Journal of Science 44, 348 (1953)
[2] A. Gal, E. V. Hungerford, and D. J. Millener, Rev. Mod. Phys. 88, 035004 (2016)
[3] M. Rufa, H. Stcker and et al., J. Phys. G: Nucl. Phys. 13, L143 (1987); M. Rufa, J. Schaffner and et al., Phys. Rev. C 42, 2469 (1990)
[4] J. Mare and J. ofka, Z. Phys. A 333, 209 (1989); 345, 47 (1993)
[5] J. Schaffner, C. Greiner, and H. Stcker, Phys. Rev. C 46, 322 (1992)
[6] J. Margueron, E. Khan, and F. Gulminelli, Phys. Rev. C 96, 054317 (2017)
[7] H. Güven, K. Bozkurt, E. Khan, and J. Margueron, Phys. Rev. C 98, 014318 (2018)
[8] A. R. Raduta, A. Sedrakian, and F. Weber, Mon. Not. R. Astron. Soc. 475, 4347 (2018)
[9] Y.-T. Rong, P. Zhao, and S.-G. Zhou, Phys. Lett. B 807, 135533 (2020)
[10] X.-R. Zhou, H.-J. Schulze, H. Sagawa and et al., Phys. Rev. C 76, 034312 (2007)
[11] M. Bender, P. H. Heenen, and P. G. Reinhard, Rev. Mod. Phys. 75, 121 (2003)
[12] M. Bender, K. Rutz, P. G. Reinhard, and J. A. Maruhn, Eur. Phys. J. A 8, 59 (2000)

How can we solve the hyperon puzzle?

—Introduction to "topical session on ΛNN three-body force"

Hirokazu Tamura[1,2,*]

[1]Department of Physics, Tohoku University, Sendai 980-8578, Japan
[2]Advanced Science Research Center, Japan Atomic Energy Agency, Tokai 319-1195, Japan

Abstract. As an introduction to the "topical session on ΛNN three-body force" in HYP2022, a scenario to solve the hyperon puzzle and to elucidate high density matter in neutron stars is proposed. It is related to the J-PARC Hadron Facility extension project, in which experimental studies of ΛNN three-body force (3BF) via high precision spectroscopy of Λ hypernuclei and high quality ΛN scattering are one of the flagship subjects in the project. Development of theoretical methods to extract 3BF effects from experimental data and to reliably extrapolate them to higher density is challenging but extremely important.

1 Introduction

Microscopic understanding of the high density matter in the core of neutron stars (NS's) is one of the most challenging subjects in hadron/nuclear physics. In particular, there is a serious contradiction called "hyperon puzzle" between observational data of massive NS's and our knowledge and frameworks of nuclear physics; huge fermi energies of neutrons in the NS core provoke their conversion to hyperons, particularly to Λ's which undergo considerable attraction in nuclear matter as demonstrated by hypernuclear data. Such hyperon appearance drastically softens the Equation-Of-State (EOS) of nuclear matter, leading to the conclusion that the three reliable observational data of NS's with twice the solar mass cannot be supported at all.

Most probably, the contradiction is caused by our naive assumption that the Λ's in-medium attraction experimentally confirmed only at the normal nuclear density (ρ_0) can be also applied to higher density nuclear matter. It is well known that microscopic calculations based on the realistic NN interaction models, which describe all the NN scattering data and light nuclei (by adding attractive NNN 3BF from pion exchange), fail to reproduce the nuclear saturation density ρ_0, indicating that a repulsive, short-range 3BF is at work. It is natural to imagine a similar repulsive ΛNN (and $\Lambda\Lambda N$, $\Lambda\Lambda\Lambda$) interaction. However, we have no experimental evidence so far [1] to support this idea.

2 How to see the ΛNN 3BF effects in experiments

Yamamoto *et al.* used the Nijmegen ESC08c models for YN interaction and phenomenologically introduced a "universal" baryon-baryon-baryon (BBB) 3BF common to all the baryon channels, of which strength was determined from the NNN 3BF strength necessary to reproduce the ^{16}O-^{16}O scattering cross sections at 70 MeV/A [2]. Using this model, they calculated

*e-mail: tamura@lambda.phys.tohoku.ac.jp

[1]In this topical session, Friedman and Gal suggested a hint of ΛNN 3BF in the B_Λ data of Λ hypernuclei [1].

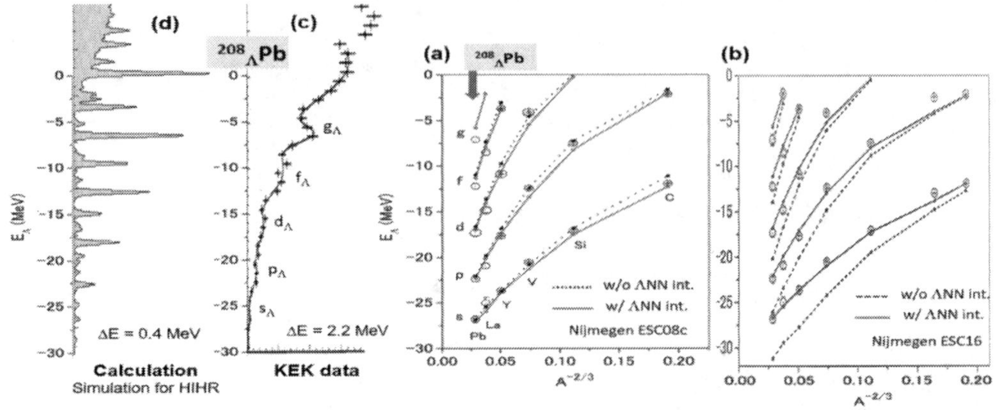

Figure 1. (a) B_Λ values of various Λ's single particle states for a wide range of hypernuclear mass number A calculated using the ESC08 interaction [2], without and with the universal 3BF for ΛNN. Experimental data are shown in inside circles. (b) The same as (a) but the ESC16 interaction is used [3]. (c) $^{208}_{\Lambda}$Pb spectrum measured at KEK [4] shown together with the figures (a) and (b). (d) Simulated $^{208}_{\Lambda}$Pb spectrum expected in the proposed experiment at HIHR [5] based on a theoretical calculation.

binding energies (B_Λ) of Λ's single-particle orbits ($0s$, $0p$, $0d$, $0f$) for a wide range of hypernuclear mass number ($A = 12 - 208$) and compared them with the data, as shown in Fig. 1 (a). When the universal 3BF is turned on, the A-dependence of the B_Λ values became slightly weaker, reflecting the density-dependent repulsive force. Thus, it was found that those B_Λ values contain a density dependent effect of the ΛN interaction, although the effect is only less than 1 MeV for $A = 12 - 208$. They also found that the EOS calculated with the universal 3BF successfully supports the two solar-mass NS's.

It is noted that in the NN case the density dependence cannot be studied in this method, because most of the single-particle nucleon hole states are too broad to be observed. On the contrary, a Λ hyperon is free from Pauli principle from nucleons and behaves as a distinguishable particle in a nucleus. For each of the single-particle Λ states, we know the local nuclear density where the Λ stays, since the spatial distribution of the Λ wave function in the hypernucleus can be theoretically calculated rather well. Thus, the A-dependence and the orbital dependence of the B_Λ values in Λ hypernuclei can be unique probes to extract the strength of the ΛNN force, if the B_Λ values are measured in ~0.1 MeV accuracy.

In order to reliably extract the ΛNN force strength, the ΛN interaction in the free space should be accurately known. Actually, as shown in Fig. 1 (b), another version of the ESC models (ESC16) gives a drastically different A-dependence between the cases without and with the universal 3BF [3]. This change between Fig. 1 (a) and (b) is caused by a difference of the p-wave component of the ΛN interaction between ESC08c and ESC16. Our present knowledge of the ΛN interaction is based on sparce Λp scattering data supplemented by Λ hypernuclear data. However, the hypernuclear data cannot be used for the present purpose, and accurate Λp scattering data are absolutely necessary.

3 Experimental prospects to solve the hyperon puzzle at J-PARC

Aiming at solving the hyperon puzzle, new experiments are proposed at J-PARC. In order to study ΛNN 3BF, we plan to perform a series of high precision Λ hypernuclear spectroscopy study via the (π^+, K^+) reaction [6], combined with high quality Λp scattering experiments [7]. It is one of the flagship programs in the J-PARC Hadron Facility extension project [5]. In this extension project, the primary proton beam line is extended and a new production target is installed, and then several new secondary beam lines are constructed. With the momentum dispersion matching technique, the High Intensity High Resolution beam line (HIHR) delivers an intense ($\sim 2.5 \times 10^8$/spill) pion beam with a high resolution of ~0.4 MeV (FWHM) for the (π^+, K^+) hypernuclear spectroscopy [6]. The K1.1 beam line is a high-purity mass-separated line for low momentum ($\leq$1.1 GeV/c) kaon beams, which will be used for Λ production in the ΛN scattering experiment [7].

4 Theoretical prospects

In our plan, the property and the strength of the ΛNN 3BF will be extracted from the high-resolution Λ hypernuclear spectroscopy and ΛN scattering experiments and then applied to higher density matter in NS's. However, how to extrapolate them to densities higher than ρ_0, strongly depends on the theoretical framework. Discussing how to develop appropriate theoretical frameworks is an important motivation of this topical session.

The chiral effective field theory (chEFT) approach for NN interaction has a great potential to describe higher density matter via systematic treatment of the NNN 3BF appearing at N^{2}LO and higher order calculations. The low energy constants (LECs) are determined from the rich NN scattering data. However, determination of the two contact terms (c_D, c_E) responsible for the NNN 3BF is not straightforward. Studies are in progress using various data of light nuclei (^{3}H, ^{4}He *etc.*), *pd* scattering, or the nuclear saturation density [8].

The chEFT approach is extended to BB interactions including hyperons and the NLO calculation has been made [9]. In order to determine the LECs, high quality YN scattering data are essential, even though the flavor SU(3) symmetry is helpful. Then, in a similar way as the NNN case, the LEC's for the ΛNN 3BF could be determined from light Λ hypernuclear data as well as Λd scattering data to be available in the future. Here, unlike the case of NNN 3BF, precise B_Λ data of single-particle Λ states of various Λ hypernuclei are expected to play significant roles particularly in determining the 3BF contact terms. Here, theoretical development for precise calculations of light to heavy hypernuclear structure is quite important.

It is noted that lattice QCD calculations also play important roles. The HAL QCD approach has succeeded in deriving the BB interaction potentials [10]. Calculations for various BB channels will be checked by correlation measurements with high energy nuclear collisions ("femtoscopy") and used to supplement the ΛN, ΣN scattering data in the construction of the SU(3)$_f$ extended chEFT interactions. In addition, efforts are being made to apply the HAL QCD method to the BBB 3BF [11]. In the future, lattice-calculated ΛNN 3BF will be directly compared with the hypernuclear data.

5 Scenario to solve the hyperon puzzle

Figure 2 summarizes our scenario to solve the hyperon puzzle. We plan to collect high-quality Λp and Σp scattering data from the K1.1 (and High-p) beam lines at J-PARC and provide YN scattering database to theorists. YN scattering data from JLab(CLAS) are also useful particularly for momenta higher than 1 GeV/c. In addition, YN/YY interaction data of other channels are also required to construct BB interaction models with a help of SU(3)$_f$ symmetry. $\Xi/\Lambda\Lambda$ hypernuclei and Ξ^- atom data from the K1.8 line in the present J-PARC Hadron Facility provide ΞN and $\Lambda\Lambda$ interactions, and femtoscopy data for various channels ($\Lambda\Lambda$, ΞN, $\Xi\Xi$, ΩN *etc.*) will be also accumulated in high energy experiments. It is to be also mentioned that $\bar{K}N$ interaction data from kaonic nuclei/atoms from J-PARC K1.8BR and DAΦNE are also necessary for constructing EOS.

Employing all these data, "realistic BB interaction" models will be constructed, for example, by upgrading the Nijmegen ESC models. In parallel, the SU(3)$_f$-extended chEFT model

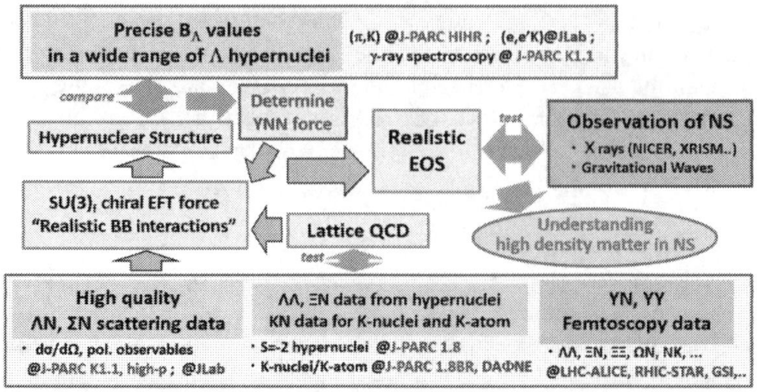

Figure 2. Scenario to solve the hyperon puzzle in the J-PARC Hadron Facility extension project.

will proceed to the N^2LO calculation and the LEC's, including those for the ΛNN 3BF, will be hopefully determined. The lattice QCD calculations confirmed by experiments will be employed to improve the realistic BB interactions. Then, Λ hypernuclei will be precisely calculated from the realistic BB interactions.

Precise hypernuclear B_Λ data to be measured at HIHR as described above, as well as $(e, e'K^+)$ spectroscopy data at JLab, will be compared with the precise calculations. To separate and assign closely-located peaks in the spectra, γ-ray spectroscopy will be applied at K1.1 for low-lying s_Λ states of the same hypernucleus. Thus, the property and strength of the ΛNN 3BF will be extracted. If the 3BF is found to be not very repulsive and cannot reproduce the observed massive NS's, it could suggest that deconfined quark matter begins to appear in NS's at rather low densities of 2–3ρ_0. If not, the 3BF should be properly incorporated into the extended chEFT interaction by determining the LEC's from those hypernuclear data, so that the chEFT interaction can be reliably applied to higher density for calculation of EOS. This would be a theoretical challenge.

Finally, the calculated EOS will be tested by mass and radius data of NS's. Such data are being accumulated via X-ray observation by NICER and then by XRISM satellite. Gravitational wave (GW) data will also provide additional information on the stiffness of NS mergers. When the BB interaction is reliably determined by our studies, more realistic calculations of GW templates will be possible and unique information will be extracted from GW data.

References

[1] E. Friedman and A. Gal, in these proceedings.
[2] Y. Yamamoto *et al.*, Phys. Rev. C **88**, 022801 (2013); Phys. Rev. C **90**, 045805 (2014)
[3] M. M. Nagels *et al.*, Phys. Rev. C **99**, 044003 (2019)
[4] T. Hasegawa *et al.*, Phys. Rev. C **53**, 1210 (1996)
[5] K. Aoki *et al.*, arXiv:2110.04462v1 [nucl-ex] (2021); F. Sakuma, in these proceedings
[6] S. N. Nakamura, in these proceedings
[7] K. Miwa, in these proceedings
[8] E. Epelbaum, H. Krebs, P. Reinert, Front. in Phys. **8**, 98 (2020)
[9] W. Weise, in these proceedings and references therein
[10] S. Aoki *et al.*, Prog. Theor. Exp. Phys. **2012**, 01A105 (2012)
[11] T. Doi *et al.*, Prog. Theor. Phys. **127**, 723 (2012)
[12] For example, ALICE collaboration, Nature **588**, 232 (2020)

EPJ Web of Conferences **271**, 06002 (2022) https://doi.org/10.1051/epjconf/202227106002
HYP2022

ΛNN content of Λ-nucleus potential

Eliahu Friedman[1,⋆] and *Avraham* Gal[1,⋆⋆]

[1] *Racah Institute of Physics, The Hebrew University, Jerusalem 9190401, Israel*

Abstract. A minimally constructed Λ-nucleus density-dependent optical potential is used to calculate binding energies of observed $1s_\Lambda$, $1p_\Lambda$ states across the periodic table, leading to a repulsive ΛNN contribution $D_\Lambda^{(3)} \approx 14$ MeV to the phenomenological Λ-nucleus potential depth $D_\Lambda \approx -30$ MeV. This value is significant in connection with the so-called 'hyperon puzzle.'

1 Introduction

The Λ-nucleus potential depth provides an important constraint in ongoing attempts to resolve the 'hyperon puzzle', i.e., whether or not dense neutron-star matter contains hyperons, primarily Λs besides nucleons [1]. Figure 1 presents compilation of most of the known Λ hypernuclear binding energies (B_Λ) across the periodic table, fitted by a three-parameter Woods-Saxon (WS) attractive potential. As $A \to \infty$, a limiting value of $B_\Lambda(A) \to 30$ MeV is obtained. Interestingly, studies of density dependent Λ-nuclear optical potentials $V_\Lambda(\rho)$ in Ref. [2], with ρ the nuclear density normalized to the number of nucleons A, conclude that a ρ^2 term motivated by three-body ΛNN interactions provides a large repulsive (positive) contribution to the Λ-nuclear potential depth D_Λ at nuclear-matter density ρ_0: $D_\Lambda^{(3)} \approx 30$ MeV. This repulsive component of D_Λ is more than just compensated at ρ_0 by a roughly twice larger attractive depth value $D_\Lambda^{(2)} \approx -60$ MeV, motivated by a two-body ΛN interaction. Note that D_Λ is defined as $V_\Lambda(\rho_0)$ in the limit $A \to \infty$ at a given nuclear-matter density ρ_0, with a value 0.17 fm^{-3} assumed here.

Most hyperon-nucleon potential models overbind Λ hypernuclei, yielding values of $D_\Lambda^{(2)}$ deeper than -30 MeV. Whereas such overbinding amounts to only few MeV in the often used Nijmegen soft-core model versions NSC97e,f [4] it is considerably stronger, by more than 10 MeV, in the recent Nijmegen extended soft-core model ESC16 [5]. A similar overbinding arises at leading order in chiral effective field theory (χEFT) [6]. The situation at next-to leading order (NLO) is less clear owing to a strong dependence of $D_\Lambda^{(2)}$ on the momentum cutoff scale λ [7]. At λ=500 MeV/c, however, it is found in Ref. [8] that both versions NLO13 [9] and NLO19 [10] overbind by a few MeV. Finally, recent Quantum Monte Carlo (QMC) calculations [11, 12], using a $\Lambda N + \Lambda NN$ interaction model designed to bind correctly ${}^5_\Lambda$He, result in a strongly attractive $D_\Lambda^{(2)}$ of order -100 MeV and a correspondingly large repulsive (positive) $D_\Lambda^{(3)}$, reproducing the overall potential depth $D_\Lambda \approx -30$ MeV.

Our aim in the present phenomenological study is to check to what extent properly chosen Λ hypernuclear binding energy data, with *minimal* extra assumptions, imply positive values of $D_\Lambda^{(3)}$, and

⋆Eliahu.Friedman@mail.huji.ac.il
⋆⋆avragal@savion.huji.ac.il

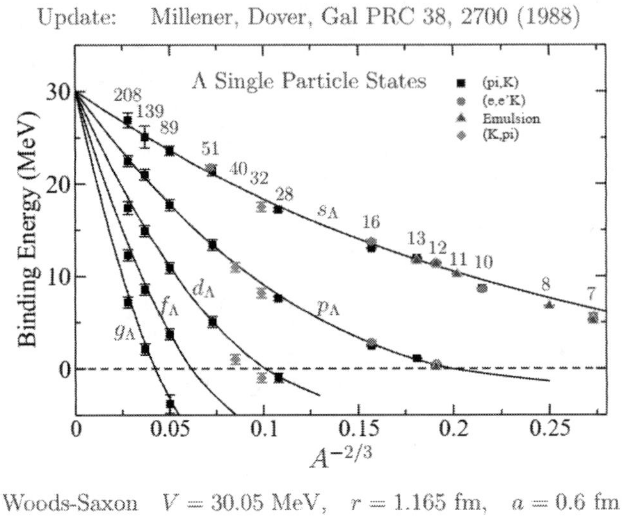

Figure 1. Compilation of Λ binding energies in ${}^{7}_{\Lambda}$Li to ${}^{208}_{\Lambda}$Pb from various sources, and as calculated using a three-parameter WS potential [2]. Figure adapted from Ref. [3]

how large it is [13]. Repulsive three-body ΛNN interactions go beyond just providing solution of the overbinding problem: as nuclear density is increased beyond nuclear matter density ρ_0, the balance between attractive $D_\Lambda^{(2)}$ and repulsive $D_\Lambda^{(3)}$ tilts towards the latter. This results in nearly total expulsion of Λ hyperons from neutron-star matter, suggesting an equation of state (EoS) sufficiently stiff to support two solar-mass neutron stars, thereby providing a possible solution to the 'hyperon puzzle'. The larger $D_\Lambda^{(3)}$ is, the more likely it is a solution [14, 15]. However, there is no guarantee that three-body ΛNN interactions are universally repulsive. For a recent discussion of this problem within an SU(3) 'decuplet dominance' approach practised in modern χEFT studies at NLO, see Ref. [8].

In this Contribution we adopt the optical potential approach as applied by Dover-Hüfner-Lemmer to pions in nuclear matter [16]. For the Λ-nucleus system, it provides expansion in powers of the nuclear density $\rho(r)$, consisting of a linear term induced by a two-body ΛN interaction plus two higher-power density terms: (i) a long-range Pauli correlations term starting at $\rho^{4/3}$, and (ii) a short-range ΛNN interaction term dominated in the present context by three-body ΛNN interactions, starting at ρ^2. As demonstrated below, the contribution of the Pauli correlations term is non negligible, propagating to higher powers of density terms than just $\rho^{4/3}$, such as the ρ^2 ΛNN interaction term. This explains why the value derived here, $D_\Lambda^{(3)} = (13.9 \pm 1.4)$ MeV, differs from any of those suggested earlier in Ref. [2] and in Skyrme Hartree Fock studies [17] where Pauli correlations are usually disregarded. Our value of $D_\Lambda^{(3)}$ strongly disagrees with the much larger value inferred in QMC calculations [12]. We comment on these discrepancies below.

2 Nuclear densities

In optical model applications aimed at establishing relations between components with different powers of density $\rho = \rho_p + \rho_n$, it is crucial to ensure that the radial extent of the densities, e.g., their r.m.s.

radii, follows closely values derived from experiment. For proton densities we used charge densities, with proton finite-size and recoil effects included. Harmonic-oscillator type densities [18] were used for the lightest elements, assuming the same radial parameters for protons and neutrons. A variation of 1% in the r.m.s. neutron radius was found to affect calculated Λ binding energies considerably less than given by most of the experimental uncertainties listed in Table 1 below. For a detailed discussion in the analogous case of light Ξ^- hypernuclei, see Ref. [19]. For species beyond the nuclear $1p$ shell we used two-parameter Fermi distributions normalized to Z for protons and $N = A - Z$ for neutrons, derived from assembled nuclear charge distributions [20]. For medium-weight and heavy nuclei, the r.m.s. radii of our neutron density distributions assume larger values than those for proton density distributions, as practiced in analyses of exotic atoms [21]. Furthermore, once neutron orbits extend beyond proton orbits, it is useful to represent the nuclear density $\rho(r)$ as

$$\rho(r) = \rho_{\text{core}}(r) + \rho_{\text{excess}}(r), \tag{1}$$

where ρ_{core} refers to the Z protons plus the charge symmetric Z neutrons occupying the same nuclear 'core' orbits, and ρ_{excess} refers to the $(N - Z)$ 'excess' neutrons associated with the nuclear periphery.

3 Optical potential

The optical potential employed in this work, $V_\Lambda^{\text{opt}}(\rho) = V_\Lambda^{(2)}(\rho) + V_\Lambda^{(3)}(\rho)$, consists of terms representing two-body ΛN and three-body ΛNN interactions, respectively:

$$V_\Lambda^{(2)}(\rho) = -\frac{4\pi}{2\mu_\Lambda} f_A\, C_{\text{Pauli}}(\rho)\, b_0 \rho, \tag{2}$$

$$V_\Lambda^{(3)}(\rho) = +\frac{4\pi}{2\mu_\Lambda} f_A\, B_0 \frac{\rho^2}{\rho_0}, \tag{3}$$

with b_0 and B_0 strength parameters in units of fm ($\hbar = c = 1$). In these expressions, $\rho(r)$ is a nuclear density distribution normalized to the number of nucleons A, $\rho_0 = 0.17$ fm^{-3} stands for nuclear-matter density, μ_Λ is the Λ-nucleus reduced mass and f_A is a kinematical factor transforming b_0 from the ΛN c.m. system to the Λ-nucleus c.m. system:

$$f_A = 1 + \frac{A-1}{A}\frac{\mu_\Lambda}{m_N}. \tag{4}$$

This form of f_A coincides with the way it is used for $V_\Lambda^{(2)}$ in atomic/nuclear hadron-nucleus bound-state problems [21] and its A dependence provides good approximation for $V_\Lambda^{(3)}$. Next is the density dependent factor $C_{\text{Pauli}}(\rho)$ in Eq. (2), standing for a Pauli correlation function:

$$C_{\text{Pauli}}(\rho) = (1 + \alpha_P \frac{3k_F}{2\pi} f_A b_0)^{-1}, \tag{5}$$

with Fermi momentum $k_F = (3\pi^2\rho/2)^{1/3}$. The parameter α_P in Eq. (5) switches off (α_P=0) or on (α_P=1) Pauli correlations in a form suggested in Ref. [22] and practised in K^- atoms studies [23]. To estimate $1/A$ correction terms, we also approximated $C_{\text{Pauli}}(\rho)$ by [19]:

$$C_{\text{Pauli}}(\rho) \approx (1 + \alpha_P \frac{3k_F}{2\pi} b_0^{\text{lab}})^{-1}, \qquad b_0^{\text{lab}} = (1 + \frac{m_\Lambda}{m_N})\, b_0. \tag{6}$$

As shown below, including $C_{\text{Pauli}}(\rho)$ in $V_\Lambda^{(2)}$ affects strongly the balance between the derived potential depths $D_\Lambda^{(2)}$ and $D_\Lambda^{(3)}$. However, introducing it also in $V_\Lambda^{(3)}$ is found to make little difference, which is why it is skipped in Eq. (3). Finally we note that the low-density limit of V_Λ^{opt} requires according to Ref. [16] that b_0 is identified with the c.m. ΛN spin-averaged scattering length (positive here).

4 Data

Table 1. Binding energies in MeV, including uncertainties, considered here; taken from Table IV of Ref. [3].

hypernucleus	$1s_\Lambda$	$\pm$	$1p_\Lambda$	$\pm$
${}^{12}_{\Lambda}$B	11.52	0.02	0.54	0.04
${}^{13}_{\Lambda}$C	12.0	0.2	1.1	0.2
${}^{16}_{\Lambda}$N	13.76	0.16	2.84	0.18
${}^{28}_{\Lambda}$Si	17.2	0.2	7.6	0.2
${}^{32}_{\Lambda}$S	17.5	0.5	8.2	0.5
${}^{51}_{\Lambda}$V	21.5	0.6	13.4	0.6
${}^{89}_{\Lambda}$Y	23.6	0.5	17.7	0.6
${}^{139}_{\Lambda}$La	25.1	1.2	21.0	0.6
${}^{208}_{\Lambda}$Pb	26.9	0.8	22.5	0.6

The present work does not attempt to reproduce the full range of B_Λ data shown in Fig. 1. It is limited to $1s_\Lambda$ and $1p_\Lambda$ states listed in Table 1. We fit to such states in *one* of the nuclear $1p$-shell hypernuclei listed in the table, where the $1s_\Lambda$ state is bound by over 10 MeV while the $1p_\Lambda$ state has just become bound. This helps resolve the density dependence of $V_\Lambda^{\rm opt}$ by setting a good balance between its two components, $V_\Lambda^{(2)}(\rho)$ and $V_\Lambda^{(3)}(\rho)$, following it all the way to ${}^{208}_{\Lambda}$Pb the heaviest hypernucleus marked in Fig. 1. We chose to fit the ${}^{16}_{\Lambda}$N precise $B_\Lambda^{\rm exp}(1s, 1p)$ values derived, respectively, from the first and third peaks to the left in Fig. 2. The extremely simple $1p$ proton hole structure of the ^{15}N nuclear core in this case removes most of the uncertainty arising from spin-dependent residual ΛN interactions [25]. The fitted optical-potential parameters b_0, Eq. (2), and B_0, Eq. (3), are then used to calculate the $B_\Lambda^{1s,1p}$ values of the other eight species listed in Table 1.

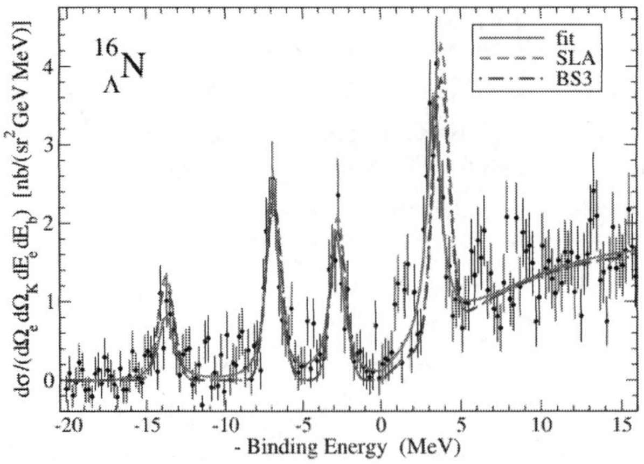

Figure 2. ^{16}O$(e, e'K^+)$ spectrum of ${}^{16}_{\Lambda}$N from JLab Hall A measurements. Figure adapted from Ref. [24].

5 Results

The two strength parameters b_0, B_0 of the optical potential terms Eqs. (2,3) were obtained by fitting to the $^{16}_{\Lambda}$N $B^{\rm exp}_{\Lambda}(1s, 1p)$ values listed in Table 1. Suppressing Pauli correlations by setting $\alpha_P = 0$ in Eqs. (5,6), the resulting Λ potential depth $D_\Lambda = -27.4$ MeV reflects a sizable cancellation between a strongly attractive two-body potential depth $D^{(2)}_{\Lambda}$ and a strongly repulsive three-body potential depth $D^{(3)}_{\Lambda}$. The overall agreement between calculations and experiment is acceptable, but some underbinding appears to develop for increasing mass numbers A, noticed clearly in the three heaviest $1s_\Lambda$ and two heaviest $1p_\Lambda$ states. The resulting b_0 is about half of the known Λp scattering length of (1.7 ± 0.1) fm [26, 27].

When the full potential Eqs. (2-6) is used (marked here as model X, including Pauli correlations through $\alpha_P = 1$) the overall picture remains unchanged regarding underbinding for the heavier elements, see Fig. 3. However, the fit parameter b_0=1.85 fm agrees now with the Λp scattering length. The other parameter, $B_0 = 0.170$ fm, is about twice smaller than for $\alpha_P = 0$.

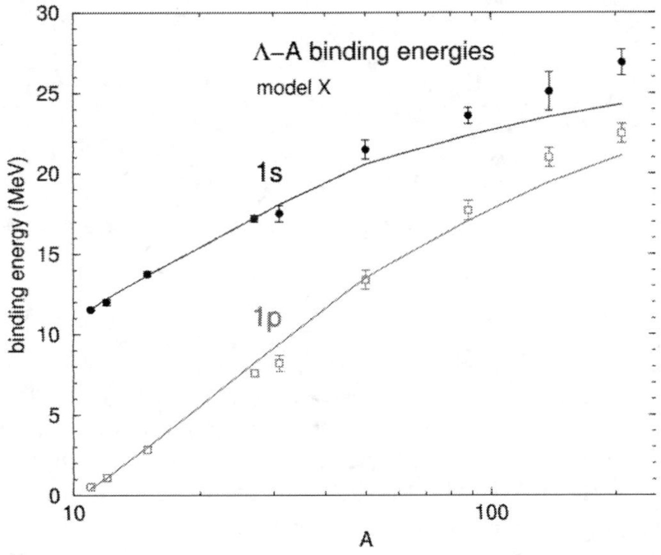

Figure 3. $B^{1s,1p}_{\Lambda}(A)$ values from model X compared with data. Continuous lines connect calculated values.

The phenomenon of underbinding associated with the optical potential Eqs. (2-6) is likely to be a result of the use of ρ^2 in nuclei where excess neutrons occupy shell-model orbits higher than those occupied by protons. This situation occurs in Fig. 3 for the four hypernuclei with $A \gtrsim 50$. Expecting that direct three-body ΛNN contributions involving one 'core' nucleon and one 'excess' nucleon vanish upon summing on the T=0 'core' closed-shell nucleons, we modify $\rho^2 = (\rho_{\rm core} + \rho_{\rm excess})^2$ by discarding the bilinear term $\rho_{\rm core}\,\rho_{\rm excess}$, thereby replacing ρ^2 in $V^{(3)}_{\Lambda}$, Eq. (3), by

$$\rho^2_{\rm core} + \rho^2_{\rm excess} \;=\; (2\rho_p)^2 + (\rho_n - \rho_p)^2 \qquad (7)$$

in terms of the input densities ρ_p and ρ_n. This ansatz is consistent with an overall isospin factor $\tau_1 \cdot \tau_2$ in two-pion exchange ΛNN forces, as first realized back in 1958 [28]. Results of applying this ansatz are shown in the lower part of Fig. 4 as model Y, where the underbinding of calculated $1s_\Lambda$ and

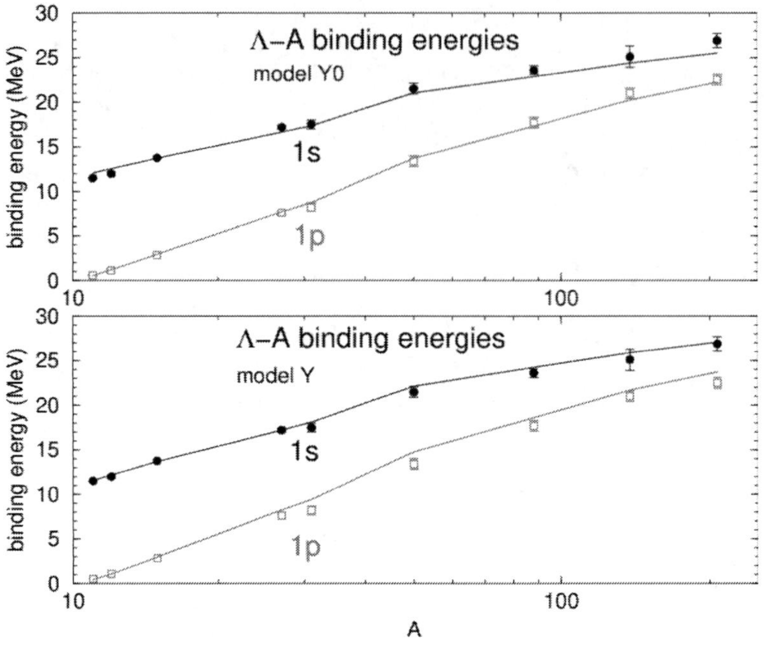

Figure 4. $B_\Lambda^{1s,1p}(A)$ values from models Y0 and Y compared with data, see text. Continuous lines connect calculated values.

$1p_\Lambda$ binding energies noticed in model X is no longer observed. The fit parameters, nevertheless, are the same as for model X above. In the upper part of Fig. 4, model Y0 shows similar results where the Pauli-correlations correction in model Y, Eq. (6), is replaced by Eq. (5). This provides a rough estimate of the impact of $1/A$ corrections typical for our optical-potential methodology. Potential depth values in model Y are $D_\Lambda^{(2)} = -41.6$ MeV, $D_\Lambda^{(3)} = 13.9$ MeV.

To estimate uncertainties, we act as follows: (i) decreasing the input value of $B_\Lambda^{1s}({}^{16}_\Lambda\text{N})$ fitted to by 0.2 MeV, thereby getting halfway to the central value of $B_\Lambda^{1s}({}^{16}_\Lambda\text{O})$=(13.4±0.4) MeV for ${}^{16}_\Lambda$O [29] the charge-symmetric partner of ${}^{16}_\Lambda$N, results in approximately 10% larger value of $D_\Lambda^{(3)}$, and (ii) applying Pauli correlations to $V_\Lambda^{(3)}$ too reduces $D_\Lambda^{(3)}$ roughly by 10%. In both cases $D_\Lambda^{(2)}$ increases moderately by $\lesssim$1 MeV. On the other hand, $D_\Lambda^{(2)}$ decreases by 1.7 MeV if Eq. (5) is used for $C_{\text{Pauli}}(\rho)$ instead of Eq. (6). Considering these uncertainties, our final values are (in MeV)

$$D_\Lambda^{(2)} = -40.6 \pm 1.0 \quad D_\Lambda^{(3)} = 13.9 \pm 1.4 \quad D_\Lambda = -26.7 \pm 1.7 \tag{8}$$

6 Discussion

The $D_\Lambda^{(2)}$ and $D_\Lambda^{(3)}$ values in Eq. (8) are considerably smaller than those deduced in QMC calculations [11, 12]. Note that the QMC nuclear densities $\rho_{\text{QMC}}(r)$ are much too compact with respect to our realistic densities, with nuclear r.m.s. radii r_N(QMC) about 0.8 of the known r.m.s. charge radii in ^{16}O and ^{40}Ca [30]. Since ρ scales as r_N^{-3}, applying it to the density dependence of our V_Λ^{opt} would transform $D_\Lambda^{(2)}$ and $D_\Lambda^{(3)}$ of Eq. (8) to as large depth values as $D_\Lambda^{(2)}$(QMC)=(−79.3±2.0) MeV and

$D_\Lambda^{(3)}$(QMC)=(53.0±5.3) MeV, their sum D_Λ(QMC)=(−26.3±5.7) MeV agreeing within uncertainties with ours.

Table 2. Λ-nuclear potential depths (in MeV) from two SHF calculations fitting B_Λ data points, and from our own $V_\Lambda^{\rm opt}(\alpha_P = 0)$ two-parameter (b_0, B_0) fit to the two $B_\Lambda^{1s,1p}({}^{16}_{\Lambda}{\rm N})$ values listed in Table 1.

Method	Data Points	$D_\Lambda^{(2)}$	$D_\Lambda^{(3)}$	D_Λ
SHF [2]	3	−57.8	31.4	−26.4
SHF [17]	35	−55.4	20.4	−35.0
$V_\Lambda^{\rm opt}(\alpha_P = 0)$ [13]	2	−57.6	30.2	−27.4

Smaller-size but still inflated values of $D_\Lambda^{(2)}$ and $D_\Lambda^{(3)}$ are obtained by applying the Skyrme Hartree Fock (SHF) methodology [2, 17]. Apart from small nonlocal potential terms and effective mass corrections, the SHF Λ-nuclear mean-field potential $V_\Lambda(\rho)$ consists of two terms: $V_\Lambda^{(2)}(\rho) \propto \rho$ and $V_\Lambda^{(3)}(\rho) \propto \rho^2$. A large-scale SHF fit [17] of the corresponding Λ potential depths to 35 B_Λ data points is listed in the middle row of Table 2. We note that the overall $D_\Lambda = -35$ MeV value becomes −31 MeV upon including a Λ effective-mass correction, a bit closer to the other D_Λ values listed in the table. Similar results, particularly for $D_\Lambda^{(2)}$, can be obtained in fact by choosing a considerably smaller number of fitted data points, as shown by the fits listed in the other two rows of the table. The 11 MeV difference between the $D_\Lambda^{(3)}$ values derived in these two SHF calculations arises mostly from nonlocal lower-power density terms, like $\rho^{5/3}$, present in [17] but absent in [2]. Interestingly, the last row lists a fit to the two $B_\Lambda^{1s,1p}({}^{16}_{\Lambda}{\rm N})$ values used here when Pauli correlations are suppressed, $\alpha_P = 0$ in Eq. (5). The sizable difference between $D_\Lambda^{(2)}$ and $D_\Lambda^{(3)}$ values listed in Table 2, all of which disregard Pauli correlations, and the $V_\Lambda^{\rm opt}$ values listed in Eq. (8) demonstrates the importance of including in $V_\Lambda^{\rm opt}$ a Pauli-correlations term ($\alpha_P = 1$) starting as $\rho^{4/3}$.

7 Summary

In summary, we have presented a straightforward optical-potential analysis of $1s_\Lambda$ and $1p_\Lambda$ binding energies across the periodic table, $12 \le A \le 208$, based on nuclear densities constrained by charge r.m.s. radii. The potential is parameterized by constants b_0 and B_0 in front of two-body ΛN and three-body ΛNN interaction terms. These parameters were fitted to precise $B_\Lambda^{\rm exp}(1s, 1p)$ values in ${}^{16}_{\Lambda}$N [31] and then used to evaluate $B_\Lambda^{1s,1p}$ values in the other hypernuclei considered here. Pauli correlations were found essential to establish a correct balance between b_0 and B_0, as judged by b_0 coming out in the final model Y analysis close to the value of the ΛN spin-averaged s-wave scattering length. Good agreement was reached in this model between the calculated $B_\Lambda^{1s,1p}$ values and their corresponding $B_\Lambda^{\rm exp}$ values, see Fig. 4.

The potential depth $D_\Lambda^{(3)}$ derived here, Eq. (8), suggests that in symmetric nuclear matter the Λ-nucleus potential becomes repulsive near three times ρ_0. Our derived depth $D_\Lambda^{(3)}$ is larger by a few MeV than the one yielding $\mu(\Lambda) > \mu(n)$ for Λ and neutron chemical potentials in purely neutron matter, respectively, under a 'decuplet dominance' construction for the underlying ΛNN interaction terms within a χEFT(NLO) model [8]. This suggests that the strength of the corresponding repulsive $V_\Lambda^{(3)}$ optical potential component, as constrained in the present work by data, is sufficient to prevent Λ hyperons from playing active role in neutron-star matter, thereby enabling a stiff EoS that supports two solar-mass neutron stars.

Acknowledgments

One of us (A.G.) thanks Jiří Mareš and other members of the HYP2022 organizing team for their generous hospitality during the Conference. The present work is part of a project funded by the European Union's Horizon 2020 research & innovation programme, grant agreement 824093.

References

[1] L. Tolos, L. Fabbietti, Prog. Part. Nucl. Phys. **112**, 103770 (2020), and past neutron-star work cited therein
[2] D.J. Millener, C.B. Dover, A. Gal, Phys. Rev. C **38**, 2700 (1988)
[3] A. Gal, E.V. Hungerford, D.J. Millener, Rev. Mod. Phys. **88**, 035004 (2016)
[4] Th.A. Rijken, V.G.J. Stoks, Y. Yamamoto, Phys. Rev. C **59**, 21 (1999)
[5] M.M. Nagels, Th.A. Rijken, Y. Yamamoto, Phys. Rev. C **99**, 044003 (2019)
[6] H. Polinder, J. Haidenbauer, U.-G. Meißner, Nucl. Phys. A **779**, 244 (2006)
[7] J. Haidenbauer, I. Vidaña, Eur. Phys. J. A **56**, 55 (2020)
[8] D. Gerstung, N. Kaiser, W. Weise, Eur. Phys. J. A **56**, 175 (2020), and references cited therein to earlier works on ΛNN interactions in χEFT.
[9] J. Haidenbauer, S. Petschauer, N. Kaiser, U.-G. Meißner, A. Nogga, W. Weise, Nucl. Phys. A **915**, 24 (2013)
[10] J. Haidenbauer, U.-G. Meißner, A. Nogga, Eur. Phys. J. A **56**, 91 (2020)
[11] D. Lonardoni, S. Gandolfi, F. Pederiva, Phys. Rev. C **87**, 041303(R) (2013)
[12] D. Lonardoni, F. Pederiva, S. Gandolfi, Phys. Rev. C **89**, 014314 (2014)
[13] E. Friedman, A. Gal, *Constraints from Λ hypernuclei on the ΛNN content of the Λ-nucleus potential*, arXiv:2204.02264
[14] D. Lonardoni, A. Lovato, S. Gandolfi, F. Pederiva, Phys. Rev. Lett. **114**, 092301 (2015)
[15] D. Logoteta, I. Vidaña, I. Bombaci, Eur. Phys. J. A **55**, 207 (2019)
[16] C.B. Dover, J. Hüfner, R.H. Lemmer, Ann. Phys. (NY) **66**, 248 (1971)
[17] H.-J. Schulze, E. Hiyama, Phys. Rev. C **90**, 047301 (2014), and past SHF work cited therein
[18] L.R.B. Elton, *N*uclear Sizes (Oxford University Press, Oxford, 1961)
[19] E. Friedman, A. Gal, Phys. Lett. B **820**, 136555 (2021)
[20] I. Angeli, K.P. Marinova, At. Data Nucl. Data Tables **99**, 69 (2013)
[21] E. Friedman, A. Gal, Phys. Rep. **452**, 89 (2007)
[22] T. Waas, M. Rho, W. Weise, Nucl. Phys. A **617**, 449 (1997)
[23] E. Friedman, A. Gal, Nucl. Phys. A **959**, 66 (2017), and past K^- atoms work cited therein
[24] F. Garibaldi, *et al.* (Jefferson Lab Hall A Collaboration), Phys. Rev. C **99**, 054309 (2019)
[25] D.J. Millener, Nucl. Phys. A **804**, 84 (2008), **881**, 298 (2012), **914**, 109 (2013)
[26] G. Alexander, *et al.*, Phys. Rev. **173**, 1452 (1968)
[27] A. Budzanowski, *et al.* (HIRES Collaboration), Phys. Lett. B **687**, 31 (2010)
[28] R. Spitzer, Phys. Rev. **110**, 1190 (1958)
[29] M. Agnello, *et al.* (FINUDA Collaboration), Phys. Lett. B **698**, 219 (2011)
[30] D. Lonardoni, Ph.D. thesis, University of Trento, Italy, arXiv:1311.6672; see Table 4.7
[31] F. Cusanno, *et al.* (Jefferson Lab Hall A Collaboration), Phys. Rev. Lett. **103**, 202501 (2009)

Equation of state and strangeness in neutron stars

- role of hyperon-nuclear three-body forces -

Wolfram Weise[1,*]

[1]Physics Department, Technical University of Munich, 85748 Garching, Germany

Abstract. A brief survey is presented of our present understanding of the equation-of-state of cold, dense matter and the speed of sound in the interior of neutron stars, based on the constraints inferred from observational data. The second part focuses on strangeness in baryonic matter and the role of hyperon-nuclear two- and three-body forces, with reference to the "hyperon puzzle" in neutron stars and possible scenarios for its solution.

1 Introduction

The structure of highly compressed baryonic matter at zero temperature is still largely unknown. Neutron stars are the extreme systems of choice to learn about the properties of strongly interacting cold matter at high baryon densities. Their detailed composition, ranging from nuclear degrees of freedom at low densities to various possible forms of quark matter at high densities, is subject to ongoing speculations.

Within the last decade, the observational data base for neutron stars has expanded substantially. Its detailed analysis sharpened the empirical constraints on the equation-of-state (EoS) of the matter deep inside the core of such objects. In particular, the discovery of two-solar-mass neutron stars implied that the EoS, i.e. pressure $P(\varepsilon)$ as function of energy density ε, must be sufficiently stiff in order to support such heavy compact stars against gravitational collapse. Some previously discussed simple forms of exotic matter (quark matter, kaon condensation, ...) were thus excluded when their corresponding EoS's turned out to be too soft and unable to satisfy the stability conditions.

Viewing neutron stars as systems primarily composed of neutrons in beta equlibrium with a small fraction of protons, it has long been considered that the formation of hyperons through weak processes might become energetically favourable at sufficiently high densities. However, it was soon realized that when adding these strangeness degrees of freedom, the EoS would again become too soft and unable to support $2M_\odot$ neutron stars. This started a discussion under the keyword *hyperon puzzle*, which is still continuing and will be a main theme in this presentation. To frame this discussion in a broader context, it is useful first to summarize the present state of knowledge about the neutron star EoS as inferred from the observational data. Of particular interest is the speed of sound in the core of neutron stars. Its behaviour as a function of energy density is a sensitive indicator for phase transitions or continuous changes of degrees of freedom (crossovers) in the star's composition.

*e-mail: weise@tum.de

Table 1. Masses of two neutron stars together with radii inferred from NICER measurements at 68% credible level.

	Mass M [$M_\odot$]	Radius R [km]	Ref.
PSR J0030+0451	$1.44^{+0.15}_{-0.14}$	$13.02^{+1.24}_{-1.06}$	[7]
	$1.34^{+0.15}_{-0.16}$	$12.71^{+1.14}_{-1.19}$	[9]
PSR J0740+6620	2.08 ± 0.07	$13.7^{+2.6}_{-1.5}$	[8]
	$2.072^{+0.067}_{-0.066}$	$12.39^{+1.30}_{-0.98}$	[10]

2 Equation-of-state of neutron star matter

2.1 Observational constraints

Neutron star masses and radii. The information about the masses of the heaviest neutron stars derives primarily from precise Shapiro time delay measurements of pulsars orbiting in binary systems with white dwarfs. Three such massive objects, PSR J1614–2230 [1–3], PSR J0348+0432 [4] and PSR J0740+6620 [5, 6], have been established:

$$\text{PSR J1614–2230} \qquad M = 1.908 \pm 0.016\, M_\odot\,, \tag{1}$$

$$\text{PSR J0348+0432} \qquad M = 2.01 \pm 0.04\, M_\odot\,, \tag{2}$$

$$\text{PSR J0740+6620} \qquad M = 2.08 \pm 0.07\, M_\odot\,. \tag{3}$$

Radii of neutron stars together with their masses can be inferred from X-ray profiles of rotating hot spot patterns measured with the NICER telescope stationed at the ISS, also in combination with other multimessenger data. Two representative neutron stars have been investigated in this way and analysed by two independent groups using different methods. Results are summarized in Table 1. These combined mass and radius data are now included as prime input in setting constraints for the neutron star EoS.

Binary neutron star mergers and tidal deformabillities. Gravitational wave signals produced by the merger of two neutron stars in a binary have been detected by the LIGO and Virgo collaborations. These signals are interpreted using theoretical waveform models that depend on the mass ratio of the two neutron stars, M_2/M_1, and a mass-weighted combination of their tidal deformabilities, Λ_1 and Λ_2:

$$\bar{\Lambda} = \frac{16}{13}\frac{(M_1 + 12M_2)M_1^4\Lambda_1 + (M_2 + 12M_1)M_2^4\Lambda_2}{(M_1 + M_2)^5}\,. \tag{4}$$

So far, two binary neutron star merger events, GW170817 [11] and GW190425 [12], were observed, yielding the following constraints on $\bar{\Lambda}$ at the 90% level:

$$\text{GW170817}: \quad \bar{\Lambda} = 320^{+420}_{-230}\,, \qquad \text{GW190425}: \quad \bar{\Lambda} \leq 600\,. \tag{5}$$

The first one of these events (GW170817) was further evaluated together with electromagnetic signals [13]. The following masses and tidal deformabilities of the individual neutron stars in the binary were reported:

$$M_1 = 1.46^{+0.13}_{-0.09}\, M_\odot \qquad \Lambda_1 = 255^{+416}_{-171}\,,$$

$$M_2 = 1.26^{+0.09}_{-0.12}\, M_\odot \qquad \Lambda_2 = 661^{+858}_{-375}\,. \tag{6}$$

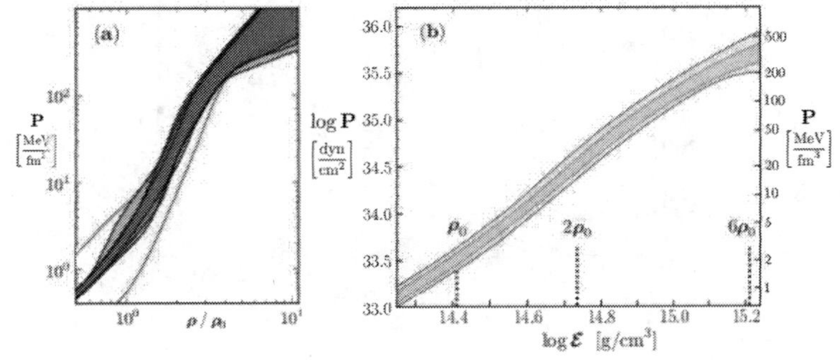

Figure 1. Two recent examples of deduced EoS of neutron star matter using Bayes inference methods with data sets described in Section 2.1 as input: (a) pressure P as function of baryon density ρ (in units of $\rho_0 = 0.16$ fm^{-3}) [14] and (b) as function of the energy density ε [15].

The tidal deformabilities of the individual neutron stars are evidently still subject to large uncertainties, but they nonetheless provide further independent data for statistical inference procedures.

2.2 Inference of the sound velocity and the EoS in neutron stars

In recent years the data listed in the previous subsection have been incorporated in a variety of studies using Bayesian statistics to infer the equation-of-state of neutron star matter and related properties. Two representative examples of such results are shown in Figure 1. Constraints at low baryon densities are commonly imposed by implementing results from nuclear chiral effective field theory (ChEFT) in the range $\rho \simeq 1-2\,\rho_0$ around the equilibrium density, $\rho_0 = 0.16$ fm^{-3}, of normal nuclear matter. Notably, the density profiles computed for neutron stars in the mass range $M \sim 1.4$ - $2.1\ M_\odot$ reach baryon densities that typically do not exceed $\rho \sim 6\,\rho_0$ in the central cores of even the most massive stars, where central pressures $P \gtrsim 300$ MeV/fm^3 are required to ensure stability.

An advanced Bayes inference procedure has recently been performed [16] with special focus on the (squared) sound velocity, $c_s^2(\varepsilon) = \partial P(\varepsilon)/\partial\varepsilon$, from which the EoS is determined as

$$P(\varepsilon) = \int_0^\varepsilon d\varepsilon' c_s^2(\varepsilon')\,. \tag{7}$$

Results are shown in Figure 2. At 68% credibility level, it can be concluded that the speed of sound in neutron stars systematically exceeds the canonical value for an ultrarelativistic Fermi gas (the conformal limit), $c_s^2 = 1/3$, at energy densities $\varepsilon \gtrsim 0.5$ GeV/fm^{-3} corresponding to baryon densities $\rho \gtrsim 3\,\rho_0$. This observation is of some significance, recalling that a first-order phase transition in the core of a neutron star would be signaled by a rapid drop of c_s to zero. Quantifying likelihood categories in terms of Bayes factors, one finds that there is strong evidence [16] that c_s^2 stays at values larger than 0.1 for neutron stars with $M \lesssim 2\,M_\odot$. This indicates that a first-order phase transition in the core of even the heaviest observed neutron stars is unlikely. On the other hand, a continuous crossover at high baryon density is not ruled out.

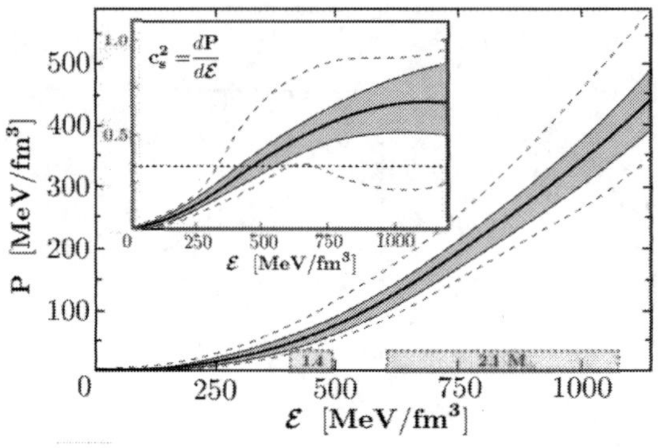

Figure 2. Pressure $P(\varepsilon)$ as function of energy density and squared sound speed $c_s^2(\varepsilon)$ of neutron star matter using Bayesian inference with data listed in Section 2.1. Displayed are medians (solid curves), 68% (grey) and 95% (dashed) credible bands. The sectors on the lower axis mark the ranges of central energy densities realised in the cores of 1.4 and 2.1 $M_\odot$ neutron stars, respectively. Adapted from [16].

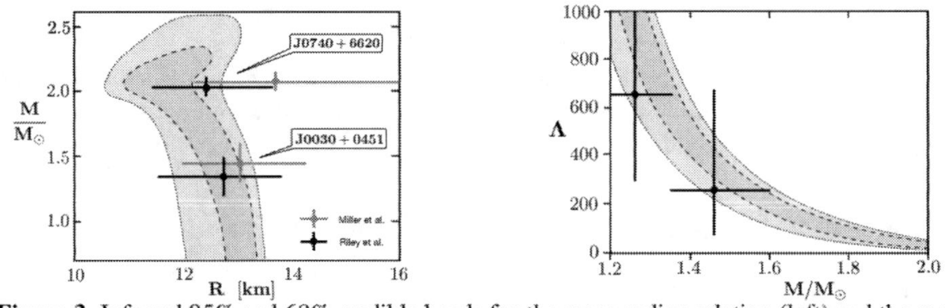

Figure 3. Inferred 95% and 68% credible bands for the mass-radius relation (left) and the mass dependence of the tidal deformabilty Λ (right) of neutron stars [16] based on the empirical EoS of Fig. 2. Data for masses and radii from [7–10]. Tidal deformabilty data (GW170817) from [13].

Figure 3 shows the mass-radius relation and tidal deformabilties for neutron stars as deduced from the inferred EoS and in comparison with data. The most probable radii are in the range 11-13 km. A tendency towards slightly smaller inferred radii for the heaviest 2 $M_\odot$ objects as compared to those listed in Table 1 appears to be favoured when the GW merger events are incorporated in the Bayesian analysis.

2.3 Theoretical frameworks and models

While the advanced statistical methods applied to the expanded observational neutron star data base sharpen the constraints on the equation-of-state of dense matter, no information is provided in this way about the physical degrees of freedom that govern the microscopic composition and dynamics of the neutron star. A variety of models exists that can roughly be grouped into three categories:

a) Models based on *nuclear degrees of freedom* (nucleons, pions and short-range correlations) using many-body theory with realistic two- and three-nucleon interactions. A time-honored example is the APR equation-of-state [17]. Spontaneously broken chiral symmetry of low-energy QCD is a guiding principle, realised through chiral EFT calculations [18–20] applicable up to baryon densities $\rho \lesssim 2\rho_0$. A similar scheme in combination with non-perturbative functional renormalisation group methods can be further extended to higher densities [21]. An important point to be aware of when working with such chiral approaches is the following: the frequently used mean-field approximation typically leads to an unphysical first-order chiral phase transition at baryon densities $\rho \lesssim 3\,\rho_0$, well within the density range realised in neutron stars. However, fluctuations beyond mean-field tend to convert this phase transition into a continuous crossover [22] that is shifted to densities quite far beyond even those expected in the centers of heavy $2\,M_\odot$ stars.

b) Models operating with *nucleons and hyperons*, using advanced many-body methods and including interactions of the baryon SU(3) octet with nucleons as input [23, 24]. These models are commonly confronted with the *hyperon puzzle* in neutron star matter unless additional repulsive mechanisms are introduced, e.g. through hyperon-nuclear many-body forces, in order to maintain the necessary stiffness of the EoS. An interesting variant in this class of approaches is a chiral quark-meson model [25] in which the members of the baryon octet are coupled to meson fields with quark bag boundary conditions. Effective in-medium baryon masses are determined by the non-linear density dependence of a scalar (sigma) field coupled to the baryons. A key quantity is the scalar polarizability of the baryons that encodes effects of repulsive three-body forces. This model can accomodate Λ hyperons in neutron star matter from an onset density $\rho \gtrsim 3\,\rho_0$ upward. Charged and neutral Ξ's would appear at higher densities, still maintaining the required stiffness of the EoS in the presence of sufficiently repulsive many-body correlations. However, the predicted pattern of the resulting sound speed in the presence of hyperons displays a rapid drop below the conformal bound, $c_s = \sqrt{1/3}$, which raises another issue in view of the empirical constraints discussed previously.

c) Hybrid models characterised by a *hadron-to-quark crossover* connecting nucleonic or baryonic matter at intermediate densities with quark matter at high densities. Such scenarios have gained significance in recent years under the keywords *quark-hadron continuity* or *quarkyonic matter* [26, 27]. Percolation mechanisms are thought to be at the origin of such a continuous transition. Quark-antiquark pairs forming meson clouds at lower densities increase their mobility upon compression. At high densities, the compact valence quark cores of the baryons begin to overlap. An example of an EoS constructed within such a picture is given in [28]. The quark matter regime at high densities is described by a Nambu - Jona-Lasinio model including pairing interactions. Strong universal repulsion between vector currents of the quarks is usually introduced to keep the EoS sufficiently stiff.

Figure 4 may be taken as a pictorial summary of this brief survey, displaying the EoS with built-in constraints from nuclear physics at low densities, the perturbative QCD limit at asymptotically high energy densities, and interpolating between these extremes. An apparent change of slope in $P(\varepsilon)$ around typical core densities of heavy neutron stars is sometimes interpreted as a possible signature of a hadron-to-quarks phase transition. But its statistical weight, judging from the previous inference analyses, is so far limited. In fact several other inference results for $P(\varepsilon)$ do not show a pronounced kink structure.

As emphasized previously, the sound velocity in neutron stars is a key quantity in order to further reduce the freedom of choice in models describing the intrinsic structure and composition of dense matter in neutron stars. This is apparent in Figure 5. Shown there is the Bayesian inference posterior of c_s^2 from [16] together with a reconstruction of the sound speed from neutron star data using a deep learning network scheme [30]. These results demonstrate once again the rise of c_s^2 beyond the conformal boundary of 1/3 within the range of neutron star densities. A chiral FRG calculation [21] based on nucleon and meson degrees of freedom follows this trend (though perhaps with a slightly too soft EoS at the highest densities). The quark-meson coupling model [25] including hyperons indicates a pattern that falls out of this

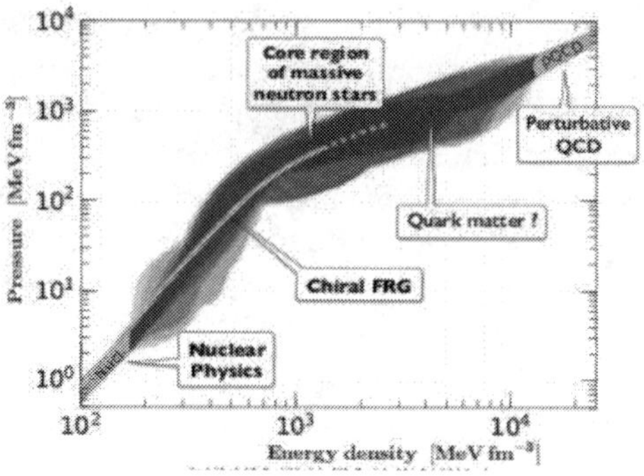

Figure 4. Equation-of-state $P(\varepsilon)$ covering neutron star matter and beyond, with low-density constraints from nuclear physics [18–20], asymptotic constraints from perturbative QCD and uncertainty bands interpolating between these extremes (adapted from [29]). Also shown is an EoS curve derived from chiral nucleon-meson field theory combined with functional renormalization group methods (Chiral FRG) [21].

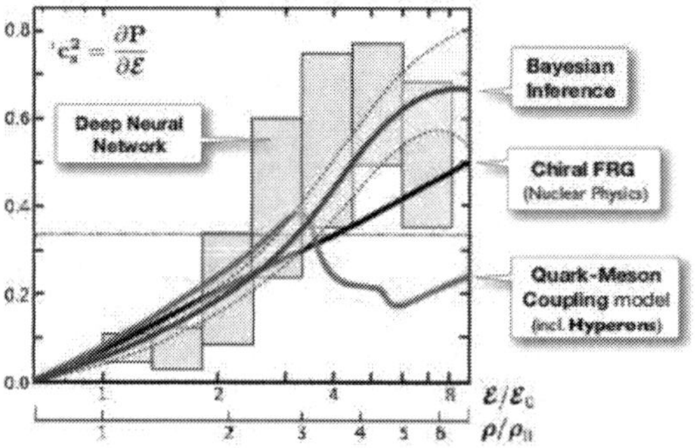

Figure 5. Squared speed of sound in neutron star matter as function of energy density (in units of $\varepsilon_0 = 150$ MeV/fm^3) and of baryon density (in units of $\rho_0 = 0.16$/fm^3). Bayes inference result (68% credible band) [16] is shown side-by-side with data-based mapping of c_s^2 using a deep learning network approach [30]. For comparison: chiral nucleon-meson field theory combined with functional renormalization group (Chiral FRG) [21], and quark-meson coupling model calculation including hyperons [25].

scheme. Obviously there is a strong quest for a further reduction of uncertainties in the sound speed along with future extensions of the neutron star data base.

3 Strangeness and baryonic matter

3.1 Hyperon-nucleon interactions from SU(3) chiral effective field theory

Guided by the symmetry breaking pattern of three-flavor QCD, chiral SU(3) effective field theory (EFT) is a systematic low-energy expansion, with controlled uncertainties, of interactions involving the baryon and pseudoscalar meson octets. Unresolved short-distance dynamics is encoded in contact terms and derivatives thereof. Our present (still limited) quantitative understanding of hyperon-nucleon interactions derives primarily from this framework at next-to-leading order (NLO) [31–33]. Steps beyond NLO are possible in principle but not yet feasible in practice because of the large number of low-energy constants that need to be determined at higher orders. This requires a much broader empirical data base compared to the one presently available. Nonetheless the updated (NLO19) version of the interaction [32] has been checked successfully against an increasing amount of data. It now serves as a prototype force for ΛN, ΣN and ΞN two-body processes, including important coupled-channel dynamics such as $\Lambda N \leftrightarrow \Sigma N$ and $\Xi N \leftrightarrow \Lambda\Lambda$.

Hyperon-nucleon scattering and correlation functions. Hyperon-nucleon cross sections measured decades ago at BNL and CERN served as a first step in constraining the theory. In the meantime interesting new experimental developments provide further tests. Examples are the $\Sigma^- p \to \Sigma^- p$ elastic scattering and $\Sigma^- p \to \Lambda n$ charge exchange scattering measurements at J-PARC (E40) [34], and the femtoscopy studies of hyperon-nucleon correlation functions from pp collisions with ALICE at LHC [35]. In particular, the accurately determined Λp correlation function [36] is well reproduced by the NLO19 interaction, especially at the lowest center-of-mass momenta ($\lesssim$ 100 MeV/c) which are not covered by scattering experiments.

Constraints from hypernuclei. Hypernuclear physics has provided a systematic set of binding energies for ground states and excited states of a Λ in nuclei from A = 5 to 208. A phenomenological Λ-nuclear single-particle potential works well to reproduce the data with a central potential depth $U_\Lambda(r = 0) \simeq -30$ MeV [37]. Hypernuclear many-body calculations, using e.g. Brueckner-Hartree-Fock methods [38, 39], have henceforth addressed the question to what extent such a shell-model potential can be understood in terms of underlying ΛN two-body forces. The results of those computations point to a trend, namely that a realistic two-body interaction produces too much binding in the lower shell-model levels of heavy hypernuclei such as ${}_\Lambda Pb$ [38]. Similarly, the Λ potential in nuclear matter calculated with the NLO19 two-body force alone is more attractive by almost 20% than the empirical potential and has its minimum located at a density 25% higher than $\rho_0 \simeq 0.16$ fm^{-3} [39].

This opens the question about the role of ΛNN three-body forces. The issue is further underlined by a recent phenomenological study [40] using an ansatz that combines contributions of Λ-nucleon two- and three-body forces to the Λ-nuclear potential: $U_\Lambda(\rho) = U_0^{(2)}[\rho(r)/\rho_0] + U_0^{(3)}[\rho(r)/\rho_0]^2$, in an optimal fit to the systematics of hypernuclear binding energies of the lowest (s and p shell) orbits. The best-fit result features a total potential depth of $U_\Lambda(\rho_0) = -(26.5 \pm 1.6)$ MeV, separated into a ΛN two-body contribution, $U_0^{(2)} = -(40.4 \pm 0.6)$ MeV, and a repulsive ΛNN three-body piece, $U_0^{(3)} = (13.9 \pm 1.4)$ MeV. Systematic uncertainties may be underestimated in this work, but the presence of a substantial repulsive hypernuclear three-body force is nonetheless suggestive.

Hyperon-nuclear three-body forces. Within the chiral EFT hierarchy, genuine three-baryon interactions first appear at next-to-next-to-leading order (NNLO) [41]. A complete calculation would involve a large number of parameters, too large for a reliable determination given the limited amount of data. However, an approximation known to work well in the three-nucleon sector, namely $\Delta(1232)$ dominance in intermediate states of the three-body mechanism, can easily be generalized to decuplet dominance in SU(3) chiral EFT, now with Σ^* and Δ intermediate virtual excitations. In this way the three-body terms are promoted from NNLO to NLO in the chiral counting, in line with the NLO treatment of the two-body forces. This leaves three coupling parameters of the three-body interaction terms. Assuming SU(3) symmetry, one of these coupling constants is already fixed by the $\Delta \to \pi N$ decay width. The remaining two parameters can be constrained by comparison with the empirical Λ-nuclear single-particle potential [39].

3.2 Hyperons in dense baryonic matter and neutron stars

Equiped with a ΛNN three-body force from chiral SU(3) EFT, one can now proceed to investigate its role in nuclear and neutron star matter.

Density dependence of the Λ-nuclear potential. With increasing baryon density beyond $\rho \sim \rho_0$ the repulsive nature of the ΛNN three-body interaction is expected to become progressively more prominent relative to the contribution of the ΛN two-body interaction, because it enters the energy density with a higher power of density. This is shown in Figure 6(a) for the Λ-nuclear single-particle potential, $U_\Lambda(\rho)$, in symmetric nuclear matter. A smilar behaviour is observed for a Λ hyperon in neutron matter. At high densities, $\rho \gtrsim 3\,\rho_0$, the three-body contribution to U_Λ begins to be dominant over the two-body part with its linear density dependence characteristic of a Hartree potential generated by the short-distance ΛN interaction.

Hyperon puzzle in neutron stars. Given the energy density $\varepsilon(\rho_i)$ of neutron star matter as a function of the densities ρ_i of baryonic constituents (neutrons, protons and possibly Λ hyperons), the individual chemical potentials are determined by $\mu_i = \partial\varepsilon/\partial\rho_i$. The onset for the appearance of hyperons in a neutron star through weak interaction processes converting neutrons into Λ's is marked by the condition that the Λ chemical potential equals that of the neutrons: $\mu_\Lambda = \mu_n$.

The density dependence of the potential U_Λ is reflected in μ_Λ. Hence the repulsive ΛNN three-body interaction shows up prominently in the Λ chemical potential at high baryon densities. Figur 6(b) displays results of a Brueckner-Hartree-Fock calculation of μ_Λ for neutron star matter in comparison to μ_n [39]. If only two-body ΛN interactions (NLO19) are included, hyperons start appearing already at densities $\rho \simeq 2 - 3\,\rho_0$. The EoS becomes too soft and unable to support two-solar-mass neutron stars. However, repulsive ΛNN three-body forces raise μ_Λ to a level such that it does not match the neutron chemical potential any more: the appearance of Λ's in the core of the neutron star is prohibited. At the same time the neutron sector (with inclusion of many-body correlations) features sufficiently strong repulsion by itself to satisfy the stabilty conditions for heavy neutron stars.

Of course, uncertainties still exist concerning the detailed quantitative behaviour of the three-body terms in the energy density, related to approximations involved e.g. in the simplified octet dominance ansatz. Hence the hyperon puzzle is not yet solved, but the results so far obtained point into a qualitatively promising direction.

4 Conclusions and outlook

Significant advances have been achieved in the physics of pulsars and the analysis of observational data, resulting in improved constraints on the equation-of-state and the sound speed at high densities in neutron star interiors. Progress has also been made concerning the role of strangeness in neutron star matter, with emphasis on hypernuclear three-body forces and their possible role in addressing the *hyperon puzzle*. The repulsion provided by the ΛNN three-body interaction at high baryon densities is potentially capable of blocking the appearance of Λ hyperons in the density range of even the heaviest observed ($\sim 2\,M_\odot$) neutron stars.

In this context, as well as in the general quest for identifying the active degrees of freedom in dense baryonic matter, an important issue is the further detailed exploration of the sound velocity in the center of neutron stars. Presently available constraints, though still with relatively large uncertainties, indicate that the sound speed systematically exceeds the conformal bound, $c_s = \sqrt{1/3}$, in the stellar core. This sets stringent limits on a possible softening of the equation-of-state through the appearance of new phases or additional degrees of freedom in the core.

Further insights into the role of hyperons in neutron star matter will require yet a more quantitative evaluation of the subtle balance between two- and three-body forces involving hyperons. Much improved and expanded sets of high-statistics two-body hyperon-nucleon scattering and reaction data, together with correlation functions from precise femtoscopy

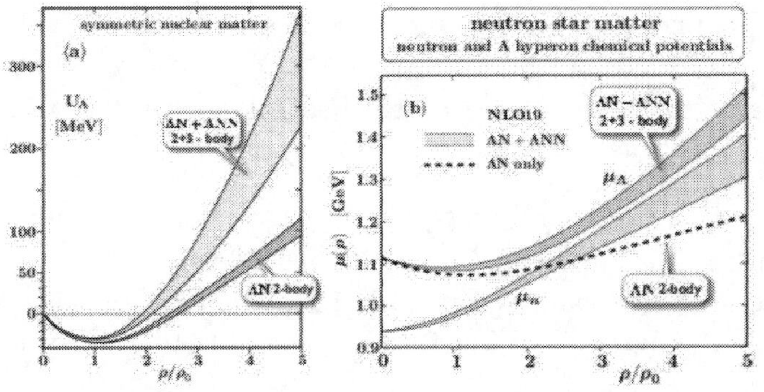

Figure 6. (a) Single-particle potential U_Λ of a Λ hyperon in symmetric nuclear matter as a function of baryon density in units of $\rho_0 = 0.16$ fm^{-3}. Results of Brueckner-Hartree-Fock calculations using NLO19 two-body ΛN interaction, in comparison with additional ΛNN three-body forces as input. Three-body parameters have been fixed to the empirical $U_\Lambda(\rho_0) \simeq -30$ MeV. (b) Chemical potentials of Λ and neutron, μ_Λ and μ_n in beta-equlibrated neutron star matter calculated with the same input as in (a). The neutron chemical potential is computed using the chiral nucleon-meson model combined with functional renormalization group methods [21]. In both Figures the bands give an impression of estimated uncertainties. Adapted from [39].

measuremets, will provide further quantitative constraints on the theory (e.g. enabling extensions of the SU(3) chiral EFT approach beyond NLO).

This is in turn necessary to define an accurate baseline for a more detailed investigation of three-body effects and beyond in hypernuclear few- and many-body systems. Future developments in high-resolution hypernuclear spectroscopy, combined with advanced methods in many-body theory, will be mandatory for further progress in specifying the input at normal nuclear densities and at the same time reducing uncertainties in the extrapolation to higher densities. A foreseen mass resolution of better than 0.5 MeV in a next generation of hypernuclear spectroscopy experiments with (π^+, K^+) reactions at J-PARC's projected HIHR beamline [42], and with $(e, e'K^+)$ measurements at JLab and MAMI, will be important steps towards these goals.

Acknowledgements: Special thanks to Len Brandes, Dominik Gerstung and Norbert Kaiser for collaborations and discussions on the topics of this report. Work supported in part by the DFG Cluster of Excellence ORIGINS.

References

[1] P.B. Demorest *et al.*, Nature **467**, 1081 (2010)
[2] E. Fonseca *et al.*, Astrophys. J. **832**, 167 (2016)
[3] Z. Arzoumanina *et al.*, Astrophys. J. Suppl. **235**, 37 (2018)
[4] J. Antoniadis *et al.*, Science **340**, 1233232 (2013)
[5] H.T. Cromartie *et al.*, Nature Astron. **4**, 72 (2020)
[6] E. Fonseca *et al.*, Astrophys. J. Lett. **915**, L12 (2021)
[7] M.C. Miller *et al.*, Astrophys. J. Lett. **887**, L24 (2019)
[8] M.C. Miller *et al.*, Astrophys. J. Lett. **918**, L28 (2021)

[9] T.E. Riley *et al.*, Astrophys. J. Lett. **887**, L21 (2019)
[10] T.E. Riley *et al.*, Astrophys. J. Lett. **918**, L27 (2021)
[11] B. Abbot *et al.*, (LIGO and Virgo Collaborations), Phys. Rev. **X9**, 011001 (2019)
[12] B. Abbot *et al.*, (LIGO and Virgo Collaborations), Astrophys. J. Lett. **892**, L3 (2020)
[13] M. Fasano, T. Abdelsalhin, A. Maselli, and V. Ferrari, Phys. Rev. Lett. **123**, 141101 (2019)
[14] R. Essick, I. Tews, P. Landry, S. Reddy, and D.E. Holz, Phys. Rev. **C102**, 055803 (2020)
[15] G. Raaijmakers *et al.*, Astrophys. J. Lett. **918**, L29 (2021)
[16] L. Brandes, W. Weise, and N. Kaiser, arXiv:2208.03026 [nucl-th] (2022)
[17] A. Akmal, V.R. Pandharipande, and D.G. Ravenhall, Phys. Rev. **C58**, 1804 (1998)
[18] K. Hebeler, J.M. Lattimer, C.J. Pethick, and A. Schwenk, Astrophys. J. **773**, 11 (2013)
[19] J.W. Holt and N. Kaiser, Phys. Rev. **C95**, 034326 (2017)
[20] S. Huth, C. Wellenhofer, and A. Schwenk, Phys. Rev. **C103**, 025803 (2021)
[21] M. Drews and W. Weise, Prog. Part. Nucl. Phys. **93**, 69 (2017)
[22] L. Brandes, N. Kaiser, and W. Weise, Eur. Phys. J. **A57**, 243 (2021)
[23] H. Djapo, B.-J. Schaefer, and J. Wambach, Phys. Rev. **C81**, 035803 (2010)
[24] D. Lonardoni, A.Lovato, S. Gandolfi, and F. Pederiva, Phys. Rev. Lett. **114**, 092301 (2015)
[25] T. Motta, P. Guichon, and A.W. Thomas, Nucl. Phys. **A1009**, 122157 (2021)
[26] L. McLerran and S. Reddy, Phys. Rev. Lett. **122**, 122701 (2019)
[27] K. Fukushima, T. Kojo, and W. Weise, Phys. Rev. **D102**, 096017 (2020)
[28] G. Baym, S. Furusawa, T. Hatsuda, T. Kojo, and H. Togashi, Astrophys. J. **885**, 42 (2019)
[29] E. Annala, T. Gorda, A. Kurkela, J. Nättilä, and A. Vuorinen, Nature Phys. **16**, 907 (2020)
[30] Y. Fujimoto, K. Fukushima, and K. Murase, Phys. Rev. **D101**, 054016 (2020); JHEP **03**, 273 (2021)
[31] J. Haidenbauer, S. Petschauer, N. Kaiser, U.-G. Meißner, A. Nogga, and W. Weise, Nucl. Phys. **A915**, 24 (2013)
[32] J. Haidenbauer, U.-G. Meißner, and A. Nogga, Eur. Phys. J. **A56**, 91 (2020)
[33] S. Petschauer, J. Haidenbauer, N. Kaiser, U.-G. Meißner, and W. Weise, Front. in Phys. **8**, 12 (2020)
[34] K. Miwa *et al.* (J-PARC E40 Collaboration), Phys. Rev. **C104**, 045204 (2021); Phys. Rev. Lett. **128**, 072501 (2022)
[35] L. Fabbietti, V. Mantovani Sarti, and O. Vazquez Doce, Ann. Rev. Nucl. Part. Sci. **71**, 377 (2021)
[36] S. Acharya *et al.* (ALICE Collaboration), arXiv:2104.04427 [nucl-ex] (2021)
[37] A. Gal, E. Hungerford, and D. J. Millener, Rev. Mod. Phys. **88**, 035004 (2016)
[38] J. Haidenbauer and I. Vidaña, Eur. Phys. J. **A56**, 56 (2020)
[39] D. Gerstung, N. Kaiser, and W. Weise, Eur. Phys. J. **A56**, 175 (2020)
[40] E. Friedman and A. Gal, arXiv:2204.02264 [nucl-th] (2022)
[41] S. Petschauer, N. Kaiser, J. Haidenbauer, U.-G. Meißner, and W. Weise, Phys. Rev. **C93**, 014001 (2016)
[42] K. Aoki *et al.*, arXiv:2110.04462 [nucl-ex] (2021), Chapter 2.3

Experimental study of $\bar{K}NN$ and future experiments for kaonic nuclei

Takumi Yamaga[1,*] for the J-PARC E15, E80, and P89 collaborations

[1]RIKEN Cluster for Pioneering Research, RIKEN, Wako, 351-0198, Japan

Abstract. Quasi-bound systems of antikaon and nucleus, $\bar{K}$-nuclei, have been considered as a natural extension of the interpretation that $\Lambda(1405)$ is the $\bar{K}$-N hadronic molecule. The lightest $\bar{K}$-nucleus, $\bar{K}NN$ quasi-bound state, is particularly important to be investigated for confirmation of the existence of such an exotic nuclear state containing a meson. We conducted an experiment, J-PARC E15, to search for $\bar{K}NN_{(I_z=+1/2)}$ by the in-flight K^- induced reaction on ^{3}He, and finally observed a clear signal of $\bar{K}NN_{(I_z=+1/2)}$ for the first time. To robustly confirm the existence of $\bar{K}$-nuclei, we have planned systematic measurements for various $\bar{K}$-nuclei at J-PARC.

1 Introduction

The kaonic nucleus is a quasi-bound system containing antikaon ($\bar{K}$) and nucleons (N). The existence of such an exotic nuclear state, which has a meson as a building block, was predicted as a natural extension of the interpretation of $\Lambda(1405)$ being $\bar{K}$-N hadronic molecule due to the strong attractive $\bar{K}N$ interaction in isospin $I_{\bar{K}N} = 0$ channel rather than simple three quark state. Many theoretical studies for kaonic nuclei have been done since the $\bar{K}N$ interaction was established to be strongly attractive by measuring X-ray from kaonic hydrogen. All those theoretical works support the existence of kaonic nuclei, although predicted binding energies and decay widths are widely distributed which is due to the energy dependence of $\bar{K}N$ interaction in their models.

After the first detailed theoretical calculation for kaonic nuclei [1, 2], several experiments have been performed to search for kaonic nuclei, especially for the lightest system, $\bar{K}NN$. Considering the probability of $I_{\bar{K}N} = 0$ component in the system, $\bar{K}NN$ would be isospin $I = 1/2$ with spin and parity of $J^\pi = 0^-$. Thus, most experiments searched for $\bar{K}NN$ by its non-mesonic decay mode of $\bar{K}NN \rightarrow \Lambda N$. Because of experimental difficulty in the detection of neutral particles, experimental searches focused on $\bar{K}NN$ with $I_z = +1/2$ state, which has a positive charge so decays as $\bar{K}NN_{(I_z=+1/2)} \rightarrow \Lambda p \rightarrow (\pi^- p)p$.

Searches for $\bar{K}NN_{(I_z=+1/2)}$ were performed with various reaction processes: pp collisions [3, 4], γ induced [5], π^+ induced [6], and stopped-K^- induced [7, 8] reactions. In the stopped-K^- induced reactions, the K^- is bound to NN in a light nucleus target, such as Li or C, by releasing extra energy, corresponding to binding energy, into residual nucleons. In the other reaction processes not using K^--beam, the K^+K^- pair creation process is expected to be the source of $\bar{K}$. Some of those works observed candidates of $\bar{K}NN_{(I_z=+1/2)}$ signal [3, 6, 7],

*e-mail: takumi.yamaga@riken.jp

while some of those reported null results [4, 5, 8]. Thus, the existence of $\bar{K}NN$ was not established with those works.

The most severe problem in those experiments is that there are many background channels to contaminate the spectrum. For instance, in the stopped-K^- induced experiment, the multi-nucleon absorption of K^- is the major background that overlaps with the $\bar{K}NN$ signal. In the K^+K^- pair creation experiments, the main background is the $K^+\Lambda$ or $K^+\Sigma$ production by converting a proton to the hyperon that needs smaller energy than the K^+K^- pair. Thus, although they tagged the K^+ production, the existence of intermediate-$\bar{K}$ is not secured.

To search for $\bar{K}NN_{(I_z=+1/2)}$, we performed the J-PARC E15 experiment. We utilized the in-flight (K^-, n) reaction to effectively produce $\bar{K}NN_{(I_z=+1/2)}$, suppress background processes, and easily isolate the signal from background contaminations. The produced $\bar{K}NN_{(I_z=+1/2)}$ state is reconstructed by its non-mesonic decay mode, $\bar{K}NN_{(I_z=+1/2)} \to \Lambda p$. From the exclusive measurement for the $K^-\ {}^3\mathrm{He} \to \Lambda pn$ reaction, we finally observed a distinct peak in the Λp invariant-mass spectrum which is interpreted as a signal of $\bar{K}NN_{(I_z=+1/2)}$ [9, 10].

In this paper, we review the result of J-PARC E15 and overview future projects to investigate $\bar{K}$ nuclei at J-PARC. In Sec. 2, we briefly introduce the J-PARC E15 experiment and the production of $\bar{K}NN_{(I_z=+1/2)}$ in the (K^-, n) reaction. In Sec. 3, we show the analysis result of the $K^-\ {}^3\mathrm{He} \to \Lambda pn$ reaction. The future projects for futher investigation of $\bar{K}$-nuclei is described in Sec. 4.

2 J-PARC E15

We measured an in-flight K^- induced reaction on the helium-three target, the $K^- + {}^3\mathrm{He} \to \Lambda pn$ reaction, to search for $\bar{K}NN_{(I_z=+1/2)}$ by its non-mesonic decay mode, $\bar{K}NN_{(I_z=+1/2)} \to \Lambda p$. Thus, we selected the Λpn final state and measured the invariant mass of the Λp system. We chose the incident K^--beam momentum of 1 GeV/c, at which the reaction cross-section of elementary reactions, $K^-N \to \bar{K}n$, is maximized corresponding Y^*-resonances located $\sim$ 1.8 GeV. Because of the large $K^-N \to \bar{K}n$ cross-section, the reaction process could be considered as a cascade reaction:

$$\begin{aligned} K^- + N &\to \bar{K} + n, \\ \bar{K} + (NN)_{\mathrm{res}} &\to \bar{K}NN_{(I_z=+1/2)}, \end{aligned} \tag{1}$$

where $(NN)_{\mathrm{res}}$ denotes residual nucleons in the first reaction. In the first step reaction, the induced K^--beam knocks out a neutron and is recoiled back by decreasing its momentum to a minimum $\sim$ 0.2 GeV/c at the small neutron emission angle. Due to the small recoiling momentum, it is expected that the intermediate-$\bar{K}$ is effectively absorbed by the two residual nucleons. Therefore, the production cross-section of $\bar{K}NN_{(I_z=+1/2)}$ would be rather large although the induced K^--beam originally has a rather large momentum.

The reaction process,

$$\begin{aligned} K^- + {}^3\mathrm{He} &\to X + n, \\ X &\to \Lambda p, \end{aligned} \tag{2}$$

can be characterized by two kinematical parameters, the invariant mass of X (m_X) and the momentum transfer to X (q_X). If we assume that the reaction process is the cascade reaction described in Eq. 1, m_X and q_X correspond to the invariant-mass of $\bar{K}+(NN)_{\mathrm{res}}$ and the momentum of the intermediate-$\bar{K}$, respectively. Thus, we analyzed a two-dimensional distribution on the (m_X, q_X) plane.

The experiment was performed at the K1.8BR beamline of the hadron experimental facility at J-PARC in 2015. A high-intence secondary K^--beam is produced by bombarding a gold target with a 30 GeV primary proton beam. The K^- beam is identified and measured its momentum by a beamline spectrometer system at the K1.8BR beamline. An experimental target, liquid ^{3}He, is located at the final focal plane of the beamline. The K^--beam is induced into the ^{3}He target, and particles generated in the reaction are detected by a cylindrical detector system (CDS), which surrounds the target. CDS is composed of a cylindrical drift chamber for tracking and a cylindrical detector hodoscope for a time-of-flight measurement. The detectors are installed inside a solenoid magnet providing a magnetic field along the K^- beam direction of which field strength is 0.7 T. More detailed information on the detector system can be found in Ref. [11].

In the analysis, we selected events of the Λpn final state and measured a two-dimensional distribution of the Λp invariant mass and the momentum transfer to Λp for the selected events. To select the Λpn final state, we required that three charged particles, $pp\pi^-$, are detected with CDS. Among the two $p\pi^-$ pairs in the detected $pp\pi^-$, the $\Lambda \to p\pi^-$ candidate is chosen by the invariant mass of a pair and distances of the closest approach between charged tracks (DCA). The neutron in the final state is not detected with CDS but is kinematically identified by the kinematical fit. The final events of the Λpn final state are selected based on a log-likelihood function containing χ^2 of the kinematical fit and DCA information. The analysis procedure is described in more detail in Ref. [10]. The number of selected events is ~ 6000 with the integrated luminosity of 2.9 nb^{-1} while the contamination ratio is evaluated to be $\sim 20\%$ which mainly comes from the wrong identification of $\Sigma^0 pn$ and $\Sigma^- pp$ final states as the Λpn final state.

3 Observation of $\bar{K}NN_{(I_z=+1/2)}$

For the selected Λpn final state events, we measured a two-dimensional distribution on the invariant mass of Λp pair (m_X) and the momentum transfer to the pair (q_X) as shown in Fig. 1. To obtain the double differential cross-section $d^2\sigma/(dm_X\, dq_X)$, we corrected the acceptance containing the geometrical acceptance, the detection efficiency of CDS, and the analysis efficiency. In the correction, we ignored small sensitivity regions having an acceptance of less than 5%, which is indicated by gray hatched in the figure, to ensure the reliable correction.

There are three clear events-concentrations in the figure. One is located at m X below the mass threshold of an antikaon and two nucleons system ($m_{\bar{K}} + 2m_N$), indicated by the gray vertical dotted line in the figure. The two events-concentrations are situated above the threshold; they are distributed along the blue dotted curve in the figure, which is described by

$$M_F(q_X) = \sqrt{4m_N^2 + m_{\bar{K}}^2 + 4m_N \sqrt{m_{\bar{K}}^2 + q_X^2}}. \quad (3)$$

Equation 3 is nothing but the invariant mass of antikaon having the momentum of q_X and two nucleons at rest. Therefore, the two events-concentrations can be interpreted as quasi-free $\bar{K}$ absorption into two nucleons, in which the intermediate-$\bar{K}$ having almost on-shell mass is absorbed by the residual two nucleons having Fermi momentum. On the other hand, the events-concentration below the mass threshold is well separated from the quasi-free events, and its m_X centroid does not change in different q_X-region. Thus, the events concentration should be produced by any resonance state, so it could be a signal of $\bar{K}NN_{(I_z=+1/2)}$.

To isolate a signal of $\bar{K}NN_{(I_z=+1/2)}$, we conducted a model fitting for the obtained two-dimensional distribution. We considered the following three processes in our model: (K) the $\bar{K}NN_{(I_z=+1/2)}$ production process, (F) the quasi-free $\bar{K}$ absorption process, and (B) a broad component. The two dimensional distributions of the processes (K) and (F) are shown in Fig. 2(a) and (b). The broad distribution (B) is produced by a quasi-free $\bar{K}$ absorption process

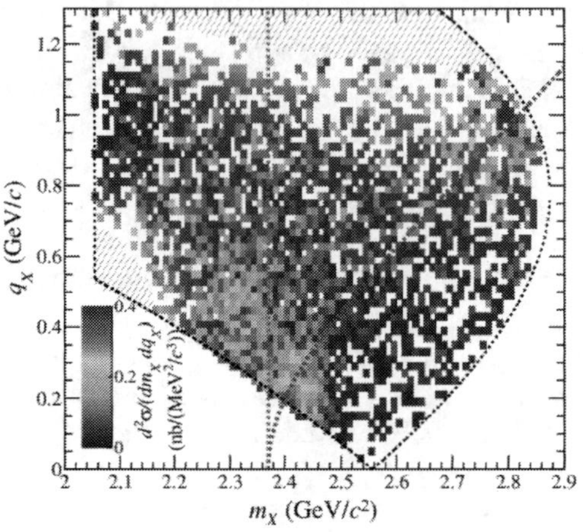

Figure 1. Two-dimensional distribution of the invariant mass of Λp pair (m_X) and the momentum transfer to the pair (q_X). The black dotted lines show the kinematical limit in the reaction. The gray dotted line is the mass threshold of $m_{\bar{K}} + 2m_N$. The blue dotted curve shows the kinematical line of the quasi-free reaction. (See the text for more details.) The figure was taken from Ref. [10].

if a proton (not a neutron) is knocked out in the first step elementary reaction described in Eq. 1. In the quasi-free process with a proton knocked out, the second step reaction, the $\bar{K}$ absorption process, is described as,

$$K^- + (pn)_{\text{res}} \to \Lambda n, \tag{4}$$

thus the same two-dimensional distribution as the process (F) (Fig. 2(b)) is observed in a two-dimensional distribution for the Λn system. Instead, the Λp system should not have any correlations between m_X and q_X, so it results in a broad distribution on the present two-dimensional distribution. We introduced a model function for the process (B) rather phenomenologically as the Breit-Wigner formula with a large width as shown in Fig. 2(c). More detailed information for the model functions is described in Ref. [10].

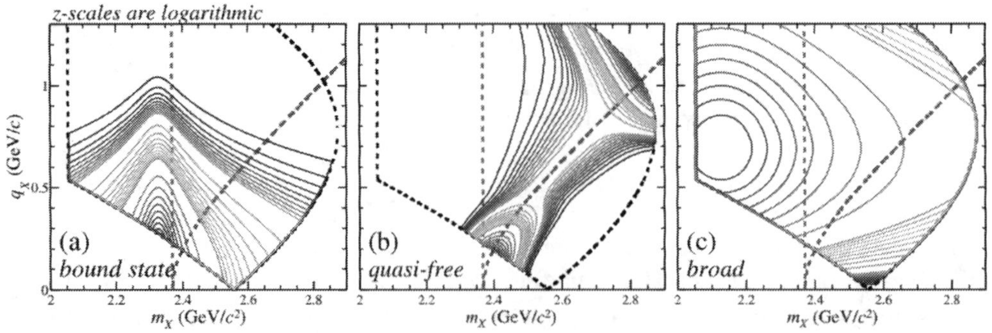

Figure 2. Two-dimensional distribution of model functions for the fit: (a) the $\bar{K}NN_{(I_z=+1/2)}$ production process, (b) the quasi-free $\bar{K}$ absorption process, and (c) the broad component. See the text for more details. The figure was taken from Ref. [10].

Figure 3 shows the fit result on the m_X distributions by selecting various q_X regions. As shown in the figure, the overall distributions are well reproduced by the model functions and background contaminations from $\Sigma^0 pn$ and $\Sigma^- pp$ final states. We obtained the mass and width of the observed state to be $2.328 \pm 0.003(\text{stat.})^{+0.004}_{-0.003}(\text{syst.})$ GeV/c^2 and $100 \pm 7(\text{stat.})^{+19}_{-9}(\text{syst.})$ MeV, respectively, so the observed state has a smaller mass than the $(\bar{K} + 2N)$ mass threshold. Additionally, the existence of intermediate-$\bar{K}$ during the reaction is clear because the quasi-free $\bar{K}$ absorption process is seen. Therefore, we interpreted the observed state as the $\bar{K}NN_{(I_z=+1/2)}$ quasi-bound state. The binding energy of $\bar{K}NN_{(I_z=+1/2)}$, mass reduction from the $(\bar{K} + 2N)$ mass threshold, was obtained to be $42 \pm 3(\text{stat.})^{+3}_{-4}(\text{syst.})$ MeV, which is consistent with some theoretical predictions [12–16].

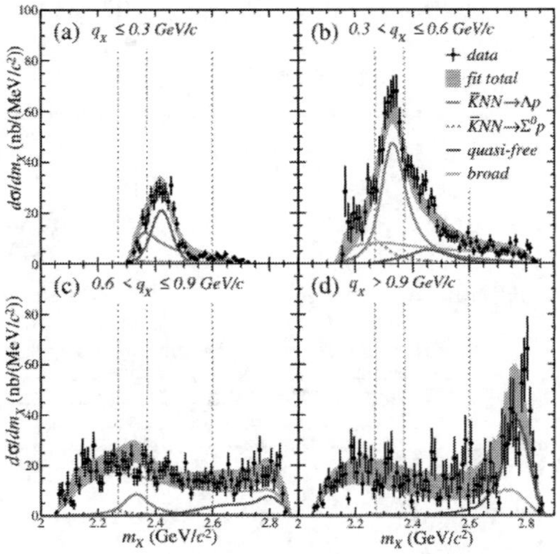

Figure 3. m_X spectra with various q_X intervals with the fit result. The figure was taken from Ref. [10].

On the other hand, the obtained decay width is relatively larger than theoretical predictions which are around 50 MeV. It is also larger than the decay width of $\Lambda(1405)$, which is considered as $\bar{K}$-N hadronic molecule decaying into $\pi\Sigma$ with 100%. A possible explanation of the large decay width of observed $\bar{K}NN_{(I_z=+1/2)}$ is the multi-nucleon absorption of $\bar{K}$. Indeed, the multi-nucleon absorption is not taken into account in the theoretical calculations. A theoretical calculation suggested that such multi-nucleon absorption could increase the total width by ~ 30 MeV [17]. Thus, we need to measure mesonic decay channels to clarify that possibility. Another possibility is the existence of sub-structure(s) around the $\bar{K}NN_{(I_z=+1/2)}$ signal region, *e.g.*, $\bar{K}NN_{(I_z=+1/2)}$ having $J^\pi = 1^-$, which is predicted as a shallow bound system by a theoretical calculation [18]. To conclude such possibilities, higher statistical data for the same reaction is essential which allows us to decompose the spectra with more detailed model functions.

We interpreted the observed resonance located at m_X below the $(\bar{K} + 2N)$ mass threshold as $\bar{K}NN_{(I_z=+1/2)}$; however, there are other candidates which could contribute to the obtained spectrum, *i.e.*, resonances which satisfy conservations, namely having strangeness $S = -1$ and baryon number $B = 2$, such as Y^*N. Even if those candidates can be excluded, we should clarify whether $\bar{K}$ keeps its particle identity in the system. For those confirmations, systematic measurements for $\bar{K}$-nuclei are desired, thus we have planned future projects of $\bar{K}$-nuclei experiments at J-PARC.

4 Future projects

The main goals of the future projects of $\bar{K}$-nuclei experiments at J-PARC are to robustly confirm the existence of $\bar{K}$-nuclei and to clarify their internal structures. For those purposes, we consider two directions of the experimental approach. One direction is a more precise measurement for $\bar{K}NN$. Another direction is a systematic measurement for several $\bar{K}$-nuclei systems having different A.

In the E15 experiment, we found that the in-flight K^- induced reaction is promising to produce $\bar{K}$-nuclei and isolate its signal from the background contaminations. Thus, the J-PARC hadron experimental facility is a unique and suitable place to investigate $\bar{K}$-nuclei further. We will construct a new large cylindrical detector system, the conceptual design of which is as shown in Fig. 4, and replace the CDS used in the E15 experiment with the new CDS to conduct future experiments.

The new CDS has approximately twice larger solid angle coverage than the previous CDS, so it can reconstruct many body decay channels of $\bar{K}$-nuclei with sufficient reconstruction efficiency. Neutrons can also be detected by the new CDS with a plastic scintillator array of which thickness is 15 cm in total. By installing trackers in between layers of the plastic scintillator array, we can measure the proton polarization from proton scattering inside the plastic scintillator. We can perform the systematic measurements for various $\bar{K}$-nuclei just by replacing the target material with the suitable one, *e.g.*, ^{4}He to measure $\bar{K}NNN$ system.

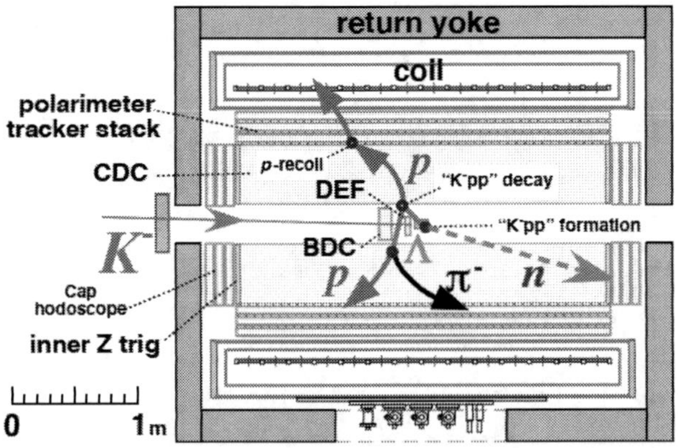

Figure 4. Conceptual design of the new cylindrical detector system.

As the first-step experiments of future projects, we proposed two experiments for the J-PARC PAC: the J-PARC E80 to measure $\bar{K}NNN$ [20], and the J-PARC P89 to measure $\bar{K}NN$ precisely [21]. After conducting those experiments, we will perform further measurements not only for heavier $\bar{K}$-nuclei than $\bar{K}NNN$ but also for the lighter system, $\Lambda(1405) \equiv \bar{K}N$. We describe those future experiments briefly in the following sections.

4.1 J-PARC E80

The experiment aims to search for the second lightest $\bar{K}$-nuclue, $\bar{K}NNN$, by the ^{4}He(K^-, n) reaction. $\bar{K}NNN$ is considered to be $I = 0$ with $J^\pi = 1/2^-$, so it is essential to measure $\bar{K}NNN \rightarrow \Lambda d$ non-mesnoic two-body decay to establish its isospin state. We will also measure three-body decay, $\bar{K}NNN \rightarrow \Lambda pn$, by detecting a neutron with new CDS. From the decay ratio between two- and three-body decay modes, we may extract information on the internal configuration of $\bar{K}NNN$. We can also measure mesonic decay modes, $\bar{K}NNN \to \pi YNN$.

4.2 J-PARC P89

The primary purpose of the experiment is to determine J^π of $\bar{K}NN$. The determination of J^π is essential for establishing the observed state as a quantum state and clarifying its internal configuration. To this end, we will measure the spin-spin correlation between Λ and p from $\bar{K}NN_{(I_z=+1/2)} \to \Lambda p$ decay. Because J^π of initial $\bar{K}NN_{(I_z=+1/2)}$ are conserved during the strong decay, the initial J^π information should appear as spin alignment of Λp system. Therefore, we can determine J^π of $\bar{K}NN$ from the spin alignment measurement in a model-independent way. For this purpose, the proton polarimeter system of the new CDS is essential to measure the proton spin direction while the Λ spin direction can be obtained rather easily from decay asymmetry in the $\Lambda \to p\pi^-$ decay.

Another purpose is to search for $\bar{K}NN_{(I_z=-1/2)}$, which must exist if $\bar{K}NN_{(I_z=+1/2)}$ exists, by its non-mesonic decay mode $\bar{K}NN_{(I_z=-1/2)} \to \Lambda n$. Thus, we will measure the ^{3}He(K^-, p) reaction going to the $\Lambda n + p$ final state.

4.3 Other perspectives

We intend to measure $d(K^-, n)$ reaction and reconstruct $\Lambda(1405)$ from $\Lambda(1405) \to \pi^\pm\Sigma^\mp$ decay by detecting a neutron from $\Sigma^\mp \to n\pi^\mp$ decay. If we replace the plastic scintillator arrays with an electromagnetic calorimeter to detect γ-rays, it will be possible to reconstruct $\Lambda(1405) \to \pi^0\Sigma^0$ decay mode, which is a particularly interesting channel because it contains only the $I = 0$ contribution. However, this measurement should be a long-term project because of the difficulty to construct a large electromagnetic calorimeter.

On the other hand, for heavier systems than $\bar{K}NNN$, experimental difficulties are expected to arise. One of the difficulties is that decay products of heavier $\bar{K}$-nuclei contain light nuclei, α or heavier, and will need a dedicated detector setup. Another difficulty is that natural width may be getting larger in heavier $\bar{K}$-nuclei since multinucleon-absorption of $\bar{K}$ may increase in the many-body system. Nevertheless we have planned to measure some heavier $\bar{K}$-nuclei, *e.g.*, $\bar{K}NNNN$.

$\bar{K}NNNN$ is considered to be $I = 1/2$ with $J^\pi = 0^-$ state, ineternal structure of which would be K^--α for $I_z = -1/2$ and $\bar{K}^0$-α for $I_z = +1/2$ [19]. Because there is no stable nuclear target with $A = 5$, we will utilize the ^{6}Li(K^-, d) reaction to produce $\bar{K}NNNN$, so the $I_z = -1/2$ state will be measured. Though the in-flight K^- induced reaction is an efficient reaction to measure $\bar{K}$-nuclei, it would be challenging to measure $\bar{K}NNNN$ because the (K^-, d) reaction has a smaller cross-section than the (K^-, N) reactions, and the momentum transfer in the (K^-, d) reaction is larger than that in the (K^-, N) reaction. Therefore, it is essential for realization of $\bar{K}NNNN$ search to measure production cross sections of $\Lambda(1405)$ and $\bar{K}NN$ with ^{3}He(K^-, d) and ^{4}He(K^-, d) reactions, respectively, in the preceding experiments. Sensitivities for other $\bar{K}$-nuclei with the in-flight K^- induced reaction and the new CDS are under consideration.

5 Summary

We reviewed the result obtained by the J-PARC E15 experiment, the observation of $\bar{K}NN_{(I_z=+1/2)}$ in the in-flight K^- reaction on ^{3}He. A clear signal of $\bar{K}NN_{(I_z=+1/2)}$ was observed in the non-mesonic decay channel, $\bar{K}NN_{(I_z=+1/2)} \to \Lambda p$ for the first time, and the binding energy and decay width were obtained from the analysis. However, there are remaining questions, especially for the internal structure of $\bar{K}$-nuclei. Thus, we have planned further experiments to robustly confirm the existence of $\bar{K}$-nuclei and to clarify their internal structure. We will conduct systematic measurements for various $\bar{K}$-nuclei, from $\Lambda(1405) \equiv \bar{K}N$ to $\bar{K}NNNN$, with a newly constructred large cylindrical detector system. The construction of the new detector system has been started in 2022 and will be completed in 2025. Thus, we expect the first measurement to start in 2026.

References

[1] T. Yamazaki and Y. Akaishi, Phys. Lett. B **535**, 70 (2002)
[2] Y. Akaishi and T. Yamazaki, Phys. Rev. C **65**, 044005 (2002)
[3] T. Yamazaki *et al.*, Phys. Rev. Lett. **104**, 132502 (2010)
[4] G. Agakishiev *et al.*, Phys. Lett. B **742**, 242 (2015)
[5] A. O. Tokiyasu *et al.*, Phys. Lett. B **728**, 616 (2014)
[6] Y. Ichikawa *et al.*, Prog. Theor. Exp. Phys. **2015**, 021D01 (2015)
[7] M. Agnello *et al.*, Nucl. Phys. A **914**, 310 (2013)
[8] O. Vázquez Doce *et al.*, Phys. Lett. B **758**, 134 (2016)
[9] S. Ajimura *et al.*, Phys. Lett. B **789**, 620 (2019)
[10] T. Yamaga *et al.*, Phys. Rev. C **102**, 044002 (2020)
[11] K. Agari *et al.*, Prog. Theor. Exp. Phys. **2012**, 02B011 (2012)
[12] T. Yamazaki and Y. Akaishi, Phys. Rev. C **76**, 045201 (2007)
[13] K. Ikeda and T. Saito, Phys. Rev. C **76**, 035203 (2007)
[14] K. Ikeda and T. Saito, Phys. Rev. C **79**, 035201 (2009)
[15] S. Wycech and A. M. Green, Phys. Rev. C **79**, 035201 (2009)
[16] J. Révai and N. V. Shevchenko, Phys. Rev. C **90**, 034004 (2014)
[17] M. Bayar and E. Oset, Phys. Rev. C **88**, 044003 (2013)
[18] N. V. Shevchenko, Few-Body Syst. **61**, 27 (2020)
[19] S. Ohnishi *et al.*, Phys. Rev. C **95**, 065202 (2017)
[20] F. Sakuma *et al.*, Proposal for J-PARC E80, https://j-parc.jp/researcher/Hadron/en/pac_2007/pdf/P80_2020-10.pdf
[21] T. Yamaga *et al.*, Proposal for J-PARC P89, https://j-parc.jp/researcher/Hadron/en/pac_2107/pdf/P89_2021-09.pdf

$\bar{K}N$ interaction, p-wave terms

Albert Feijoo[1,2,*], *Daniel* Gazda[2,**], *Volodymyr* Magas[3,***], and *Àngels* Ramos[3,****]

[1]Departamento de Física Teórica and IFIC, Centro Mixto Universidad de Valencia-CSIC, Institutos de Investigación de Paterna, Aptdo. 22085, 46071 Valencia, Spain
[2]Nuclear Physics Institute, 25068 Rez, Czech Republic
[3]Departament de Física Quàntica i Astrofísica and Institut de Ciències del Cosmos (ICCUB), Universitat de Barcelona, Martí i Franquès 1, 08028 Barcelona, Spain

Abstract. The inclusion of new ingredients that are expected to be specially relevant at higher energies could reveal more information about the physics behind the NLO terms of the chiral Lagrangian. In the present work, we explore the relevance of including partial waves higher than the $L = 0$, which is usually the only component considered in the literature to study the $\bar{K}N$ scattering phenomenology. In particular, we focus on the p-wave contribution, the effect of which is expected to be non-negligible, as we aim at obtaining the K^-p scattering amplitudes at higher energies, necessary to describe the $\eta\Lambda$, $\eta\Sigma^0$, $K^0\Xi^0$ and $K^+\Xi^-$ production reactions. Extending the $\bar{K}N$ interaction to p-wave components is also relevant for studies of bound K^- mesons in nuclei since their local momentum can acquire sizable values.

1 Introduction

It is already more than two decades that Unitarized Chiral Perturbation Theory (ChPT) has been shown as a powerful approach to describe low-energy hadron interction. This non-perturbative treatment of chiral perturbation theory (ChPT) allows one to dynamically generate resonances and bound states. ChPT is characterized by employing hadrons as degrees of freedom, instead of quarks and and gluons, and by respecting the symmetries of the underlying theory of the strong interaction, Quantum Chromodynamics (QCD), particularly its spontaneously broken chiral symmetry [1].

Already from the early stages of the $\bar{K}N$ interaction study, such a theoretical scheme provided clear evidences of the molecular nature of $\Lambda(1405)$ (see Refs [2–7]), whose mass was systematically overestimated by quark models. The interpretation of the $\Lambda(1405)$ as a $\bar{K}N$ quasi-bound state was already anticipated by the authors of Refs [8, 9], but the confirmation of the $\Lambda(1405)$ being essentially a meson-baryon bound state came after establishing its double-pole nature [4, 10] from comparing different experimental line shapes [11], which unambiguously indicates the different coupling strength to the meson-baryon components of the $\Lambda(1405)$ wave-function.

*e-mail: edfeijoo@ific.uv.es
**e-mail: gazda@ujf.cas.cz
***e-mail: vladimir@fqa.ub.edu
****e-mail: ramos@fqa.ub.edu

https://doi.org/10.1051/epjconf/202227107002

More recently, the irruption of several experiments triggered a renewed interest in improving the chiral unitary theories [19–26] for a better description of the $\bar{K}N$ interaction and related phenomenology. On the one hand, the COSY [12] and HADES [13] collaborations employed pp reactions aiming at establishing the shape of the $\Lambda(1405)$. On the other hand, with same purpose, photo- or electro-production processes were carried out at LEPS [14] and CLAS [15–17]. Additionally, the precise determination of the energy shift and width of the $1s$ state in kaonic hydrogen was measured by SIDDHARTA [18].

In our previous works [24–26], the relevance of the terms next in the hierarchy after the lowest order Weinberg–Tomozawa (WT) term was studied, in connection with the inclusion of higher-energy experimental data. An especial attention was paid on the $K\Xi$ production reactions because they do not proceed directly via the lowest order WT contribution [24]. Our further studies [25, 26] indeed demonstrated that the so-called Born diagrams as well as the next-to-leading order (NLO) terms are far from being mere corrections when it comes to the reproduction of the $K^-p \to \eta\Lambda, \eta\Sigma^0, K^+\Xi^-, K^0\Xi^0$ reaction cross sections. In particular, the inclusion of isospin filtering reactions in [26, 27] served to emphasize their relevance for avoiding potential ambiguities in the isospin components of the scattering amplitude. All this translates into stronger constraints on the models by means of which one can derive more reliable values of the low-energy constants of the chiral Lagrangian.

In Ref. [28], as we expected, the inclusion of partial waves higher than the $L = 0$ reveal more information about the physics behind the NLO terms of the chiral Lagrangian. In particular, the focus was put on p-wave contributions whose effect is proved to be non-negligible in order to describe the $\eta\Lambda$, $\eta\Sigma^0$, $K^0\Xi^0$ and $K^+\Xi^-$ production reactions. As additional ingredients, new NLO contributions were taken into account, such terms have usually been ignored for having a low impact at lower energies. We show in the present manuscript that the new terms of the NLO Lagrangian are relevant, being in fact strongly intertwined with the p-wave components of the $\bar{K}N$ interaction.

2 Formalism

As already mentioned, UChPT has been proved as a proper tool to treat the meson–baryon scattering at energies around resonances. The key point lies in the fact that the nonperturbative schemes prevent plain ChPT from non converging and, at the same time, the unitarity and analyticity of the scattering amplitude are ensured. In the present work, the Bethe–Salpeter (BS) equation solved in coupled channels guarantees unitarity. The method followed is the same employed in Refs. [3, 29] and allows factorizing the interaction kernel and the scattering amplitude out of the integral equation thereby transforming a complex system of coupled integral equations into a simple system of algebraic equations which, in matrix form, reads:

$$T_{ij} = (1 - V_{il}G_l)^{-1} V_{lj}, \qquad (1)$$

where V_{ij} is the driving kernel derived from the chiral Lagrangian, T_{ij} is the corresponding scattering amplitude for the transition from an i channel to a j one, and G_l is the loop function

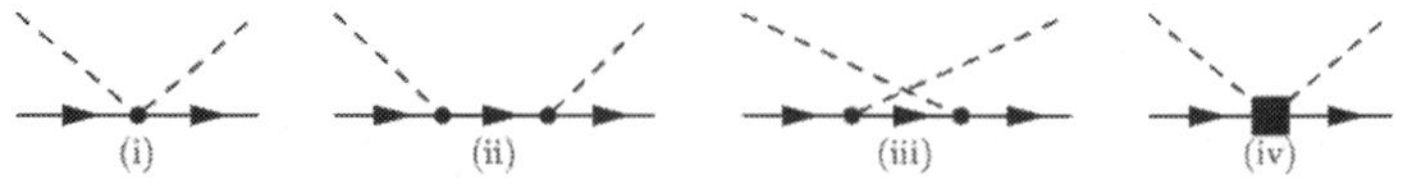

Figure 1. Diagrammatic representation of the meson-baryon interaction kernels: Weinberg–Tomozawa term (i), direct and crossed Born terms (ii) and (iii), and NLO terms (iv). Dashed (solid) lines represent the pseudoscalar octet mesons (octet baryons).

of the intermediate channel l, which reads:

$$G_l = \frac{2M_l}{(4\pi)^2}\left\{a_l(\mu) + \ln\frac{M_l^2}{\mu^2} + \frac{m_l^2 - M_l^2 + s}{2s}\ln\frac{m_l^2}{M_l^2} + \frac{q_{\rm cm}}{\sqrt{s}}\ln\left[\frac{(s+2\sqrt{s}q_{\rm cm})^2 - (M_l^2 - m_l^2)^2}{(s-2\sqrt{s}q_{\rm cm})^2 - (M_l^2 - m_l^2)^2}\right]\right\}, \quad (2)$$

with M_l and m_l being the baryon and meson masses of the channel. The subtraction constants a_l replace the divergence for a given dimensional regularization scale $\mu = 1000$ MeV. These constants are unknown parameters to be fitted to the experimental data. There are ten channels in the $S = -1$ sector, therefore one expects to have the same number of subtraction constants, although they can be reduced to six by taking into account isospin symmetry arguments.

Roughly speaking, the $SU(3)$ effective chiral Lagrangian consist of an expansion in powers of momentum over a characterizing parameter of the system. These building blocks preserve the symmetries of the fundamental interaction and are arranged in the expansion by order of relevance following a power counting scheme (see for instance [30] for a more detailed explanation). At leading order (LO), the most general Lagrangian can be expressed as:

$$\mathcal{L}_{\phi B}^{(1)} = i\langle\bar{B}\gamma_\mu[D^\mu, B]\rangle - M_0\langle\bar{B}B\rangle - \frac{1}{2}D\langle\bar{B}\gamma_\mu\gamma_5\{u^\mu, B\}\rangle - \frac{1}{2}F\langle\bar{B}\gamma_\mu\gamma_5[u^\mu, B]\rangle, \quad (3)$$

where M_0 is the common baryon octet mass in the chiral limit, the constants D, F denote the axial vector couplings of the baryons to the mesons, and the symbol $\langle\cdot\rangle$ stands for the trace in flavour space. The baryon octet field $(N, \Lambda, \Sigma, \Xi)$ is denoted by B, while the pseudoscalar meson octet field ϕ (π, K, η) enters in $u_\mu = iu^\dagger\partial_\mu U u^\dagger$, where $U(\phi) = u^2(\phi) = \exp\left(\sqrt{2}i\phi/f\right)$ (see [24]) with f being the pseudoscalar decay constant that acts as typifying scale factor in the expansion in powers of momentum. Finally, $[D_\mu, B]$ contains the covariant derivative that accounts for the local character of the chiral transformation of u and it is defined as:

$$[D_\mu, B] = \partial_\mu B + [\Gamma_\mu, B] \quad (4)$$

with $\Gamma_\mu = [u^\dagger, \partial_\mu u]/2$ being the chiral connection. The next-to-leading order (NLO) contributions are given by:

$$\begin{aligned}
\mathcal{L}_{\phi B}^{(2)} &= b_D\langle\bar{B}\{\chi_+, B\}\rangle + b_F\langle\bar{B}[\chi_+, B]\rangle + b_0\langle\bar{B}B\rangle\langle\chi_+\rangle + d_1\langle\bar{B}\{u_\mu, [u^\mu, B]\}\rangle \\
&+ d_2\langle\bar{B}[u_\mu, [u^\mu, B]]\rangle + d_3\langle\bar{B}u_\mu\rangle\langle u^\mu B\rangle + d_4\langle\bar{B}B\rangle\langle u^\mu u_\mu\rangle \\
&- \frac{g_1}{8M_N^2}\langle\bar{B}\{u_\mu, [u_\nu, \{D^\mu, D^\nu\}B]\}\rangle - \frac{g_2}{8M_N^2}\langle\bar{B}[u_\mu, [u_\nu, \{D^\mu, D^\nu\}B]]\rangle \\
&- \frac{g_3}{8M_N^2}\langle\bar{B}u_\mu\rangle\langle u_\nu, \{D^\mu, D^\nu\}B\rangle - \frac{g_4}{8M_N^2}\langle\bar{B}\{D^\mu, D^\nu\}B\rangle\langle u_\mu u_\nu\rangle \\
&- \frac{h_1}{4}\langle\bar{B}[\gamma^\mu, \gamma^\nu]Bu_\mu u_\nu\rangle - \frac{h_2}{4}\langle\bar{B}[\gamma^\mu, \gamma^\nu]u_\mu[u_\nu, B]\rangle - \frac{h_3}{4}\langle\bar{B}[\gamma^\mu, \gamma^\nu]u_\mu\{u_\nu, B\}\rangle \\
&- \frac{h_4}{4}\langle\bar{B}[\gamma^\mu, \gamma^\nu]u_\mu\rangle\langle u_\nu, B\rangle + \text{h.c.},
\end{aligned} \quad (5)$$

where M_N stands for the nucleon mass, while the coefficients b_D, b_F, b_0, d_i $(i = 1, \ldots, 4)$, g_i $(i = 1, \ldots, 4)$ and h_i $(i = 1, \ldots, 4)$ are the low-energy constants (LECs) at this order. The LECs are unknown parameters that contain all the physics of the problem. Despite the symmetries of the underlying theory cannot fix these constants, some of them can be constrained by other observables, namely: the mass splitting of baryons, the pion-Nucleon sigma term or the strangeness content of the proton. In this case, we treat the LECs as free parameters in the fitting procedure as it is usually done in the literature. The quantity $\chi_+ = 2B_0(u^\dagger\mathcal{M}u^\dagger + u\mathcal{M}u)$ explicitly breaks chiral symmetry via the quark mass matrix $\mathcal{M} =$

diag(m_u, m_d, m_s), while $B_0 = -\langle 0|\bar{q}q|0\rangle / f^2$ relates to the order parameter of spontaneously broken chiral symmetry.

The NLO Lagrangian of Eq. (5), taken from [31], differs from the one we employed in our previous works [24–26] in the terms accompanied by g_i ($i = 1, \ldots, 4$) and h_i ($i = 1, \ldots, 4$) coefficients (from now on g-terms and h-terms). The non-significant role of these terms at low energies has been the main reason to discard them in the study of the $S = -1$ meson–baryon scattering. By contrary, since our models explore higher energies, it is reasonable to consider these additional contributions. Actually, as we show in [28] and in the present manuscript, they play a relevant role mainly in the $\bar{K}N$ transitions to the $\eta\Lambda$, $\eta\Sigma^0$, and $K\Xi$ channels. It should be mentioned that there is a reduction of the number of LECs since the the g_3 monomial cancells after the hermitian conjugation (h.c.) transformation in Eq. (5) as well as the appearance of new structures coming from h.c. of the g-terms.

All the interaction kernels for the $\phi_i B_i^s \to \phi_j B_j^{s'}$ processes (with the incoming and outgoing spins s, s', respectively) can be derived from the previous Lagrangians. They can be found schematically represented in Fig. 1. Diagram (i) corresponds to the Weinberg–Tomozawa (V_{ij}^{WT}), while the Born contributions come in diagrams (ii) (direct Born term V_{ij}^{BD}) and (iii) (crossed Born term V_{ij}^{BC}), the vertices of which are obtained from the D and F terms of Eq. (3), and, finally, diagram (iv) accounts for the tree-level contributions at NLO (V_{ij}^{NLO}). The analytical form of all the contributions is given by Eqs. (7), (8), (9) and (10) in Ref. [28].

Since the total interaction kernel ($V_{ij} = V_{ij}^{WT} + V_{ij}^{BD} + V_{ij}^{BC} + V_{ij}^{NLO}$) is composed of a mixture of contributions with different angular momenta, it is not possible to incorporate V_{ij} into Eq. (1) directly. Thus, it is convenient to express the T-matrix in terms of the spin-nonflip and spin-flip parts.

$$T_{ij}(s, s') = \chi_j^{\dagger s'} \left[f(\sqrt{s}, \theta) - i(\vec{\sigma} \cdot \hat{n}) g(\sqrt{s}, \theta) \right] \chi_i^s, \tag{6}$$

where θ is the CM angle between the inital and final meson momenta and $\hat{n} = \vec{q}_j \times \vec{q}_i / |\vec{q}_j \times \vec{q}_i|$ is the normal vector to the scattering plane (being $\vec{q}_j$ and $\vec{q}_i$ the outgoing and incoming three-momentum in the CM, respectively) . The functions $f(\sqrt{s}, \theta)$ and $g(\sqrt{s}, \theta)$ can be expanded in Legendre polynomials as:

$$\begin{aligned} f(\sqrt{s}, \theta) &= \sum_{l=0}^{\infty} f_l(\sqrt{s})\, P_l(\cos\theta), \\ g(\sqrt{s}, \theta) &= \sum_{l=1}^{\infty} g_l(\sqrt{s})\, \sin\theta\, \frac{\mathrm{d}P_l(\cos\theta)}{\mathrm{d}\cos\theta}, \end{aligned} \tag{7}$$

with f_l and g_l being the projections of the sum of all the above kernels onto $P_l(\cos\theta)$ and $\sin\theta \frac{\mathrm{d}P_l(\cos\theta)}{\mathrm{d}\cos\theta}$, respectively.

The amplitudes can be redefined as:

$$\begin{aligned} f_{l+}(\sqrt{s}) &= \frac{1}{2l+1} \left(f_l(\sqrt{s}) + l\, g_l(\sqrt{s}) \right) \text{for } J = l + \frac{1}{2}, \\ f_{l-}(\sqrt{s}) &= \frac{1}{2l+1} \left(f_l(\sqrt{s}) - (l+1)\, g_l(\sqrt{s}) \right) \text{for } J = l - \frac{1}{2}, \end{aligned} \tag{8}$$

to ensure that the quantum numbers of spin ($\frac{1}{2}$), orbital angular momentum (l), and total angular momentum (J) are preserved in the unitarization procedure implemented by the BS equations. Following the notation of Ref. [32], each unitarized J-scattering amplitude should be calculated by a new version of Eq. (1), which in matrix form reads:

$$f_{l\pm} = \left[1 - f_{l\pm}^{\text{tree}} G \right]^{-1} f_{l\pm}^{\text{tree}}, \tag{9}$$

where the amplitudes $f^{\text{tree}}_{l\pm}$ are obtained from equations (6)–(8).

The aim of this work is the study of the effects caused by the inclusion of higher partial waves on the physical observables in this sector, particularly on the K^-p cross sections. The most general expression of the differential cross sections for a given $\phi_i B_i^{\chi_z} \to \phi_j B_j^{\chi'_z}$ process is

$$\frac{d\sigma_{ij}}{d\Omega} = \frac{M_i M_j q_j}{16\pi^2 s q_i} S_{ij}, \tag{10}$$

with

$$S_{ij} = \frac{1}{2\chi+1} \sum_{\chi_z, \chi'_z} |T_{ij}(\chi_z, \chi'_z)|^2 = |f(\sqrt{s}, \theta)|^2 + |g(\sqrt{s}, \theta)|^2,$$

where the first factor averages over the initial baryon spin projections, giving 1/2 for this particular case, and where we have also summed over all possible final baryon spin projections.

Since the scope of the present study does not go beyond d-wave contributions, focusing mostly on p-wave effects, the expression for the total cross section incorporating such partial waves can be written as:

$$\sigma_{ij} = \frac{M_i M_j q_j}{4\pi s q_i} \left[|f_{0+}|^2 + 2|f_{1+}|^2 + |f_{1-}|^2 + 3|f_{2+}|^2 + 2|f_{2-}|^2 \right]. \tag{11}$$

3 Results and discussion

The $S = -1$ sector is an interesting benchmark to test meson-baryon Effective Field Theories given the existence of large sets of scattering data. For this very reason, it cannot only serve for checking their predictive power but also to extract information about the LECs, especially those beyond LO. This collection of experimental data has been employed to constrain the models developed in the study, the technical details of the data treatment as well as the fitting procedure can be found in Ref. [28]. Actually, three models were introduced in [28] to discuss the effects of the new elements, two new and a third one already developed in [26]. Being the two new models the natural extension of the old one in two steps.

- **s-wave (old)**: This first model corresponds to the fit called **WT+Born+NLO** carried out in [26]. It was constructed by adding the interaction kernels derived from the Lagrangian up to NLO and neglecting the h- and g-terms. We limited this model to the s-wave contribution.
- **s-wave**: The second model improves upon the first one by incorporating the novel h- and g-terms that come from the NLO interaction kernel and, as in the first model, only the s-wave contribution is taken into account.
- **s+p-waves**: This model employs the same Lagrangian as the **s-wave** fit, but it also incorporates the p-wave contributions.

As first comment on the values of the parameters present in the models, the most eye-catching feature is the difference observed when comparing these two new models (see Table 2 in Ref. [28]), being the set of the **s+p-waves** model closer to that of our earlier **s-wave (old)** model. This makes us believe that the **s+p-waves** model developed in the present work constitutes a very good starting point for subsequent implementations of higher partial waves. Both new models, **s-wave** and **s+p-waves**, reproduces experimental data very well and achieve very low $\chi^2_{d.o.f.}$ (0.77 and 0.86 respectively). This fact is clearly appreciated in Fig. 2 of [28], where total cross sections of K^-p scattering to all channels of the $S = -1$ sector are compared to experimental data. It should be commented that one can only notice larger differences among the models in the cross sections of the $K^-p \to \eta\Lambda, \eta\Sigma^0, K^+\Xi^-, K^0\Xi^0$

Table 1. Threshold observables obtained from our fits (branching ratios and the energy shift and width of the 1s state of kaonic hydrogen). Experimental data is taken from [18, 33, 34].

	γ	R_n	R_c	ΔE_{1s}	Γ_{1s}
s+p-waves	2.36	0.188	0.662	297	532
s-wave	2.40	0.179	0.665	280	560
s-wave (old)	$2.36^{+0.03}_{-0.03}$	$0.188^{+0.010}_{-0.011}$	$0.659^{+0.005}_{-0.002}$	288^{+23}_{-8}	588^{+9}_{-40}
Exp.	2.36±0.04	0.189±0.015	0.664±0.011	283±36	541±92

processes, specially at energies above 1900 MeV (see discussion in [28]). The results of the threshold observables for the previous models are collected in Table 1 as well as the corresponding experimental values. No substantial change in the reproduction of the experimental values can be appreciated when comparing the new models to the **s-wave (old)** one, which is completely aligned with the previous comment.

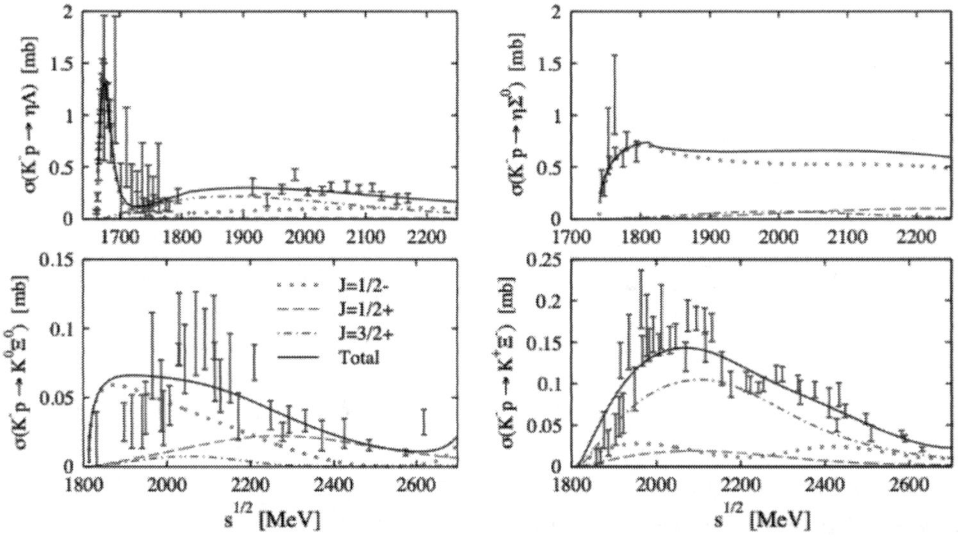

Figure 2. Total cross sections of the $K^-p \to \eta\Lambda, \eta\Sigma^0, K^+\Xi^-, K^0\Xi^0$ reactions obtained for **s+p-waves** (solid black line), and the corresponding contributions for $J^P = \frac{1}{2}^-$ (dotted red line), $J^P = \frac{1}{2}^+$ (dashed green line) and $J^P = \frac{3}{2}^+$ (dash-dotted violet line). Experimental data has been taken from [35–50].

In order to aid better understanding of the role of the p-waves, in Fig. 2 we present the individual partial-wave contributions to the $K^-p \to \eta\Lambda, \eta\Sigma^0, K^+\Xi^-, K^0\Xi^0$ processes included in the **s+p-waves** model, namely the $J^P = \frac{1}{2}^-$ (s-wave), $\frac{1}{2}^+$ and $\frac{3}{2}^+$ (p-wave) channels. We directly focus on these reactions since p-wave effects are barely noticeable in the classical processes.

Focusing first on the $K^-p \to \eta\Lambda$ cross section, it can be seen that the s-wave $J^P = \frac{1}{2}^-$ contribution dominates just above threshold due to the $\Lambda(1670)$ resonance, which is generated dynamically in this partial wave. Then, as the energy increases, its role is moderately losing relevance against the $J^P = \frac{3}{2}^+$ component which is the main contribution in the region ranging from 1750 to 2100 MeV. Now, we examine the $K^-p \to \eta\Sigma^0$ reaction. The s-wave $J^P = \frac{1}{2}^-$ contribution is clearly the dominant one over the whole range of energies explored. The p-wave contributions start being noticeable from 1850 MeV, however, one should be very cautious about drawing conclusions because of the lack of experimental data in this region.

The partial-wave decomposition of the $K\Xi$ total cross section provides very valuable information about the relevance of higher partial waves in our model. On the one hand, the $K^0\Xi^0$ production discloses the fundamental role played by the $J^P = \frac{1}{2}^+$ contribution above 2200 MeV and how the low-energy regime is dominated by the $J^P = \frac{1}{2}^-$ wave. The $J^P = \frac{3}{2}^+$ component provides a non-negligible strength around 2100 MeV, which is however not enough to reproduce the sizable experimental structure. On the other hand, in the $K^-p \to K^+\Xi^-$ process, the $J^P = \frac{1}{2}^-$ and $J^P = \frac{1}{2}^+$ contributions are moderate and roughly constant, being the $J^P = \frac{3}{2}^+$ component the one that governs the description of the experimental data.

The main conclusion of this study is the need of higher partial-wave contributions to properly describe the K^-p inelastic scattering amplitudes opening up at higher energies. This is in contrast to our previous works, where the experimental data of such processes was reasonably reproduced employing pure s-wave scattering amplitudes. In other words, our previous models have effectively overestimated the lowest partial-wave contributions, thereby masking the physics of higher partial waves behind the values of the NLO parameters. As for the relevance of the new g- and h-terms considered in this work, compared to the rest of NLO contributions, we cannot say anything conclusive, as we have obtained very different values of the parameters in the two new fits presented in [28]. Given the fact that these terms provide a non-negligible contribution to the scattering amplitudes, especially far enough from thresholds, it is natural to consider the g- and h-terms when implementing higher partial waves in the scattering amplitudes for their strong dependence on momenta.

References

[1] J. Gasser and H. Leutwyler, Annals Phys. **158**, 142 (1984)
[2] N. Kaiser, P. B. Siegel and W. Weise, Nucl. Phys. A **594**, 325-345 (1995)
[3] E. Oset and A. Ramos, Nucl. Phys. A **635**, 99-120 (1998)
[4] J. A. Oller and U. G. Meissner, Phys. Lett. B **500**, 263-272 (2001)
[5] M. F. M. Lutz and E. E. Kolomeitsev, Nucl. Phys. A **700**, 193-308 (2002)
[6] C. Garcia-Recio, J. Nieves, E. Ruiz Arriola and M. J. Vicente Vacas, Phys. Rev. D **67**, 076009 (2003)
[7] B. Borasoy, R. Nissler and W. Weise, Eur. Phys. J. A **25**, 79-96 (2005)
[8] R. H. Dalitz and S. F. Tuan, Phys. Rev. Lett. **2**, 425-428 (1959)
[9] R. H. Dalitz and S. F. Tuan, Annals Phys. **10**, 307-351 (1960)
[10] D. Jido, J. A. Oller, E. Oset, A. Ramos and U. G. Meissner, Nucl. Phys. A **725**, 181-200 (2003)
[11] V. K. Magas, E. Oset and A. Ramos, Phys. Rev. Lett. **95**, 052301 (2005)
[12] I. Zychor, M. Buscher, M. Hartmann, A. Kacharava, I. Keshelashvili, A. Khoukaz, V. Kleber, V. Koptev, Y. Maeda and T. Mersmann, *et al.* Phys. Lett. B **660**, 167-171 (2008)
[13] G. Agakishiev *et al.* [HADES], Phys. Rev. C **87**, 025201 (2013)
[14] M. Niiyama, H. Fujimura, D. S. Ahn, J. K. Ahn, S. Ajimura, H. C. Bhang, T. H. Chang, W. C. Chang, J. Y. Chen and S. Date, *et al.* Phys. Rev. C **78**, 035202 (2008)
[15] K. Moriya *et al.* [CLAS], Phys. Rev. C **87**, no.3, 035206 (2013)
[16] K. Moriya *et al.* [CLAS], Phys. Rev. C **88**, 045201 (2013)
[17] H. Y. Lu *et al.* [CLAS], Phys. Rev. C **88**, 045202 (2013)
[18] M. Bazzi *et al.* [SIDDHARTA], Phys. Lett. B **704**, 113-117 (2011)

[19] Z. H. Guo and J. A. Oller, Phys. Rev. C **87**, no.3, 035202 (2013)
[20] Y. Ikeda, T. Hyodo and W. Weise, Nucl. Phys. A **881**, 98-114 (2012)
[21] L. Roca and E. Oset, Phys. Rev. C **88**, no.5, 055206 (2013)
[22] M. Mai and U. G. Meißner, Eur. Phys. J. A **51**, no.3, 30 (2015)
[23] A. Cieplý, M. Mai, U. G. Meißner and J. Smejkal, Nucl. Phys. A **954**, 17-40 (2016)
[24] A. Feijoo, V. K. Magas and A. Ramos, Phys. Rev. C **92**, no.1, 015206 (2015)
[25] A. Ramos, A. Feijoo and V. K. Magas, Nucl. Phys. A **954**, 58-74 (2016)
[26] A. Feijoo, V. Magas and A. Ramos, Phys. Rev. C **99**, no.3, 035211 (2019)
[27] A. Feijoo, V. K. Magas, A. Ramos and E. Oset, Phys. Rev. D **92**, no.7, 076015 (2015) [erratum: Phys. Rev. D **95**, no.3, 039905 (2017)]
[28] A. Feijoo, D. Gazda, V. Magas and A. Ramos, Symmetry **13**, no.8, 1434 (2021)
[29] T. Hyodo and D. Jido, Prog. Part. Nucl. Phys. **67**, 55-98 (2012)
[30] S. Scherer, Adv. Nucl. Phys. **27**, 277 (2003)
[31] K. Aoki and D. Jido, PTEP **2019**, no.1, 013D01 (2019)
[32] D. Jido, E. Oset and A. Ramos, Phys. Rev. C **66**, 055203 (2002)
[33] R. J. Nowak, J. Armstrong, D. H. Davis, *et al.*, Nucl. Phys. B **139**, 61-71 (1978)
[34] D. N. Tovee, D. H. Davis, J. Simonovic, *et al.*, Nucl. Phys. B **33**, 493-504 (1971)
[35] J. K. Kim, Phys. Rev. Lett. **14**, 29 (1965)
[36] T. S. Mast, M. Alston-Garnjost, R. O. Bangerter, *et al.*, Phys. Rev. D **14**, 13 (1976)
[37] R. O. Bangerter, M. Alston-Garnjost, A. Barbaro-Galtieri, *et al.*, Phys. Rev. D **23**, 1484 (1981)
[38] J. Ciborowski, J. Gwizdz, D. Kielczewska, *et al.*, J. Phys. G **8**, 13-32 (1982)
[39] G. Burgun, J. Meyer, E. Pauli, *et al.*, Nucl. Phys. B **8**, 447-459 (1968)
[40] J. R. Carlson, H. F. Davis, D. E. Jauch, N. D. Sossong, R. Ellsworth, Phys. Rev. D **7**, 2533-2537 (1973)
[41] P. M. Dauber, J. P. Berge, J. R. Hubbard, D. W. Merrill, R. A. Muller, Phys. Rev. **179**, 1262-1285 (1969)
[42] M. Haque, *et al.* (Birmingham-Glasgow-London(I.C.)-Oxford-Rutherford), Phys. Rev. **152**, 1148-1161 (1966)
[43] G. W. London, R. R. Rau, N. P. Samios, S. S. Yamamoto, M. Goldberg, S. Lichtman, M. Prime, J. Leitner, Phys. Rev. **143**, 1034-1091 (1966)
[44] T. G. Trippe, P. E. Schlein, Phys. Rev. **158**, 1334-1337 (1967)
[45] W. P. Trower, J. R. Ficenec, R. I. Hulsizer, J. Lathrop, J. N. Snyder, W. P. Swanson, Phys. Rev. **170**, 1207-1222 (1968)
[46] A. Starostin, B. M. K. Nefkens, E. Berger, *et al.* (Crystal Ball Collaboration), Phys. Rev. C **64**, 055205 (2001)
[47] D. F. Baxter, I. D. Buckingham, I. F. Corbett, *et al.*, Nucl. Phys. B **67**, 125-156 (1973)
[48] M. Jones, R. Levi Setti, D. Merrill, R. D. Tripp, Nucl. Phys. B **90**, 349-383 (1975)
[49] A. Berthon, G. Tristram, J. Vrana, *et al.*, Nuovo Cim. A **21**, 146-167 (1974)
[50] M. D. Jones, Nucl. Phys. B **73**, 141-165 (1974)

On K^--nuclear interaction, K^--nuclear quasibound states and K^- atoms

Jaroslava Óbertová[1,*], *Eliahu* Friedman[2], *Jiří* Mareš[3], and *Àngels* Ramos[4]

[1]Faculty of Nuclear Sciences and Physical Engineering, Czech Technical University in Prague, Břehová 7, 11519 Prague, Czech Republic
[2]Racah Institute of Physics, The Hebrew University, 91904 Jerusalem, Israel
[3]Nuclear Physics Institute of the Czech Academy of Sciences, 25068 Řež, Czech Republic
[4]Departament de Física Quàntica i Astrofísica and Institut de Ciències del Cosmos, Universitat de Barcelona, Barcelona, Spain

Abstract. The K^-N potentials derived from SU(3) chiral coupled-channel meson-baryon interaction models fail to describe kaonic atom data. It was shown that K^- single-nucleon potentials have to be supplemented by a phenomenological K^- multi-nucleon optical potential in order to achieve good description of the data. Moreover, only the Prague, Kyoto-Munich and Barcelona models are able to account for the experimental data on K^- atoms and K^- single-nucleon absorption ratios from bubble chamber experiments. A considerable imaginary part coming from the multi-nucleon absorption ruled out existence of narrow K^--nuclear quasibound states in nuclei with $A \geq 6$.
We have developed a microscopic model for the K^-NN absorption in nuclear matter. The absorption was described as a meson-exchange process and the primary K^-N interaction strength was derived from the state-of-the-art chiral models. The medium modifications of the K^-N scattering amplitudes due to the Pauli exclusion principle were taken into account. The model was applied in calculations of kaonic atoms for the first time. The description of the data improved significantly when the two nucleon absorption was considered.

1 K^-N interaction models

The description of low-energy meson-baryon interactions is currently provided by SU(3) chiral coupled-channel meson-baryon interaction models which represent a synergy between the chiral perturbation theory and the coupled channel T-matrix resummation techniques. The chiral models were successfully applied to the description of the K^-N interaction near threshold. The parameters of the models are fitted to K^-p scattering data [1–3], threshold branching ratios [4, 5] and strong-interaction energy shifts and widths in kaonic hydrogen [6]. In Figure 1, there is a comparison of real and imaginary parts of the s-wave free-space K^-p (top panel) and K^-n (bottom panel) scattering amplitudes in the state-of-the-art chiral models. There is a good agreement of the Prague (P) [7], Kyoto-Munich (KM) [8], Murcia (M1 and M2) [9], and Barcelona (BCN) [11] models on the K^-p amplitude at and above threshold since their parameters are fitted to the experimental data in this region. The Bonn

*e-mail: jaroslava.obertova@fjfi.cvut.cz

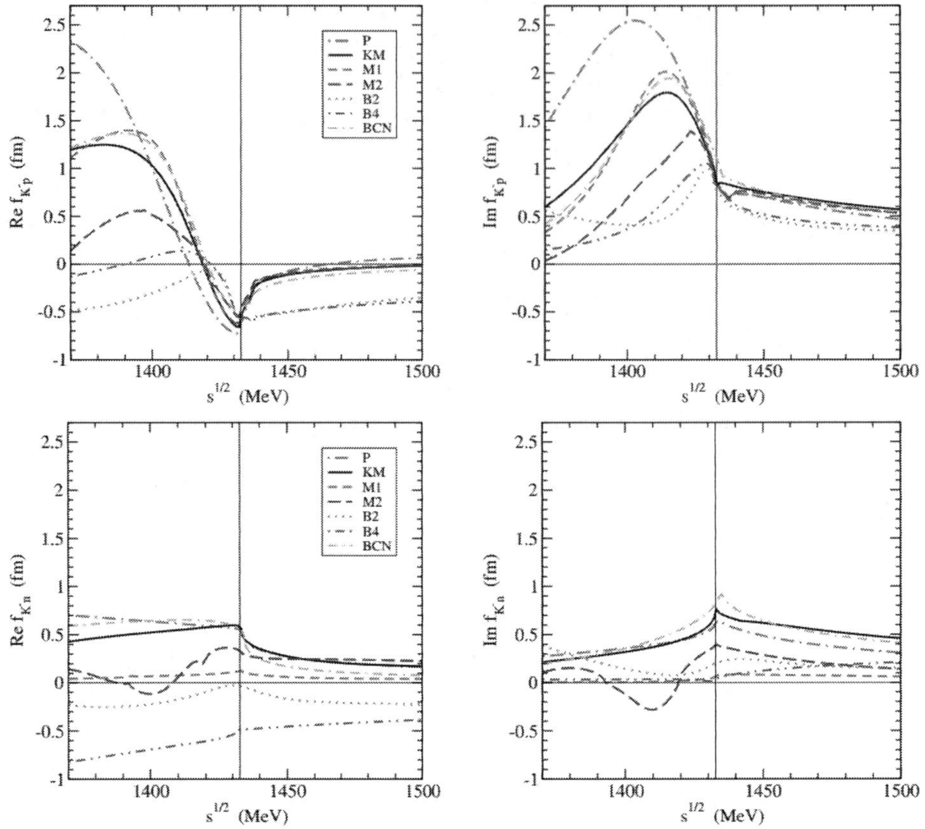

Figure 1. Real (left) and imaginary (right) parts of the *s*-wave free-space K^-p (top panel) and K^-n (bottom panel) amplitude in the state-of-the-art chiral models.

models (B2 and B4) [10] do not agree with the rest of the models above threshold since higher partial waves were included in the fit. Below threshold the models differ significantly from each other. In the case of the K^-n amplitude, the models do not match each other even at and above threshold due to the lack of sufficiently accurate data. As a consequence, various chiral models yield considerably different K^- optical potentials below threshold.

2 Kaonic atoms and multi-nucleon absorption

A valuable source of information about the K^--nuclear interaction below threshold are kaonic atoms. Their study gives us unique information about the K^--nuclear potential. A sensitive test of any K^-N interaction model is its ability to fit the kaonic atoms data. In Table 1, there are values of $\chi^2(65)$ stemming from the comparison of the results for K^- potentials with the data for all considered chiral models [12]. The values of $\chi^2(65)$ are exceedingly large, implying that K^- optical potentials based on K^- single-nucleon chiral amplitudes are generally unable to fit the kaonic atoms data. The reason for this discrepancy lies in the fact that the considered chiral models do not address the K^- interactions with two and more nucleons. The K^- multi-nucleon absorption represents about 20% of all K^- absorptions in the surface region of atomic nuclei and its role increases with density; its consideration is thus crucial for

Table 1. Values of $\chi^2(65)$ resulting from confrontation of selected chiral K^- optical potentials with kaonic atoms data.

model	B2	B4	M1	M2	P	KM
$\chi^2(65)$	1174	2358	2544	3548	2300	1806

a realistic description of the K^--nucleus interaction. Friedman and Gal [12] supplemented the optical potential constructed from chiral K^-N amplitudes by a phenomenological potential describing the K^- multi-nucleon processes:

$$2\mu_{K^-} V^{\rm phen}_{K^-{\rm multiN}} = -4\pi B(\frac{\rho}{\rho_0})^\alpha \rho\,, \tag{1}$$

where μ_{K^-} is the K^--nucleus reduced mass, ρ is the nuclear density distribution, ρ_0 is the saturation density, and complex amplitude B together with parameter α are fitted to the data. The K^- single-nucleon potentials supplemented by a phenomenological density dependent multi-nucleon term provided considerably better description of kaonic atoms. Equally good fits were achieved for all considered chiral models, with values of $\chi^2(65)$ for all models in the range from 105 to 125 [12].

The analysis of Ref. [12] revealed in addition that only three models supplemented by the phenomenological K^- multi-nucleon potential were able to reproduce simultaneously kaonic atoms data and K^- single-nucleon absorption fraction from old bubble chamber experiments, $\mathrm{BR}_{K^-N} = 0.75 \pm 0.05$ [13–15]. Figure 2 shows the calculated values of K^- single-nucleon absorption fraction for lower (solid lines) and upper (dashed lines) atomic states as a function of the proton number Z for considered chiral models. The gray band denotes the experimental value including the error bar. Only the P, KM, and BCN models are compatible with the data. The compatibility of the three models with the data was confirmed later by the measurement of the $K^-n \to \pi\Lambda$ amplitude by the AMADEUS collaboration [16].

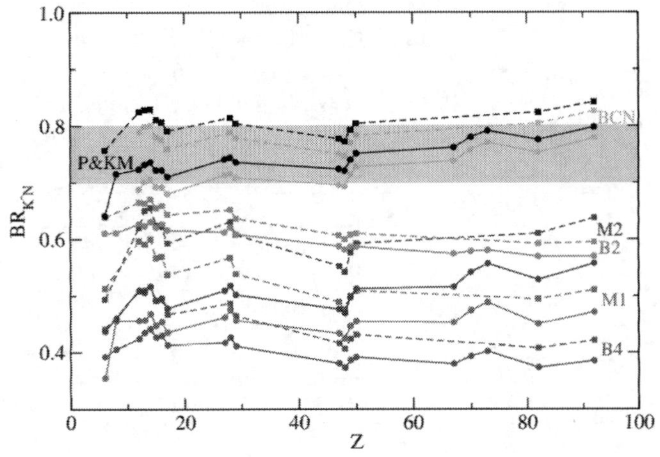

Figure 2. K^- single-nucleon absorption fraction, BR_{K^-N}, calculated for lower (solid lines) and upper (dashed lines) atomic levels using the state-of-the-art chiral models as a function of the proton number Z. The gray band denotes the experimental value including the error bar deduced from bubble chamber experiments [13–15].

3 K^- quasibound states in many-body systems

The attractive nature of K^-N interaction led to conjectures about the existence of K^- bound states in atomic nuclei (see, e.g., [17]). We explored nuclear K^- quasibound states in many-body systems and calculated their binding energies and widths employing the chiral K^- single-nucleon + phen. multi-nucleon optical potential [18, 19]. The K^- single-nucleon potential was derived from the P and KM model amplitudes. In-medium modifications of the amplitudes due to the Pauli correlations were taken into account with the help of the multiple scattering approach by Wass, Rho, and Weise (WRW) [20].

The K^- in a nuclear bound state probes the K^- optical potential around the saturation density ρ_0, whereas experiments with kaonic atoms constrain reliably the K^- potential up to $\sim 50\%$ of ρ_0 [12]. Beyond that limit the shape of the phenomenological multi-nucleon K^- potential in Eq. (1) results from extrapolation of the empirical formula to higher densities. Therefore, we considered two limiting options in our calculations. First, we took the full form of the phenomenological potential as is in Eq. (1) - full density option (FD). Then, we fixed the potential $V^{\rm phen}_{K^-\rm multiN}$ at constant value $V^{\rm phen}_{K^-\rm multiN}(0.5\rho_0)$ for $\rho(r) \geq 0.5\rho_0$ - half density limit (HD). In Table 2, we present $1s$ K^- binding energies B_{K^-} and corresponding widths Γ_{K^-}, calculated in the KM and P models for the two versions of the K^- multi-nucleon potential. For comparison, there are also binding energies and widths calculated only with the underlying K^- single-nucleon potential. When the K^- multi-nucleon processes are included the widths increase considerably while the binding energies change only slightly. For the FD option, the antikaon is not even bound in most of the nuclei due to very strong multi-nucleon absorption.

Table 2. 1s K^- binding energies B_{K^-} and corresponding widths Γ_{K^-} (in MeV) in various nuclei calculated using the single nucleon K^-N amplitudes (denoted KN); plus K^- multi-nucleon amplitude $B(\rho/\rho_0)^\alpha$, where $\alpha = 1$ and 2, for the HD and FD options (see text for details).

KM model			$\alpha = 1$		$\alpha = 2$	
		KN	HD	FD	HD	FD
^{16}O	B_{K^-}	45	34	not	48	not
	Γ_{K^-}	40	109	bound	121	bound
^{40}Ca	B_{K^-}	59	50	not	64	not
	Γ_{K^-}	37	113	bound	126	bound
^{208}Pb	B_{K^-}	78	64	33	80	53
	Γ_{K^-}	38	108	273	122	429
P model			$\alpha = 1$		$\alpha = 2$	
^{16}O	B_{K^-}	64	49	not	63	not
	Γ_{K^-}	25	94	bound	117	bound
^{40}Ca	B_{K^-}	81	67	not	82	not
	Γ_{K^-}	14	95	bound	120	bound
^{208}Pb	B_{K^-}	99	82	36	96	47
	Γ_{K^-}	14	92	302	117	412

4 Microscopic K^-NN absorption model and its application in kaonic atoms calculations

The absorption of K^- on two or more nucleons was found to be essential for a proper description of kaonic atoms data. So far, it was described mainly by a phenomenological optical potential which had to be extrapolated to the saturation density. Sekihara et al. [21]

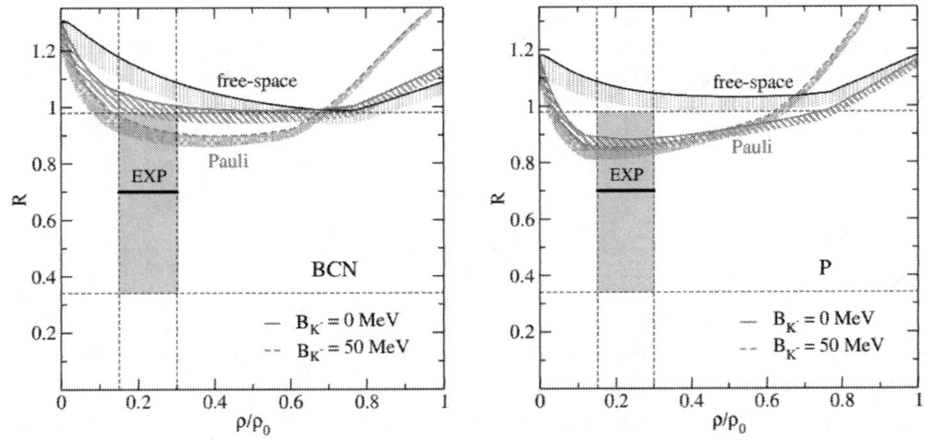

Figure 3. The ratio R as a function of relative density, calculated using the free-space and Pauli blocked amplitudes for $B_{K^-} = 0$ MeV and $B_{K^-} = 50$ MeV. Color bands denote the uncertainty due to different cut-off values $\Lambda_c = 800 - 1200$ MeV. Figure adapted from Ref. [23].

connected the two-nucleon antikaon absorption to chiral meson-baryon interaction models and developed the first microscopic model for K^-NN absorption in nuclear matter. Inspired by the evaluation of an η'-nucleus optical potential including η'-$2N$ absorption [22], we have developed a microscopic model for K^-NN processes in symmetric nuclear matter [23]. The description of K^- two-nucleon absorption was based on the meson-exchange picture. In the model, the scattering amplitudes derived from the Prague and Barcelona models were employed. We took into account the medium modifications of the free-space amplitudes due to the Pauli exclusion principle. We derived the real part of the K^-NN potential as well. Moreover, we derived the K^-N potential within the same approach. We built the total K^-N and K^-NN optical potentials as a sum of potentials coming from all possible absorption channels. For more details regarding the model see Ref. [23].

In order to test the validity of the K^-NN model, we calculated the ratio

$$R = \frac{\mathrm{BR}(K^-pp \to \Lambda p)}{\mathrm{BR}(K^-pp \to \Sigma^0 p)} = 0.7 \pm 0.2(\mathrm{stat.})^{+0.2}_{-0.3}(\mathrm{syst.})\,, \tag{2}$$

measured recently by the AMADEUS collaboration in reactions of low-energy K^- with carbon [24]. Figure 3 shows the ratio R as a function of relative density, calculated using free-space (black) and Pauli blocked (red) amplitudes derived from the BCN (left) and P (right) models. We performed calculations for two values of the kaon binding energy, $B_{K^-} = 0$ MeV (solid lines) and $B_{K^-} = 50$ MeV (dashed line). The AMADEUS experimental value including the error bar is denoted by the gray box and spans over the density area relevant for kaonic carbon. First, we evaluated the ratio using the free-space BCN and P model amplitudes. It lies above 1 and out of the gray box in the relevant density region. When the Pauli blocked amplitudes are taken into account, the value of the ratio drops down within the error bar of the experimental value. This result clearly illustrates the importance of in-medium modifications of the free-space amplitudes. More results of the K^-NN model can be found in Ref. [23].

As the next step, we applied the microscopic K^-NN model in kaonic atoms calculations to test further its validity [25]. We calculated the microscopic $K^-N + K^-NN$ potentials for 23 nuclear targets and confronted them with kaonic atom data. Next, we added a phenomeno-

Table 3. Values of $\chi^2(65)$ resulting from the comparison of the calculations of kaonic atoms using K^-N, $K^-N + K^-NN$, and $K^-N + K^-NN$+phen. multi-nucleon potentials based on the Pauli blocked and WRW modified BCN amplitudes with the experimental data. Values of complex amplitude B and parameter α are presented as well.

	K^-N	$K^-N + K^-NN$	+ phen.	ReB (fm)	ImB (fm)	α
Pauli	825	565	105	-1.97(13)	-0.93(11)	1.4
WRW	2378	1123	116	-0.90(9)	0.72(10)	0.6

logical term $\sim B(\frac{\rho}{\rho_0})^\alpha \rho$ to the $K^-N + K^-NN$ potentials to quantify the effect of K^--3N(4N) processes which are missing in the microscopic model. In the calculations, we employed the BCN and P model amplitudes and considered their in-medium modification due to the Pauli exclusion principle within two approaches: 1) the Pauli blocked amplitudes were evaluated directly in the chiral model (denoted by 'Pauli') 2) we employed the multiple scattering approach WRW [20].

In Table 3, the values of $\chi^2(65)$ resulting from the comparison of predictions of K^-N, $K^-N + K^-NN$, and $K^-N + K^-NN$+phen. potentials with kaonic atoms data are presented. The description of the data improves significantly when the K^-NN absorption is taken into account, the value of $\chi^2(65)$ decreases to about one half, however, it is still considerable. When the additional phenomenological term is added to the $K^-N + K^-NN$ potentials, values of $\chi^2(65)$ further decrease to $\sim$ 100 and become comparable with the best fit K^-N+phenomenological potential, ReB = -1.3 fm, ImB = 1.9 fm, α = 1, $\chi^2(65)$ = 112.3. The values of the complex amplitude B and the parameter α fitted to the data are presented as well. For the Pauli blocked amplitudes, the value of ImB of the additional phenomenological term is negative. The $K^-N + K^-NN$ potentials thus seem to be too absorptive with no space for additional absorption from 3N(4N) processes. The potentials based on the WRW amplitudes yield positive ImB = 0.72 fm, being about half of the value of the amplitude obtained with the best fit K^-N+ phen. multi-nucleon potential, which seems reasonable. The parameter α controls the density dependence of the additional phenomenological term. For the description of 3N(4N) processes, it is expected $\alpha \geq 2$. The fit yields values of α lower than 2 in both cases. The above results indicate some deficiencies in the microscopic K^- potentials which need to be resolved. A possible solution could be a proper inclusion of the hadron self-energies, which are part of the medium modifications, in the K^-NN model as well as in the K^-N chiral amplitudes. They were not considered in the present calculations, however, an extension of the model to involve the hadron self-energies is currently under investigation.

5 Summary

In this contribution, we reviewed the current status of the K^-N and K^--nuclear interactions derived from the SU(3) chiral meson-baryon interaction models. The K^- single-nucleon potentials based on chiral K^-N scattering amplitudes are unable to describe kaonic atoms data on their own. It turned out that an indispensable ingredient of the K^--nuclear interaction is the K^- absorption on two and more nucleons. After supplementing the K^-N chiral potentials by a phenomenological density-dependent multi-nucleon term, the description of the data significantly improves, yielding the value χ^2/d.p.$\leq$ 2 for all state-of-the-art chiral models. Moreover, only the P, KM, and BCN models were found compatible with kaonic atoms data and the value of the K^- single-nucleon absorption fraction deduced from bubble chamber experiments. The K^-N chiral + phen. multi-nucleon potentials based on the P and KM models were applied in the calculations of K^--nuclear quasibound states in many-body systems

($A \geq 6$). The strong multi-nucleon absorption caused the K^- widths to be up to order of magnitude larger than the corresponding binding energies. In some nuclei, depending on the form of the phenomenological multi-nucleon potential applied at saturation density, the K^- was found even unbound.

Next, we developed a microscopic model for the K^-NN absorption in nuclear matter. The model was applied in the kaonic atoms calculations for the first time. The description of the data significantly improved when the microscopic K^-NN potential was added to the K^- single-nucleon potential derived from the BCN chiral amplitudes. However, the values of $\chi^2(65)$ were still sizeable suggesting that some processes of K^--$3N(4N)$ absorption are missing. Therefore, we added a phenomenological term to the $K^-N + K^-NN$ microscopic potentials to quantify the $3N(4N)$ processes. The unexpected density dependence of the additional phenomenological term revealed some deficiencies in the microscopic potentials. Further improvements, i.e. proper inclusion of hadron self-energies in the K^-N chiral amplitudes, of the microscopic $K^-N + K^-NN$ potentials are thus desirable.

Acknowledgements

J.O. and J.M. acknowledge support from the GACR Grant No. 19-19640S. J.O. acknowledge partial support from the project of the European Regional Development Fund No. CZ.02.1.01/0.0/0.0/16 019/0000778. A.R. acknowledges support from the grants CEX2019-00098-M and PID2020-118758GB-100 of the AEI (Spain). This work is also part of a project funded by the European Union's Horizon 2020 research & innovation programme, grant agreement 824093.

References

[1] A. D. Martin, Nucl. Phys. **B 179**, 33 (1981)
[2] J. Ciborowski *et al.*, J. Phys. **G 8**, 13 (1982)
[3] D. Evans *et al.*, J. Phys. **G 9**, 885 (1983)
[4] D. N. Tovee *et al.*, Nucl. Phys. **B 33**, 493 (1971)
[5] R. J. Nowak *et al.*, Nucl. Phys. **B 139**, 61 (1978)
[6] M. Bazzi *et al.* (SIDDHARTA collaboration), Phys. Lett. **B 704**, 113 (2011); Nucl. Phys. **A 881**, 88 (2012)
[7] A. Cieplý, J. Smejkal, Nucl. Phys. **A 881**, 115 (2012)
[8] Y. Ikeda, T. Hyodo, W. Weise, Nucl. Phys. **A 881**, 98 (2012)
[9] Z. H. Guo, J. A. Oller, Phys. Rev. **C 87**, 035202 (2013)
[10] M. Mai, U.-G. Meißner, Nucl. Phys. **A 900**, 51 (2013)
[11] A. Feijoo, V. Magas, À. Ramos, Phys. Rev. **C 99**, 035211 (2019)
[12] E. Friedman, A. Gal, Nucl. Phys. **A 959**, 66 (2017)
[13] H. Davis *et al.*, Nuovo Cimento **53 A**, 313 (1968)
[14] J. W. Moulder *et al.*, Nucl. Phys. **B 35**, 332 (1971)
[15] C. Vander Velde-Wilquet *et al.*, Nuovo Cimento **39 A**, 538 (1977)
[16] K. Piscicchia, S. Wycech, L. Fabbietti *et al.*, Phys. Lett. **B 782**, 339 (2018)
[17] W. Weise, Nucl. Phys. **A 835**, 51 (2010), and references therein
[18] J. Hrtánková, J. Mareš, Phys. Lett. **B 770**, 342 (2017)
[19] J. Hrtánková, J. Mareš, Phys. Rev. **C 96**, 015205 (2017)
[20] T. Wass, M. Rho, W. Weise, Nucl. Phys. **A 617**, 449 (1997)
[21] T. Sekihara *et al.*, Phys. Rev. **C 86**, 065205 (2012)

[22] H. Nagahiro *et al.*, Phys. Lett. **B 709**, 87 (2012)
[23] J. Hrtánková, À. Ramos, Phys. Rev. **C 101**, 035204 (2020)
[24] R. Del Grande *et al.*, Eur. Phys. J. **C 79**, 190 (2019)
[25] J. Óbertová, E. Friedman, J. Mareš, preprint submitted to Phys. Rev. **C**, (2022), arXiv:2208.14946 [nucl-th]

A novel approach to the measurement of the hyperon nucleon/s interaction by AMADEUS

Kristian Piscicchia[1,2*], *Michael* Cargnelli[3], *Raffaele* Del Grande[4,2], *Laura* Fabbietti[5,4], *Johann* Marton[3], *Pawel* Moskal[6], *Àngels* Ramos[7], *Alessandro* Scordo[2], *Diana* Sirghi[2], *Magdalena* Skurzok[2,6], *Oton* Vazquez Doce[2], *Slavomir* Wycech[8], *Johann* Zmeskal[3], *Paolo* Branchini[9], *Eryk* Czerwinski[6], *Veronica* De Leo[10,11], *Erika* De Lucia[2], *Alessandro* Di Cicco[12,9], *Eleonora* Diociaiuti[2], *Paolo* Fermani[2], *Salvatore* Fiore[13,11], *Matteo* Martini[2,14], *Elena* Perez Del Rio[6], *Andrea* Selce[12,9], *Michal* Silarski[6], and *Catalina* Curceanu[2]

[1]Centro Ricerche Enrico Fermi - Museo Storico della Fisica e Centro Studi e Ricerche "Enrico Fermi", 00184 Rome, Italy
[2]INFN - Laboratori Nazionali di Frascati, 00044 Frascati, Italy
[3]Stefan-Meyer-Institut für Subatomare Physik, 1030 Wien, Austria
[4]Physik Department E12, Technische Universität München, 85748 Garching, Germany
[5]Excellence Cluster "Origin and Structure of the Universe", 85748 Garching, Germany
[6]Institute of Physics, Jagiellonian University, 30-348 Cracow, Poland
[7]Facultat de Física, Universitat de Barcelona, 08028 Barcelona, Spain
[8]National Centre for Nuclear Research, 02-093 Warsaw, Poland
[9]INFN Sezione di Roma Tre, Roma, Italy
[10]Dipartimento di Fisica dell'Università "Sapienza", Roma, Italy
[11]INFN Sezione di Roma, Roma, Italy
[12]Dipartimento di Matematica e Fisica dell'Università "Roma Tre", Roma, Italy
[13]ENEA, Department of Fusion and Technology for Nuclear Safety and Security, Frascati (RM), Italy
[14]Dipartimento di Scienze e Tecnologie applicate, Università "Guglielmo Marconi", Roma, Italy

Abstract. The AMADEUS collaboration is investigating the low-energy antikaon interactions with nucleons and nuclei, taking advantage of the low-momentum antikaons beam provided by the DAΦNE collider at LNF-INFN.
In this work a novel technique is outlined for the measurement of the hyperon-nucleon two and three body scattering cross sections. The method consists in producing hyperons by antikaons atomic captures in light nuclear targets, and extrapolating the cross sections from the measurement of the yields of the corresponding elastic final state interactions of the hyperons.
The feasibility of this kind of analyses is shown by comparison of calculated Σ^0 production in ^{4}He by K^- absorption on three nucleons, with a sample of $K^-{}^{12}$C absorption measured by AMADEUS in collaboration with KLOE. The feasibility of a dedicated high statistics measurement is discussed.

1 Introduction

The AMADEUS collaboration is investigating the low-energy antikaon interaction with nucleons and nuclei, exploiting the unique low-momentum beam of kaons ($p_K \sim 127$ MeV/c)

*email: kristian.piscicchia@cref.it

produced by the DAΦNE collider at LNF-INFN [1], to constrain hadronic nuclear physics models in the strangeness -1 sector.

As a first step, the data collected in 2004/2005 by the KLOE collaboration [2] (corresponding to an integrated luminosity of about 1.74 fb^{-1}) consisting in a complex of K^- absorptions in H, ^{4}He, ^{9}Be and ^{12}C, was analyzed, with the aim to measure e.g. K^- multi-nucleon absorption branching fractions and cross sections [3–5], yields and spectral shapes of the non-resonant antikaon-nucleon absorption transition amplitudes below the $\overline{K}N$ threshold [6] and to perform for the first time invariant mass spectroscopic studies with very low momentum (about 100 MeV/c) in-flight K^- captures [7].

A second dedicated measurement was performed in 2012, with a high purity carbon (graphite) target, which was installed between the beam pipe and the KLOE DC inner wall. Aim of this measurement was to collect a reference sample of pure K^- absorption at-rest in ^{12}C. The geometry of the target was optimized to maximize the kaon stopping power (technical details can be found in Ref. [8]). The total reconstructed integrated luminosity amounts to about 37 pb^{-1}. Details of the event selection and particle identification procedure are reported in Ref. [8].

The investigation of the hyperons dynamics in nuclear matter and of the hyperon-nucleon/multi-nucleons interaction is a subject of utmost interest, with profound implications on the features of possible strange phases in compact astrophysical objects, such as Neutron Stars (NSs) (see e.g. Ref. [9] for a recent review). NSs are among the most compact objects in the Universe, thus providing a perfect testbed for the investigation of the matter properties in extreme conditions of density, isospin asymmetry and temperature.

The study of strange baryons behavior in nuclear matter was pursued by exploiting several techniques: via scattering experiments with secondary hyperon beams [10–12] interacting with a hydrogen target, through the analysis of exotic nuclei containing hyperons (the so called hypernuclei, see Refs. [13–15] and references therein) and, more recently, using femtoscopic techniques [16–21].

In this work we describe a novel approach, which could supply the measurement of the hyperon-1N (interaction with one nucleon) and hyperon-2N (interaction with two nucleons) cross sections, also for channels which are not accessible in scattering experiments (as in the case of Σ^0-1/2N), in a hyperon momentum range which is still unexplored. The idea consists in producing hyperons, as a consequence of K^--multi nucleon interaction processes, through K^- induced reactions on light nuclear targets, and extract the hyperon-1/2N cross sections from the measurement of the yields of the elastic Final State Interaction (FSI) processes. The explicative example of Σ^0 production due to K^--3NA (absorption on three nucleons) in ^{4}He is discussed in Sections 2 and 3.

2 Measurement of the Σ^0-N two and three body interactions

In the following Sections the aim of the proposed measurement will be outlined using the example of the Σ^0-N two and three body interaction cross section. The idea consists in exploiting the K^- three nucleon absorption at rest, in a ^{4}He gaseous target, to produce a primary Σ^0d pair, and then to disentangle and measure the absolute yields of the Σ^0 elastic FSI reactions with the deuteron or with the residual neutron to obtain the Σ^0-1/2N interaction cross sections. ^{4}He is the most appropriate nuclear target, since the abovementioned *signal reaction* can not involve fragmentation of the residual nucleus, which reduces in this case to a neutron. Moreover, an exclusive analysis, with the identification of the Σ^0, d and n, would allow to close the kinematics, for a clean discrimination of the involved background processes, and to get rid of the dominant inelastic FSI conversion processes. The signal reaction under study is initiated by the atomic capture of a negatively charged kaon followed by the nuclear K^--3NA capture at-rest:

$$K^{-}\,{}^{4}\text{He} \rightarrow (K^{-}\,ppn) + n \rightarrow \Sigma^{0}\,d + n. \quad (1)$$

The K^--3NA can be followed by two kinds of elastic FSIs:

- $n\,d \rightarrow n\,d$, – to disentangle this effect on the deformation of the final state kinematic distributions we can take advantage of the well known scattering cross sections,
- $\Sigma^0\,n/d \rightarrow \Sigma^0\,n/d$, – from which we aim to extract information on the $\Sigma^0 - 1/2N$ interaction cross section.

The AMADEUS collaboration has a track record experience in the investigation of K^- induced reactions on light nuclear targets (both at rest and in-flight) aimed to study the K^- single- [6, 7] and multi-Nucleon capture processes [3–5]. In particular in Refs. [3–5] the branching ratios and the low-energy cross sections (for a K^- momentum of about 100 MeV/c) were measured for the K^--2/3NA, in the $\Lambda\,p$ and $\Sigma^0 p$ channels. It was shown that the Quasi-Free (QF) K^--multiN absorption (without FSI) can be well disentangled from the same K^--multiN reaction, when this is followed by elastic FSI, namely the re-scattering of the hyperon or the nucleon, which are produced in the primary hadronic interaction, with the residual nucleus. For the generic multi-nucleon capture

$$K^{-}\,AN \rightarrow Y\,(A-1)N, \quad (2)$$

which occurs in the field of the residual nucleus, the hyperon-nucleon(s) pair $(Y\,(A-1))$, which emerges from the primary interaction, always populate the most energetic part of the invariant mass distribution (just below the kinematc threshold) and is characterized by the strongest angular correlation. A clear example is shown in Figure 1 (adapted from Ref. [4]), where the invariant mass (panel a) and angular correlation (panel b) distributions are shown for the $\Lambda\,p$ channel, produced in K^--multiN captures on a ^{12}C target. The elastic FSI degrades such sharp correlation, shifting the corresponding invariant mass to lower values. This is among the major kinematical properties which will be exploited in the analysis we aim to perform.

The proposal for the realization of a dedicated high statistics measurement at the DAΦNE collider, which would fulfill all the requirements of this method, was recently put forward (see Ref. [24])

The carbon target is not ideal for the purpose of our measurement, since many processes can contribute to form the final state under study, such as fragmentation of the residual nucleus and charge exchange reactions which greatly complicate the data interpretation. A dedicated measurement with a high density gaseous ^{4}He target is necessary to pin down the searched signal, and maximizes the probability of K^- nuclear captures at rest, with respect to the in-flight competitor reactions. Nevertheless, to demonstrate the potentiality of the proposed measurement, in Section 3 the calculated distributions of the signal channel are compared to the measured distributions from a sample of $\Sigma^0\,d$ events, reconstructed by using partial statistics from the 2012 data taking campaign. The data correspond to K^- absorption events on a solid ^{12}C target, almost purely at rest.

3 Kinematics of the K^--3NA

In Figure 2 the calculated invariant mass (left) of the $\Sigma^0\,d$ pairs produced in the reaction Eq. (1) is compared with the measured invariant mass (right), of the $\Sigma^0\,d$ events reconstructed by using partial statistics from the 2012 data taking campaign. Calculations are performed

https://doi.org/10.1051/epjconf/202227107004

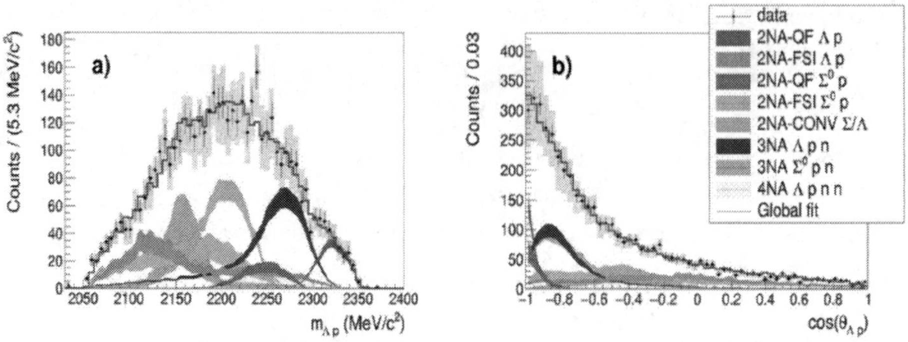

Figure 1. The $\Lambda\, p$ invariant mass (panel a), $\Lambda\, p$ angular correlation (panel b) distributions are shown for K^- absorption on ^{12}C. Black points represent the data, black error bars correspond to the statistical errors, cyan error bars correspond to the systematic errors. The grey line distributions represent the global fit function, the coloured distributions represent the contributing processes according to the colour code reported in the legend and the widths correspond to the statistical error. The figure is adapted from Ref. [4].

according to a phenomelogical model of the K^- capture (see Refs. [4, 6, 22, 23]). The figure evidences that the QF $\Sigma^0\, d$ events, produced in the K^--3NA reaction, populate a narrow invariant mass distribution at the higher edge of the invariant mass spectrum, limited by the kinematical threshold which is located at about 3290 MeV/c^2. The few events in the measured spectrum which exceed the threshold are due the experimental resolution. The invariant mass region lying immediately below the QF distribution is expected to be populated by the K^--3NA reaction, followed by elastic FSI of the primary products of the hadronic interaction, namely the $\Sigma^0\, n/d \rightarrow \Sigma^0\, n/d$ two and three body interactions. The goal of our proposal is to measure the yields of these reactions, from which we aim to extract the corresponding cross sections.

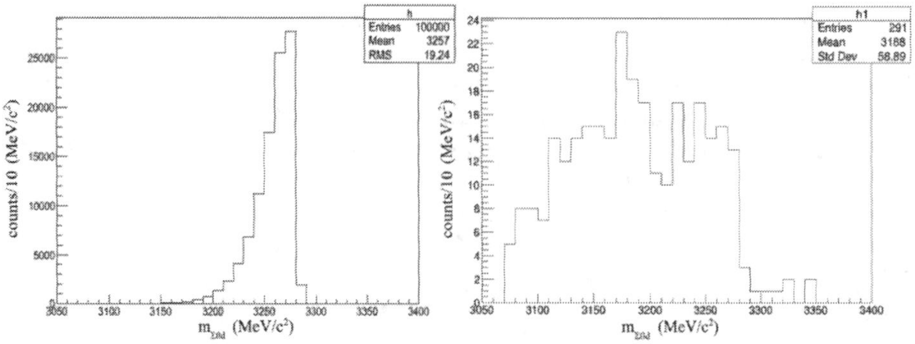

Figure 2. The panel on the left shows the calculated $\Sigma^0\, d$ invariant mass spectrum for the process in Eq. (1). This is compared with the measured invariant mass (right), of the $\Sigma^0\, d$ events reconstructed by using partial statistics from the 2012 data taking campaign.

The calculated Σ^0 momentum (p_{Σ^0}), for the reaction in Eq. (1), is shown in Figure 3 (left) and compared with the corresponding measured distribution (right). The QF K$^-$-3NA process corresponds to the higher part of the hyperon momentum spectrum. Moreover the narrow p_{Σ^0} distribution allows to pin down the cross sections of the $\Sigma^0 - 1/2$N interactions at a Σ^0 momentum of about $p_{\Sigma^0} = 550 \pm 50$ MeV/c.

The main background process is represented by the K$^-$-2NA nuclear capture at-rest:

$$K^- \, {}^4\mathrm{He} \rightarrow (K^- \, pn) + d \rightarrow \Sigma^0 \, n + d, \tag{3}$$

which can, as well, be followed by elastic FSIs of Σ^0, neutron and deuteron. It is straightforward to shown that the $\Sigma^0 \, d$ invariant mass spectrum pertaining to the reaction in Eq. (3) is much broader than the distribution in Figure 2 and centered at about 3180 MeV/c^2. Moreover, the detection of the neutron would allow a clean distinction of the K$^-$-3NA signal from the K$^-$-2NA. Energetic neutrons emerging from the latter process, whose momentum distribution is peaked at about 450 MeV/c, could indeed be well disentangled from the Fermi neutrons produced in a K$^-$-3NA reaction, whose momentum distribution is centered at about 170 MeV/c.

To conclude the discussion of the involved background processes, the copious contribution of K$^-$-1NA reactions should be easily discriminated, as the production of a pion would necessarily lead to uncorrelated and low-energy $\Sigma^0 \, d$ pairs (see also Refs. [3–5]).

It is worth noticing that the measurement proposed in this work depends, as a fundamental ingredient, on a reliable calculation of the elastic final state interaction reactions of the hyperon with the residual nucleus. Presently, the phenomenological model adopted in our calculation does not involve energy dependence of the scattering amplitude for this process, which requires further investigation.

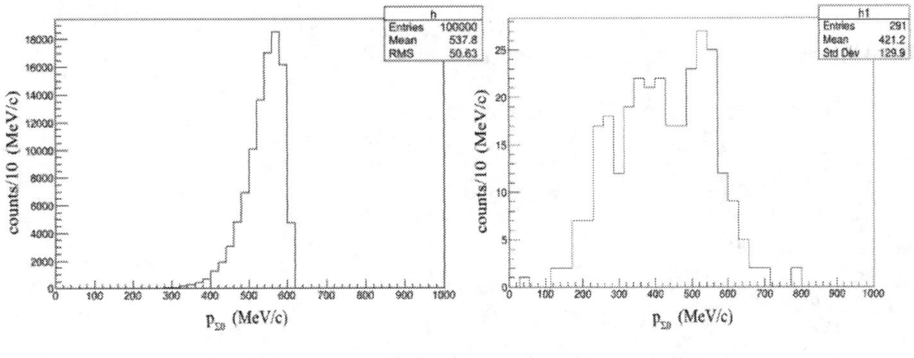

Figure 3. The figure shows comparison among the calculated Σ^0 momentum spectrum (left) for the process in Eq. (1) and the measured distribution (right) for the $\Sigma^0 \, d$ events reconstructed by using partial statistics from the 2012 data taking campaign.

4 Discussion

In this work a novel approach is presented which could provide experimental information on the cross section of hyperon-1/2N interaction (hyperon interaction with one/two nucleons) also for channels which are not accessible in scattering experiments (as in the case of Σ^0-1/2N). The idea consists in producing hyperons in K$^-$ multi-nucleon interaction processes, through K$^-$ atomic captures on light nuclear targets, and then evaluate the cross sections from the corresponding yields of the elastic final state interaction processes.

The feasibility of the proposed measurement is demonstrated by comparing the calculated spectra for the $K^- \, {}^4He \rightarrow \Sigma^0 \, d + n$ atomic absorption with a preliminarily analyzed sample of $K^- \, {}^{12}C \rightarrow \Sigma^0 \, d +$ Residual events collected by AMADEUS in collaboration with KLOE.

High statistics samples could be collected in the optimal experimental conditions, by stopping antikaons in high-density light nuclear gaseous targets, exploiting a future dedicated apparatus working on the DAΦNE collider (see Ref. [24]).

Refined calculations are demanded, for a precise discrimination of the hyperons final state interactions, accounting for the eventual energy dependence of the corresponding scattering amplitudes.

ACKNOWLEDGMENTS

We acknowledge the KLOE/KLOE-2 Collaboration for their support and for having provided us with the data and the tools to perform the analysis presented in this paper. We thank as well the DAΦNE staff for the excellent working conditions and permanent support. Part of this work was supported by Minstero degli Affari Esteri e della Cooperazione Internazionale, EXOTICA project, PO21MO03; the Austrian Science Fund (FWF): [P24756-N20 and P33037-N]; the Croatian Science Foundation under the project IP-2018-01-8570; the EU STRONG-2020 project (Grant Agreement No. 824093), the EU Horizon 2020 project under the MSCA (Grant Agreement 754496); Japan Society for the Promotion of Science JSPS KAKENHI Grant No. JP18H05402; the Polish Ministry of Science and Higher Education grant No. 7150/E-338/M/2018 and the Polish National Agency for Academic Exchange (grant no PPN/BIT/2021/1/00037).

References

[1] A. Gallo *et al.*, Conf. Proc. C **060626**, 604 (2006)
[2] F. Bossi, E. De Lucia, J. Lee-Franzini, S. Miscetti, M. Palutan and KLOE coll., Riv. Nuovo Cim. **31**, 531 (2008)
[3] O. Vazquez Doce et al., Phys. Lett. B **758**, 134 (2016)
[4] R. Del Grande et al., Eur. Phys. J. C **79** no. 3, 190 (2019)
[5] R. Del Grande et al, Phys. Scr. **95**, 084012 (2020)
[6] K. Piscicchia et al., Phys. Lett. B **782**, 339 (2018)
[7] K. Piscicchia et al., EPJ Web of Conferences **166**, 00020 (2018)
[8] C. Curceanu et al., Acta Phys. Polon. B **46**, 203 (2015)
[9] L. Tolos, L. Fabbietti, Prog. Part. Nucl. Phys. **112**, 103770 (2020)
[10] F. Eisele, H. Filthuth, W. Foehlisch, V. Hepp, G. Zech, Phys. Lett. B**37**, 204 (1971)
[11] G. Alexander, et al., Phys. Rev. **173**, 1452 (1968)
[12] B. Sechi-Zorn, B. Kehoe, J. Twitty, R. A. Burnstein, Phys. Rev. **175**, 1735 (1968)
[13] A. Gal, E. V. Hungerford, D. J. Millener, Rev. Mod. Phys. **88**, 035004 (2016)
[14] H. Tamura, et al., Nucl. Phys. A**914**, 99 (2013)
[15] A. Feliciello, T. Nagae, Rept. Prog. Phys. **78**, 096301 (2015)
[16] S. Acharya, et al., Phys. Rev. Lett. **123**, 112002 (2019)
[17] D. L. Mihaylov, V. Mantovani Sarti, O. W. Arnold, L. Fabbietti, B. Hohlweger, A. M. Mathis, Eur. Phys. J. C**78**, 394 (2018)
[18] S. Acharya, et al., Phys. Rev. C**99**, 024001 (2019)

[19] J. Adamczewski-Musch, et al., Phys. Rev. C**94**, 025201 (2016)
[20] S. Acharya, et al., PLB **805**, 135419 (2020)
[21] S. Acharya, et al., Phys. Lett. B**797**, 134822 (2019)
[22] R. Del Grande, K. Piscicchia, S. Wycech, Acta Phys. Polon. B **48**, 1881 (2017)
[23] K. Piscicchia, S. Wycech, C. Curceanu, Nucl. Phys. A **954**, 75 (2016)
[24] C. Curceanu, et al., "Fundamental physics at the strangeness frontier at DAΦNE. Outline of a proposal for future measurements," [arXiv:2104.06076 [nucl-ex]]

Decay of the $\Lambda(1405)$ hyperon to $\Sigma^0\pi^0$ measured at GlueX

Nilanga Wickramaarachchi[1,*], *Reinhard A.* Schumacher[2], and *Grzegorz* Kalicy[1] for the GlueX Collaboration

[1]The Catholic University of America, Washington, D.C. 20064, USA
[2]Carnegie Mellon University, Pittsburgh, PA 15213, USA

Abstract. Among the light baryons, the $J^\pi = \frac{1}{2}^-$ $\Lambda(1405)$ hyperon is an important special case as it sits just below the $N\bar{K}$ threshold and decays almost exclusively to $\Sigma\pi$. Some long-standing hypotheses for $\Lambda(1405)$ are that it could be a $N\bar{K}$ bound state or a $\Sigma\pi$ continuum resonance. It may also be considered as a simple quark-model resonance, the P-wave companion of the $\Lambda(1520)$. In recent years chiral unitary models have suggested that there are two isospin zero poles present in this mass region, and that the "line shape" of the $\Lambda(1405)$ depends to what extent each of the two poles are stimulated in a given reaction. Below the $N\bar{K}$ threshold, the $\Lambda(1405)$ decays to the three $\Sigma\pi$ charge combinations. The $\Sigma^0\pi^0$ mode is purely $I = 0$ and it is uncontaminated by complications arising from $I = 1$ scattering processes contributing to the reaction mechanism in the $\Sigma^+\pi^-$ and $\Sigma^-\pi^+$ decays. It is also not affected from production and decay of the nearby $\Sigma^0(1385)$ hyperon. The GlueX experiment at Jefferson Lab has been used to study the $\Lambda(1405) \to \Sigma^0\pi^0$ decay mode. We focus on the preliminary results of $d\sigma/dM_{\Sigma^0\pi^0}$ and fits to the line shape of the $\Lambda(1405)$ region in the $-(t - t_{min})$ range 0 - 1.5 GeV^2 from analyzing the reaction $\gamma p \to K^+\Lambda^*$ using the data collected during the first phase of the GlueX program.

1 Introduction

The $\Lambda(1405)$ is a hyperon with $J^\pi = \frac{1}{2}^-$ and located just below the $N\bar{K}$ threshold. Observation of this state dates back to 1961 where it was produced by K^-p interactions using a 1.15 GeV/c momentum K^- beam in bubble chamber experiments [1]. Because of its proximity to the $N\bar{K}$ threshold, it was predicted on the basis of hadronic scattering data to be a lightly bound "molecule" of these particles [2]. Since then the $\Lambda(1405)$ has been widely discussed to understand it's nature. In the quark model the $\Lambda(1405)$ can be considered as the spin-orbit partner of $\Lambda(1520)$. One of the interesting features of the $\Lambda(1405)$ is its invariant mass spectrum, also called its "line shape". The line shape in a given production reaction may give clues to the internal structure of the state. It has been measured by experiments including CLAS [3], ANKE [4] and BGO-OD [5], which show distortions from a simple resonance form. The cause for this behavior is not fully understood, but new measurements at GlueX may help to clarify the picture.

*e-mail: wickramaarachchi@cua.edu

Below the $N\bar{K}$ threshold, the $\Lambda(1405)$ decays 100% into $\Sigma\pi$, with each mode $\Sigma^+\pi^-$, $\Sigma^-\pi^+$ and $\Sigma^0\pi^0$ expected to have close to 1/3 branching fraction. Above the $N\bar{K}$ threshold, the $\Lambda(1405)$ is assumed to couple strongly to this channel. Various chiral unitary models suggest that the $\Lambda(1405)$ is a compound state consisting of two $I = 0$ poles; for a recent discussion, see Ref. [6]. The line shape can depend on how each of the two poles are contributing to the reaction. In fact, the most recent PDG has added a $\Lambda(1380)$ as a two-star resonance to account for this two-pole picture of $\Lambda(1405)$ [7]. Among the decay modes of the $\Lambda(1405)$, $\Sigma^0\pi^0$ is a special case since it is purely $I = 0$. This is shown in Eq. 1-3 where the differential cross section for each mode is written in terms of contributions from different isospin amplitudes [8].

$$\frac{d\sigma(\pi^+\Sigma^-)}{dM_I} \propto \frac{1}{3}|T^{(0)}|^2 + \frac{1}{2}|T^{(1)}|^2 + \frac{2}{\sqrt{6}}\mathrm{Re}(T^{(0)}T^{(1)*}) \tag{1}$$

$$\frac{d\sigma(\pi^-\Sigma^+)}{dM_I} \propto \frac{1}{3}|T^{(0)}|^2 + \frac{1}{2}|T^{(1)}|^2 - \frac{2}{\sqrt{6}}\mathrm{Re}(T^{(0)}T^{(1)*}) \tag{2}$$

$$\frac{d\sigma(\pi^0\Sigma^0)}{dM_I} \propto \frac{1}{3}|T^{(0)}|^2 \tag{3}$$

Unlike the two charged modes there is no contamination from the $I = 1$ $\Sigma(1385)$ to the $\Sigma^0\pi^0$ mode. So reconstructing $\Lambda(1405)$ in the $\Sigma^0\pi^0$ mode is very useful to study its line shape. Studying the $\Lambda(1405)$ line shape would provide more information on how the $\Sigma\pi$ and $N\bar{K}$ channels contribute to its production.

2 Experimental procedures

2.1 GlueX detector

The GlueX detector is located in Hall D at Jefferson Lab, Newport News, Virginia, USA. It uses an 11.6 GeV electron beam provided by the Continuous Electron Beam Accelerator Facility (CEBAF). The electron beam is incident on a thin diamond radiator and produces bremsstrahlung photons. The scattered electron is measured and used to tag the energy of the beam photon in the energy range 3.0-11.6 GeV. A fraction of the photons which are ~9 GeV are linearly polarized and produced by coherent bremsstrahlung. While this is important for the main goals of the GlueX experiment, particularly the search for hybrid mesons, the present study does not exploit the linearly polarized beam. The tagged photon beam is incident on a 30 cm long liquid hydrogen target within a superconducting solenoid magnet that provides a 2 T magnetic field. The GlueX detector includes forward and cylindrical drift chambers to track charged particles. Forward and cylindrical electromagnetic calorimeters are used to detect neutral showers and to provide triggering information. Precision timing information from the forward time-of-flight system and the barrel calorimeter is used for particle identification. GlueX detector provides nearly 4π angular coverage and allows the exclusive reconstruction of many reactions. Complete information on the GlueX detector can be found in Ref. [9]. Phase I of the GlueX experiment ran during 2017-2018. The total luminosity for the photon beam energy range 6.5-11.6 GeV is 423 pb^{-1}.

2.2 Analysis method

The $\Lambda(1405) \to \Sigma^0\pi^0$ decay was reconstructed using the exclusive reaction $\gamma p \to K^+\Sigma^0\pi^0$. The decay chain is $\Sigma^0 \to \Lambda\gamma$, $\pi^0 \to \gamma\gamma$ and $\Lambda \to \pi^- p$ as shown in Fig. 1. The final state for this reaction contains two positively charged tracks (K^+ and proton), one negatively charged track (π^-) and three neutral showers (3γ). A kinematic fit was applied conserving four-momentum of the reaction and with a vertex constraint to allow for displaced vertices of the hyperons. The π^0 and Σ^0 masses were constrained during the kinematic fit to improve mass resolution for $\Sigma^0\pi^0$. The full beam energy range $E_{beam} > 6.5$ GeV was used for the analysis in order to apply the measured photon flux to produce cross section $d\sigma/dM_{\Sigma^0\pi^0}$. The Mandelstam t was calculated using four-momenta of the photon beam and the K^+, $t = (p_{beam} - p_{K^+})^2$. Events were selected in the range $0 < -(t - t_{min}) < 1.5$ GeV^2 focusing on the t-channel production of $\Lambda(1405)$.

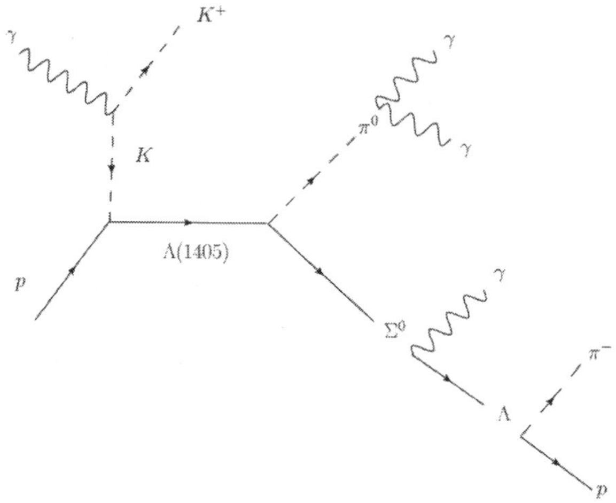

Figure 1: The decay chain of the reaction $\gamma p \to K^+\Lambda(1405)$.

Figure 2(a) shows the invariant mass of $\Sigma^0\pi^0$ after applying all event selection criteria. Clear peaks of $\Lambda(1405)$ and $\Lambda(1520)$ are seen. There are 13351 ± 139 counts in the $\Lambda(1405)$ mass region ($M_{\Sigma^0\pi^0} < 1.47$ GeV). A sharp drop of yield for $\Lambda(1405)$ is seen at the $N\bar{K}$ threshold. This indicates the $\Sigma^0\pi^0$ mass resolution is good to resolve such threshold effects. From detector simulation it was found that the $\Sigma^0\pi^0$ mass resolution is ~5 MeV in the $\Lambda(1405)$ mass region.

2.3 Results

In order to extract differential cross section $d\sigma/dM_{\Sigma^0\pi^0}$, the acceptance in the $\Sigma^0\pi^0$ mass range from threshold up to 1.7 GeV was found using simulated events which were analyzed in the same way as data. Figure 2(b) shows the distribution for $d\sigma/dM_{\Sigma^0\pi^0}$ for $0 < -(t - t_{min}) < 1.5$ GeV^2. The uncertainties shown are statistical. The absolute normalization of this measurement has not been finalized so the differential cross section is given without a numerical scale on the vertical axis. Therefore we report here on the line shape of the mass spectrum but not the absolute scale of the cross section. It can be seen that the $\Lambda(1405)$ line shape appears to deviate from a simple Breit-Wigner form while the $\Lambda(1520)$ could be fit with a Breit-Wigner. To test this hypothesis, we consider two fits to the $\Lambda(1405)$ line

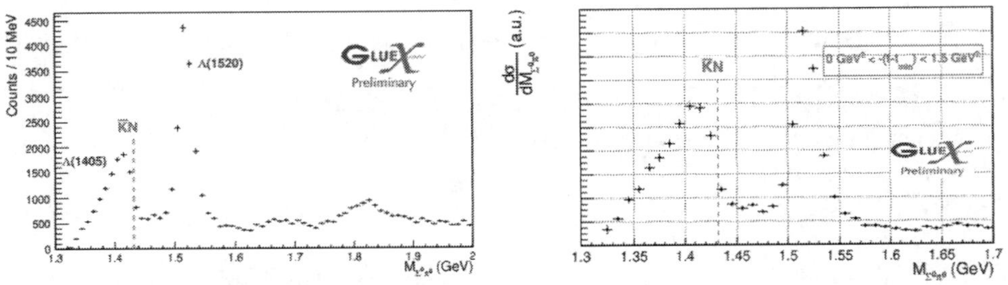

(a) (b)

Figure 2: Invariant mass spectrum of $\Sigma^0\pi^0$ (a) for selected $\gamma p \rightarrow K^+\Sigma^0\pi^0$ events and (b) the differential cross section $d\sigma/dM_{\Sigma^0\pi^0}$ in the lower-mass region shown without a numerical scale since the absolute normalization has not been finalized. Red vertical dashed lines indicate the $N\bar{K}$ threshold.

shape considering the $\Lambda(1405)$ to be described by a single Flatté amplitude or by two such amplitudes (compound coherent $\Lambda(1405)$'s) as shown in Fig. 3. Table 1 shows a summary of the different fitting procedures and results. The background parameterization takes into account the $Y^*(1670)$ contribution in addition to the linear background under $\Lambda(1405)$ and $\Lambda(1520)$ peaks. The two Flatté amplitudes used in the $\Lambda(1405)$ two-state hypothesis were assumed to have a common Flatté factor. Based on the χ^2/d.o.f. of the fits the two-state hypothesis for $\Lambda(1405)$ is preferred to the one-state hypothesis. The centroids of two states are around $\approx$1387 MeV and $\approx$1409 MeV.

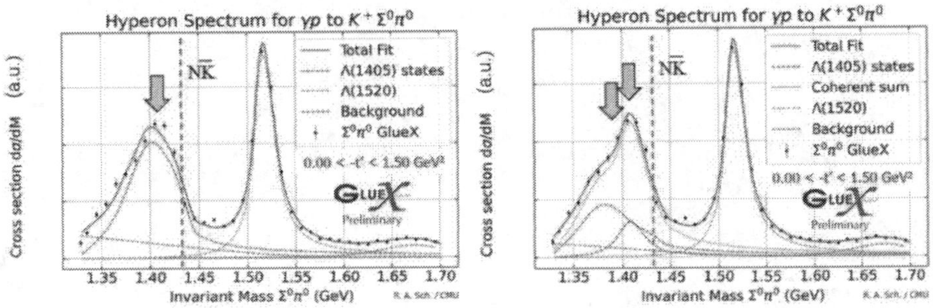

Figure 3: Comparison of two hypotheses for the $\Lambda(1405)$ line shape, differing by having either one or two coherent Flatté amplitudes in the fits. Green arrows indicate the centroids.

In this simple model approach it is expected that unitarity is not preserved. A more proper K-matrix approach to the problem is under development. It enforces unitarity and analyticity of the resulting T-matrix and will result in proper identification of the T-matrix poles of the compound-state $\Lambda(1405)$ region.

Differential cross sections $d\sigma/dM_{\Sigma^0\pi^0}$ were found for 3 different $-(t - t_{min})$ bins 0-0.35, 0.35-0.6 and 0.6-1.5 GeV^2. Figure 4 shows these distributions fit with the two-$\Lambda(1405)$ states hypothesis. Each of the fits seem to describe the data well and reproduce the drop of cross section at $N\bar{K}$ threshold. A notable feature is that the $\Lambda(1405)$ line shape seems to depend on

Table 1: Summary of two fit hypotheses used to fit $d\sigma/dM_{\Sigma^0\pi^0}$

One Λ(1405) state	Two Λ(1405) states
Parameterized with one Flatté amplitude	Two Flatté amplitudes
Incoherent Λ(1520)	Incoherent Λ(1520)
Incoherent background	Incoherent background
χ^2/d.o.f. of the fit = 5.1	χ^2/d.o.f. of the fit = 3.5
Centroid at 1407 MeV	Two centroids at ≈1387 MeV and ≈1409 MeV

$-(t-t_{min})$. The relative intensities of two Λ(1405) states also seem to change with $-(t-t_{min})$. This possibly means that the contributions of the $N\bar{K}$ and $\Sigma\pi$ poles to Λ(1405) production depend on t. Since the Λ(1405) couples to both $\Sigma\pi$ and $N\bar{K}$ channels, a full description of the state should include both of these final states. GlueX has data for both $\gamma p \to K^+\Sigma^0\pi^0$ and $\gamma p \to K^+K^-p$ channels. A coupled-channel fit approach is in progress with using additional measurements from K^-p data.

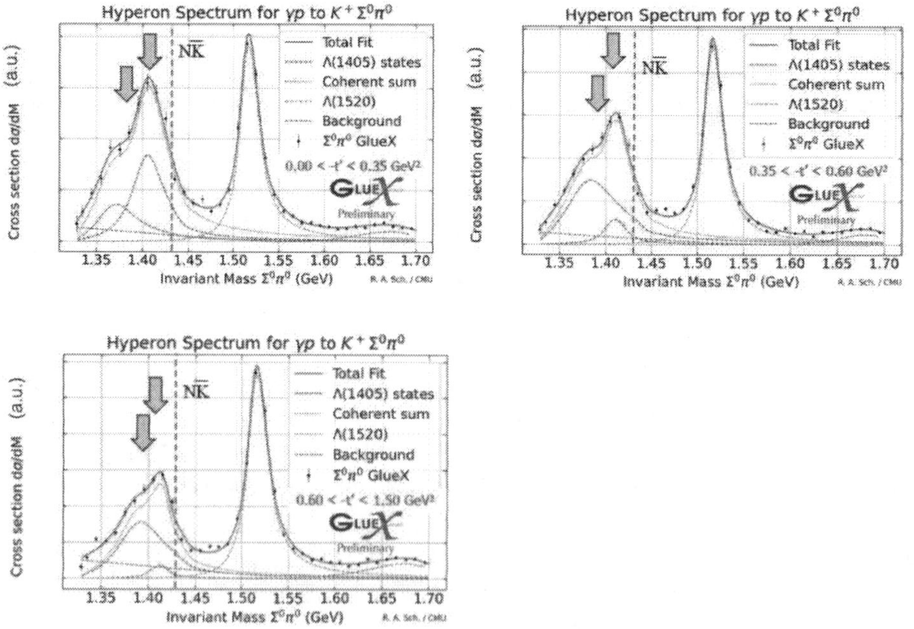

Figure 4: The differential cross section $d\sigma/dM_{\Sigma^0\pi^0}$ in three bins of $-(t-t_{min})$ shown without a numerical scale since the absolute normalization has not been finalized. Red vertical dashed lines indicate the $N\bar{K}$ threshold. Green arrows indicate centroid positions.

3 Conclusion

The GlueX detector is well suited for exclusive reconstruction of $\Lambda(1405) \rightarrow \Sigma^0\pi^0$. GlueX-I data provide the highest statistics obtained for this decay mode with ~5 MeV resolution for $\Sigma^0\pi^0$ mass in the $\Lambda(1405)$ region. The extracted $\Lambda(1405)$ line shape clearly shows deviation from a single Flatté-type Breit-Wigner amplitude. Our preliminary fits to the line shape favor a picture with two coherent states for the $\Lambda(1405)$ region, and support previous theoretical and experimental evidence suggesting the $\Lambda(1405)$ is a composite baryon state. The differential cross sections $d\sigma/dM_{\Sigma^0\pi^0}$ indicate a possible t-dependence of the $\Lambda(1405)$ line shape. More elaborate fits using a K-matrix ansatz and data from $N\bar{K}$ final states are ongoing.

4 Acknowledgements

We would like to acknowledge the outstanding efforts of the staff of the Accelerator and the Physics Divisions at Jefferson Lab that made the experiment possible. This work was supported in part by the U.S. Department of Energy, the U.S. National Science Foundation, the German Research Foundation, Forschungszentrum Jülich GmbH, GSI Helmholtzzentrum für Schwerionenforschung GmbH, the Natural Sciences and Engineering Research Council of Canada, the Russian Foundation for Basic Research, the UK Science and Technology Facilities Council, the Chilean Comisión Nacional de Investigación Científica y Tecnológica, the National Natural Science Foundation of China and the China Scholarship Council. This material is based upon work supported by the U.S. Department of Energy, Office of Science, Office of Nuclear Physics under contract DE-AC05-06OR23177.

References

[1] M. Alston *et al.*, Phys. Rev. Lett. **6**, 698 (1961)

[2] R. H. Dalitz and S. F. Tuan, Ann. Phys. 8, 100 (1959); Phys. Rev. Lett. 2, 425 (1959); Ann. Phys. 10, 307 (1960)

[3] K. Moriya *et al.* [CLAS Collaboration], Phys. Rev. **C 87**, 035206 (2013)

[4] I. Zychor *et al.*, Phys. Lett. **B660**, 167-171 (2008)

[5] G. Scheluchin *et al.*, Phys. Lett. **B833**, 137375 (2022)

[6] M. Mai, U.-G. Meissner, Eur. Phys. J. **A51**, 30 (2015)

[7] R.L. Workman *et al.*, [Particle Data Group], Prog. Theor. Exp. Phys. 2022, 083C01 (2022); See in particular V. D. Burkert *et al.* "82. Λ and Σ Resonances", and T. Hyodo *et al.* "83. Pole Structure of the Λ(1405) Region".

[8] J. C. Nacher, E. Oset, H. Toki, A. Ramos, Phys. Lett. **B455**, 55 (1999)

[9] S. Adhikari *et al.* [GlueX Collaboration], Nucl. Instrum. Meth. **A987**, 164807 (2021)

EPJ Web of Conferences **271**, 07006 (2022) https://doi.org/10.1051/epjconf/202227107006
HYP2022

Triangle singularity in the $K^- d \to p\Sigma^-$ reaction and its relevance to find the $\bar{K}N$ amplitude below threshold

Eulogio Oset[1,*], *Raquel* Molina[1,**], *Lian Rong* Dai[2,***], and *Albert* Feijoo[1,3,****]

[1]Departamento de Física Teórica and IFIC, Centro Mixto Universidad de Valencia-CSIC, Institutos de Investigación de Paterna, Aptdo. 22085, 46071 Valencia, Spain
[2]Nuclear Physics Institute, 25068 Rez, Czech Republic
[3]School of Science, Huzhou University, Huzhou 313000, Zhejiang, China

Abstract. We study for the first time the $p\Sigma^- \to K^- d$ and its time reversal reaction close to threshold and show that they are driven by a triangle mechanism, with the $\Lambda(1405)$, a proton and a neutron as intermediate states, which develops a triangle singularity close to the $\bar{K}d$ threshold. As the main result, we find that the cross section, well within measurable range, is very sensitive to different models that, while reproducing $\bar{K}N$ observables above threshold, provide different extrapolations of the $\bar{K}N$ amplitudes below threshold. The observables of this reaction will provide new constraints on the theoretical models, leading to more reliable extrapolations of the $\bar{K}N$ amplitudes below threshold and to more accurate predictions of the $\Lambda(1405)$ state of lower mass.

1 Introduction

The interaction of kaons with nucleons and nuclei has attracted much attention in the last decades and continues to draw attention. While the advent of chiral unitary theory brought much light into the issue [1–3], the present situation is such that different models agree above the $\bar{K}N$ threshold but give rise to very different $\bar{K}N$ amplitudes below threshold [4]. In this talk we present results from [5] where we suggest the use of the $p\Sigma^- \to K^- d$ ($K^- d \to p\Sigma^-$) reaction to pin down information from the $\bar{K}N$ amplitude below threshold. This reaction proceeds via the mechanism shown in figure 1.

In complete analogy with the $pp \to \pi^+ d$ reaction studied in [6–10], we study the triangle mechanism shown in figure 1, where a $\Lambda(1405)$ resonance substitutes a $\Delta(1232)$ in the loop. The interesting thing observed in Ref. [10] is that the triangle mechanism for the $pp \to \pi^+ d$ reaction was driven by a triangle singularity (TS), which made its strength very large and was responsible for some peculiar characteristics which were in agreement with experiment [11]. The diagrams shown in figure 1 also develop a triangle singularity which makes it stand over other possible mechanisms. TS were studied by Landau [12] and it was shown that a triangle mechanism of the type of figure 1 develops a singularity in the amplitude (in the limit of zero

*e-mail: Eulogio.Oset@ific.uv.es
**e-mail: Raquel.Molina@ific.uv.es
***e-mail: dailianrong@zjhu.edu.cn
****e-mail: edfeijoo@ific.uv.es

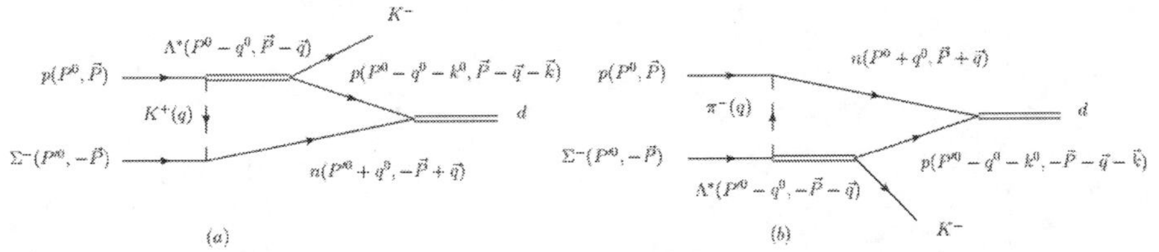

Figure 1. Feynman diagrams for the $p\Sigma^- \to K^-d$ reaction. The momenta of the lines are shown in brackets. Here, Λ^* stands for $\Lambda(1405)$.

width of the resonance) if the $\Lambda(1405)$ and the two nucleons are placed simultaneously on shell and the $\Lambda(1405)$ goes in the same direction as the K^-, together with the condition that the nucleon produced from the $\Lambda(1405)$ decay goes faster than the other nucleon and can catch up to fuse into the deuteron. The possibility that this process occurs at the classical level to give the singularity is the content of Coleman-Norton theorem [13].

2 Discussion and results

In the present case this reaction offers information on the $\bar{K}N$ amplitude below threshold, as we show below. Details on the calculations are shown in [5] and here we just present results. In figure 2 we show the contributions of the different spin transitions to the total cross section of the $K^-d \to p\Sigma^-$ reaction (σ),. We see that the most important is $\uparrow\uparrow\to\uparrow\uparrow$, or $\downarrow\downarrow\to\downarrow\downarrow$ which has the same strength, where the arrows indicate the spin of the nucleons in the deuteron for the initial K^-d state, or the p or Σ in the final state.

In figure 3 we show the contribution of the higher pole of the $\Lambda(1405)$ to the $K^-d \to p\Sigma^-$ total cross section, around 1420 MeV, and the lower pole, around 1380 MeV, [14] for both diagrams of figure 1. We observe two features, the higher pole always gives the largest contribution, and the diagram (b) of figure 1 is the dominant one, because it relies upon pion exchange, contrary to diagram (a) that relies on kaon exchange. We also observe that there is a destructive interference between the two Λ^* states.

Finally, in order to show the sensitivity of the reaction cross section to different models that have the same $\bar{K}N$ amplitude above threshold, but different below threshold, we show the results with four models , namely

Oset-Ramos [1]: This first model uses a Weinberg-Tomozawa (WT) contribution as driving term in the interaction kernel. The authors took into account, for the first time, the full $S = -1$ meson-baryon basis for the regularization of the loop integral in coupled channels.

Roca-Oset [15]: This model takes as building block the previous one, yet reducing the basis to $\pi\Lambda, \pi\Sigma, \bar{K}N$ channels. It has special interest for the present study because, apart from the ordinary $K^-p \to \pi\Lambda, \pi\Sigma, \bar{K}N$ total cross sections and threshold observables, the authors incorporated the CLAS data for the $\Lambda(1405)$ photoproduction [16] in the fits to constrain the model parameters.

Cieplý-Smejkal: The third one is the model called NLO30 in Ref. [17]. This is a model based on a chirally motivated potential, written in a separable form, whose central piece is derived from the Lagrangian up to next-to-leading order (NLO). The authors took into

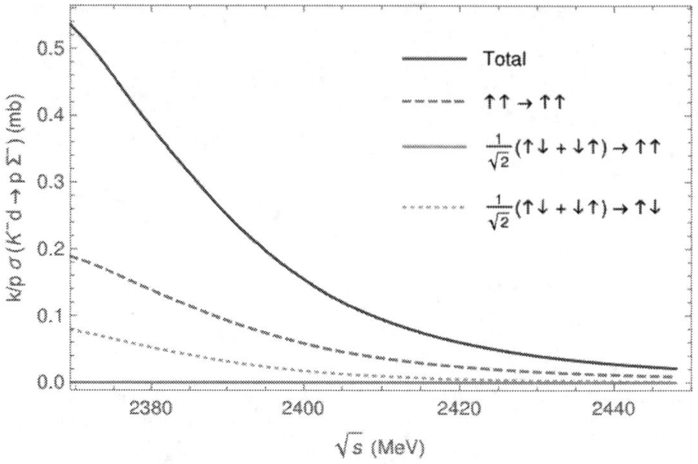

Figure 2. Contribution from several spin transitions in $\frac{k}{p}\sigma$. Here, $\sqrt{s}$ and p are the total CM energy and the momentum modulus, respectively, of the incoming $p\Sigma^-$ system, while k represents the momentum modulus of the outgoing K^-.

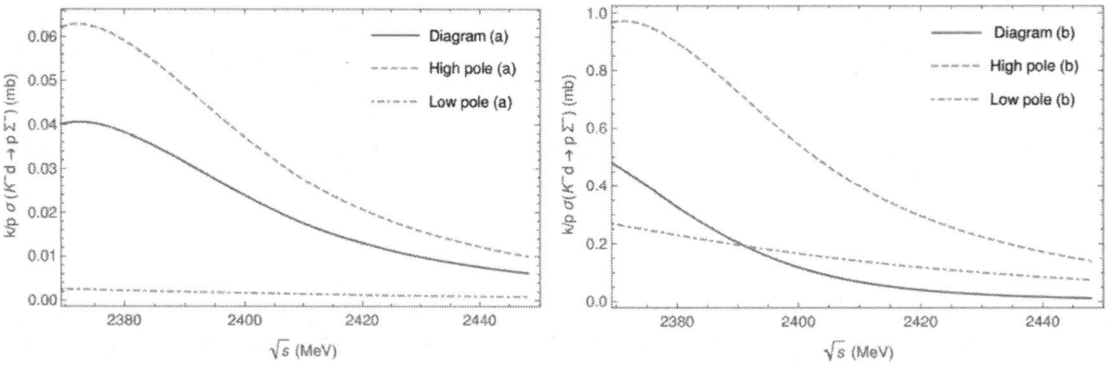

Figure 3. Contributions of the high and low mass poles to the diagrams (a) and (b) in $\frac{k}{p}\sigma$.

account in the fitting procedure the very precise measurements of the shift and width of the 1S state in kaonic hydrogen carried out by SIDDHARTA Collaboration [18].

Feijoo-Magas-Ramos [19] This model corresponds to the fit called WT+Born+NLO carried out in [19]. It was constructed by adding the interaction kernels derived from the Lagrangian up to NLO, and including additional experimental data at higher energies in the fits. This model is the natural extension of Oset-Ramos.

The results obtained with the four different models are shown in figure 4. As we can see, the cross sections can differ by a factor of two sometimes and this is due to the different amplitudes below threshold since all these amplitudes have about the same amplitude above threshold. This clearly tells us that we can use this reaction to discriminate among different models, or what is the same to pin down the $\bar{K}N$ amplitude below threshold, which is the object of present discussion and is essential to make good predictions of $\bar{K}$ bound states in nuclei, another subject of much interest in the field [20–22].

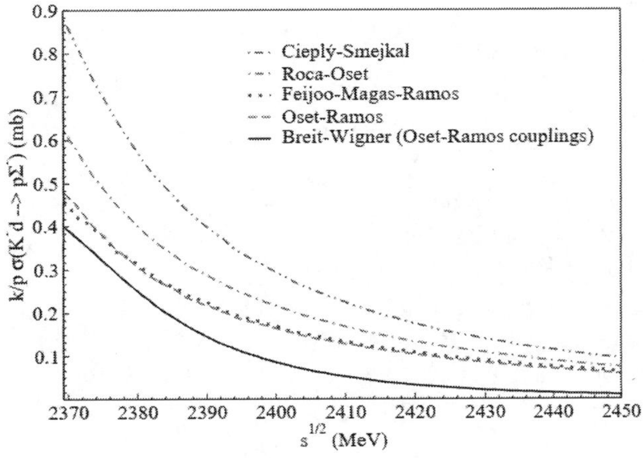

Figure 4. $K^{-}d \to p\Sigma^{-}$ cross sections ($\frac{k}{p}\sigma$) for the considered models. More details in the text.

Acknowledgments

We would like to thank Aleš Cieplý for providing us with the $\bar{K}N$ scattering amplitudes calculated from NLO30 model as well as the corresponding $\Lambda(1405)$ pole positions and their couplings. R. M. acknowledges support from the CIDEGENT program with Ref. CIDEGENT/2019/015 and from the spanish national grant PID2019-106080GB-C21. This work is also partly supported by the Spanish Ministerio de Economia y Competitividad and European FEDER funds under Contracts No. FIS2017-84038-C2-1-P B and No. FIS2017-84038-C2-2-P B. This project has also received funding from the European Union's Horizon 2020 programme No. 824093 for the STRONG-2020 project and by Generalitat Valenciana under PROMETEO/2020/023 contract . The work of A. F. was partially supported by the Czech Science Foundation, GAČR Grant No. 19-19640S, and the Generalitat Valenciana and European Social Fund APOSTD- 2021-112. L. R. D. acknowledges the support from the National Natural Science Foundation of China (Grant Nos. 11975009, 12175066, 12147219).

References

[1] E. Oset and A. Ramos, Nucl. Phys. A **635**, 99 (1998)
[2] J. A. Oller and U. G. Meissner, Phys. Lett. B **500**, 263 (2001)
[3] A. Cieplý, J. Hrtánková, J. Mareš, E. Friedman, A. Gal, and À. Ramos, AIP Conf. Proc. **2249**, 030014 (2020)
[4] C. Curceanu, C. Guaraldo, D. Sirghi, A. Amirkhani, A. Baniahmad, M. Bazzi, G. Bellotti, D. Bosnar, M. Bragadireanu, and M. Cargnelli, *et al.* Symmetry **12**, 547 (2020)
[5] A. Feijoo, R. Molina, L. R. Dai, and E. Oset, arXiv:2105.09654 [nucl-th]
[6] D. O. Riska, M. Brack and W. Weise, Phys. Lett. B **61**, 41 (1976)
[7] A. M. Green and J. A. Niskanen, Nucl. Phys. A **271**, 503 (1976)
[8] M. Brack, D. O. Riska and W. Weise, Nucl. Phys. A **287**, 425 (1977)
[9] D. Schiff and J. Tran Thanh Van, Nucl. Phys. B **5**, 529 (1968)

[10] N. Ikeno, R. Molina and E. Oset, Phys. Rev. C **104**, 014614 (2021)
[11] R. A. Arndt, I. I. Strakovsky, R. L. Workman, and D. V. Bugg, Phys. Rev. C **48**, 1926 (1993)
[12] L. D. Landau, Nucl. Phys. **13**, 181 (1959)
[13] S. Coleman and R. E. Norton, Nuovo Cim. **38**, 438-442 (1965)
[14] D. Jido, J. A. Oller, E. Oset, A. Ramos, and U. G. Meissner, Nucl. Phys. A **725**, 181 (2003)
[15] L. Roca and E. Oset, Phys. Rev. C **88**, 055206 (2013)
[16] K. Moriya *et al.* [CLAS], Phys. Rev. C **87**, 035206 (2013)
[17] A. Cieply and J. Smejkal, Nucl. Phys. A **881**, 115 (2012)
[18] M. Bazzi *et al.* [SIDDHARTA], Phys. Lett. B **704**, 113 (2011)
[19] A. Feijoo, V. Magas and A. Ramos, Phys. Rev. C **99**, 035211 (2019)
[20] J. Mares, E. Friedman and A. Gal, Nucl. Phys. A **770**, 84 (2006)
[21] S. Hirenzaki, Y. Okumura, H. Toki, E. Oset, and A. Ramos, Phys. Rev. C **61**, 055205 (2000)
[22] A. Baca, C. Garcia-Recio and J. Nieves, Nucl. Phys. A **673**, 335 (2000)

Four-body Faddeev-type calculation of the $\bar{K}NNN$ system

Nina V. Shevchenko[1,*]

[1]Nuclear Physics Institute, 25068 Řež, Czech Republic

Abstract. A quasi-bound state in the $\bar{K}NNN$ system caused by strong interactions was studied using dynamically exact four-body Faddeev-type equations. Binding energy and width of the $K^- ppn$ state were calculated using three antikaon-nucleon and three nucleon-nucleon interaction models.

The attractive nature of $\bar{K}N$ interaction has stimulated theoretical and experimental searches for K^- bound states in different systems. In particular, many efforts were devoted to the lightest possible system $\bar{K}NN$, see e.g. [1]. All theoretical results agree that a quasi-bound state in the $K^- pp$ system exists, but the predicted binding energies and widths are quite diverse. Several experimental evidences of the $K^- pp$ quasi-bound state existence were already reported. The estimated characteristics of the state are different from each other and also from the theoretical predictions.

Some theoretical works were devoted to the question of the quasi-bound state caused by strong interactions in the heavier $\bar{K}NNN$ system with different quantum numbers [2–5]. However, more accurate calculations within Faddeev-type equations are needed. Only these dynamically exact equations written in momentum representation can treat energy dependent $\bar{K}N$ potentials, necessary for study of this system, exactly.

We solved four-body Faddeev equations in the Grassberger-Sandhas (GS) form [6]. If separable potentials leading to the corresponding separable T-matrices

$$T_\alpha(z) = |g_\alpha\rangle \tau_\alpha(z) \langle g_\alpha| \tag{1}$$

are used as an input, and the three-body T-matrices $\bar{T}^\tau_{\alpha\beta}(z)$ are presented in a separable form

$$\bar{T}^\tau_{\alpha\beta}(z) = |\bar{g}^\tau_\alpha\rangle \bar{\tau}^\tau_{\alpha\beta}(z) \langle \bar{g}^\tau_\beta|, \tag{2}$$

the four-body equations can be written as

$$\bar{X}^{\sigma\rho}_{\alpha\beta}(z) = \bar{Z}^{\sigma\rho}_{\alpha\beta}(z) + \sum_{\tau,\gamma,\delta} \bar{Z}^{\sigma\tau}_{\alpha\gamma}(z) \bar{\tau}^\tau_{\gamma\delta}(z) \bar{X}^{\tau\rho}_{\delta\beta}(z) \tag{3}$$

with four-body transition $\bar{X}^{\sigma\rho}_{\alpha\beta}$ and kernel $\bar{Z}^{\sigma\rho}_{\alpha\beta}$ operators

$$\bar{X}^{\sigma\rho}_{\alpha\beta}(z) = \langle \bar{g}^\sigma_\alpha | (\bar{G}_0)_{\alpha\alpha}(z) \bar{U}^{\sigma\rho}_{\alpha\beta}(z) (\bar{G}_0)_{\beta\beta}(z) | \bar{g}^\rho_\beta \rangle, \tag{4}$$

$$\bar{Z}^{\sigma\rho}_{\alpha\beta}(z) = (1 - \delta_{\sigma\rho}) \langle \bar{g}^\sigma_\alpha | (\bar{G}_0)_{\alpha\beta}(z) | \bar{g}^\rho_\beta \rangle. \tag{5}$$

The four-body operators $\bar{U}^{\sigma\rho}_{\alpha\beta}(z)$ and $(\bar{G}_0)_{\alpha\beta}(z)$ in Eqs.(4,5) enter the Faddeev-type GS equations with separable two-body potentials and three-body amplitudes in an arbitrary form.

*e-mail: shevchenko@ujf.cas.cz

They are defined as

$$\bar{U}^{\sigma\rho}_{\alpha\beta}(z) = \langle g_\alpha | G_0(z) U^{\sigma\rho}_{\alpha\beta}(z) G_0(z) | g_\beta \rangle, \tag{6}$$

$$(\bar{G}_0)_{\alpha\beta}(z) = \delta_{\alpha\beta}\tau_\alpha(z), \tag{7}$$

where $G_0(z)$ is a free Green function in four-body space, and $U^{\sigma\rho}_{\alpha\beta}(z)$ is a four-body transition operator. The high indices σ, ρ, τ define a partition, which could be 3 + 1 or 2 + 2 type, while the low indices α, β define two-body subsystems of the particular three-body subsystem or the 2 + 2 partition, denoted by the high index.

The input for the system Eq.(3) are $\bar{K}N$ and NN potentials. We used two-body interaction models constructed for our three-body calculations [1]. Namely, three separable models of the $\bar{K}N$ interaction were used: two phenomenological potentials with one- $V^{1,\mathrm{SIDD}}_{\bar{K}N}$ or two-pole $V^{2,\mathrm{SIDD}}_{\bar{K}N}$ structure of the $\Lambda(1405)$ resonance and a chirally motivated potential $V^{\mathrm{Chiral}}_{\bar{K}N}$ with two-pole structure. All three potentials describe low-energy $K^- p$ scattering and the $1s$ level shift of kaonic hydrogen with equally high accuracy.

Two-term separable nucleon-nucleon potentials $V^{\mathrm{TSA-A}}_{NN}$ and $V^{\mathrm{TSA-B}}_{NN}$ from our three-body calculations [1] were used here together with the new version of the NN potential V^{TSN}_{NN}, described in [7]. All three potentials reproduce the Argonne v18 NN phase shifts at low energies up to 500 MeV with the change of sign, so they are repulsive at short distances. All three models of nucleon-nucleon interaction give proper singlet and triplet NN scattering lengths and deuteron binding energy.

Separable versions of the three-body $\bar{K}NN$, NNN and the 2 + 2 $\bar{K}N + NN$ amplitudes are not an input for the four-body equations, they should be calculated when the four-body system of equations Eq.(3) is being solved. We used Energy Dependent Pole Expansion/Approximation (EDPE/EDPA) method, suggested in [8] specially for the four-body Faddeev-type equations. It needs a solution of eigenequations for a fixed energy z_{fix} and an evaluation of the corresponding eigenvalues λ_n and eigenfunctions $g_{n\alpha}(p; z)$ first. Usually, z_{fix} is chosen to be equal to the binding energy $z_{\mathrm{fix}} = E_B$ if a bound state in the system exists or $z_{\mathrm{fix}} = 0$ if not. After that, the energy-dependent form-factors $g_{n\alpha}(p; z)$ and propagators $\Theta_{mn}(z)$ are calculated. Finally, the separable three-body amplitude is written as

$$X^{\mathrm{EDPE}}_{\alpha\beta}(p, p'; z) = \sum_{m,n=1}^{\infty} g_{m\alpha}(p; z)\, \Theta_{mn}(z)\, g_{n\beta}(p'; z). \tag{8}$$

Therefore, the EDPE method needs only one solution of eigenvalue equations and calculations of integrals after that. The series Eq.(8) converges quite fast, it is accurate already with one term [8].

One of the three-body subsystems of the $\bar{K}NNN$ system, namely $\bar{K}NN$, was studied in our previous works [1]. The binding energy and width of the quasi-bound state in the $K^- pp$ system, which is $\bar{K}NN$ with zero spin, were calculated. Our most recent values of the $K^- pp$ quasi-bound state characteristics can be found in [7]. Another, spin one state of the $\bar{K}NN$ system, which is $K^- np$, can have two different quasi-bound states. One of them, kaonic deuterium, is an atomic state mainly caused by the Coulomb interaction between the antikaon and the proton. Another state, which is similar to the one in $K^- pp$, is caused by strong interactions. Since the pole corresponding to the "strong" $K^- np$ state is situated very close to the $K^- d$ threshold, theoretical prediction of its existence depends on the interaction models. In particular, calculations with the new NN potential V^{TSN}_{NN} together with the two-pole $\bar{K}N$ interaction models predict an existence of the strong quasi-bound state, while the use of V^{TSA}_{NN} leads to the negative result [7].

Three-body equations for the NNN subsystem, also necessary for the four-body calculations, were solved additionally. The remaining object to study is the amplitude of the $\bar{K}N + NN$ partition. It is a special state consisting of a pair of non-interacting pairs of particles. It is described by a system of three-body equations, which were written down and solved separately. Partition $\bar{K}N + NN$ with zero isospin in each pair of particles has a quasi-bound state since $\bar{K}N$ with zero isospin has a quasi-bound state, the $\Lambda(1405)$ resonance, while the nucleon-nucleon pair can be bound into deuteron. The $2 + 2$ partition with isospin of the pairs equal to one does not have a quasi-bound state, so the fixed energy $z_{\rm fix} = 0$ was set for this case. All three-body program codes were rewritten for evaluation of the separable versions of the three-body $\bar{K}NN$, NNN and "three-body" $\bar{K}N + NN$ amplitudes using the EDPE method.

The Faddeev-type system of equations Eq.(3) consists of 18 equations while all four particles are different. In the case of the $\bar{K}NNN$ system, the number of equations after antisymmetrization is reduced to five. However, our $\bar{K}N$ and NN potentials are spin- and isospin-dependent models, moreover, nucleon-nucleon interaction is described by the two-term potentials. Since only one term in the series Eq.(8) was used at the first step, the final number of the coupled integral equations to be solved for the $\bar{K}NNN$ system is equal to 18.

In contrast to our three-body calculations, here we did not take coupling between the $\bar{K}N$ and $\pi\Sigma$, $\pi\Lambda$ channels into account directly. We used exact optical $\bar{K}N$ potentials instead, which reproduce the elastic part of the coupled-channel amplitudes exactly. The inelasticity is taken into account through the energy-dependent imaginary parts of the potentials. We also did not take the Coulomb interaction into consideration assuming that it plays a minor role.

The $\bar{K}NNN$ system with zero isospin and spin one half ($K^-ppn - \bar{K}^0nnp$ in particle representation) was investigated, and binding energy and width of the quasi-bound state in the system were calculated. Three versions of the antikaon-nucleon interaction: $V^{1,\rm SIDD}_{\bar{K}N}$, $V^{2,\rm SIDD}_{\bar{K}N}$ and $V^{\rm Chiral}_{\bar{K}N}$, - together with the three versions of the NN potentials: $V^{\rm TSA-A}_{NN}$, $V^{\rm TSA-B}_{NN}$ and $V^{\rm TSN}_{NN}$, - used in the calculations allowed us to investigate the dependence of the results on the two-body input. The results are presented in Fig.1. It is seen that the four-body binding energies and widths, in the same way as those of the three-body $\bar{K}NN$ or $\bar{K}\bar{K}N$ systems, strongly depend on the model of the $\bar{K}N$ interaction. If the chirally motivated antikaon-nucleon potential $V^{\rm Chiral}_{\bar{K}N}$ is used, the binding energy B_{K^-ppn} and the width Γ_{K^-ppn} of the $\bar{K}NNN$ quasi-bound state also significantly depend on the nucleon-nucleon potential. While one-pole $V^{1,\rm SIDD}_{\bar{K}N}$ or two-pole $V^{2,\rm SIDD}_{\bar{K}N}$ phenomenological antikaon-nucleon model is used, dependence on the NN potential is weak.

The previously calculated characteristics of the $\bar{K}NNN$ quasi-bound state with the same quantum numbers are also shown in the figure. The prediction [2], denoted as AY in the figure, with the largest binding energy and the smallest width was done using the G-matrix approach and an antikaon-nucleon interaction model, which does not reproduce experimental data. Variational calculations were performed by two groups, and the results are denoted as BGL [3] and OHHMH [4]. Both used a chirally motivated antikaon-nucleon potential, which is a problem for variational calculations in coordinate space since it is not possible to take the energy dependence of the potentials into account. The authors used different ways to deal with the energy dependence of the $\bar{K}N$ interaction model. The second group [4] fixed the energy at two different values, and the resulting widths differ drastically one from another. The authors of [5] (ME in Fig.1) used the same Faddeev-type four-body equations as we did. Moreover, two of the three antikaon-nucleon potentials they used are our phenomenological $V^{1,\rm SIDD}_{\bar{K}N}$ and $V^{2,\rm SIDD}_{\bar{K}N}$ models. The third one is a chiral potential. The results of [5] differ drastically from ours, namely, their binding energies are larger, while the widths are much smaller than ours for the same $V_{\bar{K}N}$. The nucleon-nucleon potentials used by us and by the authors of [5] are different, but it hardly can explain such differences. The reason can be an inappropriate treatment of the $2 + 2$ partition, which is not described in [5], or inaccuracy

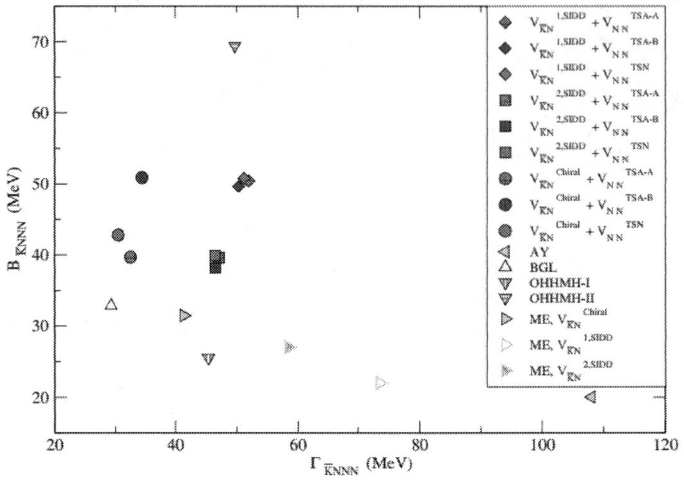

Figure 1. Binding energies $B_{\bar{K}NNN}$ (MeV) and widths $\Gamma_{\bar{K}NNN}$ (MeV) of the quasi-bound state of the $\bar{K}NNN$ state calculated using three different models of antikaon-nucleon interaction and three nucleon-nucleon models. The results of other authors are also presented.

during construction of the separable versions of the three-body subamplitudes and the $\bar{K}N + NN$ partition.

The binding energies B_{K^-ppn} in the K^-ppn system turned out to be quite close to those evaluated previously for the K^-pp system with the same two-body potentials. Namely $B_{K^-ppn} \sim 30.5 - 34.5$ MeV was obtained with the chirally motivated $\bar{K}N$ potential and $B_{K^-ppn} \sim 46.4 - 52.0$ MeV with phenomenological models of antikaon-nucleon interaction. The four-body widths $\Gamma_{K^-ppn} \sim 38.2 - 50.9$ MeV are smaller than the three-body ones.

A more detailed description of the calculation and discussion of the results can be found in [9].

Asknowledgements. The work was supported by GACR grant 19-19640S.

References

[1] N.V. Shevchenko, Few Body Syst. **58**, 6 (2017)
[2] Y. Akaishi, T. Yamazaki, Phys. Rev. C **65**, 044005 (2002)
[3] N. Barnea, A. Gal, E.Z. Liverts, Phys. Lett. B **712**, 132 (2012)
[4] S. Ohnishi et.al, Phys. Rev C **95**, 065202 (2017)
[5] S. Marri, J. Esmaili, Eur. Phys. J. A **55**, 43 (2019)
[6] P. Grassberger, W. Sandhas, Nucl. Phys. B **2**, 181 (1967)
[7] N.V. Shevchenko, Few Body Syst. **61**, 27 (2020)
[8] S. Sofianos, N.J. McGurk, H. Fiedeldeldey, Nucl. Phys. A **318**, 295 (1979)
[9] N.V. Shevchenko, arXiv:nucl-th/2206.14514 (2022)

Meson-baryon scattering and $\Lambda(1405)$ in chiral effective field theory

Xiu-Lei Ren[1,2,*]

[1]Institut für Kernphysik & PRISMA[+] Cluster of Excellence, Johannes Gutenberg Universität, D-55099 Mainz, Germany
[2]Helmholtz Institut Mainz, D-55099 Mainz, Germany

Abstract. We investigated the meson-baryon scattering using time-order perturbation theory (TOPT) based on covariant chiral effective field theory. The effective potential is defined as the sum of two-particle irreducible contributions of time-ordered diagrams, and the renormalized scattering amplitude is obtained by solving the integral equation, which is derived self-consistently in TOPT. Our developed formalism has been successfully applied to the πN scattering at leading order, and it has been extended to the meson-baryon scattering in $S = -1$ sector and found the two-pole structure of the $\Lambda(1405)$ resonance.

1 Introduction

Meson-baryon scattering contains very interesting phenomena, such as the πN scattering data can be used to extract the sigma term [1], $\sigma_{\pi N}$, which is a key input in the evaluation of the neutralino-nucleon cross section [2]; the $\bar{K}N$ interaction, due to its strong attractive nature, is important in strangeness nuclear physics, e.g. the $\Lambda(1405)$ resonance, $\bar{K}NN$ and multi-$\bar{K}$ nuclei, and the kaon-condensate in neutron stars [3]. Furthermore, the study of meson-baryon scattering can deepen our understanding of SU(3) dynamics in nonperturbative QCD.

Chiral unitary approach [4], which extends the applicable range of low-energy chiral effective field theory to the resonance region by iterating the interaction kernel in the Lippmann-Schwinger or Bethe-Salpeter equations ($T = V + VGT$), has been extensively applied to the studies of meson-baryon scattering processes. However, in practice, the on-shell factorization is often used in solving the scattering equation by neglecting the off-shell effects. Furthermore, the finite momentum cutoff or subtraction constant is introduced to renormalize the one-loop Green function [5], which results in the cutoff or model dependence of the scattering amplitude.

To avoid these approximations or obstacles, in this work, we tentatively propose a renormalized framework for meson-baryon scattering using TOPT with the covariant chiral Lagrangians [6, 7], which accompanies the research line of baryon-baryon scattering in Refs. [8–10]. Such a framework allows us to obtain the effective potential and the scattering equation on an equal footing. By including the off-shell effects of potential and utilizing the subtractive renormalization, we obtain a renormalized scattering amplitude at leading order (LO). We have successfully applied the developed formalism to the pion-nucleon and $S = -1$ meson-baryon scatterings.

*e-mail: xiulei.ren@uni-mainz.de

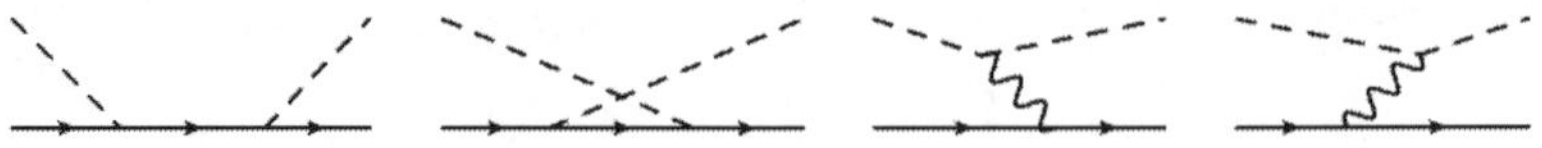

Figure 1. Time-ordered diagrams contributing to the LO meson-baryon potential.

2 Theoretical framework

Here we briefly present the major ingredients of our framework, the details can be found in Refs.[6, 7]. First is to obtain the coupled-channel scattering equations in TOPT,

$$\begin{aligned} T^{M_fB_f,M_iB_i}(E;\boldsymbol{p}',\boldsymbol{p}) =& V^{M_fB_f,M_iB_i}(E;\boldsymbol{p}',\boldsymbol{p}) \\ &+\sum_{M,B}\int\frac{d^3\boldsymbol{k}}{(2\pi)^3}V^{M_fB_f,MB}(E;\boldsymbol{p}',\boldsymbol{k})\,G^{MB}(E)\,T^{MB,M_iB_i}(E;\boldsymbol{k},\boldsymbol{p}), \end{aligned} \tag{1}$$

where M_iB_i, M_fB_f and MB denote initial, final and intermediate particle channels. The two-body Green function, which is obtained according to the diagrammatic rules of TOPT [8], has the form

$$G^{MB}(E) = \frac{1}{2\,\omega_M\omega_B}\,\frac{-m_B}{E-\omega_M-\omega_B+i\epsilon}\,, \tag{2}$$

where $\omega_i \equiv \sqrt{\boldsymbol{p}_i^2+m_i^2}$ is the energy of the $i_{\rm th}$ hadron with the mass m_i and momentum $\boldsymbol{p}_i$.

The effective potential $V(E;\boldsymbol{p}',\boldsymbol{p})$ is defined as the sum of two-particle irreducible time ordered diagrams of the meson-baryon scattering. At leading order, using the covariant Lagrangians,

$$\begin{aligned} \mathcal{L}_{\rm LO} &= \frac{F_0^2}{4}\left\langle u_\mu u^\mu+\chi_+\right\rangle+\left\langle\bar{B}\left(i\gamma_\mu\partial^\mu-m\right)B\right\rangle+\frac{D/F}{2}\left\langle\bar{B}\gamma_\mu\gamma_5[u^\mu,B]_\pm\right\rangle \\ &\quad -\frac{1}{4}\left\langle V_{\mu\nu}V^{\mu\nu}-2\mathring{M}_V^2\left(V_\mu-\frac{i}{g}\Gamma_\mu\right)\left(V^\mu-\frac{i}{g}\Gamma^\mu\right)\right\rangle+g\left\langle\bar{B}\gamma_\mu[V^\mu,B]\right\rangle, \end{aligned} \tag{3}$$

where the definition of symbols can be found in Ref. [7], one can calculate the time-ordered diagrams shown in Fig. 1 to obtain the effective potential. It is worthy to note that the lowest-lying vector mesons are involved as the dynamical degrees of freedom in the effective Lagrangians. Because the vector-meson exchange contribution has a better ultraviolet behavior than the standard Weinberg-Tomozawa term, without changing the low-energy physics.

Next, we can divide the leading order potential into the one-baryon reducible part and irreducible part,

$$V_{\rm LO} = V_R + V_I,\quad \text{with}\quad V_R = V_a,\ \ V_I = V_{b+c+d}. \tag{4}$$

This allows us to rewrite the integral equation $T = V + VGT$ (the symbolic form of Eq. (1)) as follows,

$$\begin{aligned} T &= T_I + (1+T_I\,G)\,T_R\,(1+G\,T_I), \\ T_I &= V_I + V_I\,G\,T_I,\quad T_R = V_R + V_R\,G\,(1+T_I\,G)\,T_R. \end{aligned}$$

It is found that the irreducible part T_I is finite in the removed regulator limit, while the reducible part T_R is divergent. The divergence terms in T_R can be systematically removed by using the subtractive renormalization [11]: replacing the meson-baryon propagator $G_{MB}(E)$ with the subtracted one $G^S_{MB}(E) = G_{MB}(E) - G_{MB}(m_B)$. Note that such subtraction, which is an analogy to the extended-on-mass-shell scheme [12], corresponds to the expansion around the threshold and can systematically remove the chiral power-counting breaking terms in the iterated amplitudes. Finally, we obtain a renormalized finite T-matrix.

Table 1. Pole positions of $\Lambda(1405)$ in the $S = -1$ sector (units are MeV).

		lower pole	higher pole
This work	$F_0 = F_\pi$	$1337.7 - i\,79.1$	$1430.9 - i\,8.0$
(LO)	$F_0 = 103.4$	$1348.2 - i\,120.2$	$1436.3 - i\,0.7$
NLO	Ref. [13]	$1381^{+18}_{-6} - i81^{+19}_{-8}$	$1424^{+7}_{-23} - i26^{+3}_{-14}$
	Ref. [14], Fit II	$1388^{+9}_{-9} - i114^{+24}_{-25}$	$1421^{+3}_{-2} - i19^{+8}_{-5}$
	Ref. [15], sol-2	$1330^{+4}_{-5} - i56^{+17}_{-11}$	$1434^{+2}_{-2} - i10^{+2}_{-1}$
	Ref. [15], sol-4	$1325^{+15}_{-15} - i90^{+12}_{-18}$	$1429^{+8}_{-7} - i12^{+2}_{-3}$

3 Results and discussion

As a simple application, we first studied the πN scattering up to LO using the proposed formalism in Ref. [6]. We found that phase shifts based on the non-perturbative renormalized amplitude are only slightly different from the ones obtained from the perturbative tree-level calculation. This indicates that our non-perturbative treatment is valid, since the chiral expansion has good convergence in the SU(2) sector.

Next, in Ref. [7] we extended the TOPT framework to the $S = -1$ meson-baryon scattering, which has four coupled channels: $\bar{K}N$, $\pi\Sigma$, $\eta\Lambda$, and $K\Xi$. Using the LO effective potential, we exactly solved the coupled-channel scattering equation (Eq. (1)) in the isospin basis by taking into account the off-shell effects. In order to obtain the renormalized T-matrix, we utilized the subtractive renormalization which allows us to take the momentum cutoff to infinity. We would like to emphasize that no free parameters are introduced in the whole process. After performing the analytic continuation of the renormalized T-matrix into the complex s-plane, we find the two-pole structure of $\Lambda(1405)$ in the second Riemann sheet, and the corresponding pole positions are presented in Table 1. We can see that by changing the meson-decay constant, the width of lower pole is increasing, and the higher pole lies close and moves beyond the threshold of $\bar{K}N$ channel and its width decreases. In comparison with the previous NLO studies [13–15], we found that our LO results are consistent with Mai's results [15], in particular for the lower pole. The $\pi\Sigma$ invariant mass spectrum is presented in Ref. [7], which is in good agreement with experimental data. Furthermore, we calculated the coupling strengths of the initial/final transition channels, and found that the two poles of $\Lambda(1405)$ have different coupling natures: the lower pole couples predominantly to the $\pi\Sigma$ channel, and the higher pole couples strongly to the $\bar{K}N$ channel.

Finally, using the $I = 0, 1$ amplitudes of $\bar{K}N$ scattering in the coupled channels, we calculated the total cross section of K^-p scattering to various channels, as shown in Fig. 2. Our LO prediction covers well the data of $K^-p \to \pi\Sigma$ cross section, and is slightly larger than the data of $K^-p \to K^-p, \pi^0\Lambda$. This will be improved by introducing the isospin-breaking effects and beyond LO studies.

4 Conclusion

In this work, we studied the meson-baryon scattering using the time-ordered perturbation theory based on covariant chiral effective field theory. The effective potential and the scattering equation are formulated within the same framework. By taking the vector mesons as the explicit degrees of freedom in the chiral effective Lagrangians, we achieved the renormalized scattering amplitude at leading order by solving the integral equations of meson-baryon scattering. The proposed formalism has been successfully applied to the πN scattering and the $S = -1$ meson-baryon scattering, and found the two-pole structure of the $\Lambda(1405)$ state.

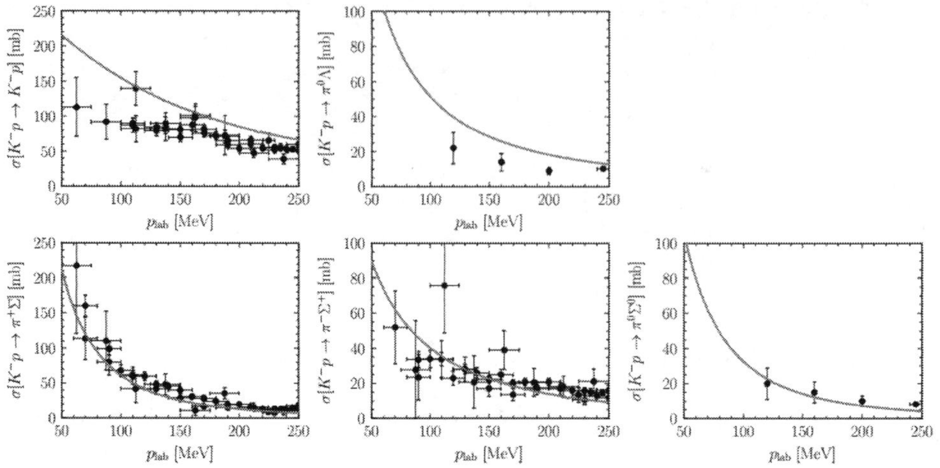

Figure 2. Total cross section of $K^- p$ scattering to various channels as the function of incident K^- laboratory momentum p_{lab}.

Acknowledgments

This work was supported by the Deutsche Forschungsgemeinschaft (DFG, German Research Foundation), in part through the Research Unit (Project ID 458854507 - FOR 5327), and in part through the Cluster of Excellence (PRISMA$^+$ EXC 2118/1) within the German Excellence Strategy (Project ID 39083149), and was supported by the European Union's Horizon 2020 research and innovation programme under grant agreement No 824093.

References

[1] M. Hoferichter, J. Ruiz de Elvira, B. Kubis and U.-G. Meißner, Phys. Rev. Lett. **115**, 092301 (2015)
[2] J. R. Ellis, K. A. Olive and C. Savage, Phys. Rev. D **77**, 065026 (2008)
[3] L. Tolos and L. Fabbietti, Prog. Part. Nucl. Phys. **112**, 103770 (2020)
[4] J. A. Oller, E. Oset and A. Ramos, Prog. Part. Nucl. Phys. **45**, 157-242 (2000)
[5] T. Hyodo and D. Jido, Prog. Part. Nucl. Phys. **67**, 55-98 (2012)
[6] X.-L. Ren, E. Epelbaum, J. Gegelia and U.-G. Meißner, Eur. Phys. J. C **80**, 406 (2020)
[7] X.-L. Ren, E. Epelbaum, J. Gegelia and U.-G. Meißner, Eur. Phys. J. C **81**, 582 (2021)
[8] V. Baru, E. Epelbaum, J. Gegelia and X.-L. Ren, Phys. Lett. B **798**, 134987 (2019)
[9] X.-L. Ren, E. Epelbaum and J. Gegelia, Phys. Rev. C **101**, 034001 (2020)
[10] X. L. Ren, E. Epelbaum and J. Gegelia, Phys. Rev. C **106**, 034001 (2022)
[11] E. Epelbaum, A. M. Gasparyan, J. Gegelia, U.-G. Meißner and X.-L. Ren, Eur. Phys. J. A **56**, 152 (2020)
[12] T. Fuchs, J. Gegelia, G. Japaridze and S. Scherer, Phys. Rev. D **68**, 056005 (2003)
[13] Y. Ikeda, T. Hyodo and W. Weise, Nucl. Phys. A **881**, 98-114 (2012)
[14] Z. H. Guo and J. A. Oller, Phys. Rev. C **87**, 035202 (2013)
[15] M. Mai and U.-G. Meißner, Eur. Phys. J. A **51**, 30 (2015)

Study of strange matter in STAR with express analysis

Ivan Kisel[1,2,3,4,*], for the STAR Collaboration

[1]Goethe University, Frankfurt am Main, Germany
[2]FIAS, Frankfurt Institute for Advanced Studies, Frankfurt am Main, Germany
[3]HFHF, Helmholtz Research Academy Hesse, Frankfurt am Main, Germany
[4]GSI, Helmholtz Center for Heavy Ion Research, Darmstadt, Germany

Abstract. The STAR experiment provides a perfect machinery for studying strange matter for more than two decades. Recently, we developed the express procedure, which allows online monitoring of the collected physics data. The high quality of express calibration and reconstruction provides a unique possibility to run the express production and observe almost in real time strange particles including mesons, hyperons, resonances and even hypernuclei.
The STAR Beam Energy Scan II program, including fixed target Au+Au collisions taken in 2018–2021, is particularly suited to study hypernuclei. Light hypernuclei are expected to be abundantly produced in low energy heavy-ion collisions. Measurements of hypernuclei production and their properties will provide information on the hyperon-nucleon interactions, which are essential ingredients for understanding nuclear matter equation of state at high net-baryon densities, such as inside neutron stars.
With the heavy fragment trigger introduced for the 2021 data taking, we were able to run the express production at the STAR High Level Trigger farm. The collected data were sufficient to observe the decay process of ${}^{5}_{\Lambda}\mathrm{He} \rightarrow {}^{4}\mathrm{He} p \pi^{-}$ with more than 11σ significance, measure binding energy as a function of hypernuclei mass, and study hypernuclei decay properties with the Dalitz plot technique.

1 Introduction

Since hyperons are elementary particles containing at least one s-quark, but do not contain the heavier c and b quarks, i.e., they consist of three light quarks, at least one of which is an s-quark, hyperons are formed in large amounts during collisions of heavy ions, which allows one to study their properties in detail with heavy ion experiments.

The formation of hyperons in large amounts also makes it possible to study hypernuclei, which are a system of strongly interacting particles consisting of nucleons (protons and neutrons) and one or more hyperons. The structure of hypernuclei is determined by nuclear interaction between nucleons and hyperons.

Hypernuclei are excellent experimental probes for studying hyperon-nucleon (Y-N) interactions. The Y-N interaction is an important component in the description of the hadronic phase of heavy ion collisions, as well as in the equation of state (EoS) of astrophysical objects, such as neutron stars [1]. Therefore, heavy ion collisions at RHIC provide a unique

*e-mail: I.Kisel@compeng.uni-frankfurt.de

opportunity to study also the Y-N interaction in finite temperature and density regions by measuring the hypernuclear lifetime, production yield, etc.

In this publication, we present preliminary results from a real-time search for hyperons and hypernuclei using an express data stream on a computer cluster of the High Level Trigger (HLT) in the STAR experiment within the Beam Energy Scan phase-II (BES-II) program (2018–2021) [2].

2 Extended functionality of the High Level Trigger for BES-II

In order to create a package of algorithms for full processing and analysis of data in real time within the BES-II physics program, the functionality of the HLT computer cluster was significantly extended. This was done within the FAIR Phase-0 program, which allowed adapting the package of fast algorithms for processing and analysis FLES (First Level Event Selection) [3] of the CBM experiment (FAIR/GSI) data to work with real data of the STAR experiment.

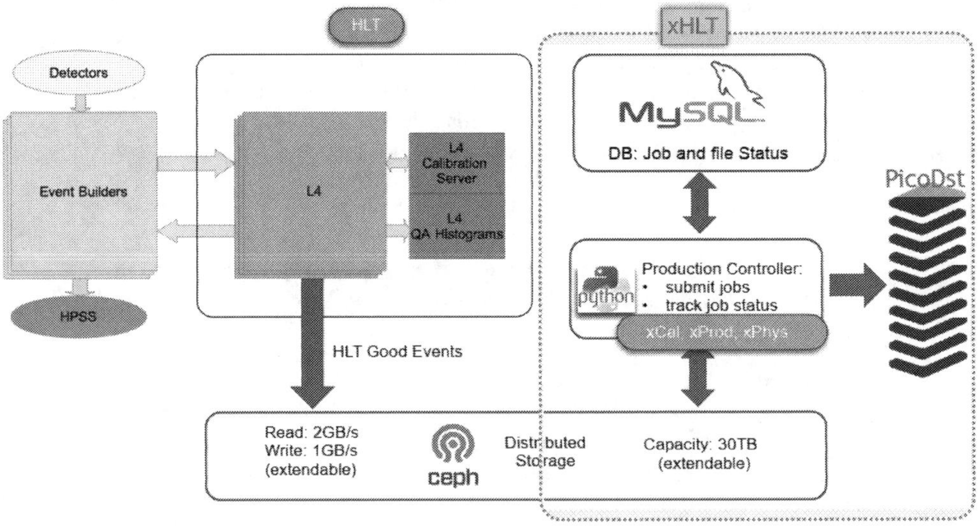

Figure 1. Full chain of express production and analysis has been running on STAR HLT since 2019.

Basic elements of data processing and analysis are search of particle trajectories in the detector system based on the Cellular Automaton (CA Track Finder) and search of short-lived particles based on the Kalman filter method (KF Particle Finder) [3]. These algorithms, after careful adaptation and detailed testing on simulated data, were checked with the Au+Au collisions recorded in 2014, 2016, and BES-I. It was shown that using the CA Track Finder provides 25% more D^0 and 20% more W when processing a sample of pp collisions at 510 GeV collected in 2013. Also, the KF Particle Finder package provides twice as many signal particles with the same background level as the standard approach used in STAR. The reliability and high performance of both algorithms allowed them to be included in the real-time express physics analysis chain at the HLT computer cluster during the BES-II (2018-2021) runs.

It is also essential that a data calibration package, optimized for real-time operation, has been added to the HLT operation. This package has shown an exceptionally high performance with BES-II data, which resulted in high quality of the CA Track Finder, as well as in the KF Particle Finder's accurate search for short-lived particles, in particular hyperons and hypernuclei.

3 BES-II: Search for mesons and hyperons

The quality of the express chain of data processing and analysis was continuously tested in real time using meson reconstructions as an example.

Table 1. Results of the search for mesons and hyperons in the STAR HLT express data stream.

Decay	Mass, MeV/c^2	σ, MeV/c^2	S	S/B	$S/\sqrt{S+B}$
$\pi^0 \to \gamma_{e^+e^-}\gamma_{e^+e^-}$	135.9	4.2	$14.0 \cdot 10^3$	0.2	48
$K_s^0 \to \pi^+\pi^-$	497.3	4.1	$67.1 \cdot 10^6$	6.5	7629
$K_s^+ \to \pi^+\pi^+\pi^-$	493.9	2.6	$2.4 \cdot 10^6$	24.3	1524
$K_s^- \to \pi^+\pi^-\pi^-$	493.9	2.4	$0.7 \cdot 10^6$	8.3	839
$K_s^+ \to \pi^+\pi^+\pi^- + K$ track	493.8	2.0	$35.7 \cdot 10^3$	inf	189
$K_s^- \to \pi^+\pi^-\pi^- + K$ track	493.8	2.0	$12.9 \cdot 10^3$	inf	114
$\pi^+ \to \mu^+\nu_\mu$	138.2	2.2	$2.1 \cdot 10^6$	75.9	1443
$\pi^- \to \mu^-\bar{\nu}_\mu$	138.2	2.2	$2.4 \cdot 10^6$	78.6	1546
$K^+ \to \mu^+\nu_\mu$	493.8	9.1	$3.1 \cdot 10^6$	4.7	1606
$K^- \to \mu^-\bar{\nu}_\mu$	493.6	9.0	$1.1 \cdot 10^6$	4.5	956
$K^+ \to \pi^+\pi^0$	493.2	6.7	$1.0 \cdot 10^6$	2.6	830
$K^- \to \pi^-\pi^0$	493.1	6.6	$0.3 \cdot 10^6$	2.4	489
$\Lambda \to p\pi^-$	1115.7	1.5	$60.1 \cdot 10^6$	24.6	7601
$\bar{\Lambda} \to \bar{p}\pi^+$	1115.7	1.4	$0.9 \cdot 10^6$	7.1	931
$\Xi^- \to \Lambda\pi^-$	1321.9	2.1	$0.8 \cdot 10^6$	21.8	890
$\bar{\Xi}^+ \to \bar{\Lambda}\pi^+$	1321.9	2.1	$45.8 \cdot 10^3$	36.7	211
$\Omega^- \to \Lambda K^-$	1672.4	2.2	$9.2 \cdot 10^3$	3.9	86
$\bar{\Omega}^+ \to \bar{\Lambda}K^+$	1672.4	2.4	$2.2 \cdot 10^3$	12.0	46

The upper part of table 1 shows the results of reconstruction of decay channels $\pi^0 \to \gamma_{e^+e^-}\gamma_{e^+e^-}$, $K_s^0 \to \pi^+\pi^-$, $K_s^+ \to \pi^+\pi^+\pi^-$ and $K_s^- \to \pi^+\pi^-\pi^-$ after processing 140M Au+Au events at 7.7 GeV, collected in 2021. Due to the high quality of online calibration and processing, strange mesons are reconstructed with high significance and S/B ratio, and even π^0 is observed with a significant of 48σ. Reconstruction of π^0 relies on a rather complex search of photons, as electron and positron are parallel at the conversion point, and requires high efficiency of track finding, since 4 tracks are produced in the decay tree.

Also there is an example of processing 32.5M Au+Au events at 7.7 GeV to search for decays $K_s^+ \to \pi^+\pi^+\pi^-$ and $K_s^- \to \pi^+\pi^-\pi^-$ when all four tracks are registered in the detector system and reconstructed. STAR with its perfectly working TPC detector allows to identify charged kaons without background by full topological reconstruction with all 4 tracks including kaon. Reconstruction of such full decay topologies provides additional technical opportunities to study quality of detector performance and reconstruction algorithms.

Decay channels of pions and kaons with a neutral daughter particle can also be found using the missing mass method. The middle part of table 1 shows the results of the reconstruction of decay channels $\pi^+ \to \mu^+\nu_\mu$, $\pi^- \to \mu^-\bar{\nu}_\mu$, $K^+ \to \mu^+\nu_\mu$, $K^- \to \mu^-\bar{\nu}_\mu$, $K^+ \to \pi^+\pi^0$ and $K^- \to \pi^-\pi^0$ after processing 32.5M Au+Au events at 7.7 GeV, collected in 2021. The missing mass method provides various opportunities in the study of different decay channels with a neutral daughter particle.

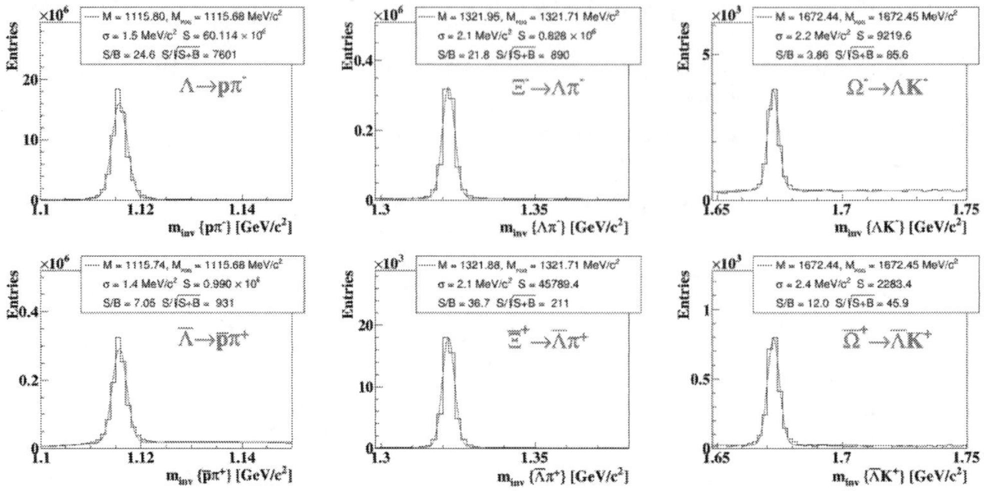

Figure 2. With express calibration and production we observe in real time all hyperons with high significance and S/B ratio (140M AuAu events at 7.7 GeV, 2021).

STAR has recently upgraded the inner part of the TPC, which, together with the enhanced CA track finder, has improved the efficiency of hyperon reconstruction. The bottom part of table 1 and Fig. 2 show the results of reconstruction of the hyperon decay channels after processing 140M Au+Au events at 7.7 GeV, collected in 2021. It can be seen that the high quality of the new BES-II experimental data provides an excellent opportunity to study hyperons.

4 BES-II: Search for hypernuclei

In order to increase statistics of the experimental data set for the search of hypernuclei at the lowest energy of 3.0 GeV in the BES-II program, intensity of the beam collision in the fixed-target mode was significantly increased. This led to that more than half of the events in the STAR detector consist of at least two closely overlapping heavy ion collisions, i.e., after reconstruction the events had two or more reconstructed primary vertices. This required a more detailed data analysis within the KF Particle Finder package to detect and carefully clean up interactions of beam particles with the beam pipe material and support structures. Such a cleaning procedure involved, among other things, reconstruction of multiple primary vertices from interactions with the beam pipe and support structures as well as the pileup events, and then discarding such found primary tracks. This cleaning procedure reduced significantly the background in hypernuclei spectra, especially in three-body channels, compared to the standard procedure used at higher energies and lower intensities.

Another feature of the search for hypernuclei was that as the intensity of the beam increased in the fixed-target mode, the HLT computer cluster did not have enough resources to process and analyze all the collected data online. Therefore, in order to fully process the data with the goal of searching for hypernuclei, a trigger was introduced, requiring the presence of a He nucleus in the event, for subsequent real-time analysis. A set of 437M triggered Au+Au collisions at $\sqrt{s_{NN}} = 3.0$ GeV in the fixed-target mode recorded in 2021 on HTL proved to be sufficient to measure the yield, lifetime, and spectra of the hypernuclei.

Using the same procedure, the 7.7 GeV data in the collider mode collected in 2021 were also analyzed within the express chain, as well as data sets of different energies in the fixed-

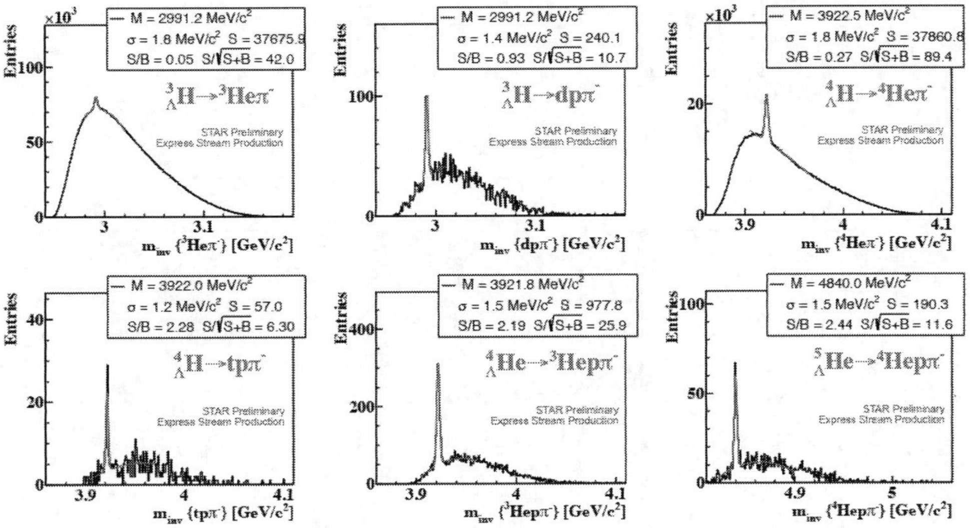

Figure 3. Standard and express reconstruction of hypernuclei using 2018, 2019, 2020, and 2021 data collected at different energies in the collider and fixed-target modes. The signal of $^{5}_{\Lambda}$He is visible with a significance of 11.6σ.

target mode collected in 2018, 2019, and 2020 and processed with the STAR standard production chain after the final calibration. After such (express and standard) processing of all the data, the signal of the $^{5}_{\Lambda}$He hypernucleus is clearly visible with a significance of 11.6σ (see Fig. 3).

Collected statistics are also sufficient to study Dalitz plots in the 3-body decay channels. Thus, Figs. 4 and 5 show the Dalitz plots for the decay $^{4}_{\Lambda}$He $\rightarrow$ ^{3}He+p+π^- which has the largest number of signals (978 decays) found. The background was estimated using the side-band method and subtracted under the peak. As can be seen, the background is smooth and no structures are observed. The complex structure in the pπ projection can be explained as a possible spin effect. In the nucleus-p combination, the signal and the background show different behavior. There can be seen a hint of two-body ^{4}Li π^- decay (^{4}Li ($J^P = 2^-$): M = 3.75 GeV/c^2, σ = 8.7 MeV/c^2). Similar behavior is observed in the Dalitz plots for the decays of hypernuclei $^{5}_{\Lambda}$He $\rightarrow$ ^{4}He+p+π^- and $^{3}_{\Lambda}$He $\rightarrow$ d+p+π^-, but with lower statistics, 190 and 240 signal particles, respectively.

As for the binding energy of the hypernuclei (Fig. 6), one can see that it increases for heavier hypernuclei up to $^{5}_{\Lambda}$He. When comparing binding energies per nucleon, hypernuclei behave similarly to ordinary nuclei (see Table 2). Thus, the value of the binding energy per nucleon in the hypernuclei, together with hints of their two-body resonance decays, may be an indication that Λ in the hypernuclei behaves similarly to other nucleons, n and p.

Table 2. Hypernuclei behave similarly to ordinary nuclei with respect to the binding energy per nucleon.

$^{5}_{\Lambda}$He	6.30 MeV/c^2	^{4}He	7.07 MeV/c^2	^{5}He	5.48 MeV/c^2	^{5}Li	5.27 MeV/c^2
$^{4}_{\Lambda}$He	2.55 MeV/c^2	^{3}He	2.57 MeV/c^2	^{4}He	7.07 MeV/c^2	^{4}Li	1.15 MeV/c^2
$^{4}_{\Lambda}$H	2.65 MeV/c^2	^{3}H	2.83 MeV/c^2	^{4}H	1.40 MeV/c^2		
$^{3}_{\Lambda}$H	0.8-0.9 MeV/c^2	^{2}H	1.11 MeV/c^2	^{3}H	2.83 MeV/c^2		

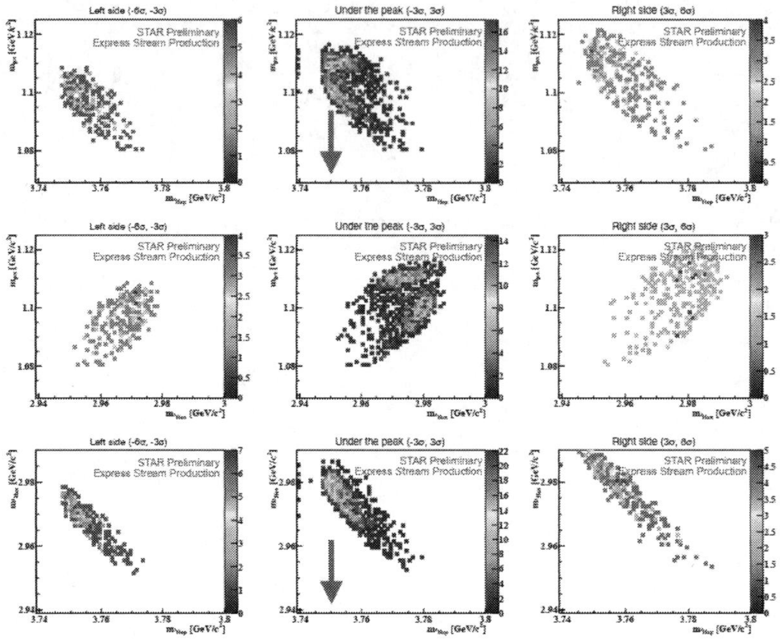

Figure 4. The Dalitz plots for the decay ${}^{4}_{\Lambda}\mathrm{He} \to {}^{3}\mathrm{He}+\mathrm{p}+\pi^{-}$ which has the largest number of signals (978 decays) found using 2018, 2019, 2020, and 2021 data collected at different energies in the collider and fixed-target modes.

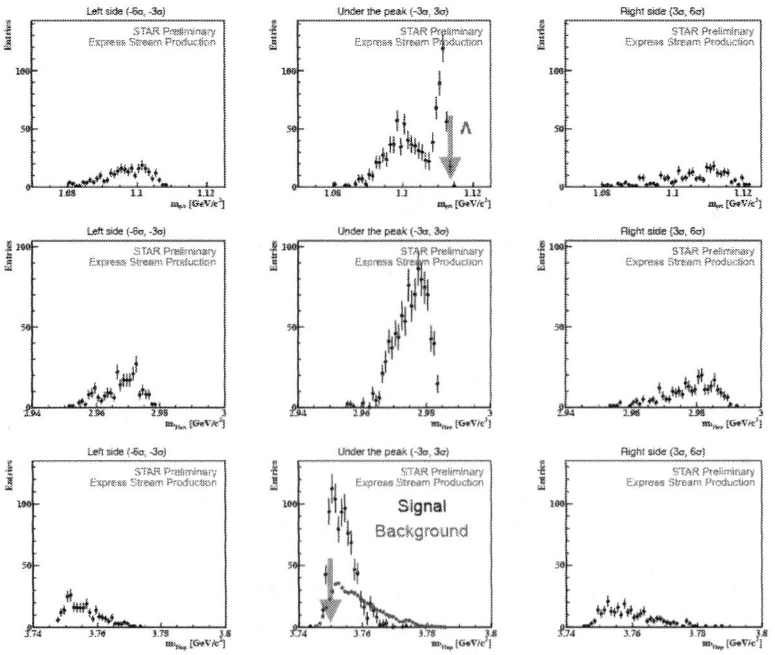

Figure 5. Projections of the Dalitz plots for the decay ${}^{4}_{\Lambda}\mathrm{He} \to {}^{3}\mathrm{He}+\mathrm{p}+\pi^{-}$. The complex structure in the signal in the pπ projection can be explained as a possible spin effect. In the nucleus-p combination, the signal and the background show different behavior. There can be seen a hint of two-body ${}^{4}\mathrm{Li}\,\pi^{-}$ decay (${}^{4}\mathrm{Li}$ ($J^{P} = 2^{-}$): M = 3.75 GeV/c^{2}, σ = 8.7 MeV/c^{2}).

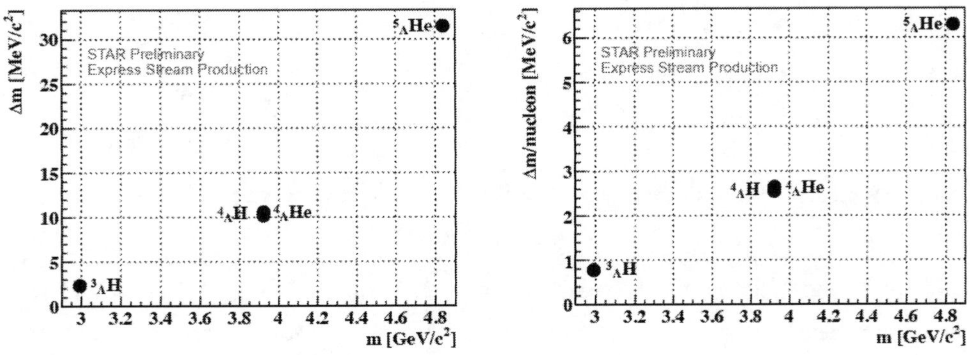

Figure 6. The binding energy of the hypernuclei increases for heavier hypernuclei up to ${}^{5}_{\Lambda}$He. Only statistical errors are shown, which are smaller than the symbol size.

5 Conclusion

During processing and analysis of express data stream on the STAR HLT within the BES-II program in 2018–2021 the reconstruction of particle trajectories was done in real time by the track finder based on the Cellular Automaton, and the search for short-lived particles and hypernuclei by the KF Particle Finder package based on the Kalman filter.

Throughout the data acquisition of the BES-II physics program operation, real-time data analysis showed high quality of the experimental data collected by the STAR detector. The online processing of the express data stream on the STAR HLT farm showed excellent signatures ranging from mesons to hypernuclei. High quality of the collected experimental data and online calibration, and reliable performance of data processing and analysis algorithms allowed us to observe and investigate various hypernuclei up to ${}^{5}_{\Lambda}$He with a significance of 11.6σ. The Dalitz plots of three-body decays of hypernuclei show complex structures with the possible presence of spin effects. There are also hints that a significant fraction of such three-body decays happen via nuclear resonances.

The binding energy per nucleon in hypernuclei increases with increasing A up to ${}^{5}_{\Lambda}$He. There are indications that Λ behaves similarly to other nucleons, n and p, in hypernuclei. Theoretical models of the hypernuclei decays are needed to make a more detailed analysis of the efficiency of three-body decay channels and branching ratios.

6 Acknowledgements

The author considers it his duty to express his gratitude to the entire STAR Collaboration, and especially to the members of STAR HLT/TFG Group Y. Fisyak, H. Ke, S. Margetis, A. Tang, I. Vassiliev and M. Zyzak for their enormous work in obtaining the results given in the paper.

References

[1] J. Steinheimer et al., Phys. Lett. **B 714**, 85 (2012)
[2] Q. Yang (for the STAR Collaboration), Nucl. Phys. **A 982**, 951 (2019)
[3] I. Kisel, (CBM, STAR collaborations), J. Phys. Conf. Ser. **1602**, 012006 (2020)

EPJ Web of Conferences **271**, 08002 (2022) https://doi.org/10.1051/epjconf/202227108002
HYP2022

Recent studies on hypernuclei lifetimes from STAR

Yue Hang Leung[1,*], for the STAR collaboration

[1]University of Heidelberg

Abstract. The hyperon-nucleon (Y-N) interaction is an essential ingredient in the description of the equation-of-state of high-baryon-density matter. Light hypernuclei ($A = 3,4$), being simple Y-N bound states, serve as cornerstones of our understanding of the Y-N interaction. Thus, precise measurements of their lifetimes are important, as they provide stringent tests to hyperon-nucleon interaction models.

The yields of light hypernuclei are expected to increase from high to low energy heavy-ion collisions due to the increase in baryon density. As a result, the STAR Beam Energy Scan II program, which spans an energy range of $\sqrt{s_{NN}} = 3.0 - 27.0$ GeV, is particularly suited for hypernuclei studies. In these proceedings, recent results on the lifetimes of light hypernuclei (${}^3_\Lambda$H, ${}^4_\Lambda$H, ${}^4_\Lambda$He) measured in $\sqrt{s_{NN}} = 3.0$ and 7.2 GeV Au+Au collisions are presented. The relative branching ratio R_3 of the ${}^3_\Lambda$H is intimately related to its lifetime. A new R_3 measurement using data from $\sqrt{s_{NN}} = 3.0$ GeV Au+Au collisions is reported. These results will be compared to previous measurements and theoretical calculations, and the physics implications will be discussed.

1 Introduction

Hypernuclei are nuclei containing at least one hyperon. They serve as important experimental probes to access the hyperon-nucleon (Y-N) interaction, which is an important component in the equation-of-state of high-baryon-density matter, such as neutron stars. Measurements of the lifetimes and binding energies of hypernuclei provide tests for hyperon-nucleon interaction models. In particular, the hypertriton, ${}^3_\Lambda$H, the lightest known hypernuclei, has a small binding energy ($B_\Lambda \sim O(0.1$ MeV)). Due to its loosely bounded nature, it is believed that the ${}^3_\Lambda$H lifetime $\tau({}^3_\Lambda\text{H})$ is very close to the free Λ lifetime $\tau(\Lambda)$, 263 ± 2 ps [1]. Recently, STAR, ALICE and HypHI have reported ${}^3_\Lambda$H lifetimes ranging from $\sim 50\%$ to $\sim 100\%$ of $\tau(\Lambda)$ with large uncertainties. This situation calls for more precise measurements of the ${}^3_\Lambda$H lifetime to clarify the situation.

2 Hypernuclei reconstruction with the STAR detector

In heavy-ion collisions, hypernuclei yields are expected to increase towards lower beam energies due to the increasing baryon density [2]. The STAR Beam Energy Scan II program

*e-mail: leung@physi.uni-heidelberg.de

(BES-II), which covers collision energies from $\sqrt{s_{NN}}$ = 3.0 to 27.0 GeV, provides a great opportunity for studies of light hypernuclei.

At STAR, hypernuclei are reconstructed using their mesonic decay channels, e.g. ${}^{3}_{\Lambda}\mathrm{H} \rightarrow {}^{3}\mathrm{He} + \pi^{-}$, ${}^{4}_{\Lambda}\mathrm{He} \rightarrow {}^{3}\mathrm{He} + \mathrm{p} + \pi^{-}$. Particle identification of the daughter tracks is achieved by the measured ionization energy loss in the Time Projection Chamber (TPC). The KFParticle package [3], based on the Kalman Filter method, is utilized to reconstruct hypernuclei candidates from their daughter tracks. The combinatorial background is estimated via event mixing or rotating all daughter pion candidates within one event [4]. In the following, we will discuss recent hypernuclei lifetime and relative branching ratio results from STAR.

3 Hypernuclei lifetimes

3.1 Measurements from BES-II

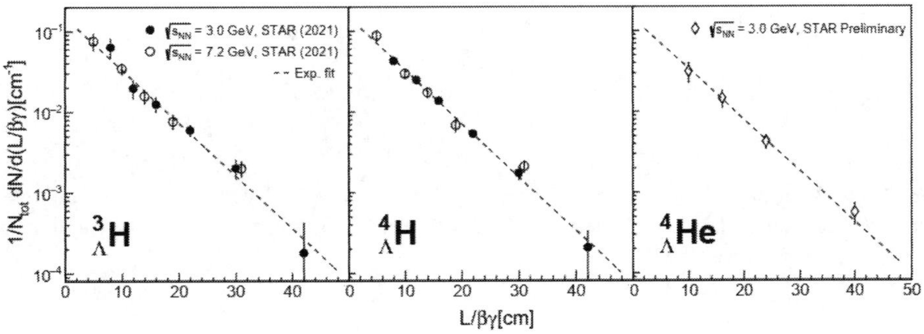

Figure 1. The normalized yield versus the proper decay length $L/\beta\gamma$ for ${}^{3}_{\Lambda}$H (left panel), ${}^{4}_{\Lambda}$H (middle panel), and ${}^{4}_{\Lambda}$He (right panel). The dotted lines represent exponential fits to the data.

In 2018, data from Au+Au collisions at $\sqrt{s_{NN}}$ = 3.0 GeV and 7.2 GeV have been collected, with 258 and 155 million recorded events respectively. The lifetime analyses were carried out for ${}^{3}_{\Lambda}$H and ${}^{4}_{\Lambda}$H using both data sets, while the ${}^{4}_{\Lambda}$He analysis utilized the $\sqrt{s_{NN}}$ = 3.0 GeV dataset only. In order to extract the lifetime, the hypernuclei yields are measured as a function of proper decay length. As shown in Fig. 1, the resultant distributions are well described by exponential functions.

The lifetimes are extracted via χ^2 fits with exponential functions. As a cross-check, the lifetime of the Λ is extracted using the same method and the measured lifetime, $267 \pm 1(stat) \pm 4(syst)$ ps, is consistent with the PDG value [1]. Four major sources of systematic uncertainties are considered. They include imperfect description of topological variables in the GEANT simulations for efficiency estimation, imperfect knowledge of the true kinematic distribution of the hypernuclei, the simulated TPC tracking efficiency, and the signal extraction technique. Their contributions are estimated by varying topological cuts for hypernuclei candidate selection, the hypernuclei transverse montentum p_T and rapidity y distributions in the GEANT simulations, the TPC track quality selection criteria, and the background subtraction method. Other effects, such as particle misidentification, contamination from three-body decays, and coulomb dissociation through target material, have been quantified via Monte-Carlo simulations and are negligible compared to other sources of uncertainty. Different sources of systematic uncertainties are assumed to be uncorrelated and

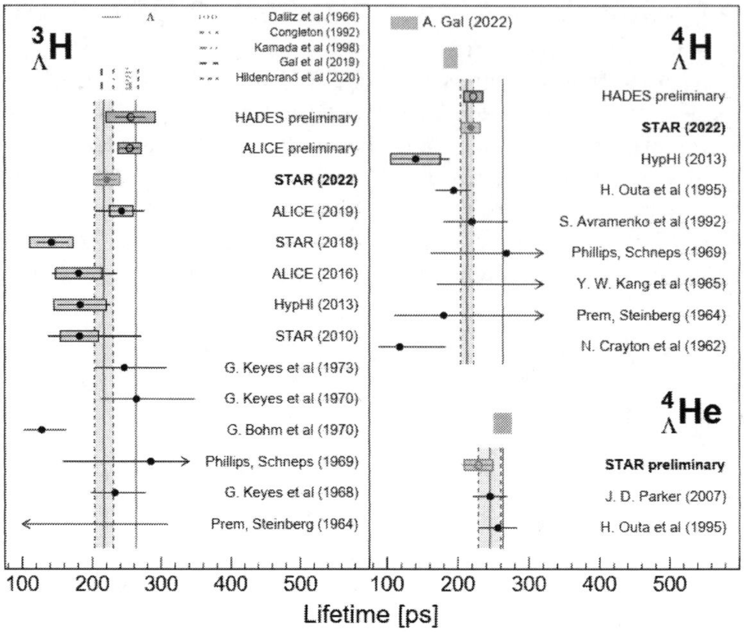

Figure 2. Compilation of $^{3}_{\Lambda}$H, $^{4}_{\Lambda}$H and $^{4}_{\Lambda}$He lifetimes. The experimental average lifetimes are indicated by blue shaded bands. The short dashed lines and solid bands represent theoretical calculations while the solid grey line indicates the free Λ lifetime.

added in quadrature. The total systematic uncertainty amounts to 8.2%, 6.0%, and 8.7% for $^{3}_{\Lambda}$H, $^{4}_{\Lambda}$H, and $^{4}_{\Lambda}$He respectively.

The $^{3}_{\Lambda}$H, $^{4}_{\Lambda}$H, and $^{4}_{\Lambda}$He lifetimes are measured to be 221 ± 15(*stat*) ± 19(*syst*) ps, 218 ± 6(*stat*) ± 13(*syst*) ps [5], and 229 ± 23(*stat*) ± 20(*syst*) ps respectively. In Fig. 2, the results are compared to published measurements, preliminary results from ALICE [6] and HADES [7], and theoretical calculations. The experimentally averaged $^{3}_{\Lambda}$H lifetime is (82 ± 5)% of the Λ lifetime, consistent with theoretical calculations incorporating pion final-state interactions [8]. Similar to the $^{3}_{\Lambda}$H, the experimental averages of $^{4}_{\Lambda}$H and $^{4}_{\Lambda}$He lie below the Λ lifetime. Their ratio $\tau_{avg}(^{4}_{\Lambda}\mathrm{H})/\tau_{\mathrm{avg}}(^{4}_{\Lambda}\mathrm{He})$ is equal to 0.85 ± 0.07, compatible with theoretical estimates invoking the isospin rule [9], which is based on the experimentally measured ratio, $\Gamma(\Lambda \to n + \pi^{0})/\Gamma(\Lambda \to p + \pi^{-}) \approx 0.5$. The new $^{3}_{\Lambda}$H and $^{4}_{\Lambda}$H results from STAR have an improved precision compared to previous measurements, providing stronger constraints to hyperon-nucleon interaction models.

4 Relative branching ratio of the hypertriton

The $^{3}_{\Lambda}$H relative branching ratio R_3, defined as:

$$R_3 = \frac{BR(^{3}_{\Lambda}\mathrm{H} \to {}^{3}\mathrm{He} + \pi^{-})}{BR(^{3}_{\Lambda}\mathrm{H} \to {}^{3}\mathrm{He} + \pi^{-}) + \mathrm{BR}(^{3}_{\Lambda}\mathrm{H} \to \mathrm{d} + \mathrm{p} + \pi^{-})}, \quad (1)$$

where *BR* stands for branching ratio, is an important input to theoretical computations of the $^{3}_{\Lambda}$H lifetime. Calculations has shown that the two-body and three-body mesonic decay channels of the $^{3}_{\Lambda}$H contribute ~ 97% of the total decay rate [10], while the remaining ~ 3%

stems from four-body mesonic decays and non-mesonic decays. Since the π^- decay rate and the π^0 decay rate are related to each other via the isospin rule, the lifetime of the ${}^3_\Lambda$H can be estimated via a hybrid method: theoretically computing the ${}^3_\Lambda\mathrm{H} \rightarrow {}^3\mathrm{He} + \pi^-$ decay rate and combining with the experimentally determined R_3 [11]. Thus, the ${}^3_\Lambda$H R_3 provides an additional handle to access its lifetime.

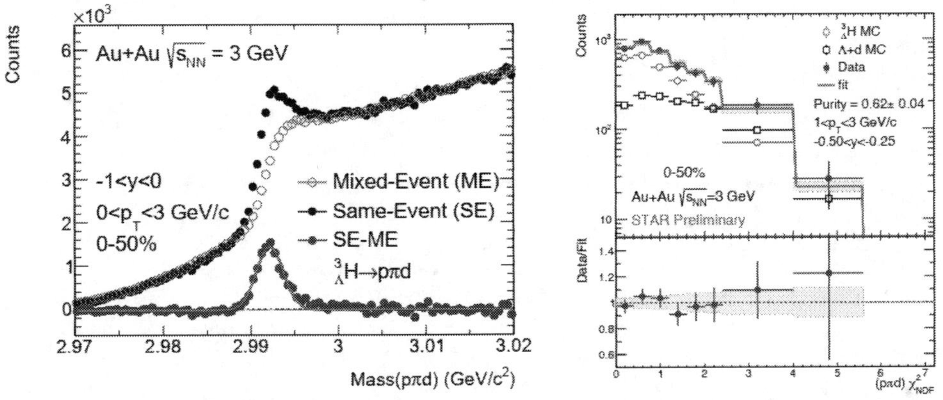

Figure 3. (left) Invariant mass of $d - p - \pi^-$ triplets from Au+Au collisions at $\sqrt{s_{NN}}$ =3.0 GeV. Data are shown as solid black markers, the combinatorial background estimated via event mixing is shown as the open red markers, and the background subtracted distribution is shown as the solid blue markers. (right, upper panel) Template fit to χ^2 of the secondary vertex fit. The solid blue markers represent the data after combinatorial background subtraction, red and black open markers represent contributions from signal and correlated background respectively. The solid red line indicates the sum of the components. (right, bottom panel) Ratio of the data after combinatorial background subtraction to the sum of the components.

The ${}^3_\Lambda$H yields are measured in $\sqrt{s_{NN}}$ = 3.0 GeV Au+Au collisions via both two-body and three-body decay channels. The extraction of ${}^3_\Lambda$H signal via its three-body decay is more complicated compared to the two-body decay due to significant contributions of correlated background in its invariant mass spectrum. As demonstrated in Fig. 3, after subtracting the combinatorial background which is estimated via event-mixing, a template fit is applied to the data to statistically separate contributions from correlated background and signal. The fit exploits the fact that, for true ${}^3_\Lambda$H signal, all three daughter tracks point to the same secondary vertex, which gives rise to smaller χ^2 values in the secondary vertex fit, while correlated background does not, and lead to larger χ^2 values.

By comparing the corrected yields from the two decay channels, R_3 can be determined. The preliminary result $R_3 = 0.27 \pm 0.03(stat) \pm 0.04(syst)$, as shown in Fig. 4, is consistent with previous measurements. Our new measurements provide improved precision to the ${}^3_\Lambda$H R_3, which, aside from its connection to the lifetime, may also provide constraints on its binding energy [11].

5 Summary and outlook

In summary, recent studies on hypernuclei lifetimes and branching ratios from STAR have been discussed. New lifetime measurements of ${}^3_\Lambda$H, ${}^4_\Lambda$H, and ${}^4_\Lambda$He from the Beam Energy Scan II progam have been presented. In particular, the new ${}^3_\Lambda$H and ${}^4_\Lambda$H lifetime measurements are the most precise published results to date, providing strong constraints to hyperon-nucleon

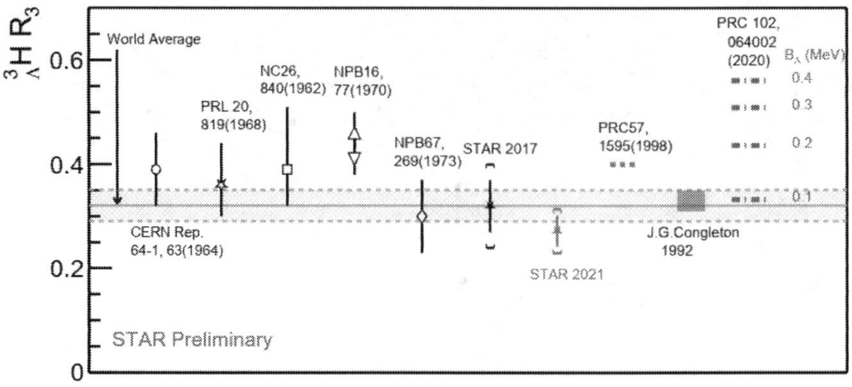

Figure 4. Compilation of $^{3}_{\Lambda}$H relative branching ratio R_3. The experimental average R_3 is indicated by the blue shaded band. The magenta box and dashed red and orange lines represent theoretical calculations.

interaction models. In addition, the $^{3}_{\Lambda}$H relative branching ratio R_3 has been extracted in $\sqrt{s_{NN}}$ = 3.0 GeV Au+Au collisions. The improved precision on R_3 provides the necessary input for connecting theoretically computed two-body mesonic decay rates and the $^{3}_{\Lambda}$H lifetime.

References

[1] P. Zyla et al. (Particle Data Group), Prog. Theor. Exp. Phys. 2020, 083C01 (2020)
[2] A. Andronic et al., Phys. Lett. B **697**, 203 (2011), (update, preliminary)
[3] I. Kisel (CBM, STAR collaborations), J. Phys. Conf. Ser. **1602**, 012006 (2020)
[4] J. Adam et al. (STAR collaboration), Phys. Rev. C **102**, 034909 (2020)
[5] M. S. Abdallah et al. (STAR collaboration), Phys. Rev. Lett. **128**, 202301 (2022)
[6] E. Bartsch (ALICE collaboration), Nucl. Phys. A **1005**, 121791 (2021)
[7] S. Spies (HADES collaboration), these proceedings
[8] A. Gal and H. Garcilazo, Phys. Lett. B **791**, 48 (2019)
[9] A. Gal, EPJ Web Conf. **259**, 08002 (2022)
[10] H. Kamada et al., Phys. Rev. C **57**, 1595 (1998)
[11] F. Hildenbrand and H.-W. Hammer, Phys. Rev. C **102**, 064002 (2020)

Studies of hypernuclei with heavy-ion beams, nuclear emulsions and machine learning

Takehiko R. Saito[1,2,3,*]
for the WASA-FRS collaboration and the Emulsion-ML collaboration [**]

[1]High Energy Nuclear Physics Laboratory, RIKEN, Wako, Japan
[2]GSI Helmholtz Centre for Heavy Ion Research, Darmstadt, Germany
[3]School of Nuclear Physics and Technology, Lanzhou University, China

Abstract. The lightest hypernucleus, the hypertriton, has been a benchmark in the field of hypernuclear physics. However, some of recent experiments employing energetic heavy-ion beams have revealed that the hypertriton lifetime is significantly shorter than 263 ps which is expected by considering the known weakly binding nature of the hypertriton. The STAR collaboration has also measured the hypertriton binding energy, and the deduced value is contradicting to its formerly known small binding energy. These measurements have indicated that the fundamental physics quantities of the hypertriton such as its lifetime and binding energy have not been understood, therefore, they have to be measured very precisely. Furthermore, an unprecedented Λnn bound state observed by the HypHI collaboration has to be studied in order to draw a conclusion whether or not such a bound state exists. These three-body hypernuclear states are studied by the heavy-ion beam data in the WASA-FRS experiment and by analysing J-PARC E07 nuclear emulsion data with machine learning.

1 Introduction

The hypertriton lifetime and binding energy have recently become hot topics in the field of hypernuclear and few-body physics. The hypertriton was extensively studied until the 1970s by using nuclear emulsions and bubble chambers. Those experiments concluded that a Λ hyperon is very weakly bound to a deuteron core with a small binding energy of 0.13 ± 0.05 MeV [1, 2]. The lifetime of the hypertriton was also measured [3–8] but did not deliver a firm conclusion on the lifetime due to inaccuracies in the measurement. Therefore, the lifetime of the hypertriton has been considered to be very close to the lifetime of a free Λ hyperons, i.e., 263 ps [9], without having precise experimental data.

Recently, studies of light hypernuclei by means of induced reactions [10] of and collisions of heavy-ion beams have provided new precise data on the hypertriton lifetime. The HypHI experiment at GSI studied hypertritons produced in the ^{6}Li+^{12}C reactions at 2 *A* GeV and showed that the deduced lifetime of the hypertriton, 183^{+42}_{-32}(stat.)±37(syst.) ps [10], is significantly shorter than the presently considered value [9]. Ultra-relativistic heavy-ion collisions have also become a powerful tool for studying the hypertriton, and the lifetime of the

*e-mail: takehiko.saito@riken.jp
**Members of the Emulsion-ML collaboration can be found in the author list of one of these proceedings by Manami Nakagawa

hypertriton was also measured by the STAR collaboration at the RHIC and the ALICE collaboration at the LHC. All recently published values are summarised in Table 1. Additionally, the HADES collaboration measured the lifetime of hypertritons produced in ^{197}Au+^{197}Au at 1.23 *A* GeV, and the tentative deduced value is $256^{+22}_{-36} \pm 27$ ps [18]. The STAR collaboration performed the combined analysis for all the published hypertriton lifetime values, and they have concluded that the averaged value is 200 ± 13 ps and that it is considerably lower than that of the Λ hyperon [15]. However, as shown in Table 1, the measured values have a large span even for measurements obtained by one collaboration. Furthermore, the precision of the current measurements does not allows us to conclude about the value of the hypertriton lifetime. Therefore, at least one measurement with an excellent accuracy is needed to clarify the situation. The binding energy of the hypertriton is also of great interest since it is expected to be strongly correlated to its lifetime. After the binding energy was deduced by employing nuclear emulsions in 1968 [1] and 1973 [2], additional experimental studies have not been performed for he hypertriton binding energy. However recently, the STAR collaboration deduced the binding energy of the hypertriton B_Λ to be 0.41 ± 0.12(stat.) $\pm$ 0.11(syst.) MeV [19]. This value is significantly larger than the previously known binding energy of 0.13 ± 0.05 MeV from the nuclear emulsions, however, the accuracy of the STAR result is not sufficient to reach a firm conclusion.

Another type of three-body hypernuclear states have also become to be of special interest. The HypHI collaboration observed an enhancement in the invariant mass distributions of the d+π^- and t+π^- final states [20]. Although the significance of the possible signals is not large, approximately 5 σ, those results may indicate the existence of an unprecedented neutral bound state formed by a Λ hyperon and two neutrons, i.e., Λnn [20]. This observation has initiated theoreticians and experimentalists to study this state. A large theoretical effort has been carried out using various approaches to check the existence of the Λnn bound state [21–24], however, the obtained results did not show that the system Λnn is bound. On the other hand, pionless effective field theory studying the Λnn state with $I = 0$ has not ruled out the bound state [25, 26]. Possible resonance states associated with Λnn have also been studied theoretically [24, 27]. Very recently, the E12-17-003 experiment was performed at JLab to search for bound and resonance states with Λnn by employing electron beams to bombard a tritium target, and an indication of the existence of the Λnn state has been observed with small significance [28]. It should be, however, noted that the same collaboration published different interpretations of the data, which suggest non-existence of Λnn [29], therefore, a conclusion has not been reached yet. The existence of the Λnn should be thus further studied experimentally with better precision and more data samples.

As discussed above, there are puzzles associated with three-body hypernuclear states, namely on the hypertriton and Λnn. We are aiming to solve these puzzles by performing precise measurements. We will exploit data from heavy-ion collisions collected by the WASA-FRS experiment to address the issues of the hypertriton lifetime and the Λnn bound state, and we will apply machine learning techniques to identify hypertritons produced in nuclear emulsions use them to determine hypertriton binding energy [30].

2 The WASA-FRS experiment with heavy-ion beams

We have developed a novel technique for producing and measuring hypernuclei by utilising the GSI fragment separator (FRS) [31] as a high momentum-resolution forward magnetic spectrometer. A schematic layout of the FRS is presented in Figure 1. Inside of the mid focal plane (F2) of the FRS indicated by a red circle in the figure, the WASA central detector [32] is mounted. The WASA central detector that is used in our experiment consists of a superconducting solenoid magnet (up to 1 T magnetic field) with its associated iron yokes and

Table 1. Summary of the measured hypertriton lifetimes obtained by the HypHI, the STAR and the ALICE collaborations. The values are shown with their statistical and systematic uncertainties.

Exp.	Reaction and energy	Prod. meth.	Lifetime [ps]	Year of pub.
HypHI at GSI	^{6}Li+^{12}C at 2 *A* GeV	Projectile fragmentation	$183^{+42}_{-32} \pm 37$	2013 [10]
STAR at RHIC	^{197}Au+^{197}Au at $\sqrt{s_{NN}} = 200$ GeV	Central collision	$182^{+89}_{-45} \pm 27$	2010 [11]
			$155^{+25}_{-22} \pm 31$	2017 [12]
			$142^{+24}_{-21} \pm 29$	2018 [13, 14]
	^{197}Au+^{197}Au at $\sqrt{s_{NN}} = 3$ & 7.2 GeV		$221^{+15}_{-19} \pm 27$	2022 [15]
ALICE at the LHC	^{208}Pb+^{208}Pb at $\sqrt{s_{NN}} = 2.76$ TeV	Central collision	$181^{+54}_{-39} \pm 33$	2016 [16]
	^{208}Pb+^{208}Pb at $\sqrt{s_{NN}} = 5.02$ TeV		$242^{+34}_{-38} \pm 17$	2019 [17]

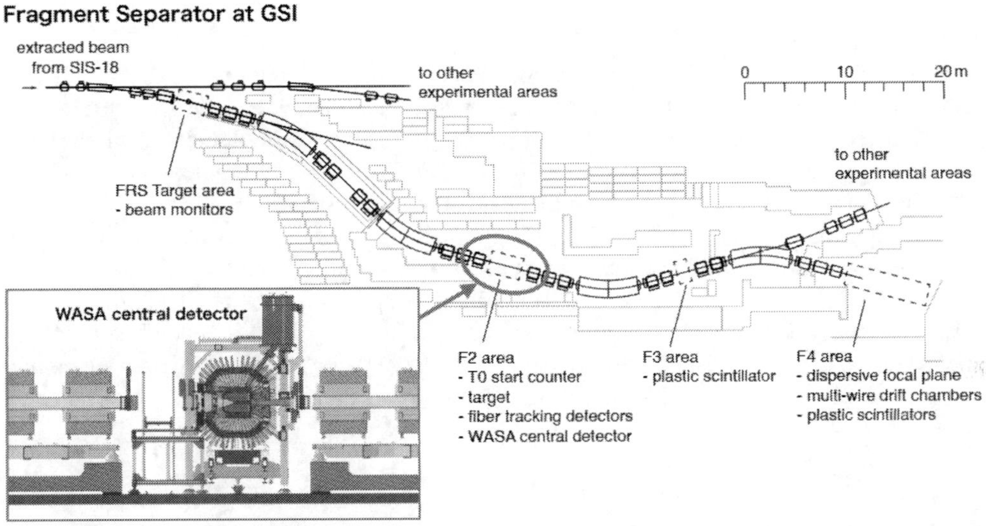

Figure 1. The Fragment Separator (FRS) and the WASA detector.

a cryogenic system, a calorimeter with CsI crystals, inner drift chambers with straw tubes, and newly developed arrays of plastic barrel hodoscopes [33] and end-cap hodoscopes [34]. Additionally, we have supplemented the detector with six scintillating fibre detectors and a start counter for Time-of-Flight measurements dedicated to the WASA-FRS hypernuclear experiments [34]. The experiments which are using the WASA detector and the FRS are called "the WASA-FRS experiments" [30]. The WASA-FRS experiments studying the hypertriton and Λnn production use beams of ^{6}Li at 2 *A* GeV provided by the SIS-18 synchrotron. These beams are delivered to the mid-focal plane of the FRS, and hypernuclei of interest are produced on the production target installed in front of the WASA detector. Residual nuclei after the emission of a π^{-} meson are forward boosted since they are produced from the decay of hypernuclei that emerge as projectile fragments. Therefore, approximately 10% of the residual nuclei from the hypernuclei of interest fall within the acceptance of the FRS behind the F2 and are further transported from F2 through the final focal plane (F4). The particle rate

Figure 2. Photographs of the WASA setup with a side view (left) and the iron-yoke opened (right). The photos are provided by Jan Hosan and GSI/FAIR.

at F4 is estimated to be only a few hundreds Hz even with 10 times larger beam intensity than that in the HypHI, which is sufficiently small to take data only with a requirement of particle detection at F4. Therefore, it provides an opportunity to implement a simpler data-acquisition system without a complicated displaced vertex trigger, as employed in the HypHI [10]. The use of the FRS for measuring the residual nuclei produced by the hypernuclear decay of interest without the displaced vertex trigger will increase the size of the data sample by approximately a factor 100 in comparison to the HypHI. Furthermore, the standard resolving power of the FRS is 10^4 for the measurement of the momentum of the residual nuclei, which is approximately three orders of magnitude better than the HypHI. Decays of hypernuclei of interest take place inside the F2 area and π^- mesons from the hypernuclear decay are emitted widely, and they are measured by the WASA detector with additional detectors at F2 of the FRS.

According to Monte Carlo simulations [30, 34], the signal integral is approximately 5.8×10^3, and the significance of the peak was calculated with the amount of the signal and background to be approximately 120 σ. In comparison to the HypHI, the peak integral has been increased by a factor of approximately 38, and the significance is approximately 25 times higher [10]. The peak width is $\sim$ 3.18 MeV which is 1.5 times narrower than that observed in the HypHI [10]. The accuracies for the lifetime have also been estimated for different lifetime values and they are 5 ps for the lifetime value of 150 ps, 8 ps for 200 ps and 13 ps for 250 ps. Additionally, we have estimated that the performance of the measurements should be similar for the case when one searches for the existence of the Λnn bound state. This measurement should reach the same level of significance as for the hypertriton case, which is approximately 120 σ.

The WASA-FRS experiment to study the hypertriton, ${}^4_\Lambda$H and Λnn was already performed at GSI in the first quarter of 2022 with ^{6}Li projectiles at 2 A GeV bombarding a diamond target with a thickness of 9.87 g/cm^2, and the data analysis is in progress. Developments of new data analysis algorithms for track recognition with the Graph Neural Network (GNN) [35] have already been completed and are being prepared for publication [36]. Details of the performed WASA-FRS experiment and its performance are discussed by Ekawa in these proceedings [34].

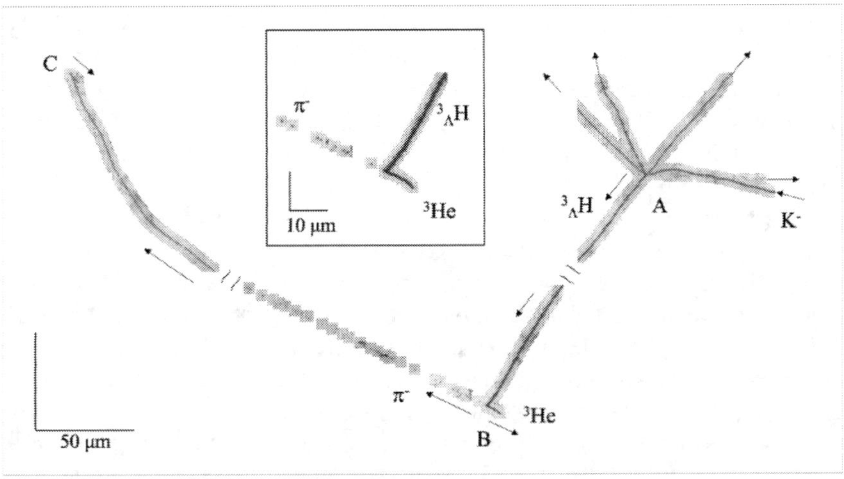

Figure 3. The first discovered hypertriton event in the nuclear emulsion of the E07 experiment at J-PARC [30]. A K^- meson enters the emulsion from outside the emulsion modules and then propagates to the point A. Here the K^- meson is stopped and it is absorbed by one of the nuclei in the emulsion to produce five visible nuclear fragments and $^3_\Lambda$H. The hypertriton ($^3_\Lambda$H) proceeds towards the point B, where it stops. The hypertriton decays to ^{3}He and π^-. The inset of the figure shows a magnification around the point B.

3 Precise measurement of the hypertriton binding energy with nuclear emulsion and machine learning

In order to solve the hypertriton puzzle, in addition we plan to determine the hypertriton binding energy very precisely, in addition to improving the accuracy of the hypertriton lifetime with the WASA-FRS experiment discussed in the previous subsection. We are currently re-analysing the existing J-PARC E07 nuclear emulsions by means of machine learning techniques to deduce the binding energy of the hypertriton to exploit this large data sample and to achieve the best possible precision.

The E07 experiment [37] at J-PARC, aims to search for double-strangeness hypernuclei such as double-Λ- and Ξ-hypernuclei. The experiment was already performed in 2016–2017, and it employed dedicated nuclear emulsion sheets. Approximately 1300 nuclear emulsion sheets were irradiated by intense K^- beams. Searches for double-strangeness hypernuclei were performed by using the so-called hybrid method [37]. In this method, a Ξ^- hyperon produced by the (K^-, K^+) reaction is identified and tracked by other detectors when the outgoing K^+ mesons are also detected, and the position of the stopped Ξ^- hyperon that induces the production of double-strangeness hypernuclei is estimated. Visual inspections by an optical microscope were performed only around the estimated stop position. Therefore, a very tiny portion of all the irradiated emulsions was analysed, which greatly reduced the human workload. Hypertritons were also produced by induced reactions of K^- beams on nuclei inside the E07 nuclear emulsions. Tracks of produced hypertritons and charged particles originated from their decays are also recorded in the emulsions. However, in order to identify these hypertriton decays, a complete scan of the entire nuclear emulsion sheets is necessary since the information associated with the hypertriton production is not recorded by the hybrid method.

We are aiming at detecting hypertriton decay at rest via the ^{3}He+π^- two-body decay, which can be used to measure the invariant mass which is necessary to deduce the binding energy. The advantage of using this hammer-like event with decay into ^{3}He+π^- is that the track length of the π^- meson is well defined only by the two-body decay kinematics, being approximately 28 mm. This length is unique and very different for the other hypernuclear decays (for example, the π^- track length for the case of $^4_\Lambda$H is approximately 42 mm). Therefore, the hypertriton event can be firmly identified. However, since the shape of the hypertriton decay event is too simple and is associated with a very sparse π^- track, analytical filters can hardly classify these shapes from other backgrounds with high efficiency; therefore, candidates should be sorted out by human visual inspection with the current technology for emulsion analysis. There are approximately 1.4 billion images per emulsion sheet to be inspected, hence an analysis of one emulsion sheet would require about 560 years [30]. Furthermore, the ratio of the number of candidates to be found by visual inspection over the number of images without candidates is expected to be only $\sim 10^{-7}$. Because of the huge human load and the extremely small signal-to-background ratio, hypertritons were never observed in the data of the E07 [37] experiment and the former experiments at KEK, and only a small amount of other single-strangeness hypernuclei were observed in the same data sets.

For overcoming the difficulties discussed above, we have developed machine learning models to detect candidates for hypertriton events in the E07 nuclear emulsion data [30]. One of difficulties in the development is that there is no real training data with hypertriton events since no hypertriton event was observed in the E07 nuclear emulsion. Therefore, we employed Monte Carlo simulations to produce data emulating production and decay of the hypertriton in the emulsion material. The Monte Carlo simulations calculate particle trajectories as sharp lines. These lines have to be converted to defused line shapes like track images in the emulsion by considering the size of grains in the emulsion. This is achieved by employing one of the machine learning techniques, i.e., Generative Adversarial Networks (GANs) [38]. Background images are taken randomly from the real emulsion data and mixed with the converted defused lines to imitate hypertriton events in the nuclear emulsion. With this technique, we have produced a sufficient amount of training data for developing the machine learning models. We have already developed an object detector for hypertriton events with one machine learning model called Mask R-CNN [39]. We employ the surrogated images for training the object detection model. For the analysis, the emulsion sheets of the E07 experiment are scanned by the dedicated automated microscope scanning devices and all the raw images are recorded on a storage disk. The trained machine learning models are applied to the stored data to detect tracks of interest within real emulsion images. It has already been shown that the current model improves the signal-to-background ratio from $\sim 10^{-7}$ to 2×10^{-4}. The human load decreased by a factor of 2000. This improvement has already allowed us to mine hypertriton events with the support of visual inspection. After the detection of candidates of the images that may be associated to the hypertriton events, they are again visually inspected, and clear candidates are sorted out. Then, we perform a detailed analysis of the tracks by using the microscopes. The procedure is already established and the number of the discovered hypertriton decays increases day by day. The first hypertriton event discovered on February 2nd in 2021 [30, 40] is displayed in Figure 3. We have already analysed approximately 0.9% of the total emulsion data (1300 sheets) until August 25th in 2022, and we have already discovered few tens of hypertriton events. Additionally, four times more $^4_\Lambda$H events have been discovered and identified. The discovery of these hypertriton events in the E07 emulsion data already guarantees the improvement in the accuracy for the determination of the hypertriton binding energy. The systematic uncertainty for the determination of the hypertriton binding energy has been estimated to be 28 keV by revisiting the former emulsion data and by employing Monte Carlo simulations [41], and this uncertainty will be improved by introducing a new method for calibrating the density of nuclear emulsions, which will be completed soon [40]. The developed machine learning technique will be applied to search for three-body decay events, therefore, the binding energy of, for example, $^{4}_{\Lambda}$He and $^{5}_{\Lambda}$He will be precisely determined [42]. The current status of the precise determination of the hypertriton binding energy in this project and its perspective are discussed by Nakagawa in these proceedings [42].

4 Summary and outlook

Physicists dealing with few-body hypernuclei systems are highly interested to improve precision of the lifetime and the binding energy of the hypertriton. In addition, it needs to be clarified whether or not the Λnn bound state exists. Recently, experiments with heavy-ion beams such as ALICE, HADES, HypHI and STAR measured the hypertriton lifetime, however, the obtained results vary a lot thus a conclusion has not been yet drawn. The recent measurement of the hypertriton binding energy by STAR has indicated that the hypertriton binding energy could be significantly larger than the formerly known value of 130 ± 50 keV, however, the precision of the deduced binding energy by STAR does not allow us to come to a firm conclusion. Therefore, the precise measurements on the both lifetime and binding energy of the hypertriton are awaited. We will provide the most accurate data on both the lifetime and binding energy of the hypertriton for improving our understanding of the real nature of the hypertriton drastically. Furthermore, the existence of the Λnn state has to be experimentally studied with high precision. We have already performed the WASA-FRS experiment at GSI with the ^{6}Li+^{12}C reaction at 2 *A* GeV for measuring the hypertriton lifetime with an accuracy of approximately 10 ps and for studying whether or not the Λnn state can exist. The analysis is in progress and the results will be presented in near future. For the hypertriton binding energy, we are analysing the J-PARC E07 nuclear emulsions by developing the machine learning models, and the detection of the hypertriton events have already been accomplished. Determination of the hypertriton binding energy with an accuracy of 28 keV or better is foreseen in near future.

Similar experiments like the WASA-FRS will be continued with the Super-FRS [43] at FAIR (Facility of Anti-proton and Ion Research) [44] in Germany. The upgrade of the superconducting solenoid magnet of the WASA detector is planed, and we are aiming at achieving a maximum central magnetic field of 2 T by using a cryocooler. The upgrade of the inner drift chamber is also in progress. With the upgraded WASA at Super-FRS, lifetime and binding nature of hypernuclei, especially for proton rich hypernuclei [30, 45], will be investigated. Furthermore, very neutron rich hypernuclei will be studied by employing charge-exchange-reactions combined with strangeness production [46]. We are also preparing hypernuclear experiments with heavy-ion beams at HIAF (High-Intensity Heavy Ion Accelerator Facility) [47] in China, and it is planed to extend hypernuclear studies to double-strangeness [30]. The analysis of the E07 nuclear emulsions will be continued, and the analysis for single-strangeness hypernuclei will be completed in near future. The development of machine learning models for detecting events associated with double-strangeness hypernuclei has already been started, and it is expected to detect approximately 1000 double-strangeness hypernuclear events by analysing the entire E07 emulsion data. Design studies for future experiments with nuclear emulsions at J-PARC and elsewhere have also been started.

References

[1] G. Bohm, et al., Nucl. Phys. **B 4**, 511–526 (1968)

[2] M. Juric, et al., Nucl. Phys. **B 52**, 1–30 (1973)
[3] R. Prem, et al., Phys. Rev. **136**, B1803–B1806 (1964)
[4] G. Keyes, et al., Phys. Rev. Lett. **20**, 819–821 (1968)
[5] R.E. Phillips, et al., Phys. Rev. **180**, 1307–1318 (1969)
[6] G. Bohm, et al., Nucl. Phys. **B 16**, 46–52 (1970)
[7] G. Keyes, et al., Phys. Rev. **D 1**, 66–77 (1970)
[8] G. Keyes, et al., Nucl. Phys. **B 67**, 269–283 (1973)
[9] Particle Data Group, et al., Prog. Theor. Exp. Phys. 2020, (2020)
[10] C. Rappold, et al., Nucl. Phys. **A 913**, 170–184 (2013)
[11] The STAR collaboration, Science **328**, 58–62 (2010)
[12] Xu, Y. for the STAR Collaboration, Proc. 12th Int. Conf. on Hypernucl. Strange Part. Phys. (HYP2015) 021005 (2017)
[13] L. Adamczyk, et al., Phys. Rev. **C 97**, 054909 (2018)
[14] J. Chen, et al., Phys. Reports **760**, 1–39 (2018)
[15] M.S. Abdallah, et al., Phys. Rev. Lett. **128**, 202301 (2022)
[16] J. Adam, et al., Phys. Lett. **B 754**, 360–372 (2016)
[17] S. Acharya, et al., Phys. Lett. **B 797**, 134905 (2019)
[18] Simon Spies for the HADES collaboration, in these proceedings
[19] J. Adam, et al., Nat. Phys. 16, 409–412 (2020)
[20] C. Rappold et al., Phys. Rev. **C 88**, 041001 (2013)
[21] E. Hiyama, et al., Phys. Rev. **C 89**, 061302 (2014)
[22] A. Gal, et al., Phys. Lett. **B 736**, 93–97 (2014)
[23] H. Garcilazo, et al., Phys. Rev. **C 89**, 057001 (2014)
[24] M. Schäfer, et al., Phys. Lett. **B 808**, 135614 (2020)
[25] S. -I. Ando, et al., Phys. Rev. **C 92**, 024325 (2015)
[26] F. Hildenbrand, et al., Phys. Rev. **C 100**, 034002 (2019)
[27] I.R. Afnan, et al., Phys. Rev. **C 92**, 054608 (2015)
[28] B. Pandey, et al., Phys. Rev. **C 105**, L051001 (2022). L. Tang, et al., these proceedings
[29] K. Itabashi, et al., Few Body Syst. **63**, 16 (2022). K.N. Suzuki, et al., PTEP **2022**, Issue 1, 013D01 (2022)
[30] T.R. Saito et al., Nat. Rev. Phys. **3**, 803–813 (2021)
[31] H. Geissel, et al., Nucl. Instruments Methods Phys. Res. **B 70**, 286–297 (1992)
[32] C. Bargholtz, et al., Nucl. Instruments Methods Phys. Res. **A 594**, 339–350 (2008)
[33] R. Sekiya et al., Nucl. Instruments Methods Phys. Res. **A 1034**, 166745 (2022)
[34] H. Ekawa for the WASA-FRS collaboration, these proceedings
[35] F. Drielsma, et al., arXiv preprint (2020), https://arxiv.org/abs/2007.01335. J. Zhou, et al., arXiv preprint (2019), https://arxiv.org/ abs/1812.08434
[36] H. Ekawa, et al, submitted to Eur. Phys. J. A
[37] E. Ekawa, et al., Prog. Theor. Exp. Phys. **2019** , 021D02 (2019). S.H. Hayakawa, et al., Phys. Rev. Lett. **126**, 062501 (2021)
[38] I.J. Goodfellow, et al., arXiv preprint (2014), https://arxiv.org/abs/1406.2661. T. Wang, et al., arXiv preprint (2017), http://arxiv.org/abs/1711.11585
[39] K. He, et al., arXiv preprint (2017), https://arxiv.org/abs/1703.06870
[40] A. Kasagi, Ph.D. thesis, Gifu University, 2023
[41] E. Liu et al., Eur. Phys. J. **A 57**, 327 (2021)
[42] M. Nakagawa, et al., these proceedings

[43] Webpage of Super-FRS. https://fair-center.eu/en/for-users/experiments/nustar/super-frs.html

[44] Webpage of Facility for Antiproton and Ion Research in Europe (FAIR). https://fair-center.eu

[45] C. Rappold, et al., Phys. Rev. **C 94**, 044616 (2016)

[46] T.R. Saito, et al., Eur. Phys. J. **A 57**, 159 (2021)

[47] Webpage of High Intensity Heavy-ion Accelerator Facility (HIAF). http://hiaf.impcas.ac.cn/hiaf_en/public/c/news.html

New results on light nuclei, hyperons and hypernuclei from HADES

(HADES collaboration)

Rayane Abou Yassine[6,14], *Jörn* Adamczewski-Musch[5], *Marten* Becker[10], *Philip* Bergmann[5], *Alberto* Blanco[1], *Christoph* Blume[8], *Lukas* Chlad[15,21], *Petr* Chudoba[15], *Izabela* Ciepał[3], *Malte* Cordts[6], *Jörn* Dreyer[7], *Waleed Ahmed* Esmail[5], *Miroslaw* Firlej[2], *Tomasz* Fiutowski[2], *Henrik* Floersheimer[6], *Paulo* Fonte[1,19], *Jürgen* Friese[9], *Ingo* Fröhlich[8], *Jörg* Förtsch[18], *Tetyana* Galatyuk[6,5], *Tomasz* Gniazdowski[17], *Robert* Greifenhagen[7,20], *Mateusz* Grunwald[17], *Dieter* Grzonka[11], *Malgorzata* Gumberidze[5], *Szymon* Harabasz[6], *Thorsten* Heinz[5], *Claudia* Höhne[10,5], *Fatima* Hojeij[14], *Romain* Holzmann[5], *Holger* Huck[8], *Marek* Idzik[2], *Burkhard* Kämpfer[7,20], *Karl-Heinz* Kampert[18], *Behruz* Kardan[8], *Vadym* Kedych[6], *Ilse* Koenig[5], *Wolfgang* Koenig[5], *Marvin* Kohls[8], *Jedrzej* Kolas[17], *Grzegorz* Korcyl[4], *Georgy* Kornakov[17], *Frederic* Kornas[6,5], *Roland* Kotte[7], *Wilhelm* Krueger[6], *Andrej* Kugler[15], *Pawel* Kulessa[11], *Rafal* Lalik[4], *Semen* Lebedev[10], *Sergey* Linev[5], *Luís* Lopes[1], *Manuel* Lorenz[8], *Akshay* Malige[4], *Jochen* Markert[5], *Tomasz* Matulewicz[16], *Johan* Messchendorp[5], *Volker* Metag[10], *Jan* Michel[8], *Aleksandra* Molenda[2], *Jakub* Moron[2], *Christian* Müntz[8], *Marvin* Nabroth[8], *Lothar* Naumann[7], *Jan* Orliński[16], *Jan-Hendrik* Otto[10], *Yannis* Parpottas[12], *Mirco* Parschau[8], *Christian* Pauly[18], *Vladimir* Pechenov[5], *Olga* Pechenova[5], *Gabriela* Perez Andrade[11], *Dennis* Pfeifer[18], *Krzysztof* Piasecki[16], *Jerzy* Pietraszko[5], *Tetiana* Povar[18], *Alexandr* Prozorov[15,21], *Witold* Przygoda[4], *Krzysztof* Pysz[3], *Béatrice* Ramstein[14], *Narendra* Rathod[4], *James* Ritman[5], *Adrian* Rost[6,5], *Anar* Rustamov[5], *Piotr* Salabura[4], *Joao* Saraiva[1], *Susan* Schadmand[5], *Niklas* Schild[6], *Erwin* Schwab[5], *Florian* Seck[6], *Ilya* Selyuzhenkov[5], *Udai* Singh[4], *Leon* Skorpil[8], *Jerzy* Smyrski[4], *Manfred* Sobiella[7], *Stefano* Spataro[22], *Simon* Spies[8,*], *Maria* Stefaniak[17], *Herbert* Ströbele[8], *Joachim* Stroth[8,5], *Konrad* Sumara[4], *Ondřej* Svoboda[15], *Krzysztof* Swientek[2], *Melanie* Szala[8], *Pavel* Tlusty[15], *Michael* Traxler[5], *Haralabos* Tsertos[12,13], *Vladimir* Wagner[15], *Mateusz* Wasiluk[17], *Adrian Amatus* Weber[10], *Christian* Wendisch[5], *Peter* Wintz[11], *Hanna* Zbroszczyk[17], *Elizaveta* Zherebtsova[5,23], *Marcin* Zielinski[4], and *Peter* Zumbruch[5]

[1]LIP-Laboratório de Instrumentação e Física Experimental de Partículas , 3004-516 Coimbra, Portugal
[2]AGH University of Science and Technology, Faculty of Physics and Applied Computer Science, 30-059 Kraków, Poland
[3]Institute of Nuclear Physics, Polish Academy of Sciences, 31342 Kraków, Poland
[4]Smoluchowski Institute of Physics, Jagiellonian University of Cracow, 30-059 Kraków, Poland
[5]GSI Helmholtzzentrum für Schwerionenforschung GmbH, 64291 Darmstadt, Germany
[6]Technische Universität Darmstadt, 64289 Darmstadt, Germany
[7]Institut für Strahlenphysik, Helmholtz-Zentrum Dresden-Rossendorf, 01314 Dresden, Germany
[8]Institut für Kernphysik, Goethe-Universität, 60438 Frankfurt, Germany
[9]Physik Department E62, Technische Universität München, 85748 Garching, Germany
[10]II.Physikalisches Institut, Justus Liebig Universität Giessen, 35392 Giessen, Germany
[11]Forschungszentrum Juelich, 52428 Juelich, Germany

[*]e-mail: s.spies@gsi.de

[12]Frederick University, 1036 Nicosia, Cyprus
[13]Department of Physics, University of Cyprus, 1678 Nicosia, Cyprus
[14]Laboratoire de Physique des 2 infinis Irène Joliot-Curie, Université Paris-Saclay, CNRS-IN2P3. , F-91405 Orsay , France
[15]Nuclear Physics Institute, The Czech Academy of Sciences, 25068 Rez, Czech Republic
[16]Uniwersytet Warszawski - Instytut Fizyki Doświadczalnej, 02-093 Warszawa, Poland
[17]Warsaw University of Technology, 00-662 Warsaw, Poland
[18]Bergische Universität Wuppertal, 42119 Wuppertal, Germany
[19]also at Coimbra Polytechnic - ISEC, Coimbra, Portugal
[20]also at Technische Universität Dresden, 01062 Dresden, Germany
[21]also at Charles University, Faculty of Mathematics and Physics, 12116 Prague, Czech Republic
[22]also at Dipartimento di Fisica and INFN, Università di Torino, 10125 Torino, Italy
[23]also at University of Wrocław, 50-204 Wrocław, Poland

Abstract. In March 2019 the HADES experiment recorded 14 billion Ag+Ag collisions at $\sqrt{s_{NN}} = 2.55$ GeV as a part of the FAIR phase-0 physics program. In this contribution, we present and investigate our capabilities to reconstruct and analyze weakly decaying strange hadrons and hypernuclei emerging from these collisions. The focus is put on measuring the mean lifetimes of these particles.

1 Introduction

In heavy-ion collisions in the energy regime of few GeV per nucleon, which are explored by HADES, the nuclear matter is heated to temperatures of 70 MeV deduced from dielectron spectroscopy [1]. At the same time, it is compressed to densities above its ground state density. Densities and temperatures of the same order of magnitude are also expected in merging neutron stars [2, 3]. For this reason, the study of these few GeV heavy-ion collisions allows us to gain information on the microscopic composition and also deduce macroscopic properties like the equation of state of astronomic objects like merging neutron stars.

1.1 Strange hadron and hypernuclei production

At collision energies of few GeV, the lightest strange mesons, the Kaons, as well as the lightest hyperons, the Λ hyperons, are produced close to their free nucleon nucleon threshold energy. Therefore, their excitation function experiences a steep rise with increasing $\sqrt{s_{NN}}$, while the production rates are at the order of $\lesssim 0.1$ per collision [4–6]. At the same time, nucleons from the colliding nuclei are stopped noticeably in the collision zone and dominate the fireball. This offers excellent conditions for the formation and emission of light nuclei as well as hypernuclei involving one (or more) hyperons [7]. Hypernuclei allow us to probe the underlying hyperon-nucleon interactions which are highly relevant for the equation of state at high densities [8, 9]. Furthermore, the production rates and kinematic distributions of hypernuclei are investigated to determine their dominant formation mechanism (e.g. thermal production, coalescence [10]).

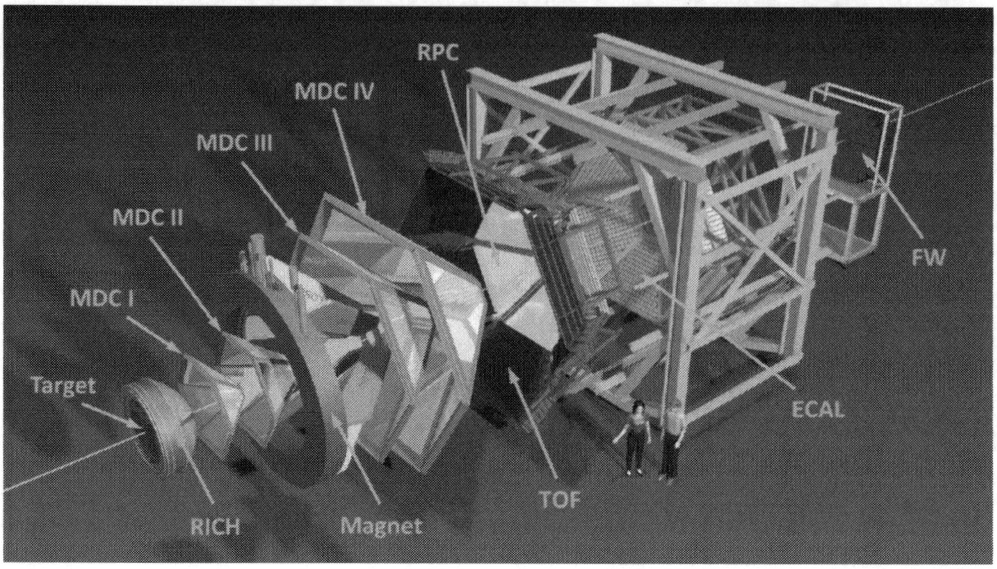

Figure 1. Schematic expanded view of the HADES detector setup.

2 HADES experiment

The HADES experiment, which is schematically depicted in an expanded view in Fig. 1, is a multi-purpose high-acceptance compact magnet spectrometer. It is operated in a fixed-target mode using beams from the SIS18 accelerator at GSI, Darmstadt, Germany. Besides heavy-ion A+A collisions, also elementary p+A and p+p as well as pion induced π+A collisions are measured, e.g. [11–13]. Starting from the segmented 15-fold silver target, the particles traverse a Ring-Imaging CHerenkov (RICH) detector for e^+/e^- identification, two low-mass Mini Drift Chamber (MDC) tracking stations, a toroidal magnetic field, followed by two further MDC tracking stations and depending on the polar angle either a Resistive Plate Chamber (RPC) or a Time of Flight (ToF) scintillation detector for a time of flight measurement. The setup is completed by an Electromagnetic CALorimeter (ECAL) for photon detection, a Forward Wall (FW) hodoscope for measurement of the event plane and a diamond reaction time and trigger detector (START) located in front of the target. Besides the RICH, the FW and the START detector, all detectors are split into six independent sectors covering almost the entire azimuthal angle. More details on the HADES experiment are given in [14].

3 Weak decay recognition

While the charged kaons have a mean flight length $c\tau$ of a few meters and are thereby measured directly in the detectors, the Λ hyperon, K^0_S meson and all known hypernuclei have mean flight lengths $c\tau$ of only a few centimeters and thereby need to be reconstructed from their decay daughters. The large amount of protons, pions and light nuclei, not associated with weak decays, makes a method to effectively separate the signals from the combinatorial background inevitable. For this purpose, the macroscopic mean flight lengths $c\tau$ of weakly decaying particles are utilized, resulting in a spatial displacement between the primary event vertex and the secondary decay vertex which can be detected. This displacement further leads to a so-called *Off-Vertex-Decay-Topology* which is quantified using the following parameters defined in the laboratory reference system:

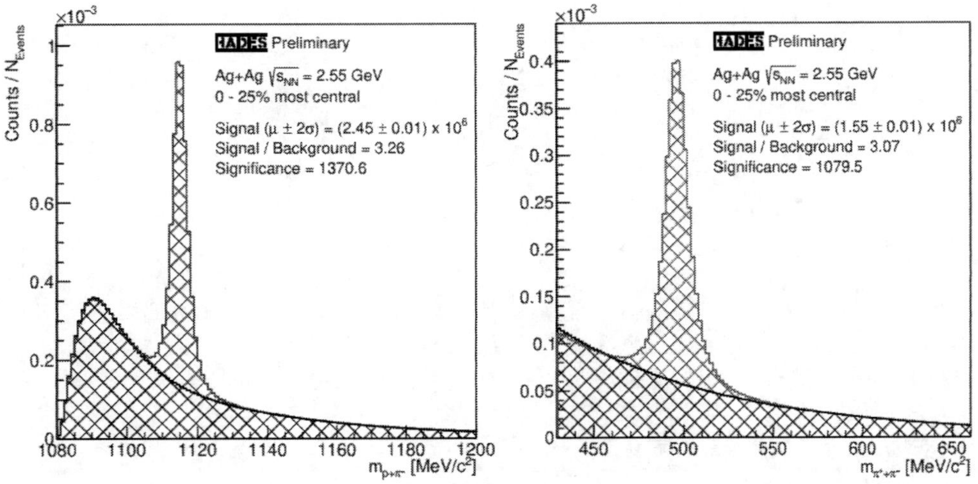

Figure 2. Invariant mass spectra of $\Lambda \rightarrow p + \pi^-$ decays (left) and $K^0_S \rightarrow \pi^+ + \pi^-$ decays (right) from the 0-25% most central Ag(1.58A GeV)+Ag events measured by HADES.

- Distance between the primary and secondary vertices.
- Distance of closest approach (DCA) between the daughter tracks and the primary vertex.
- DCA between reconstructed mother track and primary vertex.
- DCA between the two daughter tracks.
- Opening angle between the two daugther tracks.

In addition, the method proposed by R. Armenteros and J. Podolanski [15] is used to take the kinematic constraints of two-body decays into account as described in [6, 16]. Finally, the procedure is completed by using a Multi-Layer-Perceptron (MLP) artificial Neural Network (aNN) from the Toolkit for Multivariate Data Analysis (TMVA) [17] to further enhance the rejection of combinatorial background. For this purpose, the Off-Vertex-Decay-Topology parameters except for the opening angle as well as the azimuthal angle of the Armenteros-Podolanski ellipse (cf. [6, 16]) are used as input parameters of the aNN which is trained using simulated signal samples and background samples obtained by the mixed-event technique.

In previous analyses [5, 18] it is demonstrated that the use of an aNN to reject the combinatorial background in the recognition procedure of weak decays significantly improves the statistical significance of the reconstructed signals compared to pure hard-cut analyses. In total, the reconstruction method allows the reconstruction of 2.45 million Λ hyperons and 1.55 million K^0_S mesons in the relevant two-body decay channels as shown in the invariant mass distributions in Fig. 2.

4 Lifetime measurements

To achieve proper measurements of the mean lifetimes of $^3_\Lambda$H and $^4_\Lambda$H hypernuclei, first, the method has to be assessed to prevent systematic effects coming from e.g. the required correction factors for acceptance and efficiency effects. Therefore, the mean lifetimes of Λ hyperons and K^0_S mesons are ideally suited, since they are measured with high precision by many experiments and do not depend on the production mechanism of the particle. Besides that, the

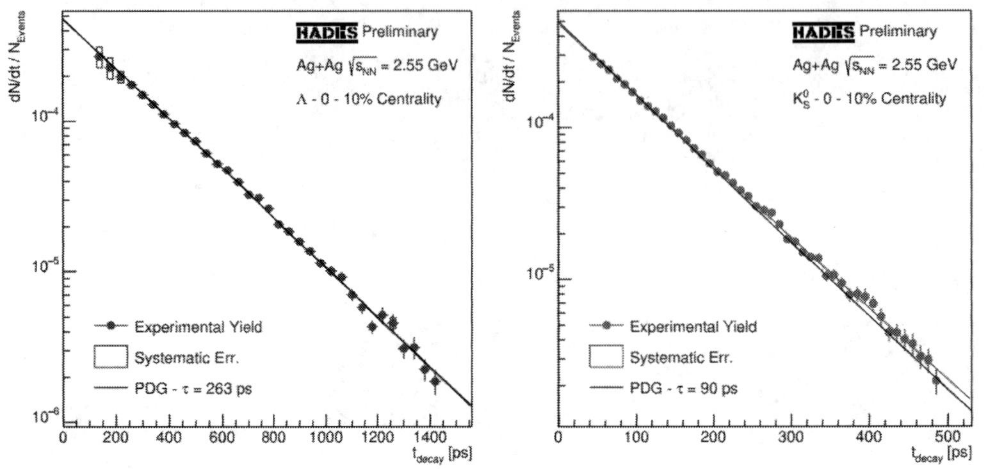

Figure 3. Decay curves of Λ hyperons (left) and K^0_S mesons (right) from the 0-10% most central Ag(1.58A GeV)+Ag events measured by HADES. The colored lines correspond to the exponential fit functions and the black lines to the PDG literature lifetimes [19].

highly significant signals of these particles allow for studies with vanishing statistical uncertainties. In the following, the measurement technique for mean lifetimes of weakly decaying hadrons is described.

The decay time t of a particle is calculated from its decay length l, its velocity β and its Lorentz-factor γ via equation 1:

$$t = \frac{l}{\beta\gamma c} \tag{1}$$

Using the calculated decay times, the reconstructed Λ hyperons are distributed to 37 equally sized intervals of 40 ps width ranging from 120 to 1600 ps, and the reconstructed K^0_S mesons are distributed to 46 equally sized intervals of 10 ps width ranging from 40 to 500 ps. In each interval, the amount of reconstructed particles is extracted from the corresponding invariant mass spectrum in a $\pm 2\sigma$ range determined by Gaussian fit functions.

These extracted signal counts are corrected for acceptance and efficiency losses using simulated Λ hyperons and K^0_S mesons generated according to the kinematic distributions determined in the multi-differential analyses described in [6]. For a given decay time, the kinematic distributions strongly influence the distribution of decay lengths which further influences the integrated acceptance and efficiency correction factors. Thus, proper kinematic distributions are mandatory. The simulated particles are processed by the HGeant package, based on the GEometry And TRacking (GEANT) 3.21 package [20], to emulate their interactions with the detector materials and the corresponding detector responses. Afterward, they are merged with actually measured events and processed by the full event reconstruction procedure to determine their acceptance and efficiency correction factors (cf. [6]).

Using this procedure, the decay curves shown in figure 3 are obtained. They are fitted with an exponential function representing the decay law according to equation 2 to extract the mean lifetime of Λ hyperons and K^0_S mesons:

$$N(t) = N_0 \cdot \exp\left(-\frac{t}{\tau}\right) \Rightarrow \frac{dN}{dt} = -\frac{N_0}{\tau} \cdot \exp\left(-\frac{t}{\tau}\right) \tag{2}$$

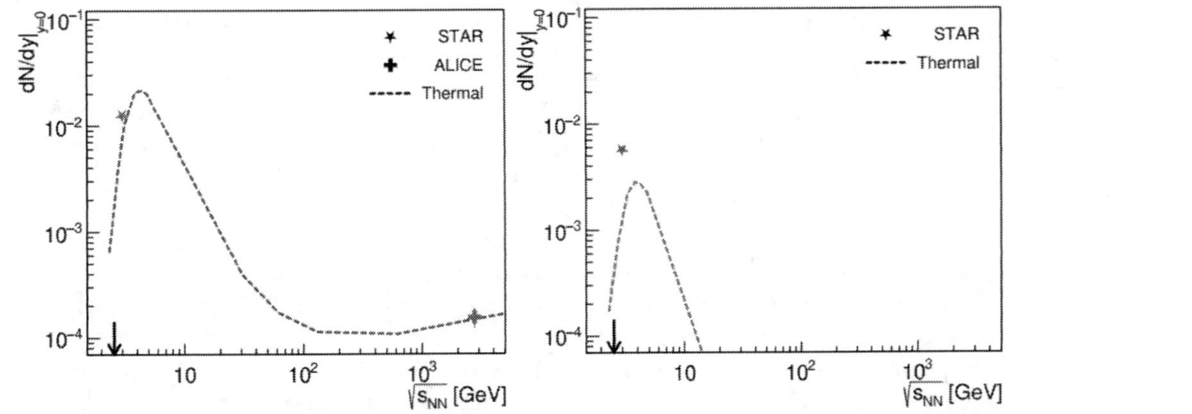

Figure 4. Energy excitation functions of the ${}^{3}_{\Lambda}$H (left) and ${}^{4}_{\Lambda}$H (right) production rates at mid-rapidity. The data are taken from [7, 21]. The dashed lines correspond to thermal model predictions [22] taken from [7] as well.

The obtained mean lifetimes τ of $(262 \pm 2 \pm 3)$ ps (Λ) and $(92 \pm 1 \pm 1)$ ps (K^0_S) are both in accordance with the corresponding literature lifetimes of 263 ps and 90 ps [19]. The first given uncertainty is of statistical nature, while the second one is of systematic nature and obtained by altering selection criteria used for the identification of the corresponding decay candidates. These alternative selection criteria correspond to pure hard-cut analyses, which is why any systematic effect that might arise from using the aNN is included in the systematic uncertainties. This proves that the HADES spectrometer as well as the applied procedures are suited to measure the lifetimes of weakly decaying hadrons.

5 Production rates

Fig. 4 shows the energy excitation functions of ${}^{3}_{\Lambda}$H and ${}^{4}_{\Lambda}$H production rates at mid-rapidity. At the moment only very few data on the production of hypernuclei in heavy-ion collisions at mid-rapidity are available. Therefore, the currently available data set leaves space for various different model predictions. Yet, the general trend is clear. At collision energies at or below the energy threshold of hyperon production in binary NN collisions, the production of hypernuclei is limited by the availability of produced hyperons, and their energy excitation function shows an exponential increase. In the energy region $\sqrt{s_{NN}} \gtrsim 10$ GeV, the number of nuclei present in the collision system ceases as discussed in Sec. 1.1. Therefore, the production rates of hypernuclei are decreasing again resulting in a maximum of hypernuclei production in the energy region of $\sqrt{s_{NN}} \approx 3$-6 GeV. Heavier hypernuclei are stronger affected by the ceasing amount of nucleons in the collision system which is why they develop maximal production rates at lower energies than lighter hypernuclei.

The thermal predictions depicted by the dashed lines in Fig. 4 which are supported by the available data points show these discussed effects. However, deeper and more quantitative investigations of the mechanisms relevant to the formation of Hypernuclei in heavy-ion collisions require further measurements from different energies and collision systems.

6 Summary and outlook

In this contribution, we have presented benchmark measurements of Λ hyperons and K^0_S mesons demonstrating the capability of HADES for high-precision lifetime measurements of weakly decaying particles. In the next step, the lifetimes of the currently accessible hypernuclei, the $^3_\Lambda$H and the $^4_\Lambda$H, are going to be measured which might provide further constraints to models describing the highly relevant hyperon-nucleon interactions.

Furthermore, we discussed the energy excitation function of hypernuclei production and the requirement for more independent measurements to constrain it. In particular, the upcoming CBM experiment, which will operate in the energy region of $\sqrt{s_{NN}} \approx$ 3-6 GeV where maximal hypernuclei production rates are expected, is going to provide very valuable data on hypernuclei. Also, the final results from the most recent STAR fixed-target measurement campaign will significantly extend the available data set on hypernuclei production in heavy-ion collisions.

In parallel to the ongoing analysis of the Ag(1.58A GeV)+Ag data, the HADES collaboration is implementing the Kalman-Filter based KF-Particle-Finder [23] package to reconstruct decaying particles. This approach is already employed by STAR, ALICE and CBM to increase the reconstruction efficiencies of decaying particles. Thereby, it will strongly enhance the reconstruction and analysis of hypernuclei, in particular with respect to three-body decay channels to reconstruct for example the $^4_\Lambda$He.

We gratefully acknowledge the support by SIP JUC Cracow, Cracow (Poland), 2017/26/M/ST2/00600; TU Darmstadt, Darmstadt (Germany), VH-NG-823, DFG GRK 2128, DFG CRC-TR 211, BMBF:05P18RDFC1; Goethe-University, Frankfurt (Germany) and TU Darmstadt, Darmstadt (Germany), DFG GRK 2128, DFG CRC-TR 211, BMBF:05P18RDFC1, HFHF, ELEMENTS:500/10.006, VH-NG-823, GSI F&E, ExtreMe Matter Institute EMMI at GSI Darmstadt; JLU Giessen, Giessen (Germany), BMBF:05P12RGGHM; IJCLab Orsay, Orsay (France), CNRS/IN2P3; NPI CAS, Rez, Rez (Czech Republic), MSMT LTT17003, MSMT LM2018112, MSMT OP VVV CZ.02.1.01/0.0/0.0/18_046/0016066.

References

[1] J. Adamczewski-Musch *et al.* HADES Collaboration, Nature Phys. **15** 1040-1045 (2019).

[2] M. Hanauske, J. Steinheimer, L. Bovard, A. Mukherjee, S. Schramm *et al.*, J. Phys. Conf. Ser. **878** 012031 (2017)

[3] C. Ecker, M. Järvinen, G. Nijs and W. van der Schee, Phys. Rev. D **101** 103006 (2020)

[4] V. Steinberg, J. Staudenmaier, D. Oliinychenko, F. Li, Ö. Erkiner and H. Elfner, Phys. Rev. C **99**, 064908 (2019)

[5] J. Adamczewski-Musch *et al.* HADES Collaboration, Phys. Lett. B **793** 457-463 (2019)

[6] S. Spies, PhD Thesis Goethe Universität Frankfurt (2022)

[7] M. Abdallah *et al.* STAR Collaboration, Phys.Rev.Lett. **128** 20, 202301 (2022)

[8] T. Saito *et al.*, Nature Rev.Phys. **3** 12, 803-813 (2021)

[9] A. Gal, E. V. Hungerford, and D. J. Millener, Rev.Mod.Phys. **88**, 035004 (2016)

[10] J. Steinheimer, K. Gudima, A. Botvina, I. Mishustin, M. Bleicher and H. Stöcker, Phys. Lett. B **714** 85-91 (2012)

[11] G. Agakishiev *et al.* HADES Collaboration, Eur. Phys. J. A **50** 81 (2014)

[12] J. Adamczewski-Musch *et al.* HADES Collaboration, Phys.Rev.C **95** 015207 (2017)

[13] J. Adamczewski-Musch *et al.* HADES Collaboration, Phys. Rev. C **102** 024001 (2020)

[14] G. Agakishiev *et al.* HADES Collaboration, Eur. Phys. J. A **41** 243-277 (2009)

[15] J. Podolanski and R. Armenteros, Phil. Mag. **45** 13-30 (1954)

[16] S. Spies, EPJ Web Conf. **259** 01007 (2022)

[17] A. Hocker, P. Speckmayer, J. Stelzer, J. Therhaag, E. von Toerne *et al.*, TMVA - Toolkit for Multivariate Data Analysis, CERN-OPEN-2007-007 (2007)

[18] S. Spies, J. Phys. Conf. Ser. **1667** 012041 (2020)

[19] R.L. Workman *et al.* Particle Data Group, Prog.Theor.Exp.Phys. **2022**, 083C01 (2022)

[20] R. Brun, F. Bruyant, M. Maire, A.C. McPherson and P. Zanarini, CERN-DD-EE-84-1 (1987)

[21] J. Adam *et al.* ALICE Collaboration, Phys. Lett. B **754** 360-372 (2016)

[22] A. Andronic, P. Braun-Munzinger, J. Stachel and H. Stöcker, Phys. Lett. B **697** 203-207 (2011)

[23] S. Gorbunov and I. Kisel, CBM-SOFT-note-2007-003 (2007)

Strange hadron correlations in heavy-ion collisions at RHIC energies and below

Hanna Zbroszczyk[1,*]

[1]Warsaw University of Technology, Faculty of Physics, Koszykowa 75, 00-662 Warsaw, POLAND

Abstract. Geometry and dynamics of the particle-emitting source in heavy-ion collisions can be inferred via the femtoscopy method. Two-particle correlations at small relative momentum exploit Quantum Statistics and the Final State Interactions (Coulomb and strong), which allow one to study the space-time characteristics of the source of the order of 10^{-15} m and 10^{-23} s. Femtoscopic measurements allow one also to study FSI, especially the strong one, which is unknown for many two-particle systems. The RHIC program covers a significant part of the QCD phase diagram using collisions of Au nuclei for several beam energies from 7.7 to 200 GeV for which baryon-rich region is studied via femtoscopy. These measurements complement those obtained from AGS, SPS, and SIS experiments. Strange hadron measurements together with non-strange ones provide complementary information about source characteristics. Strangeness is a significant observable for many regimes of collision energies providing a complete view of strangeness production in heavy-ion collisions. These proceedings shows the femtoscopic measurements of various strange particle combinations at different collision energies and centralities of the collision.

1 Introduction

For the last decades, many theoretical and experimental efforts have been devoted to study properties of strongly interacting matter described by the QCD phase diagram. At the Large Hadron Collider (LHC) and top Relativistic Heavy Ion Collider (RHIC) energies, the explored matter is described by high temperature and low baryon chemical potential and exists in the state of Quark-Gluon-Plasma (QGP). Hadron Gas (HG) transition to QGP is found to be of a smooth cross-over type. Many other areas of the QCD phase diagram, for significantly lower temperature and higher baryon chemical potential values are currently being explored by the HADES experiment conducted at Heavy Ion Synchrotron SIS18 at GSI and will be studied in the future by the CBM experiment at FAIR, currently under construction. Two experiments nowadays study areas of the phase diagram for intermediate temperature and baryon chemical potential: STAR conducted at RHIC with its Beam Energy Scan (BES) program [1], [2] and NA61/SHINE at the Super Proton Synchrotron (SPS). In this region, the phase transition from HG to QGP could be the first-order and ends at a Critical Point (CP). However, the CP has not been discovered yet, and its location is still unknown.
Understanding the QCD phase diagram is one of the most essential goals in the field of relativistic heavy-ion physics. Several methods are proposed to study such baryon densities; two-particle correlations being one of them. Changes in the particle spectra shape as a function of

*e-mail: hanna.zbroszczyk@pw.edu.pl

collision energy can be studied to reveal changes in the kinetic freeze-out temperature of the hadrons from the medium and their mean transverse velocity. Momentum spectra can give partial information only. The bulk response of the produced system has a non-trivial structure in both space and time. By the measurement of particle correlations in the region of small relative velocities, the space-time information can be extracted. This method, called correlation femtoscopy, is widely used in heavy-ion collisions studies. Its name: femtoscopy is related to femto-scale (10^{-15} m = 1 fm), which can not be accessed by any other experimental technique, is a method originating from astronomy - HBT. By studying the space-time sizes of particle emission region as a function of beam energy, it is possible to infer the medium's energy density at the last re-scattering of the hadrons. These measurements can be sensitive to a softening in the equation of state, related to a first-order phase transition. STAR, as an example of a currently operating experiment for non-zero baryon chemical potential, performed detailed studies of two-pions femtoscopy for data collected in the frame of the BES-I program. With the relations between thermal and collective motions, between chemical and thermal freeze-out, with the effects of resonance production and secondary re-scattering, the system's final image of space-time evolution represents a very complex phenomenon, difficult to describe quantitatively.
RHIC, one of the largest accelerators in the world, is located at Brookhaven National Laboratory (BNL), Upton. Four facilities were located at RHIC: STAR, PHENIX, PHOBOS, and BRAHMS. Currently, STAR is the only running experiment. Solenoidal Tracker At RHIC (STAR) is an experiment, which primary goal is to study the properties of matter created under extreme conditions (such as pressure, density) and learn about interactions between hadrons. For almost 20 years, STAR has been collecting data from various collisions of different ions at different energies: 7.7, 11.5, 19.6, 27, 39, 62.4, 130, and 200 GeV. The highest collision energy of Au nuclei is $\sqrt{s_{NN}}$ = 200 GeV, and it enables measurement of hot and dense matter after QGP formation. The BES program which first part includes colliding energies below $\sqrt{s_{NN}}$ = 62.4 GeV, is mainly focused on the exploration of the QCD phase transition between QGP and a hadron gas and also to search for the CP between first-order phase-transition and cross-over transition described by QCD.
The High-Acceptance Di-Electron Spectrometer (HADES) was originally designed to measure low mass di-electrons originating from the decay of vector mesons, HADES is also well suited to measure charged hadrons with good efficiency. The synchrotron SIS18 provides the beams with energies between 1 - 2 A GeV for heavy ions and up to 4.5 GeV for protons. Secondary pion beams with momenta up to 2 GeV/c are also available at this facility.

2 The femtoscopy method

Femtoscopy is a powerful tool which allows one to study through the effects of quantum statistics (QS) and the FSI, geometrical (source sizes), and the dynamical (space-time asymmetries in the emission process) properties of the expanding source. As a result of the heavy-ion collision, a hot and dense system produces different types of particles. Using the femtoscopy method, it is possible to learn about the source's geometrical and dynamical properties. Femtoscopy enables one to study space characteristics of source emitting pairs of particles. Thus, it is impossible to explore the whole source's information but rather to its part, emitting considered pairs of particles. It explains why source sizes extracted from various studies are different depending on the considered system. Therefore the geometry of heavy-ion collision can be studied via femtoscopic parameters describing source sizes of emission areas. In order to define a femtoscopic correlation function [3], [4] in the case of identical particle combinations a Lonitudinally Co-Moving System (LCMS) is used

which is defined as $p_{L,1} + p_{L,2} = 0$. $p_{L,1}$ and $p_{L,2}$ are longitudinal components of momentum for single particles. Using the LCMS system the correlation function is defined using $Q_{\text{inv}} = \sqrt{(p_1 - p_2)^2 - (E_1 - E_2)^2}$ variable. In the case of nonidentical particle combinations Pair Rest Frame (PRF) reference is used which defined $k^* = p_1 = -p_2$ variable. The PRF reference can be also used for identical pairs of particles and then $Q_{\text{inv}} = 2k^*$.
The definition of the correlation functions is as follows: $C(k^*) = \frac{A(k^*)}{B(k^*)}$ or $C(Q_{\text{inv}}) = \frac{A(Q_{\text{inv}})}{B(Q_{\text{inv}})}$. Correlated pairs of particles (from the same event enter) enter the numerator $A(k^*)$ or $A(Q_{\text{inv}})$ and uncorrelated pairs of particles (from different events) enter the denominator $B(k^*)$ or $B(Q_{\text{inv}})$ of the correlation function. If two particles are identical, the wave function must be adequately (anti-) symmetrized for pairs with even (odd) total pair spin. It is the QS effect. It is referred to as the Hanbury - Brown Twiss (HBT) effect, originating from the two astrophysicists' names that performed similar measurements for the correlation between photons from stars. Therefore, a correlation function in identical particles is sensitive to the QS effects (if there are considered bosons, they obey Bose-Einstein description; if they are fermions, they obey a Fermi-Dirac description). For pairs of hadrons, it will also contain the contribution from FSI. If both particles are charged ones, it will contain a contribution from Coulomb FSI. For pairs of charged hadrons, both types of FSI act at the same time. In traditional femtoscopy, considered usually for pions and kaons, the interaction is well known. For these pairs, the QS symmetrization is a dominating effect, the strong interaction is negligible, while the Coulomb FSI can be treated as a correction and can be considered with simplified methods. Besides, pions are the most abundantly produced particles in heavy-ion collisions, and the multiplicity of kaons is also large. The resulting experimental correlation function has excellent statistical precision. It can be fitted by a known theoretical formula, providing exact information about source sizes. A traditional femtoscopy method provides precise and detailed information relating to the source's space-time structure.

3 Results of nucleon-nucleon interactions for Au + Au collisions at $\sqrt{s_{NN}}$ = 200 GeV

The study of nucleon-nucleon (N-N), nucleon-hyperon (N-Y), and hyperon-hyperon (Y-Y) interactions are fundamental to understand the physics of relativistic heavy-ion collisions, neutron stars and the existence of various exotic hadrons. A significant amount of N-N scattering data allows us to construct precise N-N potential models. However, the limited availability of N-Y scattering data and no scattering data for the Y-Y systems makes understanding the N-Y and Y-Y potentials complicated and challenging. Commonly, the experimental information on the bound states of strange baryons and nucleons (hypernuclei) is used to provide information on N-Y interactions. However, the extraction of strong interactions' parameters becomes difficult due to contamination by many-body effects. STAR has measured correlation functions of proton-proton, antiproton-antiproton, and the ratio of proton-proton and antiproton-antiproton for the collision energy $\sqrt{s_{NN}} = 200$ GeV. Correlation functions for baryon-baryon and antibaryon-antibaryon pairs are assumed to be consistent with each other within uncertainties [5], see Figure 1. The scattering length, f_0 is a parameter that describes low-energy scattering, and d_0 is defined as the effective range of strong interaction between two particles. From these studies, parameters f_0 and d_0 are estimated for antiproton-antiproton pairs, and they are found to be consistent with those for proton-proton pairs (Figure 2). Source sizes in the case of two protons or two antiprotons were found as $R_{pp} = 2.75 \pm 0.01(stat) \pm 0.04(syst.)$ fm and $R_{\bar{p}\bar{p}} = 2.80 \pm 0.02(stat) \pm 0.03(syst.)$ fm.

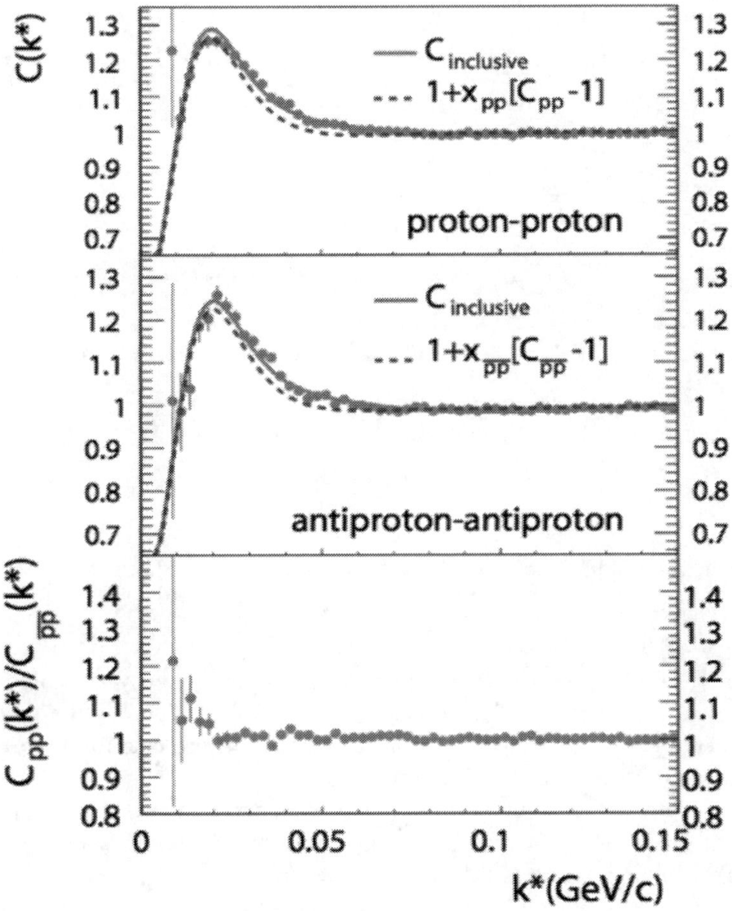

Figure 1. Proton-proton (top), antiproton-antiproton (middle) correlation functions and their ratio (bottom) for Au+Au collisions at $\sqrt{s_{NN}}$ = 200 GeV. [5]

4 Strange baryon correlations

High-energy heavy-ion collisions provide a significant number of hyperons in each collision, which gives an excellent opportunity to study strong interactions. Using the $N-\Omega$ interaction, the shape of the two-particle correlation function at low relative momentum changes significantly with the strength of the $N-\Omega$ attraction. However, the Coulomb interaction in the $p-\Omega$ channel makes it challenging to access the strong interaction parameters directly from the measured two-particle correlation function. Therefore, a new measure, namely the ratio of the correlation functions between the peripheral (small) and central (large) collision systems, is proposed. This ratio provides direct access to the strong interaction between proton and Ω, independent of the model used for the emission source. The attractive nature of an $N-\Omega$ interaction leads to the possible existence of the $N-\Omega$ dibaryon. Such an $N-\Omega$ dibaryon is the most interesting candidate after the H-dibaryon [6]. Several attempts have been made to estimate the binding energy of the $N-\Omega$ state in different QCD-motivated

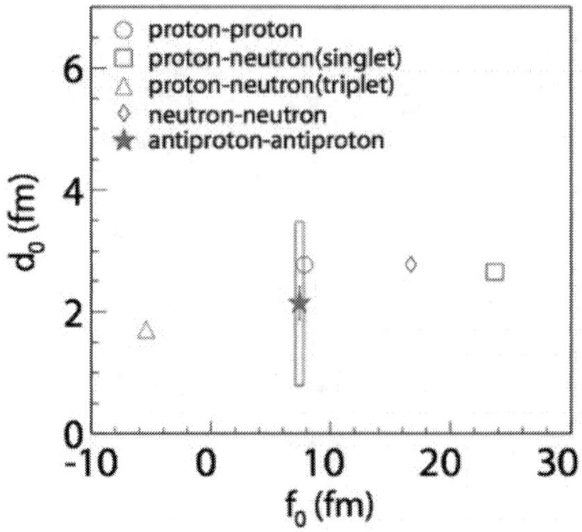

Figure 2. Parameters of strong interactions for different hadron pairs. [5]

models. The $N - \Omega$ dibaryon can be produced in high-energy heavy-ion collisions through the coalescence mechanism [7]. The measurement of the $p - \Omega$ correlation function for peripheral and central Au+Au collisions at $\sqrt{s_{NN}}$ = 200 GeV, presented in Figure 3, provides insight into the existence of an $N - \Omega$ dibaryon.

In order to estimate the parameters of the strong interaction, it is better to use data from the experiment operating at lower collision energies due to the much lower contamination of the particle sample with the products from the decays. Such an example comes from the HADES experiment, where the parameters of the strong interaction between proton and lambdas pairs are estimated. The parameters were extracted from the measured correlation function that includes both: femtoscopic correlations and long-range correlations (LRC), seen as a slope for higher momentum difference. Long-range correlations are parametrized as $C_{LRC}(k) = 1 + ak + bk^2$. Figure 4 shows example of $p - p$ (top) and $p - \Lambda$ correlations functions. The source size of the region emitting proton-lambda pairs is estimated from distributions of the proton-proton system. Then, having estimated the source, determining f_0 and d_0 parameters was possible. The values are as following: $f^{S=0}_{0,NLO}$ = 2.91 fm, $d^{S=0}_{0,NLO}$ = 2.78 fm,$f^{S=1}_{0,NLO}$ = 1.54 fm, $d^{S=1}_{0,NLO}$ = 2.72 fm, and $f^{S=0}_{0,LO}$ = 1.91 fm, $d^{S=0}_{0,LO}$ = 1.40 fm,$f^{S=1}_{0,LO}$ = 1.23 fm, $d^{S=1}_{0,LO}$ = 2.13 fm.

5 Strange meson correlations

Examination of femtoscopic correlations between even lighter strange particles (e.g., charged and neutral kaons) for various collision energies is attributed to studies of strangeness production within different phase diagram regions. Inclusion of particle correlations between hyperons may be attributed to studies of the minimum collision energy required for heavier particle species to be produced. Considering the lightest strange particles – kaons, both: charged and neutral ones can extract different elements of information. It is very well known that a significant fraction of pions come from resonance decays after freeze-out, thus complicating the identical pion femtoscopic measurements. While the direct pion source may

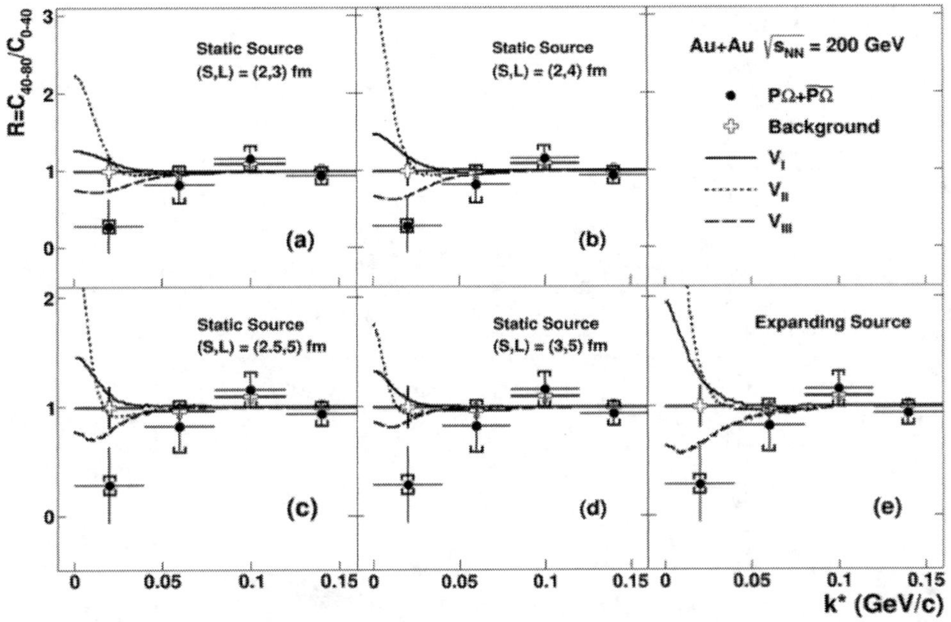

Figure 3. The solid circle represents the ratio (R) of the small system (40-80% collisions) to the large system (0-40% collisions) for proton-omega and $\bar{p}-\Omega$. The error bars correspond to the statistical errors, and caps correspond to systematic errors. The open crosses represent the ratio for background candidates from the side-band of omega invariant mass. Predictions for the ratio of the small system to the large system for $p-\Omega$ interaction potentials V_I, V_{II} and V_{III} for static source with different source sizes (S, L) = (2,3), (2,4), (2.5, 5) and (3,5) fm, where S and L are corresponding to small and large systems, are shown in (a), (b), (c) and (d) respectively. In addition, the prediction for the expanding source is shown in (e) [8].

be inherently non-Gaussian, the resonances extend the source size due to their finite life-time, introduce an additional essentially non-Gaussian distortion in the two-pion correlator, and thus they reduce the fitted correlation strength. Kaon interferometry, on the other hand, suffers less from resonance contributions and could provide a cleaner signal for correlation studies than pions. Femtoscopic correlations for two identical charged pions or kaons are described by both QS effects and FSI: Coulomb and negligible strong one. Therefore, source characteristics derived from charged kaons studies ideally complement those learned from two-pions. An example of the correlation function for two neutral kaons for the most central Au+Au collisions at $\sqrt{s_{NN}}$ = 200 GeV is shown in the Figure 5 (left), while the combination of neutral and positive kaon in the Figure 5 (right). The shape of the correlation function for identical and non-identical kaon pairs is different. The QS effects are dominant in the case of the both neutral kaons, QS (which relies on the Gaussian parameterization) is not sufficient to describe the shape of the correlation function around a dip structure around q_{inv} = 0.1 GeV/c. Only including the strong FSI (Lednicky-Lyuboshitz parametrization) allows us to describe the area of the correlation function below the unity. The parameterization of the function for neutral and positive kaons only considers the strong interaction related to the presence of a_0 resonance.

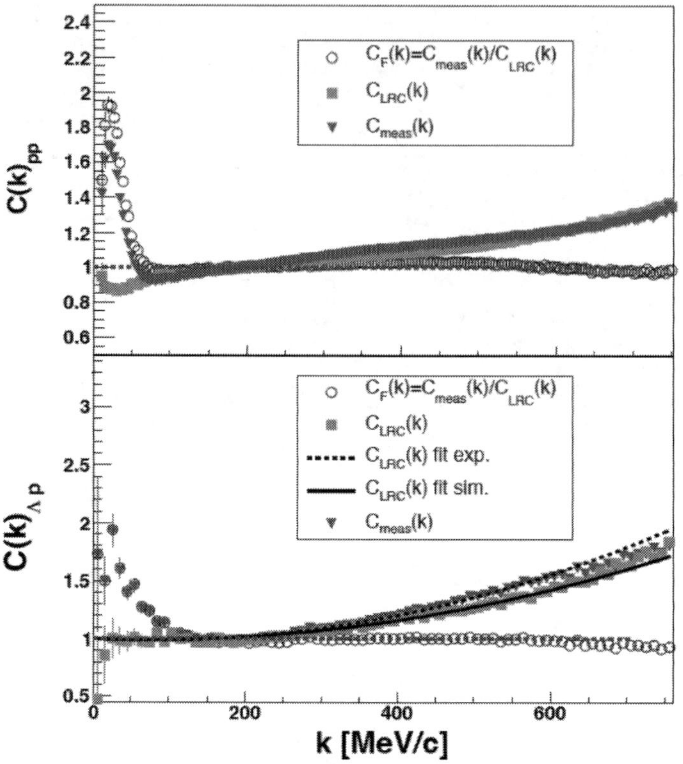

Figure 4. Proton-proton correlation function (top) used to extract source size and proton-lambda (bottom) correlation functions, femtoscopic correlations together with long-range correlation [9].

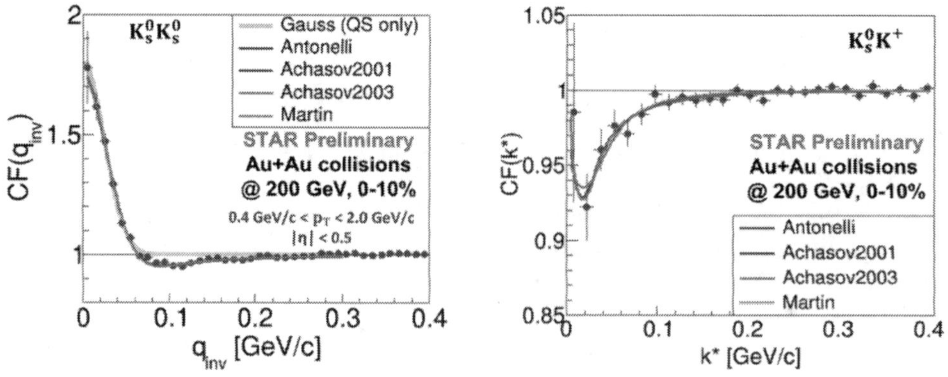

Figure 5. Two kaon correlations at Au+Au collisions at $\sqrt{s_{NN}}$ = 200 GeV for neutral kaons (left) and combination of neutral and positive kaons (right) [10].

6 Summary

These proceedings presented femtoscopy results obtained for RHIC energies and below. The knowledge about N-N, N-Y, and Y-Y interactions is crucial to understanding the nature of elementary interactions responsible for building matter components, e.g., neutron stars. In the context of N-N and $\bar{N} - \bar{N}$ interactions, results of the proton-proton, antiproton-antiproton, and proton-antiproton femtoscopy for Au + Au collisions at $\sqrt{s_{NN}}$ =200 GeV and at BES program at STAR were discussed. Except for source sizes, these studies delivered the scattering length (f_0) and effective range (d_0) parameters of the strong interaction of protons and antiprotons. Parameters of the strong interactions were also studied for $p - \Omega$ pairs; however, it was impossible to draw definitive conclusions due to insufficient statistics. It was also discussed that with the help of lower collision energies, studying elementary interactions can be conclusive and competitive with higher-energy experiments due to a much lower fraction of contaminated particles originating from decays. For example, HADES data were discussed to provide parameters of the strong interactions between $p - \Lambda$ pairs. In terms of strange meson correlations, strong interactions between neutral kaons and pairs of neutral and charged kaons were significant when the hypothesis about the possible formation of a_0 resonance as a tetraquark state.

ACKNOWLEDGMENTS

This work was supported by the Grant of National Science Centre, Poland, No: 2021/41/B/ST2/02409 and 2020/38/E/ST2/00019. Studies were funded by IDUB-POB-FWEiTE-1 and IDUB-POB-FWEiTE-3, project granted by Warsaw University of Technology under the program Excellence Initiative: Research University (ID-UB).

References

[1] G. Odyniec *et al.*, J. Phys. **G37**, 094028 (2010)
[2] M.M.Aggarwal *et al.*, arXiv:1007.2613v1
[3] V.G. Grishin, G.I. Kopylov, M.I. Podgoretskii, Sov. J. Nucl. Phys. **13** (1971) 638
[4] R. Lednicky, V. L. Lyuboshitz, Sov. Journ. Nucl. Phys. **35** (1982) 770
[5] L. Adamczyk, H. Zbroszczyk *et al.*, Nature **527** (2015) 345-348
[6] R. Jaffe, Phys. Rev. Lett. **38**, 195 (1977)
[7] N. Shah, Y. G. Ma, P. N. Chen, S. Zhang, Phys. Lett. B **754**, 6 (2016)
[8] J. Adam *et al.* (STAR Collaboration), Phys. Lett. B **790** 490 (2019)
[9] J. Adamczewski-Musch *et al.* (HADES Collaboration), Phys. Rev. C **94**, 025201
[10] D. Pawlowska (STAR Collaboration); these proceedings

Directed flow of Λ from heavy-ion collisions and hyperon puzzle of neutron stars

Akira Ohnishi[1,*], *Asanosuke* Jinno[2], *Koichi* Murase[1], and *Yasushi* Nara[3]

[1]Yukawa Institute for Theoretical Physics, Kyoto University, Kyoto 606-8502, Japan
[2]Department of Physics, Faculty of Science, Kyoto University, Kyoto 606-8502, Japan
[3]Akita International University, Yuwa, Akita-city 010-1292, Japan

Abstract. We examine the Λ potential from the chiral effective field theory (χEFT) via the Λ directed flow from heavy-ion collisions. We implement the Λ potential obtained from the χEFT in a vector potential version of relativistic quantum molecular dynamics. We find that the Λ potentials obtained from the χEFT assuming weak momentum dependence reproduce the Λ directed flow measured by the STAR collaboration in the Beam Energy Scan program. While the Λ directed flow is not very sensitive to the density dependence of the potential, the directed flow at large rapidities is susceptible to the momentum dependence. Thus understanding the directed flow of hyperons in a wide range of beam energy and rapidity is helpful in understanding hyperon potentials in dense matter.

1 Introduction

The hyperon puzzle is one of the central issues in neutron star physics in view of hypernuclear and strange particle physics [1]. When empirical two-body interactions are considered, hyperons are predicted to appear in the neutron star matter at $(2\text{-}4)\rho_0$, soften the dense nuclear matter equation of state (EoS), and reduce the maximum mass of neutron stars to be $M_{\max} = (1.3\text{-}1.6)M_\odot$. Since several neutron stars are found to have masses around $2M_\odot$ or more, most of the proposed hyperonic matter EoS were ruled out. Thus something is wrong, and it is a challenge to elucidate the mechanism to solve this hyperon puzzle.

Among several solution candidates to the hyperon puzzle, we here concentrate on the solution based on the three-body YNN repulsion [2]. Since the three-body NNN interactions are known to be necessary to describe $A = 3$ nuclear binding energies and saturation point of nuclear matter, it is natural to expect that the three-body YNN interactions play some role in hypernuclei and hyperonic matter. However, there is almost no information on three-body YNN interactions from experiments, and we have to rely on theoretical predictions. For example, the chiral effective field theory (χEFT) is a first-principles framework and is expected to give reliable results once the relevant low-energy constants are determined. There are several predictions of the Λ potential $U_\Lambda(\rho)$ at high densities obtained by using χEFT [3, 4]. In Ref. [3], Gerstung, Kaiser, and Weise (GKW) calculated U_Λ in neutron star matter using χEFT under the assumption that the diagrams relevant to the three-body forces are saturated by the decuplet baryon propagation [5]. The obtained U_Λ with 2+3-body interactions is more

*e-mail: ohnishi@yukawa.kyoto-u.ac.jp

repulsive at high densities than that only with 2-body interactions, and it suppresses Λ to appear in neutron stars: the single particle energy of Λ at zero momentum is larger than the neutron chemical potential, $M_\Lambda + U_\Lambda(\rho, \boldsymbol{p} = 0) > \mu_n(\rho)$. Confirmation of the repulsive U_Λ at high densities is a challenge in hypernuclear and strangeness particle physics. Actually, one of the main goals of the J-PARC hadron hall extension [6] is to extract the three-body YNN interaction via precision experiment of hypernuclear spectroscopy.

Together with the hypernuclear spectroscopy, hyperon observables from heavy-ion collisions should be useful, since the formed nuclear matter reaches high densities. Specifically, the anisotropic flows such as the directed flow, $v_1 = \langle \cos\phi \rangle$ with ϕ being the azimuthal angle relative to the reaction plane, are generated in the initial stage and have been utilized to constrain the equation of state (EoS). Recently, the directed flow slopes (dv_1/dy) for identified hadrons were measured in the colliding energy range of 3 GeV $\leq \sqrt{s_{NN}} \leq$ 200 GeV [7]. The transition of the slope from positive to negative was discovered for both protons and Λs at $\sqrt{s_{NN}} \approx$ 10 GeV. Theoretical models with a first-order phase transition predict this transition point at much lower beam energies [8], and the transition occurs at higher beam energies in hadron transport models without potential effects [9]. Recently, the proton directed-flow in the above colliding energy range was explained by a transport model [10] with a purely hadronic EoS. A repulsive EoS contributes positively to the slope in the early compression stage, while the tilted ellipsoid of the formed matter causes a negative contribution in the late expansion stage. With increasing colliding energy, the negative contribution takes over the positive one, and then the above transition of the slope is realized. The nuclear mean field including both the density and momentum dependence explains the observed transition energy [10].

Thus, we expect that the Λ directed flow is sensitive to the Λ potential at high densities. Provided that the above-mentioned compression/expansion mechanism also applies to Λ, one expects that the directed flow of Λs should be small since Λs are produced during the compression stage and the positive contribution would be smaller. Then the agreement of the transition energies of protons and Λs may suggest more repulsive potential for Λs than for protons at higher densities, as suggested by the χEFT.

In the Hyp2022 presentation, we discussed the effects of Λ potential on the directed flow of Λ by using an event generator JAM [10, 11]. The density and momentum dependences of the Λ potential are taken from the χEFT in Refs. [3] and [4], respectively, as discussed in Sec. 2. In Sec. 3, we discuss the results of transport model calculations using JAM performed in the relativistic quantum molecular dynamics mode with a Lorentz-vector implementation of the potential (RQMDv) [10]. Readers are referred to Ref. [12] for the details of the transport model and the calculated results.

2 Λ potential from chiral effective field theory

The Λ potential as a function of baryon density has been studied extensively in Refs. [13, 14], where the Λ separation energies of various hypernuclei have been well fitted and the EoS of the neutron-star matter has been predicted. However, most of them fail to sustain massive neutron stars [1]. While it is possible to increase the maximum mass of neutron stars above $2\,M_\odot$ by introducing repulsive three-body YNN interactions or repulsive hyperon potentials at high densities [2], most of these prescriptions introduce additional parameters which are not constrained by data or first principles.

A promising way to systematically describe many-body interactions is to invoke the χEFT, which is a first-principles theory based on the chiral symmetry of QCD. As shown in the left panel of Fig. 1, the Λ potential in nuclear matter was computed by GKW [3]. In uniform nuclear matter, the Λ potential is given as a function of the density and momentum,

$U_\Lambda(\rho, \boldsymbol{p})$. As shown in the right panel of Fig. 1, the momentum dependence of U_Λ in the χEFT is given, for example, by Kohno [4]. By including the three-body YNN interaction, the Λ potential becomes more repulsive at high densities and momenta.

We have parameterized the density and momentum dependence of the single-particle Λ potential U_Λ from χEFT in the Lorentz-vector type potential [15],

$$U^\mu_\Lambda(\rho, \boldsymbol{p}) = \frac{J^\mu}{\rho} U_{\rho\Lambda}(\rho) + U^\mu_{m\Lambda}(\rho, \boldsymbol{p}), \tag{1}$$

$$U_{\rho\Lambda}(\rho) = au + bu^{4/3} + cu^{5/3}, \tag{2}$$

$$U^\mu_{m\Lambda}(\rho, \boldsymbol{p}) = \frac{C}{\rho_0} \int d^3p' \frac{p'^{*\mu}}{p'^{*0}} \frac{f(x, \boldsymbol{p}')}{1 + [(\boldsymbol{p} - \boldsymbol{p}')/\mu]^2}, \tag{3}$$

where J^μ is the baryon current, $u = \rho/\rho_0$ is the nucleon density normalized by the saturation density $\rho_0 = 0.168\ \mathrm{fm}^{-3}$, and $p^{*\mu} = p^\mu - U^\mu$ and $p^{*0} = \sqrt{m_N^2 + \boldsymbol{p}^{*2}}$.

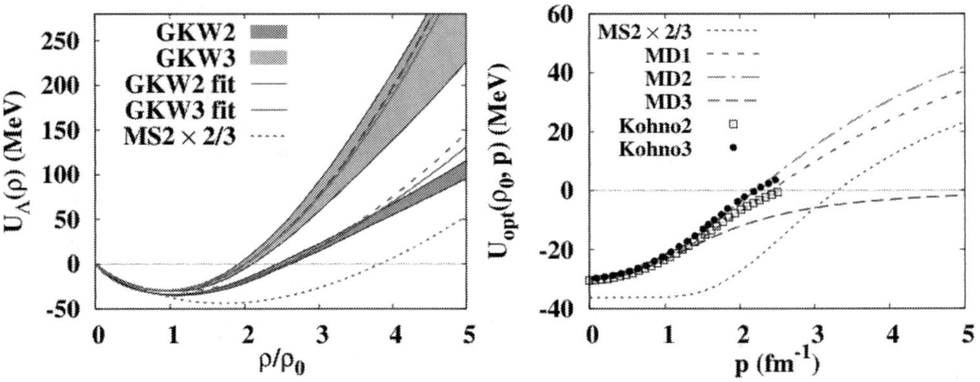

Figure 1. Baryon-density (left) and momentum (right) dependence of the single-particle potentials for Λ. GKW2 and GKW3 represent the Λ single-particle potential from χEFT with 2- and 2+3-body interactions [3]. MS2×2/3 shows the nucleon single-particle potential multiplied by 2/3. MD1, MD2, and MD3 are momentum-dependent potentials obtained by fitting the χEFT results [4].

In Fig. 1, we show the fitting results of the Λ potential. MD1 and MD2 potentials (right panel) are obtained by fitting the χEFT results by Kohno [4] at $\rho = \rho_0$ with 2- and 2+3-body interactions, respectively, up to the momentum of $2.5\ \mathrm{fm}^{-1} \simeq \lambda$ (cutoff) assuming the range parameter $\mu = 3.23\ \mathrm{fm}^{-1}$ [10]. The real part of the optical potential U_{opt} is defined as the difference between the single-particle energy and the kinetic energy. Since the momentum dependence in Ref. [4] seems to be stronger than the previous estimate [16], the potential with a weaker momentum dependence (MD3) is also prepared. For MD3, we require that the optical potential is zero at high momentum ($p \simeq 8.6\ \mathrm{fm}^{-1}$) and reproduces the result of Ref. [4] at $p = 1\ \mathrm{fm}^{-1} \simeq 0.4\lambda$. The density dependence (left panel) is determined by fitting the χEFT results by GKW [3]. GKW2 and GKW3 potentials are those with 2- and 2+3-body interactions, respectively. Since the Brückner-Hartree-Fock calculation was found to be unstable at $\rho/\rho_0 > 3.5$ [3], we fit the χEFT results in the range $\rho/\rho_0 \leq 3$. At zero Λ momentum in cold nuclear matter, we find $J^\mu = (\rho, \boldsymbol{0})$ and $\boldsymbol{U}_{m\Lambda} = \boldsymbol{0}$, and thus $U_\Lambda(\rho) \equiv U^0_\Lambda(\rho, \boldsymbol{0}) = U_{\rho\Lambda}(\rho) + U^0_{m\Lambda}(\rho, \boldsymbol{0})$. The density dependence from the momentum-dependent interaction (second term) can be absorbed by tuning parameters (a, b, c), so the density dependence of U_Λ is almost the same with and without the momentum-dependent potentials: GKW2 fit (thin solid, without momentum dependent terms) and GKW2+MD1 (dashed), and GKW3 fit (thick solid) and GKW3+MD2/MD3 (dash-dotted, long dashed) at $\rho/\rho_0 \leq 3$.

3 Directed flow of Λ from heavy-ion collisions

In order to simulate high-energy heavy-ion collisions, we use an event generator JAM2.1 [17]. In JAM2.1, there are several updates: one can utilize Pythia 8 [18] library as it is, potentials for leading baryons are included during their formation time with the reduced factor, and collision time and ordering time has been modified [19].

We implement the Λ potential described in the previous section in the form of Lorentz-vector potential in the relativistic quantum molecular dynamics (RQMDv) approach [10]. The RQMDv equations of motion for the i-th particle having the position q_i^μ and the momentum p_i^μ is given by [20]

$$\frac{dq_i^\mu}{dt} = v_i^{*\mu} - \sum_j v_j^{*\nu} \frac{\partial V_{j\nu}}{\partial p_{i\mu}}, \quad \frac{dp_i^\mu}{dt} = \sum_j v_j^{*\nu} \frac{\partial V_{j\nu}}{\partial q_{i\mu}}, \tag{4}$$

where $v_i^{*\mu} = p_i^{*\mu}/p_i^{*0}$. For the momentum-dependent part, we replace $(\boldsymbol{p}_i - \boldsymbol{p}_j)^2$ in Eq. (3) with the two-body relative momentum squared, $p_{R,ij}^2$, in the rest frame of the particle j. We use U_Λ^μ in Eq. (1) for Λ and other hyperons.[1]

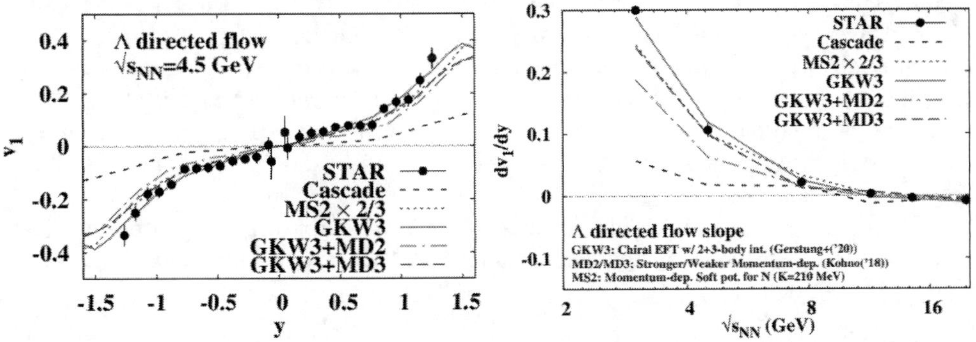

Figure 2. Left: Calculated directed flow of Λ in mid-central Au + Au collisions at $\sqrt{s_{NN}} = 4.5$ GeV. Right: The directed flow slope of Λ. Data are taken from Ref. [7].

In Fig. 2, we compare the calculated Λ directed flow $v_1 = \langle \cos\phi \rangle$ in the impact parameter range 4.6 fm $< b <$ 9.4 fm with the STAR data [7] in midcentral Au + Au collisions at $\sqrt{s_{NN}} = 4.5$ GeV (left panel), and the directed flow slope of Λ at $\sqrt{s_{NN}} = (3.0\text{-}19.6)$ GeV (right panel). The cascade results (without potential effects) underestimate the data. With the potential effects, the directed flow data of Λ are roughly explained. Specifically, GKW3 without momentum dependence explains the data well.[2]

[1] In the HYP2022 presentation, I answered to the question by Hoai Le that we do include different potentials for Σ and Ξ, but it was not correct. The results shown in HYP2022 and in this proceedings are obtained by assuming the same potential for all hyperons.

[2] In the HYP2022 presentation, we showed the calculated results with an incorrect p_T cut at $\sqrt{s_{NN}} = 3.0$ GeV, and the GKW3 results underestimated the directed flow data. By using the correct cut adopted in the experimental data analysis, 0.4 GeV $\leq p_T \leq$ 2.0 GeV [7], the results from GKW3 explain the directed flow well also at $\sqrt{s_{NN}} =$ 3.0 GeV.

Now let us discuss the sensitivity of the directed flow to the density- and momentum-dependence of the Λ potential. Unfortunately, GKW2, not shown in the figure, gives the Λ directed flow similar to that from GKW3. Thus we cannot discriminate the Λ potentials from χEFT with 2- and 2+3-body interactions by the directed flow. We also note that the strong momentum dependence of the Λ potential, e.g. GKW3+MD2, reduces the directed flow, especially at large rapidities [12]. We do not completely understand the origin of this dependence, but we guess that the cancellation of the contributions in the compression and expansion stages is one of the origins. The repulsive potential at high densities increases the directed flow in the compression stage, but a part of the increase is canceled by the negative contribution from the expansion of the tilted matter.

4 Summary and discussion

We have investigated the Λ potential via the Λ directed flow from heavy-ion collisions [12]. We implement the Λ potential from the χEFT [3, 4] parameterized as a function of density and momentum in the RQMDv mode of an event generator JAM. The calculated results using the Λ potential from χEFT do reproduce the Λ directed flow in a wide beam energy range. This is the first support by experimental data of the strongly repulsive Λ potential at high densities. It should be noted that similar results are obtained for the sideward flow $\langle p_x \rangle$ at $\sqrt{s_{NN}} = 2.7$ GeV in Ref. [21].

Yet there are several questions and problems. First, the directed flow of Λ can also be reproduced by the Λ potential only from the 2-body interactions and with a weaker repulsion at high densities. Next, the Λ directed flow depends on the momentum dependence of the potential, especially at large rapidities. Provided that the compression/expansion mechanism [10] is correct, the directed flow of Λs produced during the compression stage is affected by the expansion of tilted matter and would be more sensitive to dependence on momentum than density. More analyses of the evolution are needed to understand the sensitivity to the details of the Λ potential.

It is interesting to study Ξ and Σ flows as well. These are related to the repulsive Σ potential in dense matter [21], the dense matter EoS [22], and the formation of deconfined matter [23].

The authors would like to thank Yue-Hang Leung for telling us that the p_T cut is different at $\sqrt{s_{NN}} = 3.0$ GeV. This work was supported in part by the Grants-in-Aid for Scientific Research from JSPS (No. JP21K03577, No. JP19H01898, and No. JP21H00121).

References

[1] P. Demorest *et al.*, Nature **467**, 1081 (2010); E. Fonseca *et al.*, Astrophys. J. **832**, 167 (2016); J. Antoniadis *et al.*, Science **340**, 6131 (2013); H. T. Cromartie *et al.* [NANOGrav], Nature Astron. **4**, 72 (2019); M. C. Miller *et al.*, Astrophys. J. Lett. **918**, L28 (2021)

[2] S. Nishizaki, T. Takatsuka and Y. Yamamoto, Prog. Theor. Phys. **108**, 703 (2002); J. Rikovska-Stone, P. A. M. Guichon, H. H. Matevosyan and A. W. Thomas, Nucl. Phys. A **792**, 341 (2007); S. Weissenborn, D. Chatterjee and J. Schaffner-Bielich, Phys. Rev. C **85**, 065802 (2012) [erratum: Phys. Rev. C **90**, 019904 (2014)]; T. Miyatsu, S. Yamamuro and K. Nakazato, Astrophys. J. **777**, 4 (2013). H. Togashi, E. Hiyama, Y. Yamamoto and M. Takano, Phys. Rev. C **93**, 035808 (2016)

[3] D. Gerstung, N. Kaiser, W. Weise, Eur. Phys. J. A **56**, 175 (2020)

[4] M. Kohno, Phys. Rev. C **97**, 035206 (2018)

[5] S. Petschauer, J. Haidenbauer, N. Kaiser, U. G. Meißner and W. Weise, Nucl. Phys. A **957**, 347 (2017); J. Haidenbauer, S. Petschauer, N. Kaiser, U. G. Meißner and W. Weise, Eur. Phys. J. C **77**, 760 (2017)

[6] K. Aoki *et al.*, *Extension of the J-PARC Hadron Experimental Facility: Third White Paper*, arXiv:2110.04462 [nucl-ex]

[7] L. Adamczyk *et al.* [STAR], Phys. Rev. Lett. **112**, 162301 (2014); P. Shanmuganathan *et al.* [STAR], Nucl. Phys. A **956**, 260 (2016); L. Adamczyk *et al.* [STAR], Phys. Rev. Lett. **120**, 062301 (2018); J. Adam *et al.* [STAR], Phys. Rev. C **103**, 034908 (2021); M. S. Abdallah *et al.* [STAR], Phys. Lett. B **827**, 137003 (2022)

[8] J. Steinheimer, J. Auvinen, H. Petersen, M. Bleicher and H. Stoecker, Phys. Rev. C **89**, 054913 (2014); Y. B. Ivanov and A. A. Soldatov, Phys. Rev. C **91**, 024915 (2015); Y. B. Ivanov and A. A. Soldatov, Eur. Phys. J. A **52**, 10 (2016); Y. Nara, H. Niemi, J. Steinheimer and H. Stoecker. Phys. Lett. B **769**, 543 (2017)

[9] V. P. Konchakovski, W. Cassing, Y. B. Ivanov and V. D. Toneev, Phys. Rev. C **90**, 014903 (2014)

[10] Y. Nara and A. Ohnishi, Phys. Rev. C **105**, 014911 (2022)

[11] Y. Nara, N. Otuka, A. Ohnishi, K. Niita and S. Chiba, Phys. Rev. C **61**, 024901 (2000)

[12] Y. Nara, A. Jinno, K. Murase and A. Ohnishi, Phys. Rev. C **106**, 044902 (2022) [arXiv:2208.01297 [nucl-th]]

[13] S. Balberg and A. Gal, Nucl. Phys. A **625**, 435 (1997); D. E. Lanskoy and Y. Yamamoto, Phys. Rev. C **55**, 2330 (1997); M. Baldo, G. F. Burgio and H. J. Schulze, Phys. Rev. C **61**, 055801 (2000)

[14] N. K. Glendenning and S. A. Moszkowski, Phys. Rev. Lett. **67**, 2414 (1991); J. Schaffner and I. N. Mishustin, Phys. Rev. C **53**, 1416 (1996); C. Ishizuka, A. Ohnishi, K. Tsubakihara, K. Sumiyoshi and S. Yamada, J. Phys. G **35**, 085201 (2008); H. Shen, H. Toki, K. Oyamatsu and K. Sumiyoshi, Astrophys. J. Suppl. **197**, 20 (2011)

[15] P. Danielewicz, P. B. Gossiaux and R. A. Lacey, Fundam. Theor. Phys. **95**, 69 (1999); A. Sorensen and V. Koch, Phys. Rev. C **104**, 034904 (2021); C. Fuchs and H. H. Wolter, Nucl. Phys. A **589**, 732 (1995); K. Weber, B. Blaettel, W. Cassing, H. C. Doenges, V. Koch, A. Lang and U. Mosel, Nucl. Phys. A **539**, 713 (1992)

[16] Y. Fujiwara, Y. Suzuki and C. Nakamoto, Prog. Part. Nucl. Phys. **58**, 439 (2007)

[17] JAM2, an event generator for high-energy nuclear collisions, https://gitlab.com/transportmodel/jam2

[18] T. Sjöstrand, S. Ask, J. R. Christiansen, R. Corke, N. Desai, P. Ilten, S. Mrenna, S. Prestel, C. O. Rasmussen and P. Z. Skands, Comput. Phys. Commun. **191**, 159-177 (2015), https://pythia.org/

[19] X. L. Zhao, G. L. Ma, Y. G. Ma and Z. W. Lin, Phys. Rev. C **102**, 024904 (2020)

[20] H. Sorge, Phys. Rev. C **52**, 3291 (1995); Y. Nara and H. Stoecker, Phys. Rev. C **100**, 054902 (2019); Y. Nara, T. Maruyama and H. Stoecker, Phys. Rev. C **102**, 024913 (2020)

[21] D. C. Zhang, H. G. Cheng and Z. Q. Feng, Chin. Phys. Lett. **38**, 092501 (2021)

[22] G. C. Yong, Z. G. Xiao, Y. Gao and Z. W. Lin, Phys. Lett. B **820**, 136521 (2021)

[23] K. Nayak, S. Shi, N. Xu and Z. W. Lin, Phys. Rev. C **100**, 054903 (2019)

Centrality dependence of multistrangeness production in high-energy heavy-ion collisions

Gevorg H. Arakelyan[1,2,†], *Carlos* Merino[2,*], and *Yuli M.* Shabelski[2,3]

[1]A.Alikhanyan National Scientific Laboratory (Yerevan Physics Institute), Yerevan, Armenia
[2]Dpto. de Física de Partículas, Facultade de Física, Instit. Galego de Física de Altas Enerxías (IGFAE) Universidade de Santiago de Compostela, Galiza, Spain
[3]Petersburg Nuclear Physics Institute, NCR Kurchatov Institute, Gatchina, St.Petersburg, Russia

Abstract. We compare the experimental data on yields of protons, strange Λ's, and multistrange baryons (Ξ, Ω), and antibaryons production on nuclear targets, and the experimental ratios of multistrange to strange antibaryon production, at the energy region from SPS up to LHC, with the corresponding results of the Quark-Gluon String Model calculations. In the case of heavy nucleus collisions, the experimental dependence of the $\overline{\Xi}^+/\overline{\Lambda}$, and $\overline{\Omega}^+/\overline{\Lambda}$ ratios, on the centrality of the collision, shows a manifest violation of quark combinatorial rules.

1 Introduction

We compare [1] the results obtained in the Quark-Gluon String Model (QGSM) formalism, with the corresponding experimental data on yields of p, Λ, Ξ, and Ω baryons, and the corresponding antibaryons, in nucleus-nucleus collisions, for a wide energy region, going from SPS up to LHC ranges. We also consider the ratios of multistrange to strange antihyperon production in nucleus-nucleus collisions with different centralities, at CERN-SPS and RHIC energies.

The QGSM [2–4] is based on the Dual Topological Unitarization (DTU), Regge phenomenology [5], and nonperturbative notions of QCD [6]. In QGSM, high energy interactions are considered as proceeding via the exchange of one or several Pomerons. The cut of at least some of those Pomerons determines the inelastic scattering amplitude of the particle production processes, that occur through the production and subsequent decay of the quark-gluon strings resulting of the cut of the Pomerons.

In the case of interaction with a nuclear target, the Multiple Scattering Theory (Gribov-Glauber Theory) [7] is used. For nucleus-nucleus collisions, the Multiple Scattering Theory allows to consider these interactions as a superposition of independent nucleon-nucleon interactions.

At very high energies, the contribution of enhanced Reggeon diagrams (percolation effects) becomes important, leading to a new effect, the suppression of the inclusive density of secondaries [8] into the central (midrapidity) region. This effect corresponds to a significant fusion of the primarily produced quark-gluon strings.

†Deceased
*e-mail: carlos.merino@usc.es

The QGSM provides a succesful thorough description of multiparticle production processes in hadron-hadron [9–13], hadron-nucleus [14–16], and nucleus-nucleus [17–20] collisions, for a wide energy region. In particular, the inclusive spectra of charged pions, kaons, nucleons, and Λ hyperons, were correctly described in the cited papers.

The production of multistrange hyperons, Ξ^- (dss), and Ω^- (sss), has special interest in high energy particle and nuclear physics. Since the initial-state colliding projectiles contain no strange valence quarks, all particles in the final state with non-zero strangeness quantum number should have been created in the process of the collision. This makes multistrange baryons a valuable probe in understanding the particle production mechanisms in high energy collisions. A remarkable feature of strangeness production is that the production of each additional strange quark featuring in the secondary baryons, i.e., the production rate of secondary $B(qqs)$ over secondary $B(qqq)$, then of $B(qss)$ over $B(qqs)$, and, finally, of $B(sss)$ over $B(qss)$, is affected by one universal strangeness suppression factor, λ_s:

$$\lambda_s = \frac{B(qqs)}{B(qqq)} = \frac{B(qss)}{B(qqs)} = \frac{B(sss)}{B(qss)} , \tag{1}$$

together with some simple quark combinatorics [21, 22].

Let us define:

$$R(\overline{\Xi}^+/\overline{\Lambda}) = \frac{dn}{dy}(A + B \to \overline{\Xi}^+ + X)/\frac{dn}{dy}(A + B \to \overline{\Lambda} + X) , \tag{2}$$

$$R(\overline{\Omega}^+/\overline{\Lambda}) = \frac{dn}{dy}(A + B \to \overline{\Omega}^+ + X)/\frac{dn}{dy}(A + B \to \overline{\Lambda} + X) . \tag{3}$$

The produced antihyperons, $\overline{\Xi}^+$ and $\overline{\Omega}^+$, contain valence antiquarks newly produced during the collision. The ratios in eqs. (2) and (3) are reasonably described by QGSM when a relatively small number of incident nucleons participate in the collision (nucleon-nucleus collisions, or peripheral nucleus-nucleus collisions).

The number of quark-gluon strings (cut pomerons) in nucleus-nucleus collisions increases with the centrality of the collision, but if the secondaries are independently produced in a single quark-gluon string, the ratio of yields of different particles should not depend on the centrality.

2 Comparison of QGSM calculations with experimental data

2.1 Fixed Target Energy Data

The experimental data on proton and hyperon productiom were reasonably described in [16]. The data for p+Be and p+Pb collisions were also described in the same reference, but it appears that the value of the strange suppression parameter λ_s=0.25 would have to be slightly increased to get a better comparison with the experimental data. In the present paper we use the value λ_s=0.32, that provides good results for the description of the experimental data in references [23, 24].

In the case of Pb+Pb collisions at 158 GeV/c per nucleon, we compare the experimental data on dn/dy for central collisions ($|\ y\ | \leq 0.5$), measured by NA49 [25–30], NA57 [31, 32], and WA97 [33] collaborations, with the corresponding QGSM predictions.

In Fig. 1 we show the comparison of the QGSM prediction with the experimental data [32] on the dependence of the ratios $\overline{\Omega}^+/\overline{\Lambda}$ (left panel) and $\overline{\Xi}^+$ to $\overline{\Lambda}$ (right panel), on the number of wounded nucleons, N_ω, in Pb+Pb collisions at 158 GeV/c per nucleon. The number of

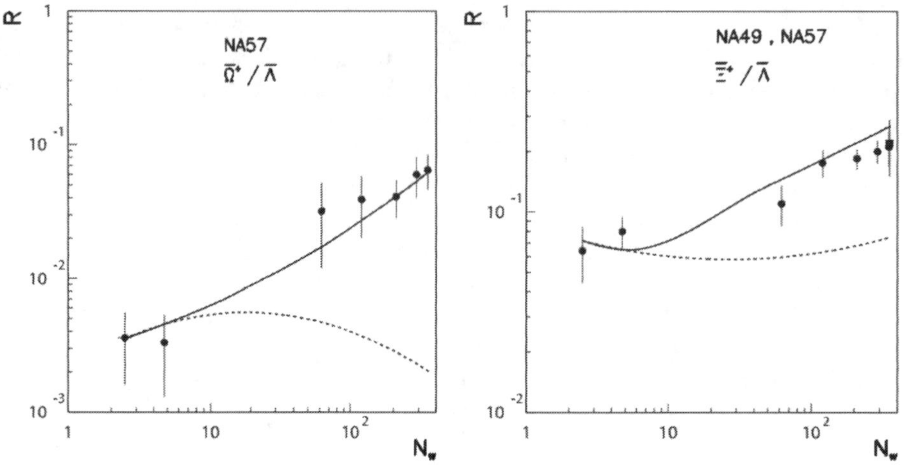

Figure 1. Ratios of $\overline{\Omega}^+/\overline{\Lambda}$ (left panel), and of $\overline{\Xi}^+/\overline{\Lambda}$ (right panel) as functions of the number of wounded nucleons, N_w. The experimental data of Pb+Pb collisions for different numbers of wounded nucleons, N_w (different centralities) measured by the NA57 Collaboration (points), and by the NA49 Collaboration (squares) are presented, and compared with the corresponding QGSM predictions.

wounded nucleons, N_ω, is directly related to the centrality of collisions. Thus, small values of N_w correspond to peripheral collisions (large impact parameter), while large values of N_w correspond to central collisions (small impact parameter).

At small values of N_w the ratio is practically equal to that in the cases of p+Be and p+Pb collisions, and all they can be correctly described by the QGSM by using a value the of the strangeness suppression parameter, λ_s=0.32. Then, the experimental ratio increases rather fast with the incresing value of N_w, i.e. when we move from peripheral to central Pb+Pb collisions. This behavior is reasonably reproduced by the full line in Fig. 1, that it has been calculated with the value λ_s=0.32 at its left end, and with larger values of λ_s at its right end (see ref. [1] for details). The results of QGSM calculations with a constant value λ_s=0.32, disregarding of the value of the number of wounded nucleons (centrality), N_ω, are shown in Fig. 1 by dashed lines.

Obviously, for small values of N_w, both curves coincide, but when N_w increases the full line also increases in agreement with the data, while the dashed line is practically constant, with exception of small corrections mainly connected to energy conservation. This dashed line shows a very significant disagreement with the experimenatl data at large values of N_ω. The difference between the full and the dashed lines for the ratio $\overline{\Omega}^+/\overline{\Lambda}$ in a very central event is of about one order of magnitude. Such a large difference comes from the fact that the cross section for $\overline{\Omega}$ production is proportional to $\lambda_s{}^3$.

This behaviour in which the value of the strangeness suppression factor λ_s increases with the value of N_w (centrality), indicates that the simple quark combinatorial rules are not valid for central collisions of heavy nuclei.

2.2 STAR Collaboration Data

Now we consider the experimental data on midrapidity densities of protons and hyperons in Au+Au measured at RHIC by the STAR [34–37] Collaboration at $\sqrt{s}$ = 62.4 GeV, and we compare them with the results of the corresponding QGSM calculations. One can appreciate that, in general, the agreement of the QGSM predictions with the experimental data in this energy region is quite reasonable.

Again, we see here that the value of λ_s for multistrange hyperon production is larger than for the case of Λ and $\overline{\Lambda}$ production. Now this difference is not so large as it is for collisions at lower fixed-target energies, in what it seems an indication that the violation of the quark combinatorial rules becomes less important for high energy collisions.

In Fig. 2 we present the comparison of the QGSM predictions with experimental data on the N_ω dependence of the ratios $\overline{\Omega}^+/\overline{\Lambda}$ (left panel), and $\overline{\Xi}^+/\overline{\Lambda}$ (right panel), measured by the STAR Collaboration in the midrapidity region, at $\sqrt{s}$ = 62.4 GeV. Similarly as in Fig. 1, the left end of the full line here was calculated with a value λ_s=0.32, while for the right end larger values of λ_s were used (see [1]). The dashed line was calculated with a constant value of λ_s=0.32.

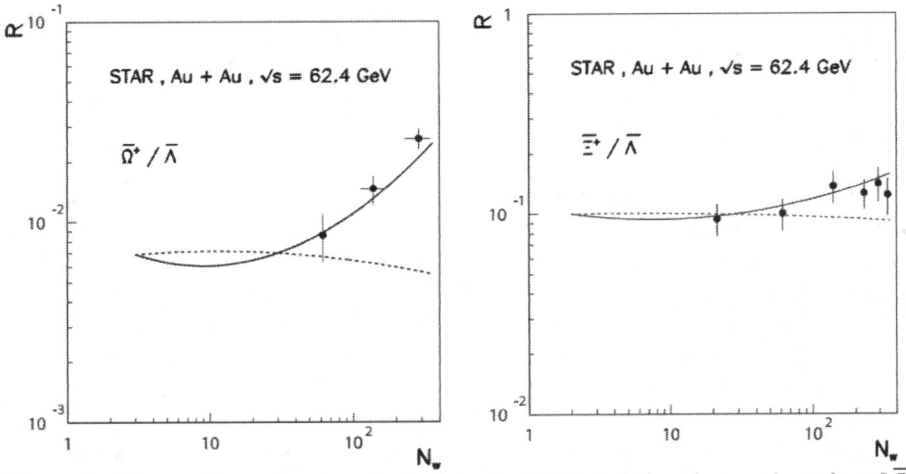

Figure 2. The experimental points obtained by the STAR Collaboration on the ratios of $\overline{\Omega}^+/\overline{\Lambda}$ (left panel), and $\overline{\Xi}^+$ to $\overline{\Lambda}$ (right panel), in Au+Au collisions at $\sqrt{s}$ = 62.4 GeV, at different centralities, as a function of the number of wounded nucleons, N_ω, together with the results of the corresponding QGSM calculations (solid curves).

The same calculations for the ratio of $\overline{\Xi}^+/\overline{\Lambda}$ in Cu+Cu collisions at $\sqrt{s}$ = 200 GeV [38] show again that the violation of the quark combinatorial rules decreses with the growth of the energy of the collision, the difference between the full and dashed lines being of the order of the experimental error bars, at $\sqrt{s}$ = 200 GeV.

2.3 LHC experimental Data

In Table 1 we consider the experimental data on p, $\overline{p}$, and Ξ^-, $\overline{\Xi}^+$, Ω^-, $\overline{\Omega}^+$ production in central Pb+Pb collisions at $\sqrt{s}$ = 2.76 Tev, and of $p + \overline{p}$ production in central Pb+Pb collisions at $\sqrt{s}$ = 5.02 Tev, measured by the ALICE Collaboration [39–41], at the CERN LHC.

The value of the strangeness suppression parameter λ_s needed for a correct description of Ξ^- and $\overline{\Xi}^+$ production at LHC is smaller than at RHIC energies, and it coincides with the standard value λ_s=0.32. In the case of Ω^- and $\overline{\Omega}^+$ production at LHC, the value of λ_s also decreases with respect to the RHIC energy range. Thus, the unusually large values of λ_s that correspond to central Pb+Pb collisions at 158 GeV/c per nucleon, monotonically decrease with the increase of the initial energy of the collision.

Thus, the experimental data on strange and multistrange particle production show that the value of the strangeness suppression factor to be taken in QGSM to correctly describe those experimental data, is lower at RHIC and LHC energy ranges, than at SPS (i.e. a smaller enhancement of strangeness production at RHIC and LHC, when compared with that at SPS).

Table 1. Experimental data on dn/dy by the ALICE Collaboration of p, $\bar{p}$ central production at $\sqrt{s}$ = 2.76 TeV [39], of p+$\bar{p}$ central production at $\sqrt{s}$ = 5.02 TeV [41], Ξ^-, and of Ξ^-, $\overline{\Xi}^+$, Ω^-, and $\overline{\Omega}^+$ production in central Pb+Pb collisions at $\sqrt{s}$ = 2.76 TeV per nucleon [40], and the comparison with the results of the corresponding QGSM calculations.

Process	$\sqrt{s}$ (TeV)	Centrality	dn/dy (Exp. Data)	dn/dy (QGSM)	λ_s
Pb+Pb $\to p$	2.76	0–5%	34 ± 3	34.604	0,32
Pb+Pb $\to \overline{p}$	2.76	0–5%	33 ± 3	33.898	
Pb+Pb $\to p + \overline{p}$	5.02	0–5%	74.56 ± 0.06 ± 3.75	77.71	0,32
Pb+Pb $\to \Xi^-$	2.76	0–10%	3.34 ± 0.06 ± 0.24	3.357	0,32
Pb+Pb $\to \overline{\Xi}^+$	2.76	0–10%	3.28 ± 0.06 ± 0.23	3.317	
Pb+Pb $\to \Omega^-$	2.76	0–10%	0, 58 ± 0.04 ± 0.09	0.606	0.38
Pb+Pb $\to \overline{\Omega}^+$	2.76	0–10%	0, 60 ± 0.05 ± 0.09	0,601	

3 Conclusion

We consider the production of hyperons in collisions on nuclear targets, in the framework of the QGSM formalism, and we find that the experimental data on multistrange hyperon and antihyperon production in p-nucleus and in peripheral nucleus-nucleus collisions can be reasonably described in the QGSM framework, by using for all baryons the same standard value of the strange suppression parameter λ_s=0.32, while to get a correct description in QGSM of multistrange hyperon and antihyperon production in central Pb+Pb collisions in the midrapidity region, a larger value of λ_s is needed.

In particular, one can see that the experimental probability of multistrange hyperon production in Pb+Pb collisions increases monotonically with the number of wounded nucleons, N_w, i.e. when going from peripheral to very central nucleus-nucleus collisions. A similar situation occurs in the case of Au+Au collisions at RHIC energies, though the effect is not so pronounced.

As a matter of fact, the experimental data on the production of hyperons in central collisions of heavy nuclei show a very significant violation, essentially large at CERN-SPS energies $\sqrt{s}$ = 17.3 GeV, of simple quark combinatorial rules, this violation decreasing with the growth of the initial energy of the collision. On the contrary, the corresponding data in proton-nucleus and in peripheral nucleus-nucleus collisions are in agreement with the quark combinatorics theoretical description, by considering one universal and constant strangeness suppression factor.

Acknowledgments
C.M. wants to thank the organizers of HYP2022 for a great conference, in which the nice enviroment thas been both personally and scientifically encouraging.

References

[1] G.H. Arakelyan, C. Merino, and Yu.M. Shabelski, Phys. Rev. **D105**, 114013 (2022)
[2] A.B. Kaidalov and K.A. Ter-Martorisyan, Phys. Lett. **B117**, 247 (1982)
[3] A.B. Kaidalov and K.A. Ter-Martirosyan, Sov. J. Nucl. Phys. **39**, 979 (1984), Yad. Fiz. **39**, 1545 (1984); Sov. J. Nucl. Phys. **40**, 135 (1984), Yad. Fiz. **40**, 211 (1984)
[4] A.B. Kaidalov, Sov. J. Nucl. Phys. **45**, 902 (1987), Yad. Fiz. **45**, 1452 (1987)
[5] G.C. Rossi and G. Veneziano, Nucl. Phys. **B123**, 507 (1977)
[6] M. Ciafalloni, G. Marchesini and G. Veneziano, Nucl. Phys. **B98**, 472 (1975)

[7] V.N. Gribov, Sov. Phys. JETP **29**, 483 (1969), Zh. Eksp. Teor. Fiz. **56**, 892 (1969)
[8] A. Capella, A. Kaidalov, and J. Tran Thanh Van, Acta Phys. Hung. **A9**, 169 (1999)
[9] A.B. Kaidalov and O.I. Piskounova, Sov. J. Nucl. Phys. **41**, 816 (1985), Yad. Fiz. **41**, 1278 (1985); Z. Phys. **C30**, 145 (1986)
[10] Yu.M. Shabelski, Sov. J. Nucl. Phys. **44**, 117 (1986), Yad. Fiz. **44**, 186 (1986)
[11] G.H. Arakelyan, A. Capella, A.B. Kaidalov, and Yu.M. Shabelski, Eur. Phys. J. **C26**, 81 (2002)
[12] G.H. Arakelyan, C. Merino, C. Pajares, and Yu.M. Shabelski, Eur. Phys. J. **C54**, 577 (2008)
[13] C. Merino, C. Pajares and Yu.M. Shabelski, Eur. Phys. J. **C71**, 1652 (2011)
[14] A.B. Kaidalov, K.A. Ter-Martirosyan, and Yu.M. Shabelski, Sov. J. Nucl. Phys. **43**, 822 (1986), Yad. Fiz. **43**, 1282 (1986)
[15] Yu.M. Shabelski, Z. Phys. **C38**, 569 (1988)
[16] G.H. Arakelyan, C. Merino, and Yu.M. Shabelski, Eur. Phys. J. **A52**, 9 (2016)
[17] Yu.M. Shabelski, Sov. J. Nucl. Phys. **50**, 149 (1989), Yad. Fiz. **50**, 239 (1989)
[18] Yu.M. Shabelski, Z. Phys. **C57**, 409 (1993)
[19] J. Dias de Deus and Yu.M. Shabelski, Phys. Atom. Nucl. **71**, 190 (2008), Yad. Fiz. **71**, 191 (2008)
[20] G.H. Arakelyan, C. Merino, and Yu.M. Shabelski, Phys. Atom. Nucl. **77**, 626 (2014), Yad. Fiz. **77**, 661 (2014)
[21] V.V. Anisovich and V.M. Shekhter, Nucl. Phys. **B55**, 455 (1973), Nucl. Phys. **B63**, 542 (1973) (erratum)
[22] G.H. Arakelyan, A.B. Kaidalov, C. Merino, and Yu.M. Shabelski, Phys. Atom. Nucl. **74**, 126 (2011)
[23] G.H. Arakelyan, C. Merino, and Yu.M. Shabelski, Int. J. Mod. Phys. **A33**, 1850202 (2018)
[24] G.H. Arakelyan, C. Merino, and Yu.M. Shabelski, Eur. Phys. J. **A55**, 151 (2019)
[25] T. Anticic *et al.*, NA49 Collaboration, Phys. Rev. **C69**, 024902 (2004)
[26] T. Anticic *et al.*, NA49 Collaboration, Phys. Rev. Lett. **93**, 022302 (2004)
[27] C. Alt *et al.*, NA49 Collaboration, Phys. Rev. Lett. **94**, 192301 (2005)
[28] C. Alt *et al.*, NA49 Collaboration, Phys. Rev. **C78**, 034918 (2008)
[29] T. Anticic *et al.*, NA49 Collaboration, Phys. Rev. **C80**, 034906 (2009)
[30] C. Blume and C. Markert, Prog. Part. Nucl. Phys. **66**, 834 (2011)
[31] F. Antinori *et al.*, NA57 Collaboration, Phys. Lett. **B595**, 68 (2004)
[32] F. Antinori *et al.*, NA57 Collaboration, J. Phys. **G32**, 427 (2006)
[33] E. Andersen *et al.*, WA97 Collaboration, Phys. Lett. **B449**, 401 (1999)
[34] J. Adams *et al.*, STAR Collaboration, Phys. Rev. Lett. **92**, 182301 (2004)
[35] J. Adams *et al.*, STAR Collaboration, Phys. Rev. Lett. **98**, 062301 (2007)
[36] B.I. Abelev *et al.*, STAR Collaboration, Phys. Rev.**C79**, 034909 (2009)
[37] M.M. Aggarwal *et al.*, STAR Collaboration, Phys. Rev. **C83**, 024901 (2011)
[38] G. Agakishiev *et al.*, STAR Collaboration, Phys. Rev. Lett. **108**, 072301 (2012)
[39] B. Abelev *et al.*, ALICE Collaboration, Phys. Rev. **C88**, 044910 (2013)
[40] B. Abelev *et al.*, ALICE Collaboration, Phys. Lett. **B728**, 216 (2014), Phys. Lett. **B734**, 409 (2014) (erratum)
[41] S. Acharya *et al.*, ALICE Collaboration, Phys. Rev. **C101**, 044907 (2020)

Recent progress on global and local polarization of hyperons in heavy-ion collisions

Takafumi Niida[1,*]

[1]University of Tsukuba, 1-1-1 Tennoudai, Tsukuba, Ibaraki 305-8571, JAPAN

Abstract. Since the observation of Λ global polarization at RHIC, indicating the most vortical fluid ever observed so far, many experimental and theoretical progresses have been made in understanding of vorticity and polarization phenomena in heavy-ion collisions. But there still exist open questions which are currently being studied. In these proceedings, recent experimental results on global and local polarization of hyperons in heavy-ion collisions are reviewed.

1 Introduction

A partial conversion of an orbital angular momentum carried by two colliding nuclei into spin angular momentum of produced particles is referred to as global polarization [1–3]. Observation of Λ global polarization by STAR experiment [4] revealed the creation of most vortical fluid ever observed. It will also help to constrain the lifetime of the initial magnetic field created in the collisions as the direction of the magnetic field coincides with the orbital angular momentum direction. While the experimental results of average global polarization are well reproduced by theoretical models, most of which are based on local thermal equilibrium, there are open questions and some of differential measurements including local polarization are not yet fully understood. Study of the polarization phenomena in heavy-ion collisions is relatively new but rapidly growing in this field, providing new information on the initial condition, medium properties, and the collision dynamics. In these proceedings, recent results of hyperon polarization in heavy-ion collisions are reviewed.

2 Global polarization

Global polarization P_H, the polarization component along the initial orbital angular momentum carried by two colliding nuclei in heavy-ion collisions, was first observed using Λ hyperons in Au+Au collisions at $\sqrt{s_{NN}}$ = 7.7-200 GeV by STAR experiment [4, 5] where the polarization signal decreases with increasing the energy. There was also an attempt to measure it at the LHC energy but the result is consistent with zero as the expected signal is at the level of current precision. Theoretical models predict that the polarization increases starting from $\sqrt{s_{NN}} \sim 2m_N$ and becomes a maximum around $\sqrt{s_{NN}}$ = 3 GeV and the peak position could be different between Λ and $\bar{\Lambda}$ [6].

Recently, the measurement has been extended to lower energies by STAR fixed-target mode and HADES experiment [7, 8]. Figure 1(left) shows a compilation of Λ global polarization as a function of the collision energy. New results at 3 GeV from STAR and at 2.4

*e-mail: niida@bnl.gov

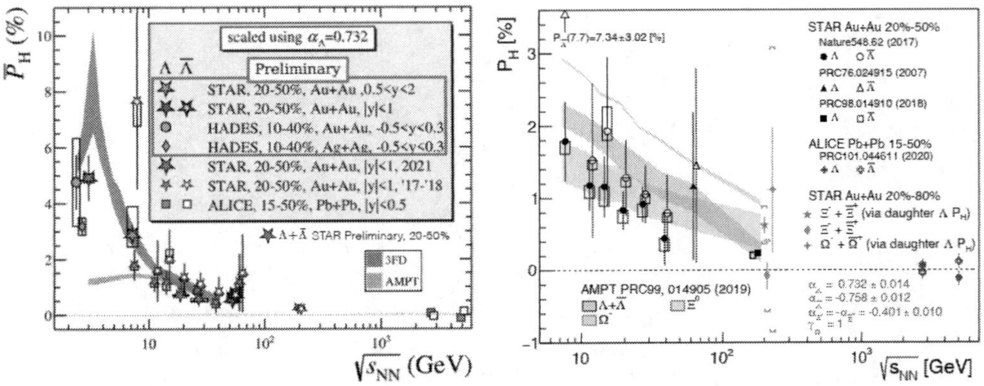

Figure 1. (Left) Global polarization of Λ and $\bar{\Lambda}$ hyperons as a function of the center-of-mass energy from STAR, ALICE, and HADES experiments [10]. (Right) global polarization of Ξ and Ω hyperons compared to that of Λ hyperons as a function of the collision energy [11].

and 2.55 GeV from HADES still show an increasing trend of the global polarization, consistent with the calculation from the hydrodynamic model based on three-fluid dynamics [9]. Note that the lower energy corresponds to the high baryon density region where the medium properties would be notably different. It should be also noted that the detector acceptance in rapidity and transverse momentum p_T affects the energy dependence since it covers only mid-rapidity at higher energies while it does a nearly whole distribution of produced particles at lower energies. Theoretical models predict the rapidity dependence of global polarization differently although the current measurements do not show any significant dependence [5, 7, 8]. Difference between Λ and $\bar{\Lambda}$, as a measure of the initial magnetic field, is not significant with large uncertainties which will be improved in the beam energy scan phase-II at RHIC-STAR.

To confirm the vorticity and global polarization picture, it is important to have the measurements for different particle species, preferably with different spins, which provides new insights into the polarization mechanism in heavy-ion collisions. Recently, the measurements with multistrangeness, i.e. Ξ (spin-1/2) and Ω (spin-3/2), have been performed by STAR experiment [11]. Unlike for Λ hyperons ($\alpha_\Lambda = 0.732$), the magnitude of decay parameter is small for Ξ ($\alpha_\Xi = -0.401$) and is close to zero for Ω ($\alpha_\Omega = 0.0157$) which makes the polarization measurement difficult. One can use the measurement of daughter Λ polarization to know its parent particle's polarization: $P^*_\Lambda = C_{X\Lambda} P^*_X$ where P^*_X and P^*_Λ are polarizations of parent particle X and its daughter Λ in parent rest frame, and $C_{X\Lambda}$ denotes polarization transfer factor. The factor $C_{X\Lambda}$ is +0.944 for the decay of $\Xi^- \to \Lambda\pi^-$ and could be either +1 or -0.6 for the decay of $\Omega^- \to \Lambda K^-$ due to the uncertainty of decay parameter $\gamma_\Omega (\approx \pm 1)$. Based on the vorticity picture, the sign of γ_Ω can be determined by the measurement of Ω global polarization.

Figure 1(right) shows global polarization of Ξ and Ω hyperons comparing to Λ global polarization as well as transport model calculations. The global polarization averaged over transverse momentum and rapidity for 20-80% Au+Au collisions at $\sqrt{s_{NN}}$ = 200 GeV is measured to be $\langle P_\Lambda \rangle = 0.24 \pm 0.03(\text{stat}) \pm 0.03(\text{stat})\%$, $\langle P_\Xi \rangle = 0.47 \pm 0.10(\text{stat}) \pm 0.23(\text{stat})\%$, and $\langle P_\Omega \rangle = 1.11 \pm 0.87(\text{stat}) \pm 1.97(\text{stat})\%$. Based on thermal model in the nonrelativistic limit [12], the polarization can be given by

$$\boldsymbol{P} = \langle \boldsymbol{s} \rangle / s \approx (s+1)\boldsymbol{\omega}/(3T), \quad (1)$$

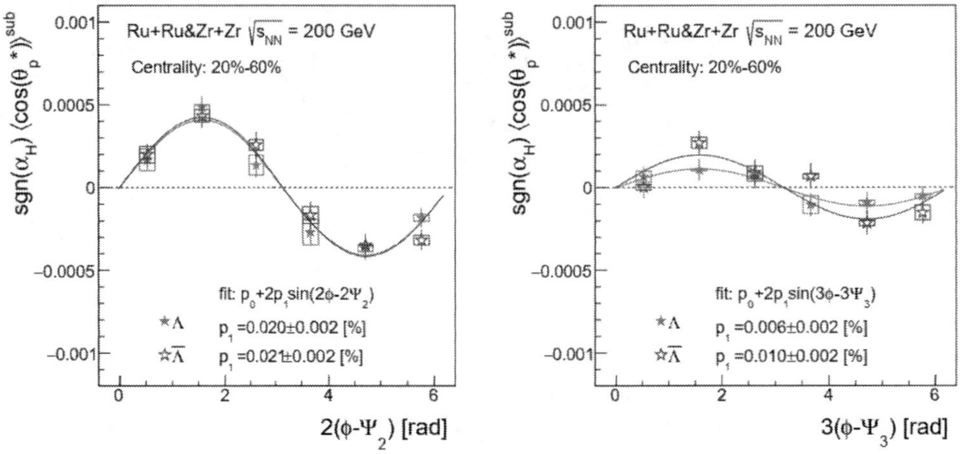

Figure 2. Longitudinal polarization of Λ and $\bar{\Lambda}$ hyperons, $\langle\cos\theta_p^*\rangle \approx P_z\alpha_H/3$, as a function of azimuthal angle with respect to the second-order event plane Ψ_2 (left) and to the third order event plane Ψ_3 (right) in 20-60% isobar (Ru+Ru and Zr+Zr) collisions at $\sqrt{s_{NN}}$ = 200 GeV [22]. The solid lines are fit functions as indicated in the figures.

where s is the spin of the particle, ω is the local vorticity of the fluid, and T is the temperature. There seems to be a hierarchy in the measured $\langle P_{\mathrm{H}}\rangle$ as expected from Eq. (1) although the uncertainty is still large to make definitive conclusion. The current measurement seems to favor the decay parameter γ_Ω to be +1. More precise measurements in the future runs at RHIC and the LHC will help to shed light on the uncertainty of γ_Ω as well as for better understanding of the polarization dependence on particle species.

3 Local polarization

In general, local velocity would depend on many other things, such as density fluctuations, the effect of jets passing through the medium, and collective flow. Thus vorticity could be created locally in heavy-ion collisions [13–15]. Vorticity along the beam direction, therefore the polarization, is expected to be induced by elliptic flow, i.e. stronger expansion (more particle emission) in the reaction plane direction than in the out-of-plane direction [16, 17]. Such a longitudinal polarization was first observed in Au+Au collisions at $\sqrt{s_{NN}}$ = 200 GeV by STAR experiment [18], and later in Pb+Pb collisions at $\sqrt{s_{NN}}$ = 5.02 TeV by ALICE experiment [19]. There have been extensive theoretical studies to understand the data since the hydrodynamic and transport models fail to explain the sign of the observed signal, although they reasonably describe the data of the global polarization. New term of shear tensor proposed in Refs. [20, 21] seems to be necessary to explain the data but further experimental inputs are needed for better understanding.

Recent high statistics data in isobar collisions at $\sqrt{s_{NN}}$ = 200 GeV taken by STAR allow to study the polarization even with higher harmonic flow plane. Figure 2 shows raw signal of the longitudinal polarization as a function of hyperon's azimuthal angle relative to the second-order (left) and third-order (right) event planes [22]. Results with the second-order event plane show a clear sine modulation as seen in Au+Au collisions. Results with the third-order event plane also exhibit a similar sinusoidal pattern relative to the azimuthal angle, indicating that vorticity along the beam direction is induced by triangular flow as well as elliptic flow. Hydrodynamic model with the shear term [23] qualitatively explain the data for both the second and third-orders in central-midcentral collisions (but not in peripheral collisions), showing sensitivity of the measurement to specific shear viscosity of quark-gluon plasma formed in the collisions.

4 Summary

In these proceedings, recent progress on the polarization measurements has been reviewed. Thanks to new experimental data at lower energies, energy dependence of global polarization has been studied in detail. Interestingly, the global polarization smoothly increases in lowering collision energy, with a possible maximum around $\sqrt{s_{NN}}$ = 3 GeV, despite the change from partonic matter to hadronic matter in studied energy range. The uncertainty is still large but it can be improved in future analyses or experiments. Measurements with multistrangeness are of great interest and future improvements will provide critical information on spin dynamics in the collisions. New measurement on local polarization with the third-order event plane in isobar collisions indicates the creation of triangular-flow-driven vorticity along the beam direction, which shed light on the sign problem of the local polarization.

Acknowledgment T. Niida is supported by JSPS KAKENHI Grant Number JP22K03648.

References

[1] Z.T. Liang, X.N. Wang, Phys. Rev. Lett. **94**, 102301 (2005), [Erratum: Phys.Rev.Lett. 96, 039901 (2006)]
[2] S.A. Voloshin (2004), `nucl-th/0410089`
[3] F. Becattini, F. Piccinini, J. Rizzo, Phys. Rev. C **77**, 024906 (2008)
[4] L. Adamczyk et al. (STAR), Nature **548**, 62 (2017)
[5] J. Adam et al. (STAR), Phys. Rev. C **98**, 014910 (2018)
[6] A. Ayala, I. Domínguez, I. Maldonado, M.E. Tejeda-Yeomans, Phys. Rev. C **105**, 034907 (2022)
[7] M.S. Abdallah et al. (STAR), Phys. Rev. C **104**, L061901 (2021)
[8] R. A. Yassine et al. (HADES) (2022), `2207.05160`
[9] Y.B. Ivanov, Phys. Rev. C **103**, L031903 (2021)
[10] J. Adams (STAR), Quark Matter 2022
[11] J. Adam et al. (STAR), Phys. Rev. Lett. **126**, 162301 (2021)
[12] F. Becattini, I. Karpenko, M. Lisa, I. Upsal, S. Voloshin, Phys. Rev. C **95**, 054902 (2017)
[13] L.G. Pang, H. Petersen, Q. Wang, X.N. Wang, Phys. Rev. Lett. **117**, 192301 (2016)
[14] B. Betz, M. Gyulassy, G. Torrieri, Phys. Rev. C **76**, 044901 (2007)
[15] X.L. Xia, H. Li, Z.B. Tang, Q. Wang, Phys. Rev. C **98**, 024905 (2018)
[16] S.A. Voloshin, EPJ Web Conf. **171**, 07002 (2018)
[17] F. Becattini, I. Karpenko, Phys. Rev. Lett. **120**, 012302 (2018)
[18] J. Adam et al. (STAR), Phys. Rev. Lett. **123**, 132301 (2019)
[19] S. Acharya et al. (ALICE), Phys. Rev. Lett. **128**, 172005 (2022)
[20] F. Becattini, M. Buzzegoli, G. Inghirami, I. Karpenko, A. Palermo, Phys. Rev. Lett. **127**, 272302 (2021)
[21] B. Fu, S.Y.F. Liu, L. Pang, H. Song, Y. Yin, Phys. Rev. Lett. **127**, 142301 (2021)
[22] T. Niida (STAR), Quark Matter 2022
[23] S. Alzhrani, S. Ryu, C. Shen, Phys. Rev. C **106**, 014905 (2022)

EPJ Web of Conferences **271**, 08009 (2022) https://doi.org/10.1051/epjconf/202227108009
HYP2022

Femtoscopic measurements of two-kaon combinations in Au+Au collisions at the RHIC energies

Diana Pawłowska[1,*] for the STAR Collaboration

[1]Faculty of Physics, Warsaw University of Technology, Koszykowa 75, 00-662 Warsaw, Poland

Abstract. Femtoscopic measurements allow one to study the geometric characteristics of the homogeneity region created during the heavy-ion collisions. The lightest strange hadrons are charged and neutral kaons, which enhancement of the productions is one of the Quark-Gluon Plasma signatures. Analyzing high statistic data taken by the STAR experiment, it is possible to study the short-lived neutral kaons identified through their decay products: positive and negative pions. Thanks to the $K_S^0 K^{\pm}$ analysis, one can also study the production of a_0 resonance, which might be a tetraquark. These proceedings will present the femtoscopic analysis of charged and neutral kaons correlations in Au+Au collisions at $\sqrt{s_{NN}}$ = 39 GeV and 200 GeV measured by the STAR experiment.

1 Introduction

The Solenoidal Tracker at RHIC (STAR) is currently the only experiment at the Relativistic Heavy Ion Collider (RHIC). The main reason for designing this experiment was to measure the properties of hot and dense matter at high collision energy and to find the Quark-Gluon Plasma (QGP). Thanks to data taken by the STAR experiment, it is possible to measure the femtoscopic characteristic between two strange particles (like kaons), which could provide different information about the collision interaction region in addition to what was deduced from pion studies. This is due to the collective effects that form a strongly interacting source. An example of such behavior is that kaons are heavier than pions, so they have smaller thermal velocities, and the collective flow more strongly confines their source volumes. When selecting different particles, we analyze only some parts of this source - Figure 1.

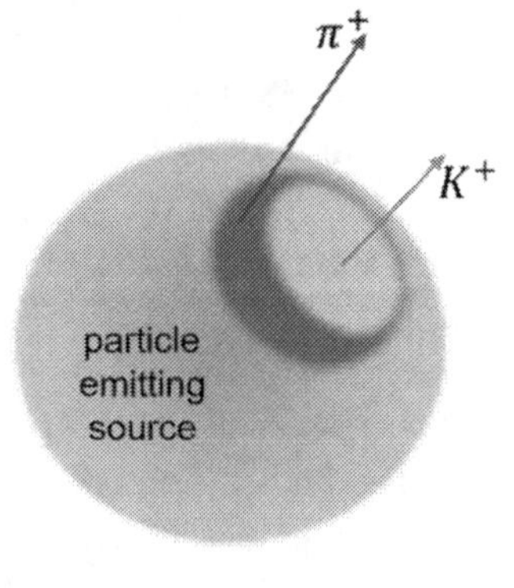

Figure 1. A schematic view of a particle emitting source.

Kaon femtoscopy uses data from the Time Projection Chamber (TPC) and Time of Flight (TOF) to identify particles. TPC is used to register the charged particle tracks in three-dimensional space and to provide information on ionization energy loss. TOF determines the particle's velocity

*e-mail: diana.pawlowska.dokt@pw.edu.pl

using a given particle's time of flight and path length. Combining information from both detectors, the full particle's characteristics regarding their momentum and mass are extracted. The situation is different when particles from weak decays are analyzed and identified based on their decay products. This is much more difficult because one must reconstruct daughters and pair them up. In the case of neutral kaons, decay products are positively and negatively charged pions.

2 Femtoscopy technique

Femtoscopy measures correlations arising from quantum statistical (QS) effects and final state interaction (FSI), probing space-time characteristics of the source emitting given particle pairs. This technique allows one for measurements on the scale of the femtometers and time of the order of 10^{-23} s. The correlation function is expressed by the ratio of the probability of measuring two particles with specific momentum $P_2(\vec{p_1}, \vec{p_2})$ difference to the product of the possibilities of finding these particles independently $P_1(\vec{p_1})P_1(\vec{p_2})$:

$$CF(\vec{p_1}, \vec{p_2}) = \frac{P_2(\vec{p_1}, \vec{p_2})}{P_1(\vec{p_1})P_1(\vec{p_2})}. \quad (1)$$

The theoretical $K_S^0K_S^0$ correlation function is described by the Lednicky-Lyuboshitz model [1], which takes into account strong FSI:

$$CF(q_{inv}) = 1 + \lambda\left(e^{[-R_{inv}^2 q_{inv}^2]} + \alpha\left[\left|\frac{f(k^*)}{R_{inv}}\right|^2 + \frac{4\Re f(k^*)}{\sqrt{\pi}R_{inv}}F_1(q_{inv}R_{inv}) - \frac{2\Im f(k^*)}{\sqrt{\pi}R_{inv}}F_2(q_{inv}R_{inv})\right]\right), \quad (2)$$

where $\alpha = 0.5$ for $K_S^0K_S^0$ correlation, $F_1(z) = \int_0^z dx e^{x^2-z^2}/z$ and $F_2(z) = (1-e^{-z^2})/z$ are analytic forms, $f(k^*)$ is s-wave $K_S^0K_S^0$ scattering amplitude depending on near-threshold resonances $f_0(980)$ and $a_0(980)$, written in terms by theirs masses and decay couplings (the most popular parametrizations are described in [2–5]), $q_{inv} = \sqrt{(\vec{p_1} - \vec{p_2})^2 - (E_1 - E_2)^2}$ is pair relative momentum (k^* is a first particle momentum in pair rest frame), R_{inv} is the size of the particle-emitting source and λ is the chaoticity parameter.
In the case of $K_S^0K^{\pm}$, the Lednicky-Lyuboshitz model is also used. However, a component related to QS effects does not exist; only the strong interaction is present (see Eq. 2). In the case of $K_S^0K^{\pm}$, the interaction depends only on the properties of one resonance - a_0.

3 Results

$K_S^0K_S^0$ measurements were performed for Au+Au collisions at $\sqrt{s_{NN}} = 39$ GeV and 200 GeV and for $K_S^0K^{\pm}$ combination only for the top RHIC energy.

3.1 $K_S^0K_S^0$ and $K_S^0K^+$ correlation functions

An example of the correlation function for two neutral kaons for central Au+Au collisions at $\sqrt{s_{NN}} = 200$ GeV is shown in Figure 2a, while the $K_S^0K^{\pm}$ combination in Figure 2b. The shape of the functions for identical and non-identical pairs of particles is different. The QS effects are dominant in the case of the $K_S^0K_S^0$ system. However, QS (which relies on the Gaussian parameterization) is not sufficient to describe the shape of the correlation function around $q_{inv} = 0.1$ GeV/c, i.e. a dip structure. Only including the strong FSI (Lednicky-Lyuboshitz parametrization) allows us to describe the area of the correlation function below the unity. The parameterization of the function for $K_S^0K^{\pm}$ only considers the strong interaction related to the a_0 resonance.

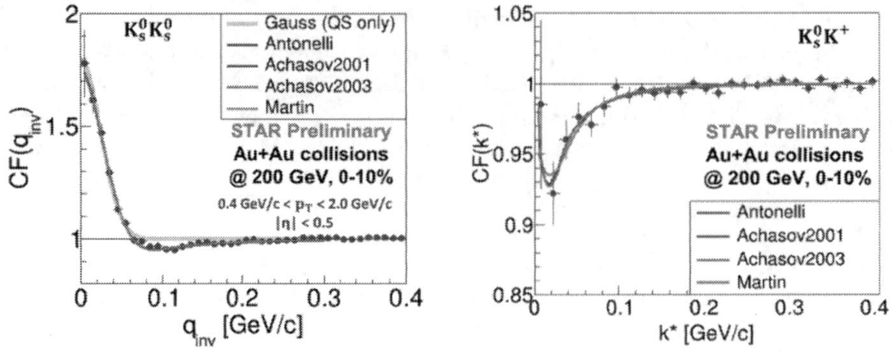

(a) (b)

Figure 2. $K_S^0 K_S^0$ (a) and $K_S^0 K^+$ (b) correlation functions for Au+Au collisions at $\sqrt{s_{NN}} = 200$ GeV with theoretical fits with the Lednicky-Lyuboshitz parametrization.

3.2 Centrality dependence for all kaon combinations

Figure 3 shows the comparison of the source sizes from all K_S^0 and $K^\pm$ combinations for the top RHIC energy as a function of the centrality, where decreasing dependence with increasing collision centrality is observed. The obtained results of the $K^\pm K^\pm$ [6], $K_S^0 K_S^0$, and the $K_S^0 K^+$

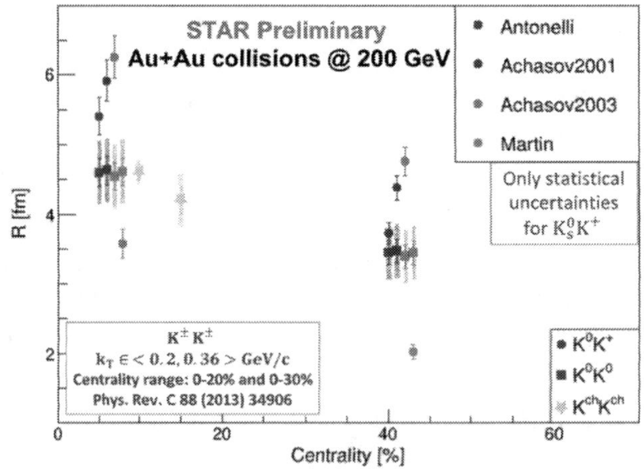

Figure 3. Centrality dependence of the particle emitting source sizes for the $K^\pm K^\pm$ [6], $K_S^0 K_S^0$, and $K_S^0 K^\pm$ systems for Au+Au collisions at $\sqrt{s_{NN}} = 200$ GeV. The circle markers represent the results from $K_S^0 K^+$, the squared markers correspond to $K_S^0 K_S^0$ radii, and the star markers are from the $K^\pm K^\pm$ analysis. Different colors describe different Lednicky-Lyuboshitz parametrizations. For $K_S^0 K^\pm$ only statistical uncertainties are shown.

results favor the Antonelli parameterization [2], which suggests a_0 resonance to be a four-quark state. On the other hand, based on Achasow2001 and Achasov2003 parametrizations [3, 4], the bigger the resonance mass leads to the larger source size (offer a much larger radius for $K_S^0 K^+$ system). Martin parametrization [5] provides the smallest size of the particle emitting source.

3.3 Energy dependence for $K_S^0 K_S^0$ system

The obtained source sizes also indicate a clear dependence on the collision energy. Figure 4 shows that the collisions at higher energy give larger size of the homogeneity region as expected. The extracted source radii are consistent among different parameterizations except the case with QS only, indicating the significance of including strong FSI. This again confirms how important it is to consider this impact in fits. This figure also presents calculations from two theoretical models - UrQMD [7] and Therminator2 [8]. The model calculations agree with the data within their uncertainties.

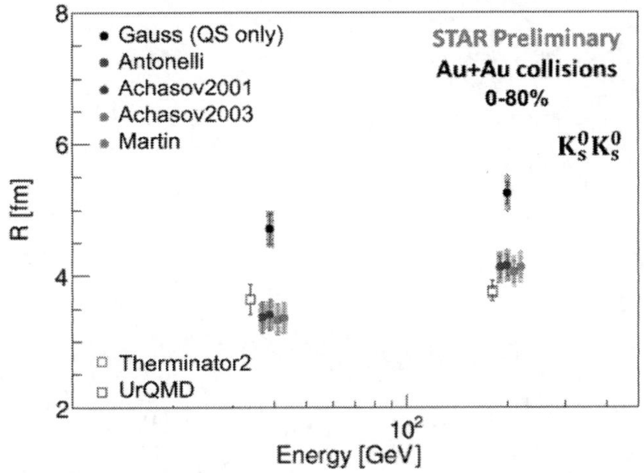

Figure 4. Energy dependence of the source sizes observed for the $K_S^0 K_S^0$ combination in Au+Au collisions. The filled markers are results from the experiment, and the empty markers correspond with model predictions.

4 Summary

In these proceedings, the results from femtoscopic measurements of all kaon combinations in Au+Au collisions at $\sqrt{s_{NN}} = 39$ GeV and 200 GeV have been presented. The obtained results show the importance of considering the strong FSI in the parameterization in the case of $K_S^0 K_S^0$ correlation functions. The comparison of the source sizes from the measurements of all systems $K^{\pm}K^{\pm}$, $K_S^0 K_S^0$, and $K_S^0 K^{\pm}$ implies that the a_0 resonance is favored to be a tetraquark because of consistent source radii for all the combinations by using the Antonelli parameterization.

Acknowledgement This work was supported by the Grant of National Science Centre, Poland, No: 2021/41/B/ST2/02409. Studies were funded by IDUB-POB-FWEiTE-1, IDUB-POB-FWEiTE-2 and IDUB-POB-FWEiTE-3 projects granted by Warsaw University of Technology under the program Excellence Initiative: Research University (ID-UB)

References

[1] R. Lednicky and V.L. Lyuboshitz, Soviet Journal of Nuclear Physics **35**, 770 (1982)
[2] A. Antonelli, eConf **C020620**, THAT06 (2002)
[3] N.N. Achasov, V.V. Gubin, Physical Review D **63**, 094007 (2001)
[4] N.N. Achasov, A.V. Kiselev, Physical Review D **68**, 014006 (2003)
[5] A.D. Martin, E.N. Ozmutlu, E.J. Squires, Nuclear Physics B **121**, 514-530 (1977)

[6] STAR Collaboration, Physical Review C **88**, 34906 (2013)
[7] M. Bleicher et al, Journal of Physics G **25**. 1859-1896 (1999)
[8] M. Chojnacki, A. Kisiel, W. Florkowski, W. Broniowski, Computer Physics Communications **183**, 746-773 (2012)

Study of charge symmetry breaking in A = 4 hypernuclei in $\sqrt{s_{NN}} = 3$ GeV Au+Au collisions at RHIC

Tianhao Shao[1] (for the STAR Collaboration),*

[1]Key Laboratory of Nuclear Physics and Ion-beam Application (MOE), Institute of Modern Physics, Fudan University, Shanghai 200433, China

Abstract. In these proceedings, we present the measurement of the charge symmetry breaking in $A = 4$ hypernuclei in Au+Au collisions at $\sqrt{s_{NN}} = 3$ GeV. The signal reconstruction and binding energy measurements of ${}^4_\Lambda$H and ${}^4_\Lambda$He, including corrections on momentum, are discussed. Our result of Λ binding energy difference for ground states is $\Delta B_\Lambda(0^+) = 0.16 \pm 0.14(\text{stat.}) \pm 0.10(\text{syst.})$ MeV. Combined with the energy levels of excited states, the difference for excited states is $\Delta B_\Lambda(1^+) = -0.16 \pm 0.14(\text{stat.}) \pm 0.10(\text{syst.})$ MeV which shows a negative sign with a magnitude comparable to the result of ground states. These results are compared with previous measurements and theoretical model calculations.

1 Introduction

The charge symmetry of strong interactions predicts that the Λ-p and Λ-n interactions should be identical as they only differ in charge. This leads to the conclusion that the Λ binding energies of a pair of mirror hypernuclei should be identical. However, in 1970s nuclear emulsion experiments measured the Λ binding energy difference in $A = 4$ mirror hypernuclei, ${}^4_\Lambda$H and ${}^4_\Lambda$He, and found a difference of $\Delta B^4_\Lambda(0^+_{\text{g.s.}}) = 350 \pm 50$ keV [1]. Such a large difference cannot be explained by the mass difference between up and down quarks in nuclear systems. In 2015, the J-PARC E13 γ-ray spectroscopy experiment measured the transition energy from the 1^+ excited state to the ground state of ${}^4_\Lambda$He to be $1406 \pm 2 \pm 2$ keV [2]. The E13 collaboration combined the Λ binding energies of ground states from emulsion experiments of 1970s [1] with a γ-ray transition energy for ${}^4_\Lambda$H measured in 1976 [3] and their new γ-ray transition energy measurement for ${}^4_\Lambda$He to determine the Λ binding energy difference in excited states to be $\Delta B^4_\Lambda(1^+_{\text{exc}}) = 30 \pm 50$ keV [2], which is much smaller than that in ground states. It was suggested that the charge symmetry breaking effect may have a large spin dependence. In 2016, the MAMI-A1 collaboration used spectrometers to provide a new measurement of the ground state Λ binding energy of ${}^4_\Lambda$H [4]. Combining their new measurement with the previous ${}^4_\Lambda$He Λ binding energy, and the measurements of the γ-ray transition energies for ${}^4_\Lambda$H [1] and ${}^4_\Lambda$He [2], they updated the estimate of the binding energy differences to be $\Delta B^4_\Lambda(0^+_{\text{g.s.}}) = 233 \pm 92$ keV and $\Delta B^4_\Lambda(1^+_{\text{exc}}) = -83 \pm 94$ keV. However, many theoretical model calculations failed to reproduce the experimental results [5–9]. In 2016, the ab initio no-core shell model calculations plus a charge symmetry breaking $\Lambda - \Sigma^0$ mixing

*e-mail: shaoth21@outlook.com

vertex of $A = 4$ hypernuclei obtained a large charge symmetry breaking in excited states and concluded that $\Delta B^4_\Lambda(1^+_{\rm exc}) \approx -\Delta B^4_\Lambda(0^+_{\rm g.s.}) < 0$ [10]. Independent measurements are crucially needed to test these calculations [11].

To study the physics of QCD matter in the high baryon density region, the STAR experiment ran fixed-target mode during the BES-II program. A stationary gold target was mounted inside the beam pipe. In the fixed-target mode, the lowest center-of-mass energy ($\sqrt{s_{\rm NN}}$) for Au+Au collisions that RHIC can effectively run is 3 GeV. In 2018, STAR took about 300 million events of Au+Au collisions at $\sqrt{s_{\rm NN}}$ = 3 GeV fixed-target mode. Model calculations predict that the production yields of hypernuclei will reach the maximum value at about $\sqrt{s_{\rm NN}}$ = 5 GeV [12]. The STAR fixed-target program gives us an opportunity to study the Λ binding energy of ${}^4_\Lambda$H and ${}^4_\Lambda$He in the same experiment to address the charge symmetry breaking effect.

2 Analysis details and results

In this analysis, signals of ${}^4_\Lambda$H and ${}^4_\Lambda$He are analyzed in Au+Au collisions at 3 GeV. The ${}^4_\Lambda$H is reconstructed via its two-body decay channel ${}^4_\Lambda{\rm H} \rightarrow {}^4{\rm He} + \pi^-$ and the ${}^4_\Lambda$He is reconstructed via its three-body decay channel ${}^4_\Lambda{\rm He} \rightarrow {}^3{\rm He} + {\rm p} + \pi^-$. The decay daughters are identified based on the particles' energy loss $\langle dE/dx \rangle$ information from the Time Projection Chamber (TPC). The identification of ^{4}He and ^{3}He is carried out also with the mass information from the Time of Flight (TOF) detector. Then the invariant mass distributions of ${}^4_\Lambda$H and ${}^4_\Lambda$He are reconstructed according to their decay topologies using the KF Particle package [13]. To enhance the signal significance, the TMVA-BDT package [14] is used. Figure 1 shows the invariant mass distributions of ${}^4_\Lambda$H and ${}^4_\Lambda$He reconstructed in 0–100% centrality and (-2, 0) rapidity range. The centroids and statistical uncertainties for masses of the ground state

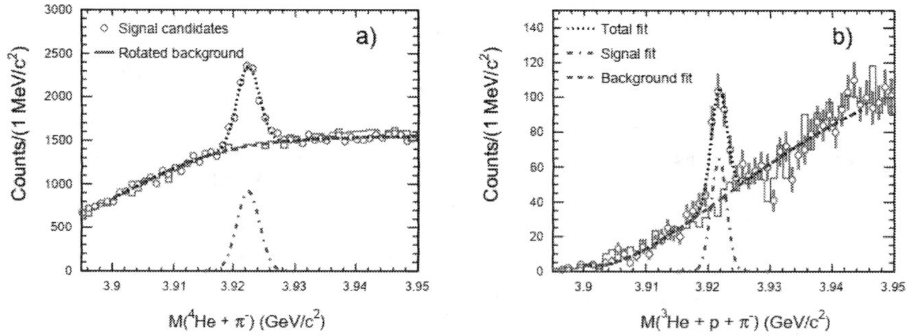

Figure 1. Invariant mass distributions for ${}^4_\Lambda$H and ${}^4_\Lambda$He reconstruction. The backgrounds are obtained by rotating ^{4}He or ^{3}He track by 180 degrees in the transverse plane. The black dashed curves represent fits with a Gaussian function plus double exponential functions. This figure is from [15].

${}^4_\Lambda$H and ${}^4_\Lambda$He are determined by fitting the invariant mass distributions with a Gaussian plus double exponential function represented by black dashed curves in Fig. 1. The extracted masses are

$$m({}^4_\Lambda{\rm H}) = 3922.38 \pm 0.06({\rm stat.}) \pm 0.14({\rm syst.})\ {\rm MeV}/c^2,$$
$$m({}^4_\Lambda{\rm He}) = 3921.69 \pm 0.13({\rm stat.}) \pm 0.12({\rm syst.})\ {\rm MeV}/c^2.$$

Due to the particle's energy loss in material before entering the TPC and the finite precision in the measured magnetic field value at STAR, the reconstructed momenta of decay daughters need to be corrected. The first correction is for the particle's energy loss. This correction is done by using the STAR embedding data. The ${}^{4}_{\Lambda}$H and ${}^{4}_{\Lambda}$He samples from Monte Carlo pass through the GEANT simulation of the STAR detector geometry and material. Then the momentum loss of particles can be determined by comparing the momentum difference between MC input and detector output. The second correction is for the used magnetic field. From previous studies of the invariant masses of known particles, it has been determined that the used magnetic field value should be scaled by 0.2%, and therefore the momenta of particles are scaled with a factor 0.998 in this analysis. The Λ invariant mass measured in Au+Au collisions at 3 GeV with these two corrections is consistent with the PDG mass. Three sources of systematic uncertainties are included: magnetic field accuracy, energy loss correction, and BDT cut.

The Λ binding energies of hypernuclei can be calculated using the masses of a given hypernucleus and its constituents. The Λ binding energies for the ground states are:

$$\mathrm{B}_{\Lambda}({}^{4}_{\Lambda}\mathrm{H}) = 2.22 \pm 0.06(\mathrm{stat.}) \pm 0.14(\mathrm{syst.})\ \mathrm{MeV},$$
$$\mathrm{B}_{\Lambda}({}^{4}_{\Lambda}\mathrm{He}) = 2.38 \pm 0.13(\mathrm{stat.}) \pm 0.12(\mathrm{syst.})\ \mathrm{MeV}.$$

The results for excited states can be obtained from the γ-ray transition energies [2, 3]. Then the Λ binding energy difference between ${}^{4}_{\Lambda}$H and ${}^{4}_{\Lambda}$He can be calculated:

$$\Delta\mathrm{B}_{\Lambda}(0^{+}) = 0.16 \pm 0.14(\mathrm{stat.}) \pm 0.10(\mathrm{syst.})\ \mathrm{MeV},$$
$$\Delta\mathrm{B}_{\Lambda}(1^{+}) = -0.16 \pm 0.14(\mathrm{stat.}) \pm 0.10(\mathrm{syst.})\ \mathrm{MeV}.$$

In this analysis, the Λ binding energy difference for excited states is negative and its magnitude is comparable to the ground states within uncertainties. Our results are consistent with the theoretical prediction, $\Delta\mathrm{B}^{4}_{\Lambda}(1^{+}_{\mathrm{exc}}) \approx -\Delta\mathrm{B}^{4}_{\Lambda}(0^{+}_{\mathrm{g.s.}}) < 0$ [10], which is from the *ab initio* calculation using the hyperon-nucleon potential from the chiral effective field theory plus a charge symmetry breaking effect.

The results in this analysis are compared to previous measurements and theoretical model calculations in Fig. 2. Due to the low statistics of ${}^{4}_{\Lambda}$He, the statistical uncertainty on the ${}^{4}_{\Lambda}$He mass drives the statistical uncertainties on the Λ binding energy differences. STAR has taken about 2 billion events in fixed-target Au+Au collisions at 3 GeV in 2021. Detector upgrades are expected to increase the tracking and particle identification acceptance. The statistical uncertainties will be reduced and their expected magnitudes are shown as green shadows in Fig. 2.

3 Conclusions

To address the charge symmetry breaking effect in A = 4 hypernuclei, we reconstructed the invariant mass distributions of ${}^{4}_{\Lambda}$H and ${}^{4}_{\Lambda}$He in Au+Au collisions at 3 GeV taken in fixed-target mode at STAR, from which the Λ binding energy difference between ${}^{4}_{\Lambda}$H and ${}^{4}_{\Lambda}$He is determined. Using our results and the γ-ray transition energies from previous measurements, we showed that the charge symmetry breaking effect in excited states has a negative value and its magnitude is comparable to that of the ground states within uncertianties. STAR has taken a factor of 7 more data for 3 GeV Au+Au collisions in 2021. The statistical uncertainties of this analysis will be reduced in the future work.

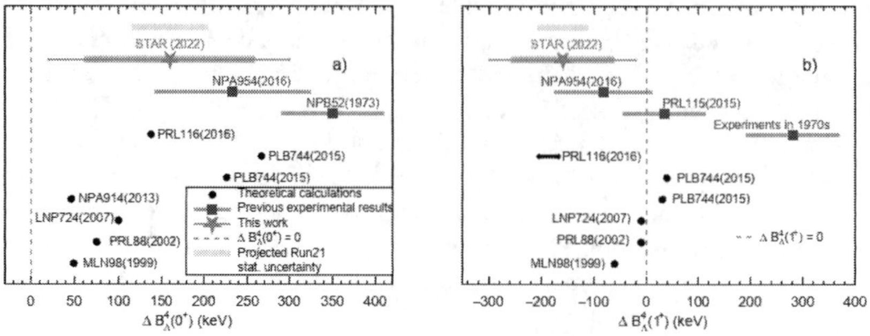

Figure 2. The Λ binding energy difference between ${}^{4}_{\Lambda}$H and ${}^{4}_{\Lambda}$He in ground states (left figure) and in excited states (right figure) compared with theoretical model calculations (black dots) [5–10] and previous measurements (blue squares) [1–4, 16]. Error bars show statistical uncertainties and shaded bands show the systematic uncertainties. The green shadows are projected statistical uncertainties for the 3 GeV data taken in 2021. This figure is from [15].

4 Acknowledegments

Finance support including the international travel expense in part by the National Natural Science Foundation of China under Contact No. 12025501 and the European Unions Horizon 2020 research and innovation programme under grant agreement No. 824093 are appreciated.

References

[1] M. Juric et al., Nucl. Phys. B **52**, 1 (1973)
[2] T.O. Yamamoto et al. (J-PARC E13), Phys. Rev. Lett. **115**, 222501 (2015)
[3] M. Bedjidian, A. Filipkowski, J.Y. Grossiord, A. Guichard, M. Gusakow, S. Majewski, H. Piekarz, J. Piekarz, J.R. Pizzi (CERN-Lyon-Warsaw), Phys. Lett. **B 62**, 467 (1976)
[4] F. Schulz et al. (A1), Nucl. Phys. **A954**, 149 (2016)
[5] A. Nogga, H. Kamada, W. Gloeckle, Phys. Rev. Lett. **88**, 172501 (2002)
[6] J. Haidenbauer, U.G. Meissner, A. Nogga, H. Polinder, Lect. Notes Phys. **724**, 113 (2007)
[7] A. Nogga, Nucl. Phys. A **914**, 140 (2013)
[8] S.A. Coon, H.K. Han, J. Carlson, B.F. Gibson, *Proceedings of 7th International Conference on Mesons and Light Nuclei 98*, 407-413, `arXiv:nucl-th/9903034`
[9] A. Gal, Phys. Lett. **B 744**, 352 (2015)
[10] D. Gazda, A. Gal, Phys. Rev. Lett. **116**, 122501 (2016)
[11] J. Chen, D. Keane, Y.G. Ma, A. Tang, Z. Xu, Phys. Rept. **760**, 1 (2018)
[12] J. Steinheimer, K. Gudima, A. Botvina, I. Mishustin, M. Bleicher, H. Stocker, Phys. Lett. **B 714**, 85 (2012)
[13] I. Kisel (CBM, STAR), J. Phys. Conf. Ser. **1602**, 012006 (2020)
[14] A. Hocker, P. Speckmayer, J. Stelzer, J. Therhaag, E. von Toerne et al., `arXiv:physics/0703039`
[15] M.S. Abdallah et al. (STAR), Phys. Lett. **B**, 137449 (2022)
[16] M. Bedjidian et al. (CERN-Lyon-Warsaw), Phys. Lett. **B 83**, 252 (1979)

Kaon flow in Au+Au collisions at 1.23AGeV measured with HADES

Lukáš Chlad[1,2,*] for the HADES collaboration

[1]Nuclear Physics Institute of the CAS, Řež 130, 25068 Husinec, Czech Republic
[2]Faculty of Nuclear Sciences and Physical Engineering, CTU in Prague, Břehová 7, 11519 Praha, Czech Republic

Abstract. We present results on the anisotropic transverse flow of kaons (K^+, K^0_S and K^-) in Au+Au collisions at $\sqrt{s_{NN}} = 2.42$ GeV measured with HADES. It was proposed already in the mid-nineties that kaon flow close to its production threshold might be a good probe for the kaon-nucleon potential and, consequently, for the nuclear equation-of-state. The presented analysis was performed on more than 2 billion events of the 40% most central collisions, which opened the possibility of analyzing the kaon flow differentially as a function of transverse momentum, rapidity, and centrality, even in this low-energy regime. The measurements are compared to microscopic transport model predictions and to other data at similar collision energies. Implications on the properties of compressed nuclear matter will be discussed.

1 Introduction

To understand the properties of hot and dense matter that exists during merger of neutron stars [1], a study of strange hadrons and namely kaons was suggested [2]. A particular interest to study transveral flow was proposed in [3] which resulted in many experiments like KaoS [4] and FOPI [5]. The lack of detailed differential measurements below and near the kaon production threshold demands further investigation, which is partially presented in this contribution.

The presented analysis was performed on data collected by the HADES Collaboration during spring 2012 when accelerated gold ions with kinetic energy $1.23A$ GeV interacted with a fixed segmented gold target. From the acquired data set, more than 2 billion events with centrality up to 40% were selected. HADES spectrometer consists of diamond based t_0 detector followed by RICH hadron blind detector. Particles are futher travesing multiwire drift chambers (two in front and two behind a toroidal magnetic field) and are finaly detected by time-of-flight wall (made of RPC detector at lower polar angles and scintillator based TOF array for larger θ angle), for more details see [6]. Further behind this setup is a scintillator-based forward wall, which is used to determine the angle of the event plane [7].

Specific energy losses of charged kaons are used for their identification together with the track quality criteria [8]. The short-lived neutral kaons are reconstructed via their decay into charged pions (identified via velocity and momentum correlation) with pair selection based on the decay topology variables [9]. Residual backgrounds in the invariant mass spectra is described by mixed-event technique for K^0_S and by cubic fit function for charged kaons.

*e-mail: chlad@ujf.cas.cz

2 Results

Using the kaon identification described above and the transverse flow analysis technique summarized in [7] we were able to obtain the following results. There is no significant difference in observed flow of K^0_S and K^+, see figure 1a, which can be anticipated as they both contain strange antiquark. Moreover, there is indication of similar behavior of K^- as well (in figure 1b) which is rather unexpected taking into account very different potentials between nucleons and kaons versus antikaons. However, from our measurement we obtained large uncertainties for negative kaons and there is also a significant ratio of antikaons coming from ϕ meson decay [8] that can influence observed flow. We observe a strong dependence of kaon flow with respect to transverse momentum, see Figure 2a as well as decreasing absolute strength of flow in the most central collisions, see Figure 2.

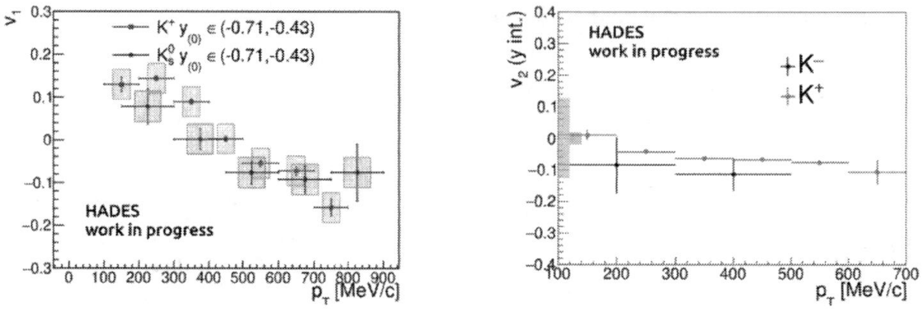

(a) directed flow of K^+ and K^0_S

(b) elliptic flow of K^+ and K^-

Figure 1: Comparison of transverse flow for different kaon's charges.

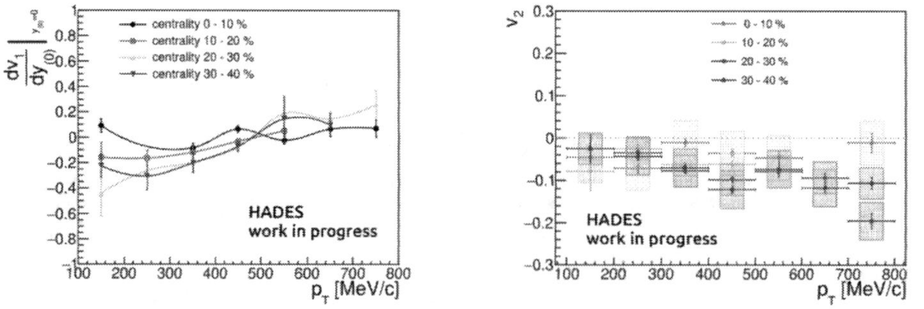

(a) slope of v_1 at midrapidity

(b) rapidity integrated v_2

Figure 2: Centrality dependence of transverse flow of K^+.

3 Comparison with predictions of kinetic transport models

Multiple kinetic transport model predictions were studied to have a first insight into which aspects of kaon creation and propagation in dense matter influence their directed and elliptic flow. Only a few selected results are presented here. We observed that pure so-called cascade models do not predict non-zero v_1 or v_2 which is in contrast to our measurements. Better agreement was achieved with models that are using some form of (anti-)kaon nucleon potential as one can see from figure 3. It is clearly visible that our measurements prefer the PHSD model with the inclusion of V_{KN} over the option without potentials. It is clear that more detailed comparison with kinetic transport models as well as further precise measurements (for example HADES Ag+Ag at 1.56*A* GeV performed in spring 2019) are necessary to make any strong conclusions.

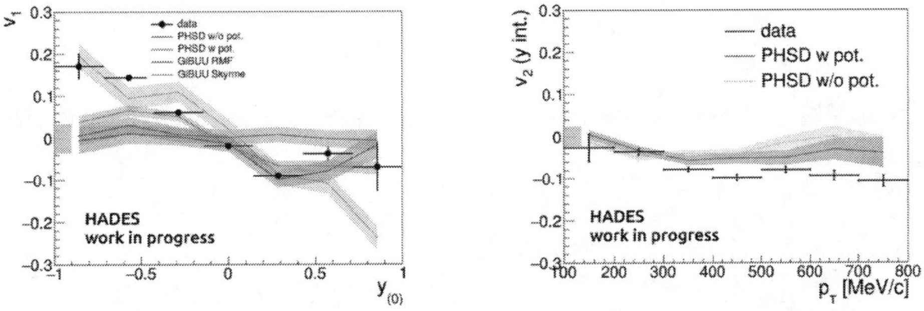

(a) v_1 for low p_T

(b) rapidity integrated v_2

Figure 3: Comparison of model predictions for transverse flow of K$^+$ for 20 – 30% most central collisions.

Acknowledgement

Work supported by: NPI CAS, Rez, Rez (Czech Republic), MSMT LTT17003, MSMT LM2018112, MSMT OP VVV CZ.02.1.01/0.0/0.0/18_046/0016066.

References

[1] J. Adamczewski-Musch, O. Arnold, C. Behnke, A. Belounnas, A. Belyaev, J.C. Berger-Chen, J. Biernat, A. Blanco, C. Blume, M. Böhmer et al. (HADES collaboration), Nature Physics **15**, 1040 (2019)

[2] V. Thorsson, M. Prakash, J.M. Lattimer, Nuclear Physics A **572**, 693 (1994)

[3] G.Q. Li, C.M. Ko, B.A. Li, Physical Review Letters **74**, 235 (1995), 9410017

[4] A. Förster, I. Böttcher, F. Dohrmann, A. Förster, E. Grosse, P. Koczoń, B. Kohlmeyer, S. Lang, F. Laue, M. Menzel et al. (KaoS collaboration), Journal of Physics G **31**, S693 (2005), 0411045

[5] V. Zinyuk, T.I. Kang, Y. Leifels, N. Herrmann, B. Hong, R. Averbeck, A. Andronic, V. Barret, Z. Basrak, N. Bastid et al. (FOPI collaboration), Physical Review C **90**, 025210 (2014), 1403.1504

[6] G. Agakichiev, C. Agodi, H. Alvarez-Pol, E. Atkin, E. Badura, A. Balanda, A. Bassi, R. Bassini, G. Bellia, D. Belver et al. (HADES collaboration), European Physical Journal A **41**, 243 (2009)

[7] J. Adamczewski-Musch, O. Arnold, C. Behnke, A. Belounnas, A. Belyaev, J.C. Berger-Chen, A. Blanco, C. Blume, M. Böhmer, P. Bordalo et al. (HADES Collaboration), Physical Review Letters **125**, 262301 (2020)

[8] J. Adamczewski-Musch, O. Arnold, E.T. Atomssa, C. Behnke, A. Belounnas, A. Belyaev, J.C. Berger-Chen, J. Biernat, A. Blanco, C. Blume et al. (HADES Collaboration), Physical Review Letters **123**, 022002 (2019)

[9] J. Adamczewski-Musch, O. Arnold, C. Behnke, A. Belounnas, A. Belyaev, J. Berger-Chen, J. Biernat, A. Blanco, C. Blume, M. Böhmer et al. (HADES collaboration), Physics Letters B **793**, 457 (2019)

EPJ Web of Conferences **271**, 08012 (2022)
HYP2022 https://doi.org/10.1051/epjconf/202227108012

WASA-FRS HypHI experiment at GSI for studying light hypernuclei

Hiroyuki Ekawa[1,*] for the WASA-FRS Collaboration

[1]High Energy Nuclear Physics Laboratory, Cluster for Pioneering Research, RIKEN, Wako, Japan.

Abstract. From January till March 2022, the WASA-FRS HypHI experiment performed a precise measurement of the hypertriton and the ${}^{4}_{\Lambda}$H hypernucleus lifetime at GSI. The data collected should also confirm whether or not the $nn\Lambda$ bound state can exist. The experiment were carried out with the WASA central detector with a complex of additional dedicated detectors mounted together at the mid-focal plane of the high-momentum-resolution forward spectrometer, the so-called fragment separator FRS. Hypernuclei of interest were produced by induced reactions of ^{6}Li projectiles at 1.96 A GeV on a diamond target with a thickness of 9.87 g/cm^2. Negatively charged π mesons from two-body decays of the hypernuclei of interest were measured by the WASA and the other detectors, and the residual nuclei after the π^- decay were measured by the FRS with a momentum resolving power of 10^4.

1 Introduction

Hypernuclei have been studied to investigate the baryon-baryon interaction [1]. In order to produce hypernuclei, secondary meson beams such as pion or kaon, or primary electron beams are mainly used. Hypernuclei can be also produced by using heavy-ion beams. The HypHI Phase 0 experiment was carried out at GSI in 2009 for the first time with heavy-ion beams [2–4]. In this experiment, ^{6}Li beams with 2 A GeV were injected into a ^{12}C target. Since the energy of the projectile is higher than the threshold to produce a Λ hyperon, the projectile fragments form a hypernuceus by capturing the Λ emitted from a hot participant zone. The hypernuclei can be identified from the invariant mass spectrum by detecting the π^- and decay residues emitted from their two-body decays. This experiment shows the feasibility of this method and reveals two major puzzles.

The first puzzle is induced by observing an indication of $nn\Lambda$ bound state [2]. The invariant mass distributions of deuteron + π^- and triton + π^- show bump structures on the top of background. In [2] it was concluded that the bump structures may indicate the possible $nn\Lambda$ bound state. However, these distributions are broad, and such a bound state is hardly reproduced from theoretical calculations. Therefore, the existence of $nn\Lambda$ is still unclear.

The second puzzle is initiated by observing of the short lifetime of hypertriton [3]. The binding energy of hypertriton is very small, 0.13 MeV, which was measured in old emulsion experiments [5, 6]. Therefore, hypertriton is regarded as a very loosely bound state between deuteron and Λ, and its lifetime is considered to be as almost the same as that of a free Λ, 263 ps. However, the HypHI Phase 0 experiment showed a shorter lifetime of hypertriton,

*e-mail: hiroyuki.ekawa@riken.jp

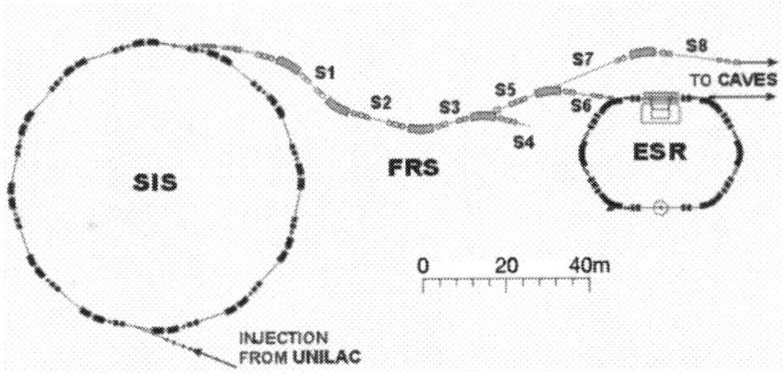

Figure 1. Schematic drawing of the SIS18 synchrotron and the fragment separator (FRS) at GSI.

$183^{+42}_{-32}(stat.) \pm 37(syst.)$ ps. Other heavy-ion experiments, ALICE and STAR, also reported short lifetimes, $142^{+24}_{-21}(stat.) \pm 29(syst.)$ ps [7] and $181^{+54}_{-39}(stat.) \pm 33(syst.)$ ps [8]. Later, their values were updated to be $221 \pm 15(stat.) \pm 19(syst.)$ ps [9] and $242^{+34}_{-38}(stat.) \pm 17(syst.)$ ps [10], respectively. Thus, the obtained average lifetime of hypertriton is 200 ± 13 ps [9], which is shorter than that of free Λ. Moreover, the STAR collaboration reported the binding energy of the hypertriton. Their deduced value is $0.41 \pm 0.12(stat.) \pm 0.11(syst.)$ MeV [11], which is much larger than the former measured value. Consequently, both the lifetime and the binding energy are not well confirmed for the hypertriton, which is even the lightest Λ-hypernuceus. In order to clarify this situation, a new and more precise measurement is necessary.

2 WASA-FRS HypHI experiment

2.1 Overview

The WASA-FRS HypHI experiment aims at studying light hypernuclei produced with heavy-ion beams. This experiment was carried out as the WASA-FRS campaign together with the η' mesonic nuclei experiment. These have been conducted at the fragment separator, so-called FRS [12] with the WASA central detector [13] at GSI. The WASA (Wide Angle Shower Apparatus), which was used at COSY-Jülich, consists of a superconducting solenoid magnet and associated detectors. The WASA-FRS HypHI experiment aims to produce and detect hypertriton, ${}^{4}_{\Lambda}$H and $nn\Lambda$. Heavy-ion beams accelerated up to 1.96 *A* GeV by the SIS18 synchrotron were injected into the FRS (Fig.1). The hypernuclei were produced on a diamond target with a thickness of 9.87 g/cm^2 at the mid-focal plane (S2) in the FRS. The second half of the FRS was used to identify their decay residues with high momentum resolution ($\delta p/p = 10^{-4}$). The π^- mesons were detected by the WASA detectors at the S2 with reasonably large acceptance.

Figure 2 shows the setup around the WASA solenoid. The WASA solenoid is located at the center surrounded by CsI calorimeters to measure energy deposit of charged particles and electromagnetic showers from photons. The magnetic field at the center of the solenoid magnet was 1 T in the experiment. Mini Drift Chamber (MDC) and plastic scintillators (PSB[14], PSBE and PSFE) are located inside of the solenoid magnet to measure the momentum of charged particles and time-of-flight (TOF), respectively. These scintillators have barrel and end-cap shapes to surround the MDC. Fiber trackers (UFTs, DFTs and MFT) are located outside of the solenoid to measure the trajectory of incoming and outgoing charged particles. One of the fiber trackers, Mini Fiber Tracker (MFT), is located inside of the iron yoke to measure the hypernuclear decay vertex precisely.

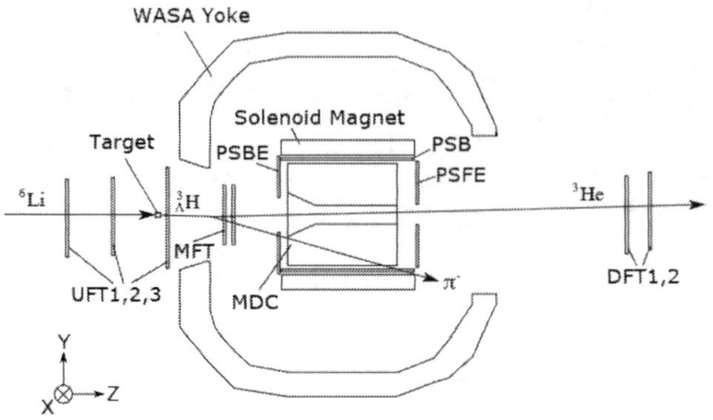

Figure 2. Experimental setup of the WASA-FRS HypHI experiment.

2.2 Monte Carlo simulation

A Monte Carlo (MC) simulation was carried out to evaluate the performance of the WASA detector system before performing the experiment. In the MC simulation, the detectors were implemented with realistic detector resolutions that were obtained by test measurements. The particles were generated by the transport model UrQMD [15] with an assumption of coalescence factor to produce hypernuclei and ordinary nuclei that can reproduce the cross sections measured in the HypHI phase 0 experiment [4]. The amount of observed hypertriton in four days was estimated to be 5800 with a peak width of 3.2 MeV(σ) in the invariant mass distribution [16]. The significance is estimated to be 120 σ. From this result, the accuracy of the lifetime measurement is estimated to be 8 ps. The simulation shows that the values have become several times more accurate compared to the previous measurement by the HypHI Phase 0 experiment.

Additionally, a track finding analysis with the graph neural network, so-called GNN, has been developed based on data obtained by the MC simulation. Since the heavy-ion collision induces many charged particles, track finding is an important issue in this experiment. The GNN can treat graph data structure, which can express data nodes and their relations. Because of this feature, by using the machine learning technique, detector hits can be clusterized by learning true connection in which the hits belong to the same track. We have confirmed more than 98 % of π^- tracks can be found with the GNN analysis [17].

2.3 Data taking

The WASA-FRS experiment was carried out from January to March in 2022 as summarized in Table 1. After a detector commissioning, we have taken data for the η' mesonic nuclei experiment and the WASA-FRS HypHI experiment. The acquired fragments at the S4 focal plane in the WASA-FRS HypHI experiment are summarized in Table 2, which includes ordinal fragments from heavy-ion collisions and the decay residues from hypernuclei. For ^{6}Li beams, data have been taken in three settings to detect different fragments. A fragment of ^{3}He was used to reconstruct hypertriton production events, and ^{4}He and deuteron were taken for $^4_\Lambda$H and $nn\Lambda$, respectively. Since ^{4}He and deuteron have the same A/Z, they can be obtained in the same magnetic setting. Additionally, protons in the mid-rapidity region were collected to detect the Λ decay as a calibration measurement. Moreover, ^{12}C beams were used to produce hypernuclei. This measurement is a first feasibility study for proton-rich

Table 1. Data obtained in the WASA-FRS experimnet.

Run	Period	Data size
Commissioning	28th Jan.– 7th Feb.	7 TB
Physics run for η' nuclei	22th Feb.– 28th Feb.	40 TB
Physics run for WASA-FRS HypHI	10th Mar.– 19th Mar.	48 TB

Table 2. Acquired fragments at the S4 focal plane in the WASA-FRS HypHI experiment. Decay channels of the objective hypernuclei are listed for each fragment.

Beam	Fragment at S4	Decay channel	Amount of fragments	Time [h]	Accepted trigger rate [Hz]
^{6}Li	^{3}He	$^{3}_{\Lambda}\mathrm{H} \to \pi^- + {}^{3}\mathrm{He}$	3.3×10^8	40.9	2600
	^{4}He	$^{4}_{\Lambda}\mathrm{H} \to \pi^- + {}^{4}\mathrm{He}$	0.9×10^8	43.9	1800
	deuteron	$nn\Lambda \to \pi^- + d + n$	1.8×10^8		
	proton (mid-rapidity)	$\Lambda \to \pi^- + p$	5.3×10^6	3.2	680
^{12}C	^{3}He	$^{3}_{\Lambda}\mathrm{H} \to \pi^- + {}^{3}\mathrm{He}$	1.0×10^8	13.5	2400
	^{9}C	$^{9}_{\Lambda}\mathrm{B} \to \pi^- + {}^{9}\mathrm{C}$	2.4×10^5		

hypernuclei such as $^{9}_{\Lambda}$B in future experiments. In this case, fragments of ^{3}He and ^{9}C were collected.

3 Data analysis

The accumulated data are being analyzed. A fragment that passes through the FRS behind the S2 focal plane is identified by measuring the TOF and energy deposit obtained by scintillators located at the S3 and S4 focal plane (Fig.1). Figure 3 (a) and (b) show the correlations between the TOF between S3–S4 and the energy deposit of the S4 scintillator for ^{3}He and ^{4}He (deuteron), respectively. The fragments are clearly observed confirming that fragment identification in the FRS performed reasonably well.

Charged particles emitted to the WASA detector system are analyzed by the tracking and TOF detectors. As a first step, particle tracks are found by requiring hits in all layers of MFT, which has six layers. The tracks are extrapolated to MDC, and associated MDC hits are chosen based on their residuals. Figure 4 (a) shows a difference of ϕ angles between hit wires of MDC and the tracks obtained by MFT. The MDC hits with a residual within $\pm$ 0.2 radian are selected as the associated hits. Associated PSB hits are also chosen based on similar criteria. Momenta of reconstructed particles are obtained by applying the Kalman filter with a tracking tool Genfit [18] using hits in MFT, MDC and PSB. Figure 4 (b) shows the correlation between particle velocity β and momentum. Three types of particles, proton, π^+ and π^-, are identified. In the current analysis, the tracking efficiency is reduced because hits on all layers of MFT are required to find tracks. Track finding is going to be replaced by the developed GNN analysis to increase the number of reconstructed tracks.

4 Summary

The WASA-FRS HypHI experiment aims at studying light hypernuclei produced with heavy-ion beams. The lifetimes of hypertriton and $^{4}_{\Lambda}$H are going to be measured precisely from their decay lengths, and whether or not the $nn\Lambda$ bound state exists is going to be confirmed. Data of the WASA-FRS experiment were successfully taken in January–March 2022. The analysis of the obtained data is in progress.

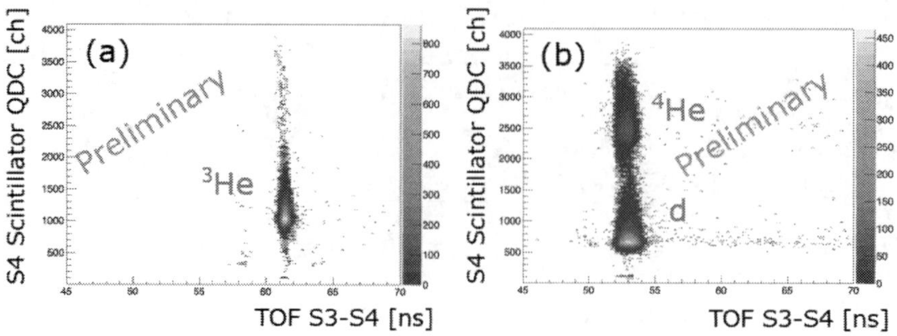

Figure 3. Correlation between S3–S4 TOF and the energy deposit on the S4 scintillator for ^{3}He (a) and ^{4}He/d (b).

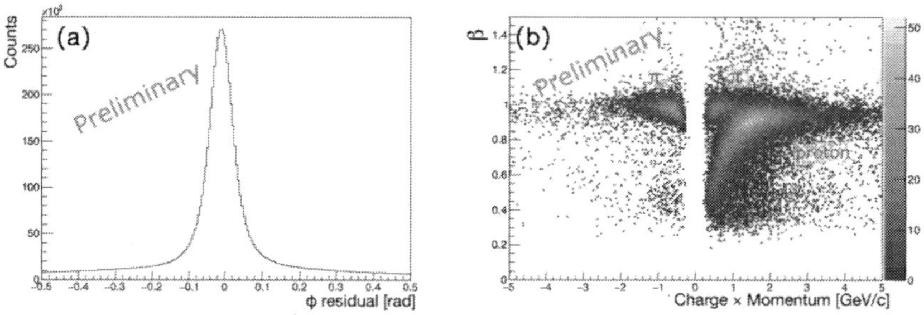

Figure 4. Analysis of the WASA detectors. (a) Difference of the ϕ angle between MDC hit wire and track obtained by MFT. (b) Correlation between β and momentum.

References

[1] O. Hashimoto, H. Tamura, Progress in Particle and Nuclear Physics **57**, 564 (2006)
[2] C. Rappold et al. (HypHI), Phys. Rev. C **88**, 041001 (2013)
[3] C. Rappold et al., Nucl. Phys. A **913**, 170 (2013)
[4] C. Rappold et al., Phys. Lett. B **747**, 129 (2015)
[5] M. Juric et al., Nucl. Phys. B **52**, 1 (1973)
[6] G. Bohm et al., Nucl. Phys. B **4**, 511 (1968)
[7] L. Adamczyk et al. (STAR Collaboration), Phys. Rev. C **97**, 054909 (2018)
[8] J. Adam et al. (ALICE Collaboration), Phys. Lett. B **754**, 360 (2016)
[9] M.S. Abdallah et al. (STAR Collaboration), Phys. Rev. Lett. **128**, 202301 (2022)
[10] S. Acharya et al. (ALICE Collaboration), Phys. Lett. B **797**, 134905 (2019)
[11] J. Adam et al. (STAR Collaboration), Nat. Phys. **16**, 409–412 (2020)
[12] H. Geissel et al., Nucl. Instrum. Meth. B **70**, 286 (1992)
[13] C. Bargholtz et al., Nucl. Instrum. Meth. A **594**, 339 (2008)
[14] R. Sekiya et al., Nucl. Instrum. Meth. A **1034**, 166745 (2022)
[15] M. Bleicher et al., J. Phys. G: Nucl. Part. Phys. **25**, 1859 (1999)
[16] T. Saito et al., Nat. Rev. Phys. **3**, 803–813 (2021)
[17] H. Ekawa et al., submitted to Eur. Phys. J. A
[18] T. Bilka et al., arXiv (2019), `1902.04405`

Investigation of the Σ^0 hyperon production in p(3.5GeV)+p collisions

Waleed Esmail[1,*] for the HADES Collaboration

[1]GSI Helmholtzzentrum für Schwerionenforschung GmbH, 64291 Darmstadt, Germany

Abstract. The production mechanism of the Σ^0 was investigated via the exclusive reaction pp $\rightarrow$ pK$^+\Sigma^0$ at a beam energy 3.5 GeV with the HADES detector. The preliminary value of the extracted total production cross section is found to be $\sigma(\text{pK}^+\Sigma^0)[\mu\text{b}] = 17.7 \pm 1.7(\text{stat}) \pm 1.6(\text{syst})$. The dynamics of the reaction pp $\rightarrow$ pK$^+\Sigma^0$ was investigated by studying the angular distributions in the center-of-mass, Gottfried-Jackson and helicity frames. The angular distributions in the center-of-mass frame supports the pion exchange mechanism. Furthermore, the helicity angular distributions are highly non-isotropic, which is a clear indication of resonant production component.

1 Introduction

Hyperons are of particular interest to the field of hadron physics. A good understanding of hyperon production in proton-proton (p+p) or proton-nucleus (p+A) reactions provides a way to study the strong interaction at the non-perturbative energy regime. Furthermore, it serves as a tool to understand hot and dense nuclear matter in heavy ion collisions, since strangeness enhancement is one of the key signatures proposed for the formation of the quark-gluon plasma [5, 8].

Although there are numerous experimental results and countless dedicated theoretical studies, hyperon production near threshold is not yet well understood. Within the framework of the one-boson exchange models [1], the initial p+p system exchanges a virtual meson, which results in the production of the final state particles. This process can proceed directly or via an intermediate resonance. Hyperon production can be put into one of two categories depending on the mediator properties. The first category, a pion exchange, represents a scenario where the initial p+p system exchange a non-strange meson. In the second category, a kaon exchange mechanism, the p+p system exchange a strange meson [7]. This formalism simplify the exclusive reaction pp $\rightarrow$ pK$^+\Sigma^0$ to pion or kaon induced reactions πp $\rightarrow$K$^+$ Σ^0 or K$^+$ p $\rightarrow$K$^+$ p, respectively. Since no resonances in the K$^+$ p system are known, the pion exchange mechanism is more likely to involve resonances, such that the hyperon is produced via an intermediate resonance πp $\rightarrow$N*/Δ^* $\rightarrow$K$^+$ Σ^0 [2].

2 Experimental details

The High Acceptance Di-Electron Spectrometer (HADES) located at the GSI Helmholtzzentrum für Schwerionenforschung in Darmstadt is a fixed target experiment designed to

*e-mail: w.esmail@gsi.de

measure low-mass di-electrons originating from the decay of vector mesons. However, it has excellent hadron detection capabilities as well, which allows various measurements including hyperon physics. HADES acceptance covers almost the full azimuthal range and polar angles from $\vartheta = 18°$ to $\vartheta = 85°$. The detector setup consists of a hadron-blind Ring Imaging CHerenkov (RICH), two sets of Mult-Wire Drift (MDC) chambers and Time-Of-Flight (TOF) system. Effective particle identification (PID) across a wide momentum range is possible by combining the information from the sub-detectors. In addition, a Forward Wall (FW) is placed 7 meters downstream of the target, which consists of a scintillating hodoscope that covers the very forward polar angles $\vartheta = 0.33°$ to $\vartheta = 7°$. More details on the experimental setup can be found in [3].

The Σ^0 decays electromagnetically to the less massive Λ hyperon with $BR(\Sigma^0 \rightarrow \Lambda\gamma) \sim 100\%$. The Λ decays weakly and reconstructed with the decay mode $\Lambda \rightarrow p_{decay}\, \pi^-$ (BR = 63.9%). Therefore, the final state contains 2 protons, 1 pion, 1 kaon and 1 photon. The Σ^0 is reconstructed via the missing mass technique since the photon is not detected directly [9]. Deep learning techniques was applied to perform the PID as explained in [9].

The strategy to select the Σ^0 signal is two fold, where two exclusive data-sets have been identified. The first, referred to as the *spectrometer data-set*, where exactly 2 p, 1 π^- and 1 K^+ are required to be within the main HADES acceptance. Since the Λ decays via the weak interaction, it travels some distance before it decays ($c\tau$ = 7.89 cm). Therefore, it was possible to apply a set of topological cuts to enhance the signal [9]. Furthermore, the signal was further purified by exploiting the kinematics of reaction $pp \rightarrow pK^+\Sigma^0$. In this context, the squared missing mass of the proton and the Λ hyperon ($MM^2(p\ p_{decay}\ \pi^-)$) are required to be > 0.2 GeV^2/c^4. This cut was introduced primarily to reject the multi-pion production $p\,p \rightarrow p\,p\,\pi^-\,\pi^+$.

The second data-set, referred to as the *wall data-set*, where exactly 1 p, 1 π^- and 1 K^+ are required to be within the main HADES acceptance. In addition to a hit registered in the FW, which is assumed to be due to the Λ decay proton. Because there is no track information since only one hit is registered, the background in this case was reduced based only the reaction kinematics. Once again, events in the range $MM^2(p\ p_{decay}\ \pi^-) > 0.2\ GeV^2/c^4$ were selected (here p_{decay} is registered in the FW). In addition, the missing mass squared of all charged particles is required to be in the range $-0.02 < MM^2(p\ K^+\ p_{decay}\ \pi^-) < 0.01\ GeV^2/c^4$ since there is a photon in the final state.

To suppress the remaining background and obtain a better mass resolution, a kinematic refit has been applied. The track parameters ($1/p$, ϑ, φ) of final state particles were fit under two kinematical constraints [9]. The first constraint: the daughter proton and the pion were constrained to the nominal Λ mass. In addition, all final state charged particles were constrained to the photon mass. In total 2613 candidate Σ^0 events were collected within the pK^+ missing mass range of 1.170-1.220 GeV/c^2.

3 Results

The dynamics of the reaction $pp \rightarrow pK^+\Sigma^0$ were investigated by analyzing the angular distributions in the Center of mass (CMS), Gottfreid-Jackson (G-J) and helicity frames. The distributions were corrected for the detector acceptance and efficiency using the inverse of the detector response matrix calculated by the singular value decomposition method and then fit with Legendre polynomials $d\sigma/d\Omega = \sum_\ell A_\ell \cdot P_\ell$, where $\ell = 0,1,2$ represent the contributing partial waves [9]. In this way the total cross section is $\sigma = 2A_0$ and the coefficients A_1 and A_2 are used to judge the asymmetries and anisotropies of the experimental distributions.

The angular distributions of the three final state particles in the CMS are shown in the top row of figure 1 alongside with the Legendre fit coefficients. Because p+p is a symmetric system, the CMS angular distributions must be symmetric with respect to $\cos\theta = 0$. The angular distributions of the hyperon 1(a) and the proton 1(b) show anisotropies, which is more pronounced in the case of the proton as quantified by the A_2 parameter. This is in contrast of to the case of the kaon 1(c), where the distribution is almost isotropic. These anisotropies indicate that there is a non-zero relative angular momentum in both the $(p - K^+\Sigma^0)$ and $(\Sigma^0 - pK^+)$ systems. In the case of a pure pion exchange mechanism, the proton is the leading particle, the small mass of the exchange pion results in a small 4-momentum transfer such that the produced proton is preferably emitted in the direction of the initial protons, which could explain the anisotropy in the proton CMS distribution.

The Gottfried-Jackson (G-J) frame [6] is defined as the rest frame of two out of the three produced particles. In the G-J frame, the G-J angle is defined as the angle between one of the rest frame particles (e.g. the kaon) and the initial proton $\theta^{RF\,K^+\Sigma^0}_{p_{p,t},\Sigma^0}$, the superscript indicates which rest frame (RF) is used and the subscript stands for the two particles, between which the angle is measured. Because the initial protons are indistinguishable, it is not known which of those protons contributes to the reaction. Therefore, the angle is calculated with respect to both protons ($p_{p,t}$ stands for the beam and target protons). The G-J angular distributions provide information about resonant production, since the internal angular momentum of the resonance is reflected in these observables.

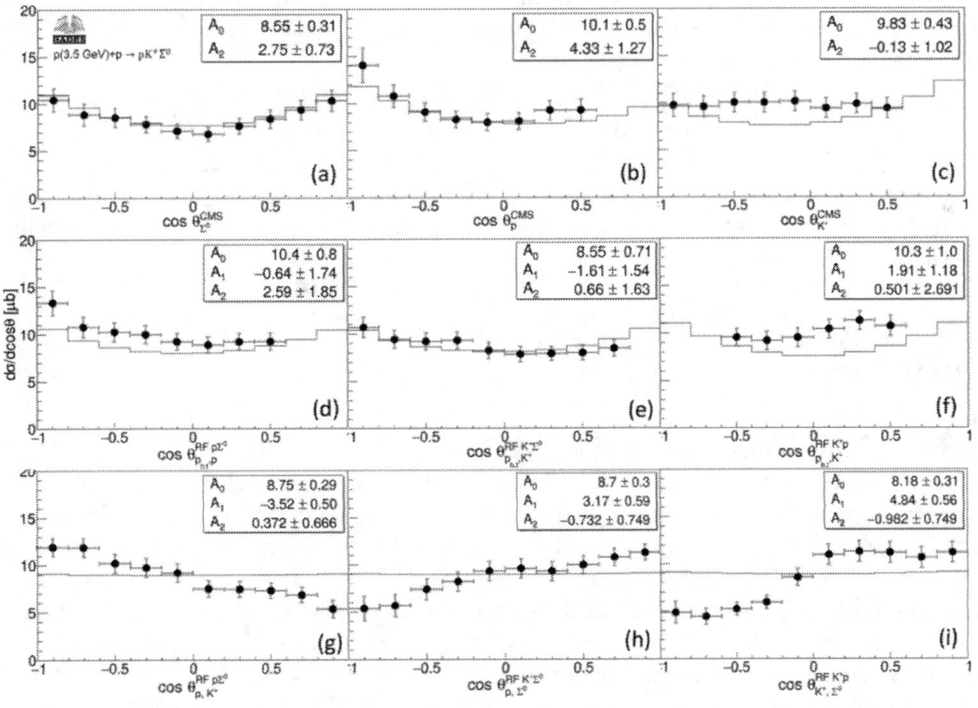

Figure 1: The angular distributions in the CMS frame (top row), G-J frames (middle row) helicity frames (bottom row). The experimental data are shown by the black points, where the error bars are the square root of the quadratic sum of the statistical and systematic uncertainties. The insets lists the Legendre fit parameters and the blue histogram represent the $pp \to pK^+\Sigma^0$ phase space simulation [1].

A pronounced anisotropy is observed in the $p\Sigma^0$ G-J frame (shown in figure 1 (d)), which could be explained as a relative angular momentum in the $p\Sigma^0$ system. This behavior is also observed in the CMS distributions of proton and the hyperon since they are related by kinematics. The angular distribution in the in $K^+\Sigma^0$ G-J frame 1(e) tends to be asymmetric, which could be caused by the excitation of nucleon resonances decaying into the $K^+\Sigma^0$ channel [7]. Because the final state proton and kaon do not originate from the same vertex, the angular distribution in the K^+p G-J frame 1(f) is expected to be isotropic if the reaction proceeds via an intermediate resonance (N^* or Δ^*). Therefore, a deviation from isotropy could be explained by a kaon exchange component.

The helicity angle is defined in the same way as the G-J angle, but instead of calculating the angle with respect to the initial proton, the angle is calculated between one of the rest frame particle and the third produced particle. In this sense, the helicity angle interrelates only the kinematics of the three output particles of the reaction. The helicity angular distribution is a special projection of the Dalitz plot, and is thus a proper observable to identify the kinematics behind the particle production. The helicity angular distributions are shown in bottom row of figure 1, all distributions are far from isotropic, which is a clear indication that the reaction proceeds in two steps via an intermediate resonance(s).

The effect of nucleon resonances on the helicity angular distributions can be tested by fitting the incoherent sum of phase space and a list of possible N^* or Δ^* resonances that might contribute at this energy to match the experimental data. The fitting was performed by means of a χ^2 minimization. A fit of phase space ($pp \rightarrow pK^+\Sigma^0$) has resulted in $\chi^2/\mathrm{ndf} \approx 4.1$, while the incoherent sum of phase space and $N^*(1710)$ ($J^P = 1/2^+$), $N^*(1880)$ $(1/2^+)$, $N^*(1895)$ $(1/2^-)$, $\Delta^*(1900)$ $(1/2^-)$ has resulted in $\chi^2/\mathrm{ndf} \approx 1.46$.

A preliminary value of the total production cross section that corresponds to an excess energy $\epsilon = 556$ MeV has been calculated by averaging the integrated yield of the different angular distributions and found to be $\sigma(pK^+\Sigma^0)[\mu b] = 17.7 \pm 1.7\,(\mathrm{stat}) \pm 1.6\,(\mathrm{syst})$. The systematic uncertainty is mainly due to the PID (5%), the signal selection cuts (2%) and the normalization to the p+p elastic cross section (7%), which is used as reference measurement at the same beam energy [4].

4 Conclusion

The exclusive reconstruction of the reaction $pp \rightarrow pK^+\Sigma^0$ at a beam kinetic energy of 3.5 GeV has been presented and the $pp \rightarrow pK^+\Sigma^0$ total production cross section was determined with competitive accuracy. The dynamics of the reaction was investigated by studying the angular distributions in the CMS, G-J and helicity frames. The CMS distributions of the hyperon and the proton show anisotropies, which is the expected behavior if the pion exchange mechanism dominates the particle production process in a simple one-boson exchange formalism. In addition, the anisotropy seen in the K^+p G-J frame indicates a non-negligible contribution of a kaon exchange mechanism. The helicity angular distributions are not isotropic, which shows that a pure phase space description of experimental data is not sufficient. The influence of different nucleon resonances has been tested by scaling the incoherent sum of phase space and nucleon resonances that might contribute at this energy to match the experimental distributions.

[1]The phase space simulation was weighted simultaneously by the Legendre fit parameters obtained from Σ^0 CMS and the $p\Sigma^0$ G-J experimental distributions. This weighting provides an acceptable description of the experimental distributions [9].

5 Acknowledgement

This research was enabled in part by support provided by Jülich Forschungszentrum and GSI Helmholtzzentrum für Schwerionenforschung. This work has been supported by the European Union's Horizon 2020 research and innovation programme under grant agreement no. 824093 (STRONG2020).

References

[1] A. M. Gasparian *et al.*, Nucl. Phys. A **684** (2001)
[2] D. Ronchen *et al.*, Eur. Phys. J. A **49** (2013)
[3] G. Agakishiev *et al.* [HADES], Eur. Phys. J. A **41** (2009)
[4] G. Agakishiev *et al.* [HADES], Eur. Phys. J. A **48** (2012)
[5] J. Rafelski and B. Muller, Phys. Rev. Lett. **48** (1982)
[6] K. Gottfried and J. D. Jackson, Nuovo Cim. **33** (1964)
[7] M. Abdel-Bary *et al.* [COSY-TOF], Eur. Phys. J. A **46** (2010)
[8] P. Koch, *et al.*, Phys. Rept. **142** (1986)
[9] W. A. M. Esmail, Ruhr-Universität Bochum, "Deep learning for track finding and the reconstruction of excited hyperons in proton induced reactions" (2022)

Neutron stars and the hyperon puzzle

Isaac Vidaña[1,*]

[1]Istituto Nazionale di Fisica Nucleare, Sezione di Catania. Dipartimento di Fisica e Astronomia "Ettore Majorana", Università di Catania, Via Santa Sofia 64, I-95123 Catania, Italy

Abstract.
In this work we shortly review the so-called "hyperon puzzle", *i.e.*, the problem of the strong softening of the equation of state of dense matter induced by the presence of hyperons which leads to maximum masses of neutron stars incompatible with the recent observations of $\sim 2\,M_{\odot}$ millisecond pulsars. In particular, we briefly go through some of the possible solutions that have been proposed to tackle this still open problem.

1 Introduction

Neutron stars are one of the possible stellar compact remnants resulting from the gravitational collapse of massive stars during a Type-II, Ib or Ic supernova event. With masses of the order of $1-2M_{\odot}$, where $M_{\odot} \simeq 2\times 10^{33}$g is the mass of the Sun, radii of about $10-12$ km, and central densities in the range of $4-8$ times the normal nuclear matter saturation density, $\epsilon_0 \sim 2.7\times 10^{14}$ g/cm^3 ($\rho_0 \sim 0.16$ fm^{-3}), neutron stars are most likely among the densest objects in the Universe [1]. These objects are excellent laboratories to test our present understanding of the theory of strongly interacting matter under extreme conditions, offering an interesting interplay between nuclear physics of dense matter and astrophysical observations. Enormous theoretical advances have been done in understanding the extreme and unique properties (magnetic fields, rotational frequencies, gravitational fields, surface temperatures) of these exotic objects. Major advances have been also achieved in their observation. Particularly important has been the observation in 2017 of the first gravitational wave signal from the merger of two neutron stars [2], inaugurating a new era in the observation of these objects.

Nowadays it is still an open question what is the true internal composition of neutron stars. Neutron stars has been traditionally modeled as a uniform fluid of neutron-rich nuclear matter in equilibrium with respect to weak interaction processes (β-stable matter). However, due to the large values reached by the density, new hadronic degrees of freedom are expected to appear in neutron star interiors in addition to nucleons. Hyperons, baryons with a strangeness content, are an example of these new degrees of freedom. Contrary to terrestrial conditions, where hyperons are unstable and decay into nucleons through weak interactions, the equilibrium conditions in neutron stars can make the inverse process happen. Hyperons may appear in the inner cores of neutron stars at densities around $(2-3)\times\rho_0$, leading their onset to a softening of the equation of state (EoS) and consequently to a reduction of the neutron star mass and, particularly, of its maximum value to values smaller than current observations of $\sim 2\,M_{\odot}$ [3–5]. The difficulty in reconciling the presence of hyperons in the neutron stars

*e-mail: isaac.vidana@ct.infn.it

interior with current mass observations is nowadays one of the most important open problem in nuclear astrophysics. This problem, known in the literature as the "hyperon puzzle", is a subject of current intensive research.

In the next sections we discuss in some detail the hyperon puzzle, and briefly present some of the possible solutions to this problem that have been proposed in the recent years. The manuscript is finished with a brief summary given at the end.

2 The hyperon puzzle

The presence of hyperons in neutron stars was considered for the first time in the pioneering work of Ambartsumyan and Saakyan in 1960 [6]. Since then, their effects on the properties of these objects have been studied by many authors using either phenomenological [7, 8] or microscopic [9–13] approaches to the EoS of neutron star matter. Phenomenological approaches, either relativistic or non-relativistic, are based on effective density-dependent interactions which typically contain a certain number of parameters adjusted to reproduce nuclear and hypernuclear observables, and neutron star properties. Relativistic mean field (RMF) models [7] and models based on the Skyrme-type interactions [8] are among the most commonly used ones. Microscopic approaches, on the other hand, are based on realistic two- and three-body baryon interactions that describe the scattering data in free space. These realistic interactions have been mainly constructed within the framework of a meson-exchange theory [14, 15], although recently a new approach based on chiral perturbation theory has emerged as a powerful tool [16]. In order to obtain the EoS one has to solve then a very complicated many-body problem [17]. A great difficulty of this problem lies in the treatment of the repulsive core, which dominates the short-range behavior of the interaction. Although many different microscopic many-body methods have been extensively used to study nuclear matter, very few of these have been extended to hypernuclear sector. To our knowledge, these many-body methods include the Brueckner–Hartree–Fock (BHF) approximation [9] of the Brueckner–Bethe–Goldstone theory, the Hartree–Fock theory based on the soft $V_{low\,k}$ interactions [10] and the Dirac–Brueckner–Hartree–Fock theory [11, 12]. Very recently the Auxiliary Field Diffusion Monte Carlo method [13] was also extended to the hyperonic sector.

All these approaches agree that hyperons may appear in the inner core of neutron stars at densities around $\sim (2-3)\times\rho_0$. At such densities, the nucleon chemical potential is large enough to make the conversion of nucleons into hyperons energetically favorable. This conversion relieves the Fermi pressure exerted by the nucleons and makes the EoS softer. This is a common feature of all models including hyperons, and it is illustrated in panel (a) of Fig. 1 for the particular example of the GM3 parametrization of the Glendenning–Moszkowski RMF model (see Ref. [7]) with (red dashed line) and without (black solid line) hyperons. As a consequence (see panel (b)) the masses of the stars with central densities beyond the onset density of the hyperons are lower and the maximum mass of hyperonic stars is substantially reduced.

Although the presence of hyperons in neutron stars seems to be energetically unavoidable, the strong softening of the EoS associated with the onset of hyperons (notably in microscopic models) leads to maximum masses not compatible with observations. The discrepancy between the theory and observations became more dramatic after the recent measurements of $\sim 2\,M_\odot$ millisecond pulsars [3–5] which rule out most of all currently proposed EoS with hyperons.

To solve this problem it is needed a mechanism that could provide the additional repulsion that makes the EoS stiffer at high densities and, therefore, the maximum mass compatible with the current observational limits. Four of the possible mechanisms that could provide

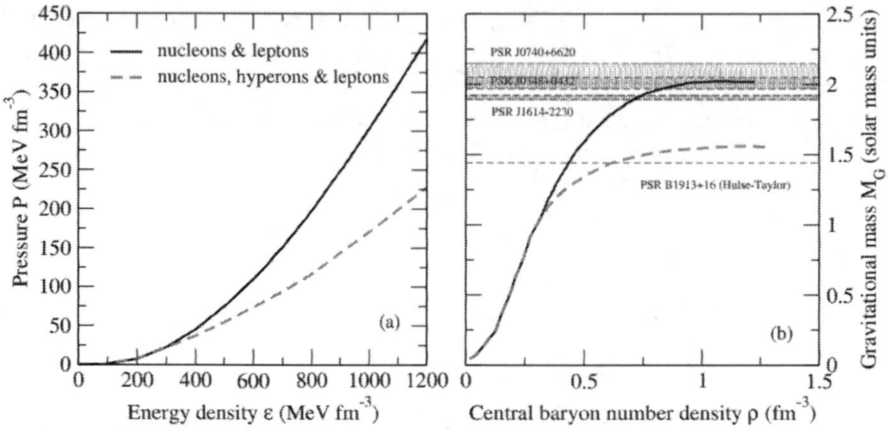

Figure 1. Illustration of the effect of the presence of hyperons on the EoS (panel (a)) and mass of a neutron star (panel (b)). The GM3 parametrization of the Glendenning–Moszkowski RMF model (see Ref. [7]) with (red dashed line) and without (black solid line) hyperons has been considered as an example. The bands show the observational masses and their corresponding errors of the PSR B1913+16 (Hulse–Taylor) [18] pulsar, and the millisecond pulsars PSR J1614-2230 [3], PSR J0348+0432 [4] and PSR J0740+6620 [5].

such additional repulsion are: (i) more repulsive hyperon-hyperon interactions in relativistic density functional methods driven by either repulsive vector mesons exchanges [19–22] or less-attractive scalar σ meson exchange [23], (ii) repulsive hyperonic three-body forces [24–28, 30], (iii) phase transition to deconfined quark matter at densities below the hyperon threshold [34–38] and (iv) appearance of other hadronic degrees of freedom, such as the Δ isobar [41] or meson condensates [42–47] that shift the onset of hyperons to larger densities. Each one of these possible solutions is briefly reviewed in the following sections.

3 Hyperon-hyperon repulsion

This solution, which has been mainly explored in the context of RMF models (see *e.g.*, Refs. [19–23]), relies on the well-known fact that in meson-exchange models of nuclear forces vector mesons generate repulsion at short distances while σ-meson is responsible for intermediate range attraction. If the interaction of hyperons with vector mesons is repulsive enough or the attraction driven by the σ-meson is weak enough then the neutron star matter EoS could be sufficiently stiff to explain the current pulsar mass observations. However, the modifications of the strength of meson exchanges must be consistent with the hypernuclear data which requires, at least, the ΛN interaction to be attractive and suitable tuned to the hypernuclear data [39]. Such tuning is not required if the repulsive vector meson interactions acts only among the hyperons through the exchange of the strange ϕ vector meson (which couples only to hyperons). In this way, the onset of hyperons would be shifted to higher densities and neutron stars with maximum masses larger than $2M_\odot$ and a significant hyperon fraction could be successfully obtained. It is also possible to tune the interactions of non-strange mesons to accommodate the hypernuclear data and astrophysical constraints on maximum masses of neutron stars in the framework of relativistic density functional theories (for such an approach see *e.g.* Ref. [23]). For more details the interested reader is referred to the original recent work quoted above.

4 Hyperonic three-body forces

It is well known that the three-nucleon forces in the nuclear Hamiltonian are fundamental ingredients that are needed to reproduce properly the properties of few-nucleon systems as well as the empirical saturation point of symmetric nuclear matter in non-relativistic many-body approaches. Therefore, it seems natural to suggest that three-body forces involving one or more hyperons (*i.e.,* NNY, NYY and YYY) could provide additional repulsion at high densities (already established in the case of three-nucleon forces) that can solve the hyperon puzzle. Indeed if three-body forces involving hyperons are repulsive enough they can make the EoS stiffer at high densities and, therefore, make the maximum mass of the star compatible with the recent observations. This idea was suggested even before the observation of neutron stars with $\sim 2M_\odot$ (see *e.g.,* Ref. [24]), and it has been explored by a number of authors in the last years [25–28, 30]. However, no general consensus has been reached regarding the role played by the hyperonic three-body forces in solving the hyperon puzzle. The authors of Refs. [24, 26] found that the inclusion of hyperonic three-body forces is sufficient to obtain hyperon stars with masses of the order of $2M_\odot$, whereas in Ref. [25] it was fund that the largest maximum mass attainable within their model is $1.6M_\odot$. The authors of [27] concluded that, with the model they considered, the presence of hyperons in the core of neutron stars could not be satisfactory established and, conse- quently, according to these authors, there is no clear incompatibility with astrophysical observations when Λs are included. However, one should note, that the presence of protons, necessary to establish the correct β-equilibrium inside neutron stars and thus a proper treatment of neutron star matter EoS, was neglected in their calculation. On the other hand, the results of Ref. [28] showed that the inclusion of a NNΛ force, derived by the Jülich–Bonn–Munich group within the framework of chiral effective field theory (χEFT) at next-to-next-to-leading order, leads to an EoS stiff enough such that the resulting hyperon star maximum mass is compatible with the largest masses currently observed. Using a perturbative many-body approach these authors calculated also the separation energy of the Λ hyperon in some hypernuclei finding that the agreement with the experimental data improves for the heavier ones when the effect of a repulsive NNΛ force of the order of $\sim$ 10 MeV at saturarion is taken into account. Recently, Friedman and Gal [29] found that a repulsive contribution of about 14 MeV to the Λ-nucleus potential in symmetric matter at saturation could potentially solve the hyperon puzzle. This is in complete agreement with the results of Ref. [28] just mentioned. Finally, the recent microscopic ab-inito calculations of Ref. [30] of the Λ hyperon properties in symmetric nuclear matter and pure neutron matter, performed with NΛ and NNΛ potentials derived within the framework of χEFT, indicate that the appearance of hyperons in neutron stars could be energetically unfavorable. However, a complete calculation under conditions of chemical equilibrium found in neutron stars made with χEFT is still missing. To conclude, although it seems that hyperonic three-body forces offer an interesting microscopic solution to the hyperon puzzle, however the uncertainties involved in the physics of hyperonic three-body forces are too large at the moment to make a definite conclusion possible.

5 Phase transition to deconfined quark matter

Several authors have suggested that an early phase transition from hadronic mater to deconfined quark matter at densities below the hyperon threshold could provide a solution to the hyperon puzzle. Therefore, massive neutron stars could actually be hybrid stars with a stiff quark matter core. The question that arises then is whether quarks can provide sufficient repulsion to produce a $2M_\odot$ neutron star. To yield maximum masses larger than $2M_\odot$, quark matter should have two important and necessary features: (i) a significant overall quark-quark repulsion to maintain stiff EoS, for example, in vector channels and (ii) a strong enough attraction in certain channels which leads to color superconductivity needed to make the deconfined quark matter phase energetically favorable over the hadronic one [40]. Several models of hybrid stars with the necessary properties to generate $2M_{\odot}$ neutron stars have been proposed [31–38]. Conversely, the observation of $2M_{\odot}$ neutron stars may also helped to impose important constraints on the models of hybrid and strange stars with a quark matter core, and improve our present understanding of the hadron-quark phase transition.

6 Δ isobar and kaon condensation in neutron stars

An alternative way to circumvent the hyperon puzzle is to invoke the appearance of other hadronic degrees of freedom such as for instance the Δ isobar or meson condensates that shift the onset of hyperons to higher densities.

The Δ isobar is often neglected in the studies of neutron stars because its threshold density was found to be higher than the typical densities prevalent in cores of neutron stars. Nevertheless, it has been recently shown by Drago *et al.*, [41] that the onset of the Δ depends crucially on the density-dependence of the slope of the nuclear symmetry energy, *i. e.*, the parameter $L = 3\rho_0(\partial E_{sym}(\rho)/\partial\rho)_{\rho_0}$. By using a state-of-the-art EoS and recent experimental constraints on L, these authors showed that the Δ isobar could actually appear before the hyperons in neutron star interiors. However, they found also that, as soon as the Δ is present the EoS, as in the case of hyperons, becomes considerably softer and, consequently, the maximum mass is reduced to values below the current observational limit. Thus, the hyperon puzzle is effectively replaced with the so-called Δ puzzle.

The possible existence of a Bose–Einstein condensate of negative kaons in the inner core of neutron stars has also been extensively considered in the literature (see *e.g.*, [42–47] and references therein). As the density of stellar matter increases, the K^- chemical potential, μ_{K^-}, is lowered by the attractive vector meson field originating from dense nucleonic mater. When μ_{K^-} becomes smaller than the electron chemical potential μ_e the process $e^- \rightarrow K^- + \nu_e$ becomes energetically possible. The critical density for this process was calculated to be in the range $2.5 - 5\rho_0$ [45, 46]. However, as in the case of the Δ, the appearance of the kaon condensation induces also a strong softening of the EoS and consequently leads to a reduction of the maximum mass to values below the current observational limits. The interested reader is referred to the original work on this subject [42–47] for a comprehensive description of the implications of kaon condensation on the structure and evolution of neutron stars.

7 Summary

In this paper we briefly reviewed the so-called "hyperon puzzle", *i.e.*, the problem of the strong softening of the EoS of dense matter due to the appearance of hyperons which leads to maximum masses of compact stars that are not compatible with the recent observations of $\sim 2M_{\odot}$ millisecond pulsars. We have discussed four of the different solutions that have been proposed to tackle this problem: (i) more repulsion in hyperon-hyperon interactions within the density functional theories of hypernuclear matter in the vector and/or scalar mesons exchange channels; (ii) repulsive hyperonic three-body forces in the ab initio microscopic calculations, (iii) a phase transition to deconfined quark matter at densities below the hyperon threshold, and (iv) the appearance of other hadronic degrees of freedom that could shift the onset of hyperons to larger densities.

Acknowledgements

The author is very grateful to the organizers of HYP2022 for their invitation. This work has received funding from the Helmholtz Institute Mainz and the European Union's Horizon 2020 research and innovation programme under grant agreement No. 824093.

References

[1] S. L. Shapiro and S. A. Teukolsky, *Black Holes, White Dwarf and Neutron Stars* (Wiley, New York) 1983; N. K. Glendenning, *Compact Stars: Nuclear Physics, Particle Physics and General Relativity*, 2nd edition (Springer, Berlin) 2000; P. Haensel, A. Y. Potekhin and D. G. Yakovlev, *Neutron Stars 1: Equation of State and Structure, Astrophysics and Space Science Library Series*, Vol. **326** (Springer, New York) 2007

[2] B. P. Abbott *et al.* (LIGO Scientific Collaboration and Virgo Collaboration), Phys. Rev. Lett. **116**, 061102

[3] P. Demorest *et al.,* Nature **467**, 1081 (2010); Z. Arzoumanian *et al.,* Astro. Phys. J. Suppl. Ser. **235** (2), 37 (2018)

[4] J. Antoniadis *et al.,* Science **340** 6131 (2013)

[5] H. T. Cromartie *et al.,* Nature Astronomy **10**, 1038 (2019); E. Fonseca *et al.,* Astro. Phys. J. Lett. **915**, L12 (2021)

[6] V. A. Ambartsumyan and G. S. Saakyan, Sov. Astron. **4**, 187 (1960)

[7] N. K. Glendenning, Phys. Lett. B **114**, 392 (1982); N. K. Glendenning, Astrophys. J. **293**, 470 (1985); N. K. Glendenning, Z. Phys. A **326**, 57 (1987); N. K. Glendening and S. A. Moszkowski, Phys. Rev. Lett. **67**, 2414 (1991); F. Weber and M. K. Weigel, Nucl. Phys. A **505**, 779 (1989). R. Knorren, M. Prakash and P. J. Ellis, Phys. Rev. C **52**, 3470 (1995); J. Schaffner and I. Mishustin, Phys. Rev. C **53**, 1416 (1996); H. Huber, F. Weber, M. K. Weigel and Ch. Schaab, Int. J. Mod. Phys. E **7**, 310 (1998)

[8] S. Balberg and A. Gal, Nucl. Phys. A **625**, 435 (1997); S. Balberg, I. Lichtenstadt and G. B. Cook, Astrophys. J. Suppl. Ser. **121**, 515 (1999). D. E. Lanskoy and Y. Yamamoto, Phys. Rev. C **55**, 2330 (1997); T. Y. Tretyakova and D. E. Lanskoy, Eur. Phys. J. A **5**, 391 (1999); J. Cugnon, A. Lejeune and H.-J. Schulze, Phys. Rev. C **62**, 064308 (2000); I. Vidaña, A. Polls, A. Ramos and H.-J. Schulze, Phys. Rev. C **64**, 044301 (2001); X.-R. Zhou , H.-J. Schulze, H. Sagawa, C.-X. Wu and E.-G. Zhao, Phys. Rev. C **76**, 034312 (2007); X.-R. Zhou, A. Polls, H.-J. Schulze and I. Vidaña, Phys. Rev. C **78**, 054306 (2008)

[9] H.-J. Schulze, M. Baldo, U. Lombardo, J. Cugnon and A. Lejeune, Phys. Lett. B **355** 21 (1995); H.-J. Schulze, M. Baldo, U. Lombardo, J. Cugnon and A. Lejeune, Phys. Rev. C **57**, 704 (1998); M. Baldo, G. F. Burgio and H.-J. Schulze, Phys. Rev. C **58**, 3688 (1998) M. Baldo, G. F. Burgio and H.-J. Schulze, Phys. Rev. C **61**, 055801 (2000); I. Vidaña, A. Polls, A. Ramos, M. Hjorth-Jensen and V. G. J. Stoks, Phys. Rev. C **61**, 025802 (2000); I. Vidaña, A. Polls, A. Ramos, L. Engvik and M. Hjorth-Jensen, Phys. Rev. C **62**, 035801 (2000); H-J. Schulze, A. Polls, A. Ramos and I. Vidaña, Phys. Rev. C **73**, 058801 (2006); H.-J. Shulze and T. Rijken, Phys. Rev. C **84**, 035801 (2011)

[10] H. Dapo, B.-J. Schaefer and J. Wambach, Phys. Rev. C **81**, 035803 (2010)

[11] F. Sammarruca, Phys. Rev. C **79** 034301 (2009)

[12] T. Katayama and K. Saito, arXiv:1410.7166 (2014); arXiv:1501.05419 (2015)

[13] D. Lonardoni, F. Pederiva and S. Gandolfi, Phys. Rev. C **89** 014314 (2014)

[14] P. M. M. Maessen, T. A. Rijken and J. J. de Swart, Phys. Rev. C **40**, 2226 (1989); Th. A. Rijken, V. G. J. Stoks and Y. Yamamoto, Phys. Rev. C **59**, 21 (1999); V. G. J. Stoks and

Th. Rijken, Phys. Rev C **59**, 3009 (1999); Th. A. Rijken, Phys. Rev. C **73**, 044007 (2006); Th. A. Rijken and Y. Yamamoto, Phys. Rev. C **73**, 044008 (2006)

[15] B. Holzenkamp, K. Holinde and J. Speth, Nucl. Phys. A **500**, 485 (1989); J. Haidenbauer and U.-G. Meissnes, Phys. Rev. C **72**, 044005 (2005)

[16] H. Polinder, J. Haidenbauer and U.- G. Meissner, Nucl. Phys. A **779**, 244 (2006); J. Haidenbauer, S. Petschauer, N. Kaiser, U.-G. Meissner, A. Nogga and W. Weise W 2013 Nucl. Phys. A **915** 24

[17] *Nuclear Methods and Nuclear Equation of State* (International Review of Nuclear Physics, Vol 8. Edited by Marcello Baldo. World Scientific Publishing Company 1999); H. Müther and A. Polls, Prog. Part. Nucl. Phys. **45**, 243 (2000)

[18] R. A. Hulse and J. H. Taylor, Astrophys. J. Lett. **195** L51 (1975)

[19] I. Bednarek, P. Haensel, J. L. Zdunik, M. Bejger and R. Mańka, Astron. and Astrophys. **543**, A157 (2012)

[20] S. Weissenborn, D. Chatterjee, and J. Schaffner–Bielich, Phys. Rev. C **85**, 065802 (2012)

[21] M. Oertel, C. Providência, F. Gulminelli and Ad. R. Raduta, J. Phys. G. **42** 075202 (2015)

[22] K. A. Maslov, E. E. Kolomeitsev and D. N. Voskresensky, Phys. Lett. B **748** 369 (2015)

[23] E. N. E. van Dalen, G. Colucci and A. Sedrakian, Phys. Lett. B **734**, 383 (2014)

[24] T. Takatsuka *et al.,* Eur. Phys. J. A **13**, 213 (2002); Prog. Theor. Phys. Suppl. **174**, 80 (2008)

[25] I. Vidaña, D. Logoteta, C. Providência, A. Polls and I. Bombaci, Eur. Phys. Lett. **94**, 11002 (2011)

[26] Y. Yamamoto, T. Furumoto, B. Yasutake and Th. A. Rijken, Phys. Rev. C **88** 022801 (2013); Phys. Rev. C **90** 045805 (2014)

[27] D. Lonardoni, A. Lovato, S. Gandolfi and F. Pederiva Phys. Rev. Lett. **114** 092301 (2015)

[28] D. Logoteta, I.Vidaña, and I. Bombaci, Eur. Phys. J. A **55**, 207 (2019)

[29] E. Friedman and A. Gal, arXiv:2204.02264

[30] D. Gerstung, N. Kaiser, and W. Weise, Eur. Phys. J. A **56**, 175 (2020)

[31] M. Baldo, M. Buballa, G. F. Burgio, F. Neumann, M. Oertel and H.-J. Schulze, Phys. Lett. B **562**, 153 (2003)

[32] M. Baldo, G. F. Burgio, P. Castorina, S. Plumari and Z. Zappalà, Phys. Rev. C **75**, 035804 (2007)

[33] A. Kurkela, P. Romatschke and A. Vuorinen, Phys. Rev. D **81**, 105021 (2010)

[34] F. Özel, D. Psaltis, S. Ransom, P. Demorest and M. Alford, Astrophys. J. Lett. **724** L199 (2010)

[35] S. Weissenborn, I. Sagert, G. Pagliara, M. Hempel and J. Schaeffner-Bielich, Astophys. J. Lett. **740** L14 (2011)

[36] T. Klähn, D. Blaschke and D. Lastowiecki, Phys. Rev. D **88** 085001 (2013)

[37] L. Bonanno and A. Sedrakian, Astron. and Astrophys. **539** 416 (2012)

[38] R. Lastowiecki, D. Blaschke, H. Grigorian and S. Typel Acta Phys. Polon. Suppl., **5** 535 (2012)

[39] O. Hashimoto and H. Tamura, Prog. Part. Nucl. Phys. **57** 564 (2006)

[40] J. L. Zdunik and P. Haensel, Astron. and Astrophys. **551** A61 (2013)

[41] A. Drago, A. Lavagno, G. Pagliara and D. Pigato, Phys. Rev. C **90**, 065809 (2014); EPJ Web Conf. **95**, 01011 (2015)

[42] D. B. Kaplan and A. E. Nelson, Phys. Lett. B **175**, 57 (1986)
[43] D. B. Kaplan and A. E. Nelson, Phys. Lett. B **179**, 409, Erratum (1986)
[44] G. E. Brown, C.-H. Lee, M. Rho and V. Thorsson, Nucl. Phys. A **567**, 937 (1994)
[45] V. Thorsson, M. Prakash and J. M. Lattimer, Nucl. Phys. A **572**, 693 (1994)
[46] C.-H. Lee, Phys. Rep. **275**, 255 (1996)
[47] N. K. Glendenning and J. Schaffner-Bielich, Phys. Rev. Lett. **81**, 4564 (1998)

Strange mesons in nuclei and neutron stars

Laura Tolos[1,2,3,*]

[1]Institute of Space Sciences (ICE, CSIC), Campus UAB, Carrer de Can Magrans, 08193, Barcelona, Spain
[2]Institut d'Estudis Espacials de Catalunya (IEEC), 08034 Barcelona, Spain
[3]Frankfurt Institute for Advanced Studies, Ruth-Moufang-Str. 1, 60438 Frankfurt am Main, Germany

Abstract. The present status in the field of strange mesons in nuclei and neutron stars is reviewed. In particular, the $\bar{K}N$ interaction, that is governed by the presence of the $\Lambda(1405)$, is analyzed and the formation of the $\bar{K}NN$ bound state is discussed. Moreover, the properties of $\bar{K}$ in dense nuclear matter are studied, in connection with strangeness production in nuclear collisions and kaon condensation in neutron stars.

1 Introduction

Understanding the dynamics of hadrons with strangeness has received a lot attention over the past decades in connection with the study of exotic atoms, the analysis of strangeness production in particle and nuclear research facilities, and the investigation of different strange phases in the interior of neutron stars (for a recent review see Ref. [1]). In particular, one venue of interest in the field of strangeness is the study of strange mesons, such as the $\bar{K}$ meson, and its dynamics with nucleons and nuclear matter.

It is known that the behaviour of the $\bar{K}N$ scattering amplitude close to threshold is governed by the presence of the $\Lambda(1405)$ resonance below threshold. The nature of this baryonic resonant state has been carefully studied, in particular after the theoretical predictions that describe this state as the superposition of two poles [2–4]. Furthermore, the two-pole nature of the $\Lambda(1405)$ shows that the $\bar{K}N$ interaction could be attractive enough to produce bound states, such as the $\bar{K}NN$ state (see Ref. [5] and references therein).

Also, the properties of $\bar{K}$ in dense nuclear matter have prompted the interest of the scientific community, in particular after the theoretical predictions of kaon condensation in neutron stars [6], and the analysis of strangeness creation and propagation in nuclear collisions (see, for example, [7, 8]).

In this paper we review the status in the field of strange mesons in nuclei and neutron stars, paying a special attention to the formation of the $\bar{K}NN$ state, the propagation of strange mesons in nuclear collisions and the phenomena of kaon condensation in neutron stars.

2 The $\bar{K}N$ interaction: the $\Lambda(1405)$

The $\bar{K}N$ scattering in the $I = 0$ channel is governed by the presence of the $J^P = 1/2^- \; S = -1$ $\Lambda(1405)$. The origin of the $\Lambda(1405)$ as a molecule was predicted more than 50 years ago

*e-mail: tolos@ice.csic.es

Reference	B [MeV]	Γ [MeV]	Method	Potential
Barnea et al. (2012)	16	41	Variational	chiral
Dote et al. (2009)	17-23	40-70	Variational	chiral
Dote et al. (2018)	14-50	16-38	ccCSM	chiral
Ikeda et al. (2010)	9-16	34-46	Faddeev	chiral
Bayar et al. (2013)	15-30	75-80	Faddeev	chiral
Sekihara et al. (2016)	15-20	70-80	Faddeev	chiral
Yamazaki et al. (2002)	48	61	Variational	phenomenological
Shevchenko et al. (2007)	50-70	90-110	Faddeev	phenomenological
Ikeda et al. (2007)	60-95	45-80	Faddeev	phenomenological
Wycech et al. (2009)	40-80	40-85	Variational	phenomenological
Dote et al. (2017)	51	32	ccCSM	phenomenological
Revai et al. (2014)	32/ 47-54	50-65	Faddeev	chiral/phenomenological

Table 1. Binding energy and width of K^-pp for different chiral and phenomenological calculations using variational, Faddeev or ccCSM+Feshbach methods. Table modified from Table 2 of Ref. [1]

by Dalitz and Tuan [9, 10]. And, since the end of the last century, a lot of effort has been invested in understanding the $\bar{K}N$ interaction and the nature of the $\Lambda(1405)$ using coupled-channel unitarized theories within meson-exchange models [11, 12] or meson-baryon chiral effective theories (χEFT) [2, 3, 13–19]. A very interesting conclusion of these works is that the dynamics of the $\Lambda(1405)$ is described by the superposition of two poles of the scattering matrix, between the $\bar{K}N$ and $\pi\Sigma$ thresholds [2–4]. Whereas experimentally the $\Lambda(1405)$ is seen as one resonance, the existence of two poles is supported in reaction-dependent line shapes [3].

3 The $\bar{K}NN$ bound state

The two-pole structure of the $\Lambda(1405)$ indicates that the $\bar{K}N$ interaction might be attractive enough to produce bound states. As a consequence, $\bar{K}$-nuclear clusters may form, such as the $\bar{K}NN$ in isospin $I = 1/2$ and $J^{\pi} = 0^-$. Indeed, the $I = 1/2$ $\bar{K}NN$ state has been extensively studied over the last decades, both theoretically and experimentally (see Ref. [5] and references therein).

Several theoretical works have addressed the existence of this state. In Table 1 the K^-pp binding energies (B) and widths (Γ) resulting from different models are shown (see references in Ref. [1]). In this table, we separate between variational, three-body Faddeev calculations or the coupled-channel Complex Scaling Method (ccCSM), whereas, at the same time, we distinguish between those models where the two-body interactions are based on χEFT or energy-independent phenomenological models. We observe that the binding energies range between 9 and 95 MeV, while the decay widths cover values from 16 to 110 MeV.

The variety of values for the binding energy and width is due to different sources: a) the uncertainties in the subthreshold extrapolation of the $\bar{K}N$ interaction. The $\bar{K}N$ interaction from χEFT is less attractive than the phenomenological potentials in the subthreshold region. Hence, the chiral interactions give binding energies considerably lower than those from phenomenological interactions; b) the use of variational or Faddeev calculations introduces certain approximations. Full three-body interactions are not accounted for in variational methods, whereas Faddeev calculations deal with separable two-body interactions. As

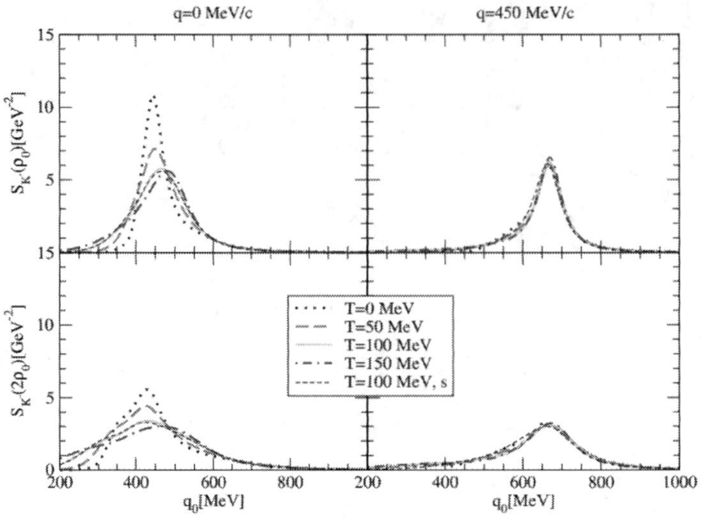

Figure 1. $\bar{K}$ spectral function for different momenta, densities and temperatures from the chiral model of Ref. [28].

for ccCSM, it combines the merits of variational and Faddeev but with a high computational cost.

4 Antikaons in matter

Antikaons in matter have been extensively studied over the last decades. Early works on relativistic mean-field models (RMF) [20] or quark-meson coupling (QMC) schemes [21] obtained potentials of a few hundreds of MeVs at saturation density ρ_0. However, most of these models assumed that the low-density theorem is valid for the calculation of the $\bar{K}$ optical potential in dense matter, which is doubtful due to the presence of the $\Lambda(1405)$ close to threshold. Thus, it is of fundamental importance to study the properties of the $\Lambda(1405)$ in matter and the effect on the $\bar{K}N$ scattering amplitude and, hence, on the $\bar{K}$ optical potential (or self-energy).

One way to proceed is to use unitarized theories in coupled channels in dense matter, using χEFT with strangeness [22–24] or from meson-exchange models [25, 26]. Within these approaches, an attractive antikaon potential with density is obtained due to the modified S-wave $\Lambda(1405)$ resonance in the medium. This modification emerges from different effects: a) the Pauli blocking on the baryon states in the intermediate meson-baryon propagator [27]; b) the inclusion of the $\bar{K}$ self-energy [23] in the propagation of $\bar{K}$ in matter; and c) the implementation of self-energies of the other mesons and baryons in the intermediate states [24].

Once the $\bar{K}N$ effective interaction in matter and, hence, the $\bar{K}$ potential (and self-energy) are determined, it is possible to obtain the spectral representation of the $\bar{K}$ propagator as

$$S_{\bar{K}}(\omega, \vec{q}) = -\frac{1}{\pi} \mathrm{Im}\, D_{\bar{K}}(\omega, \vec{q}) = -\frac{1}{\pi} \frac{\mathrm{Im}\, \Pi_{\bar{K}}(\omega, \vec{q})}{\mid \omega^2 - \vec{q}^{\,2} - m_{\bar{K}}^2 - \Pi_{\bar{K}}(\omega, \vec{q}) \mid^2}, \quad (1)$$

where $D_{\bar{K}}(\omega, \vec{q})$ stands for the $\bar{K}$ propagator and $\Pi_{\bar{K}}(\omega, \vec{q})$ is the $\bar{K}$ self-energy. Note that ω is the energy, $\vec{q}$ the momentum and $m_{\bar{K}}$ the mass of the $\bar{K}$ meson.

The $\bar{K}$ spectral function for the chiral model of Ref. [28] is shown in Fig. 1. The $\bar{K}$ spectral function considers the S- and P-wave contributions for different temperatures (different lines) for two densities (ρ_0 in the upper panels and $2\rho_0$ in the lower panels) and two momenta (left and right). The $\bar{K}$ in matter feels attraction, that is seen by the shift of the peak of the spectral function to lower energies, while it develops a considerable width that is dominated by the S-wave $\Lambda(1405)$ close to $\bar{K}N$ threshold.

With this figure we also indicate the importance of considering higher partial waves as well as taking into account the effect of finite temperature corrections for the determination of the properties of $\bar{K}$ in dense matter [28–32]. In particular, the knowledge of higher-partial waves beyond S-wave as well as finite temperature corrections become fundamental for analyzing the results of heavy-ion collisions (HICs) at beam energies below 2 AGeV [8, 33, 34].

5 Experiments and observations: heavy-ion collisions

A possible way to determine the antikaon optical potential (or self-energy) is to study the formation and propagation of antikaons in the dense medium created in nucleus-nucleus collisions for intermediate beam kinetic energies (GeV). The analysis of HICs for the determination of the properties of antikaons in dense matter requires, however, the use of transport models. Transport models are he link between the experiments and the underlying physical processes, as they consider the production and propagation of all kind of particles, such as strange mesons (see Ref. [8] for a review on strangeness production). Those models are based on the solution of Boltzmann-type semi-classical transport equations, that can be derived from the non-equilibrium quantum field theory.

The first transport calculations for antikaons were performed neglecting the finite width of the antikaon spectral function [35, 36]. Few years later, antikaon production was studied using off-shell dynamics with in-medium spectral functions in the Hadron-String-Dynamics (HSD) transport model [33]. In this case, the Jülich meson-exchange model [25, 26] was used for the $\bar{K}N$ interaction. More recently, the in-medium effects in strangeness production in HICs at (sub-)-threshold energies of 1 - 2 A GeV based on the microscopic Parton-Hadron-String Dynamics (PHSD) transport approach have been studied, taking the in-medium antikaon properties from the χEFT scheme of Refs. [29, 32]. The manifestation of the medium effects has been analyzed for the strange mesons rapidity distributions, p_T-spectra as well as the polar and azimuthal angular distributions, directed and elliptic flow in C+C, Ni+Ni, and Au+Au collisions. By comparing to experimental data from the KaoS, FOPI and HADES Collaborations, it was found that the modifications of the strange meson properties in nuclear matter are necessary to explain the data in a consistent manner [37].

6 Experiments and observations: neutron stars

One of the possible scenarios inside the core of neutron stars is the appearance of antikaons. The composition of matter in neutron stars is found by demanding equilibrium against weak interaction processes. Thus, considering neutron star matter made of neutrons, protons and electrons, the weak interaction transitions that take place are

$$\begin{aligned} & n \rightarrow p e^- \bar{\nu}_e \\ & e^- p \rightarrow n \nu_e, \end{aligned} \tag{2}$$

so that $\mu_p = \mu_n + \mu_e$ and $\rho_p = \rho_e$, with $\rho = \rho_p + \rho_n$. Nevertheless, if the chemical potential of the electron increases dramatically with density in the interior of neutron stars, it could become energetically more favourable to produce antikaons instead of electrons, via

$$n \leftrightarrow p + \bar{K}. \tag{3}$$

For this to happen, the chemical potential of the electron for a given density should be larger than the effective mass of antikaons at that density, that is, $\mu_e > m^*_{\bar{K}}$. If this happens and due to the fact that antikaons are bosons, the phenomenon of kaon condensation would take place.

The possibility of kaon condensation has been debated since the pioneering work of Ref. [6]. The discussion is based on whether the mass of antikaons could be largely modified by the interaction with the surrounding nucleons. Some phenomenological models tend to agree with this scenario. In particular, RMF models, that include or not other strange particles such as hyperons (see, for example, the recent works [38–41]), seem to allow for this possibility. However, such large modifications in the mass of antikaons is not supported by χEFT approaches [24, 29, 32].

7 Summary

We have reviewed the status in the field of strange mesons, such as antikaons, in nuclei and neutron stars. We have analyzed the $\bar{K}N$ interaction, paying a special attention to the role of the $\Lambda(1405)$. We have also presented the theoretical efforts for determining the binding energy and width of the $\bar{K}NN$ bound state. We have then shown the properties of antikaons in dense nuclear matter, by displaying the $\bar{K}$ spectral function. We have finally discussed the propagation of strange mesons in HICs and the phenomenon of kaon condensation in neutron stars.

Acknowledgments

L.T. acknowledges support from CEX2020-001058-M (Unidad de Excelencia "María de Maeztu"), PID2019-110165GB-I00 financed by the spanish MCIN/ AEI/10.13039/501100011033/, as well as by the EU STRONG-2020 project, under the program H2020-INFRAIA-2018-1 grant agreement no. 824093, and the CRC-TR 211 'Strong-interaction matter under extreme conditions'- project Nr. 315477589 - TRR 211.

References

[1] L. Tolos, L. Fabbietti, Prog. Part. Nucl. Phys. **112**, 103770 (2020), 2002.09223
[2] J.A. Oller, U.G. Meißner, Phys. Lett. **B500**, 263 (2001), hep-ph/0011146
[3] D. Jido, J.A. Oller, E. Oset, A. Ramos, U.G. Meißner, Nucl. Phys. **A725**, 181 (2003), nucl-th/0303062
[4] T. Hyodo, W. Weise, Phys. Rev. **C77**, 035204 (2008), 0712.1613
[5] T. Nagae, Nucl. Phys. **A954**, 94 (2016)
[6] D.B. Kaplan, A.E. Nelson, Phys. Lett. **B175**, 57 (1986)
[7] C. Fuchs, Prog. Part. Nucl. Phys. **56**, 1 (2006), nucl-th/0507017
[8] C. Hartnack, H. Oeschler, Y. Leifels, E.L. Bratkovskaya, J. Aichelin, Phys. Rept. **510**, 119 (2012), 1106.2083
[9] R.H. Dalitz, S.F. Tuan, Phys. Rev. Lett. **2**, 425 (1959)
[10] R.H. Dalitz, S.F. Tuan, Annals Phys. **8**, 100 (1959)
[11] A. Müller-Groeling, K. Holinde, J. Speth, Nucl. Phys. **A513**, 557 (1990)
[12] J. Haidenbauer, G. Krein, U.G. Meißner, L. Tolos, Eur. Phys. J. **A47**, 18 (2011), 1008.3794
[13] N. Kaiser, P.B. Siegel, W. Weise, Nucl. Phys. **A594**, 325 (1995), nucl-th/9505043
[14] E. Oset, A. Ramos, Nucl. Phys. **A635**, 99 (1998), nucl-th/9711022

[15] M.F.M. Lutz, E.E. Kolomeitsev, Nucl. Phys. **A700**, 193 (2002), nucl-th/0105042
[16] C. Garcia-Recio, J. Nieves, E. Ruiz Arriola, M.J. Vicente Vacas, Phys. Rev. **D67**, 076009 (2003), hep-ph/0210311
[17] B. Borasoy, R. Nissler, W. Weise, Eur. Phys. J. **A25**, 79 (2005), hep-ph/0505239
[18] J.A. Oller, Eur. Phys. J. **A28**, 63 (2006), hep-ph/0603134
[19] A. Feijoo, V. Magas, A. Ramos, Phys. Rev. **C99**, 035211 (2019), 1810.07600
[20] J. Schaffner, J. Bondorf, I.N. Mishustin, Nucl. Phys. **A625**, 325 (1997), nucl-th/9607058
[21] K. Tsushima, K. Saito, A.W. Thomas, S.V. Wright, Phys. Lett. **B429**, 239 (1998), [Erratum: Phys. Lett.B436,453(1998)], nucl-th/9712044
[22] T. Waas, N. Kaiser, W. Weise, Phys. Lett. **B379**, 34 (1996)
[23] M. Lutz, Phys. Lett. **B426**, 12 (1998), nucl-th/9709073
[24] A. Ramos, E. Oset, Nucl. Phys. **A671**, 481 (2000), nucl-th/9906016
[25] L. Tolos, A. Ramos, A. Polls, T.T.S. Kuo, Nucl. Phys. **A690**, 547 (2001), nucl-th/0007042
[26] L. Tolos, A. Ramos, A. Polls, Phys. Rev. **C65**, 054907 (2002), nucl-th/0202057
[27] V. Koch, Phys. Lett. **B337**, 7 (1994), nucl-th/9406030
[28] L. Tolos, D. Cabrera, A. Ramos, Phys. Rev. **C78**, 045205 (2008), 0807.2947
[29] L. Tolos, A. Ramos, E. Oset, Phys. Rev. **C74**, 015203 (2006), nucl-th/0603033
[30] M.F.M. Lutz, C.L. Korpa, M. Moller, Nucl. Phys. **A808**, 124 (2008), 0707.1283
[31] D. Cabrera, A. Polls, A. Ramos, L. Tolos, Phys. Rev. **C80**, 045201 (2009), 0903.1171
[32] D. Cabrera, L. Tolos, J. Aichelin, E. Bratkovskaya, Phys. Rev. **C90**, 055207 (2014), 1406.2570
[33] W. Cassing, L. Tolos, E.L. Bratkovskaya, A. Ramos, Nucl. Phys. **A727**, 59 (2003), nucl-th/0304006
[34] L. Tolos, A. Polls, A. Ramos, J. Schaffner-Bielich, Phys. Rev. **C68**, 024903 (2003), nucl-th/0302082
[35] W. Cassing, E.L. Bratkovskaya, Phys. Rept. **308**, 65 (1999)
[36] C. Hartnack, H. Oeschler, J. Aichelin, Phys. Rev. Lett. **90**, 102302 (2003), nucl-th/0109016
[37] T. Song, L. Tolos, J. Wirth, J. Aichelin, E. Bratkovskaya, Phys. Rev. C **103**, 044901 (2021), 2012.05589
[38] N. Gupta, P. Arumugam, Phys. Rev. C **88**, 015803 (2013)
[39] T. Malik, S. Banik, D. Bandyopadhyay, Astrophys. J. **910**, 96 (2021), 2104.00775
[40] T. Muto, T. Maruyama, T. Tatsumi, Phys. Lett. B **820**, 136587 (2021), 2106.03449
[41] V.B. Thapa, M. Sinha, J.J. Li, A. Sedrakian, Phys. Rev. D **103**, 063004 (2021), 2102.08787

Exploring the role of hyperons in high density matter in the Quark-Meson-Coupling model

Jirina R. Stone[1,2,*]

[1]Department of Physics and Astronomy, University of Tennessee, Knoxville, USA
[2]Department pf Physics (Astrophysics), University of Oxford, Oxford, UK

Abstract. Microscopic composition and properties of matter at super-saturation densities have been a subject of intense investigations for decades. Experimental and observational data and fundamental laws of physics indicate that heavy strange baryons and mesons are essential components of the matter. The Quark-Meson-Coupling-Model (QMC) is well suited for such a study. The model is based on interaction between quarks in individual baryons instead between the baryons as entities without internal structure. This approach significantly increases transparency and reduces the number of variable parameters of the model, thus offering deeper insight into the physics of high density hadronic matter. In this contribution, we review the effect of hyperons on neutron star (NS) properties, the speed of sound and the symmetry energy, both at zero and finite temperature. The QMC results are contrasted with the outcome of the traditional relativistic mean field DD2Y model. The (lack of) the so-called 'hyperon puzzle' in both models is discussed.

1 Introduction

Recent advances in detection and data analysis techniques in further exploration of hypernuclei and their properties, stimulated development in theoretical interpretation of the new data. More detailed information on Λ, $\Lambda\Lambda$ and, more recently, Ξ hypernuclei, and their excitations, as well as studies of the role of kaons in heavy-ion reactions is being obtained or is in an advanced planning stage. Theoretical interpretation and reliable prediction of the outcome of experiments is provided by several classes of models, based on various physical assumptions (e.g. realistic or density dependent NN potentials) used in different mathematical frameworks such as Bruckner G-matrix theory, chiral effective field theory and (non)relativistic mean-field models. As discussed, for example, by the author in [1], the main reason for existence of such a variety of theories is that the hadronic interactions in free space and, more importantly, in nuclear medium, is not known from first principles and has to be modelled. Models based on different physics and mathematical schemes yield similarly reasonable fit to data provided they have enough variable parameters, thus preventing identification of a model closest to true microscopy of hadronic interactions. Growing popularity of statistical approaches especially in astrophysical applications, taking advantage of multimessenger data, does not help in the quest for fundamentals of hadronic interactions. The consequence of this situation that we

*e-mail: jirina.stone@physics.ox.ac.uk

can obtain a reasonable reproduction of existing data but the predictive power of the models is limited, to say the least.

In this contribution we review the latest outcome of the Quark-Meson-Coupling-Model (QMC) for hyperonic matter. This model differs from the traditional relativistic mean-field models (RMF) in that that it is based on interaction between quarks in individual baryons instead between the baryons as entities without internal structure. Such approach leads to a significant reduction of variable parameters of the model as compared to traditional models, thus aiming to a well constraint parameter space. Most of the review is based on an extensive study of hadronic matter in the framework of the QMC-A variant of the model [2, 3], where more details can be found. Here we focus on effects of the hyperons on basic properties of cold and hot high-density matter and NS. The QMC results are contrasted with the outcome of the traditional RMF DD2Y model (labeled DD2 from now on) [2, 4].

2 The QMC model

2.1 The basics

The QMC is a phenomenological relativistic mean-field model of hadronic matter with a non-relativistic extension for finite systems, not discussed here. The key ideas of the model, here outlined for nucleons but valid for any baryons, are: (i) nuclear matter is a collection of quark bags with light quarks confined by attachment to a Y shaped colour string, moving in a non-perturbative QCD vacuum [5], (ii) quarks from different nucleons can come close enough to generate vacuum fluctuations, represented by scalar and vector meson fields - the origin of the nucleon-nucleon (NN) interaction in nuclear medium (see Fig. 1). The QMC

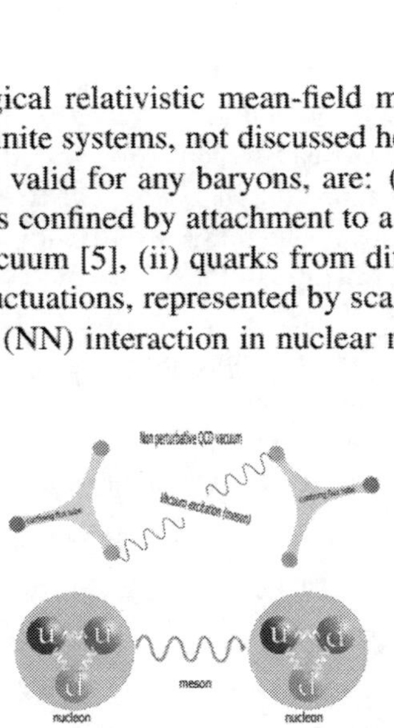

Figure 1: Mechanism of NN interaction in the QMC model (top) and in traditional RMF models which ignore the internal structure of the baryons (bottom). Adapted from Ref. [3] under Creative Commons Attribution License.

model was constructed by Guichon [6] and further developed in later years (see e.g. [7] and Refs. therein). The MIT bag is used as an effective realization of confinement for practical purposes. Although the bag strictly prevents quarks from crossing the boundary, it needs to be seen as an average representation of a more complex situation and the size and surface of the bag should not be given a deep physical meaning. The dynamical mass of the bag, immersed in a constant scalar σ field in a general case of the baryon octet (labeled by the index i for brevity) can be conveniently written as a function of the nucleon-meson coupling in free space $g_{\sigma N}$ (for details see Appendix in Ref. [8])

$$M_i(\sigma) = M_i - w_{\sigma i}\, g_{\sigma N}\, \sigma + \frac{d}{2}\tilde{w}_{\sigma i}(g_{\sigma N}\, \sigma)^2. \quad (1)$$

The coupling strength $g_{\sigma N}$ is a variable parameter of the model and is related to the quark-meson couplings as

$$g_{\sigma N} = 3g_{\sigma}^{q} \int_{Bag} d\vec{r}\bar{q}q(\vec{r}), \tag{2}$$

with q and $\bar{q}$ being the quark wavefunction and its conjugate.

The flavour dependence of the dynamical mass is controlled only by the weights $w_{\sigma i}$ and $\tilde{w_{\sigma i}}$ are fixed within the model (see Table 2 [7] and Appendix A in the Ref. [8]). The coefficient d in the third term in Equation 1, is known as the "scalar polarizability". This term is a natural consequence of the quark structure of the nucleon and is sufficient to lead to nuclear saturation. The scalar polarizability is related to the radius of the bag R_B as [7] $d = 0.0044 + 0.211 * R_B - 0.0357R_B^2$.

The total energy of a *classical* system of baryons at zero temperature, modeled as non-overlapping bags coupled to meson fields σ, ω and ρ is expressed as [7]

$$E_{QMC} = \sum_{i=1,...} \sqrt{P_i^2 + M_i^2(\sigma(\vec{R}_i))} + g_{\omega N}^{i}\omega(\vec{R}_i) + g_{\rho N}\vec{I}_i.\vec{B}(\vec{R}_i) + E_\sigma + E_{\omega,\rho}, \tag{3}$$

with $\vec{R}_i$ and $\vec{P}_i$ are the position and momentum of a baryon i and $\vec{I}$ is the isospin matrix. $\vec{B}$ stands here for the isovector ρ field to avoid a confusion with the baryon number density ρ [7]. The contributions of the quark-meson couplings are dependent on two more model variable parameters, $g_{\omega N} = 3g_{\omega}^{q}$ and $g_{\rho N} = g_{\rho}^{q}$. The meson contributions, E_σ and $E_{\omega,\rho}$ are obtained by solution, in the mean field approximation, of the meson field equations of motion. The model is then quantized by replacement $\vec{P}_i \rightarrow -i\vec{\nabla}_i$. The three variable parameters, conveniently written as effective coupling constants $G_\sigma = g_{\sigma N}^2/m_\sigma^2$, $G_\omega = g_{\omega N}^2/m_\omega^2$ and $G_\rho = g_{\rho N}^2/m_\rho^2$, are fitted to properties of nuclear matter at saturation. The fixed parameters of the model, used in this work, are the free ω and ρ meson masses, the σ mass equal to 700 MeV, the bag radius R_B = 1 fm, and σ self-interaction parameter, λ_3, which appears to be needed in finite nuclei [9], equal to zero.

2.2 Applications

Starting with the equation of state (EoS), shown in Fig. 2, we observe that there is a threshold density for appearance of hyperons at T=0; at non-zero temperature hyperons appear at all densities. The onset of hyperons at T=0 MeV, manifests itself by separation of the EoS for nucleon only (np) matter from the EoS for hyperonic matter (npY) which softens at densities above the threshold (left panel). At finite T, shown for β-equilibrated matter with entropy density S/A=2 k_B, labeled by NS(β), corresponding to temperatures between about 20 - 50 MeV [2], this kink disappears. As expected, the pressure increases with temperature, more markedly in the np matter. Comparing the results of the QMC and DD2 models at T=0 MeV, they differ significantly both in the hyperon onset density and hyperon content (see Fig. 2). The Λ hyperons appear in both models but at different thresholds; in addition, in the QMC model only both cascades are present. In the DD2 model, only the negatively charged Σ and Ξ hyperons appear.

Figure 3 illustrates the NS masses vs radius and central density of cold and hot stars, in comparison of the latest observational results. It demonstrates that neither model suffers from the so called 'hyperon puzzle' despite the sensitivity of the EoS to the hyperonic content. Naturally, the only np matter stars achieve higher maximum mass but the maximum mass of the npY stars are still within the observational limits. In the non-zero temperature case, shown for two proto-neutron stars scenarios, one with the fixed lepton fraction, S/A=1 k_B

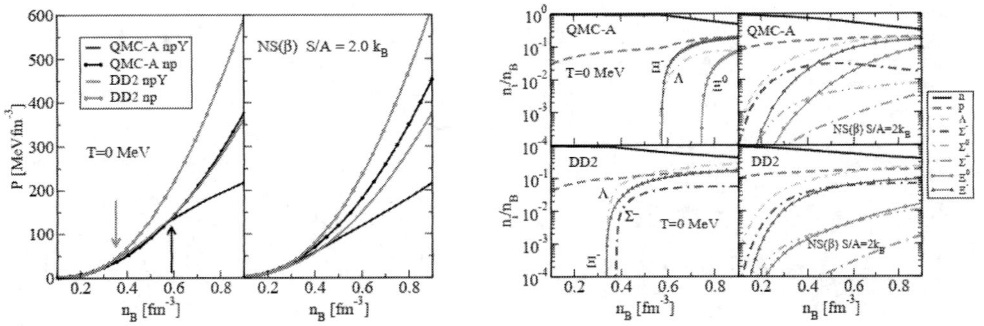

(a) *(b)*

Figure 2: (a) Equations of state of the QMC and DD2 models at T=0 MeV (left) and warm NS(β) matter (right), for full baryon octet (solid) and for only nucleons (dashed). (b) Relative population of nucleons and hyperons in units of the total baryon number density at T=0 MeV (left) and NS(β) matter (right). Only populations higher than 0.01% in the region of 0.1 - 1 fm^{-3} are shown (right). The figure is adapted from [2].

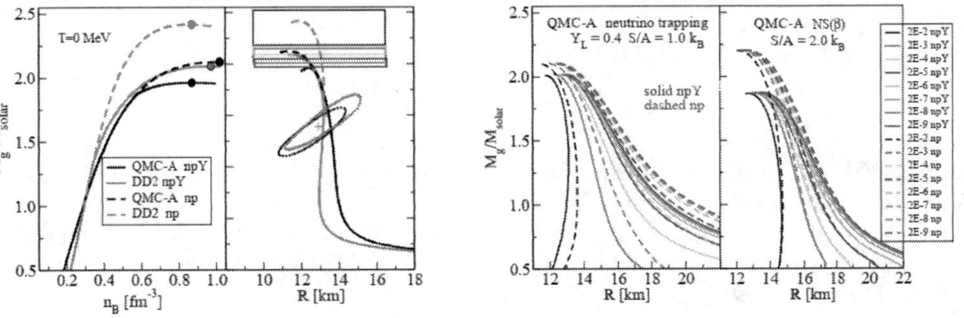

(a) *(b)*

Figure 3: (a) NS masses vs the central baryon density (left) and radius (right) computed with the QMC-A and DD2 models for hyperonic (solid line) and nucleonic (dashed) matter. Observational limits on the maximum mass configuration are illustrated by black dashed [10], green dot-double dash [11], red dot-single dash [12] and solid blue [13] rectangles. The NICER limits on the radius of a ~1.4 $M_\odot$ PSR J0030-0451 [14] (magenta full triangles) and [15] (dark green full circles) are added for completeness. The coloured full circles in the left panel indicate the maximum mass. (b) Illustration of search for the surface of a hot star with fixed Y_p and S/A=1 k_B (left) and the for NS in β (right). The figure is adapted from [2].

and trapped neutrinos, and the other with the neutrinoless NS(β) matter, the maximum mass is somewhat higher, as expected. Note that at finite T the surface of the star is diffused and has to be determined by search for a close-to-zero surface pressure as a function of radius [2].

Finally, we show in Fig. 4 two other examples, not directly related, of applications of the models. The single-particle potentials of hyperons in high-density matter determine their ap-

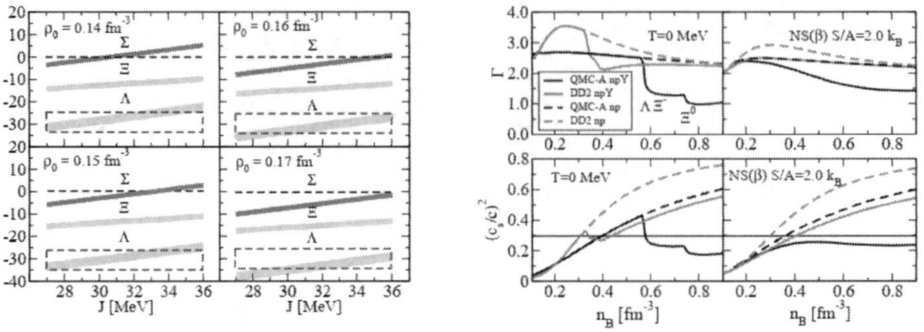

(a) *(b)*

Figure 4: (a) Single-particle potentials U_Λ (grey), U_Σ (red) and U_Ξ (green) as a function of the symmetry energy coefficient J for the saturation densities ρ_0 between 0.14 and 0.17 fm^{-3}. The filled width of the rectangles depicting the potentials represents the uncertainty due to the spread in the saturation energy E_0/A between -15.0 and -18 MeV. The dashed blue rectangle around U_Λ represents the region of experimental values reported in the literature. The horizontal black dashed line guides the eye to zero, dividing positive and negative values of U_Σ. No established experimental constraint on U_Ξ is available. Taken from Ref. [3] under Creative Commons Attribution License. (b) The adiabatic index Γ and the square of speed of sound in units of c for T=0 MeV (left top and bottom) and for NS(β) (top and bottom right). Predictions of the QMC-A model (black) and the DD2 model (green) are shown, both for npY (solid) and np (dashed) matter. The blue horizontal line indicates the conformal limit $(c_s/c)^2 = 1/3$. The figure is adopted from [2].

pearance in NS. These potentials are dependent on the nucleon-hyperon and hyperon-hyperon interactions, calculated within the QMC model, but treated as variable parameters in traditional RMF models. Specifically, the appearance of Σ hyperons in T=0 MeV models has been an issue for many years and both positive and negative U_Σ values have been considered [2, 3]. Here we show, in the left panel of Fig. 4, the relation of the potentials to saturation properties of nuclear matter as calculated in QMC-A. The positive U_Σ, consistent with the absence of Σ hyperons below ∼1 fm^{-3}, and unbound Σ hypernuclei, together with U_Λ being within a better known limit, places a strong constraint on the symmetry energy coefficient *J*. For this result being confirmed, data on U_Ξ are urgently needed, offering compelling support for experiments at JPARC.

In the right panel of Fig. 4 we show the density and temperature dependence of the adiabatic index Γ and the square of speed of sound as calculated in both QMC-A and DD2 models. Both quantities are directly related to the hyperonic content of the matter (see Fig. 2). The large drop in Γ at density 0.5-0.6 fm^{-3} for QMC-A takes place because the Λ and Ξ^- hyperons appear at almost the same density and the next drop is clearly related to appearance of the Ξ^0. The DD2 model predicts very close density thresholds for the Λ , Σ^- and Ξ^- hyperons, which can be associated with the large unresolved drop below 0.4 fm^{-3} (for more detail see [2]). At finite T, In the density dependence of Γ is smooth in both models, reflecting the absence of the thresholds, present in the T=0 MeV case. The speed of sound exhibits sensitivity to hyperon presence as well. In cold matter, $(c_s/c)^2$ decreases below the conformal limit with the onset of hyperons in the QMC-A model, but increases with density in DD2 (with a minimal glitch around 0.3 fm^{-3}). In warm matter, $(c_s/c)^2$ remains below the limit at all densities in hyperonic matter in QMC-A; in the DD2 model, it increases monotonically with density above the conformal limit for densities higher than about 0.4 fm^{-3}. Naturally, there is no structure in density dependence of either Γ or $(c_s/c)^2$ in np matter.

3 Summary and outlook

In summary, we have explored the complex effects of hyperon presence in high-density matter in the cores of cold and warm NS. Starting with the EoS and the composition of the core, following through the NS gravitational mass vs radius and central energy density dependences, to the relation between of the hyperon single-particle potentials and nuclear matter saturation properties and to the adiabatic index and the speed of sound, we conclude that the hyperons play a multifaceted role in the physics of high density matter and NS. The QMC model is one of the possible avenues to finding answers to the problem of hyperonic interactions in medium. However, it is also a phenomenological model and does not provide a full answer to the microscopy of baryon interactions which is still waiting to be discovered.

References

[1] J.R. Stone, Universe **7**, 257 (2021)
[2] J.R. Stone, V. Dexheimer, P.A.M. Guichon, A.W. Thomas, S. Typel, MNRAS **502**, 3476 (2021)
[3] J.R. Stone, P.A.M. Guichon, A.W. Thomas, Frontiers in Astronomy and Space Sciences **9** (2022)
[4] S. Typel, G. Röpke, T. Klähn, D. Blaschke, H.H. Wolter, Physical Review C **81**, 015803 (2010)
[5] F. Bissey, F.G. Cao, A.R. Kitson, A.I. Signal, D.B. Leinweber, B.G. Lasscock, A.G. Williams, Physical Review D **76**, 114512 (2007)
[6] P.A.M. Guichon, Phys. Letts. B **200**, 235 (1988)
[7] P.A.M. Guichon, J.R. Stone, A.W. Thomas, Progress in Particle and Nuclear Physics **100**, 262 (2018), 1802.08368
[8] J. Rikovska-Stone, P.A.M. Guichon, H.H. Matevosyan, A.W. Thomas, Nucl. Phys. A **792**, 341 (2007), nucl-th/0611030
[9] K.L. Martinez, A.W. Thomas, J.R. Stone, P.A.M. Guichon, Physical Review C **100**, 024333 (2019)
[10] R.W. Romani, D. Kandel, A.V. Filippenko, T.G. Brink, W. Zheng, The Astrophysical Journal Letters **934**, L17 (2022)
[11] R.W. Romani, D. Kandel, A.V. Filippenko, T.G. Brink, W. Zheng, The Astrophysical Journal Letters **908**, L46 (2021)
[12] E. Fonseca, H.T. Cromartie, T.T. Pennucci, P.S. Ray, A.Y. Kirichenko, S.M. Ransom, P.B. Demorest, I.H. Stairs, Z. Arzoumanian, L. Guillemot et al., The Astrophysical Journal Letters **915**, L12 (2021)
[13] J. Antoniadis, P.C.C. Freire, N. Wex, T.M. Tauris, R.S. Lynch, M.H. van Kerkwijk, M. Kramer, C. Bassa, V.S. Dhillon, T. Driebe et al., Science **340** (2013)
[14] M.C. Miller, F.K. Lamb, A.J. Dittmann, S. Bogdanov, Z. Arzoumanian, K.C. Gendreau, S. Guillot, A.K. Harding, W.C.G. Ho, J.M. Lattimer et al., ApJ **887**, L24 (2019)
[15] T.E. Riley, A.L. Watts, S. Bogdanov, P.S. Ray, R.M. Ludlam, S. Guillot, Z. Arzoumanian, C.L. Baker, A.V. Bilous, D. Chakrabarty et al., ApJ **887**, L21 (2019)

EPJ Web of Conferences **271**, 09004 (2022) https://doi.org/10.1051/epjconf/202227109004
HYP2022

Studying ϕ meson properties in nuclear matter from dilepton and K^+K^- decays

Philipp Gubler[1,*] and *Elena* Bratkovskaya[2,3,4,**]

[1]Advanced Science Research Center, Japan Atomic Energy Agency, Tokai, Ibaraki 319-1195, Japan
[2]GSI Helmholtzzentrum für Schwerionenforschung GmbH, Planckstrasse 1, 64291 Darmstadt, Germany
[3]Institute for Theoretical Physics, Johann Wolfgang Goethe Universität, Frankfurt am Main, Germany
[4]Helmholtz Research Academy Hesse for FAIR (HFHF), GSI Helmholtz Center for Heavy Ion Physics, Campus Frankfurt, 60438 Frankfurt, Germany

Abstract. After briefly reviewing the state of theoretical knowledge related to the behavior of the ϕ meson in nuclear matter, preliminary results of transport simulations of pA reactions corresponding to the KEK E325 experiment are presented. Finally, an outlook to current and future developments in the field is given.

1 Introduction

The behavior of the ϕ meson in hot and/or dense matter has for a long time been of interest from the point of view of probing the in-medium strange quark condensate [1, 2] and strangeness production in heavy-ion collisions [5, 6]. Recently, it has also attracted attention because its degree of its polarization [7] and the behavior of its polarization modes can provide information about the vorticity [8, 9] and the potential breaking of Lorentz symmetry of the medium [10, 11].

In these proceedings, we will particularly focus on the finite density behavior of the ϕ meson, by first reviewing some recent theoretical results obtained within a QCD sum rule approach in Section 2 and next by discussing preliminary results of numerical transport simulations of the pA reactions studied at the KEK E325 experiment in Section 3. A summary and outlook follow in Section 4.

2 Recent theory results

The QCD sum rule method is one of the few tools available to study finite density systems directly from QCD [1]. The method provides relations between the various QCD condensates and the spectral properties of hadrons, both in vacuum and in hot and/or dense matter. It was for instance shown in Ref. [2] that the mass of the ϕ meson in nuclear matter can be related to the strange sigma term $\sigma_{sN} = m_s\langle N|\bar{s}s|N\rangle$, which governs the finite density behavior of the

*e-mail: gubler@post.j-parc.jp
**e-mail: brat@fias.uni-frankfurt.de

strange quark condensate $\langle\bar{s}s\rangle_\rho$ within the linear density approximation: $\langle\bar{s}s\rangle_\rho \simeq \langle 0|\bar{s}s|0\rangle + \frac{\sigma_{sN}}{m_s}\rho$. Specifically, the relation between σ_{sN} and the ϕ meson mass $m_\phi(\rho)$ was determined as

$$\frac{m_\phi(\rho)}{m_\phi(0)} - 1 = \left[b_0 - b_1\Big(\frac{\sigma_{sN}}{1\,\text{MeV}}\Big)\right]\frac{\rho}{\rho_0}, \tag{1}$$

ρ_0 being the normal nuclear matter density and $b_0 = (1.00 \pm 0.34)\cdot 10^{-2}$, $b_1 = (2.86 \pm 0.48)\cdot 10^{-4}$.

The above result shows that an experimental measurement of $m_\phi(\rho)$ would allow to constrain the value of σ_{sN}. Equation (1), however, does not take finite momentum effects into account, which cannot simply be ignored in an actual experimental situation. Such effects were considered in Ref. [11], where it was found that the longitudinal and transverse modes, which cannot be distinguished at zero momentum, will split with increasing momentum. The origin of this splitting is the breaking of Lorentz symmetry in nuclear matter, which in QCD sum rules leads to the emergence of non-scalar condensates, such as $\mathcal{ST}\langle\bar{s}\gamma^\mu iD^\nu s\rangle_\rho$ ($\mathcal{S}$ and $\mathcal{T}$ make the following structures symmetric and traceless with respect to open Lorentz indices). These, in turn, cause the two polarizations modes to behave differently. A specific calculation leads to [11]

$$\frac{m_\phi^{\text{L/T}}(\rho)}{m_\phi(0)} - 1 = \left[a + b^{\text{L/T}}\Big(\frac{|\vec{q}|^2}{1\,\text{GeV}^2}\Big)\right]\frac{\rho}{\rho_0}, \tag{2}$$

where a is the zero momentum mass shift [which depends on σ_{sN}, as shown in Eq. (1)]. The superscripts L and T stand for longitudinal and transverse modes, respectively. The numerical values for $b^{\text{L/T}}$ are calculated as $b^{\text{L}} = (-4.8 \pm 0.8)\cdot 10^{-3}$, $b^{\text{T}} = (6.7 \pm 3.4)\cdot 10^{-3}$. It will be interesting to see whether the E16 experiment at J-PARC [12] will in the future be able to measure these two polarization modes independently.

3 Numerical pA reaction simulations

In this work, we use the PHSD transport approach [3, 4] (in the HSD version [5], i.e. ignoring the possible QGP droplet formation) which is a covariant, microscopic off-shell framework, in which the vector meson (and recently also kaon [13]) spectral functions and their density dependence can be consistently incorporated into the numerical simulation. We simulate 12 GeV pA reactions with C and Cu targets, that were studied at the E325 experiment [14] at KEK. For parametrizing the ϕ meson spectral function and its density dependence, we use a relativistic Breit-Wigner form

$$A_V(M,\rho) = C\frac{2}{\pi}\frac{M^2\Gamma_V^*(M,\rho)}{[M^2 - M_0^{*2}(\rho)]^2 + M^2\Gamma_V^{*2}(M,\rho)}, \tag{3}$$

where C is a renormalization constant and the in-medium mass $M_0^*(\rho)$ and width $\Gamma_V^*(M,\rho)$ are assumed to depend linearly on density ρ,

$$M_0^*(\rho) = M_0\Big(1 - \alpha\frac{\rho}{\rho_0}\Big), \tag{4}$$

$$\Gamma_V^*(M,\rho) = \Gamma_V(M) + \alpha_{\text{coll}}\frac{\rho}{\rho_0}, \tag{5}$$

M_0 ($\Gamma_V(M)$) are the vacuum mass (width).

The simulations were carried out with multiple ϕ meson modification scenarios, that is, with changing values of α and α_{coll} in the above parametrization. A few preliminary examples

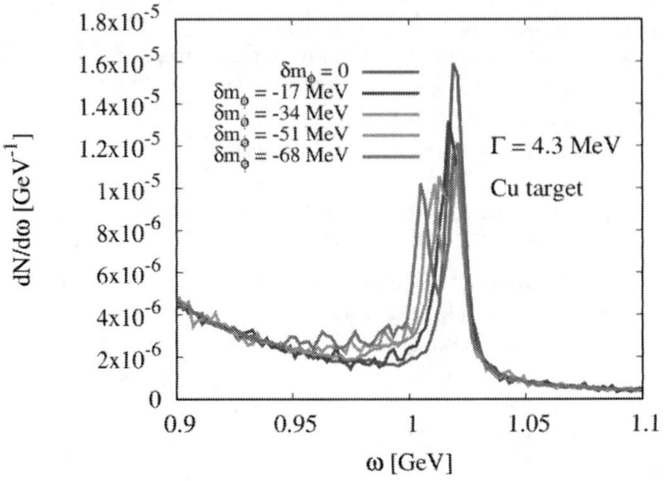

Figure 1. The di-lepton spectra in the ϕ meson mass region, obtained in simulations of 12 GeV pCu reactions for scenarios of different negative mass shift magnitudes for the ϕ meson at finite density.

of the di-lepton spectra in the ϕ meson mass region, obtained for a Cu target and scenarios with different finite density mass shifts (and no broadening) are given in Fig. 1. Experimental effects such as acceptance, limited resolution and rescattering effects of the dileptons emerging from the reaction, are not included in this plot. It is furthermore important to include final state QED radiation effects, which can deform the spectrum in a similar manner as one would expect for a mass shift case [15]. Only after taking all these effects properly into account, can a realistic comparison with experimental data become possible. Such an analysis, with the final goal of determining which mass shift and/or broadening scenarios can best reproduce the experimental data of Ref. [14], is presently ongoing.

4 Conclusions and outlook

In these proceedings, recent theoretical results related to the behavior of the ϕ meson in nuclear matter were reviewed, emphasizing especially the effects of Lorentz symmetry breaking in medium, which leads to a splitting of the longitudinal and transverse polarization modes and to a modified dispersion relation. First results based on numerical simulations of pA reactions using the microscopic PHSD transport approach were also discussed and the importance of considering all experimental and theoretical (QED) effects than can distort the dilepton spectrum before making a direct comparison with experimental data, was highlighted.

As an outlook, let us mention here some interesting directions of research, that are currently being pursued by multiple theoretical and experimental groups. Very recently, two new results regarding the ϕN scattering length have been reported from a femtoscopy measurement by ALICE at LHC [16] and a lattice QCD calculation by the HAL QCD Collaboration [17]. The scattering length can at low density can be related to a first order nuclear optical potential and hence to the nuclear matter ϕ meson mass shift and width. These new results suggest for the ϕ meson a rather strong attraction of several tens of MeV at normal nuclear matter density. It remains to be seen whether this conclusion holds in future more precise measurements and calculations. Another new development is the J-PARC E16 experiment [12], which is scheduled to start with its first physics data taking in 2024 and will, similar to the KEK E325 experiment, measure dilepton spectra generated by pA reactions to study potential in-medium modification effects of vector mesons. At a later stage, the K^+K^- decay channel will also be studied in the same experiment, which in combination with the dilepton measurement could be especially useful to disentangle the two polarization modes discussed in Section 2.

P.G. thanks HyungJoo Kim and Keisuke Ohtani for the fruitful collaborations that are mentioned in these proceedings, and Taesoo Song for his help in preparing the PHSD code. P.G. is supported by KAKENHI under Contract No. JP20K03940. E.B. acknowledges support by the Deutsche Forschungsgemeinschaft (DFG, German Research Foundation) through the grant CRC-TR 211 "Strong-interaction matter under extreme conditions" - project number 315477589 - TRR 211.

References

[1] T. Hatsuda and S. H. Lee, Phys. Rev. C **46**, no.1, R34 (1992)
[2] P. Gubler and K. Ohtani, Phys. Rev. D **90**, no.9, 094002 (2014)
[3] W. Cassing and E. L. Bratkovskaya, Phys. Rev. C **78**, 034919 (2008)
[4] W. Cassing and E. L. Bratkovskaya, Nucl. Phys. A **831**, 215-242 (2009)
[5] W. Cassing and E. L. Bratkovskaya, Phys. Rept. **308**, 65-233 (1999)
[6] T. Song, J. Aichelin and E. Bratkovskaya, Phys. Rev. C **106**, 024903 (2022)
[7] M. Abdallah *et al.* [STAR], [arXiv:2204.02302 [hep-ph]].
[8] Z. T. Liang and X. N. Wang, Phys. Rev. Lett. **94**, 102301 (2005) [erratum: Phys. Rev. Lett. **96**, 039901 (2006)].
[9] H. Taya *et al.* [ExHIC-P], Phys. Rev. C **102**, no.2, 021901 (2020)
[10] S. H. Lee, Phys. Rev. C **57**, 927-930 (1998) [erratum: Phys. Rev. C **58**, 3771 (1998)].
[11] H. Kim and P. Gubler, Phys. Lett. B **805**, 135412 (2020)
[12] S. Ashikaga, *et al.* JPS Conf. Proc. **26**, 024005 (2019)
[13] T. Song, L. Tolos, J. Wirth, J. Aichelin and E. Bratkovskaya, Phys. Rev. C **103**, no.4, 044901 (2021)
[14] R. Muto *et al.* [KEK-PS-E325], Phys. Rev. Lett. **98**, 042501 (2007)
[15] A. Spiridonov, [arXiv:hep-ex/0510076 [hep-ex]].
[16] S. Acharya *et al.* [ALICE], Phys. Rev. Lett. **127**, no.17, 172301 (2021)
[17] Y. Lyu, T. Doi, T. Hatsuda, Y. Ikeda, J. Meng, K. Sasaki and T. Sugiura, [arXiv:2205.10544 [hep-lat]].

Hyperonic equation of state at finite temperature for neutron stars

Hristijan Kochankovski[1,2,*], *Angels* Ramos[1], and *Laura* Tolos[3,4,5]

[1]Departament de Física Quàntica i Astrofísica and Institut de Ciències del Cosmos, Universitat de Barcelona, Martí i Franquès 1, 08028, Barcelona, Spain

[2]Faculty of Natural Sciences and Mathematics-Skopje, Ss. Cyril and Methodius University in Skopje, Arhimedova, 1000 Skopje, North Macedonia

[3]Institute of Space Sciences (ICE, CSIC), Campus UAB, Carrer de Can Magrans, 08193 Barcelona, Spain

[4]Institut d'Estudis Espacials de Catalunya (IEEC), 08034 Barcelona, Spain

[5]Frankfurt Institute for Advanced Studies, Ruth-Moufang-Str. 1, 60438 Frankfurt am Main, Germany

Abstract. We review the composition and the equation of state of the hyperonic core of neutron stars at finite temperature within a relativistic mean-field approach. We make use of the new FSU2H* model, which is built upon the FSU2H scheme by improving on the Ξ potential according to the recent analysis on the Ξ atoms, and we extend it to include finite temperature corrections. The calculations are done for a wide range of densities, temperatures and charge fractions, thus exploring the different conditions that can be found in proto-neutron stars, binary mergers remnants and supernovae explosions. The inclusion of hyperons has a strong effect on the composition and the equation of state at finite temperature, which consequently would lead to significant changes in the properties and evolution of hot neutron stars.

1 Introduction

The large densities in the neutron star core make neutron stars perfect candidates for testing different models of nuclear matter under extreme conditions. Since there is no experimental data that covers the supranuclear region, many uncertainties are still present. One of them is the composition of the inner core, that is strongly model dependent. Numerous models predict the appearance of exotic particles, such as hyperons, as it is energetically favourable to produce them inside the core. Although evolved neutron stars are cold objects, some of the information we have from them come from neutron star merger events or from their early evolution. In both scenarios, the matter in the star is hot, which means that a finite temperature treatment of matter is essential. In addition, one cannot always assume that star matter satisfies the beta equilibrium condition. Due to the difficulty in dealing with finite temperature nuclear models, many groups that perform complicated relativistic simulations for the evolution of hot neutron stars use approximated models for the finite-temperature equation of state (EoS) [1]. However, we have shown in Ref. [2] that those approaches are particularly inaccurate when one considers hyperons in the core of the neutron star. Thus,

*e-mail: hriskoch@fqa.ub.edu

it is of fundamental importance to make use of a baryonic EoS at high densities without further approximations when dealing with neutron star mergers and/or the early stages in the evolution of a neutron star. In this work we briefly present the FSU2H* model that was recently constructed in Ref. [2]. Within this framework we build an hyperonic EoS that covers a wide range of densities, temperatures and charge fractions, making this EoS available for relativistic neutron star simulations. We show the composition pattern and the pressure-density relation for matter at different temperatures and charge fractions.

2 Theoretical framework

We consider matter made of baryons at a given temperature T, baryon density ρ_B and fixed charged fraction $Y_Q = \sum_i q_i\rho_i/\rho_B$, where q_i and ρ_i are the charge and density of the i-th baryon. The interaction between the baryons is modeled through the exchange of different mesons, which leads to the following Lagrangian density $\mathcal{L}$ of the system that can be split into the baryonic contributions $\mathcal{L}_b$ ($b = n, p, \Lambda, \Sigma, \Xi$), and a mesonic term $\mathcal{L}_m$, which includes the contributions from the σ , ω, ρ, ϕ and σ^* mesons:

$$\mathcal{L} = \sum_b \mathcal{L}_b + \mathcal{L}_m; \tag{1}$$

$$\begin{aligned}
\mathcal{L}_b &= \bar{\Psi}_b(i\gamma_\mu\partial^\mu - q_b\gamma_\mu A^\mu - m_b + g_{\sigma b}\sigma + g_{\sigma^* b}\sigma^* - g_{\omega b}\gamma_\mu\omega^\mu - g_{\phi b}\gamma_\mu\phi^\mu - g_{\rho,b}\gamma_\mu\vec{I}_b\vec{\rho}^\mu)\Psi_b,\\
\mathcal{L}_m &= \frac{1}{2}\partial_\mu\sigma\partial^\mu\sigma - \frac{1}{2}m_\sigma^2\sigma^2 - \frac{\kappa}{3!}(g_{\sigma b}\sigma)^3 - \frac{\lambda}{4!}(g_{\sigma b})^4\\
&+ \frac{1}{2}\partial_\mu\sigma^*\partial^\mu\sigma^* - \frac{1}{2}m_{\sigma^*}^2\sigma^{*2} - \frac{1}{4}\Omega^{\mu\nu}\Omega_{\mu\nu} + \frac{1}{2}m_\omega^2\omega_\mu\omega^\mu + \frac{\zeta}{4!}g_{\omega b}^4(\omega_\mu\omega^\mu)^2\\
&- \frac{1}{4}\vec{R}^{\mu\nu}\vec{R}_{\mu\nu} + \frac{1}{2}m_\rho^2\vec{\rho}_\mu\vec{\rho}^\mu + \Lambda_\omega g_{\rho b}^2\vec{\rho}_\mu\vec{\rho}^\mu g_{\omega b}^2\omega_\mu\omega^\mu - \frac{1}{4}P^{\mu\nu}P_{\mu\nu}\\
&+ \frac{1}{2}m_\phi^2\phi_\mu\phi^\mu - \frac{1}{4}F^{\mu\nu}F_{\mu\nu};
\end{aligned} \tag{2}$$

with m_i being the mass of i-th particle, Ψ_b indicating the baryon Dirac fields, and $\Omega_{\mu\nu} = \partial_\mu\omega_\nu - \partial_\nu\omega_\mu$, $\vec{R}_{\mu\nu} = \partial_\mu\vec{\rho_\nu} - \partial_\nu\vec{\rho_\mu}$, $P_{\mu\nu} = \partial_\mu\phi_\nu - \partial_\nu\phi_\mu$ and $F_{\mu\nu} = \partial_\mu A_\nu - \partial_\nu A_\mu$ being the mesonic and electromagnetic strength tensors. The quantity $\vec{I}_b$ represents the isospin operator, γ^μ are the Dirac matrices and g_{mb} labels the couplings of the different baryons to the mesons.

Between the different baryonic species a weak interaction equilibrium is assumed that give rise to the following relations between their chemical potentials (μ_i):

$$\begin{aligned}
\mu_{b^0} &= \mu_n,\\
\mu_{b^-} &= 2\mu_n - \mu_p,\\
\mu_{b^+} &= \mu_p,
\end{aligned} \tag{3}$$

where b^0 is any neutral baryon, b^+ is any positive and b^- is any negative baryon.

In order to obtain the composition and all thermodynamical variables of interest, one needs to obtain the Euler-Lagrange equations of motion and solve them in relativistic mean field approximation. Then, from the energy-momentum tensor it is easy to obtain all thermodynamical quantities. For details see Ref. [2]. The coupling constants are chosen in order for the model to reproduce the saturation properties of nuclear matter and some properties of finite nuclei, as well as to satisfy the constraints on the high-density nuclear pressure coming from heavy-ion collisions together with the 2 $M_\odot$ neutron star observations and radii smaller than 13 km [3–10]. We list the values of the parameters in Tables (1,2), that are obtained in Ref. [2, 11]. The hyperonic couplings to the mesonic fields are given in terms of

Table 1. Parameters of the model FSU2H*. The mass of the nucleon is equal to m_N = 939 MeV.

m_σ (MeV)	m_ω (MeV)	m_ρ (MeV)	m_{σ^*} (MeV)	m_ϕ (MeV)	$g^2_{\sigma N}$	$g^2_{\omega N}$	$g^2_{\rho N}$	κ (MeV)	λ	ζ	Λ_ω
498	783	763	980	1020	103	170	197	4.00014	-0.0133	0.008	0.045

Table 2. The ratios of the couplings of hyperons to mesons with respect to the nucleonic ones.

Y	$R_{\sigma Y}$	$R_{\omega Y}$	$R_{\rho Y}$	$R_{\sigma^* Y}$	$R_{\phi Y}$
Λ	0.6613	2/3	0	0.2812	$-\sqrt{2}/3$
Σ	0.4673	2/3	1	0.2812	$-\sqrt{2}/3$
Ξ	0.3305	1/3	1	0.5624	$-2\sqrt{2}/3$

their ratios to the corresponding couplings of nucleons: $R_{iY} = g_{iY}/g_{iN}$ for $i = (\sigma, \omega, \rho)$, and $R_{\sigma^* Y} = g_{\sigma^* Y}/g_{\sigma N}$ and $R_{\phi Y} = g_{\phi Y}/g_{\omega N}$, since $g_{\sigma^* N} = 0$ and $g_{\phi N} = 0$ due to the OZI rule.

3 Composition and EoS at fixed Y_Q

As mentioned before, we show the composition and pressure-density relation [$P(\rho_B)$] for matter at fixed T and Y_Q. Given that simulations need a wide range of values of these parameters, we display our results for two temperatures (T = 25 MeV and T = 75 MeV) and two charged fractions (Y_Q = 0.01 and Y_Q = 0.5) in order to give a clear picture about what one might expect for baryonic neutron star matter at any arbitrary condition.

In the upper plots of Figure 1 we show the composition of the core for two charge fractions (Y_Q = 0.01 (left panel) and Y_Q = 0.5 (right panel)), and two different temperatures (solid lines for T = 25 MeV and dashed lines for T = 75 MeV). We observe that when the temperature is large, hyperons are present at any point of the core. All hyperons of the baryon octet can be found in the hyperonic core for certain conditions of temperature and charge fraction. The appearance of the Λ hyperon significantly lowers the neutron abundance, an effect which is even more enhanced with the appearance of the negatively charged hyperons. The presence of negatively charged hyperons then leads to an increase of the proton fraction due to the fixed charge fraction. This mechanism is more clear at low and moderate temperatures, such as T = 25 MeV. In the lower plots of the Figure 1 we display the pressure versus the baryonic density of the core for the same charge fractions and temperatures. One clearly sees that the relative difference between the pressure-density curves for different temperatures is larger for low densities in the core. This is due to the fact that temperature effects are manifest more clearly when matter is non-degenerate. However, it is interesting to note that when the charge fraction is high, as in Y_Q = 0.5, the difference between the curves for different temperatures becomes small at densities around ρ_B = 0.5 fm^{-3}. The reason lies in the fact that, due to the fixed charged fraction, the proton abundance cannot be significantly lowered, making the protons degenerate, even at high temperature.

4 Conclusions

We have briefly reviewed the EoS of hot dense baryonic matter at different charge fractions within the new FSU2H* model. All hyperons from the baryon octet can appear in the core of neutron stars in significant abundances under certain conditions of temperature and charge

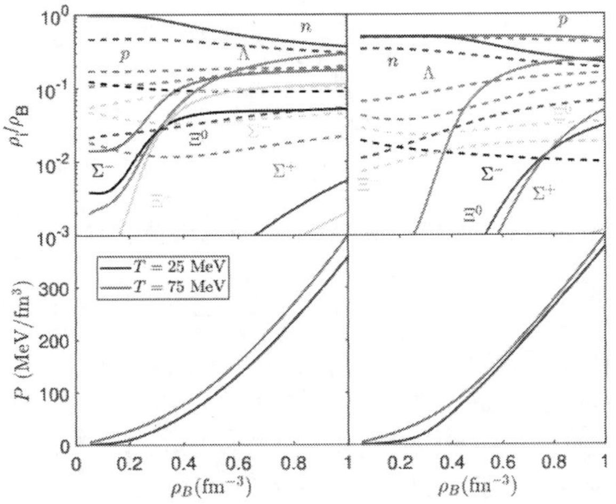

Figure 1. Composition and pressure as a function of baryonic density of the hyperonic core of neutron stars at $Y_Q = 0.01$ (left panels) and $Y_Q = 0.5$ (right panels). The solid lines in the upper plots are the compositions obtained at $T = 25$ MeV, while the dashed lines are the compositions at $T = 75$ MeV.

fraction. At sufficiently high temperature, hyperons populate the dense matter even at densities below the saturation one. This composition pattern affects the EoS, making it softer with respect to the pure nucleonic one. The correct treatment of the hyperonic EoS at finite temperature is of special importance for the correct description of the early phases in the evolution of neutron stars or for the numerical simulations of binary neutron stars mergers.

References

[1] K. Hotokezaka, K. Kiuchi, K. Kyutoku, T. Muranushi, Y.I. Sekiguchi, M. Shibata, K. Taniguchi, Physical Review D - Particles, Fields, Gravitation and Cosmology **88**, 044026 (2013)

[2] H. Kochankovski, A. Ramos, L. Tolos, *Equation of state for hot hyperonic neutron star matter* (2022), `https://arxiv.org/abs/2206.11266v1`

[3] P. Demorest, T. Pennucci, S. Ransom, M. Roberts, J. Hessels, Nature **467**, 1081 (2010)

[4] J. Antoniadis et al., Science **340**, 6131 (2013), `1304.6875`

[5] E. Fonseca, T.T. Pennucci, J.A. Ellis, I.H. Stairs, D.J. Nice, S.M. Ransom, P.B. Demorest, Z. Arzoumanian, K. Crowter, T. Dolch et al., The Astrophysical Journal **832**, 167 (2016)

[6] H.T. Cromartie, E. Fonseca, S.M. Ransom, P.B. Demorest, Z. Arzoumanian, H. Blumer, P.R. Brook, M.E. DeCesar, T. Dolch, J.A. Ellis et al., *Relativistic Shapiro delay measurements of an extremely massive millisecond pulsar* (2020), `https://arxiv.org/abs/1904.06759v2`

[7] T.E. Riley et al., Astrophys. J. Lett. **887**, L21 (2019), `1912.05702`

[8] M.C. Miller et al., Astrophys. J. Lett. **887**, L24 (2019), `1912.05705`

[9] T.E. Riley et al., Astrophys. J. Lett. **918**, L27 (2021), `2105.06980`

[10] M.C. Miller et al., Astrophys. J. Lett. **918**, L28 (2021), `2105.06979`

[11] L. Tolos, M. Centelles, A. Ramos, Publications of the Astronomical Society of Australia **34** (2017)

The T_{cc}^{+} exotic state and its doubly bottom $B_{(s)}^{(*)}B_{(s)}^{(*)}$ counterparts

Eulogio Oset[1,*], *Wei Hong* Liang[2,3,**], *Lian Rong* Dai[4,***], *Raquel* Molina[1,****], *Luís* Roca[5,†], *Alberto* Martínez Torres[6,‡], *Kanchan* Khemchandani[7,§], and *Albert* Feijoo[1,8,¶]

[1]Departamento de Física Teórica and IFIC, Centro Mixto Universidad de Valencia-CSIC, Institutos de Investigación de Paterna, Aptdo. 22085, 46071 Valencia, Spain
[2]Department of Physics, Guangxi Normal University, Guilin 541004, China
[3]Guangxi Key Laboratory of Nuclear Physics and Technology, Guangxi Normal University, Guilin 541004, China
[4]School of Science, Huzhou University, Huzhou 313000, Zhejiang, China
[5]Departamento de Física, Universidad de Murcia, E-30100 Murcia, Spain
[6]Universidade de Sao Paulo, Instituto de Fisica, C.P. 05389-970 Sao Paulo, Brazil
[7]Universidade Federal de Sao Paulo, C.P. 01302-907 Sao Paulo, Brazil
[8]Nuclear Physics Institute, 25068 Rez, Czech Republic

Abstract. We have studied the exotic doubly charmed D^*D system providing a natural explanation for the peak recently observed by LHCb, in terms of $D^{*+}D^0$ and $D^{*0}D^+$ with isospin $I = 0$. The width has been evaluated accurately based on the decay widths of the D^* states. The $D^0D^0\pi^+$ decay mode of the bound state formed is studied in detail, showing a narrow peak below the $D^{*+}D^0$ threshold and some strength above it, as observed in the experiment. The remarkable agreement of this approach and the latter experimental analysis supports strongly the molecular nature of this state, the first example of a meson with two open charmed quarks. This study can be naturally extended to the bottom sector giving interesting features for the $B_{(s)}^{(*)}B_{(s)}^{(*)}$ counterparts found there. In addition, our previous study on the D^*D^* molecular state has been revisited taking into account the insights extracted from the T_{cc} state and the new results are included in the present manuscript.

1 Introduction

The interesting discovery of the two-open-charm meson state, the T_{cc} state, by the LHCb collaboration [1, 2] has brought new excitement into the field of hadron spectroscopy, with yet another state that does not have the standard $q\bar{q}$ nature of ordinary mesons. The characterizing features of the T_{cc} state are a mass very close to the $D^{*+}D^0$ and $D^{*0}D^+$ threshold as well as

*e-mail: Eulogio.Oset@ific.uv.es
**e-mail: liangwh@gxnu.edu.cn
***e-mail: dailianrong@zjhu.edu.cn
****e-mail: Raquel.Molina@ific.uv.es
†e-mail: luisroca@um.es
‡e-mail: amartine@if.usp.br
§e-mail: kanchan.khemchandani@unifesp.br
¶e-mail: edfeijoo@ific.uv.es

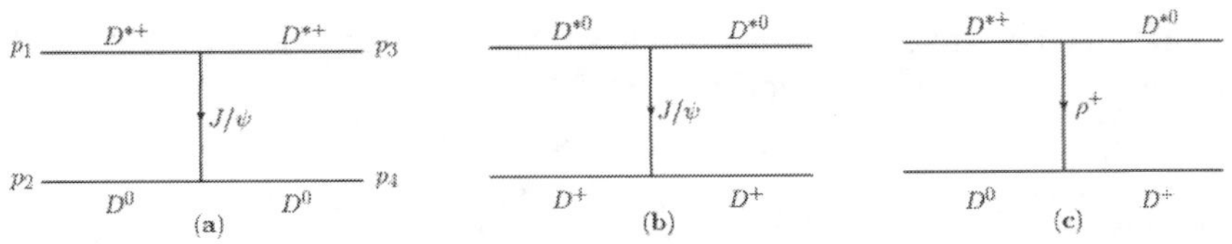

Figure 1. Diagrams considered for the interaction vector-pseudoscalar (VP).

a very small width. Being more precise, its mass is slightly below the $D^{*+}D^0$ threshold. Actually, the primary results of Ref. [1] were reanalysed by the LHCb Collaboration [2], using a unitary Breit-Wigner amplitude considering the experimental resolution. The results obtained differ from those reported in Ref. [1]. The new values obtained from the pole position of the state are

$$\delta m_{\text{exp}} = -360 \pm 40^{+4}_{-0} \text{ keV}, \tag{1}$$

$$\Gamma = 48 \pm 2^{+0}_{-14} \text{ keV}, \tag{2}$$

where δm_{exp} is the mass difference between the T_{cc} state and the $D^{*+}D^0$ threshold, while Γ represents the correspondig width. The closeness of this mass to the D^*D threshold makes one think about the possibility that this state could be a molecular state of D^*D. In fact, this structure was anticipated in Refs. [3–5]. The possible D^*D bound state would be analogous to the D^*D^* molecular state studied earlier in Ref. [6]. In that later work, predictions were made for another exotic state of $D^*\bar{K}^*$ nature that can be associated to the $X_0(2866)$ state reported in Ref. [7] (see update in Ref. [8]).

2 Formalism and results

Following [6, 8] we use the extension of the local hidden gauge approach to study the D^*D interaction, mediated by exchanges of ρ^+ and J/ψ vector mesons (diagrammatically represented in Fig. 1), and whose interaction kernel is given by

$$\begin{aligned} V_{ij} &= C_{ij}\, g^2 (p_1 + p_3) \cdot (p_2 + p_4)\, \vec{\epsilon} \cdot \vec{\epsilon}\,' \\ &\rightarrow C_{ij}\, g^2 \frac{1}{2}[3s - (M^2 + m^2 + M'^2 + m'^2) \\ &\quad - \frac{1}{s}(M^2 - m^2)(M'^2 - m'^2)]\, \vec{\epsilon} \cdot \vec{\epsilon}\,', \end{aligned} \tag{3}$$

where M, m are the initial vector, pseudoscalar masses and M', m' the corresponding final ones and, as can be seen in in Fig. 1, p_i are the corresponding incoming and outgoing momenta of the vector- and pseudoscalar-mesons in the CM framework with a total energy $\sqrt{s}$. The second expression in Eq. (3) is obtained by projection onto s-wave contribution. Also, since we are close to the D^*D threshold we neglect the ϵ^0 components of the vectors and work with the vector polarizations $\vec{\epsilon}, \vec{\epsilon}\,'$. The matrix C is a 2×2 matrix, where each element C_{ij} (with i and j covering the two channels considered in the basis) contains the information of the atractive or repulsive character as well as the strength of each transition, and it is given by

$$C = \begin{pmatrix} \frac{1}{M^2_{J/\psi}} & \frac{1}{m^2_\rho} \\ \frac{1}{m^2_\rho} & \frac{1}{M^2_{J/\psi}} \end{pmatrix}. \tag{4}$$

 https://doi.org/10.1051/epjconf/202227110001

This interaction is attractive for the D^*D system in $I = 0$ and repulsive for $I = 1$. With this potential we solve the Bethe Salpeter equation

$$T = [1 - VG]^{-1} V, \quad (5)$$

in order to treat the interaction nonperturbatively.

As for the $D^0D^0\pi^+$ production we use the decay mechanism shown in Fig. 2, for a more detailed explanation see Ref [9], and we obtain the results of Fig. 3. These results are in good agreement with the analysis of LHCb whose results were obtained by means of the unitary amplitude and the values of which are shown in Eqs. (1) and (2).

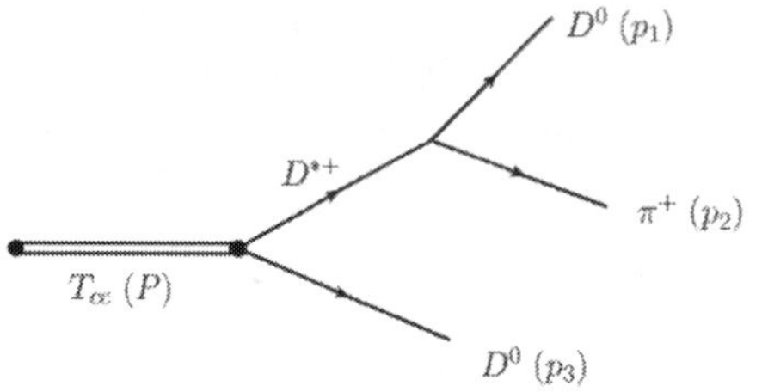

Figure 2. Mechanism for $D^0\pi^+D^0$ decay of the T_{cc} state. The diagram with $D^{*0}D^+$ decay does not lead to final $D^0D^0\pi^+$.

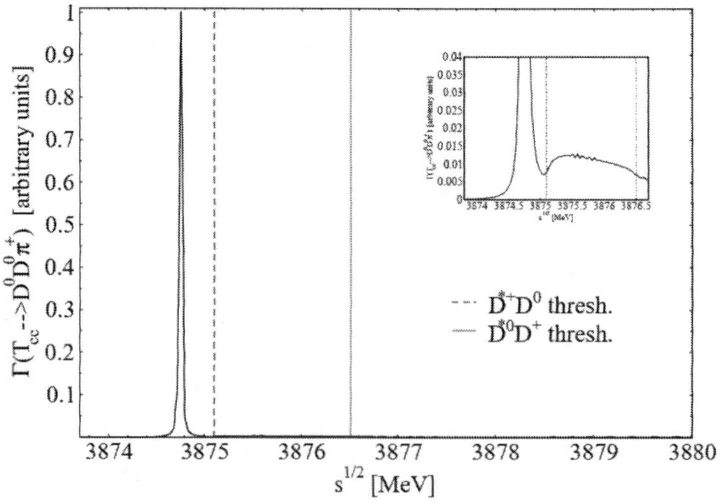

Figure 3. $\Gamma(\sqrt{s})$ for the decay of the T_{cc} into $D^0D^0\pi^+$.The inset in the figure is a zoom to illustrate the mass distribution between $D^{*+}D^0$ and $D^{*0}D^+$ thresholds.

The results obtained in this work have been extended to the study of the D^*D^* system in [10] where we find also bound states for $I = 0$ and $J = 1$, more bound that the T_{cc} and with a larger width, as shown in Table 1. The results are also extended to the study of the $B^{(*)}_{(s)}B^{(*)}_{(s)}$ states in [11], where we also find bound states as shown in Table 2.

Table 1. The predictions assuming 420 or 450 MeV cutoff values, as used for the T_{cc} state. The threshold masses for D^*D and $D_s^*D^*$ are 4017.1 MeV and 4122.46 MeV, respectively. The binding energy B is referred to the closest threshold and M and Γ stand for the mass and width of the predicted states, respectively.

	q_{max} = 450 MeV	q_{max} = 420 MeV
$M_{D^*D^*}$	4014.08 MeV	4015.54 MeV
$B_{D^*D^*}$	3.23 MeV	1.56 MeV
$\Gamma_{D^*D^*}$	2.3 MeV	1.5 MeV
$M_{D_s^*D^*}$	4122.46 MeV (cusp)	4122.46 MeV (cusp)
$\Gamma_{D_s^*D^*}$	70 – 100 KeV	70 – 100 KeV

Table 2. States of $J^P = 1^+$ obtained from the different sectors considered. The binding energy B is referred to the closest threshold and M and Γ stand for the mass and width of the predicted states, respectively..

States	M (MeV)	B (MeV)	Γ
B^*B ($I = 0$)	10583	21	14 eV
$B_s^*B - B^*B_s$ ($I = \frac{1}{2}$)	10681	11	45 eV
B^*B^* ($I = 0$)	10630	19	8 MeV
$B_s^*B^*$ ($I = \frac{1}{2}$)	10728	12	0.5 MeV

Acknowledgements

This work is partly supported by the National Natural Science Foundation of China under Grants No. 11975083 and No. 12047567. This work is also partly supported by the Spanish Ministerio de Economia y Competitividad (MINECO) and European FEDER funds under Contracts No. FIS2017-84038-C2-1-P B, PID2020-112777GB-I00, and by Generalitat Valenciana under contract PROMETEO/2020/023. This project has received funding from the European Union Horizon 2020 research and innovation programme under the program H2020-INFRAIA-2018-1, grant agreement No. 824093 of the "STRONG-2020" project. The work of A. F. was partially supported by the Czech Science Foundation, GAČR Grant No. 19-19640S, and the Generalitat Valenciana and European Social Fund APOSTD- 2021-112.

References

[1] R. Aaij *et al.* [LHCb], Nature Phys. **18**, 751-754 (2022)
[2] R. Aaij *et al.* [LHCb], Nature Commun. **13**, 3351 (2022)
[3] N. Li, Z. F. Sun, X. Liu and S. L. Zhu, Phys. Rev. D **88**, 114008 (2013)
[4] M.-Z. Liu, T.-W. Wu, M. Pavon Valderrama, J.-J. Xie, and L.-S. Geng, Phys. Rev. D **99**, 094018 (2019)
[5] Z. M. Ding, H. Y. Jiang and J. He, Eur. Phys. J. C **80**, no.12, 1179 (2020)
[6] R. Molina, T. Branz, and E. Oset, Phys. Rev. D **82**, 014010 (2010)
[7] R. Aaij et al. (LHCb), Phys. Rev. Lett. **125**, 242001 (2020)
[8] R. Molina and E. Oset, Phys. Lett. B **811**, 135870 (2020)
[9] A. Feijoo, W. H. Liang and E. Oset, Phys. Rev. D **104**, 114015 (2021)
[10] L. R. Dai, R. Molina and E. Oset, Phys. Rev. D **105**, 016029 (2022)
[11] L. R. Dai, E. Oset, A. Feijoo, R. Molina, L. Roca, A. M. Torres and K. P. Khemchandani, Phys. Rev. D **105**, 074017 (2022)

https://doi.org/10.1051/epjconf/202227110002

Nature of excited Ξ baryons with threshold effects

Takuma Nishibuchi[1,*] and *Tetsuo* Hyodo[1,**]

[1]Department of Physics, Tokyo Metropolitan University, Hachioji 192-0397, Japan

Abstract. Spectroscopy of excited baryons with strangeness $S = -2$ is stimulated by recent experimental developments. Here we focus on the $\Xi(1620)$ which locates close to the $\bar{K}\Lambda$ threshold. To take into account the threshold effects, we construct the coupled-channels meson-baryon scattering amplitude where the $\Xi(1620)$ appears as a resonance. We demonstrate that the $\bar{K}\Lambda$ threshold effects distort the peak of the $\Xi(1620)$ resonance from the simple Breit-Wigner distribution.

1 Introduction

Recent progress in hadron spectroscopy enables us to find new resonances in the strange baryon sector. In fact, $\Lambda(1380)$, $\Lambda(2070)$, $\Lambda(2080)$ $\Lambda(2085)$, $\Sigma(2010)$, $\Omega(2012)$ are newly added in the listings by the Particle Data Group [1]. At the same time, theoretical frameworks have been elaborated to describe the excited hadrons as resonances in the scattering of hadrons [2]. Thus, we are now in a position to study the strange baryon resonances systematically based on the experimental data.

In the strangeness $S = -2$ sector, the Belle collaboration reported a new precise measurement of the $\pi\Xi$ spectrum in the $\Xi_c \to \pi\pi\Xi$ decay [3]. The obtained spectrum shows clear peaks of the $\Xi(1620)$ and $\Xi(1690)$ resonances. In Ref. [3], the mass and width of the $\Xi(1620)$ are found to be $M_R = 1610.4 \pm 6.0^{+6.1}_{-4.2}$ MeV and $\Gamma_R = 59.9 \pm 4.8^{+2.8}_{-7.1}$ MeV, respectively. The properties of the neutral $\Xi(1690)$ was determined by the WA89 collaboration [4] as $M_R = 1686 \pm 4$ MeV and $\Gamma_R = 10 \pm 6$ MeV which is adopted by the current PDG [1]. We however note that the peaks of the $\Xi(1620)$ and $\Xi(1690)$ appear just on top of the $\bar{K}\Lambda$ threshold ($\sim$ 1611 MeV) and $\bar{K}\Sigma$ threshold ($\sim$ 1689 MeV), respectively. It is known that the spectrum of a near-threshold resonance can be distorted by various kinematical effects at the threshold [5]. Thus, the properties of these Ξ resonances should be determined with proper treatment of the meson-baryon thresholds.

In this work, we examine the spectrum of the $\Xi(1620)$ resonance by employing the chiral unitary approach [6–9], which describes the coupled-channels meson-baryon scattering including the threshold effects. Based on the previous work on the $\Xi(1620)$ in Ref. [10], we construct a model which has a resonance pole of the $\Xi(1620)$ in the energy region reported in Ref. [3]. Through the comparison with the Breit-Wigner distribution, we discuss the effect of the $\bar{K}\Lambda$ threshold for the spectrum of the $\Xi(1620)$.

*e-mail: nishibuchi-takuma@ed.tmu.ac.jp
**e-mail: hyodo@tmu.ac.jp

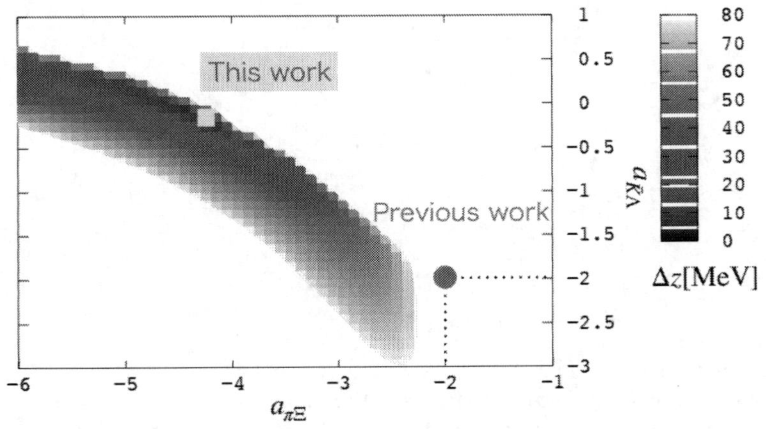

Figure 1. The density plot of the deviation of the pole position Δz in the $a_{\pi\Xi}$-$a_{\bar{K}\Lambda}$ plane.

2 Formulation

In the chiral unitary approach [6–8, 10], the coupled-channels s-wave meson-baryon scattering amplitude $T_{ij}(W)$ at total energy W is given by the solution of the scattering equation

$$T_{ij}(W) = V_{ij}(W) + \sum_k V_{ik}(W)G_k(W, a_k)T_{kj}(W), \tag{1}$$

where the indices i, j, k represent the meson-baryon channels. For the $\Xi(1620)$ with the strangeness $S = -2$ and isospin $I = 1/2$, there are four channels in the isospin basis, $\pi\Xi$, $\bar{K}\Lambda$, $\bar{K}\Sigma$, and $\eta\Xi$. For the interaction kernel $V_{ij}(W)$, we use the Weinberg-Tomozawa interaction which satisfies the chiral low-energy theorem. The divergence of the loop function $G_i(W, a_i)$ is tamed by the dimensional regularization, and the finite part is specified by the subtraction constant a_i in each channel.

Because the structure of the Weinberg-Tomozawa interaction is model-independently determined by chiral symmetry, four subtraction constants a_i in the loop function are the free parameters in this model. In Ref. [10], the subtraction constants are determined as $a_i = -2$ in all channels at the regularization scale $\mu = 630$ MeV, based on the natural size argument in Ref. [8]. The pole of the $\Xi(1620)$ is then found at $W = 1607 - 140i$ MeV, which corresponds to $M_R = 1607$ MeV and $\Gamma_R = 280$ MeV. In view of the new Belle result [3], which indicates the narrower decay width, it is necessary to update the model.

3 Numerical results

Here we aim at constructing a model which is consistent with the Belle data [3]. As a first trial, we assume that the pole of the $\Xi(1620)$ locates at $z_{\rm ex} = 1610 - 30i$ MeV motivated by the Belle result, and try to reproduce $z_{\rm ex}$ in the chiral unitary approach. The pole position of the scattering amplitude of chiral unitary approach $z_{\rm th}$ depends on the subtraction constants a_i. Starting from the model of Ref. [10], we vary the subtraction constants in the $\pi\Xi$ and $\bar{K}\Lambda$ channels, to which the pole position $z_{\rm th}$ is sensitive. To quantify the deviation of the pole

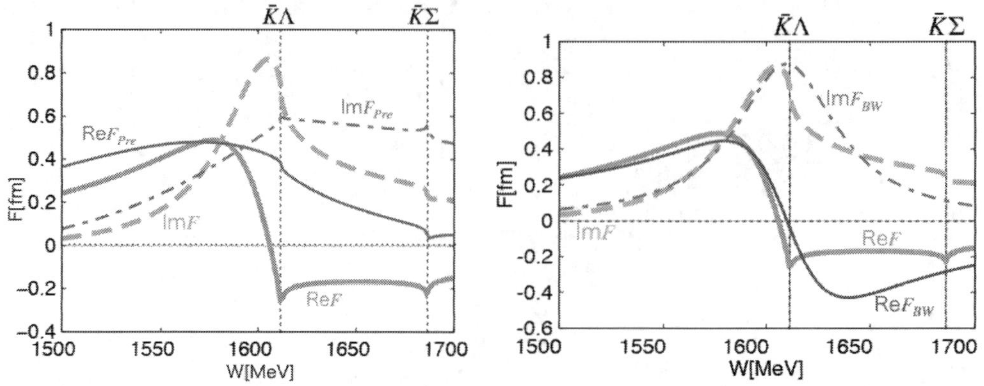

Figure 2. Left: A comparison of the $\pi\Xi$ scattering amplitude in this work F (thick lines) with the one from Ref. [10] F_{Pre} (thin lines). The real (imaginary) part of the scattering amplitude is denoted by the solid (dashed) lines. Vertical dotted lines indicate the $\bar{K}\Lambda$ and $\bar{K}\Sigma$ thresholds. Right: A comparison with the Breit-Wigner amplitude F_{BW}. The legends are the same as in the left panel.

positions, we define Δz as

$$\Delta z = |z_{\rm th} - z_{\rm ex}|. \tag{2}$$

In Fig. 1, we show the density plot of Δz in the $a_{\pi\Xi}$-$a_{\bar{K}\Lambda}$ plane. We find that Δz is minimized by $(a_{\pi\Xi}, a_{\bar{K}\Lambda}) = (-4.19, -0.14)$, which gives $z_{\rm th} = 1610 - 30i$ MeV in agreement with $z_{\rm ex}$.

Let us examine the behavior of the scattering amplitude on the real energy axis. In the left panel of Fig. 2, we show the real and imaginary parts of the scattering amplitude F in the diagonal $\pi\Xi$ channel as functions of the energy W, in comparison with the model of Ref. [10]. The amplitude of Ref. [10] (thin lines) does not exhibit a resonance behaviour (peak of the imaginary part and zero crossing of the real part) in this energy region, because of the broad decay width $\sim$ 280 MeV. In contrast, the amplitude in this work shows a clear resonance behaviour around 1600 MeV. Thus, by reducing the decay width, the $\pi\Xi$ spectrum (Im F) can exhibit a resonance peak, as observed by the Belle collaboration.

As mentioned in the introduction, the peak of the $\Xi(1620)$ is very close to the $\bar{K}\Lambda$ threshold. To examine the threshold effect, we compare the the scattering amplitude of our model with the amplitude generated by the Breit-Wigner distribution,

$$T^{\rm BW}(W) = \frac{Z}{W - M_R + i\Gamma_R/2}. \tag{3}$$

We set $M_R = 1610$ MeV and $\Gamma_R = 60$ MeV to have the same pole position as in our model. The residue Z is adjusted to scale the magnitude of the spectrum. In the right panel of Fig. 2, we present the comparison of our results with the Breit-Wigner amplitude. It can be seen from the figure that the peak of the imaginary part does not coincide with the Breit-Wigner distribution, even though the pole positions are the same. In fact, the peak of the imaginary part of our model is at $W \sim 1604$ MeV, which is shifted from the real part of the pole position ($W = 1610$ MeV). In this way, we quantitatively show that the $\bar{K}\Lambda$ threshold effect distorts the spectrum of the $\Xi(1620)$ resonance.

4 Summary

In this work, we study the spectrum of the $\Xi(1620)$ resonance by taking the nearby $\bar{K}\Lambda$ threshold into account. Within the framework of the chiral unitary approach, we construct a model of the $\Xi(1620)$ whose resonance pole appears in the energy region indicated by the recent Belle data. Through the comparison with the pure Breit-Wigner distribution, we find that the peak structure of the $\Xi(1620)$ is highly affected by the existence of the $\bar{K}\Lambda$ threshold. We conclude that the determination of the pole position of the $\Xi(1620)$ should be performed by properly taking into account the $\bar{K}\Lambda$ threshold.

As a future perspective, it will be interesting to study the origin of the $\Xi(1620)$ resonance in our model. In Ref. [11], it was shown that the change of the subtraction constants in the chiral unitary approach can be related to the inclusion of the genuine quark states. By applying the natural renormalization scheme, it is possible to extract the genuine quark component. A large deviation of the subtraction constants (Fig. 1) may indicate the importance of the genuine component. The $\bar{K}\Sigma$ threshold effect for the $\Xi(1690)$ resonance can also be studied in the same manner, by referring to the models in Refs. [12, 13] which reproduce the $\Xi(1690)$. For the $\Xi(1690)$, it should be noted that the central value of the $\Xi(1690)$ determined by the WA89 collaboration is just in between the $K^-\Sigma^+$ ($\sim$ 1683 MeV) and $\bar{K}^0\Sigma^0$ ($\sim$ 1690 MeV) thresholds. For the quantitative discussion of the $\Xi(1690)$, therefore, the isospin symmetry breaking effects should play an important role. To analyze the $\pi\Xi$ spectrum of the experimental data by the Belle collaboration, we need theoretical framework to calculate the $\Xi_c \to \pi\pi\Xi$ decay, as developed in Ref. [14]. These subjects will be investigated to understand the nature of the Ξ resonances in future.

References

[1] R.L. Workman (Particle Data Group), PTEP **2022**, 083C01 (2022)
[2] T. Hyodo, M. Niiyama, Prog. Part. Nucl. Phys. **120**, 103868 (2021)
[3] M. Sumihama et al. (Belle), Phys. Rev. Lett. **122**, 072501 (2019)
[4] M.I. Adamovich et al. (WA89), Eur. Phys. J. C **5**, 621 (1998)
[5] F.K. Guo, X.H. Liu, S. Sakai, Prog. Part. Nucl. Phys. **112**, 103757 (2020)
[6] N. Kaiser, P.B. Siegel, W. Weise, Nucl. Phys. A **594**, 325 (1995)
[7] E. Oset, A. Ramos, Nucl. Phys. A **635**, 99 (1998)
[8] J.A. Oller, U.G. Meissner, Phys. Lett. B **500**, 263 (2001)
[9] T. Hyodo, D. Jido, Prog. Part. Nucl. Phys. **67**, 55 (2012)
[10] A. Ramos, E. Oset, C. Bennhold, Phys. Rev. Lett. **89**, 252001 (2002)
[11] T. Hyodo, D. Jido, A. Hosaka, Phys. Rev. C **78**, 025203 (2008)
[12] C. Garcia-Recio, M.F.M. Lutz, J. Nieves, Phys. Lett. B **582**, 49 (2004)
[13] T. Sekihara, PTEP **2015**, 091D01 (2015)
[14] K. Miyahara, T. Hyodo, M. Oka, J. Nieves, E. Oset, Phys. Rev. C **95**, 035212 (2017)

EPJ Web of Conferences **271**, 10003 (2022) https://doi.org/10.1051/epjconf/202227110003
HYP2022

Estimation of compositeness with correction terms

Tomona Kinugawa[1,*] and *Tetsuo* Hyodo[1,**]

[1]Department of Physics, Tokyo Metropolitan University, Hachioji 192-0397, Japan

Abstract. The compositeness X is defined as the probability to observe the composite structure such as the hadronic molecule component in a bound state. One of the model-independent approaches to calculate X is the weak-binding relation. However, when the scattering length a_0 is larger than the radius of the bound state R, the central value of the compositeness X becomes larger than unity, which cannot be interpreted as a probability. For the systems with $a_0 > R$, we need to estimate the compositeness with the correction terms. For the reasonable determination of the compositeness, we first present the quantitative estimation of the correction terms. Because the exact value of the compositeness should be contained in its definition domain $0 \leq X \leq 1$, we propose the reasonable estimation method with the uncertainty band by excluding the region outside of the definition domain of the compositeness. We finally estimate the compositeness of physical systems, and obtain the result which we can interpret as the fraction of the composite component.

1 Introduction

Almost all hadrons are considered to be qqq or $q\bar{q}$ states in the constituent quark models. However, some hadrons are expected to have an extraordinary structure, and called exotic hadrons. In recent experiments in the heavy quark sector, candidates for exotic hadrons have been observed, as represented by the $X(3872)$ [1]. One possible component of the candidates for the exotic hadrons is the hadronic molecule, which is a weakly bound state of hadrons.

We can quantitatively characterize the internal structure of the state by compositeness whether it is a hadronic molecule dominant (composite dominant) state or not [2]. The compositeness X is defined as the probability to find the hadronic molecule component in the normalized wavefunction of the bound states $|\Psi\rangle$, $X = |\langle\text{molecule}|\Psi\rangle|^2$. Here $|\text{molecule}\rangle$ is the schematic notation of the hadronic molecule component. We can determine X by using the weak-binding relation [3, 4]:

$$a_0 = R\left\{\frac{2X}{1+X} + O\left(\frac{R_{\text{typ}}}{R}\right)\right\}, \tag{1}$$

where a_0 is the scattering length and $R \equiv 1/\sqrt{2\mu B}$ is the radius of the bound state, determined by the binding energy B and the reduced mass μ. Taking into account the range correction to

*e-mail: kinugawa-tomona@ed.tmu.ac.jp
**e-mail: hyodo@tmu.ac.jp

the weak-binding relation [5], we define $R_{\rm typ}$ as the largest one among the length scale of the interaction $R_{\rm int}$ and those in the effective range expansion except for a_0:

$$R_{\rm typ} = \max\{R_{\rm int}, |r_e|, |P_s/R^2|, \cdots\}, \tag{2}$$

where r_e is the effective range and P_s is the shape parameter (for more details, see Sec. III in Ref. [5]).

When we consider sufficiently shallow bound states with $R \gg R_{\rm typ}$, the correction terms of the weak-binding relation $O(R_{\rm typ}/R)$ are negligible, and the compositeness X is determined only from the observables a_0 and R. Thanks to this universal feature, the weak-binding relation has been utilized as a model-independent approach to calculate X. However, naive application of Eq. (1) without the correction terms sometimes contradicts the definition domain of the compositeness, $0 \leq X \leq 1$. For example, the compositeness of the deuteron d is given as $X = 1.68$ with $a_0 = 5.42$ fm (taken from CD-Bonn potential [6]) and $B = 2.22$ MeV (taken from PDG [7]). This problem is discussed in Refs. [8–10] in connection with the effective range. To avoid this contradiction, here we propose a reasonable estimation method of the compositeness with the uncertainty which arises from the correction terms $O(R_{\rm typ}/R)$ in Eq. (1).

2 Estimation of compositeness with correction terms

2.1 Importance of correction terms

Let us consider the relation of the scattering length a_0 and the radius R by neglecting the correction terms $O(R_{\rm typ}/R)$. Because X is the probability to find the composite component in a bound state, it is defined within $0 \leq X \leq 1$. It follows from this relation that $2X/(1 + X) \leq 1$. Therefore, to satisfy the weak-binding relation (1) without the correction terms $a_0 = R[2X/(1 + X)]$, R should be larger than a_0. However, there are some systems with $a_0 > R$ which give $X > 1$ as mentioned above. In such cases, we cannot interpret X as the probability. This problem originates in the assumption of neglecting the correction terms. For the systems with $a_0 > R$, it is expected that the weak-binding relation holds by taking into account the correction terms $O(R_{\rm typ}/R)$, because $2X/(1 + X) + O(R_{\rm typ}/R) > 1$ can be realized for $0 \leq X \leq 1$. Therefore, it is necessary to develop a quantitative estimation method of the correction terms to obtain $X \leq 1$ for the systems with $a_0 > R$.

2.2 Estimation of uncertainty band

From the discussion in Sec. 2.1, we propose the estimation method of the compositeness X with introducing the contribution from the correction terms $O(R_{\rm typ}/R)$. As discussed in Ref. [4], the correction terms $O(R_{\rm typ}/R)$ can be estimated quantitatively as the dimensionless quantity ξ:

$$\xi = \frac{R_{\rm typ}}{R}. \tag{3}$$

We then determine the upper and lower boundaries of the estimated compositeness X_u (X_l) as

$$X_u(\xi) = \frac{a_0/R + \xi}{2 - a_0/R - \xi}, \tag{4}$$

$$X_l(\xi) = \frac{a_0/R - \xi}{2 - a_0/R + \xi}, \tag{5}$$

for $0 \leq \xi \leq 1$. It is expected that the exact value of X is contained within $X_l \leq X \leq X_u$.

Numerically, X_u and X_l can go beyond the definition domain of the compositeness $0 \leq X \leq 1$, depending on the values of a_0, R and ξ. However, the results $X \geq 1$ and $X \leq 0$ do not make sense, because the exact value of X is not contained there. Therefore, we define

$$\bar{X}_u = \min\{X_u, 1\}, \quad \bar{X}_l = \max\{X_l, 0\}, \tag{6}$$

to restrict the uncertainty band of the compositeness within the definition domain of X:

$$\bar{X}_l \leq X \leq \bar{X}_u, \tag{7}$$

as illustrated in Fig. 1. We regard this uncertainty band (7) as the estimated compositeness and discuss the internal structure of the bound state with it. It is clear that the estimated compositeness with the uncertainty band (7) is restricted within $0 \leq X \leq 1$, and we can interpret X as the probability. More details about the estimation of X are discussed in Sec. III and IV in Ref. [5].

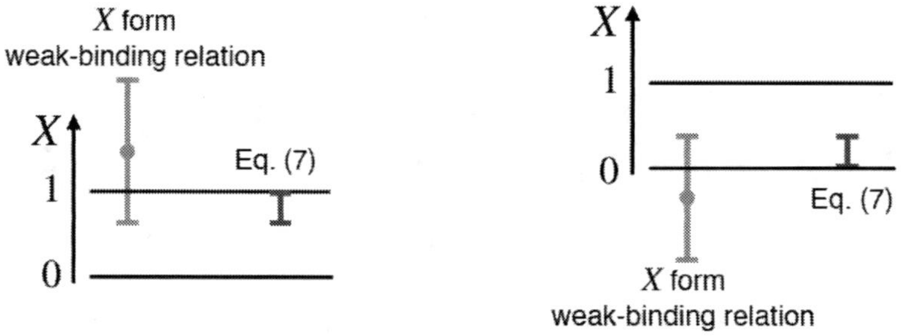

Figure 1. Schematic illustration of the definition of the uncertainty band (7). The left panel shows the case for $X_u > 1$ ($\bar{X}_u = 1$), and the right shows that for $X_l < 0$ ($\bar{X}_l = 0$).

3 Application to physical systems

Now we estimate the compositeness X of the actual physical systems with the uncertainty estimation discussed in Sec. 2.2. We consider the deuteron, $X(3872)$, $D^*_{s0}(2317)$, $D_{s1}(2460)$, $N\Omega$ dibaryon, $\Omega\Omega$ dibaryon, $^3_\Lambda$H, and ^{4}He dimer. The deuteron d in the p-n scattering is chosen as the typical observed hadron. $X(3872)$ in the D^0-$\bar{D}^{*0}$ scattering, $D^*_{s0}(2317)$ in the D-K scattering, and $D_{s1}(2460)$ in the D^*-K scattering are the candidates for the exotic hadrons which are experimentally observed [7]. $N\Omega$ and $\Omega\Omega$ dibaryons are the states obtained by the lattice QCD calculation [11, 12]. We can apply the weak-binding relation not only to the hadron systems but also to the nuclei and atomic systems. $^3_\Lambda$H in the Λ-d scattering is an example of nuclei, and ^{4}He dimer which is the weakly bound state of ^{4}He atoms is an example in the atomic systems.

For the estimation of X from the weak-binding relation, we need the scattering length a_0, the reduced mass μ, the binding energy B, the effective range r_e, and the interaction range $R_{\rm int}$. The radius of the bound state is calculated by $R = 1/\sqrt{2\mu B}$. We tabulate relevant quantities in Tab. 1. We note that $R_{\rm int}$ is not an observable, and therefore it is determined from the theoretical consideration. The procedure to determine these physical quantities is explained in Ref. [5].

Table 1. The physical quantities and the compositeness X with the uncertainty band (7). u, mK and B.R. stand for the atomic mass unit, millikelvin and the Bohr radius.

bound state	B	a_0	r_e	$R_{\rm int}$	Compositeness X
d	2.22 MeV	5.42 fm	1.75 fm	1.43 fm	$0.74 \leq X \leq 1$
$X(3872)$	0.018 MeV	28.5 fm	−5.34 fm	1.43 fm	$0.53 \leq X \leq 1$
$D^*_{s0}(2317)$	44.8 MeV	1.3 fm	−0.1 fm	0.359 fm	$0.81 \leq X \leq 1$
$D_{s1}(2460)$	45.1 MeV	1.1 fm	−0.2 fm	0.359 fm	$0.55 \leq X \leq 1$
$N\Omega$ dibaryon	1.54 MeV	5.30 fm	1.26 fm	0.676 fm	$0.80 \leq X \leq 1$
$\Omega\Omega$ dibaryon	1.6 MeV	4.6 fm	1.27 fm	0.949 fm	$0.79 \leq X \leq 1$
${}^3_\Lambda$H	0.13 MeV	16.8 fm	2.3 fm	4.32 fm	$0.74 \leq X \leq 1$
^{4}He dimer	1.30 mK	189 B.R.	13.8 B.R.	10.2 B.R.	$0.93 \leq X \leq 1$

The results of the estimated compositeness with the uncertainty band (7) are shown in the right column in Tab. 1. It is found that the range correction is important for the application to the $X(3872)$ and the $N\Omega$ dibaryon [5]. We find that those bound states are dominated by the composite component because the lower boundaries $\bar{X}_l$ are larger than 0.5.

4 Summary

The compositeness X characterizes the internal structure of shallow bound states, especially for the candidates for exotic hadrons. The weak-binding relation is one of the approaches to estimate X. When we neglect the correction terms $O(R_{\rm typ}/R)$, the weak-binding relation becomes completely model-independent. However, if the scattering length a_0 is larger than the radius of the bound state R, the compositeness is overestimated as $X \geq 1$ without the correction terms. To avoid this problem, we discuss the method to evaluate the correction terms $O(R_{\rm typ}/R)$. We propose the estimation method of X with the uncertainty band, which includes the contribution of the correction terms $O(R_{\rm typ}/R)$. Our uncertainty estimation provides the compositeness in $0 \leq X \leq 1$ which can be interpreted as a probability. We finally perform reasonable estimations of X as shown in Tab. 1, and find that all states which we consider are composite dominant ($\bar{X}_l \geq 0.5$).

References

[1] S.K. Choi et al. (Belle), Phys. Rev. Lett. **91**, 262001 (2003)
[2] T. Hyodo, Int. J. Mod. Phys. A **28**, 1330045 (2013)
[3] S. Weinberg, Phys. Rev. **137**, B672 (1965)
[4] Y. Kamiya, T. Hyodo, PTEP **2017**, 023 (2017)
[5] T. Kinugawa, T. Hyodo, Phys. Rev. C **106**, 015205 (2022)
[6] R. Machleidt, Phys. Rev. C **63**, 024001 (2001)
[7] P.A. Zyla et al. (Particle Data Group), PTEP **2020**, 083C01 (2020)
[8] Y. Li, F.K. Guo, J.Y. Pang, J.J. Wu, Phys. Rev. D **105**, L071502 (2022)
[9] J. Song, L.R. Dai, E. Oset, Eur. Phys. J. A **58**, 133 (2022)
[10] M. Albaladejo, J. Nieves, Eur. Phys. J. C **82**, 724 (2022)
[11] T. Iritani et al. (HAL QCD), Phys. Lett. B **792**, 284 (2019)
[12] S. Gongyo et al., Phys. Rev. Lett. **120**, 212001 (2018)

EPJ Web of Conferences **271**, 10004 (2022) https://doi.org/10.1051/epjconf/202227110004
HYP2022

Analysis of coupled-channel potentials with quark and hadron degrees of freedom

Ibuki Terashima[1,*] and *Tetsuo* Hyodo[1,**]

[1]Department of Physics, Tokyo Metropolitan University, Hachioji 192-0397, Japan

Abstract. The quark-antiquark potentials are known to be confining in the absence of the $\bar{q}q$ pair creation. On the other hand, the inter-hadron potentials vanish at large distance, because the interaction range is limited by the inverse pion mass. When the $\bar{q}q$ pair creation and annihilation are switched on, the channel coupling occurs between the quark and hadron degrees of freedom, modifying the behavior of potentials. In this work, we investigate the properties of the effective potentials obtained by eliminating one of the channels. We show that the coupling to the eliminated channel induces a non-local and energy dependent effective potential, irrespective of the properties of the transition potential. In addition, when the hadron channel having continuous scattering eigenstates is eliminated, the resulting inter-quark potential contains an imaginary part above the threshold to describe the decay effect into the meson pair.

1 Introduction

Potential descriptions in the strong interaction phenomena are sometimes useful, even though the fundamental theory of QCD is formulated as a quantum field theory. For instance, quark-antiquark ($\bar{Q}Q$) potentials in the static limit can be used to study the color confinement [1], and the nuclear force potentials are the basis to study various few-body and many-body physics of the atomic nuclei [2]. Recent developments of lattice QCD calculations [3] enable us to study these potentials from the first principles.

The potentials with the quark degrees of freedom and those with hadron degrees of freedom are related with each other, through the creation/annihilation of a $\bar{q}q$ pair. In fact, the static $\bar{Q}Q$ potential shows the "string breaking" indicated by the flattening of the potential at large distance [4]. This behavior should be qualitatively modified for the case with finite quark masses, because of the existence of the continuum of the meson-meson states. The lattice QCD studies of the $\bar{Q}Q$ potential with finite quark masses have been performed in Refs. [5], but the effect of the meson-meson thresholds is not well understood.

In this work, we consider the channel coupling between the quark and hadron degrees of freedom. By constructing a coupled-channel potential model, we discuss the effect of the meson-meson states for the $\bar{Q}Q$ potential with finite quark masses. At the same time, our formulation clarifies the effect of the quark dynamics in the hadron potentials. Such analysis will be important to understand the nature of the H dibaryon in the strangeness $S = -2$ baryon-baryon scattering, where the baryon-baryon potential may contain the effect of the six-quark state generated by the quark potential.

*e-mail: terashima-ibuki@ed.tmu.ac.jp
**e-mail: hyodo@tmu.ac.jp

2 Formulation

For concreteness, we consider the coupled system of the quark-antiquark channel (denoted as $\bar{c}c$) and hadron-hadron channel ($\bar{D}D$) described by the Schrödinger equation

$$H\,|\Psi\rangle = E\,|\Psi\rangle, \quad H = \begin{pmatrix} T^{\bar{c}c} & 0 \\ 0 & T^{\bar{D}D} + \Delta \end{pmatrix} + \begin{pmatrix} V^{\rm conf} & V^t \\ V^t & V^{\bar{D}D} \end{pmatrix}, \quad |\Psi\rangle = \begin{pmatrix} |\psi^{\bar{c}c}\rangle \\ |\psi^{\bar{D}D}\rangle \end{pmatrix}, \tag{1}$$

where $T^{\bar{c}c}$ ($T^{\bar{D}D}$) is the kinetic energy of the $\bar{c}c$ ($\bar{D}D$) channel and Δ is the threshold energy of $\bar{D}D$. Here all the potentials are assumed to be local ones, i.e., $\langle \boldsymbol{r}'|\,V\,|\boldsymbol{r}\rangle = V(\boldsymbol{r})\delta(\boldsymbol{r}'-\boldsymbol{r})$. The diagonal potential in the $\bar{c}c$ channel $V^{\rm conf}$ is confining, and the $\bar{D}D$ potential $V^{\bar{D}D}$ vanishes at large distance. The channel coupling is induced by the transition potential V^t.

Following the Feshbach method [6], we can construct the single-channel effective Hamiltonian for the $\bar{c}c$ channel $H^{\bar{c}c}_{\rm eff}(E)$ as

$$H^{\bar{c}c}_{\rm eff}(E) = T^{\bar{c}c} + V^{\rm conf} + V^t(E - T^{\bar{D}D} - \Delta - V^{\bar{D}D} + i0^+)^{-1}V^t \equiv T^{\bar{c}c} + V^{\bar{c}c}_{\rm eff}(E), \tag{2}$$

with $H^{\bar{c}c}_{\rm eff}(E)\,|\psi^{\bar{c}c}\rangle = E\,|\psi^{\bar{c}c}\rangle$. The coupling with the $\bar{D}D$ channel is renormalized in the effective potential $V^{\bar{c}c}_{\rm eff}(E)$. We derive, in the same way, the effective Hamiltonian for $\bar{D}D$ as

$$H^{\bar{D}D}_{\rm eff}(E) = T^{\bar{D}D} + \Delta + V^{\bar{D}D} + V^t(E - T^{\bar{c}c} - V^{\rm conf} + i0^+)^{-1}V^t \equiv T^{\bar{D}D} + \Delta + V^{\bar{D}D}_{\rm eff}(E), \tag{3}$$

with $H^{\bar{D}D}_{\rm eff}(E)\,|\psi^{\bar{D}D}\rangle = E\,|\psi^{\bar{D}D}\rangle$. By solving the single-channel effective Schrödinger equations in a self-consistent manner for the energy E, we obtain the results equivalent to the original coupled-channel equation (1).

3 Results

We first investigate the coordinate representation of the effective potential of the $\bar{D}D$ channel,

$$\langle \boldsymbol{r}'_{\bar{D}D}|\,V^{\bar{D}D}_{\rm eff}(E)\,|\boldsymbol{r}_{\bar{D}D}\rangle = V^{\bar{D}D}(\boldsymbol{r})\delta(\boldsymbol{r}'-\boldsymbol{r}) + \sum_n \frac{\langle \boldsymbol{r}'_{\bar{D}D}|\,V^t\,|\phi_n\rangle\,\langle\phi_n|\,V^t\,|\boldsymbol{r}_{\bar{D}D}\rangle}{E - E_n}, \tag{4}$$

where $|\phi_n\rangle$ and E_n respectively represent the eigenstate and eigenenergy of the $\bar{c}c$ channel without the channel coupling, $(T^{\bar{c}c} + V^{\rm conf})\,|\phi_n\rangle = E_n\,|\phi_n\rangle$. Because of the confinement, $|\phi_n\rangle$ consists only of the discrete eigenstates. The second term shows that the effective potential is a non-local one, because the $\boldsymbol{r}$ dependence is factorized from that of $\boldsymbol{r}'$. This means that a non-local potential is induced by the channel coupling regardless of the property of the transition potential V^t. In addition, the denominator of the second term shows that the effective potential depends on the energy E. It can also be seen that the potential strength diverges at $E = E_n$ when the energy coincides with an eigenenergy of $|\phi_n\rangle$.

Next, we consider the effective $\bar{c}c$ potential. Here we assume that the $\bar{D}D$ potential $V^{\bar{D}D}$ does not support any bound states, so that the eigenstates without the channel coupling are given only by the continuum of the scattering states labeled by $\boldsymbol{p}$: $(T^{\bar{D}D} - \Delta - V^{\bar{D}D})\,|\boldsymbol{p}\rangle = E_{\boldsymbol{p}}\,|\boldsymbol{p}\rangle$. The lowest energy state with $\boldsymbol{p} = \boldsymbol{0}$ appears at the threshed, $E_{\boldsymbol{0}} = \Delta$. In this case, the coordinate representation of the effective $\bar{c}c$ potential is

$$\langle \boldsymbol{r}'_{\bar{c}c}|\,V^{\bar{c}c}_{\rm eff}(E)\,|\boldsymbol{r}_{\bar{c}c}\rangle = V^{\rm conf}(\boldsymbol{r})\delta(\boldsymbol{r}'-\boldsymbol{r}) + \int d\boldsymbol{p}\frac{\langle \boldsymbol{r}'_{\bar{c}c}|\,V^t\,|\boldsymbol{p}\rangle\,\langle\boldsymbol{p}|\,V^t\,|\boldsymbol{r}_{\bar{c}c}\rangle}{E - E_{\boldsymbol{p}} + i0^+}. \tag{5}$$

As in Eq. (4), the effective potential is non-local and energy-dependent. For the energy $E \geq \Delta$, the potential has an imaginary part due to the pole at $E = E_{\boldsymbol{p}}$, which is responsible for the opening of the $\bar{D}D$ threshold, above which the decay process $\bar{c}c \to \bar{D}D$ occurs.

In the absence of the diagonal $\bar{D}D$ potential ($V^{\bar{D}D} = 0$), we can perform the integration in Eq. (5) analytically to obtain

$$\langle \boldsymbol{r}'_{\bar{c}c} | V^{\bar{c}c}_{\text{eff}}(E) | \boldsymbol{r}_{\bar{c}c} \rangle = V^{\text{conf}}(\boldsymbol{r})\delta(\boldsymbol{r}' - \boldsymbol{r}) - V^t(\boldsymbol{r}')V^t(\boldsymbol{r}) \frac{m_D \exp[-\sqrt{m_D(\Delta - E) - i0^+}|\boldsymbol{r}' - \boldsymbol{r}|]}{4\pi|\boldsymbol{r}' - \boldsymbol{r}|}, \quad (6)$$

with m_D being the mass of the D meson. In this expression, the appearance of the imaginary part for $E \geq \Delta$ is represented as $\sqrt{m_D(\Delta - E) - i0^+} = -i\sqrt{m_D(E - \Delta)}$. Because of the $V^t(\boldsymbol{r}')V^t(\boldsymbol{r})$ factor, this is still a non-local potential. By performing the derivative expansion, we obtain the local approximation

$$\langle \boldsymbol{r}'_{\bar{c}c} | V^{\bar{c}c}_{\text{eff}}(E) | \boldsymbol{r}_{\bar{c}c} \rangle = V^{\text{conf}}(\boldsymbol{r})\delta(\boldsymbol{r}' - \boldsymbol{r}) + \frac{[V^t(\boldsymbol{r})]^2}{E - \Delta}\delta(\boldsymbol{r}' - \boldsymbol{r}) + O(\nabla^2). \quad (7)$$

This expression, however, has no imaginary part for $E \geq \Delta$, in contrast to Eq. (6). We thus find that a naive derivative expansion of the non-local potential may break the physical property of the potential.

4 Conclusion and future outlook

We have discussed the effect of the transition between the quark channel with a confining potential and the hadron channel with a scattering potential. It is shown that the channel coupling introduces the non-locality and energy-dependence in the effective single-channel potentials. We find that the coupling to the hadron channel with continuous eigenstates generates the imaginary part of the effective potential above the threshold, which however, disappears when the formal derivative expansion is applied to obtain a local potential. This indicates that the construction of a local effective potential should be performed with care.

As a future outlook, it will be interesting to examine various local approximations other than the formal derivative expansion. For instance, an alternative method to obtain a local potential is proposed in Ref. [7]. Another direction is to apply the present formulation to the actual hadron systems. The charmonium like $X(3872)$ can be described as the $D\bar{D}^*$ scattering coupled with the $\bar{c}c$ core, which is a suitable system for an application.

References

[1] K.G. Wilson, Phys. Rev. D **10**, 2445 (1974)
[2] E. Epelbaum, H.W. Hammer, U.G. Meissner, Rev. Mod. Phys. **81**, 1773 (2009)
[3] G.S. Bali, Phys. Rept. **343**, 1 (2001); N. Ishii, S. Aoki, T. Hatsuda, Phys. Rev. Lett. **99**, 022001 (2007)
[4] G.S. Bali, H. Neff, T. Duessel, T. Lippert, K. Schilling (SESAM), Phys. Rev. D **71**, 114513 (2005)
[5] Y. Ikeda, H. Iida, Prog. Theor. Phys. **128**, 941 (2012); T. Kawanai, S. Sasaki, Phys. Rev. Lett. **107**, 091601 (2011); K. Nochi, T. Kawanai, S. Sasaki, Phys. Rev. D **94**, 114514 (2016)
[6] H. Feshbach, Ann. Phys. **5**, 357 (1958); ibid, **19**, 287 (1962)
[7] S. Aoki, K. Yazaki, PTEP **2022**, 033B04 (2022)

J-PARC hadron experimental facility extension project *

Taskforce on the extension of the Hadron Experimental Facility, *Fuminori* Sakuma[1,**], *Kazuya* Aoki[2], *Hiroyuki* Fujioka[3], *Toshiyuki* Gogami[4], *Yoshimasa* Hidaka[2,5,6], *Emiko* Hiyama[7], *Ryotaro* Honda[2], *Atsushi* Hosaka[8,9,10], *Yudai* Ichikawa[9], *Masaharu* Ieiri[2], *Masahiro* Isaka[11], *Noriyoshi* Ishii[8], *Takatsugu* Ishikawa[12], *Yusuke* Komatsu[2], *Takeshi* Komatsubara[2], *GeiYoub* Lim[2], *Koji* Miwa[7], *Yuhei* Morino[2], *Tomofumi* Nagae[4], *Sho* Nagao[7], *Satoshi N.* Nakamura[13], *Hajime* Nanjo[14], *Megumi* Naruki[4], *Hidekatsu* Nemura[8], *Tadashi* Nomura[2], *Hiroyuki* Noumi[8,2], *Hiroaki* Ohnishi[12], *Kyoichiro* Ozawa[2], *Shinya* Sawada[2], *Takayasu* Sekihara[15], *Sang-In* Shim[8], *Koji* Shiomi[2], *Kotaro* Shirotori[8], *Yasuhisa* Tajima[16], *Hitoshi* Takahashi[2], *Toshiyuki* Takahashi[2], *Sachiko* Takeuchi[10,17], *Makoto* Takizawa[18,2,10], *Hirokazu* Tamura[7,9], *Kiyoshi* Tanida[9], *Mifuyu* Ukai[2], *Takeshi O.* Yamamoto[9], and *Yasuo* Yamamoto[10]

[1]RIKEN Cluster for Pioneering Research, RIKEN, Wako 351-0198, Japan
[2]Institute of Particle and Nuclear Studies(IPNS), High Energy Accelerator Research Organization (KEK), Tsukuba 305-0801, Japan
[3]Tokyo Institute of Technology, Tokyo 152-8551, Japan
[4]Kyoto University, Kyoto 606-8502, Japan
[5]Graduate University for Advanced Studies (Sokendai), Tsukuba 305-0801, Japan
[6]RIKEN iTHEMS, RIKEN, Wako 351-0198, Japan
[7]Tohoku University, Sendai 980-8578, Japan
[8]Research Center for Nuclear Physics (RCNP), Osaka University, Ibaraki 567-0047, Japan
[9]Advanced Science Research Center (ASRC), Japan Atomic Energy Agency (JAEA), Tokai 319-1195, Japan
[10]RIKEN Nishina Center, RIKEN, Wako 351-0198, Japan
[11]Hosei University, Tokyo 102-8160, Japan
[12]Research Center for Electron Photon Science (ELPH), Tohoku University, Sendai 982-0826, Japan
[13]The University of Tokyo, Tokyo 113-0033, Japan
[14]Osaka University, Toyonaka 560-0043, Japan
[15]Kyoto Prefectural University, Kyoto 606-8522, Japan
[16]Yamagata University, Yamagata 990-8560, Japan
[17]Japan College of Social Work, Kiyose 204-8555, Japan
[18]Showa Pharmaceutical University, Machida 194-8543, Japan

Abstract. The J-PARC Hadron Experimental Facility was constructed with an aim to explore the origin and evolution of matter in the universe through experiments with intense particle beams. In the past decade, many results from particle and nuclear physics experiments have been obtained at the present facility. To expand the physics programs to as yet unexplored regions, the extension project of the Hadron Experimental Facility has been extensively discussed. This contribution presents the physics of the extension of the Hadron Experimental Facility to resolve issues related to strangeness nuclear physics, hadron physics, and flavor physics.

*Excerpted from "Extension of the J-PARC Hadron Experimental Facility - Third White Paper -" in arXiv:2110.04462 [nucl-ex] [1]

**e-mail: sakuma@ribf.riken.jp

1 Introduction

The Japan Proton Accelerator Research Complex (J-PARC) is a multi-purpose accelerator facility located in Tokai village, Japan. The aim of J-PARC is to promote a variety of scientific research programs ranging from the basic science of particle, nuclear, atomic, and condensed-matter physics and life science to industrial application and future nuclear transmutation using various types of intense particle beams.

The Hadron Experimental Facility focuses on particle and nuclear physics to explore the origin and evolution of matter in the universe, using the primary 30 GeV proton beam and secondary beams of pions, kaons, and muons. With the intense hadron beams, three core lines of research – strangeness nuclear physics, hadron physics, and flavor physics – have been conducted to approach the open questions in the universe.

2 Present status of J-PARC hadron experimental facility

A low-momentum charged-kaon beam line (K1.8/K1.8BR), a neutral-kaon beam line (KL), and a primary proton beam line (high-p) are being operated at the present Hadron Experimental Facility. A new primary beam line for a muon-to-electron conversion experiment (COMET) will also be ready for operation in 2023.

Primary protons are slowly extracted from the main ring accelerator (MR) and transported to the experimental hall (MR-SX operation) [2]. Kaons produced at the primary target (T1) are extracted to each secondary kaon beam line. The primary protons are also delivered to the high-p and COMET beam lines by branching off the protons in the switchyard upstream of the T1 target. As of June 2021, a beam power of 64.5 kW was achieved with a 2.0 s beam duration in a 5.2 repetition cycle, which corresponds to $7.0{\times}10^{13}$ protons per pulse.

Since the first delivery of the proton beam to the Hadron Experimental Facility in January 2009, many experiments have been conducted and fruitful results have been obtained. The nuclear physics programs at the Hadron Experimental Facility are summarized in Ref [3].

3 Extension project at J-PARC hadron experimental facility

To expand programs for particle and nuclear physics to as yet unexplored regions, more beam lines are necessary. In the present Hadron Experimental Facility, however, a single production target is in place and is shared with a limited number of secondary beam lines due to the limited space of the present hall. The extension of the facility by the addition of more production targets and the installation of new secondary beam lines will substantially expand research opportunities.

In the extension project, the construction of the following new beam lines will expand the potential for particle and nuclear physics programs at J-PARC.

- **HIHR beam line**: High-intensity, high-resolution (HIHR) charged π meson beam line for high-precision spectroscopy of Λ-hypernuclei. The beam line will adopt state-of-the-art technology using dispersion-matching, and will enable operation at more than 10^8 pions / spill up to 2 GeV/c with a beam momentum resolution of $\delta p/p \sim 10^{-4}$. A $(\pi^{\pm}, K^{+})$ missing-mass resolution of a few 100 keV (FWHM) will be achieved, corresponding to Λ-hypernuclei mass determination with several tens keV accuracy, which has never been realized to date.

- **K10 beam line**: High-momentum (2–10 GeV/c) charged K meson and anti-proton beam line for investigations of $S = -2$ and $S = -3$ strangeness physics and charm physics. The beam line will provide separated secondary beams with higher momentum than any existing beam lines. High-momentum particle separation will be realized by two-stage RF separators.
- **K1.1 beam line**: Low-momentum (<1.2 GeV/c) charged K meson beam line optimized for investigations of $S = -1$ strangeness physics. High-purity, high-intensity secondary particles will be available by the use of two-stage electrostatic separators. A branched beam line (K1.1BR) focusing on experiments using stopped kaons will also be prepared.
- **KL2 beam line**: High-intensity neutral K_L meson beam line dedicated to measuring the branching ratio of the rare decay $K_L \to \pi^0 \nu\bar{\nu}$. An extraction angle of 5 degrees will be adopted, instead of the value of 16 degrees used in the existing beam line to increase the K_L yield while maintaining the ratio of kaons and neutrons which could become a source of background events.

Figure 1 shows the layout of the extended Hadron Experimental Facility together with the present facility. The size of the experimental area will be twice as large and a new production target (T2) will be implemented. The new beam lines will be connected from the T2 target station. Construction of a test beam line is planned in the experimental area that has been used for the KL experiment. In the extended experimental hall, five beam lines (plus the test beam line) will be operated simultaneously. For further details, please refer to Ref [1].

Construction of the extended hadron hall will take 6 years in total, including a period of 2.5 years of suspending beam delivery to the existing beam lines. The project was selected as top priority, to be budgeted in the mid-term plan (FY2022-26) of of The High Energy Accelerator Research Organization (KEK) in KEK-PIP2022 (Project Implementation Plan) [4]. Funding of the extension project will start in FY2024 at the earliest, and then the expansion programs with new beam lines will start from around the end of 2029.

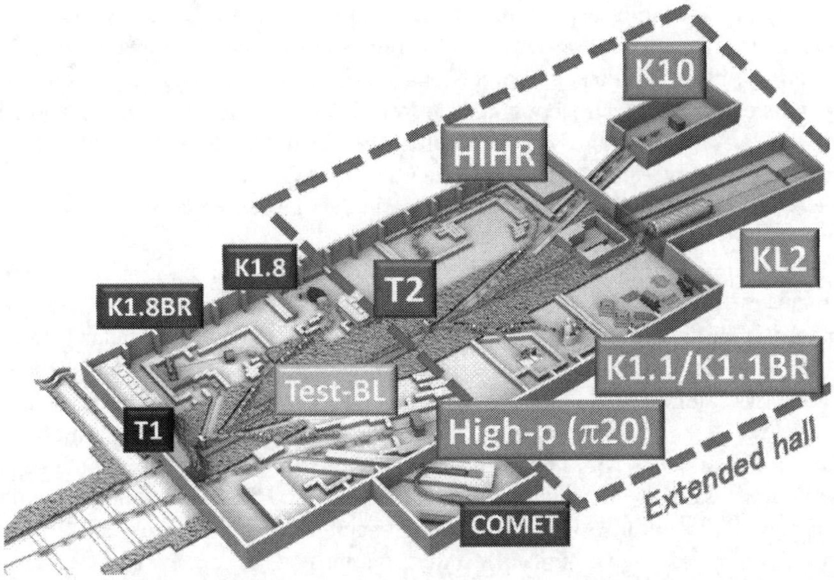

Figure 1. Layout of the extended Hadron Experimental Facility. The T2 production target and the HIHR, K1.1/K1.1BR, KL2, and K10 beam lines will be newly constructed. In addition, the π20 beam line and the test beam line are expected to be realized in the early stage of the project.

4 Scientific goals for extension project

The scientific goals and expected achievements for the extension project are summarized for each subject in this section together with the present situation: what is the problem to be solved, what has been achieved at the present facility to date, and how to approach the goal in the new project.

4.1 Strangeness nuclear physics - microscopic elucidation of neutron star matter through solving the hyperon puzzle

The so-called "hyperon puzzle" refers to the difficulty in reconciling astronomical observations of two-solar-mass neutron stars [5, 6] with the presence of hyperons in their interiors as predicted by nuclear physics; the presence of hyperons makes the equation of state (EOS) softer, so that the maximum mass of neutron stars is incompatible with observations. The solution to this problem requires a mechanism that provides an additional repulsion between baryons to make the EOS stiffer. Such additional repulsion could be described by two-body baryon-baryon interactions and three-body baryon-baryon-baryon interactions including hyperons, which produce a strong effect in dense nuclear matter. Much more comprehensive information on hyperon-nucleon (YN) and hyperon-hyperon (YY) interactions both in free space and in the nuclear medium are required. To determine the strength of hyperonic three-body repulsive forces, it is vital to precisely measure the Λ binding energy (B_Λ) for Λ-hypernuclei in a wide mass-number region. Information on the density-dependent ΛN interaction can be obtained from this binding energy.

Experiments on hypernuclei have been proposed and performed at the Hadron Experimental Facility to determine the strength of baryon-baryon interactions extended to the strangeness sector. The $\Lambda\Lambda$ and ΞN interactions have been and will be continued to be investigated using the world's most accurate measurements of $\Xi/\Lambda\Lambda$ hypernuclei at the existing **K1.8** beam line based on the (K^-, K^+) reaction. ΛN interactions in free space will be precisely determined by Λp scattering experiments planned at the new **K1.1** and **high-p** beam lines. The strength of ΛN-ΣN coupling will also be precisely obtained by spectroscopic studies of light neutron-rich hypernuclei and precision measurements of charge symmetry-breaking in Λ hypernuclei. γ-ray spectroscopy of Λ hypernuclei is a powerful tool to precisely determine low-lying level structure of hypernuclei. Differential cross-section measurements of $\Sigma^- p \to \Lambda n$ and $\Lambda p \to \Sigma^0 p$ scattering are essential for determining the ΛN-ΣN coupling in free space because this coupling could be different in the nuclear medium. These experiments have been and will be performed at K1.8 and K1.1.

However, to obtain information on the hyperonic three-body repulsive force, innovative measurements of Λ hypernuclei with unprecedented precision in a wide mass-number region are required. Such three-body effects affect the Λ binding energy, particularly in heavy Λ hypernuclei. The effect is expected to be from a few hundred keV to a few MeV, depending on the two-body ΛN interaction. In heavy Λ hypernuclei, Λ couples to core nuclei with various one-hole excited states with a small energy difference between each other, which results in several hypernuclear states with narrow energy spacings of typically a few hundred keV. A past experiment at KEK could not separate these states due to the limited energy resolution of ~2 MeV (FWHM). Therefore, a new spectroscopic method should be introduced to determine the Λ binding energies by separating each state and to provide accurate information including the ΛNN three-body effect. New experiments proposed at the newly constructed **HIHR** beam line will resolve this problem. The binding energy will be determined with a resolution of a few hundred keV (FWHM) for light ($^4_\Lambda$He) to heavy ($^{209}_{\Lambda}$Pb) hypernuclei, using (π, K^+) missing-mass spectroscopy with high-intensity pion beams and thin nuclear targets.

Systematic high-precision spectroscopy of light to heavy Λ hypernuclei provides essential data for determining the repulsive hyperonic three-body force effect through theoretical studies based on realistic ΛN interactions determined from Λp scattering experiments at K1.1. γ-ray spectroscopy provides additional information for determining the level structure of Λ hypernuclei, particularly for the heavy Λ hypernuclei where the level spacing could be narrower than the HIHR resolution. The γ-ray information will also be used to interpret the binding energy spectrum obtained by HIHR measurements. Density-dependent ΛN interaction effects also appear in the mass-number dependence of the level spacing between the s_Λ and p_Λ states because the average nuclear density in each orbital is expected to be different for different mass numbers. The level spacing between s_Λ and p_Λ will be determined with an accuracy of a few keV from γ-ray spectroscopy using germanium detectors.

Only after combining realistic YN and YY interactions with the hyperonic three-body force strength obtained from comprehensive studies in hypernuclear physics, a realistic EOS can be established to solve the hyperon puzzle.

4.2 Hadron physics - revelation of baryon structure built by quarks and gluons through spectroscopic studies of strange and charmed baryons

Quantum chromodynamics (QCD) has been successful at describing the interactions between quarks and gluons, and hadron properties. However, low-energy phenomena such as the formation of hadrons have not yet been clearly explained, because perturbative calculations do not work in the low-energy regime.

At the present Hadron Experimental Facility, experimental approaches to investigate the in-medium properties of mesons are proceeding. These approaches are expected to provide crucial information on the spontaneous breaking of chiral symmetry associated with the mass generation of hadrons, which is the most important feature of QCD at low energy. Various investigations of in-medium meson properties have been and will be performed, such as precise measurements of the spectral functions of vector mesons in media at **High-p** and systematic measurements of light kaonic nuclei at **K1.8BR**.

Baryon spectroscopy is another important approach for investigating how QCD works at low energy, and will be explored at J-PARC. In particular, the fundamental question "how quarks build hadrons" has not yet been answered clearly. It is vital to understand the dynamics of the effective degrees of freedom, which are constituent quarks and Nambu-Goldstone bosons, brought by the non-trivial gluon field at low energy. The spectroscopy of baryons with heavy flavors provides a good opportunity to investigate interactions and correlations of the effective degrees of freedom in hadrons. In excited baryons containing a heavy quark, the correlation between a quark pair inside the baryon is expected to be enhanced, and this is referred to as diquark correlation. There are longstanding arguments about whether diquark correlations play a role in the baryon structure and are related to diquark condensation in highly dense quark matter.

At the energy of J-PARC, excited baryons with a charm quark or multi-strange quarks are appropriate targets for study. An experiment to investigate diquark correlation in charmed baryons (denoted as Y_c^*) is planned at the **π20** beam line with secondary high-momentum pions. Spectroscopy of Ξ baryons is planned at the new **K10** beam line with intense, separated kaons to investigate diquark correlation in the strangeness sector, and to explore the unknown field of excited Ξ baryons (denoted as Ξ^*). The relative motion of two quarks in a diquark and the collective motion of a diquark to the other quark are kinematically separated by the introduction of quarks with different flavors and with different masses in a baryon. The $\pi^- p \to D^{*-} Y_c^{*+}$ and $K^- p \to \Xi^* K^{(*)}$ reactions will be employed because the production cross sections reflect such diquark motions [7, 8]. It is noted that Y_c^{*+} and Ξ^* can be systematically populated from the ground states up to highly excited states because they are identified in the missing mass spectra. The branching ratio (decay partial width), which carries information on the internal structure of the excited baryon, can be easily determined by measuring decay particles in coincidence with a populated state. This is an advantage of fixed-target experiments.

Further investigations of quark correlations can be performed at the K10 beam line, via the simplest sss systems - Ω baryons. The Ω baryons are unique because the pion coupling is rather weak compared to other hadrons, including $u/\bar{u}$ and $d/\bar{d}$ quarks, and the quark-gluon dynamics will thereby be directly reflected to the exited state Ω^*. However, the production cross sections for Ω^{*-} are expected to be small in the case of the $K^- p \to K^+ K^{(*)0} \Omega^{*-}$ reaction; therefore, investigations can only be performed using the intense, separated kaon beams available at K10.

4.3 Flavor physics - investigation of new physics beyond the Standard Model through rare kaon decays

The Standard Model (SM) in particle physics successfully describes how elementary particles interact with each other. The last missing particle in the SM, the Higgs particle, was discovered in 2012 by the ATLAS and CMS experiments at the LHC [9, 10]. However, many questions that cannot be answered using the SM still remain, such as the matter-dominant universe and the low mass of the Higgs boson, and therefore the search for new physics beyond the SM has been pursued intensively. The direct production of new particles would not be discovered so easily in the LHC; therefore, the role of flavor physics experiments, which investigate new physics at a high energy scale through rare phenomena using intensity frontier machines, has become more important.

At the **KL** beam line in the present hall, the KOTO experiment searches for the rare kaon decay $K_L \to \pi^0 \nu\bar{\nu}$. This decay directly violates CP symmetry. The branching ratio is highly suppressed and is predicted to be 3.0×10^{-11} with a small theoretical uncertainty of 2% [11]. This decay is thus one of the most sensitive probes to search for new physics. The KOTO experiment set the most stringent upper limit on the $K_L \to \pi^0 \nu\bar{\nu}$ branching ratio to be 3.0×10^{-9} at a 90% confidence level with the data-set taken in 2015 [12]. A sensitivity of 7.2×10^{-10} was achieved with the data-set taken in 2016-2018. Three events were observed in the signal region, which was consistent with the number of expected background events, 1.22 ± 0.26 [13]. With newly installed counters to reduce background events, the KOTO experiment continues to take physics data and will achieve a sensitivity of better than 10^{-10} in 3-4 years. However, the achievable sensitivity will eventually become saturated. It is thus desirable to design a new experiment that can observe a sufficient number of SM-predicted events for measurement of the $K_L \to \pi^0 \nu\bar{\nu}$ decay.

In the extended facility, the KOTO step-2 experiment plans to build a new neutral kaon beam line (**KL2**) with a smaller extraction angle than that for the KOTO experiment to increase the K_L flux, and prepare a longer detector to extend the signal region and improve the signal acceptance. The KOTO step-2 experiment aims to measure the branching ratio for $K_L \to \pi^0 \nu\bar{\nu}$ decay with a beam intensity of 100 kW for three snow-mass-years and an accuracy of 27% by observing 35 SM events with a signal-to-background ratio of 0.63.

The breaking of time-reversal (T) symmetry, which is connected to CP violation through the CPT theorem, is an important key to solve the matter-antimatter asymmetry in the universe. T violation will be investigated at the **K1.1BR** beam line by measuring the transverse polarization of muons from the $K^+ \to \pi^0 \mu^+ \nu$ decay.

Charged lepton flavor violation is a definite sign for new physics beyond the SM. The **COMET** experiment, which aims to search for coherent neutrino-less conversion of a muon to an electron ($\mu^- + N(A, Z) \to e^- + N(A, Z)$) in muonic atoms ($\mu - e$ conversion), provides a window on new physics with the world's highest sensitivity.

The observables associated with kaon rare decays and $\mu - e$ conversion, together with measurements in B factories such as the Belle II experiment, play an important role in the investigation of the flavor structure in new physics.

References

[1] Taskforce on the extension of the Hadron Experimental Facility, (2021), `arXiv:2110.04462 [nucl-ex]`
[2] T. Koseki et al., Prog. Theor. Exp. Phys. **2012**, 02B004 (2012)
[3] H. Ohnishi, F. Sakuma, T. Takahashi, Prog. Part. Nucl. Phys. **113**, 103773 (2020)
[4] https://www.kek.jp/wp-content/uploads/2022/07/KEK-PIP2022.pdf
[5] P.B. Demorest, T. Pennucci, S.M. Ransom, M.S.E. Roberts, J.W.T. Hessels, Nature **467**, 1081 (2010)
[6] J. Antoniadis et al., Science **340**, 1233232 (2013)
[7] S.H. Kim, A. Hosaka, H.C. Kim, H. Noumi, K. Shirotori, Prog. Theor. Exp. Phys. **2014**, 103D01 (2014)
[8] S.I. Shim, A. Hosaka, H.C. Kim, Prog. Theor. Exp. Phys. **2020**, 053D01 (2020)
[9] G. Aad et al., Phys. Lett. B **716**, 1 (2012)
[10] S. Chatrchyan et al., Phys. Lett. B **716**, 30 (2012)
[11] A.J. Buras, D. Buttazzo, J. Girrbach-Noe, R. Knegjens, J. High Energy Phys. p. 033 (2015)
[12] J.K. Ahn et al., Phys. Rev. Lett. **122**, 021802 (2019)
[13] J.K. Ahn et al., Phys. Rev. Lett. **126**, 121801 (2021)

EPJ Web of Conferences **271**, 11002 (2022) https://doi.org/10.1051/epjconf/202227111002
HYP2022

Strangeness physics programs by S-2S at J-PARC

Toshiyuki Gogami[1,*], *Patrick* Achenbach[2], *Jung Keun* Ahn[3], *Darko* Androić[4], *Kanae* Aoki[5], *Arshak* Asaturyan[6], *Elena* Botta[7], *Masroor. H.* Bukhari[8], *Alexandre* Camsonne[9], *Silviu. C.* Covrig[9], *Kengo* Ebata[1,10], *Hiroyuki* Ekawa[11], *Petr* Evtoukhovitch[12], *Alessandro* Feliciello[7], *Hiroyuki* Fujioka[13], *Manami* Fujita[10], *Franco* Garibaldi[14], *Takeshi. K.* Harada[1,10], *Shoichi* Hasegawa[10], *Tomoyuki* Hasegawa[15], *Shuhei. H.* Hayakawa[16], *Erina* Hirose[5], *Ryotaro* Honda[5], *Kenji* Hosomi[10], *Yudai* Ichikawa[10], *Kenichi* Imai[10], *Tatsuhiro* Ishige[16], *Wooseung* Jung[3], *Kento* Kamada[16], *Shunsuke* Kanatsuki[1], *Seigo* Kato[17], *Shinhyung. H.* Kim[10], *Takumi* Kobori[13], *Simonetta* Marcello[7], *Pete* Markowitz[18], *Koji* Miwa[16], *Arthur* Mkrtchyan[19], *Hamlet* Mkrtchyan[19], *Manabu* Moritsu[20], *Tomofumi* Nagae[1], *Sho* Nagao[21], *Manami* Nakagawa[11], *Satoshi. N.* Nakamura[21], *Takuya* Nanamura[1,10], *Megumi* Naruki[1], *Ryosuke* Negishi[13], *Kazuki* Okuyama[16], *Fumiya* Oura[16], *Bishnu* Pandey[22], *Josef* Pochodzalla[2], *Atsushi* Sakaguchi[23], *Tamao* Sakao[16], *Hiroyuki* Sako[10], *Chhanda* Samanta[22], *Susumu* Sato[10], *Mitra. H.* Shabestari[24], *Albert* Shahinyan[19], *Kotaro* Shirotori[25], *Simon* Širca[26], *C.* Son[16], *Hitoshi* Sugimura[5], *Hitoshi* Takahashi[5], *Shuji* Takahashi[1], *Tomonori* Takahashi[25], *Toshiyuki* Takahashi[5], *Hirokazu* Tamura[16], *Kiyoshi* Tanida[10], *Atsushi* Tokiyasu[27], *Zviadi* Tsamalaidze[12,28], *Makoto* Uchida[13], *Mifuyu* Ukai[5], *Chiaki* Une[13], *Guido. M.* Urciuoli[14], and *Takeshi. O.* Yamamoto[10] for the S-2S Collaboration

[1]Department of Physics, Graduate School of Science, Kyoto University, Kyoto 606-8502 Japan
[2]Institute for Nuclear Physics, Johannes Gutenberg-University, D-55099 Mainz, Germany
[3]Department of Physics, Korea University, Seoul 136-713, Korea
[4]Department of Physics & Department of Applied Physics, University of Zagreb, HR-10000 Zagreb, Croatia
[5]Institute of Particle and Nuclear Studies, KEK, Tsukuba, Ibaraki 305-0801, Japan
[6]University of North Carolina Wilmington (UNCW), NC 28403, USA
[7]Istituto Nazionale di Fisica Nucleare-Sezione di Torino, Via P. Giuria 1, I-10125 Torino, Italy
[8]Jazan University, Jazan 45142, Saudi Arabia
[9]Thomas Jefferson National Accelerator Facility, Newport News, Virginia 23606, USA
[10]ASRC, Japan Atomic Energy Agency, Tokai, Ibaraki 319-1195, Japan
[11]RIKEN, Wako, Saitama 351-0198, Japan
[12]Joint Institute for Nuclear Research, Dubna, Moscow Region 141980, Russia
[13]Department of Physics, Tokyo Institute of Technology, Meguro, Tokyo 152-8551, Japan
[14]INFN, Sezione di Roma, 00185, Rome, Italy
[15]School of Allied Health Sciences, Kitasato University, Sagamihara, Kanagawa 252-0373, Japan
[16]Department of Physics, Tohoku University, Sendai, Miyagi 980-8578, Japan
[17]Department of Physics, Yamagata University, Yamagata, Yamagata 990-8560, Japan
[18]Department of Physics, Florida International University, Miami, Florida 27411, USA
[19]A.I. Alikhanyan National Science Laboratory (Yerevan Physics Institute), Yerevan 0036, Armenia
[20]Department of Physics, Kyushu University, Fukuoka, Fukuoka 819-0395, Japan
[21]Graduate School of Science, The University of Tokyo, Tokyo 113-0033, Japan
[22]Department of Physics & Astronomy, Virginia Military Institute, Lexington, Virginia 24450, USA
[23]Department of Physics, Osaka University, Osaka 560-0043, Japan

*e-mail: gogami.toshiyuki.4a@kyoto-u.ac.jp

[24]Department of Physics, University of West Florida, FL 32514, USA
[25]Research Center for Nuclear Physics, Osaka University, Ibaraki, Osaka 567-0047, Japan
[26]Faculty of Mathematics and Physics, University of Ljubljana, 1000 Ljubljana, Slovenia
[27]Research Center for Electron Photon Science, Tohoku University, Sendai, Miyagi 982-0826, Japan
[28]Georgian Technical University (GTU), Tbilisi, Georgia

Abstract. In the K1.8 beam-line at Hadron Experimental Facility of J-PARC, a new magnetic spectrometer S-2S is being installed. S-2S was designed to achieve a high momentum resolution of $\Delta p/p = 6 \times 10^{-4}$ in FWHM. Several strangeness-physics programs which require the high resolution will be realized by S-2S. The present article introduces J-PARC E70 (missing-mass spectroscopy of $^{12}_{\Xi}$Be) and E94 (missing-mass spectroscopy of $^{7}_{\Lambda}$Li, $^{10}_{\Lambda}$B, and $^{12}_{\Lambda}$C) experiments.

1 Introduction

Hypernuclear spectroscopy is a significant tool to study the baryon-baryon (BB) interaction with strangeness degrees of freedom (S). The interaction between Λ and nucleon (ΛN), which is the $S = -1$ system, has been being explored by Λ hypernuclear spectroscopy [1–3] and ΛN-scattering experiments [4–6]. Despite the $S = -1$ hypernuclei have been investigated for about 70 years since the first hypernuclear event was reported in 1953 [7], the number of observed hypernuclear species and data qualities are still limited. Moreover, the BB interaction for $S = -2$ baryons is quite unknown because experimental data of such as double-Λ or Ξ hypernuclei are much fewer than those for the $S = -1$ system. Nuclear emulsion experiments that found potential events of the $S = -2$ hypernuclei [8–13] allow us to consider the $\Lambda\Lambda$ and ΞN interactions are weakly attractive. In addition, recent data of femtoscopy by the ALICE Collaboration show similar attractive behaviors for the ΞN and $\Lambda\Lambda$ interactions [14–17]. These successful findings lead to a prediction of the existence of bound hypernuclei with $S = -2$, and they strongly motivate us to go forward to a systematic investigation of Ξ (double-Λ) hypernuclei to study the ΞN ($\Lambda\Lambda$) interaction.

At the beginning of 2022, an installation of a new magnetic spectrometer, Strangeness −2 Spectrometer (S-2S), started at the K1.8 beam-line at Hadron Experimental Facility of J-PARC [18, 19]. S-2S was designed for a missing-mass spectroscopy of Ξ hypernuclei with the world's best energy resolution of 2 MeV in the full width at half maximum (FWHM). One of the important features of S-2S is the high momentum resolution of 6×10^{-4} in FWHM, maintaining a reasonable solid-angle acceptance of about 55 msr at the central momentum. The installation of S-2S magnets and detectors is planned to be completed by the end of 2022. The first S-2S commissioning by using the beam is going to be carried out in the beginning of 2023. A new era to study the $S = -2$ hypernuclei with the high-precision reaction-spectroscopy is about to be begun. There are several strangeness-physics programs with S-2S, which are labeled E70 (missing-mass spectroscopy of $^{12}_{\Xi}$Be) [20], E75 [missing-mass spectroscopy of $^{7}_{\Xi}$H (phase-1 experiment); decay-pion spectroscopy of $^{5}_{\Lambda\Lambda}$H (phase-2 experiment)] [21, 22], E90 (ΣN-cusp measurement) [23], E94 (missing-mass spectroscopy of $^{7}_{\Lambda}$Li, $^{10}_{\Lambda}$B and $^{12}_{\Lambda}$C) [24] and E96 (X-ray measurement from ^{12}C-Ξ^- atom) [25] so far. Other experimental ideas which we are examining can be found in Ref. [26]. In the present article, the E70 and E94 experiments are described since some details of the others may be shown elsewhere in the same volume of the present journal.

2 Ξ hypernuclear study in J-PARC E70

2.1 Experimental setup

The S-2S spectrometer consists of two quadrupole (Q1 and Q2) magnets, one dipole (D) magnet, and particle detectors. The magnetic flux density of the D magnet is 1.5 T at the maximum coil current of 2500 A, and it bends charged particles at about 1.37 GeV/c by 70°. The particle detectors are composed of plastic-scintillation counters for a data-taking trigger, five layers of drift chambers (SDC1–5) for a particle tracking, and two types of Cherenkov counters (radiation media of aerogel and water) for a particle identification [27, 28]. Figure 1 shows a picture of S-2S as of the end of June 2022. The back face of SDC5 and a common frame which holds SDC3–5 are seen in the downstream of the D magnet in the picture. The other detectors are now being installed. A K^--meson beam transported by the K1.8 beam-

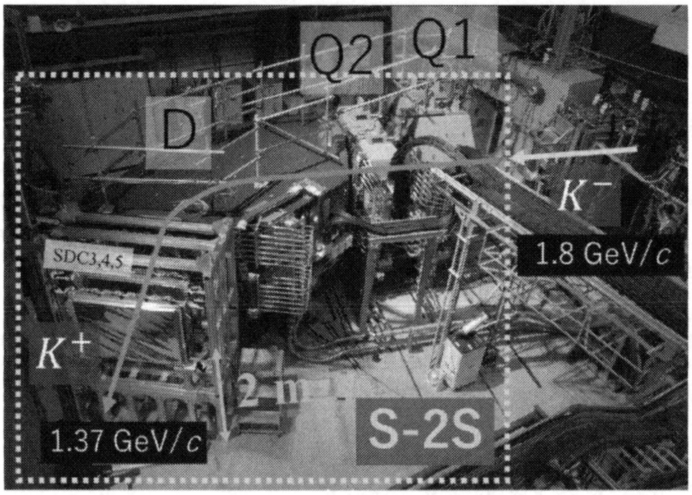

Figure 1. A photograph of S-2S which is installed in the K1.8 beam-line at Hadron Experimental Facility of J-PARC as of the end of June, 2022. SDC3–5 are drift chambers for the particle tracking, and the back face of SDC5 is seen in the photograph. Plastic scintillation counters, aerogel-Cherenkov counters, and water-Cherenkov counters are installed in the downstream of SDC5 in this order.

line spectrometer is impinged on an experimental target to produce Ξ hypernuclei through the reaction of (K^-, K^+) at $p_{K^-} = 1.8$ GeV/c. It is worth noting that the reaction rate for the $p(K^-, K^+)\Xi^-$ reaction was found to be the maximum at $p_{K^-} \simeq 1.8$ GeV/c within the range of $1.5 \leq p_{K^-} \leq 1.9$ GeV/c [29] in the previous experiment (J-PARC E05) [28, 30]. The generated K^+ is detected and momentum-analyzed by S-2S. The central momentum of S-2S is set to 1.37 GeV/c to cover the kinematics for the $p(K^-, K^+)\Xi^-$ reaction, which is used for an energy calibration, as well as the $^{12}\mathrm{C}(K^-, K^+)^{12}_{\Xi}\mathrm{Be}$ reaction. The experimental target (AFT: active fiber target) is composed of scintillation fibers with a core of polystyrene (C : H = 1 : 1). Therefore, data of the energy calibration and Ξ hypernuclei are taken simultaneously. AFT measures the energy loss of charged particles event by event. The energy-loss information in AFT is used when the missing-mass is reconstructed to suppress the mass-resolution deterioration due to the energy straggling. The goal of missing-mass resolution is about 2 MeV/c^2 in FWHM, and is expected to be realized by applying the event-by-event correction for the energy loss.

2.2 Expected results

In the J-PARC E70 experiment, the expected number of events in the bound region is about 100 counts for the $^{12}C(K^-, K^+)^{12}_{\Xi}Be$ reaction. Assumptions for the yield estimation are shown below. The beam intensity is assumed to be 8×10^5 K^-'s per spill with the spill cycle of 4.2 seconds. It is noted that, in a beam test in 2021, the ratio of K^- to π^- was achieved to be 1.67 and 0.38 at the K^--beam intensity of 6.5×10^5 and 8.6×10^5 per spill, respectively. A typical path length from the target to the end of the S-2S particle detector (water-Cherenkov counter) is 8 m, and thus the fraction of survived K^+'s is 0.46. The particle-detection efficiency which takes into account various efficiencies of such as the detectors, data acquisition system, and data analysis is assumed to be 0.5. The 9-g/cm^2 thick AFT, which is mostly composed of polystyrene (CH), is exposed to the beam for 480 hours. The assumed differential cross section is 60 nb/sr for our angular acceptance of $0° \leq \theta_{K^+} \leq 10°$. Here, the cross-section assumption is based on a result from the past experiment at BNL [31] in which the mass resolution was 14 MeV$/c^2$ (FWHM).

Figure. 2 shows expected binding-energy (B_{Ξ}) spectra for the $^{12}C(K^-, K^+)^{12}_{\Xi}Be$ reaction by a simple Monte-Carlo simulation. The assumed energy resolutions for the left and right

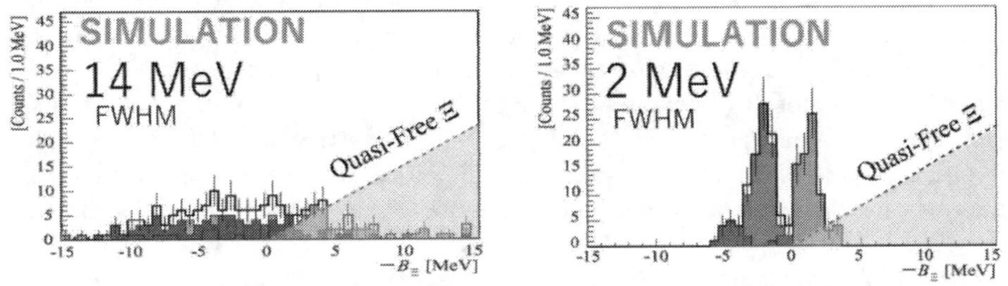

Figure 2. Expected spectra of the Ξ binding energy taken with the negative sign ($-B_{\Xi}$) for the reaction of $^{12}C(K^-, K^+)^{12}_{\Lambda}Be$. Assumed energy resolutions in the left and right panels are 14 and 2 MeV in FWHM, respectively. J-PARC E70 aims to achieve the 2-MeV resolution in FWHM (right panel). The peak positions and relative peak strengths for the assumed three states ($J^\pi = 1^-_1$, 1^-_2 and 1^-_3), which are shown in different colors, are taken from a prediction by DWIA with the interaction model of ESC08a [32].

panels in the figure are 14 and 2 MeV in FWHM, respectively. The left and right panels correspond to the energy resolutions for the cases of the past BNL experiment [31] and J-PARC E70. There are three of assumed states, $J^\pi = 1^-_1$, 1^-_2 and 1^-_3, which are shown in different colors. The energy levels and their relative cross sections are taken from a prediction by DWIA calculation with the interaction model of ESC08a [32]. J-PARC E70 would be able to observe the peak structures separately thanks to the high resolution, and provide significant information of the ΞN interaction.

3 Λ hypernuclear study in J-PARC E94

3.1 Overview

J-PARC E94 aims to establish a high resolution spectroscopy of Λ hypernuclei with the (π^+, K^+) reaction [24]. The highest energy resolution in the missing-mass spectroscopy of Λ hypernuclei with the (π^+, K^+) reaction is about 1.5 MeV in FWHM, which was realized at KEK [33]. On the other hand, the J-PARC E94 aims to achieve 1 MeV (FWHM) or a little better. Such a high resolution is expected to be realized thanks to the high momentum resolution of the new magnetic spectrometer S-2S. A similar or better energy resolution, which is 0.5–1 MeV in FWHM, was already realized by the missing-mass spectroscopy with the $(e, e'K^+)$ reaction at the Thomas Jefferson Lab (JLab) [34–37]. However, the improvement of the energy resolution for the spectroscopy with the (π^+, K^+) reaction gives a large impact to the study of the ΛN interaction because hypernuclear species which can be studied by the (π^+, K^+) and $(e, e'K^+)$ reactions is different. Therefore, complementary investigation of the ΛN interaction through the Λ hypernuclear spectroscopy would be dramatically strengthened.

J-PARC E94 aims to determine the absolute binding energy with the accuracy of $|\Delta B_\Lambda| \simeq$ 100 keV. Such a high accuracy could not be realized in the past (π^+, K^+) experiments because a source of the energy calibration was the ground-state binding energy of $^{12}_{\Lambda}$C [$B_\Lambda^{\rm g.s.}(^{12}_{\Lambda}{\rm C})$]. The measured $B_\Lambda^{\rm g.s.}(^{12}_{\Lambda}{\rm C})$ is known to be $10.76 \pm 0.19 \pm 0.04$ MeV [38]. The error on $B_\Lambda^{\rm g.s.}(^{12}_{\Lambda}{\rm C})$, which is about 200 keV, propagates to the uncertainty of measured Λ binding energies when it is used for the calibration. In addition, it is pointed out that $B_\Lambda^{\rm g.s.}(^{12}_{\Lambda}{\rm C})$ is shifted by about 500–600 keV [39, 40], which could lead to an additional systematic error on the binding-energy measurement. Therefore, J-PARC E94 tries to use different calibration sources, the binding energies of $^{7}_{\Lambda}{\rm Li}(1/2^+, 5/2^+)$. The binding energies of these states were well determined by the emulsion experiment [38] and γ-ray spectroscopy [41, 42] with high statistics. Their energy uncertainties are only about 50 keV. The small uncertainties of the calibration sources allow us to reach the high accuracy of $|\Delta B_\Lambda| \simeq$ 100 keV in J-PARC E94. The binding energies of the $1/2^+$ and $5/2^+$ states of $^{7}_{\Lambda}$Li were not be able to be used before because the energy separation between the states is only about 2 MeV and could not be separated with the energy resolutions in the past experiments. The use of $^{7}_{\Lambda}$Li for the calibration purpose is expected to be realized for the first time thanks to the high energy resolution of about 1 MeV (FWHM) in J-PARC E94.

J-PARC E94 aims to prove that the missing-mass spectroscopy with the 1-MeV resolution and the 100-keV accuracy is possible via the (π^+, K^+) reaction by measuring $^{7}_{\Lambda}$Li, $^{10}_{\Lambda}$B, and $^{12}_{\Lambda}$C. The accurate measurement of $^{12}_{\Lambda}$C would clarify the issue of the energy shift of about 500–600 keV in the existing data of $B_\Lambda^{\rm g.s.}(^{12}_{\Lambda}{\rm C})$ [39, 40]. In addition, the accurate data of $^{10}_{\Lambda}$B would lead to better understanding of the charge-symmetry breaking [43, 44] by comparing with its mirror hypernucleus $^{10}_{\Lambda}$Be [39].

3.2 Experimental conditions and status

The experimental setup for the J-PARC E94 is the same as that for J-PARC E70 except for the beam polarity, momentum settings of the spectrometers (both the beam-line spectrometer and S-2S), and experimental targets. The π^+ beam at $p_{\pi^+} = 1.05$ GeV/c is impinged on the experimental target. Target materials of ^{7}Li, ^{10}B, and ^{12}C with the areal density of 1 g/cm^2 are alternately installed to measure $^{7}_{\Lambda}$Li, $^{10}_{\Lambda}$B, and $^{12}_{\Lambda}$C, respectively, by the (π^+, K^+) reaction. The central momentum of S-2S is set to 0.72 GeV/c.

J-PARC E94 was approved to be at the Stage 1 in 2022. We are now allowed to apply to the Stage 2 which is the final status to perform the experiment. A systematic investigation with the high resolution and high accuracy for the other Λ hypernuclei, which may be up to the medium-heavy mass region (A $\sim$ 50), would be promising by S-2S in the future if J-PARC E94 is successfully completed.

4 Summary

The new magnetic spectrometer S-2S is being installed in the K1.8 beam-line at Hadron Experimental Facility of J-PARC. S-2S was designed to achieve the high momentum resolution of $\Delta p/p = 6 \times 10^{-4}$ (FWHM). Several strangeness-physics programs in which the high resolution measurements are required will be realized by S-2S. In the present article, J-PARC E70 (missing-mass spectroscopy of $^{12}_{\Xi}$Be) and E94 (missing-mass spectroscopy of $^{7}_{\Lambda}$Li, $^{10}_{\Lambda}$B, and $^{12}_{\Lambda}$C) experiments are described. The S-2S commissioning with the hadron beam is planned to be performed in the beginning of 2023, followed by the calibration and physics runs of J-PARC E70.

Acknowledgments

We thank J-PARC staff for supporting our experimental programs. This work was supported by JSPS KAKENHI Grant Numbers 17070006, 17070005, 22244025, 23000003, 13J01075, 16H02186, 18H05403, 18H01219. This work was supported by SPIRITS 2020 of Kyoto University.

References

[1] O. Hashimoto and H. Tamura, Prog. Par. Nucl. Phys. **57**, 2, 564–653 (2006)
[2] A. Feliciello and T. Nagae, Rep. Prog. Phys. **78**, 9 (2015)
[3] A. Gal, E. V. Hangerford, and D. J. Millener, Rev. Mod. Phys. **88**, 035004 (2016)
[4] T. H. Groves, Phys. Rev. **129**, 1372 (1963)
[5] G. Alexander et al., Phys. Rev. **173**, 1452 (1968)
[6] J. Rowley et al., Phys. Rev. Lett. **127**, 272303 (2021)
[7] M. Danysz and J. Pniewski, The London, Edinburgh, and Dublin Phil. Mag. Jour. Sci. **44**:350, 345–350 (1953)
[8] H. Takahashi et al., Phys. Rev. Lett. **87**, 212502 (2001)
[9] J. K. Ahn et al., Phys. Rev. C **88**, 014003 (2013)
[10] K. Nakazawa et al., Prog. Theor. Exp. Phys. **2015**, 033D02 (2015)
[11] H. Ekawa et al., Prog. Theor. Exp. Phys. **2019**, 021D02 (2019)
[12] M. Yoshimoto et al., Prog. Theor. Exp. Phys. **2021**, 073D02 (2021)
[13] S. H. Hayakawa et al., Phys. Rev. Lett. **126**, 062501 (2021)
[14] S. Acharya et al. (ALICE Collaboration), Phys. Lett. B **797**,134822 (2019)
[15] S. Acharya et al. (ALICE Collaboration), Phys. Rev. Lett. **123**, 112002 (2019)
[16] S. Acharya et al. (ALICE Collaboration), Nature **588**, 232 (2020)
[17] Y. Kamiya et al., Phys. Rev. C **105**, 014915 (2022)
[18] K. Agari et al., Prog. Theor. Exp. Phys. **2012**, 02B009 (2012)
[19] T. Takahashi et al., Prog. Theor. Exp. Phys. **2012**, 02B010 (2012)
[20] T. Nagae et al., *Proposal for the next E05 run with the S-2S spectrometer*, Proposal to J-PARC, E70 (2018)
[21] H. Fujioka et al., *Decay Pion Spectroscopy* $^{5}_{\Lambda\Lambda}H$ *produced by Ξ-hypernuclear decay*, Proposal to J-PARC, E75 (2019)
[22] H. Fujioka et al., *Phase-1 of the P75 experiment: Measurement of the formation cross section of* $^{7}_{\Xi}H$ *in the* ^{7}Li(K^-, K^+) *reaction*, Proposal to J-PARC, E75 Phase-1 (2020)
[23] Y. Ichikawa et al., *High resolution spectroscopy of the ΣN cusp by using the* $d(K^-, \pi^-)$ *reaction*, Proposal to J-PARC, E90 (2022)
[24] T. Gogami et al., *New generation Λ hypernuclear spectroscopy with the* (π^+, K^+) *reaction by S-2S*, Proposal to J-PARC, E94 (2022)
[25] T. O. Yamamoto et al., *Measurement of X rays from* Ξ^-C *atom with an active fiber target system*, Proposal to J-PARC, E96 (2022)

[26] S-2S workshop 2021, https://kds.kek.jp/event/39644/
[27] T. Gogami et al., Nucl. Instrum. Methods, Phys. Res. Sect. A **817**, 70–84 (2016)
[28] T. Gogami et al., JPS Conf. Proc. **18**, 011031 (2017)
[29] T. Gogami et al., J. Phys.:Conf. Ser. **1643**, 012133 (2020)
[30] T. Nagae et al., AIP Conf. Proc. **2130**, 020015 (2019)
[31] P. Khaustov et al., Phys. Rev. C **61**, 054603 (2000)
[32] T. Motoba and S. Sugimoto, Nucl. Phys. A **835**, 223–230 (2010)
[33] H. Hotchi et al., Phys. Rev. C **64**, 044302 (2001)
[34] S. N. Nakamura et al., Phys. Rev. Lett. **110**, 012502 (2013)
[35] L. Tang et al., Phys. Rev. C **90**, 034320 (2014)
[36] T. Gogami et al., Nucl. Instrum. Methods, Phys. Res. Sect. A **900**, 69–83 (2018)
[37] T. Gogami et al., Phys. Rev. C **103**, L041301 (2021)
[38] D. H. Davis, Nucl. Phys. A **754**, 3c–13c (2005)
[39] T. Gogami et al., Phys. Rev. C **93**, 034314 (2016)
[40] E. Botta et al., Nucl. Phys. A **960**, 165–179 (2017)
[41] K. Tanida et al., Phys. Rev. Lett. **86**, 10 (2001)
[42] M. Ukai et al., Phys. Rev. C **73**, 012501(R) (2006)
[43] E. Hiyama and Y. Yamamoto, Prog. Theor. Phys. **128**, 1, 105–124 (2012)
[44] A. Gal and D. Gazda, J. Phys.:Conf. Ser. **966**, 012006 (2018)

Future prospects of spectroscopic study of Lambda hypernuclei at JLab and J-PARC HIHR

Satoshi N. Nakamura[1,*] for the JLab hypernuclear collaboration and SπK collaboration

[1]Department of Physics, Graduate School of Science, the University of Tokyo, 113-0033, JAPAN

Abstract. The $(e, e'K^+)$ reaction spectroscopy opened a door to high resolution spectroscopy of Λ hypernuclei at JLab and it is currently the only technique to give sub-MeV energy resolution for reaction spectroscopy of wide-mass range Λ hypernuclei. New experiments from light to heavy hypernuclei are under preparation at JLab to solve hypertriton puzzle, to clarify charge symmetry breaking of Λ hypernuclei and to give a clue to solve the hyperon puzzle or the puzzle of heavy neutron stars.
As a hypernuclear precision spectroscopy experiment with other than electron beams, there is a newly proposed experiment using the (π^+, K^+) reaction at the new HIHR beamline, which is a key facility in the J-PARC hadron experimental facility extension project. The HIHR beamline adopts the momentum dispersion match technique and will enable us to perform sub-MeV resolution spectroscopy for isospin partners to the Λ hypernuclei studied with the electron beams at JLab.

1 Introduction

Neutron star is a quite interesting object which is most dense material in our Universe. It is a compact star with a radius of ~10 km and mass of 1-2 solar masses bound by gravitational force and the strong force prevents it from collapsing to a black hole. Neutron star mergers are now believed to play an important role in production of heavy elements. Astronomical observations of neutron stars by gravitational waves [1, 2] and X-ray hot spot measurement [3, 4] made significant progresses to give information about macroscopic understanding of characteristics of neutron stars. Therefore, microscopic understanding of neutron stars becomes more important than ever and collaborative works between experiments and theories are mandatory.

One of great mysteries of neutron star is the so-called "hyperon puzzle." Naïve discussion about the nucleon Fermi energy and the attractive ΛN two-body interaction, Λ hyperons are expected to appear in the core of neutron star. Inclusion of hyperon such as Λ particle softens the Equation of State (EoS) of neutron stars. Solutions of the Tolman-Oppenheimer-Volkoff equation [5] with the conventional baryon potential models predict maximum masses of neutron stars are limited to be less than about 1.5 solar masses and heavier neutron stars than the limit are considered to collapse into black holes due to their too heavy masses. However, recent astronomical observations clarified the existence of massive neutron stars as heavy as two solar masses [6–8]. The masses of these neutron star are measured by quite

*e-mail: nue@nex.phys.s.u-tokyo.ac.jp

reliable techniques such as relativistic Shapiro delay and thus, there is no doubt about their existence. The conflict between theoretical predictions of maximum masses of neutron stars with hyperons and astronomical observations of heavy neutron stars is called as the "hyperon puzzle" and it is one of most important problems to be solved for the nuclear physics.

It is common understanding that there is an additional mechanism to make the neutron star EoS stiffer to solve the hyperon puzzle and a promising scenario is inclusion of the three-body repulsive ΛNN force. The repulsive ΛNN force plays significant role in high density environment such as neutron star core. Inclusion of it makes neutron star's EoS stiffer and futhermore it may be so strong that it prevents Λ's appearance [9–12]. Inclusion of the repulsive ΛNN force gives drastic effect to the EoS of neutron stars, however, its gives small effects as small as < 1 MeV to Λ's binding energies of hypernuclei which are able to be studied by the terrestrial experiments with particle accelerators. Therefore, high resolution spectroscopic study of Λ hypernuclei is quite important.

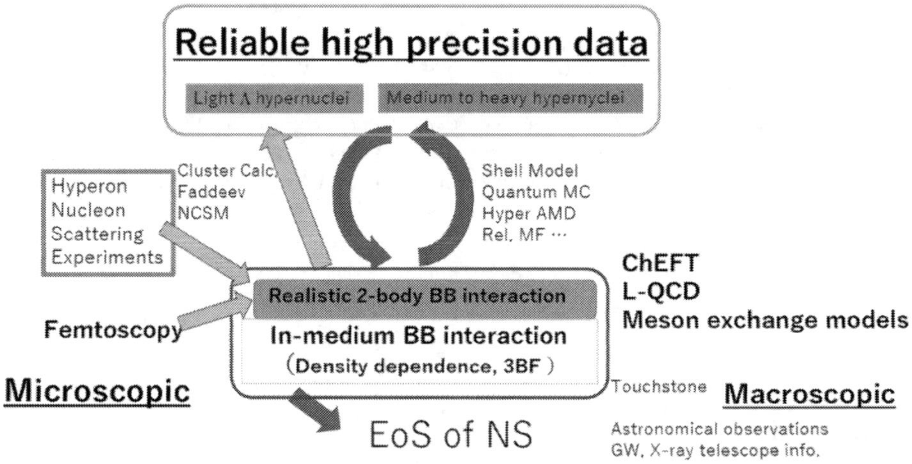

Figure 1. Strategy to solve the hyperon puzzle.

Figure 1 shows the our strategy to solve the hyperon puzzle. Starting point is reliable 2-body baryonic interaction models. Previously these baryonic interaction models have been constructed with SU(3) flavor symmetry and the limited binding energy data of hypenuclei, however, high quality hyperon nucleon scattering experiments was successfully performed recently for ΣN [13] and femotoscopy experiments [14] provides precious information about the p-Σ^0 interaction. In near future, combination of high quality experimental data and progresses of theoretical studies such as chiral effective field theory and lattice QCD will improve greatly our understanding of the 2-body baryonic interaction. The established 2-body baryonic interaction potentials will be converted to the effective in-medium baryonic potentials which include the multi-body forces or density dependences. The validity of such in-medium interaction models will be tested with high precision hypenuclear data. High precision data of hypernuclei provide also precious experimental inputs, for example, low energy constant of the chiral effective field theory. EoS of neutron stars will be deduced from such in-medium baryoic interaction models. Macroscopic information of neutron stars obtained by astronomical observations will serve as a touch stone to test our microscopic understanding of the baryon interaction. Consistent understanding of microscopic as well as macroscopic understandings of neutron stars means the successful solution of the hyperon puzzle.

High precision hypernuclear experiments play an important role in the above scenario. In the following sections, I will explain about future experiments which provide high precision spectroscopy of Λ hypernuclei, namely, a spectroscopic study of electro-produced hypernuclei at Thomas Jefferson National Accelerator Facility (JLab) and a new (π, K^+) reaction spectroscopy of hypernuclei at the HIHR beamline which will be constructed at the extended hadron experimental facility of the Japan Proton Accelerator Research Complex (J-PARC).

2 Spectroscopic study of electro-produced Λ hypernuclei at JLab

2.1 Spectroscopy of Λ hypernuclei with meson beams and electron beams

Conventionally, meson beams have been used for study of Λ hypernuclei. Figures 2 (left) show $n(K^-, \pi^-)\Lambda$ and $n(\pi^+, K^+)\Lambda$ reactions which were used for spectroscopic study of Λ hypernuclei with K^- beams and π^+ beams. The $n(K^-, \pi^-)\Lambda$ reaction is the quark exchange reaction and the $n(\pi^+, K^+)\Lambda$ reaction is the $s\bar{s}$ pair production to convert a neutron to a Λ particle. As shown in figure 2 (right), the $p(e, e'K^+)\Lambda$ reaction is quite similar reaction to the $n(\pi^+, K^+)\Lambda$ reaction while a virtual photon produces $s\bar{s}$ pair in the $p(e, e'K^+)\Lambda$ reaction and a proton is converted to a Λ. Even with the same targets, $p(e, e'K^+)\Lambda$ reaction and $n(K^-, \pi^-)\Lambda$, $n(\pi^+, K^+)\Lambda$ reactions produces different hypernuclei. They are complementary and experiments with both electron beams and meson beams are important to study isospin dependence or charge symmetry breaking of hypernuclei.

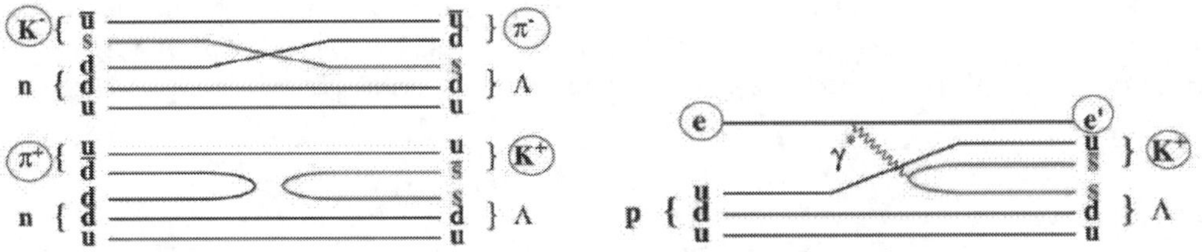

Figure 2. $n(K^-, \pi^-)\Lambda$ and $n(\pi^+, K^+)\Lambda$ reactions (left), and $p(e, e'K^+)\Lambda$ reaction (right)

Though the production cross section of electroproduction of Λ is a few orders of magnitude smaller than that for (π^+, K^+), (K^-, π^-) reactions, the energy resolution of the $p(e, e'K^+)$ experiments is 0.5-0.7 MeV (FWHM) while the resolutions are limited to be a few MeV for the experiments with meson beams due to quality of secondary meson beams. Electrons can be directly accelerated by particle accelerators but meson beams should be produced as secondary beams with proton accelerators. Therefore, momentum information and beam position of electron beams are not necessary to be measured event by event but beam momentum, position measurements are essential for meson beams. Beam detectors limits maximum intensity of meson beams but there is no limitation from beam detectors for electron beams. Therefore meson beam intensity is limited to be several 10^6/s while electron beam intensity can be as high as 100 μA which corresponds to 6×10^{14}/s.

Overcoming lots of experimental difficulties due to high electron background, the $(e, e'K^+)$ reaction spectroscopy of Λ hypernuclei was established in JLab Hall-C [15–23] and Hall-A [24–26].

Table 1. Approved JLab experiments on Λ hypernuclear spectroscopy

Experimental Number	Reaction	Comments on target
E12-15-008 [27]	$^{40,48}\mathrm{Ca}(e,e'K^+)^{40,48}_{\Lambda}\mathrm{K}$	isotopically enriched
E12-19-002 [28]	$^{3,4}\mathrm{He}(e,e'K^+)^{3,4}_{\Lambda}\mathrm{H}$	gaseous cryogenic
E12-20-013 [29]	$^{208}\mathrm{Pb}(e,e'K^+)^{208}_{\Lambda}\mathrm{Tl}$	low melting point

2.2 New JLab experiments of Λ-hypernuclear spectroscopy

Three new experiments on Λ hypernuclei with the $(e, e'K^+)$ reaction were proposed and approved at JLab. Table 1 summarizes these three experiments' information. Neutron star is a nucleon (baryon) many-body system characterized by large mass numbers and very high neutron abundances, *ie.* $A \gg 1$ and asymmetric parameter is close to one ($\delta = (N-Z)/A \sim 1$). So far, most of spectroscopic studies of Λ hypernuclei have been performed with solid targets of stable nuclei. The $(e, e'K^+)$ reaction converts ^AZ target to $^A_\Lambda(Z-1)$ hypernucleus and the (π^+, K^+) reaction does ^AZ to $^A_\Lambda Z$. Therefore A and δ of produced hypernuclei cannot be far from the original targets' A and δ as shown in figure 3. Solid line shows that $A^{-2/3}-\delta$ relation of stable targets and so far produced hypernuclei locate along this line. The newly proposed experiments try to widen our region of hypernuclear study: E12-19-002 tries to study light hypernuclei with cryogenic gaseous targets (^{3,4}He) to study the Charge Symmetry Breaking effects of hypernuclei and to solve the hypertriton puzzle which is the discrepancy between observed short lifetime and shallow binding energy of $^3_\Lambda$H; E12-20-013 aims to study $^{208}_{\Lambda}$Tl with ^{208}Pb target of which properties can be associated with the uniform nuclear matter; the E12-15-008 experiment investigates the isospin dependence of medium-heavy hypernuclei $^{40,48}_{\Lambda}$K to study isospin dependence of the ΛNN three-body repulsive force. Detail of physics motivations of these three experiments can be found in the experimental proposals and this HYP2022 proceedings [27–29].

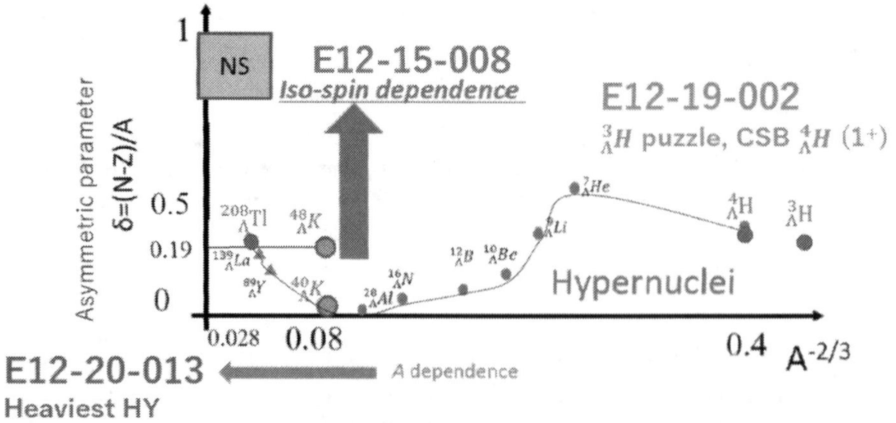

Figure 3. Mass number ($A^{-2/3}$) and asymmetric of parameter ($\delta = (N-Z)/A$) of Λ hypernuclei which are able to be produced from the stable targets by the $(e, e'K^+)$ reaction.

2.3 Experimental setup of JLab hypernuclear experiments

Basic principle of the ${}^{A}Z(e, e'K^{+})^{A}_{\Lambda}(Z-1)$ spectroscopy of hypernuclei is missing mass spectroscopy with the measurement of e' momentum, K^{+} momentum with the known incoming electron beam energy. The cross section of the elementary process, $p(\gamma, K^{+})\Lambda$, becomes maximum in the region of $E_{\gamma} = 1.2 - 1.5$ GeV and recoil momentum of the hypernucleus becomes smaller for higher beam energy in the (γ, K^{+}) reaction which results in larger capture probability of Λ to produce a hypernucleus. Therefore, we chose the virtual photon energy of $E_{\gamma^{*}} = 1.5$ GeV and the central momentum of the kaon spectrometer (HKS) was set at 1.2 GeV/c for the new experiments as we used it in the previous Hall-C hypernuclear experiments.

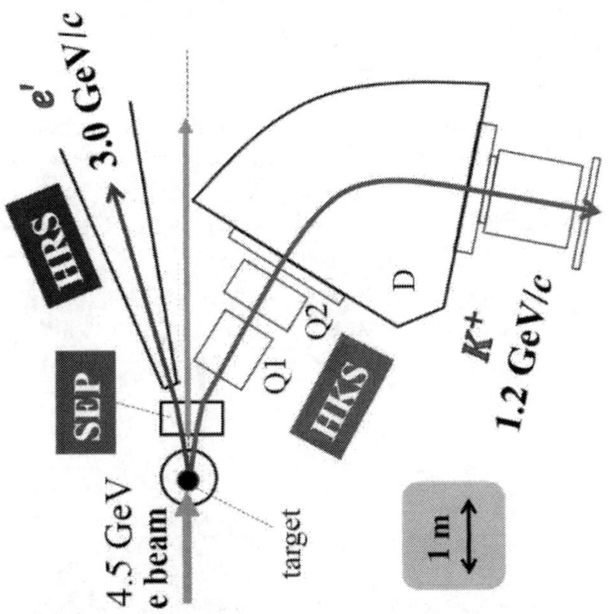

Figure 4. Setup in Hall A. HKS which was used in previous Hall-C experiments will be moved to Hall-A and permanently installed HRS in Hall-A will be used in this option.

There are two possibilities for the experimental site at JLab, namely Hall-A and Hall-C. In Hall-A, we can use the High Resolution Spectrometer (HRS) as the spectrometer for the scattered electrons. It is a vertical bending type spectrometer and it can measure high momentum electrons with an excellent momentum resolution of $\Delta p/p \sim 3 \times 10^{-4}$. Therefore we set the beam energy of $E_{e} = 4.5$ GeV and planned to use HRS with a central momentum of $E_{e'} = 3.0$ GeV which makes the virtual photon energy as $E_{\gamma^{*}} = 1.5$ GeV (Fig. 4). Higher beam energy allows us to make the bremsstrahlung electron background Lorentz boosted to forward angles out of spectrometer's acceptances. Therefore, we can expect good signal-to-noise ratio, though absolute e' momentum resolution which contributes to the missing mass resolution, becomes worse.

Another possibility is the experiment in Hall-C with the scattered electron spectrometer (HES) which was used for the previous hypernuclear experiments in Hall-C. It is designed to measure low energy electrons less than 1 GeV. Therefore, electron beam energy is limited to be less than 2.5 GeV if $E_{\gamma^{*}} = 1.5$ GeV is kept. Lower energy of scattered electron is favored for better energy resolution since smaller momentum (p) gives better absolute resolution for the fixed relative momentum resolution $\Delta p/p$, however, it is disfavored in terms of the signal-to-noise ratio.

In order to avoid physical interference between the forward located kaon and electron spectrometers, we will install a new pair of charge separation magnets (PCS) in both Hall-A and Hall-C options (Fig. 5). Introduction of PCS minimizes effects from the magnetic field to the unused electron beam while previously used a single large dipole (SPLitter) bent the electron beam as well as separated positively charged K^+ and negative e'. Furthermore, the single dipole magnet coupled the kaon spectrometer and the electron spectrometer beam-optically. It resulted in very complicated beam tune and analysis.

Another challenge in Hall-C option is that both HES and HKS are horizontal bending spectrometers which have no vertex resolution along the beamline. It was not a disadvantage for thin solid targets which were used in previous Hall-C experiments since these solid targets' (material) thicknesses were $\sim$ 100 mg/cm^2 or less than $<$ 1 mm. Therefore, we know the reaction point along the beamline with an accuracy of $<$ 1 mm. However, we will use a gaseous target with a length of 20 cm in the new experiment and the reaction point information is necessary for the high-resolution spectroscopy. Therefore, we plan to modify HES to be a vertical bending type spectrometer as shown in Fig. 6. Based on the preliminary simulations, solid angle of vertical HES with the PCS is slightly reduced but optical natures will not be affected much.

Considering the advantages and disadvantages of the spectrometers' performance and the JLab's experimental allocation plan in Halls A and C, we chose to perform the experiments in Hall-C. Detailed simulation and modification design of HES for the safe rotation of 100 ton spectrometer are in progress.

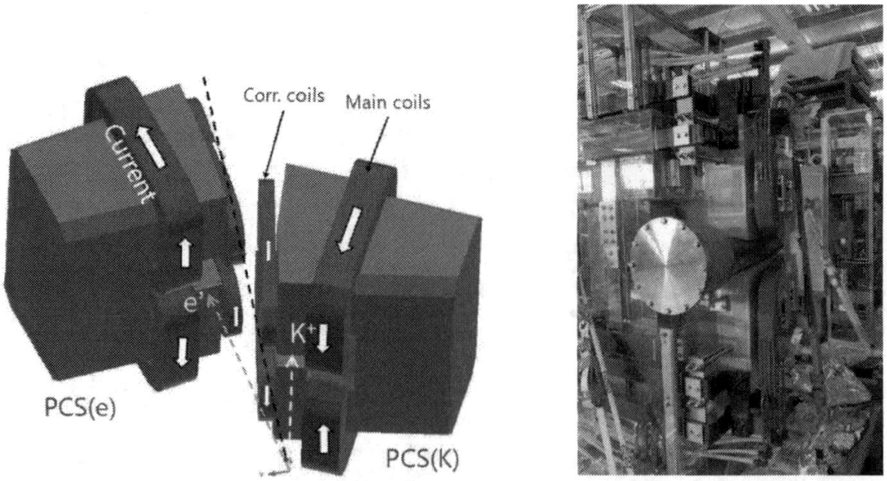

Figure 5. Particle Charge Separator (PCS) consists of a pair of dipole magnets (left). PCS(e) can bend 3 GeV/c electron by 6 degrees and PCS(K) can do 1.2GeV/c K^+ by 18.4 degrees. They were constructed in Japan and shipped to JLab in 2022 (right).

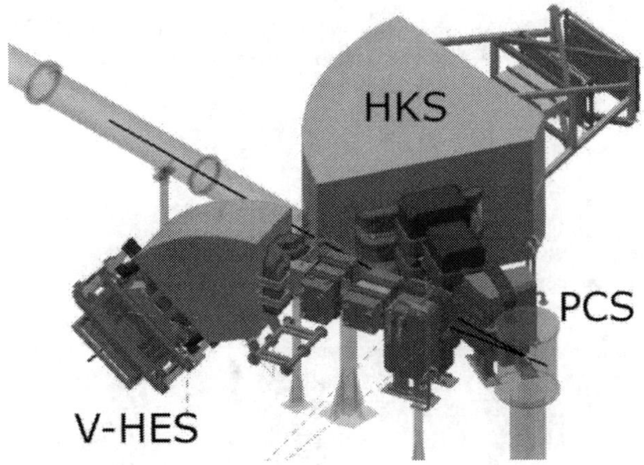

Figure 6. Setup in Hall C. Both HKS and HES were used in previous Hall-C experiments, however, HES will be modified to be a vertical bending spectrometer.

3 Supra-precision (π^+, K^+) reaction spectroscopy of Λ hypernuclei (SπK) at J-PARC HIHR beamline

In the previous section, it was mentioned that high-quality primary electron beam at JLab allows us to avoid measuring momenta and positions of beam electrons. High-resolution kaon and scattered electron spectrometers make it possible to perform the $(e, e'K^+)$ spectroscopy of hypernulei with a resolution of 0.5 MeV.

Until now, the (π^+, K^+) reaction spectroscopy of hypernuclei has been limited not only in yield but also in resolution. Pion beams are produced as secondary particles and their momentum distribution is large; therefore beam momentum analysis is inevitable. The π^+ beam intensity is limited by the operational limits of the beamline detectors. For these reasons, the energy resolution and beam intensity of the (π^+, K^+) hyperuclear spectroscopy have been limited to a few MeV and a few millions of π^+s/sec, respectively. The (π^+, K^+) and $(e, e'K^+)$ spectroscopies, which produce different Λ hypernuclei even with the same target, should provide complementary high-quality data for the detailed isospin dependence or the charge symmetry breaking studies. However, until now, the resolution of (π^+, K^+) spectroscopy has not matched with that of the $(e, e'K^+)$ reaction.

In order to avoid these limitations, a new experiment of supra-precision (π^+, K^+) reaction spectroscopy of Λ hypernuclei (SπK; J-PARC P84) was proposed [30] at the new High Intensity High Resolution beamline (HIHR) which is planned at the extended hadron hall of J-PARC [31, 32].

Introducing HIHR, these limitations of (π^+, K^+) reaction spectroscopy are removed and an excellent energy resolution of 0.4 MeV (FWHM), which is comparable to or even better than the $(e, e'K^+)$ spectroscopy, will be achieved. It enables us to determine the binding energies of Λ for various hypernuclei with a few 10 keV accuracy.

3.1 High Intensity High Resolution beamline (HIHR)

Currently the project to extend the hadron experimental facility is seriously considered and its preparation works are underway. HIHR is a key beamline designed for the Λ hypernuclear spectrscopy with the (π^+, K^+) reaction.

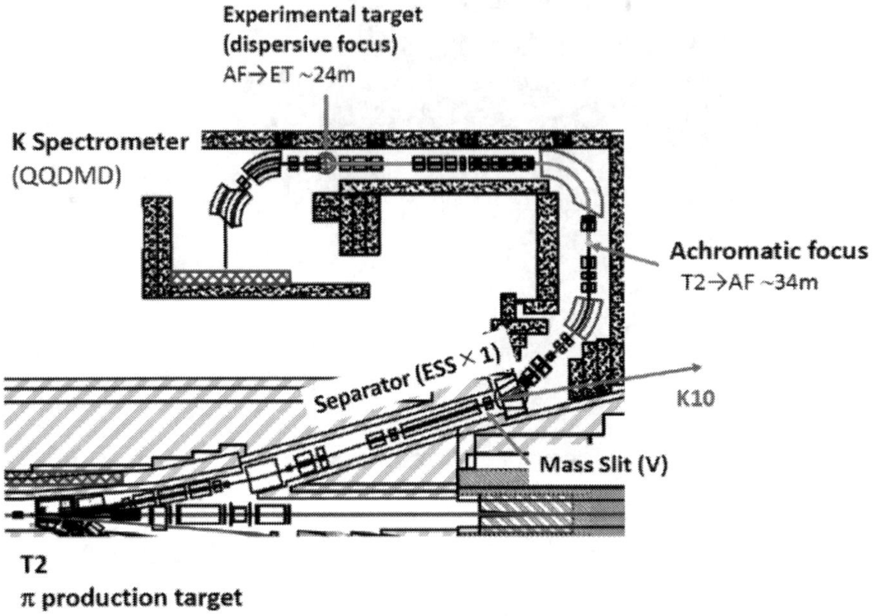

Figure 7. Schematic illustration of the HIHR beamline with a kaon spectrometer [30].

Figure 7 shows a schematic illustration the HIHR beamline with a kaon spectrometer. Primary proton beams bombard a new primary target (T2) to be installed in the hadron extension experimental facility and secondary meson beams are extracted at the production angle of 3 degrees; An Electro-Static Separator (ESS) is used for π, K separation. This beam extraction part is shared with HIHR and K10 beamlines. After this beamline branch, the π beam is focused achromatically and beam profile is redefined here with a beam slit. After the achromatic focus point, the beam is focused with a magnification of 1.1 and a dispersion of 11 cm/%. On the experimental target where the (π^+, K^+) reaction takes place, the momentum dispersion is converted to spatial dispersion and the kaon spectrometer after the experimental target is designed to cancel this spatial dispersion. This technique is called as the dispersion match (Fig. 8) and is well established technique for long-living ions [33–35], however, it is a big challenge to apply this technique to secondary GeV meson beams.

A dispersion match π beamline with an adequately designed kaon spectrometer enable us to avoid any beamline detectors. Because there are no limitations from the beamline detectors, we can use as strong as possible pion beams and can achieve excellent missing mass resolution.

Based on preliminary design of beamline and spectrometer systems, a GEANT4 simulation gives 400 keV (FWHM) resolution and 53/h yield for ${}^{12}_{\Lambda}$C hypernuclei with $2 \times 10^8 \pi^+$s/spill on the ^{12}C target of 400 mg/cm^2 [30]. Figure 9 shows an expected Λ binding energy spectrum for ${}^{208}_{\Lambda}$Pb with a beamtime of 60 days based on the above simulation with the theoretical predictions [36, 37]. Major peaks of $i_{13/2}^{-1}$ neutron holes with a Λ in $s, p, d, ..$ angular momentum states are clearly seen, and furthermore sub-major peaks which correspond to $g_{9/2}^{-1}$ neutron holes are also expected to be observed. It should be noted that sub-major peaks filled the valleys between the major peaks and they were treated as background in the previous (π^+, K^+) hypernuclear spectroscopy [38]. The improved resolution at HIHR means not just an improvement in the quality of the spectrum, but takes the study to a different stage. The first campaign of the experiments was proposed for study of $^{6,7}_{\Lambda}$Li, $^{9}_{\Lambda}$Be, $^{10,11}_{\Lambda}$B, $^{28}_{\Lambda}$Si, $^{40}_{\Lambda}$Ca, $^{51}_{\Lambda}$V, $^{89}_{\Lambda}$Y, $^{139}_{\Lambda}$La and $^{208}_{\Lambda}$Pb with a 104 days of beamtime request. This is not the entire program of supra-precision hypernuclear spectroscopy but the first step to realize the "Hypernuclear Factory" at J-PARC.

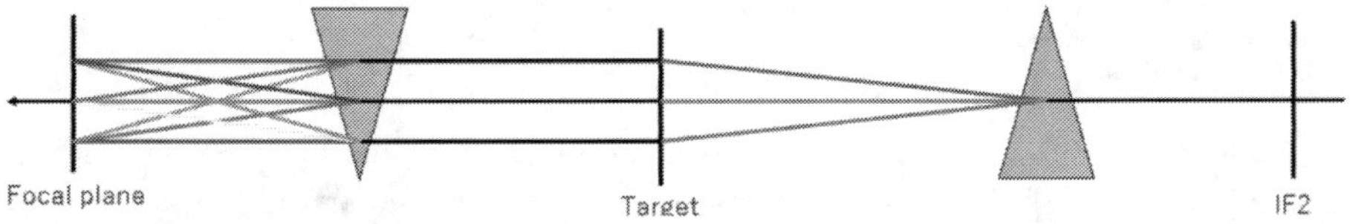

Figure 8. Conceptual illustration of the dispersion match technique. Extracted π beam has a momentum dispersion and it is converted to the spatial dispersion on the experimental target by the HIHR beamline. The kaon spectrometer is designed to cancel this spatial dispersion and the position information on the focal plane gives the excitation energies of the produced hypernuclei.

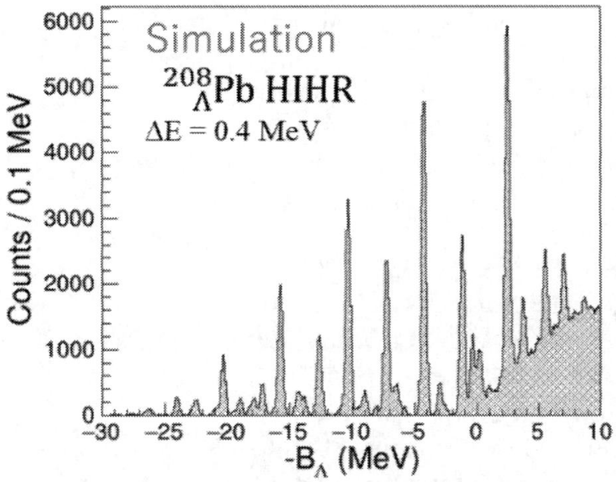

Figure 9. Expected Λ binding energy spectrum of $^{208}_{\Lambda}$Pb to be measured at HIHR. Major peaks are clearly observed, and sub-major peaks, which were previously treated as background due to the limited resolution of the experiments, can be also observed.

4 Summary

The $(e, e'K^+)$ reaction spectroscopy of Λ hypernuclei at JLab was established and new experiments from light to heavy hypernuclei are under preparation at JLab. They will provide precious data to give a clue to solve the hyperon puzzle.

At the HIHR beamline in the J-PARC hadron experimental facility extension project, a new precision spectroscopic experiment of Λ hypernuclei (SπK) was proposed using the (π^+, K^+) reaction. With the momentum dispersion match technique, HIHR with a dedicated kaon spectrometer will realize 0.4 MeV (FWHM) resolution spectroscopy to determine the Λ binding energy with an accuracy of a few 10 keV. HIHR will be a hypernuclear factory to study Λ hypernuclei in whole mass range. The high-precision $(e, e'K^+)$ experiments at JLab and new (π^+, K^+) experiments at HIHR are complementary and both of them are essentially important to solve the hyperon puzzle.

Acknowledgements

The hypernuclear experiments at JLab was partially supported by the Grant-in-Aid for Scientific Research on Innovative Areas "Toward new frontiers Encounter and synergy of state-of-the-art astronomical detectors and exotic quantum beams." This work was supported by JSPS KAKENHI grants nos. 18H05459, 18H05457, 18H01219, 17H01121, 19J22055, and 18H01220. This work was also supported by SPIRITS 2020 of Kyoto University, the Graduate Program on Physics for the Universe, Tohoku University (GP-PU), U.S. Department of Energy (DOE) grant DE-AC05-06OR23177, DE-AC02-06CH11357, DE-FG02-97ER41047 and the U.S. National Science Foundation grant no. PHY-1714809.

References

[1] B.P. Abott *et al.* (LIGO Scientific Collaboration and Virgo Collaboration), Phys. Rev. Lett. **119**, 161101 (2017)
[2] B.P. Abott *et al.* (LIGO Scientific Collaboration and Virgo Collaboration), Phys. Rev. Lett. **121**, 161101 (2018)
[3] T.E. Riley *et al.*, Astrophys. J. Lett. **887**, L21 (2019)
[4] T.E. Riley *et al.*, Astrophys. J. Lett. **918**, L27 (2021)
[5] J.R. Oppenheimer and G.M. Volkoff, Phys. Rev. **55**, 374 (1939)
[6] P.B. Demorest, T. Pennucci, S.M. Ransom, M.S.E. Roberts and J.W.T. Hessels, Nature (London) **467**, 1081 (2010)
[7] J. Antoniadis *et al.*, Science **340**, 1233232 (2013)
[8] H.T. Cromatie *et al.*, Nature Astronomy **4**, 72 (2020)
[9] D. Lonardoni *et al.*, Phys. Rev. Lett. **114**, 092301 (2015)
[10] Y. Yamamoto *et al.*, Phys. Rev. C **90**, 045805 (2014)
[11] H. Togashi *et al.*, Phys. Rev. C **93**, 035808 (2016)
[12] D. Gerstung, N. Kaiser and W. Weise, Eur. Phys. J. A **56**, 175 (2020)
[13] K. Miwa *et al.*, Phys. Rev. C **104**, 045204 (2021)
[14] S. Acharya *et al.* ALICE Collaboration, Phys. Lett. B **805**, 135419 (2020)
[15] T. Miyoshi *et al.*, Phys. Rev. Letters **90**, 232502 (2003)
[16] L. Yuan *et al.*, Phys. Rev. **C 73**, 044607 (2006)
[17] S.N. Nakamura *et al.*, Phys. Rev. Letters **110**, 012502 (2013)
[18] L. Tang *et al.*, Phys. Rev. C **90**, 034320 (2014)

[19] T. Gogami *et al.*, Phys. Rev. C **93**, 034314 (2015)
[20] T. Gogami *et al.*, Phys. Rev. C **94**, 021302(R) (2016)
[21] S.N. Nakamura, T. Gogami and L. Tang, JPS Conf. Proc., **17**, 011002 (2017)
[22] T. Gogami *et al.*, Phys. Rev. C **103**, L041301 (2021)
[23] T. Gogami *et al.*, Nucl. Inst. and Meth. A **900**, 69 (2018)
[24] M. Iodice *et al.*, Phys. Rev. Lett. **99**, 052501 (2007)
[25] F. Cusanno *et al.*, Phys. Rev. Lett. **103**, 202501 (2009)
[26] F. Garibaldi *et al.*, Phys. Rev. C **99**, 054309 (2019)
[27] S.N. Nakamura *et al.*, JLab PAC44 E12-15-008 Proposal (2016)
[28] T. Gogami *et al.*, JLab PAC49 E12-19-002 Proposal (2021); HYP2022, this proceedings
[29] F. Garibaldi *et al.*, JLab PAC48 E12-15-008 Proposal (2020); HYP2022, this proceedings
[30] S.N. Nakamura *et al.* (HIHR Hypernuclear Collaboration), J-PARC PAC32 P84 Proposal (2021)
[31] K. Aoki *et al.*, *Extension of the J-PARC Hadron Experimental Facility: Third White Paper*, ArXiv:2110.04462 [nucl-ex] (2021)
[32] F. Sakuma *et al.*, *J-PARC Hadron Experimental Facility Extension Project*, HYP2022, this proceedings
[33] BO. Sjögren, Nucl. Inst. and Meth. **7**, 76 (1960)
[34] Y. Fujita *et al.*, Nucl. Inst. and Meth. B **126**, 274 (1997)
[35] M. Fujiwara *et al.*, Nucl. Inst. and Meth. A **422**, 484 (1999)
[36] T. Motoba, Nucl. Phys. A **479** 227c (1988)
[37] T. Motoba, K. Itonaga and Y. Yamamoto, Progr. Theor. Phys. Suppl. **185** 197 (2010)
[38] T. Hasegawa *et al.*, Phys. Rev. C **53** 1210 (1996)

The J-PARC heavy ion project

Kyoichiro Ozawa[1,2,3,*], *Kazuya* Aoki[1], *Shin-ichi* Esumi[2], *Taku* Gunji[4], *Takashi* Hachiya[5], *Hiroyuki* Harada[6], *Yudai* Ichikawa[7], *Masakiyo* Kitazawa[8], *Yasuo* Miake[2], *Yuhei* Morino[1], *Shoji* Nagamiya[1,7,15], *Yasushi* Nara[9], *Megumi* Naruki[10], *Takafumi* Niida[2], *Toshihiro* Nonaka[2], *Takehiko R.* Saito[11,12], *Takao* Sakaguchi[13], *Hiroyuki* Sako[7,2], *Susumu* Sato[7], *Hitoshi* Takahashi[1], *Hirokazu* Tamura[14,7], *Kazuhiro* Tanaka[1,7], and *Satoshi* Yokkaichi[15]

[1]Institute of Particle and Nuclear Studies, KEK, Tsukuba, Ibaraki, Japan
[2]Tomonoga Center for the History of the Universe, University of Tsukuba, Tsukuba, Ibaraki, Japan
[3]Department of physics, University of Tokyo, Tokyo, Japan
[4]Center for Nuclear Study, University of Tokyo, Tokyo, Japan
[5]Department of Mathematical and Physical Sciences, Nara Women's University, Nara, Japan
[6]Accelerator division, J-PARC Center, Japan Atomic Energy Agency, Tokai, Ibaraki, Japan
[7]Advanced Science Research Center, Japan Atomic Energy Agency, Tokai, Ibaraki, Japan
[8]Yukawa Institute for Theoretical Physics, Kyoto University, Kyoto, Kyoto, Japan
[9]Faculty of International Liberal Arts Global Connectivity Program, Akita International University, Akita, Akita, Japan
[10]Division of Physics and Astronomy, Kyoto University, Kyoto, Kyoto, Japan
[11]High Energy Nuclear Physics Laboratory, Cluster for Pioneering Research, RIKEN, Wako, Saitama, Japan
[12]GSI Helmholtz Center for Heavy Ion Research, Darmstadt, Germany
[13]Brookhaven National Laboratory, Upton, NY, USA
[14]Department of Physics, Tohoku University, Sendai, Miyagi, Japan
[15]RIKEN Nishina Center for Accelerator-Based Science, RIKEN, Wako, Saitama, Japan

Abstract. A project to study high-density nuclear matter using heavy ion collisions in a beam energy range of few GeV is being prepared at J-PARC. The goal of the project is to perform experiments with beam energies of 1-12 AGeV/c and the collision rate of 10^{11} Hz. The project is divided into two phases. For the first stage, measurements with a limited beam intensity will be performed with upgraded spectrometer of an on-going experiment. Full performance will be implemented at the second phase to study in detail the high density matter and light hypernuclei. Feasibility of measurements for both phases are being evaluated.

1 Introduction

A phase diagram of Quantum Chromodynamics (QCD) is being studied intensively as shown in Fig. 1. For a high temperature region, experiments using high energy heavy ion collisions are performed at RHIC and LHC to observe and study quark gluon plasma. Recently, study of a high density region attracts considerable interest both theoretically and experimentally, since various specific phase structures, such as the first-order phase boundary, the QCD

*e-mail: ozawa@post.kek.jp

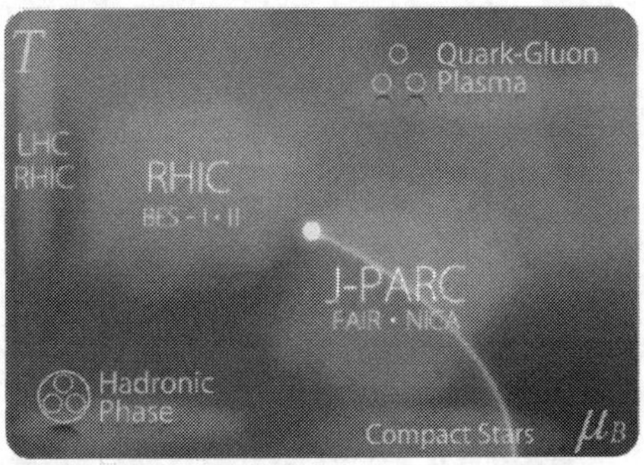

Figure 1. Schematic view of the QCD phase diagram

critical point, and color-superconducting phase, are expected to be realized in this region. In addition, recent observations of gravitational waves from neutron star mergers can give important information on a high density hadronic matter.

Experimentally, heavy ion collisions with center-of-mass energy of few GeV are suitable to generate high density matter. According to past experiments and model calculations, achieved density in such collisions is few times larger than in normal nuclear matter [1]. Several measurements have been performed to study the phase structure in the high density region. Also, measurements of various hadron productions and particle correlations are important to unveil the equation of state of the matter. For example, it is pointed out that hadron interactions in finite density matter can be measured using Lambda particle correlations in heavy ion collisions [2]. In addition, we also measure hadron properties, such as mass, width and decay rates, in the dense medium. Such measurements give important experimental information on hadron in the high-density region.

Several experiments are being prepared to study high-density matter using heavy ion collisions with center-of-mass energy of few GeV. One of the major activities is the Compressed Baryonic Matter (CBM) experiment at FAIR/GSI. At J-PARC, a project for heavy ion experiments are being prepared as well. In this manuscript, details of the heavy ion project at J-PARC are described.

2 Heavy ion program at J-PARC

A project to study high-density nuclear matter using heavy ion collisions in a beam energy range of few GeV is being prepared at J-PARC. We have designed a heavy-ion acceleration scheme and spectrometers for experiments. The goal of the project is to perform experiments with beam energies of 1-12 AGeV/c and the collision rate of 10^{11} Hz.

The project is planned using a staging approach to realize the experiment in a reasonable cost and time scale. As the first phase of the project, we will perform the experiment with upgrades of an existing spectrometer with 10^8 Hz beams. The original spectrometer was constructed for the J-PARC E16 experiment aiming to measure properties of vector mesons in a nucleus to study a chiral symmetry restoration in finite density matter [3]. For the second phase, a new dedicated spectrometer will be constructed and the maximal collision rate will be achieved.

Figure 2 shows a schematic view of the J-PARC accelerator setup in the final configuration and required upgrades for the heavy ion project.

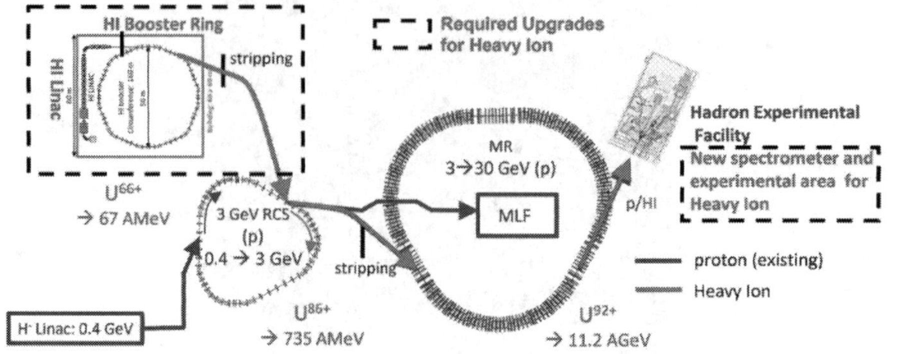

Figure 2. Schematic view of the J-PARC accelerator setup and required upgrades for the heavy ion project

To realize a high intensity heavy ion driver, a new heavy ion injector is planned in combination with existing J-PARC accelerators, which are the 3-GeV RCS and MR [4]. The driver aims to accelerate high-intensity heavy ion beams up to uranium to the energy of 11 GeV/nucleon (*A*GeV). Ions lighter than uranium can be accelerated up to 11-15 *A*GeV. The new heavy ion injector is composed of a new heavy ion LINAC and a heavy ion booster ring.

For the first phase experiment, the heavy ion injector is composed of a heavy ion LINAC and the KEK 500-MeV booster proton synchrotron (KEK-BS), since the required intensity is 10^{-3} smaller than that of the second phase experiment. The heavy ion LINAC for the first phase has the half length of the final configuration and we reuse the KEK-BS as the booster ring, which was used as the injector to the 12-GeV Proton Synchrotron at KEK. For the lower-intensity scheme, the designed maximum energy and the intensity of heavy-ions up to Au delivered from the MR to experimental beam line are 12-15 *A*GeV and 1.0×10^{8}, respectively. The various ion species are required such as deuteron, Carbon, Oxygen, Calcium and Au.

3 Experiment at phase I

The first phase experiment is proposed as an upgrade of the on-going J-PARC E16 experiment [5]. The goal of the phase-I experiment is to measure electrons and hadrons productions in heavy ion collisions with a reasonable statistics and study the quark phase in a high density region.

The J-PARC E16 experiment measures mass modifications of vector mesons in a nucleus to study partial restoration of chiral symmetry. The mass spectra are measured through an electron-positron decay mode to avoid final state interactions of hadron decays. The spectrometer of the experiment consists of tracking devices and electron identification counters as shown in Fig. 3. Three layers of Gas Electron Multiplier (GEM) Trackers and one layer of Silicon Strip Detectors (SSD) are placed for the tracking and one layer of Hadron Bind Detectors (HBD) and one layer of Lead Glass Calorimeters are placed for the electron identification. The HBD is a gas Čerenkov counter with a window-less and a mirror-less configuration [6].

The existing spectrometer focuses on the measurements of electron and positron pairs in a backward region of the center-of-mass frame, since slowly moving vector mesons are generated in this kinematical region and such mesons are expected to have a large decay

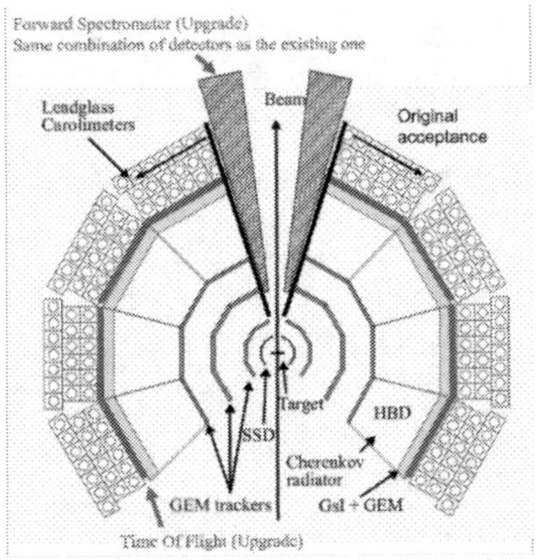

Figure 3. Schematic view of upgrades of the existing spectrometer

possibility in the nucleus. Thus, detectors in the forward region must be installed for the heavy ion experiment to cover a central rapidity region of the collision. In addition, fluctuations and correlations of particle productions are important to study thermal properties of generated matter and additional detectors for hadron measurements must be installed. We are developing a new time-of-flight counters for the hadron identifications.

Experimental feasibility is evaluated using a transport calculation code, JAM2 [7], and GEANT4 simulations [8]. Figure 4 shows evaluated acceptance of generated protons. The coverage of rapidity direction is significantly extended with the planned detector upgrade in the forward region.

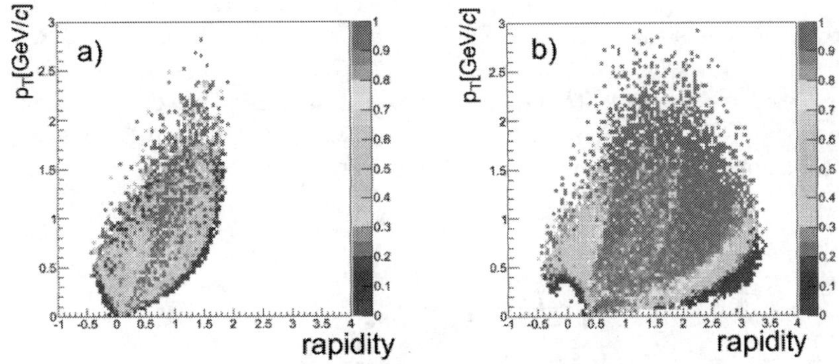

Figure 4. Acceptance of generated protons in heavy ion collisions for the existing spectrometer (a) and upgraded spectrometer (b)

One of the most important measurements using electron positron pairs is a temperature of the generated matter. Temperature can be measured as a slope of the mass spectra in the mass range higher than 1 GeV/c^2 as shown in Fig. 5. According to the recent lattice calculation [9], the energy range of the J-PARC is expected to be the region where the phase transition exists. Temperature will be measured as a function of collision energies. When obtained temperature curve exhibits a non-monotonic behavior, it indicates the existence of the first-order transition. It will be an important signature of a quark phase formation.

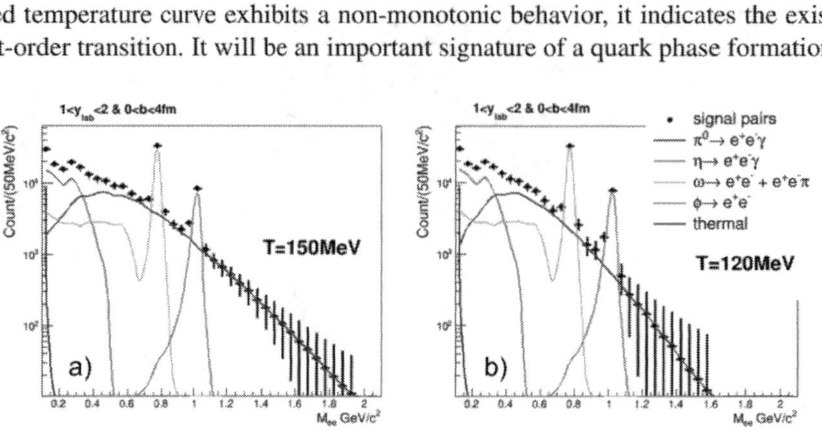

Figure 5. Expected invariant mass spectra of electron positron pairs for radiation temperature of 150 Mev (a) and 120 MeV (b)

4 Experiment at phase II

One major physics goal in the second phase is a detailed study of the high density matter. One of important measurements concerns fluctuations of conserved charges [10]. To realize such measurement, a large acceptance spectrometer to cover a wide rapidity range is required. Also, a high statistics data are needed, because signals of fluctuations are very small. Thus, we need to upgrade both the spectrometer and the accelerator.

Another physics goal is a comprehensive study of light hypernuclei. One can expect effective productions of strangeness particles at J-PARC energies, since the measured K/π ratio is maximal in this energy range [11]. Using a high intensity beam and closed geometry spectrometer, lifetimes and magnetic moments of hypernuclei can be measured precisely.

The planned spectrometer for the second phase is shown in Fig. 6. It contains detectors for hadron measurements and a sweeping magnet for closed geometry measurements. A main detector for the hadron measurements is a time projection chamber.

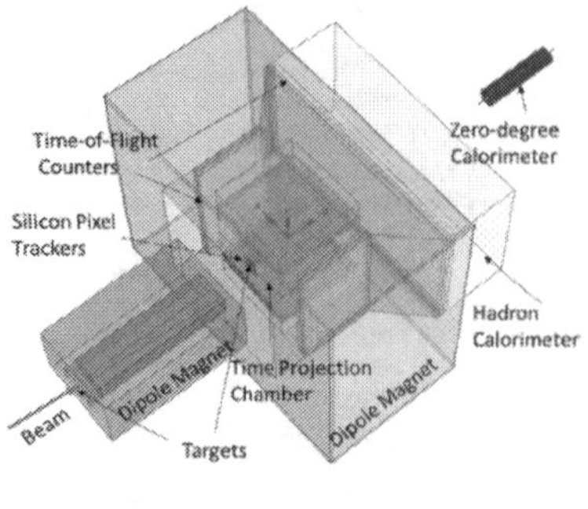

Figure 6. Schematic view of the proposed spectrometer in phase II

In addition, feasibility of other measurements is also under discussion. For example, as a signal of the phase transition between the quark phase and color-superconducting phase, a theoretical model suggests an enhance of invariant mass of electron positron pairs in a low mass region [12]. Such measurement requires extremely high statistics and it will be realized with a high intensity heavy ion beam. Furthermore, feasibility of measurements of higher order flows and correlations and heavy flavor production is being evaluated carefully.

5 Summary

A project using heavy ion beams at J-PARC are planned to study high-density nuclear matter in a beam energy range of 1-12 AGeV/c. A heavy-ion acceleration scheme and spectrometers for experiments are designed. The project has two phases. The first and second phase aims to achieve the collision rate of 10^8 Hz and 10^{11} Hz, respectively.

The accelerator needs new LINAC and booster to provide heavy ion beams. For the first phase, a new LINAC will be constructed and the KEK-PS Booster will be reused. A new large Booster will be constructed for the second phase.

The design of spectrometers are also under discussion. Upgrades of the existing spectrometer will be used for the first phase. Feasibility of temperature measurements in the first phase is evaluated and the proposal is submitted. Final design of the spectrometer for the second phase is being discussed intensively.

Acknowledgement

We express our gratitude to the members of the J-PARC Heavy Ion Program for their fruitful discussions on physics and detectors. We also thank to all accelerator physicists of J-PARC and JAEA Tandem accelerator group for designing acceleration scheme of the J-PARC Heavy Ion Program. This project was supported by MEXT/JSPS KAKENHI Grant Numbers JP19654036, JP19340075, JP21105004, JP26247048, JP15H05449, JP15K17669, JP18H05235, JP20H01935, JP20H05647 and JP21H01102.

References

[1] P. Danielewicz, R. Lacey and W. G. Lynch, Science **298**, 1592-1596 (2002)
[2] Y. Nara, A. Jinno, K. Murase and A. Ohnishi, arXiv:2208.01297 [nucl-th]
[3] S. Ashikaga *et al.*, JPS Conf. Proc. **26**, 024005 (2019)
T. N. Takahashi *et al.*, IEEE Trans. on Nucl. Sci. **68**, 1907 (2021)
[4] H. Harada *et al.*, JPS Conf. Proc. **33**, 011028 (2021)
[5] Proposal: *Proposal for dielectron measurements in heavy-ion collisions at J- PARC with E16 upgrades*, http://j-parc.jp/researcher/Hadron/en/Proposal_e.html
[6] K. Kanno *et al.*, Nucl. Instrum. and Methods. in Phys. Res. A **819**, 20 (2016)
I. Tserruya, K. Aoki, and C. Woody, Nucl. Instrum. and Methods in Phys. Res. A **970**, 163765 (2020)
[7] Y. Nara and A. Ohnishi, Phys. Rev. C **105**, no.1, 014911 (2022)
[8] J. Allison *et al.*, Nucl. Instrum. and Methods in Phys. Res. A **835**, 186-225 (2016)
[9] H. T. Ding *et al.*, Phys. Rev. Lett. **123**, 062002 (2019)
[10] M. Asakawa and M. Kitazawa, Prog. Part. Nucl. Phys. **90**, 299-342 (2016)
[11] A. Andronic *et al.* Nucl. Phys. A **837**, 65-86 (2010)
[12] T. Nishimura, M. Kitazawa and T. Kunihiro, Prog. Theor. Exp. Phys. **2022**, 093D02 (2022)

Information and perspective for doubly-strange hypernuclei with nuclear emulsion detector

Kazuma Nakazawa[1,2,*]
for the E07 collaboration (J-PARC) and the Emulsion-ML collaboration[**]

[1]Faculty of Education, Gifu University, 501-1193 Gifu, Japan
[2]Graduate School of Engineering, Gifu University, 501-1193 Gifu, Japan

Abstract. The study on doubly-strange hypernuclei has been performed with nuclear emulsion detector at KEK-PS and J-PARC for 30 years. Detected 47 candidate events of doubly-strange hypernuclei show characteristics for the Λ-Λ and Ξ-*N* interactions. The Λ-Λ interaction is weakly at-tractive and the binding energy of two Λ hyperons to nuclei shows linear dependence on the mass number of a nucleus. The Ξ^- hyperon is found to be bound by the ^{14}N nucleus, forming the $^{15}_{\Xi}$C hypernucleus, with larger energy than that given by Coulomb force. Also, the level scheme of the Ξ^- hyperon in the $^{15}_{\Xi}$C hypernucleus can be seen. With the light source of hard X-ray instead of an optical microscope, since it will enable to count grains constituting tracks in the emulsion, charge recognitions of light nuclei, especially H^+ and He^{2+}, can be possible. This would work effectively in determining the nuclides of doubly-strange hypernuclei. The '*overall-scanning method*' to scan whole volume of the emulsion sheets will allow to detect nearly 1 x 10^3 events of doubly-strange hypernuclei, which is the expected number recorded in the emulsion of the E07 experiment, with machine learning in the near future.

1 Introduction

The detections of double-Λ hypernucleus were reported for more than 50 years ago [1-3]. The interaction energy, $\Delta B_{\Lambda\Lambda}$, between two Λ hyperons was presented to be about 3 – 5 MeV. After that, no doubly-strange hypernuclei were detected, because detection by skilled personnel had reached its limits. Doubly-strange hypernuclei have been very useful for unified understanding of the baryon(*B*)-*B* interaction under SU(3)$_f$ symmetry, and for the study of the nuclear equation of state of neutron stars because of likely appearance of hyperons in them.

In the E176 experiment at KEK-PS, four candidates of doubly-strange hypernuclei in about 52 at-rest events of Ξ^- hyperon were successfully detected via $p(K^-, K^+)\Xi^-$ reactions with Emulsion-Counter '*hybrid-method*' [4-7], with which electric detectors predict the highly possible positions of production of doubly-strange hypernuclei in the emulsion sheet.

* e-mail: nakazawa.kz@gmail.com
** Members of the Emulsion-ML collaboration can be found in the author list of one of these proceedings by Manami Nakagawa

However, the nuclide of double-Λ hypernucleus was not identified uniquely, and the case close to the $\Delta B_{\Lambda\Lambda}$ in the past was preferred. Other two events were candidates of Ξ hypernucleus formed via at-rest absorption of Ξ^- hyperon in ^{12}C nucleus.

Among nearly 4 x 10^2 events of Ξ^- hyperon captured at rest by nuclei in the emulsion with the *hybrid-method*, the detection of the Nagara event, ${}^{6}_{\Lambda\Lambda}$He (^{4}He + 2Λ) in the E373 experiment at KEK-PS requested change of the $\Delta B_{\Lambda\Lambda}$ to about 1 MeV, then the Λ-Λ interaction was confirmed to be weakly attractive [8, 9]. The past three double-Λ hypernuclei showing the range of $\Delta B_{\Lambda\Lambda}$ to be 3 – 5 MeV were also re-examined. The $\Delta B_{\Lambda\Lambda}$ reported in Ref. [3] had a large discrepancy with the one given by the Nagara event, even if both events were composed by the same nuclide, so that the authenticity of one event of Ref. [3] was lost. Among 10 candidate events of doubly-strange hypernucleus, the Kiso event was the first demonstration of the presence of deeply bound Ξ hypernucleus, ${}^{15}_{\Xi}$C, where the event was detected in the development of '*overall-scanning method*' to search for typical-topology event of doubly-strange hypernuclei in whole volume of the emulsion sheets [10]. Later analysis showed that the Kinka event gave even deeper binding [11].

In the E07 experiment at J-PARC, 33 candidate events of doubly-strange hypernuclei have been detected among 2 x 10^3 at-rest captured events of Ξ^- hyperon with the *hybrid-method*. One double-Λ hypernucleus, the Mino event, has presented the $\Delta B_{\Lambda\Lambda}$ of $_{\Lambda\Lambda}$Be nucleus, where $\Delta B_{\Lambda\Lambda}$ value was not consistent with the one by the Nagara event in the most likely case as a ${}^{11}_{\Lambda\Lambda}$Be [12]. Regarding Ξ hypernuclei, the Ibuki event showed a Coulomb-assisted nuclear bound state of ${}^{15}_{\Xi}$C uniquely [13]. The Irrawaddy event was understood to be unique detection of a nuclear *s*-state of ${}^{15}_{\Xi}$C [11]. The accumulation of Ξ hypernuclei has revealed the level structure of the Ξ hypernucleus of ${}^{15}_{\Xi}$C.

Total number of candidates of doubly-strange hypernuclei is 47, and information on Λ-Λ and Ξ-*N* interactions was obtained in less than 20 events. This is because, in many events, the daughter particles emitted from production and decay of doubly-strange hypernuclei cannot be identified. To identify charges of the daughter particles, *e.g.* H^+, H^{2+}, Li^{3+}, Be^{4+} or B^{5+}, the development has been made for the measurement of track thickness in the emulsion [14]. A more promising approach is the use of X-rays as a light source for microscopy [15]. By using X-ray microscopy, developed grains consisting tracks can be counted, then the discrepancy of the grain densities of tracks allows for the identification of the charge.

In the E07 emulsion, the *hybrid-method* was applied to effectively get information of at-rest stopping Ξ^- hyperons. However, many events were lost due to mis-detection of electric detectors, especially no-detection of $n(K^-, K^0)\Xi^-$ events. If it is possible to detect the production and decay of doubly-strange hypernuclei with the *overall-scanning method*, the detectable number of doubly-strange hypernuclei is estimated to reach one thousand. The R&D of the *overall-scanning method* with a machine learning method is on-going.

2 Information for doubly-strange hypernuclei

Regarding double-Λ hypernuclei, two physical values are defined as

$$B_{\Lambda\Lambda} = M({}^{A-2}\mathrm{Z}) + 2M(\Lambda) - M({}^{A}_{\Lambda\Lambda}\mathrm{Z}), \quad (1)$$
$$\Delta B_{\Lambda\Lambda} = B_{\Lambda\Lambda} - 2B_{\Lambda}({}^{A-1}_{\Lambda}\mathrm{Z}), \quad (2)$$

where M is the symbol for the mass of the particles in brackets. The $B_{\Lambda\Lambda}$ is the binding energy of two Λ hyperons by a double-Λ hypernucleus with the mass of $M({}^{A}_{\Lambda\Lambda}\mathrm{Z})$. The Λ binding energy by a single-Λ hypernucleus of ${}^{A-1}_{\Lambda}$Z is denoted as $B_{\Lambda}({}^{A-1}_{\Lambda}\mathrm{Z})$.

The Nagara event presented the $B_{\Lambda\Lambda}$ and $\Delta B_{\Lambda\Lambda}$ values to be 7.25 ± 0.19 MeV and 1.01 ± 0.20 MeV, respectively [8]. As mentioned above, the oldest double-Λ hypernucleus event [1, 2] was then revised for the values of $B_{\Lambda\Lambda}$ and $\Delta B_{\Lambda\Lambda}$ to be 14.7 ± 0.4 MeV and 1.3 ± 0.4 MeV as the formation of ${}^{10}_{\Lambda\Lambda}$Be, respectively [9]. The $B_{\Lambda\Lambda}$ and $\Delta B_{\Lambda\Lambda}$ values of E176 double-Λ hypernucleus were also revised as 23.3 ± 0.7 MeV and 0.6 ± 0.8 MeV, respectively [9]. From the E373 and E07 experiments, three and two events have been published as double-Λ hypernuclei, respectively [9, 10, 16]. Published data for $B_{\Lambda\Lambda}$ and $\Delta B_{\Lambda\Lambda}$ of double-Λ hypernuclei are listed in Table 1. The relationship between $B_{\Lambda\Lambda}$ and mass number A for most probable case with small errors is plotted in Figure 1, where a linear dependence between them can be seen.

Table 1. Published data of $B_{\Lambda\Lambda}$ and $\Delta B_{\Lambda\Lambda}$ under assumption of Ξ^- hyperon captured in atomic 3D state in ^{12}C, ^{14}N or ^{16}O. Regarding the Nagara event, the values of the $B_{\Lambda\Lambda}$ and $\Delta B_{\Lambda\Lambda}$ have been revised in Ref. [9] as in this table due to change of the mass of Ξ^- hyperon in Ref. [17].

Event	Ξ^- captured by	nuclide	$B_{\Lambda\Lambda}$ (MeV)	$\Delta B_{\Lambda\Lambda}$ (MeV)	Comments [reference(s)]
Nagara	^{12}C	${}^{6}_{\Lambda\Lambda}$He	6.91 ± 0.16	0.67 ± 0.17	Uniquely identified [9]
Danysz, et al.	^{12}C	${}^{10}_{\Lambda\Lambda}$Be	14.7 ± 0.4	1.3 ± 0.4	${}^{10}_{\Lambda\Lambda}$Be → ${}^{9}_{\Lambda}$Be* (*Ex.* = 3.0 MeV) [9, 18]
E176	^{14}N	${}^{13}_{\Lambda\Lambda}$B	23.3 ± 0.7	0.6 ± 0.8	${}^{13}_{\Lambda\Lambda}$B → ${}^{13}_{\Lambda}$C* (*Ex.* = 4.9 MeV) [7]
Demachi-yanagi	^{14}N	${}^{10}_{\Lambda\Lambda}$Be*	11.90 ± 0.13	− 1.52 ± 0.15	most probable (topology) [9] *Ex.* ~ 2.8 MeV for ${}^{10}_{\Lambda\Lambda}$Be*
Mikage	^{12}C	${}^{6}_{\Lambda\Lambda}$He	10.01 ± 1.71	3.77 ± 1.71	most probable (mesonic decay) [9]
	^{12}C	${}^{11}_{\Lambda\Lambda}$Be	22.15 ± 2.94	3.95 ± 3.00	
	^{14}N	${}^{11}_{\Lambda\Lambda}$Be	23.05 ± 2.59	4.85 ± 2.63	
Hida	^{12}C	${}^{11}_{\Lambda\Lambda}$Be	20.83 ± 1.27	2.61 ± 1.34	
	^{14}N	${}^{12}_{\Lambda\Lambda}$Be	20.48 ± 1.21	2.00 ± 1.21	Assumed 10.24 MeV for B_{Λ}(${}^{11}_{\Lambda}$Be) [9]
Mino	^{16}O	${}^{10}_{\Lambda\Lambda}$Be	15.05 ± 0.11	1.63 ± 0.14	
		${}^{11}_{\Lambda\Lambda}$Be	19.07 ± 0.11	1.87 ± 0.37	most probable (χ^2 minimum) [12]
		${}^{12}_{\Lambda\Lambda}$Be	13.68 ± 0.11	− 2.7 ± 1.0	
D001	^{12}C	${}^{8}_{\Lambda\Lambda}$Li	17.50 ± 1.46	6.34 ± 1.46	
	^{1N}N	${}^{11}_{\Lambda\Lambda}$Be	15.05 ± 2.78	1.63 ± 2.78	likely by $B_{\Lambda\Lambda}$ [16]

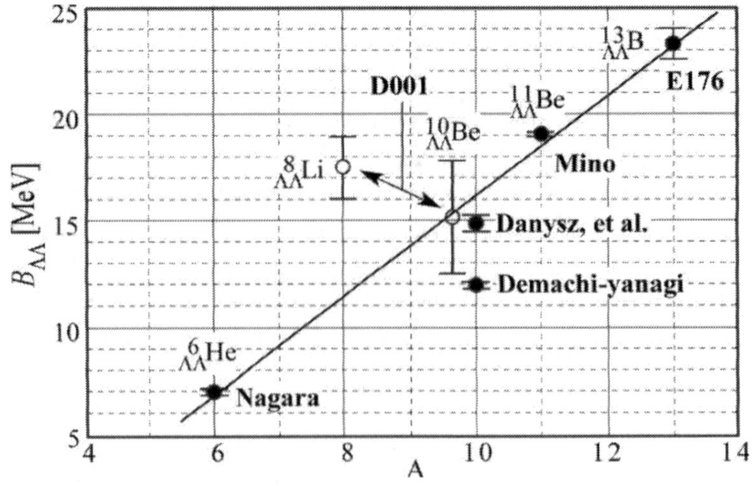

Fig. 1. Relationship between the mass number A and $B_{\Lambda\Lambda}$ for the most probable cases with small errors form Table 1.

If Ξ^- hyperon is bound by a nucleus with a much strong force than Coulomb one, a Ξ hypernucleus is formed, and it is desirable to measure the binding energy of the Ξ^- hyperon, B_{Ξ^-}, in the nucleus. This is equivalent to measuring the mass of the nucleus containing the Ξ^- hyperon by measuring the total energy obtained from all decay products. For this measurement, an event, twin hypernucleus, in which two single-Λ hypernuclei of known mass are emitted from the Ξ^- absorption point is preferred. Therefore B_{Ξ^-} is obtained as

$$M(^{A}Z) + M(\Xi^-) - B_{\Xi^-} = M(^{A1}_{\Lambda}Z_1) + M(^{A2}_{\Lambda}Z_2) + M(\textstyle\sum_{i=1}^{k} m_i) + KE, \quad (3)$$

where the $^{A1}_{\Lambda}Z_1$, $^{A2}_{\Lambda}Z_2$ and m_i are two single-Λ hypernuclei and other emitted particles, respectively. KE is the total kinetic energy of every emitted particle from the absorption point.

There have been 18 events of twin hypernuclei detected in the past three experiments. The nuclides that could be identified as absorbing Ξ^- hyperons were ^{12}C, ^{14}N and ^{16}O, with the number of events, 4, 7 and 1, although the molar ratio are 0.55 : 0.15 : 0.30, respectively. In the case of ^{14}N, the core nucleus after absorbing Ξ^- hyperon becomes carbon. It may be easy to identify the absorbing nuclide because decay daughters of the Ξ^--^{14}N system tend to be α clusters. Among the events that have not been identified, there may be events absorbed by ^{12}C and ^{16}O. However, the number of ^{14}N absorption of 7 is large compared to 18 x 0.15 = 2.7.

The experimental values of B_{Ξ^-} in the case of the ^{14}N absorption are plotted with several theoretical calculations in Figure 2. The nuclear $2p$ state is clearly seen at about – 1 MeV. There should be the nuclear $1s$ state at – 5 ~ – 8 MeV. Theoretical calculations are in good agreement with experimental data. Detail discussion is given in Ref. [11] and experimental values can be found therein.

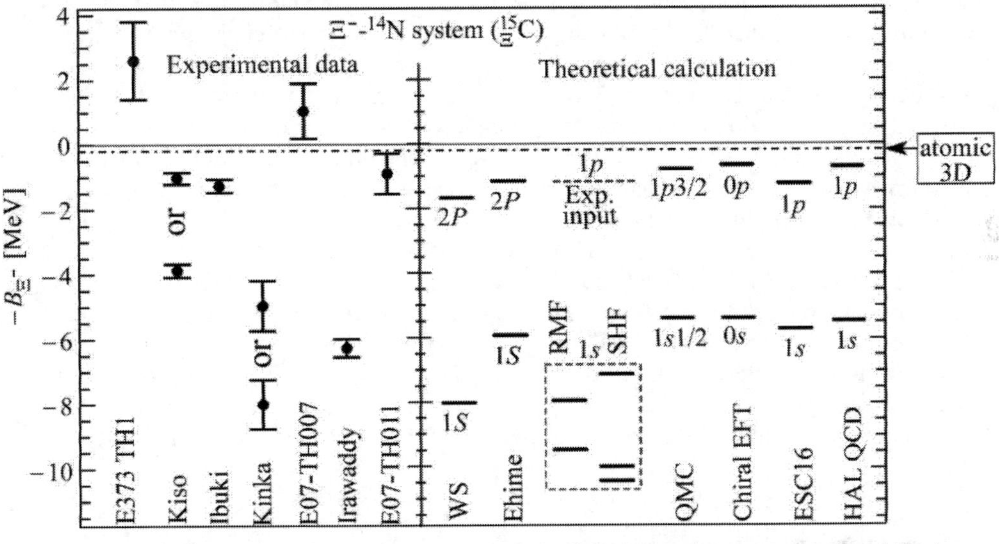

Fig. 2. Experimental data of B_{Ξ^-} and various theoretical calculations. Relevant references of both ones are given in Ref. [11] except for Chiral-EFT [19], ESC16 [20] and HAL-QCD [21].

3 R&D for systematic study of doubly-strange hypernuclei

Much research and development has been done to identify as many doubly-strange hypernuclei as possible. However, for more than 50% of the candidates, no detailed information is available at the moment, because the nuclide of decay daughters cannot be determined as introduced in Section 1.

Very recently, the use of X-rays reveals the possibility to identify the decay daughter particles by measuring the relation between the energy loss, developed grain density (G.D. : grains/0.1mm), and the length of the track with X-ray microscopy. Figure 3 shows track images of a $^{1}H^{+}$ and a $^{4}He^{2+}$ near their stopping points. The G.D. was measured for 3 and 10 tracks of $^{1}H^{+}$ and $^{4}He^{2+}$, respectively, on the range 10 ~ 80 μm from the stopping point. Very preliminary results show 231 ± 11 and 267 ± 6 [grains/0.1mm] for $^{1}H^{+}$ and $^{4}He^{2+}$, respectively. It thus seems to be possible to recognize the charge of both nuclei. The number of samples is currently small and should be increased, and then the possibility of identification would be confirmed for other nuclides of Li, Be and B.

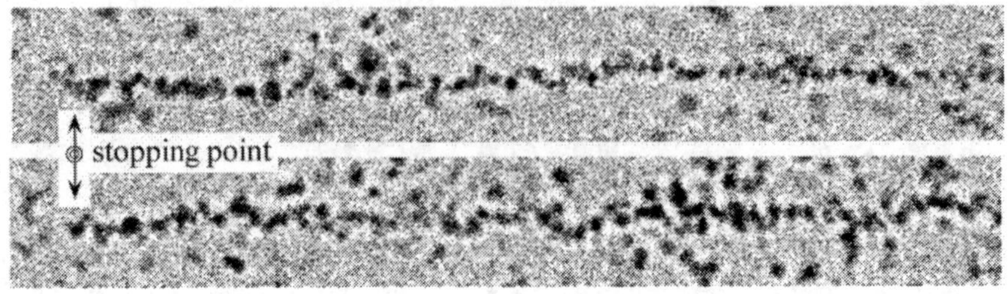

Fig. 3. Images taken by X-ray microscopy for tracks of a) $^{1}H^{+}$ and b) $^{4}He^{2+}$ around their stopping point.

As mentioned in Section 1, about 1 x 10^3 candidate events might be recorded in the emulsion sheets. Although it would be sufficient to search the whole volume of the emulsion sheets, it would take 560 years if people have to search for them. Therefore, the development of the *overall-scanning method* has been started for more than 10 years ago. Initially, line-segment detection in microscope images of tracks was attempted by means of the Hough transform. It was applied to the detection of αparticle groups associated with the decay of natural radioisotopes with some success [22], but was not put into practical use.

Recently, machine learning models have been developed to detect rare events recorded in the emulsion sheets, with positive results. Track information of rare events was generated by Monte Carlo simulation and its image style was transformed into mimetic image in the emulsion. They are employed to train a machine learning model for event detection, and the model can obtain the position and topology of events in actual microscopic images. These models were applied to search for $^{3}_{\Lambda}H$ to precisely measure the Λ binding energy, and the decay of $^{3}_{\Lambda}H$ was successfully detected [23-25]. During the search for the decay of $^{3}_{\Lambda}H$, an event with a topology of production and decay of double-Λ hypernucleus was successfully detected, although the nuclide could not be determined uniquely. The development of machine learning models has been started for selective detection of doubly-strange hypernuclei [25]. Now, a model successfully detects the Nagara event. No doubt further development will lead to the detection of numerous doubly-strange hypernuclei in the near future.

4 Summary and perspective

The *hybrid-method* with nuclear emulsion detector has opened a doorway for the study of doubly-strange hypernuclei. By the detection of double-Λ hypernuclei, the Λ-Λ interaction was found to be weakly attractive and two Λ hyperons' binding energies, $B_{\Lambda\Lambda}$, seem to linearly depend on the mass number of core nuclei. The presence of a Ξ hypernucleus being deeper bound than by Coulomb force shows the Ξ-*N* interaction to be attractive. By dedicated technics analysing the events in the emulsion and the *overall-scanning method* with machine Learning models, an order of magnitude more doubly-strange hypernuclei will be detected. Much detailed understanding will then be obtained of the interaction relevant to double strangeness sector using doubly-strange hypernuclei. By the Hadron-Hall extension at J-PARC, we will be able to step out of the doubly-strange hypernuclei and study the triple strangeness world.

We thank the KEK staff for the E176 and E373 experiment, the staff of the J-PARC accelerator and the Hadron experimental facility, FUJIFILM Corporation, KEK computing research, NII for SINET5, Kamioka Observatory, ICRR for the E07 experiment. This work was supported by Japan Society for the Promotion of Science (JSPS) KAKENHI Grant Numbers JP23224006, JP16H02180, JP20H00155, JP20J00682; Ministry of Education, Culture, Sports, Science and Technology (MEXT) KAKENHI Grant Numbers JP15001001, JP24105002, JP18H05403, JP19H05147; the National Research Foundation (NRF) of Korea with Grant Number 2018R1A2B2007757; and Deutscher Akademischer Austauschdienst (DAAD) Grant numbers PPP Japan 2017, No. 57345296 by DE/Federal Republic of Germany, and Grant agreement STRONG 2020, No. 824093 by the European Union.

References

1. M. Danysz, et al, Nucl. Phys. **49**, 121 (1963)
2. M. Danysz, et al., Phys. Rev. Lett. **11**, 29 (1963)
3. D.J. Prowse, Phys. Rev. Lett. **17**, 782 (1966)
4. S. Aoki, et al., Prog. Theor. Phys. **85**, 1287 (1991)
5. S. Aoki, et al., Prog. Theor. Phys. **89**, 493 (1993)
6. S. Aoki, et al., Phys. Lett. B **335**, 45 (1995)
7. S. Aoki, et al., Nucl. Phys. A **828**, 191 (2009)
8. H. Takahashi, et al., Phys. Rev. Lett. **87**, 212502 (2001)
9. J.K. Ahn, et al., Phys. Rev. C **88**, 014003 (2013)
10. K. Nakazawa, et al., Prog. Theor. Exp. Phys. **2015**, 033D02 (2015)
11. M. Yoshimoto, et al., Prog. Theor. Exp. Phys. **2021**, 073D02 (2021)
12. H. Ekawa, et al., Prog. Theor. Exp. Phys. **2019**, 021D02 (2019)
13. S.H. Hayakawa, et al., Phys. Rev. Lett. **126**, 062501 (2021)
14. S. Kinbara, et al., Prog. Theor. Exp. Phys. **2019**, 011D01 (2019)
15. A. Kasagi, et al., Eur. Phys. J. A **58**, 190 (2022)
16. A.N.L. Nyaw, et al., Bull. Soc. Photogr. Imag. Japan **30**, 22 (2020)
17. C. Amsler, et al. (Particle Data Group), Phys. Lett. B **667**, 1 (2008)
18. D.H. Davis, Nucl. Phys. A **754**, 3c (2005)
19. M. Khono, Phys. Rev. C **100**, 024313 (2019)
20. M.M. Nagels, et al., Phys. Rev. C **99**, 044003 (2019)
21. K. Sasaki, et al., Nucl. Phys. A **988**, 121737 (2020)
22. J. Yoshida, et al., Nucl. Instr. Meth A **847**, 86 (2017)
23. T.R. Saito et al., Nat. Rev. Phys. **3**, 803 (2021)
24. T.R. Saito, these proceedings
25. M. Nakagawa, these proceedings

Unique approach for precise determination of binding energies of hypernuclei with nuclear emulsion and machine learning

Manami Nakagawa[1,*], *Ayumi* Kasagi[1,2], *Enqiang* Liu[1,3,4], *Hiroyuki* Ekawa[1], *Junya* Yoshida[5], *Wenbo* Dou[1,6], *Yan* He[1,11], *Abdul* Muneem[1,7], *Kazuma* Nakazawa[1,8], *Christophe* Rappold[9], *Nami* Saito[1], *Takehiko R.* Saito[1,10,11], *Shohei* Sugimoto[1,6], *Masato* Taki[12], *Yoshiki K.* Tanaka[1], *He* Wang[1], *Yiming* Gao[1,3,4], *Ayari* Yanai[1,6], and *Masahiro* Yoshimoto[13]

[1]High Energy Nuclear Physics Laboratory, Cluster for Pioneering Research, RIKEN, 2-1 Hirosawa, Wako, Saitama, 351-0198, Japan

[2]Graduate School of Engineering, Gifu University, 1-1 Yanagido, Gifu 501-1193, Japan

[3]Institute of Modern Physics, Chinese Academy of Sciences, 509 Nanchang Road, Lanzhou, Gansu Province, 730000, China

[4]School of Nuclear Science and Technology, University of Chinese Academy of Sciences, No.19(A) Yuquan Road, Shijingshan District, Beijing, 100049, China

[5]Department of physics, Tohoku University, Aramaki, Aoba-ku, Sendai 980-8578, Japan

[6]Department of Physics, Saitama University, Saitama, 338-8570, Japan

[7]Faculty of Engineering Sciences, Ghulam Ishaq Khan Institute of Engineering Sciences and Technology, Topi, 23640, KP, Pakistan

[8]Faculty of Education, Gifu University, 1-1 Yanagido, Gifu 501-1193, Japan

[9]Instituto de Estructura de la Materia, CSIC, Madrid, Spain

[10]GSI Helmholtz Centre for Heavy Ion Research, Planckstrasse 1, D-64291 Darmstadt, Germany

[11]School of Nuclear Science and Technology, Lanzhou University, 222 South Tianshui Road, Lanzhou, Gansu Province, 730000, China

[12]Graduate School of Artificial Intelligence and Science, Rikkyo University, 3-34-1 Nishi Ikebukuro, Toshima-ku, Tokyo, 171-8501, Japan

[13]RIKEN Nishina Center for Accelerator-Based Science, RIKEN, 2-1 Hirosawa, Wako, Saitama 351-0198, Japan

Abstract. Hypertriton is the lightest hypernucleus and a benchmark in hypernuclear physics. However, it has recently been suggested that its lifetime and binding energy values may differ from the established values. To solve this puzzle, it is necessary to measure both values with a higher precision. For the precise measurement of the binding energy, we are aiming at developing a novel technique to measure the hypertriton binding energy with unprecedented accuracy by combining nuclear emulsion data and machine learning techniques. The analysis will be based on the J-PARC E07 nuclear emulsion data. Furthermore, a machine-learning model is being developed to identify other single and double-strangeness hypernuclei.

*e-mail: manami.nakagawa@riken.jp

1 Introduction

From the 1950s untill the 1970s, hypernuclei were studied using nuclear emulsions and bubble chambers, which recorded the tracks of the produced hypernuclei and their decay particles as photographic images [1]. In those studies, nuclides of hypernuclei were determined from production and decay kinematics, and their masses and the binding energies were obtained. Those studies have provided fruitful information on the potential and interaction associated with Lambda hyperons in hypernuclei. Recently, the Λ-Λ and Ξ-nucleon interactions have also been studied. To investigate the interactions associated with hyperons in detail, precise measurements of the binding energy of hypernuclei are important in the hypernuclear physics.

Since the hypertriton ($^{3}_{\Lambda}$H) is a benchmark in the hypernuclear physics, information on its lifetime and binding energy is a basis for various theoretical calculations of hypernuclei. The structure of hypertriton is formed by a Λ hyperon which is very weakly bound to a deuteron core with a small binding energy of 0.13 ± 0.05 MeV [1]. The distance between the Λ and deuteron is approximately 10 fm. By considering the small binding energy, it has been suggested that the lifetime of a hypertriton is comparable to the lifetime of a free Λ which is 263 ps. However, in 2013, the HypHI experiment measured hypertriton production using heavy ion beams and found a significantly smaller lifetime than that of a free Λ [2]. Subsequently, various experiments reported smaller values [4], raising questions about the binding energy of the hypertriton.

In 2020, the STAR experiment reported the binding energy of hypertriton based on measured samples of hypertritons and anti-hypertritons, and the combined value was deduced to be 0.41 ± 0.16 MeV [3]. This value differs significantly from the world average, however, the accuracy of the STAR measurement was not sufficient to draw a conclusion. The current reported values of the binding energy and lifetime are summarized in Ref. [5] and [4], respectively. Thus, even fundamental properties of a hypertriton, the simplest hypernucleus, are still not well understood, in spite of the fact that it is a benchmark in the hypernuclear physics. This situation demands that the binding energy of a hypertriton should be measured with high precision. Measuring the binding energy of other hypernuclei with high precision is also important in the hypernuclear physics.

2 Detection of Hypertriton

2.1 Nuclear emulsion

We utilize a nuclear emulsion to measure the binding energy precisely. A nuclear emulsion has the capability to record three-dimensional trajectories of charged particles as they travel through these detectors. An experiment with nuclear emulsions (the J-PARC E07 experiment) was conducted at J-PARC from 2016 untill 2017 to search for Ξ^- or $\Lambda\Lambda$ hypernuclei with two strange quarks [6]. The experiment reported the so-called MINO, IBUKI, and IRRAWADDY events [7–9].

The J-PARC E07 nuclear emulsions irradiated with K^- mesons are expected to record millions of single hypernuclear and thousands of double-strangeness hypernuclear events. Those events are recorded with a high position resolution of less than 1 μm for charged particle tracks from hypernuclear production and decay. The accuracy for measuring the binding energy of hypertriton by the nuclear emulsion can be approximately 30 keV(stat.) + 30 keV (syst.) [10]. The J-PARC E07 experiment aims to detect and sort out hypertriton events in nuclear emulsion data. Since the only hypernuclei that decay in a two-body decay with helium isotopes in a final state are hypertritons and $^{4}_{\Lambda}$H, which can be easily distinguished by the difference in the range of pion, we focused on this two-body decay mode of hypertriton. However, with the conventional image analysis techniques for nuclear emulsions, a large number of background events are misdetected [11]. Furthermore, required visual inspection by human's eyes would take 560 years to analyse all the nuclear emulsion images. Therefore, we have developed machine learning techniques to detect events efficiently.

2.2 Machine learning

A machine learning model for detecting hypertriton events of interest from images in the nuclear emulsions requires a large amount of training data. Such images with hypertriton events were missing since hypertritons were never been detected from the J-PARC E07 nuclear emulsions. Therefore, we have developed a machine learning model to generate artificial microscopic images by employing physics simulations based on Geant4 [12]. The algorithm for generation artificial nuclear emulsion images works as follows.

First, hypertriton events associated with the production and decay of hypertritons are generated by the Geant4 simulation, and real nuclear emulsion images are synthesized as a background image. For the synthesis method, an image transfer technique with pix2pix [13], a machine learning technique that utilizes the GAN technique [14], is used. These artificial microscopic images are used as training data for an object detection model with Mask R-CNN [15]. The hypertriton events that appear in the microscopic images provide the ground truth for the machine learning algorithm. The trained model was then applied to the real microscope images. As a result, hypertriton events have been successfully observed for the first time oin our nuclear emulsion [16]. Data collection, analysis, and energy calibrations still continues.

3 Development for Single- and Double-strangeness hypernuclei

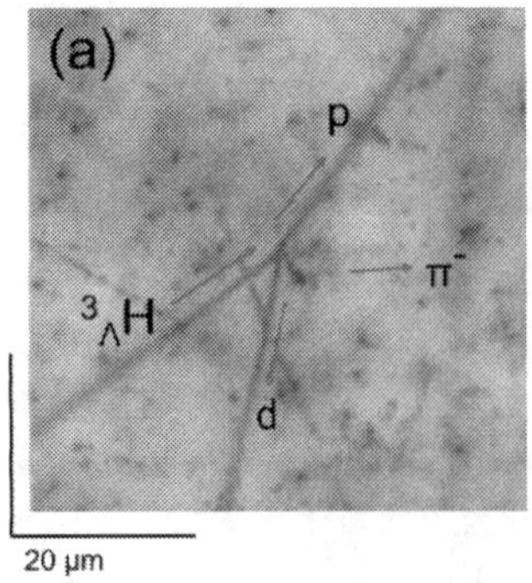

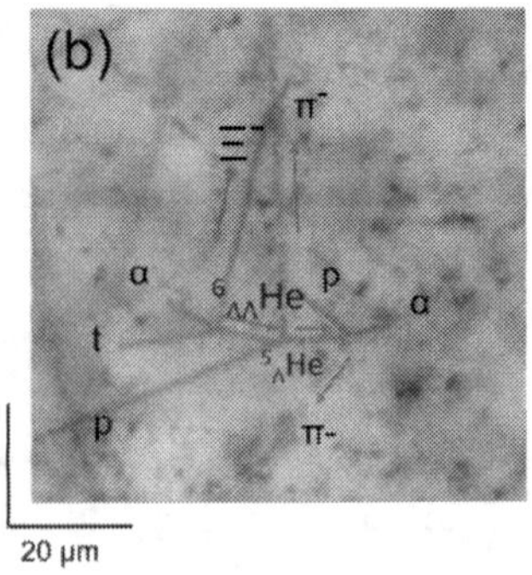

Figure 1. Artificial microscopic images of (a) a ${}^{3}_{\Lambda}\mathrm{H} \rightarrow p + d + \pi^-$ decay event and (b) a ${}^{6}_{\Lambda\Lambda}\mathrm{He}$ decay event

We are currently developing other models for hypernuclei detections. Since hypernuclei with a many-body decay are dominant, as a first step, we are developing a model for detecting hypernuclei with a three-body decay. Among these, especially important hypernuclei are hypertriton, ${}^{4}_{\Lambda}\mathrm{H}$, and ${}^{4}_{\Lambda}\mathrm{He}$. The hydrogen hypernuclei, hypertriton and ${}^{4}_{\Lambda}\mathrm{H}$ are important to cross-check the results obtained with the analysis of the two-body decay channels. Let us note that the ${}^{4}_{\Lambda}\mathrm{He}$ is a nuclide for which charge symmetry breaking is observed in the results of the γ-ray spectroscopy while the ${}^{4}_{\Lambda}$H is a mirror hypernucleus of ${}^{4}_{\Lambda}$He [17]. The recent binding energy data of the ground state of ${}^{4}_{\Lambda}$H are known to be consistent with the results of former nuclear emulsion experiments [18]. However, the binding energy data for the ground state of ${}^{4}_{\Lambda}$He is only available for former nuclear emulsions [1]. Thus, it is important to develop the detection model for hypernuclei with three-body decay modes such as hypertriton, ${}^{4}_{\Lambda}$H, and ${}^{4}_{\Lambda}$He. In addition, we are developing a detection model for double-strangeness hypernuclei. statistical and systematic uncertainties related to parameters of double-strange nuclei are too large to draw conclusions on the nature of baryonic interaction involving double strangeness. Therefore, it is necessary to discover and identify many double-strangeness hypernuclei, which would also allow to improve the accuracy of the binding energy for those double-strangeness hypernuclei that have already been observed. Panels in Figure 1 show artificial microscopic images of a three-body decay ${}^{3}_{\Lambda}\text{H} \rightarrow p + d + \pi^{-}$ event and of a ${}^{6}_{\Lambda\Lambda}$He decay event generated by our current framework.

4 Summary

A combination of a nuclear emulsion and a machine learning technique is employed to measure the hypertriton binding energy with high precision. The hypertriton events are searched for in nuclear emulsion data collected by the J-PARC E07 experiment. The machine learning technique also allows to search for many other single- and double-strangeness hypernuclei decay channels.

References

[1] M. Juric *et al.*, Nucl. Phys. B **52**, 1-30 (1973)
[2] C. Rappold *et al.* (HypHI Collaboration), Nucl. Phys. A **913**, 170-184 (2013)
[3] The STAR Collaboration, Nat. Phys. **16**, 409-412 (2020)
[4] M. S. Abdallah *et al.* (STAR Collaboration), Phys. Rev. Lett. **128**, 202301 (2022)
[5] A. Calivà (ALICE Collaboration), EPJ Web Conf. **259**, 03004 (2022)
[6] K Imai, K. Nakazawa, and H. Tamura, Proposal for Nuclear and Particle Physics Experiments at J-PARC (2006)
[7] H. Ekawa *et al.*, Prog. Theor. Exp. Phys. **2019**, 021D02 (2019)
[8] S. H. Hayakawa *et al.*, Phys. Rev. Lett. **126**, 062501 (2021)
[9] M. Yoshimoto *et al.*, Prog. Theor. Exp. Phys. **2021**, 073D02 (2021)
[10] E. Liu *et al.*, Eur. Phys. J. A **57**, 327 (2021)
[11] J. Yoshida *et al.*, Nucl. Instrum. Methods Phys. Res. A **847**, 86-92 (2017)
[12] A. Kasagi *et al.*, Supl. Rev. Mex. Fis. **3**, 0308122 (2022)
[13] T. Wang *et.al.*, arXiv:1711.11585 (2017)
[14] I. J. Goodfellow *et al.*, arXiv:1406.2661 (2014)
[15] K. He, G. Gkioxari, P. Dollar, and R. B. Girshick *et al.*, arXiv:1703.06870 (2017)
[16] T. R. Saito *et al.*, Nat. Rev. Phys. **3**, 803-813 (2021)
[17] T. O. Yamamoto *et al.* (J-PARC E13 Collaboration), Phys. Rev. Lett. **115**, 222501 (2015)
[18] A. Esser *et al.* (A1 Collaboration), Phys. Rev. Lett. **114**, 232501 (2015)

EPJ Web of Conferences **271**, 11007 (2022) https://doi.org/10.1051/epjconf/202227111007
HYP2022

Development of a triple coincidence method of reaction, gamma-ray, and weak decay in the hypernuclear gamma-ray spectroscopy at J-PARC

Fumiya Oura[1,*], *Manami* Fujita[2], *Yuji* Ishikawa[1], *Tatyana* G. Rogers[1], *Hirokazu* Tamura[1,2], *Mifuyu* Ukai[3,1], and *Takeshi* O. Yamamoto[2] for the J-PARC E63 collaborations.

[1]Tohoku University, Sendai 980-8578, Japan
[2]Japan Atomic Energy Agency (JAEA), Tokai 319-1195, Japan
[3]High Energy Accelerator Research Organization (KEK), Tsukuba 305-0801, Japan

Abstract. To understand the mechanism of the charge symmetry breaking between ${}^4_\Lambda$H and ${}^4_\Lambda$He, we plan to measure the gamma-transition energy of ${}^4_\Lambda$H ($1^+ \to 0^+$) with a Ge detector array. For identification of the hypernucleus, we will perform triple coincidence with the reaction, γ-ray, and weak decay for the first time. We measure the pion from weak decay with a range counter. This method will enable γ-ray spectroscopy of various hyperfragments which cannot be directly produced by (K^-, π^-) or (π^+, K^+) reactions.

1 Motivations

1.1 Study of the charge symmetry breaking in the Λ-N interaction

It has been suggested that there is a sizable effect of charge symmetry breaking (CSB) in the four-body system of a Λ hyperon and nucleons (N). The mechanism of the effect is still unclear but it will provide a clue to deeper understanding the baryon-baryon interaction. Recently, our group measured the γ-ray from the ${}^4_\Lambda$He ($1^+ \to 0^+$) transition and confirmed the presence of the large CSB effect (Fig. 1 (A)) [4].

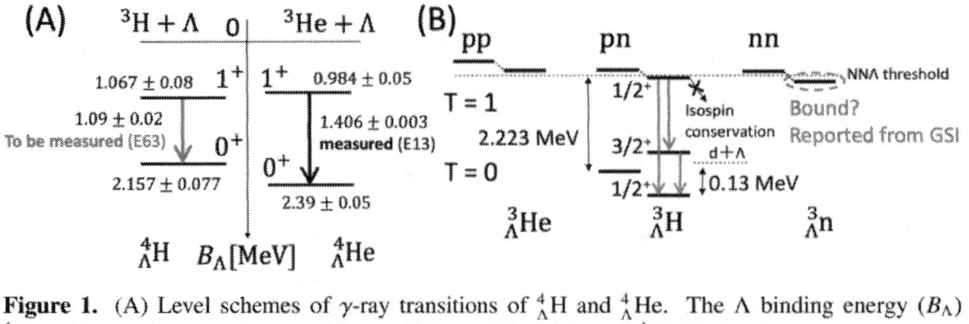

Figure 1. (A) Level schemes of γ-ray transitions of ${}^4_\Lambda$H and ${}^4_\Lambda$He. The Λ binding energy (B_Λ) of ${}^4_\Lambda$He (0^+) was measured in emulsion experiments [1] and that of ${}^4_\Lambda$H (0^+) in decay pion spectroscopy [2, 3]. (B) Level schemes of ${}^3_\Lambda$He, ${}^3_\Lambda$H, and ${}^3_\Lambda$n and possible γ-ray transitions in ${}^3_\Lambda$H.

*e-mail: ooura@lambda.phys.tohoku.ac.jp

The transition energy of ${}^{4}_{\Lambda}$He, which we measured with germanium (Ge) detectors, has good statistics and high resolution. In contrast, the data of ${}^{4}_{\Lambda}$H measured in the 1970s and 1990s is poor in statistics and resolution. Thus, we plan to precisely measure the γ-ray energy of the ${}^{4}_{\Lambda}$H ($1^{+} \to 0^{+}$) transition using a high-resolution Ge detector array, Hyperball-J, at J-PARC as the E63 experiment [5].

1.2 Search for excited states of hypertriton

Excited states of ${}^{3}_{\Lambda}$H are probably unbound, but there is a possibility to observe γ-rays from them (Fig. 1 (B)). If their energies are measured, we can approach two problems. One is whether the nnΛ state is bound or not. The energy of the $1/2^{+}$ (T=1) excited state can be very close to the NNΛ threshold. Due to the isospin conservation, the decay of ${}^{3}_{\Lambda}$H (T=1) $\to$ dΛ is suppressed and ${}^{3}_{\Lambda}$H (T=1) is expected to deexcite (partly) with γ emission. The measurement will give us important imformation on the states of ${}^{3}_{\Lambda}$H (T=1) and nnΛ. The other is related to the hypertriton lifetime puzzle. The reported values of the ${}^{3}_{\Lambda}$H binding energy are different between experiments. The energy of the excited state of $3/2^{+}$(T = 0) is possibly closer to the dΛ threshold than expected or lower than the dΛ threshold. In such cases, the gamma transition ($3/2^{+} \to 1/2^{+}$) may be observed. If the energy of the transition is measured, the information helps us understand the ΛN-ΣN coupling interaction. In either case measurement of γ-rays from the simplest hypernucleus, ${}^{3}_{\Lambda}$H, is particularly important. We also plan to measure the γ-rays from ${}^{3}_{\Lambda}$H in the E63 experiment.

2 Triple coincidence of (K^{-}, π^{-}) reaction, γ-ray, and weak decay

The E63 experiment is going to be performed at the K1.1 beam area in J-PARC (Fig. 2 left) [5]. It is impossible to produce ${}^{4}_{\Lambda}$H and ${}^{3}_{\Lambda}$H directly from (K^{-}, π^{-}) or (π^{+}, K^{+}) reactions. Thus, we will produce ${}^{4}_{\Lambda}$H and ${}^{3}_{\Lambda}$H as hyperfragments from the ^{7}Li (K^{-}, π^{-}) reaction. In the reaction, highly excited states of ${}^{7}_{\Lambda}$Li are produced and decay into ${}^{4}_{\Lambda}$H, ${}^{3}_{\Lambda}$H, and other hyperfragments. We identify a K^{-} beam particle and a π^{-} scattered particle and measure their momenta with the upstream and downstream spectrometers. Using this information, we obtain the missing mass spectrum of the ^{7}Li (K^{-}, π^{-}) reaction, and then select the excited states of ${}^{7}_{\Lambda}$Li, which are expected to decay into hyperfragments such as ${}^{4}_{\Lambda}$H and ${}^{3}_{\Lambda}$H (Fig. 2 right).

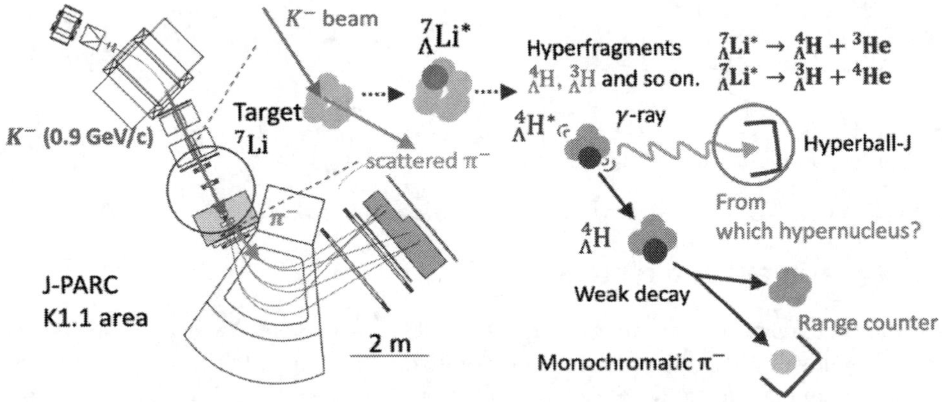

Figure 2. (Left) Experimental setup. The (K^{-}, π^{-}) reaction is identified using upstream or downstream spectorometers. (Right) The gamma-weak coincidence of ${}^{4}_{\Lambda}$H using a range counter.

When a produced hyperfragment is excited, it decays to the ground state with a γ emission. We measure the γ-rays with Hyperball-J. In this method, however, we do not know which hypernucleus emitted the γ-ray. In order to identify the hypernucleus, we measure kinetic energies of monochromatic pions from two-body weak decay of ${}^{4}_{\Lambda}\mathrm{H} \rightarrow {}^{4}\mathrm{He} + \pi^{-}$ and ${}^{3}_{\Lambda}\mathrm{H} \rightarrow {}^{3}\mathrm{He} + \pi^{-}$ by a range counter (their kinetic energies are 53 MeV and 40 MeV, respectively). The branching ratios of these two-body weak decays are 50% and 27%, respectively. For identification of the hypernucleus, we will perform triple coincidence measurement for the in-flight (K^{-}, π^{-}) reaction, a γ-ray, and a weak decay pion for the first time. The requirement to the range counter system is to separate two peaks from these two monochromatic pions in the range spectrum with a confidence level of more than 3σ.

3 Range counter system

We developed a range counter system made of 24 layers of a 6 mm-thick plastic counter. We adopted a readout method using wavelength-shifting fibers and SiPMs. Scintillation photons are converted and transported in the wavelength-shifting fibers embedded in the two sides of the scintillator, and read out with SiPMs. In front of the range counter, plastic scintillator hodoscopes as a tracking device for pions (TD) are installed. Thanks to the tracking device, angular inaccuracy is removed. Figure 3 shows the range counter system which is installed inside the Hyperball-J system.

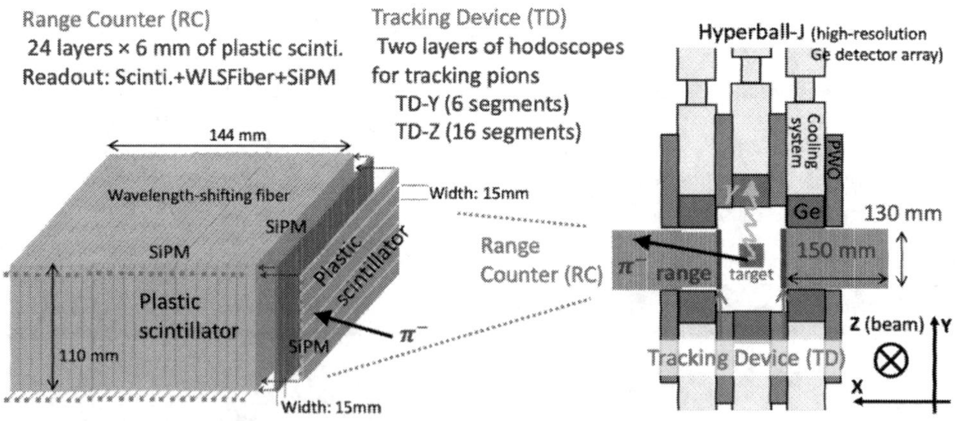

Figure 3. (Left) The range counter system, which is composed of a range counter and a tracking device. (Right) The system is installed inside the Hyperball-J.

4 Test experiment and results

We fabricated a prototype range counter made of 8 layers of a 6 mm-thick plastic counter. The total thickness of the prototype is one-third of that of the counter for the E63 experiment. We conducted an experiment at the downstream end of the J-PARC K1.8 beam line to test the prototype and evaluate its performance. The experimental setup is shown in Fig. 4 (A). Pions with momentum of 300 MeV/c were degraded down to less than 110 MeV/c to stop them in the range counter, and time of flight between two plastic hodoscope (TOF counters) located before the range counter was measured. Since the pions fly with various velocities,

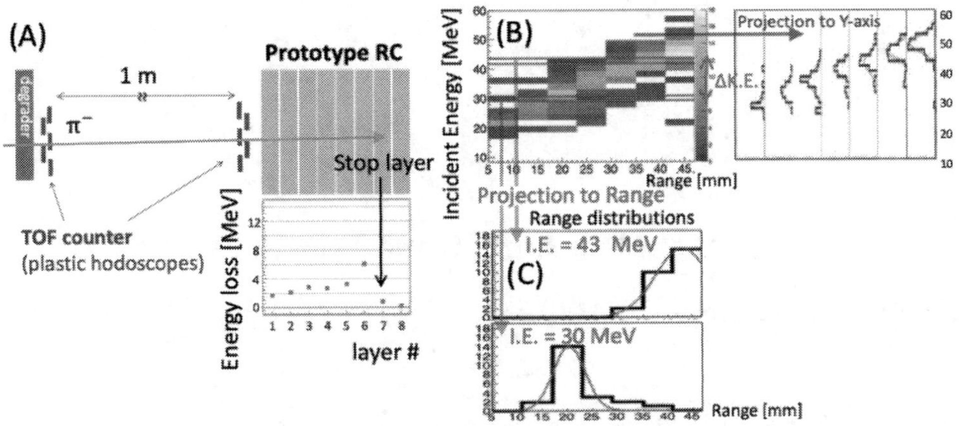

Figure 4. (A) Setup of the test experiment of a prototype range counter. The lower histogram shows the energy loss in each layer for a typical event in which a pion stopped in the prototype. (B) Correlation between incident energy and range for stop pion events. (C) Two distributions of range at energy regions around 43 and 30 MeV.

the incident energy before the range counter is not uniform. Figure 4 shows two results. First, we looked at the time of flight between TOF counters and selected slow pion events. When we plotted the energy loss in each layer of the range counter for each event, we observed Bragg curves as shown in Fig. 4 (A). In this way, we confirmed that about 300 pions stopped in the range counter. For these stop events, we plotted the correlation between the range and the incident energy (Fig. 4 (B)). We selected two energy regions around 43 MeV and 30 MeV, and then we obtained range distributions for these two regions (Fig. 4 (C)). They are found to be well separated. The difference of 43 MeV and 30 MeV (13 MeV) is equal to the difference of kinetic energies of pions from ${}^{4}_{\Lambda}$H and ${}^{3}_{\Lambda}$H. In the E63 experiment, we measure pions with higher kinetic energies than in the present test experiment, which means that the prototype can separate the pions from ${}^{4}_{\Lambda}$H and ${}^{3}_{\Lambda}$H sufficiently well.

5 Summary

We plan to measure the energy of γ-rays from ${}^{4}_{\Lambda}$H and ${}^{3}_{\Lambda}$H. For identifying the hypernucleus produced as hyperfragments, we are going to perform a triple coincidence method with a range counter which is to be newly developed. We fabricated a prototype with a thickness of one-third of the whole detector for the E63 experiment and conducted a test experiment at J-PARC using pions. As a result, we found that the prototype has the ability to measure the pion energy accurately enough in the energy region required for the E63 experiment. Based on the results, we are currently fabricating a whole set of the range counter system for the beam time coming near future.

References

[1] M. Jurič *et al.*, Nucl. Phys. B **52**, 1-30 (1973)
[2] A. Esser, S. Nagao *et al.*, Phys. Rev. Lett. **114**, 232501 (2015)
[3] F. Schulz *et al.*, Nucl. Phys. A **954**, 149-160 (2016)
[4] T.O. Yamamoto *et al.*, Phys. Rev. Lett. **115**, 222501 (2015)
[5] H. Tamura *et al.*, the proposal of the J-PARC E63 experiment, *Gamma-Ray Spectroscopy of Light* Λ *Hypernuclei* II

Overview of hypernuclear and strange particle physics

–Experimental summary of HYP2022

Hirokazu Tamura[1,2,*]

[1]Department of Physics, Tohoku University, Sendai 980-8578, Japan
[2]Advanced Science Research Center, Japan Atomic Energy Agency, Tokai 319-1195, Japan

Abstract. Recent results and future prospects in strangeness nuclear and hadron physics are summarized from an experimental point of view based on plenary presentations in the HYP2022 conference.

1 Introduction

The 14th International Conference on Hypernuclear and Strange Particle Physics was held in Prague from June 27 through July 1, 2022, hosted by Nuclear Physics Institute (NPI Řež) of the Czech Academy of Sciences and the Faculty of Nuclear Sciences and Physical Engineering of the Czech Technical University in Prague (FNSPE CTU). The conference covered various aspects in hadron and nuclear physics with strangeness. In this paper, experimental results and prospects in this field are summarized based on the experimental plenary talks given in the conference. It is to be noted that not all the talks are sufficiently covered.

2 Hypertriton and few-body hypernuclear systems

2.1 Significance and status of the hypertriton

One of the main subjects in this field is how to understand and describe the baryon-baryon interactions. The ΛN interaction, the most basic part of this subject, has been studied with various theoretical approaches, but the experimental information is still not satisfactory due to insufficient ΛN scattering data. In this situation, the hypertriton (${}^{3}_{\Lambda}$H), the lightest bound Λ hypernucleus, plays particularly important roles as the deuteron did in the study of NN interactions. However, experimental data for hypertriton are not firmly established.

Although binding energies, level structures and lifetimes of various Λ hypernuclei have been studied via counter experiments with direct reactions such as (K^-, π^-), (π^+, K^+) and $(e, e'K^+)$ since 1970s up to now, but the data for the hypertriton has not been updated due to experimental difficulties from those with large errors determined by the old emulsion experiments in 1960-70s.

"Hypertriton puzzle" arose when the heavy ion collision experiments (HypHI, then ALCIE and STAR) produced hypertritons and reported its lifetime significantly shorter than the free Λ lifetime τ_Λ = 263 ps. The reported value of the hypertriton binding energy combined for several emulsion experimental data is quite small (B_Λ = 0.13±0.05 MeV), suggesting that

*e-mail: tamura@lambda.phys.tohoku.ac.jp

the Λ wavefunction in ${}^3_\Lambda$H is widely spread as $[2\mu_{\Lambda d}B_\Lambda({}^3_\Lambda\mathrm{H})]^{-1/2} \sim 15$ fm and consequently the ${}^3_\Lambda$H lifetime is expected to be almost the same as τ_Λ.

During this decade, importance of the lifetime and the binding energy of the hypertriton is recognized and various experimental efforts have been made. Table 1 shows the recent results and the present status of experiments reported in the HYP2022 conference.

2.2 Hypertriton lifetime

In the heavy ion collision experiments, STAR, ALICE and HADES reported new values of the hypertriton lifetime. STAR once reported a quite short lifetime ($142^{+24}_{-21} \pm 29$ ps) [1] in 2018, but a new result (221±15±19) ps from Beam Energy Scan II (BES-II, 3.0 and 7.2 GeV) is significantly different from the previous one and close to the free lifetime [2]. It is noted that since hypernuclear production rates increase for lower energy collision, the Beam Enrgy Scan data contain more light hypernuclear events including ${}^5_\Lambda$He. ALICE collaboration, which reported a value of $242^{+34}_{-38} \pm 17$ ps [3] in 2019, is updating it to a new value with a smaller error close to the free lifetime [4]. In addition, HADES at GSI reported a preliminary result of 256 ± 22 ± 36 ps [5]. These recent data from heavy ion experiments seem to be converged to a value consistent or slightly lower than the free Λ lifetime.

The hypertriton lifetime puzzle was first raised by HypHI group at GSI with ^{6}Li beam at 2 GeV/A on ^{12}C target [6]. In order to improve the lifetime data for light hypernuclei as well as to confirm the suggested Λnn bound state, the group has organized a new collaboration and just conducted data-taking of the updated experiment employing FRS combined with the WASA detector moved from Juelich [7].

In the heavy ion collision experiments, the lifetime is measured as the flight length of the hypernucleus between the production point and the decay point. In order to obtain reliable lifetime value, a different type of measurement with different systematic errors is necessary. At J-PARC, the experiment E73 is being performed for a direct time measurement between

Table 1. Present status and prospects of ${}^3_\Lambda$H (and ${}^4_\Lambda$H) experiments presented in HYP2022. The data reported after the previous HYP2018 are given. $\tau({}^3_\Lambda\mathrm{H})$ and $\tau({}^4_\Lambda\mathrm{H})$ are the measured lifetimes of ${}^3_\Lambda$H and ${}^4_\Lambda$H, respectively, in ps, and $B_\Lambda({}^3_\Lambda\mathrm{H})$ is the measured Λ binding energy of ${}^3_\Lambda$H in MeV.

Experiment	Reaction	Method	$\tau({}^3_\Lambda\mathrm{H})$	$\tau({}^4_\Lambda\mathrm{H})$	$B_\Lambda({}^3_\Lambda\mathrm{H})$
STAR [2]	HI (Au+Au)	decay length	$142^{+24}_{-21} \pm 29$		0.41±0.12±0.11
	$\sqrt{s}$ = 3.0, 7.2 GeV		221±15±19	218±6±13	analysis
ALICE [4]	HI (Pb+Pb) $\sqrt{s}$ = 5 TeV	decay length	$242^{+34}_{-38} \pm 17$		(~0.05)
HADES [5]	HI (Ag+Ag) $\sqrt{s}$ =2.55 GeV	decay length	256±22±36	222±8±13	
WASA-FRS [7]	HI (^{6}Li+^{12}C)	decay length	analyzed	analyzed	analyzed
J-PARC E73 [8]	^{3}He(K^-,π^0)	decay time	run soon	(190±8±?)	-
MAMI [11]	^{7}Li(e, K^+)	decay π momentum	-	-	running
J-PARC E07 [7]	K^-+emulsion	decay time/energy	analyzed	to be analyzed	analyzed
JLab E12-19-002 [12]	^{3}He($e, e'K^+$)	missing mass	-	-	approved

production and decay. The hypertriton is produced by ^{3}He(K^-, π^0) reaction in which energetic photons from π^0 are tagged and the monochromatic pion from $^3_\Lambda$H $\to$ ^{3}He π^- is detected and used to measure the decay time. The group confirmed this method by measuring the $^4_\Lambda$H lifetime via the ^{4}He($K^-.\pi^0$) reaction and $^4_\Lambda$H $\to$ ^{4}He π^- decay, and by identifying the $^3_\Lambda$H decay from $^3_\Lambda$H $\to$ ^{3}He π^- [8]. The $^3_\Lambda$H lifetime measurement will be conducted in 2023.

2.3 Hypertriton binding energy

The experimental value of the hypertriton binding energy, $B_\Lambda(^3_\Lambda\text{H}) = 0.13 \pm 0.05$ MeV, is given as the average of rather diverged three values of old emulsion experiments, and it was not updated for a long time. Recently, STAR reported a significantly larger value of $B_\Lambda(^3_\Lambda\text{H}) = 0.41 \pm 0.12 \pm 0.11$ MeV [9], suggesting that the lifetime can be significantly shorter than the free Λ lifetime. On the other hand, ALICE reported a preliminary value of less than 0.1 MeV in the conference [4]. STAR BES-II data is also being analyzed [10].

In such invariant mass measurement in high energy experiments, however, it seems difficult to reduce the systematic error down to a few 10 keV. Two different types of approaches have also been pursued. At MAMI the $B_\Lambda(^3_\Lambda\text{H})$ value is going to be precisely determined with a precision of $\pm$20 keV via a momentum measurement of the weak decay pion in $^3_\Lambda$H$\to$^{3}He+π^- [11]. In order to calibrate the pion magnetic spectrometer using MAMI's electron beams, a novel method to precisely determine the beam energy using interference of the undulator light has been developed.

Another type of measurement is being carried out with K^--irradiated emulsion in the J-PARC E07 experiment. A machine learning method has been introduced in the emulsion analysis to search for hypertriton decay images [7]. They have found several hypertriton events and aim at obtaining the binding energy with a statistical error of 30 keV and a systematic error of 30 keV.

These various efforts in the lifetime and binding energy measurements will allow us to solve the hypertriton puzzle and precisely and reliably determine these basic quantities for the *YN* interactions.

2.4 Λnn and $\Lambda NN(T = 1)$ systems

The HypHI group reported possible existence of a weakly-decaying Λnn bound state from a ^{6}Li + ^{12}C experiment, although theoretical studies disfavor the existence of the bound state. An experiment with the same reaction was just performed by the WASA-FRS collaboration and the data analysis is in progress [7]. On the other hand, the ^{3}H($e, e'K^+$) reaction was also studied at JLab Hall A with a tritium target in order to clarify the situation. The missing mass spectrum did not exhibit a significant peak around the Λ-binding threshold and an upper limit of the Λnn production cross section was obtained [13]. Theoretically, possible existence of a Λnn resonance state is suggested, and the measured spectrum seems to exhibit a hint of an enhancement just above the threshold. Thus, the experiment had better be repeated with better statistics in the near future.

If the Λnn state is bound, the $^3_\Lambda$H(T=1) state belonging to the same T=1 isotriplet can be investigated via γ-ray spectroscopy because the $^3_\Lambda$H(T=1) $\to$ d + Λ decay is suppressed by isospin conservation. In addition, the $^3_\Lambda$H(T=0) ($3/2^+$$\to$$1/2^+$) γ transition may be also observed if the $B_\Lambda(^3_\Lambda\text{H})$ value is large enough to make the $^3_\Lambda$H(T=0) ($3/2^+$) state bound. It would provide valuable information on the ΛN-ΣN interaction, which may significantly affect the ΛN spin-spin interaction in nuclear matter. A γ-ray spectroscopy experiment for $^4_\Lambda$H and $^3_\Lambda$H is planned as J-PARC E63 [14] with the main motivation of precisely measuring the $^4_\Lambda$H ($1^+$$\to$$0^+$) γ ray for studying the charge symmetry breaking in A=4 hypernuclei. Here, $^4_\Lambda$H and $^3_\Lambda$H are simultaneously produced as hyperfragments from the ^{7}Li(K^-, π^-) reaction.

3 $S = -2$ systems

The double strangeness systems have been intensively studied at J-PARC. Three types of experiments, emulsion experiments, Ξ hypernuclear missing-mass spectroscopy, and Ξ atomic X-ray measurement, are being carried out and valuable results are coming.

3.1 $S = -2$ nuclei with emulsion

In the J-PARC E07 experiment, Ξ^- hyperons produced by the (K^-,K^+) reaction were injected into emulsion stacks and their images were analyzed with an automatic microscope scanning system with a help of the information on the measured Ξ^- injection point to the emulsion [15]. At present, 33 events of double strange hypernuclei have been observed, and among them, two events of ${}^{15}_{\Xi}$C hypernucleus decaying to twin hypernuclei (IBUKI and IRRAWADDY events) were identified. The IBUKI event gives, for the first time, an accurate value of the Ξ binding energy as $B_\Xi = 1.27 \pm 0.21$ MeV, which is in agreement with one of the two candidate values derived from the KISO event. On the other hand, the IRRAWADDY event gives a much larger value of $B_\Xi = 6.27 \pm 0.27$ MeV. Together with the reanalyzed value of the KINKA event from KEK E373 (8.00 ± 0.77 or 4.96 ± 0.77 MeV) , it was concluded that IRRAWADDY and KINKA correspond to the nuclear s_Ξ-state and IBUKI and KISO to the p_Ξ-state. This result shows a Ξ-nuclear potential with a certain depth and a quite small imaginary part. In addition, a new $\Lambda\Lambda$ hypernucleus (${}^{11}_{\Lambda\Lambda}$Be, MINO event) was observed and the Λ-Λ interaction energy was obtained as $\Delta B_{\Lambda\Lambda} = 1.87 \pm 0.37$ MeV, which should be consistently explained with the value from the NAGARA event, $\Delta B_{\Lambda\Lambda} = 0.67 \pm 0.17$ MeV for ${}^{6}_{\Lambda\Lambda}$He.

3.2 Ξ hypernuclear missing-mass spectroscopy

Since the emulsion method cannot observe exited states of hypernuclei, the missing mass spectroscopy of the (K^-, K^+) reaction is also important. Following a pilot run (J-PARC E05), an upgraded experiment for ^{12}C$(K^-, K^+){}^{12}_{\Xi}$Be reaction (J-PARC E70) with a new dedicated magnetic spectrometer (S-2S) is under preparation [16]. Employing an active target with plastic scintillator, a ${}^{12}_{\Xi}$Be spectrum will be measured with a resolution of 2 MeV(FWHM). Then, ${}^{7}_{\Xi}$H will be also studied via ^{7}Li(K^-, K^+) reaction (J-PARC E75).

3.3 Ξ-atomic X rays

Measurement of Ξ^--atomic X rays provides another means to investigate Ξ-nucleus interaction. At J-PARC, two types of experiments were carried out [17]. In the E07 experiment described above, nuclear absorption of Ξ^- in emulsion atoms was identified via image analysis of emulsion. In the E03 experiment dedicated to Ξ-atomic X rays, Ξ^- hyperons produced by the (K^-, K^+) reaction were stopped in an iron target. In both experiments, Ξ-atomic X rays are measured with a germanium detector array. Data analysis is in progress.

4 YN, YY interactions

Because of experimental difficulties in YN scattering experiments, hypernuclear data have been used to supplement our understanding of YN interactions. Employing plenty of Λ hypernuclear data, the ΛN interaction has been well restricted and the properties of various Λ hypernuclei are able to be understood. However, this approach cannot be extrapolated to high density baryonic matter in neutron stars. In order to extract information on the nuclear density dependence of the ΛN interaction in matter (in the other words, the ΛNN interaction) from Λ hypernuclear data and to properly describe high density matter in neutron stars, two-body YN interaction data in the free space is indispensable. In the case of ΣN interaction, hypernuclear data cannot be used because they are not available except for ${}^{4}_{\Sigma}$He due to repulsive nature of ΣN interaction, .

Recently, two experimental breakthroughs have been achieved. One is a new type of high statistics YN scattering experiments, and the other is so-called "femtoscopy", baryon-baryon momentum correlation measurement in high energy nuclear collision experiments.

4.1 YN scattering

Because of a short flight length of hyperons, previous YN scattering experiments employed visual detectors such as bubble chambers and scintillating fibers with image intensifier tubes. Such "slow methods" strongly restricted event rates and data acquisition rates. Recently, those difficulties have been solved by employing fast detectors and a fast data acquisition system.

The J-PARC E40 group successfully carried out a high-statistics $\Sigma^{\pm}$-p scattering experiment [18]. By employing a fast tracking detector made of scintillating fibers individually read out by SiPM sensors, they measured differential cross sections of Σ^{+}-p and Σ^{-}-p elastic scattering and $\Sigma^{-}p \rightarrow \Lambda n$ reaction at $p_{\Sigma} = 0.45 - 0.8\text{GeV}/c$. In the $\Sigma^{+}p$ channel, the obtained cross section was converted to the phase shift, showing that the $\Sigma^{+}p$ interaction has a much stronger repulsive core than the NN interaction, which is interpreted as a result of Pauli principle in the quark level.

At JLab, the CLAS collaboration derived high-statistics Λp total cross sections for $p_{\Lambda} = 0.9 - 2.0$ GeV/c, the momentum region relevant to the core of neutron stars, by using Λ hyperons which are photo-produced in a hydrogen target and scattered there [19].

4.2 Femtoscopy for YN and YY interactions

Momentum correlation of two hadrons emitted in high-energy nuclear collisions provides a unique and powerful means to investigate the strong interaction between the two hadrons. The correlation function $C(k^{*})$, the distribution of the relative momentum k^{*} for two particles in the same event divided by the one for two particles in different events, is described in terms of the source function $S(r^{*})$ and the wave function of the relative coordinate $\Psi(\mathbf{k}^{*}, \mathbf{r}^{*})$ as $C(k^{*}) = \int S(r^{*})\, |\Psi(\mathbf{k}^{*}, \mathbf{r}^{*})|^{2} d^{3}r^{*}$. Since $\Psi(\mathbf{k}^{*}, \mathbf{r}^{*})$ reflects the interaction between the two particles, the correlation function can be calculated assuming the interaction once the source function is determined from a measured correlation function of two hadrons with a known interaction such as proton-proton. Thus, information on the unknown interaction can be derived. This method called "femtoscopy" has been applied to various hadron-hadron pairs in ALICE, STAR, HADES and others. It has played significant roles in recent studies of baryon-baryon interactions. In particular, this method is powerful for low-momentum s-wave YN interactions that is difficult for scattering experiments. In addition, it is the only method to study YY interactions in the free space because YY scattering experiments are impossible.

In the series of HYP conferences, femtoscopy results on YN and YY interactions were first presented in the previous HYP2018; Λp, $\Lambda\Lambda$ and $\Xi^{-}p$ correlation data were presented from ALICE. Then, ALICE reported $\Omega^{-}p$ and Ξ^{-}-p correlation results [20], which indicate attractive $\Omega^{-}p$ and Ξ^{-}-p forces consistent with the lattice calculations by the HAL QCD group. The $\Omega^{-}p$ channel is quite interesting because HAL QCD predicts a $\Omega^{-}p$ bound state [21]. ALICE also reported a surprisingly precise ΛN correlation function with a clear ΛN-ΣN cusp structure, providing low momentum ΛN interaction data of unprecedented quality contributing to development of the chiral EFT *YN* interaction [22]. On the other hand, STAR reported $\Xi\Xi$ correlation data for the first time, as well as $p\Xi^-$ and $p\Omega^-$ data [23].

The femtoscopy method has been established as a powerful tool for investigating of baryon-baryon interaction through comparison with the correlation functions calculated with theoretical interactions from the lattice QCD and the chiral effective field theory. In near future, more data with much higher statistics and varieties of the baryon-baryon pair will be available. In order to analyze high quality data, efforts are being made to estimate various effects such as the feed-down effects from other channels and the source size effects, as well as to take into account the coupled-channel effects. In ALICE the femtoscopy method is being extended to nuclear systems such as the *p-d* system, but the measured *pd* correlation is found to be very different from the one expected from the *pd* interaction. Effects of the three-body force and the deuteron formation mechanism are investigated [24].

4.3 ΛNN three-body force and ΛN interaction in neutron stars

The observation of massive neutron stars with twice the solar mass gave a great impact to nuclear physicists, since hyperon appearance expected in the core of neutron stars cannot support such massive neutron stars. One of the scenarios to solve this "hyperon puzzle" is to clarify possible repulsive nature of the density dependence of the ΛN interaction, which is described by the ΛNN three-body force.

In the conference, several theoretical talks on Equation-Of-State (EOS) and the topical session on ΛNN three-body force were given. The topical session mainly discussed theoretical approaches to this problem, but from an experimental point of view, a femtoscopy approach by ALICE to extract the ΛNN force effect from the three-body correlation was presented [25]. To see the effect, more data with a larger statistics by two orders of magnitude is necessary, but is possible with coming RUN3 data.

At J-PARC, new experiments are proposed to study ΛNN three-body force through high precision Λ hypernuclear spectroscopy via the (π^+, K^+) reaction [12] combined with high quality Λp scattering experiments [18], as one of the main program in the J-PARC Hadron Facility extension project [26, 27]. It is noted that Ohnish pointed out a possibility to study ΛN interaction in high density matter via directed flow of Λ hyperons in high energy nuclear collision experiments [28].

In order to investigate ΛN interaction in neutron stars, the Λ binding energy in neutron-rich nuclear matter should be also measured because the ΛNN (T=1) interaction which governs the neutron star EOS can be very different from the ΛNN (T=0) interaction studied with ordinary (isospin~0) target nuclei. In JLab, a high-resolution $(e, e'K^+)$ Λ hypernuclear spectroscopy experiment on ^{40}Ca and ^{48}Ca targets (E12-15-008) is under preparation [12].

5 Kaonic systems and $\bar{K}N$ interaction

From the strongly-attractive K^-p interaction derived from K^-p atomic and low-energy scattering data, the $\Lambda(1405)$ resonance was interpreted as a $\bar{K}N$ bound state and possible existence of $\bar{K}$-nucleus quasibound states was conjectured more than two decades ago. Since then the $\bar{K}N$ interaction has attracted great attention of hadron and nuclear physicists. Various updated data on kaonic nuclei, $\bar{K}N$ femtoscopy, $\Lambda(1405)$, *etc.*, have been reported as summarized by Doce [29], and the conjecture has been confirmed and accepted through great efforts by theorists.

5.1 Kaonic nuclei

Until recently, no clear experimental evidence was given for the existence of a $\bar{K}$-nucleus quasibound state, although some experiments reported positive suggestions. The J-PARC E15 experiment recently observed a peak located 42 MeV below the K^-pp threshold in the Λp invariant mass spectrum obtained in the K^-+^3He $\rightarrow \Lambda pn$ reaction [30]. Since the peak position does not depend on the momentum transfer to the Λp system and the quasi-free K^- absorption process is also observed above the threshold, the peak is reasonably interpreted as a K^-pp bound state. Interestingly, a similar peak was also reported in the Λd invariant mass in the ^{4}He$(K^-, \Lambda p)n$ reaction [31]. Further studies are planned with a larger cylindrical spectrometer in order to assign its spin-parity and to create heavier kaonic nucleus systems.

5.2 Λ(1405)

Since the mass and the parity (−) of Λ(1405) cannot be explained by orbital excitation of quarks and the mass is close to the K^-p threshold, Λ(1405) was conjectured to have a K^-p molecular structure. This assumption has been confirmed not only by lattice QCD calculations but also by the $\bar{K}N$ amplitude analyzed with the K^-p atomic and low-energy scattering data using the chiral unitary model with the $\bar{K}N$-$\Sigma\pi$ coupled channel. The calculation predicted two-pole structure at 1380 and 1405 MeV [32], which has been supported by several experimental data for $\pi\Sigma$ mass spectra. The Glue X collaboration at JLab Hall D reported a Λ(1405) line shape recently measured via the $\Sigma^0\pi^0$ decay channel. It is well fitted with two Flatte functions at ~1387 and ~1409 MeV [33], clearly indicating the two-pole structure.

5.3 Kaonic atoms and low energy $\bar{K}N$ interaction

Although X rays from kaonic atoms have been measured for various elements including the kaonic hydrogen (K^-p), the kaonic deuterium (K^--d) X ray has not been observed yet in spite of its significance in separating the $\bar{K}N$ T=0 and 1 interactions. The SIDDHARTA-2 experiment at DAΦNE and the J-PARC E57 will soon challenge the first observation with silicon drift detectors (SDD) [35]. On the other hand, the J-PARC E62 experiment successfully employed Transition Edge Sensor (TES) with an excellent resolution of ~30 times better under a high rate beam background. This experiment determined energy shifts of the $2p$ states in K^--^{3}He and K^--^{4}He atoms with ~10 times smaller error bars than before [35].

The low-energy K^--nucleus interaction has been also studied by the AMADEUS experiment at DAΦNE, studying the K^- absorption on two, three and four nucleons, as well as $K^-p \rightarrow \Sigma^0/\Lambda\,\pi^0$ cross sections for p_{K^-}~100 MeV/c [36].

5.4 $\bar{K}N$ interaction from femtoscopy

The femtoscopy technique was also applied to the $\bar{K}N$ interaction. The ALICE collaboration reported the K^-p (and K^+p) correlation [37]. The measured K^-p correlation function from the pp collision clearly exhibits cusp structure at the $K^-p \rightarrow \bar{K}^0n$ and $K^-p \rightarrow$ Λ(1520) thresholds. It was compared with the correlation function calculated by taking into account the K^-p-$\Sigma\pi$-$\Lambda\pi$ coupled-channel effects, together with the threshold difference of different charge states such as K^-p-$\bar{K}^0n$ [38]. The data was well reproduced by the predicted curve with a significant coupling to the $\pi\Sigma$ channel. It is noted that the measured correlation function for a large source size in the Pb-Pp collision shows weaker cusp structure, and actually the calculated correlation function for a large source size is found to be less sensitive to the $\pi\Sigma$ coupling. In this way, the femtoscopy method is useful to approach the K^-p amplitude below the threshold coupled to the $\pi\Sigma$ channel.

6 Other topics with strange and heavy flavor hadrons

6.1 Other hadron-hadron interactions

The femtoscopy analysis was also performed for various channels.

Hadron-hadron correlation data including D mesons were presented for the first time by the ALICE collaboration [22]. From the π^+D^- and π^+D^+ data, the scattering lengths for the isospin 3/2 and 1/2 channels were derived and compared with lattice calculations. In addition, pD^- correlation data was also reported. Since recently-observed several exotic hadrons with charm have masses close to the threshold going to two hadrons, they are considered to be candidates of hadron-hadron molecules. The femtoscopy studies with charm will clarify the interaction between the two hadrons and reveal the nature of those exotic hadrons.

On the other hand, the STAR collaboration reported $K_s^0K_s^0$ and $K_s^0K^+$ correlation data [39]. They are reproduced well by calculations with the strong final state interaction through the $f_0(980)$ and $a_0(980)$ scalar resonances. In this way, femtoscopy could be useful in clarifying the structure of such resonances, for example, whether $f_0(980)$ and $a_0(980)$ are $qq\bar{q}\bar{q}$ states or $K\bar{K}$ molecules.

6.2 Hadron polarization in heavy ion collisions

Hyperons are good probes for investigating "global polarization" of particles in hot/dense matter created by high energy nuclear collisions because of the self-analyzing property of hyperon's weak decay. The STAR collaboration discovered a global polarization of Λ hyperons in noncentral nuclear collisions at $\sqrt{s_{NN}}$ = 7.7–62.4 GeV [40], indicating a creation of strongly vortical fluid of quark gluon plasma. Experimental studies have been made for lower collision energies down to $\sqrt{s_{NN}}$ = 7.7 GeV by STAR in Beam Energy Scan [41] and down to $\sqrt{s_{NN}}$ = 2.4 GeV by HADES [42]. The results indicates increasing trend of the polarization to lower energies. In addition, STAR also observed a surprisingly large global spin alignment of ϕ meson at low energies ($\sqrt{s_{NN}} < 62$ GeV) [43].

6.3 Exotic hadrons with heavy flavors

Studies of exotic hadrons (tetraquarks and pentaquarks) with heavy flavors have been actively conducted. In the conference, recently-observed exotic hadrons with charm and strangeness are reviewed [44]. The exotic mesons with open-charm and strangeness, $D^*_{s0}(2317)$ and $D_{s1}(2460)$, are well interpreted as DK and D^*K molecules, respectively. Recently, LHCb reported one or more new exotic resonances, $X(2900)$, decaying to D^-K^+ at the mass close to the $D^*\bar{K}^*$ threshold, which suggests a universal feature of the $\bar{K}$ and $\bar{K}^*$ bound states. In addition, at the mass close to the $\bar{D}^*_sD + \bar{D}_sD^*$ threshold, BESII reported $Z_{cs}(3985)$ in the $e^+e^- \to K^+X$ reaction, and LHCb reported $Z_{cs}(4000)$ decaying to $J/\psi K^+$. LHCb also reported a hidden charm pentaquark with strangeness, P_{cs}, in the $J/\psi\Lambda$ mass spectrum in the $\Xi_b^0 \to J/\psi\Lambda K^-$ decay. It is a strange analog of the original pentaquark, P_c, observed in $J/\psi p$ mass in the $\Lambda_b^0 \to J/\psi pK^-$ decay.

Recent results from Belle data were also reported [45]. A peak was observed in the pK^- mass spectrum at the $\Lambda\eta$ threshold, and is well explained as a threshold cusp rather than a new resonance. It may imply that some of the exotic hadrons observed at thresholds are threshold cusps rather than molecular states. From Belle data, $\Xi_c(2970)$ was assigned to $1/2^+$. It is noted that this state, located ~500 MeV above the Ξ_c ground state, is possibly an analog of the Roper resonance $N(1440)$ with $1/2^+$. It may give a hint to elucidate the Roper resonance that has not been understood yet.

7 Future Facilities

In the conference, various results on hypernuclei and strange (and heavy flavored) hadrons were reported based on recent experiments performed at a variety of facilities in the world, JLab (Hall A, Hall B (CLAS), Hall D (Glue-X)), J-PARC (K1.8, K1.8BR, High-p), GSI/FAIR (HADES, HypHI, WASA-FRS), DAΦNE (SHIDDHARTA, AMADEUS), LHC (ALICE, LHC-b), RHIC (STAR), (Super)KEKB (Belle/Belle II), and BEPCII (BESIII). Their experimental prospects were also reported in the conference. In particular, STAR is collecting lower energy data in Beam Energy Scan, which enables abundant hypernuclear production. ALICE will accumulate much larger statistics data by two orders of magnitude in the coming RUN 3, which will be used for further femtoscopy studies. Glue-X at JLab Hall D is studying hyperon photoproduction with higher energy photon beams than before [34]. Belle2 at SuperKEKB will acquire 50 times more data than Belle [45]. In addition, the following future plans of J-PARC facilities were also presented.

7.1 J-PARC Hadron Facility extension

The J-PARC Hadron Experimental Facility has played significant roles in the recent progress of nuclear and hadron physics with strangeness. In order to enhance the activity, a project is proposed to extend the area of the facility building almost twice and construct a new production target and several new secondary beam lines [27]. The main motivation of the HIHR (High Intensity High Resolution) beam line and the K1.1 beam line is to investigate YNN three-body force via high precision hypernuclear spectroscopy and hyperon-nucleon scattering experiments [26]. The π20 and K10 beam lines will deliver high momentum pions and kaons for spectroscopic studies of charm baryons and multi-strange baryons, respectively.

7.2 J-PARC Heavy Ion project

A further J-PARC project of accelerating heavy ions (J-PARC HI project) is also proposed [46]. In this project, heavy ions are accelerated by a newly-constructed superconducting linac and a booster ring, and then injected to and accelerated by the RCS ring and the Main Ring of J-PARC. Heavy ions can be accelerated up to $\sim$11 GeV/A with the world highest intensity. It will be used to produce dense nuclear matter, and the highest beam intensity will help to detect deconfined quark matter at high density via fluctuation of the particle yields. This project also aims at production of exotic' (proton/neutron rich and multi-strange) hypernuclei and exotic hadrons, as well as femtoscopy studies for hadron-hadron interactions.

8 Concluding remark

At the end, the author dared to choose the highlights of the experimental results in HYP2022.
(1) All the femtoscopy results. This new technique, being applicable to variety of hadron-hadron channels, has clearly changed the game.
(2) High statistics ΣN and ΛN scattering results at J-PARC and JLab. After half a century, the long-awaited significant update of YN scattering data has been achieved.
(3) Ξ hypernuclei in emulsion ($^{15}_{\Xi}$C) and kaonic nuclei ($K^- pp$ and $K^- ppn$) at J-PARC. Both confirmed their existence and determined their binding energies (although further studies are necessary).

All the other experimental results and plans reported in the conference were also valuable and encouraging. Considering theoretical progress as well, the author is looking forward to the next HYP conference with optimistic expectations.

In HYP2022, the participants enjoyed face-to-face presentations and discussions after difficult years of the Corona disaster. The author appreciates the organizers' decision of postponement of the conference for one year to avoid a full online meeting. The wonderful banquet as well as the fantastic concert and the beautiful city tour also delighted the participants. The author thanks the organizers for their great efforts to realize this memorable event. The HYP2022 conference was closed after the announcement by the conference chair, Jiří Mareš, that the next conference HYP2025 will be held in Japan hosted by S. N. Nakamura.

References

[1] Adamczyk *et al.*, Phys. Rev. C **97**, 054909 (2018)
[2] Y. H. Leung, presented at *HYP2022* conference
[3] Acharya *et al.*, Phys. Lett. B **797**, 134905 (2019)
[4] M. Puccio, presented at *HYP2022* conference
[5] S. Spies, presented at *HYP2022* conference
[6] C. Rappold *et al.*, Nucl. Phys. A **913**, 170 (2013)
[7] T. R. Saito, presented at *HYP2022* conference
[8] Y. Ma, presented at *HYP2022* conference
[9] STAR Collaboration, Nature Physics, **16**, 409 (2020)
[10] I. Kisel, presented at *HYP2022* conference
[11] P. Eckert, presented at *HYP2022* conference
[12] S. N. Nakamura, presented at *HYP2022* conference
[13] L. Tang, presented at *HYP2022* conference
[14] M. Ukai, presented at *HYP2022* conference
[15] K. Nakazawa, presented at *HYP2022* conference
[16] T. Gogami, presented at *HYP2022* conference
[17] T. O. Yamamoto, presented at *HYP2022* conference
[18] K. Miwa, presented at *HYP2022* conference
[19] N. Zachariou, presented at *HYP2022* conference
[20] ALICE collaboration, Nature **588**, 232 (2020)
[21] T. Iritani *et al.*, Phys. Lett. B **792**, 284 (2019)
[22] D. Mihaylov, presented at *HYP2022* conference
[23] H. P. Zbroszczyk, presented at *HYP2022* conference
[24] B. Singh, presented at *HYP2022* conference
[25] L. Šerkšnytė, presented at *HYP2022* conference
[26] H. Tamura, presented at *HYP2022* conference
[27] F. Sakuma, presented at *HYP2022* conference
[28] A. Ohnishi, presented at *HYP2022* conference
[29] O. V. Doce, presented at *HYP2022* conference
[30] S. Ajimura *et al.*, Phys. Lett. B **789**, 620 (2019); T. Yamaga, presented at *HYP2022* conference
[31] T. Hashimoto, presented at *HYP2022* conference
[32] For example, see T. Hyodo, presented at *HYP2022* conference
[33] N. Wickramaarachchi, presented at *HYP2022* conference
[34] P. Pauli, presented at *HYP2022* conference
[35] J. Zmeskal, presented at *HYP2022* conference
[36] K. Piscicchia, presented at *HYP2022* conference

[37] R. Lea, presented at *HYP2022* conference
[38] Y. Kamiya, presented at *HYP2022* conference
[39] D. Pawlowska, presented at *HYP2022* conference
[40] STAR Collaboration, Nature **548**, 62 (2017)
[41] K. Niida, presented at *HYP2022* conference
[42] R. Lalik, presented at *HYP2022* conference
[43] J. H. Chen, presented at *HYP2022* conference
[44] F. K. Guo, presented at *HYP2022* conference
[45] K. Tanida, presented at *HYP2022* conference
[46] K. Ozawa, presented at *HYP2022* conference

AUTHOR INDEX

Abou Yassine Rayane....298
Abrams D....77, 83, 96
Achenbach P....1
Achenbach Patrick....32, 117, 402
Adamczewski-Musch Jörn....298
Afnan Iraj R....89
Ahn J. K....164
Ahn Jung Keun....402
Ahn Jungkeun....121
Akaishi Takaya....15
Akazawa Y....164
Akiyama T....1, 77, 83, 96
Akiyama Takeru....32
Alef Stefan....101
Androiæ Darko....402
Androic D....1, 77, 83, 96
Aniol K....77, 83, 96
Aoki Kanae....402
Aoki Kazuya....395, 420
Arakelyan Gevorg H....320
Aramaki T....164
Asano Hidemitsu....15
Asaturyan A....1
Asaturyan Arshak....402
Ashikaga S....164
Ayerbe Gayoso C....77, 83, 96
Balkova Yuliia....126
Bane J....77, 83, 96
Barcus S....77, 83, 96
Barnea Nir....26, 58
Barrow J....77, 83, 96
Bazak Betzalel....26, 58
Beck Reinhard....101
Becker Marten....298
Bellini V....77, 83, 96
Benhar Omar....38
Bergmann Philip....298
Bhatt H....77, 83, 96
Bhetuwal D....77, 83, 96
Biswas D....77, 83, 96
Blanco Alberto....298
Blume Christoph....298
Bölting Michael....117
Botta Elena....402
Braghieri Alessandro....101
Branchini Paolo....251
Brash E....1
Bratkovskaya Elena....372
Bruns Peter C....92
Bukhari M. H....1
Bukhari Masroor. H....402
Bydžovský P....63
Bydzovsky Petr....38
Bydžovský Petr....109, 113
Callier S....164
Camsonne A....1, 77, 83, 96
Camsonne Alexandre....402
Cargnelli Michael....251
Castellanos J....77, 83, 96
Chen J....77, 83, 96
Chen Jinhui....32
Chen J-P....77, 83, 96
Chen Xurong....15
Chiga N....164, 164
Chlad Lukas....298
Chlad Lukáš....339
Choi S. W....164
Choi Sungwook....121
Chrisman D....77, 83, 96
Christiansen Martin....117
Chudoba Petr....298
Ciepa³ Izabela....298
Cieplý Aleš....92, 109, 113
Clozza Alberto....15
Cole Philip....101
Contessi Lorenzo....26
Cordts Malte....298
Covrig Dusa S....1
Covrig S....77, 83, 96
Covrig Silviu....38
Covrig Silviu. C....402
Cruz-Torres R....77, 83, 96
Curceanu Catalina....15, 251
Czerwinski Eryk....251
Dai Lian Rong....264, 380
Das R....77, 83, 96
De Gregorio G....63
De Leo Veronica....251
De Lucia Erika....251
Del Grande Raffaele....15, 251
Denisova D....63
Denisova Daria....66
Dhar Madhumita....190, 197
Di Cicco Alessandro....251
Di Salvo Rachele....101
Diociaiuti Eleonora....251
Distler Michael O....32
Dou Wenbo....432
Dreyer Jörn....298
Ebata K....1, 154
Ebata Kengo....161, 402
Eckert Philipp....32
Ekawa H....150, 164
Ekawa Hiroyuki....342, 402, 432
Elaasar M. A....1
Elsner Daniel....101
Esmail Waleed....347
Esmail Waleed Ahmed....298

AUTHOR INDEX

Esser Anselm 32
Esumi Shin-ichi 420
Evtoukhovitch P. 164
Evtoukhovitch Petr 402
Fabbietti Laura 251
Fantini Alessia 101
Feijoo Albert 235, 264, 380
Feliciello Alessandro 402
Fermani Paolo 251
Fiore Salvatore 251
Firlej Miroslaw 298
Fiutowski Tomasz 298
Floersheimer Henrik 298
Fonte Paulo 298
Förtsch Jörg 298
Freyermuth Oliver 101
Friedman Eliahu 7, 135, 209, 243
Friese Jürgen 298
Fröhlich Ingo 298
Fromberger Frank 101
Fuchey E. 77, 83, 96
Fujii Y. 1
Fujioka H. 161
Fujioka Hiroyuki 147, 395, 402
Fujioka N. 164
Fujita M. 130, 161, 164
Fujita Manami 50, 121, 150, 402, 436
Fujiwara T. 1
Furic M. 1
Gal Avraham 7, 26, 135, 209
Galatyuk Tetyana 298
Gao Yiming 432
Garibaldi F. 1, 77, 83, 96
Garibaldi Franco 38, 402
Gautam T. 77, 83, 96
Gazda Daniel 7, 235
Geratz Julian 32
Ghio Francesco 101
Gibson Benjamin F. 89
Gnanvo K. 77, 83, 96
Gniazdowski Tomasz 298
Gogami T. 1, 77, 83, 96, 130, 154, 161, 164
Gogami Toshiyuki 38, 121, 395, 402
Gomez J. 77, 83, 96
Greifenhagen Robert 298
Groß Johannes 101
Grunwald Mateusz 298
Grzonka Dieter 298
Guaraldo Carlo 15
Gubler Philipp 372
Gueye P. 1, 77, 83, 96
Gumberidze Malgorzata 298
Gunji Taku 420
Guo Jing 141, 201
Hachiya Takashi 420
Hague T. J. 77, 83
Hague T.J. 96
Haidenbauer Johann 176
Han Chengdong 15
Hansen O. 77, 83, 96
Harabasz Szymon 298
Harada Hiroyuki 420
Harada T. 164
Harada T. K. 130, 161
Harada Takeshi 121
Harada Takeshi K. 154
Harada Takeshi. K. 402
Hasegawa S. 164
Hasegawa Shoichi 121, 402
Hasegawa Tomoyuki 402
Hashimoto Tadashi 15
Hauenstein F. 77, 83, 96
Hayakawa S. H. 130, 154, 161, 164
Hayakawa S.H. 150
Hayakawa Shuhei 121
Hayakawa Shuhei H. 171
Hayakawa Shuhei. H. 402
Hayashi K. 150
He Yan 432
Heinz Thorsten 298
Helmel Christian 32
Henry W. 77, 83, 96
Hidaka Yoshimasa 395
Higinbotham D. W. 1, 77, 83
Higinbotham D.W. 96
Hirose Erina 402
Hiyama Emiko 395
Hoek Michael 32
Höhne Claudia 298
Hojeij Fatima 298
Holzmann Romain 298
Honda R. 154, 161, 164
Honda Ryotaro 105, 171, 395, 402
Hosaka Atsushi 395
Hoshino S. 164
Hosomi K. 130, 150, 164
Hosomi Kenji 402
Huck Holger 298
Hyde C. E. 77, 83
Hyde C.E. 96
Hyodo Tetsuo 384, 388, 392
Ichikawa M. 164
Ichikawa Y. 130, 154, 161, 164
Ichikawa Yudai 121, 395, 402, 420
Idzik Marek 298
Ieiri M. 164
Ieiri Masaharu 395
Ikedai M. 164

AUTHOR INDEX

Iliescu Mihai ... 15
Imai K. ... 164
Imai Kenichi ... 402
Inoue Kentaro ... 15
Isaka Masahiro ... 395
Ishige T. ... 1
Ishige Tatsuhiro ... 402
Ishii Noriyoshi ... 395
Ishikawa Takatsugu ... 395
Ishikawa Y. ... 130, 150, 164
Ishikawa Yuji ... 436
Ishimoto S. ... 164
Ishimoto Shigeru ... 15, 121
Itabashi K. ... 1, 77, 83, 96
Itabashi Kosuke ... 32
Itahashi Kenta ... 15
Itonaga Kazunori ... 54
Iwasaki Masahiko ... 15
Jinno Asanosuke ... 314
Jude Thomas ... 101
Jung W. S. ... 164
Jung Wooseung ... 402
Jung Woose-ung ... 121
Kajikawa S. ... 164
Kalicy Grzegorz ... 258
Kamada K. ... 130, 161
Kamada Kento ... 50, 402
Kamiya Yuki ... 184
Kampert Karl-Heinz ... 298
Kämpfer Burkhard ... 298
Kanatsuki Shunsuke ... 402
Kanauchi H. ... 130, 150, 164
Kanda H. ... 164
Kaneta M. ... 1, 77, 83, 96
Kaneta Masashi ... 32
Kang B. M. ... 164
Kang Byungmin ... 121
Kardan Behruz ... 298
Kasagi A. ... 150
Kasagi Ayumi ... 432
Kato Seigo ... 402
Kawai H. ... 164
Kedych Vadym ... 298
Keppel C. ... 77, 83, 96
Khemchandani Kanchan ... 380
Kim S. H. ... 164
Kim Shinhyung ... 121
Kim Shinhyung. H. ... 402
Kino R. ... 1
Kino Ryoko ... 32
Kinugawa Tomona ... 388
Kisel Ivan ... 277
Kitaoka T. ... 164
Kitazawa Masakiyo ... 420
Klag Pascal ... 32
Knapp F. ... 63
Kobayashi K. ... 164
Kobori T. ... 161
Kobori Takumi ... 402
Kochankovski Hristijan ... 376
Koenig Ilse ... 298
Koenig Wolfgang ... 298
Kohl Katrin ... 101
Kohls Marvin ... 298
Koike T. ... 130, 150, 164
Kolas Jedrzej ... 298
Komatsu Yusuke ... 395
Komatsubara Takeshi ... 395
Korcyl Grzegorz ... 298
Kornakov Georgy ... 298
Kornas Frederic ... 298
Kotte Roland ... 298
Krueger Wilhelm ... 298
Kugler Andrej ... 298
Kulessa Pawel ... 298
Kutz T. ... 77, 83, 96
Lalik Rafal ... 298
Lashley N. ... 1
Lashley-Colthirst N. ... 77, 83, 96
Le Hoai ... 18
Lebedev Semen ... 298
Lenske Horst ... 190, 197
Leung Yue Hang ... 284
Levi Sandri Paolo ... 101
Li S. ... 77, 83, 96
Liang Wei Hong ... 380
Lim GeiYoub ... 395
Linev Sergey ... 298
Liu Enqiang ... 432
Liu H. ... 77, 83, 96
Lo Iudice N. ... 63
Lopes Luís ... 298
Lorenz Manuel ... 298
Ma Yue ... 15
Ma Yu-Gang ... 32
Magas Volodymyr ... 235
Malige Akshay ... 298
Mammei J. ... 77, 83, 96
Mandaglio Guiseppe ... 101
Marcello Simonetta ... 402
Mare. Ji.i ... 243
Mareš Jiøí ... 26
Markert Jochen ... 298
Markowitz P. ... 1, 77, 83, 96
Markowitz Pete ... 402
Markowitz Pete E.C. ... 38
Martínez Torres Alberto ... 380
Martini Matteo ... 251

AUTHOR INDEX

Marton Johann....251
Matsuda K....164
Matsumoto Y....164
Matulewicz Tomasz....298
McClellan R. E....77, 83, 96
Meddi F....77, 83, 96
Meekins D....1, 77, 83, 96
Meißner Ulf-G....176
Merino Carlos....320
Merkel Harald....32
Messchendorp Johan....298
Metag Volker....298
Miake Yasuo....420
Michaels R....77, 83, 96
Michel Jan....298
Mihovilovic M....77, 83, 96
Miliucci Marco....15
Millener John....38
Miwa K....130, 154, 161
Miwa Koji....105, 121, 164, 171, 395, 402
Mizuno M....1
Mizuno Masaya....32
Mkrtchyan A. H....1
Mkrtchyan Arthur....402
Mkrtchyan H. G....1
Mkrtchyan Hamlet....402
Molenda Aleksandra....298
Molina Raquel....264, 380
Morino Yuhei....395, 420
Moritsu Manabu....402
Moron Jakub....298
Moskal Pawel....251
Motoba Toshio....38, 54
Moyer A....77, 83, 96
Müller Ulrich....32
Muneem Abdul....432
Müntz Christian....298
Murase Koichi....314
Murayama Rie....15
Nabroth Marvin....298
Nagae T....130, 154, 161
Nagae Tomofumi....121, 395, 402
Nagafusa S....1
Nagamiya Shoji....420
Nagano S....1
Nagao S....1, 77, 83, 96, 164
Nagao Sho....32, 38, 395, 402
Nagatomi R....164
Nakada Y....164
Nakada Yoshiyuki....105, 171
Nakagawa M....164
Nakagawa Manami....402, 432
Nakamura I....164
Nakamura S. N....1, 77, 83
Nakamura S.N....96
Nakamura Satoshi N....32, 38, 395, 409
Nakamura Satoshi. N....402
Nakamura Y. R....1
Nakamura Yuki R....32
Nakazawa K....150
Nakazawa Kazuma....426, 432
Nanamura T....154, 161
Nanamura Takuya....105, 121, 164, 171, 402
Nanjo Hajime....395
Nara Yasushi....314, 420
Naruki M....164
Naruki Megumi....395, 402, 420
Naumann Lothar....298
Negishi R....161
Negishi Ryosuke....402
Nemura Hidekatsu....395
Nguyen D....77, 83, 96
Niculescu G....1
Niculescu I....1
Niida Takafumi....326, 420
Nishibuchi Takuma....384
Nishimura N....150
Nomura Tadashi....395
Nonaka Toshihiro....420
Noumi Hiroyuki....15, 395
Nycz M....77, 83, 96
Obertova Jaroslava....243
Ohnishi Akira....314
Ohnishi Hiroaki....395
Ohnishi Hiroyuki....15
Okada Shinji....15
Okuyama K....1, 77, 83, 96
Okuyama Kazuki....32, 402
Orliñski Jan....298
Oset Eulogio....264, 380
Otto Jan-Hendrik....298
Oura F....130, 161
Oura Fumiya....402, 436
Outa Haruhiko....15
Owen V....77, 83, 96
Ozawa Kyoichiro....395, 420
Ozawa S....164
Palatchi C....77, 83, 96
Pandey B....1, 77, 83, 96
Pandey Bishnu....402
Park S....77, 83, 96
Parpottas Yannis....298
Parschau Mirco....298
Pauli Peter....69
Pauly Christian....298
Paw owska Diana....330
Pechenov Vladimir....298
Pechenova Olga....298

AUTHOR INDEX

Pedroni Paolo ... 101
Perez Andrade Gabriela ... 298
Perez Del Rio Elena ... 251
Pérez-Obiol Axel ... 7
Petkovic T. ... 77, 83, 96
Petrellis D. ... 63
Petrellis Dimitrios ... 109, 113
Pfeifer Dennis ... 298
Piasecki Krzysztof ... 298
Pietraszko Jerzy ... 298
Piscicchia Kristian ... 15, 251
Pochodzalla J. ... 1
Pochodzalla Josef ... 32, 117, 402
Pokorný J. ... 63
Povar Tetiana ... 298
Premathilake S. ... 77, 83, 96
Prozorov Alexandr ... 298
Przygoda Witold ... 298
Pysz Krzysztof ... 298
Ramos Angels ... 243, 376
Ramos Àngels ... 235, 251
Ramstein Béatrice ... 298
Rappold Christophe ... 432
Rathod Narendra ... 298
Raux L. ... 164
Reimer P. E. ... 77, 83
Reimer P.E. ... 96
Reinhold J. ... 1, 77, 83, 96
Reinhold Joerg ... 38
Ren Xiu-Lei ... 273
Riordan S. ... 77, 83, 96
Ritman James ... 298
Roca Luís ... 380
Rodriguez V. ... 77, 83, 96
Rodriguez V. M. ... 1
Rogers T. G. ... 164
Rogers Tatyana G. ... 436
Romaniuk Mariia ... 101
Rost Adrian ... 298
Rustamov Anar ... 298
Sada Yuta ... 15
Saito Nami ... 432
Saito Takehiko R. ... 289, 420, 432
Sakaguchi A. ... 164
Sakaguchi Atsushi ... 15, 402
Sakaguchi Takao ... 420
Sakao T. ... 161
Sakao Tamao ... 105, 164, 402
Sako H. ... 164
Sako Hiroyuki ... 121, 402, 420
Sakuma Fuminori ... 15, 395
Salabura Piotr ... 298
Samanta C. ... 1, 77, 83, 96
Samanta Chhanda ... 402
Santiesteban S. N. ... 77, 83
Santiesteban S.N. ... 96
Saraiva Joao ... 298
Sato Masaharu ... 15
Sato S. ... 164
Sato Susumu ... 121, 402, 420
Sawada Shinya ... 395
Sawatzky B. ... 1, 77, 83, 96
Schadmand Susan ... 298
Schäfer Martin ... 26, 58
Scheluchin Georg ... 101
Schild Niklas ... 298
Schlimme Björn Sören ... 32
Schmieden Hartmut ... 101
Schulze Hans-Josef ... 141
Schumacher Reinhard A. ... 258
Schupp Falk ... 117
Schwab Erwin ... 298
Scordo Alessandro ... 15, 251
Seck Florian ... 298
Sekihara Takayasu ... 395
Selce Andrea ... 251
Selyuzhenkov Ilya ... 298
Sfienti Concettina ... 32
Shabelski Yuli M. ... 320
Shabestari M. H. ... 1
Shabestari Mitra. H. ... 402
Shahinyan A. ... 1
Shahinyan Albert ... 402
Shao Tianhao ... 32, 335
Shevchenko Nina V. ... 269
Shim Sang-In ... 395
Shiomi Koji ... 395
Shiozaki T. ... 164
Shirotori K. ... 164
Shirotori Kotaro ... 15, 395, 402
Silarski Michal ... 251
Singh Udai ... 298
Sirca S. ... 1
Širca S. ... 77, 83, 96
Širca Simon ... 402
Sirghi Diana ... 15, 251
Sirghi Florin ... 15
Skorpil Leon ... 298
Skoupil D. ... 63
Skoupil Dalibor ... 109, 113
Skurzok Magdalena ... 251
Slifer K. ... 77, 83, 96
Smyrski Jerzy ... 298
Sobiella Manfred ... 298
Son C. ... 161, 402
Spataro Stefano ... 298
Spies Simon ... 298
Stefaniak Maria ... 298

AUTHOR INDEX

Steinen Marcell 32, 117
Stone Jirina R. 366
Ströbele Herbert 298
Stroth Joachim 298
Su T. 77, 83, 96
Sugimoto Shohei 432
Sugimura Hitoshi 402
Sumara Konrad 298
Suzuki K. N. 1, 77, 83, 164
Suzuki K.N. 96
Suzuki S. 164
Suzuki Shoji 15, 121
Svoboda Ondøej 298
Swientek Krzysztof 298
Szala Melanie 298
Tabata M. 164
Tachibana K. 1
Tachibana Koga 32
Taille C. d. L. 164
Tajima Yasuhisa 395
Takahashi H. 161, 164
Takahashi Hitoshi 395, 402, 420
Takahashi Shuji 402
Takahashi T. 130, 154, 161, 164
Takahashi T. N. 164
Takahashi Tomonori 402
Takahashi Toshiyuki 105, 171, 395, 402
Takeuchi Sachiko 395
Taki Masato 432
Takizawa Makoto 395
Tamura H. 130, 150, 161, 164
Tamura Hirokazu 50, 121, 205, 395, 402, 420, 436, 440
Tanaka Kazuhiro 420
Tanaka M. 164
Tanaka Yoshiki K. 432
Tang L. 1, 77, 83, 96
Tang Liguang 38
Tanida K. 130, 164
Tanida Kiyoshi 15, 121, 395, 402
Terashima Ibuki 392
Thiel Michaela 32
Tian Y. 77, 83, 96
Tlusty Pavel 298
Toda Taiga 15
Tokiyasu A. O. 161
Tokiyasu Atsushi 158, 402
Tokuda Megumu 15
Tolos Laura 360, 376
Toyama Y. 1, 77, 83, 96
Toyama Yuichi 32
Traxler Michael 298
Tsamalaidze Z. 164
Tsamalaidze Zviadi 402
Tsertos Haralabos 298
Tsuruta Masato 15
Tsutsumi K. 1
Uchida Makoto 402
Uehara K. 1, 77, 83, 96
Uehara Keita 32
Ukai M. 130, 150, 154, 161, 164
Ukai Mifuyu 44, 105, 121, 171, 395, 402, 436
Umetsu H. 164
Umeya Atsushi 54
Umezaki E. 1
Une Chiaki 402
Urciuoli G. M. 1, 77, 83
Urciuoli G.M. 96
Urciuoli Guido Maria 38
Urciuoli Guido. M. 402
Vazquez Doce Oton 251
Veselý P. 63
Vidana Isaac 38
Vidaña Isaac 352
Votaw D. 77, 83, 96
Wada S. 164
Wagner Vladimir 298
Wang He 432
Wasiluk Mateusz 298
Watanabe D. 1
Weber Adrian Amatus 298
Weise Wolfram 217
Wendisch Christian 298
Wickramaarachchi Nilanga 258
Williamson J. 77, 83, 96
Wintz Peter 298
Wojtsekhowski B. 77, 83, 96
Wood S. A. 1, 77, 83
Wood S.A. 96
Wycech Slavomir 251
Yale B. 77, 83, 96
Yamaga Takumi 15, 227
Yamamoto T. O. 154, 161, 164
Yamamoto T.O. 150
Yamamoto Takeshi 105, 121
Yamamoto Takeshi O. 171, 395, 436
Yamamoto Takeshi. O. 130, 402
Yamamoto Yasuo 395
Yanai Ayari 432
Yang Seongbae 121
Yaron Roee 58
Ye Z. 77, 83, 96
Yokkaichi Satoshi 420
Yoshida J. 150, 164
Yoshida Junya 432
Yoshimoto M. 150
Yoshimoto Masahiro 432
Yoshimura K. 164
Yuan Xiaohua 15

AUTHOR INDEX

Zachariou Nicholas 109, 113
Zbroszczyk Hanna 298, 306
Zhang Hongqiang 15
Zhang J. 77, 83, 96
Zhang Pengming 15
Zhang Yapeng...15
Zheng X. 77, 83, 96
Zherebtsova Elizaveta...298
Zhou Xian-Rong..141, 201
Zielinski Marcin 298
Zmeskal Johann 251
Zumbruch Peter 298